Antioxidants

Scrivener Publishing
100 Cummings Center, Suite 541J
Beverly, MA 01915-6106

Publishers at Scrivener
Martin Scrivener (martin@scrivenerpublishing.com)
Phillip Carmical (pcarmical@scrivenerpublishing.com)

Antioxidants

Nature's Defense Against Disease

Edited by

Rakesh K. Sindhu

School of Pharmacy, Sharda University, Greater Noida, Uttar Pradesh, India

Inderbir Singh

Chitkara College of Pharmacy, Chitkara University, Patiala, Punjab, India

and

M. Arockia Babu

Institute of Pharmaceutical Research, GLA University, Mathura, Uttar Pradesh, India

Scrivener Publishing

WILEY

This edition first published 2025 by John Wiley & Sons, Inc., 111 River Street, Hoboken, NJ 07030, USA and Scrivener Publishing LLC, 100 Cummings Center, Suite 541J, Beverly, MA 01915, USA

For more information about Scrivener publications please visit www.scrivenerpublishing.com.

Wiley Global Headquarters
111 River Street, Hoboken, NJ 07030, USA

For details of our global editorial offices, customer services, and more information about Wiley products visit us at www.wiley.com.

Library of Congress Cataloging-in-Publication Data

ISBN 978-1-394-27054-5

Front cover images courtesy of Pixabay.com
Cover design by Russell Richardson

Set in size of 11pt and Minion Pro by Manila Typesetting Company, Makati, Philippines

Printed in the USA

10 9 8 7 6 5 4 3 2 1

Contents

Preface

Free radicals, also known as reactive oxygen species (ROS), are formed in the human body as by-products of normal cellular processes. These reactive molecules, particularly capable of damaging cellular components such as DNA, lipids, and proteins, can initiate a process called oxidative stress. Leading experts from various disciplines have contributed chapters to this book, exploring a wide range of topics, including the role of antioxidants in the management of cancer, gastrointestinal disorders, skin conditions, cardiovascular diseases, and rheumatoid arthritis.

Antioxidants serve as a crucial defense mechanism against oxidative stress, acting as free radical scavengers that neutralize these harmful molecules before they can cause cellular damage. These chemicals may be synthesized endogenously within the body or derived exogenously from natural sources such as fruits, vegetables, whole grains, and nuts. This book delves into the role of antioxidants in preventing and managing a variety of diseases and critically examines current research on the effects of nutritional antioxidants on specific disease states.

This book is intended for a broad audience, including healthcare professionals seeking a deeper understanding of the relationship between antioxidants and disease prevention, nutritionists and dietitians looking to incorporate this knowledge into clinical dietary plans, and individuals interested in making informed dietary choices for better health. By providing a thorough and evidence-based exploration of this field, we aim to empower readers to adopt proactive strategies for long-term health and well-being.

As editors, we believe this book will serve as a milestone for future research and development in the study of antioxidants and their health benefits. We extend our gratitude to all the contributors, whose dedication and expertise have enriched this volume, and to Martin Scrivener and Scrivener Publishing for their support and publication.

Prof. (Dr.) Rakesh K. Sindhu
Prof. (Dr.) Inderbir Singh
Prof. (Dr.) M. Arockia Babu

1

Basics of Antioxidants and Their Importance

Shuchi Goyal[1], Divya Thirumal[2], Sumitra Singh[3], Dinesh Kumar[4], Inderbir Singh[1], Gautam Kumar[5] and Rakesh K. Sindhu[5]*

[1]*Chitkara College of Pharmacy, Chitkara University, Punjab, India*
[2]*Manipal College of Pharmaceutical Sciences, Manipal University, Manipal, Karnatka, India*
[3]*Department of Pharmaceutical Sciences, Guru Jambheshwar University of Science and Technology, Hisar, Haryana, India*
[4]*Department of Pharmaceutical Sciences, Central University of Haryana, Jant-Pali, Mahendergarh, India*
[5]*School of Pharmacy, Sharda University, Greater Noida, Uttar Pradesh, India*

Abstract
Fast living leads to an overabundance of free radicals in the body, which ultimately leads to mortality and a reduction in life expectancy by damaging cells, tissues, and organs. Consuming antioxidants aids in scavenging free radicals to ward off both acute and persistent illnesses. Antioxidants are essential that reducing the reactive mechanisms and the negative consequences of reactive oxygen species (ROS) throughout the chain supply and individual physiology. ROS are crucial for neuronal signaling, differentiation, tissue homeostasis, and longevity. In this overview, we go over the many forms of ROS, how they affect the function of cells, and whether they promote or inhibit cancer development. ROSs' detrimental impacts and their significance in the initiation of pathology are explored. A crucial part of these defense strategies is played by antioxidants. It is generally recognized that the inclusion of phenolic chemicals, particularly phenolic acids and flavonoids, is associated with antioxidative and pharmacologic effects. Antioxidants are now a crucial component of sophisticated health care. Antioxidants, whether they be organic or artificial, can help combat many diseases early on and works best when they are present in high concentrations. They affect how adequately the therapy responds. In addition to its usage in nutritious dietary supplements, emphasis is being placed on utilizing them as natural substitutes for synthetic versions to improve food durability and prevent degradation by oxidation throughout manufacturing and preservation. In purpose to support technological improvement in this domain, this overview summarizes relevant and widely recognized findings on the efficient significance of organic and synthesized antioxidants with associated therapeutic value.

***Keywords*:** Antioxidants, ROS, free radicals, pathology, importance, applications

**Corresponding author*: drrakeshksindhu@gmail.com

Rakesh K. Sindhu, Inderbir Singh and M. Arockia Babu (eds.) Antioxidants: Nature's Defense Against Disease, (1–20)

Abbreviations

RON	reactive oxygen and nitrogen species
ET	electron transfer
HTD	hydrogen atom donation
EDTA	ethylenediaminetetraacetic acid
CA	citric acid
Vit C	Vitamin C
Vit B	Vitamin B
Vit A	Vitamin A
MDA	malondialdehyde
8-OHdG	8-hydroxy-2′-deoxyguanosine levels

1.1 Introduction

Imagine your body as a bustling city, with millions of residents busily going about their daily activities. In this city, just like in any vibrant community, there is a natural process of wear and tear. As time passes, structures deteriorate, and waste accumulates. However, to maintain the city's vitality and ensure its residents' well-being, there are diligent workers known as "antioxidants." Antioxidants, pivotal in cellular protection, counteract the harm induced by unstable molecules termed free radicals. These radicals, through oxidative reactions, instigate cellular damage, potentially culminating in conditions like cancer. Antioxidants engage with free radicals, stabilizing them and thwarting potential harm. Key antioxidants encompass bella carotin, carotenoid, vit A, B, C, as well as various polyphenols [1]. An antioxidant has the ability to slow down or prevent different compounds from oxidizing, which is the chemical process by which ions move from one material to an oxidation agent. This decay procedure births liberated, setting off detrimental chain reaction within cells [2, 3]. Antioxidants step in, oxidizing themselves and removing the free radical intermediates that cause these chain reactions to stop. Interestingly, several antioxidants—such as polyphenols, thiols, and ascorbic acid—also function as reducing agents [4]. While oxidation reactions are vital for life, their immoderation can prove deleterious. Consequently, organisms, both flora and fauna, maintain intricate antioxidant defense systems. These include a range of enzymes, including catalase, superoxide dismutase, and several oxidoreductases, as well as antioxidants, including glutathione, vitamin C, and vitamin E [5]. Free radical damage can be brought on by low antioxidant levels or malfunctioning antioxidant enzymes, which may cause damage or even death to cells [6]. Given the potentials implication of Free radical damage in numerous human ailments, extensive research explores the utility of antioxidants in pharmacology, particularly in treating stroke and neurodegenerative disorder [7]. Nonetheless, whether oxidative stress acts as the cause or consequence of diseases remains unclear. Antioxidants are extensively utilized in dietary supplements, aiming to sustain health and avert conditions like cancer and coronary heart disease. While early studies suggested potential health benefits of antioxidant supplements, subsequent large clinical trials failed to validate such advantages and instead hinted at potential harm with excessive supplementation. Additionally, antioxidants find broad industrial applications, functioning as chemicals in food and Corrective and inhibiting condescension in latex and petrol [8, 9]. Chemists have long acknowledged the ability of antioxidants to mitigate oxidation caused by free radicals, essential for maintaining stability in various substances, including lubrication oils and plastics. Human biological processes, encompassing respiration, metabolism, digestion, and energy conversion, generate reactive oxygen and nitrogen species (RONs), which can manifest as free radicals or readily generate them [10]. RONs, at moderate concentrations, play pivotal roles in biological pathways but can cause considerable damage at elevated levels, leading to disruptions in cellular signaling and Free radical damage [11]. This imbalance can lead to irreversible alterations in cell compounds, affecting cellular health and contributing to major chronic ailments such as cancer, cardiovascular, liver, and neurological disorders. Antioxidant defense mechanisms encompass a spectrum of approaches, including inhibiting free radical production, scavenging free radicals, converting free radicals into less dangerous substances, postponing the emergence of more hazardous species and halting the spread of chains reactions, bolstering the endogenous antioxidant defense system through synergistic action, and chelation. These multifaceted mechanisms collectively contribute to cellular resilience against Free radical damage and its detrimental moment [12].

1.2 Generalization of Antioxidant

Antioxidants is compounds that protect cells from oxidative damage caused by free radicals and reactive oxygen species (ROS). Oxidative stress resulting from an inequality between the production of these harmful molecules and the body's ability to neutralize them has been linked to various chronic diseases, including cardiovascular diseases, cancer, neurodegenerative disorders, and aging [13]. The main types of antioxidants are those that use a single electron transfer (ET) mechanism or hydrogen atom donation (HAT) to eliminate free radicals. Antioxidant catalysts that are prooxidant are neutralized by secondary antioxidants [14]. These include compounds that deactivate reactive species like singlet oxygen (beta-carotene) or chelate prooxidant metal ions (such iron and copper), such as (EDTA) and citric acid (CA) [10]. Mechanism of antioxidants -Scavenging Free Radicals: Antioxidants, through various enzymes and molecules, can neutralize free radicals by donating electrons [15]. This process mitigates the harmful chain reactions initiated by free radicals.

Enzymatic Scavengers: as reverse fibrosis, catalase, and GPx1, act as the first line of defense against free radical damage, reverse fibrosis. For example, catalyzes the dismutation of superoxide radicals into less harmful species.

Non-enzymatic Antioxidants:

- Non-enzymatic antioxidants encompass a wide range of molecules, including vit (e.g., vitamin E,C), lignans (e.g., flavonoids and resveratrol), and trace elements (e.g., selenium and zinc). These antioxidants exert their effects by quenching ROS directly or indirectly [16].

Chelation of Metal Ions:

- Some antioxidants, like metal-binding proteins and chelators, combat oxidative stress by binding to metal ions (e.g., iron and copper) that catalyze the formation of highly reactive radicals. This prevents metal-mediated ROS generation [17].

Regulation of Transcription Factors:

- Some antioxidants activate transcription variables, like the nuclear factor erythroid 2-associated factor 2 (Nrf2), to modify how cells react to oxidative stress. Nrf2 controls the expression of numerous antioxidant genes, enhancing the body's overall antioxidant capacity [18].

Mitochondrial Protection:
Mitochondria are major sources of ROS production. Antioxidants, particularly those targeted to the mitochondria, reduce mitochondrial ROS generation and maintain mitochondrial function [19].

The mechanisms of action of natural antioxidants and oxidative processes. When polyunsaturated lipids are exposed to light, heat, ionizing radiation, metal ions, or metalloprotein catalysts, a free radical chain reaction is set off, which causes the autoxidation of the lipids in food. Oxidation can also be initiated by the enzyme lipoxygenase. Photooxidation can occur when exposed to light, and high-temperature thermal oxidation—which occurs when food is cooked, grilled, or fried—produces polar and polymeric compounds. The oxidation that happens to food most frequently is called autoxidation. The initiation (creation of lipid free radicals), propagation, and termination (generation of nonradical products) reactions are part of the traditional route of autoxidation, as shown in Figure 1.1 [20].

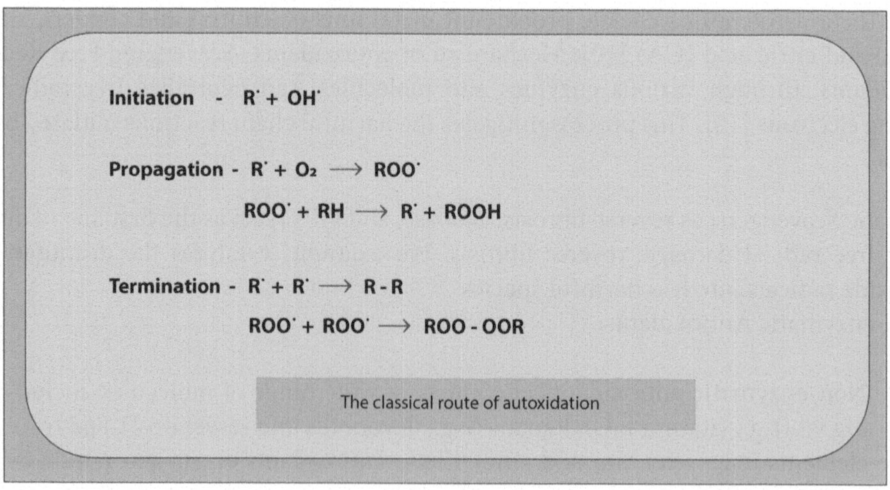

Figure 1.1 Autoxidation process [20].

1.3 Reactive Oxygen Species (ROS) and Free Radicals

Reactive oxygen species (ROS) and free radicals are fundamental components of cellular physiology, denoting chemically reactive molecular entities that harbor unpaired electrons in their outer orbits [21, 22] Reactive Oxygen Species surround a contrast array of oxygen-derived species, including superoxide anion (O2•–), hydroxyl radical (•OH), hydrogen peroxide (H2O2), and singlet oxygen (1O2), among others. On the other hand, free radicals, a subset of ROS, exhibit exceptional reactivity due to their unpaired electron, which drives them to readily engage in oxidative reactions [23].

These molecules are generated within the cell during various physiological processes, including aerobic respiration, immune responses, and cellular signaling [24, 25]. However, when their production exceeds the capacity of endogenous antioxidant defenses, it can lead to free radical damage.

The causes of ROS and free radical formation are multifaceted, encompassing both endogenous and exogenous sources [26]. Endogenously, ROS are produced during mitochondrial respiration, where electrons escape from the ETC [27]. Additionally, cellular enzyme such as nicotinamide adenine dinucleotide hydrogen phosphate oxidases and xanthine oxidoreductase contribute to ROS production as part of normal cellular functions [28]. Exogenously, exposure to outside influences like ionizing rays, pollutants, and heavy metals can increase ROS generation. The consequences of elevated ROS and free radicals are profound and multifarious. Oxidative stress, characterized by an inequality between ROS production and the body's ability to detoxify these species, is a common result [29]. This imbalance can lead to oxidative damage to biological molecules found in cells, such as proteins, fats, and DNA and RNA ultimately contributing to the pathogenesis of various diseases, such as cancer, neurodegenerative disorders, cardiovascular diseases, and inflammation-related conditions.

In response to the challenge posed by ROS and free radicals, the body has evolved a sophisticated defense system, employing endogenous antioxidants like superoxide dismutase, catalase, and glutathione peroxidase [30]. However, exogenous antioxidants from the diet, including vit E,C as well as polyphenols, play a pivotal role in bolstering the cellular defense against oxidative stress. Antioxidants function by donating electrons to free radicals without becoming destabilized themselves, thereby neutralizing these potentially harmful species By counteracting the detrimental effects of reactive oxidative stress and free radicals, antioxidants serve as an essential safeguard against the progression of diseases rooted in oxidative stress Understanding the intricate interplay between ROS, free radicals, and antioxidants is central to elucidating the molecular underpinnings of health and disease and may offer novel avenues for therapeutic intervention [31].

Pathology related to antioxidants: Pathology related to antioxidants is fundamentally centered on the intricate interplay between the protective role of antioxidants and the detrimental consequences of free radical damage [32] free radical damage, marked by an imbalance between the generation of reactive oxygen species and the bodies power to detoxify or repair the resultant damage, underpins numerous pathological conditions spanning various organ systems [33–37]. Oxidative stress stands as a linchpin in a plethora of ailments, ranging from cardiovascular diseases, neurodegenerative disorders, and cancer to diabetes, inflammatory conditions, and the process of aging itself [38]. At its core, oxidative stress is perpetuated by free radicals, which include highly reactive molecules such as superoxide anion (O2-), hydroxyl radicals (OH·), and hydrogen peroxide (H2O2) [39]. These agents wreak havoc on cellular components, inflicting harm upon DNA, proteins, and lipids [40]. In this intricate biological battle, antioxidants emerge as the valiant defenders of cellular integrity. These compounds come in two categories [41]: enzymatic antioxidants like superoxide dismutase, catalase, and glutathione peroxidase, and non-enzymatic antioxidants including vit C and E, glutathione, and flavonoids [42]. Acting in unison, these antioxidants mount a concerted effort to eliminate radicals and maintain the delicate balance of cellular redox equilibrium [43]. The pathological implications of oxidative stress are far-reaching and multi-faceted [44]. Cardiovascular diseases, encompassing conditions like atherosclerosis, hypertension, ischemia-reperfusion injury, and heart failure, bear the heavy burden of oxidative stress-induced damage [45]. Emerging evidence suggests that diets rich in antioxidants and antioxidant supplementation hold promise in mitigating cardiovascular pathology by alleviating oxidative damage [46]. Neurodegenerative disorders, such as Alzheimer's disease, Parkinson's disease, ALS, and multiple sclerosis, have their origins intertwined with oxidative stress-induced damage, contributing to their onset and progression [47]. Antioxidant therapy offers hope in delaying neurodegeneration and enhancing cognitive function. Cancer development, a multifaceted process governed by cellular and molecular alterations driven by a variety of endogenous and exogenous factors, is marked by oxidative DNA damage [48]. This damage serves as a key hallmark of carcinogenesis, instigating the initiation and promotion of cancer through chromosomal defects and the activation of oncogenes by free radicals. Diabetes and metabolic disease, characterized by free radical damage as a defining feature, underscore the potential of antioxidants in preventing diabetic complications, such as mononeuropathy, and retinopathy, by counteracting oxidative stress-induced damage. Furthermore, inflammatory disorders, particularly those of a chronic nature, are often perpetuated by oxidative stress [49]. Antioxidants, equipped with their anti-inflammatory properties, present a potential avenue for attenuating conditions like rheumatoid arthritis and inflammatory bowel disease [50]. The repercussions of antioxidant deficiency are dire, as insufficient intake or impaired antioxidant defense mechanisms significantly heighten the susceptibility to oxidative damage [51]. This exacerbates the pathology associated with various diseases, reinforcing the paramount importance of maintaining an adequate antioxidant defense system. In summary, the intricate web of oxidative stress, antioxidants, and their pathological implications underscores the critical role of these molecules in maintaining cellular health and the prevention of numerous diseases across various systems in the body [52]. Understanding this delicate balance is pivotal in the ongoing pursuit of novel therapeutic interventions and lifestyle modifications aimed at mitigating the burden of oxidative stress-related diseases [53–58].

Natural origin of Antioxidant: Reactive Oxygen Species (ROS) and free radicals are dynamic entities known for their potential to cause cellular damage, contributing to a range of diseases, including cancer and cardiovascular conditions [59]. In response to this challenge, antioxidants found in various foods play a crucial role in maintaining health. These antioxidants are abundant in fruits, vegetables, nuts, seeds, and select beverages, offering a multifaceted defense against oxidative stress. Fruits and vegetables, for instance, are rich sources of antioxidants, including vitamin C, E [60], elements such as selenium, and phytochemicals, such as flavonoids and carotenoids [61] as shown in Figure 1.2.

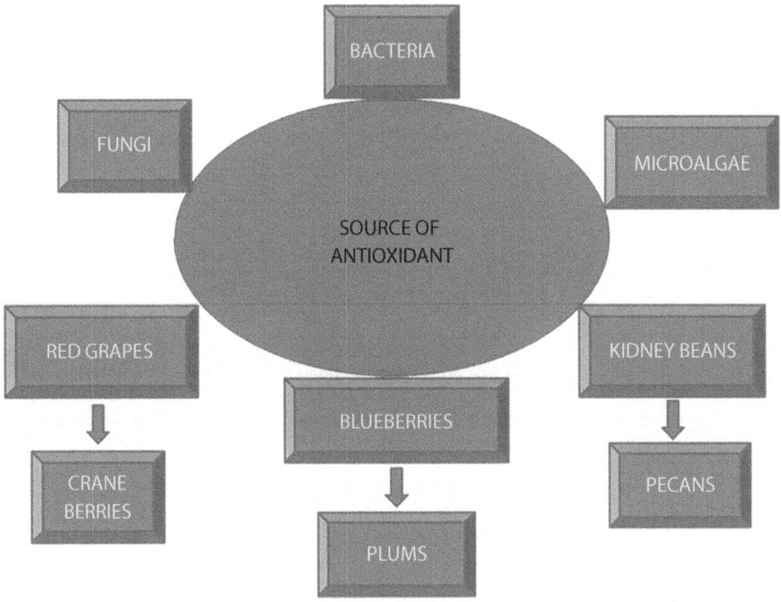

Figure 1.2 Sources of antioxidants.

These compounds are synthesized by plants as a defense mechanism against environmental stressors and as a means to attract pollinators [62]. In the realm of cereals and legumes, polyphenols, particularly phenolic acids, feature prominently. Ferulic acid, monohydroxybenzoic acid and (E)-3-(4-Hydroxyphenyl)-2-propenoic acid [63] are vital components, while buckwheat stands out with its abundant rutin content [64]. These polyphenolic compounds exist in both ester [65] and glycoside forms [66], with their chemical structure and concentration directly influencing their antioxidant activity [67, 68]. Additionally, catechins are found in cereal grains, with buckwheat, oats, and rye containing noteworthy amounts. Beverages [69] and spices also contribute significantly to antioxidant intake [70]. Coffee, tea, cocoa, red wine, and herbs and spices are rich sources of antioxidants [71]. In tea, catechins are prominent, particularly in green tea, which boasts over 2-times higher antioxidant activity compared to black or red tea. Roasted coffee, on the other hand, features chlorogenic acid as the dominant polyphenolic compound [73]. Flavonoids [72], a diverse group of naturally occurring compounds, offer strong antioxidant properties and can be found in various plant-based foods. Citrus fruits, berries, and dark chocolate are rich sources of specific flavonoids like hesperidin, naringin, anthocyanins, and epicatechin. Flavonoids combat oxidative stress through mechanisms such as free radical scavenging, metal chelation, and the enhancement of antioxidant enzymes [74]. Furthermore, marine environments are emerging as novel sources of antioxidants, with microalgae [75] and seaweed offering valuable contributions. Fucoxanthin from *Sargassum siliquastrum* [76], for example, has been found to protect human fibroblasts from UV-B-induced oxidative stress, thereby reducing intracellular [77] ROS, enhancing cell survival, and minimizing DNA damage. Additionally, compounds like fragilamide from *Martensia fragilis* exhibit moderate antioxidant activity, further expanding the scope of potential natural sources for antioxidants. Mechanism of action of various antioxidants is shown in Figure 1.3.

Antioxidant category	Mechanism of action
Carotenoids Lycopene β-Carotene	Lycopene traps $O_2^{\bullet}$; β-Carotene β-carotene scavenges $O_2^{\bullet}$ – anions and peroxyl radicals
Isothiocyanates Sulforaphane	Phase II enzyme enhancer.
Phenolic compounds BHA, t-BHQ, curcumin	Enhancers of phase II enzyme such as reductases (e.g. quinone reductase) and conjugating enzymes (e.g. glutathione S-transferase).
Micronutrients Selenium	Selenium is an essential component of several antioxidant enzymes, including glutathione peroxidase (GPx), thioredoxin reductase, phospholipid hydroperoxide glutathione peroxidase, and selenoprotein P. Its presence within these enzymes is crucial for its role in preventing lipid peroxidation, which is the process of oxidative damage to lipids or fats.
Antioxidant enzymes SOD Catalase	SOD dismutates $O_2^{\bullet} \rightarrow H_2O_2$ Catalese converts $H_2O_2 \rightarrow H_2O$

Figure 1.3 Mechanism of action of antioxidants.

Research has explored the connection between the color spectrum of fruits and vegetables and their antioxidant capacity, categorizing them into three groups based on Total Antioxidant Capacity (TAC). The first group includes high-TAC fruits like black grapes and strawberries. The second group comprises medium-TAC fruits such as blueberries and green apples. The third group consists of fruits with low-medium TAC, including mangoes and bananas, revealing the diversity of antioxidants in nature and emphasizing the health benefits of dietary variety. This classification is derived from computer vision-based image analysis of color hues related to TAC.

1.4 Importance of Antioxidant in Medicine

Antioxidants, the molecular heroes that combat the destructive forces of oxidative stress, play a pivotal role in the field of medicine. As the human body relentlessly generates reactive oxygen species (ROS) and free radicals through metabolic processes, it faces a constant battle against cellular damage, inflammation and causes the rise of various diseases. The importance of antioxidants in medicine lies in their remarkable ability to neutralize these harmful molecules and safeguard cellular health. Antioxidants have become a formidable asset in the field of medical science, presenting a potent means for not only preventing chronic ailments such as cancer, cardiovascular diseases, and neurodegenerative disorders but also for fortifying the body's immune defenses. Their emergence marks a promising pathway for interventions that encompass both preventive measures and therapeutic strategies. This introduction sets the stage to explore the multifaceted impact of antioxidants in medicine, underscoring their significance in the quest for better health and well-being. Here we are going to discuss on uses of antioxidant in various conditions and diseases [78, 79].

In Pregnancy -Pregnancy imposes heightened nutritional demands on mothers to support fetal development, with several micronutrients serving as essential cofactors or antioxidants. Normal placental development generates oxidative stress, which intensifies in the absence of sufficient antioxidant micronutrients, adversely affecting both the placenta and maternal circulation. Selenium, a crucial component of selenoproteins, including antioxidant enzymes like glutathione peroxidase, experiences reduced concentrations and activity during pregnancy (1st trimester: 65 μg/L; 3rd trimester: 50 μg/L). In pregnant women prone to postpartum thyroid dysfunction (PPTD), supplementation with 200 μg/d selenomethionine significantly lowers PPTD and permanent hypothyroidism incidence, attributed to enhanced selenoprotein activities mitigating postpartum immunological rebound [80].

Shifting to heart health, carotenoids, responsible for the vibrant hues of fruits and vegetables, showcase potent antioxidant properties. Beyond adding color, they contribute to lowering heart disease and cancer risks, boosting the immune system, and safeguarding against age-related macular degeneration – a leading cause of irreversible blindness in adults [81].

Neurological effect -A study has demonstrated Vit A neuroprotective effects against cytotoxicity associated with amyloid fibrillation. Given its capacity to interact with the Aβ-42 peptide, vitamin A emerges as a potential candidate for the therapy of Alzheimer's and Parkinson's diseases, countering systemic amyloidosis [82]. Various studies have established a correlation between vitamin C deficiencies and neurodegenerative disorders, including Parkinson's, Alzheimer's, Huntington's diseases, and amyotrophic sclerosis. Notably, a clinical study indicated that the administration of vitamin C and/or E supplements resulted in a reduced risk of cognitive decline in individuals aged ≥65 years. Furthermore, proanthocyanidins, a class of natural flavonoids, demonstrated efficacy in alleviating rotenone-induced oxidative stress in human neuroblastoma SH-SY5Y dopaminergic cells, serving as a model for Parkinson's disease [83, 84]. Co-administration of astaxanthin and fucoxanthin exhibited neuroprotective effects on pheochromocytoma neuronal cells, presenting a potential treatment approach for Alzheimer's disease.

In cancer, oleuropein has been identified as a compound that not only hampers cell proliferation but also triggers apoptosis in breast cancer cells. Its antiproliferative impact extends to human hepatocellular carcinoma cells, breast cancer cells, various human tumor cells, and human colon carcinoma cells [85]. Additionally, pectin, extracted from Citrus peels, serves as a widely employed gelling agent in diverse food industrial processes prevent cancer. In MTT assays, the exposure to 6-OHDA resulted in a concentration-dependent reduction in cell viability in PC12 cells, corroborated by apoptosis confirmation Particularly, concentrations ranging from 50 μM to 400 μM revealed the enhanced protective potential of EGCG, leading to a substantial reduction in apoptotic cells and the prevention of apoptosis-related nuclear changes. Mechanism of action of different natural antioxidants are shown in Figure 1.4.

Toxicity of antioxidant: Antioxidants are substances that help protect the body from the damaging effects of free radicals, which are unstable molecules that can damage cells and contribute to the aging process and various diseases, including cancer and cardiovascular diseases [96]. While antioxidants are generally beneficial for health, like many compounds, they can have potential toxic effects if consumed in excessive amounts. Here are some aspects to consider regarding the potential toxicity of antioxidants:

1) Excessive Supplementation: Consuming high doses of antioxidant supplements, such as vit A, C, E, and beta-carotene, beyond the recommended daily allowances (RDAs), can lead to toxicity. Excessive intake of these supplements

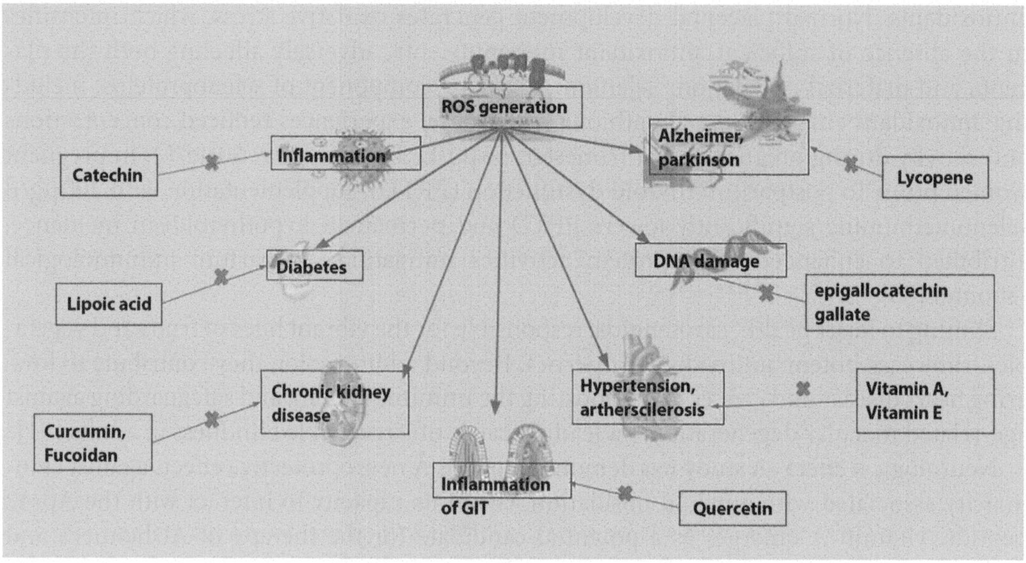

Figure 1.4 Mechanism of action of different natural antioxidants [89, 90].

may cause nausea, diarrhea, stomach cramps, and in severe cases, can lead to organ damage [97–99].

2) Vitamin A (Retinol) Toxicity: Vitamin A is an essential nutrient, but excessive intake, especially in the form of supplements, can lead to vitamin A toxicity [100]. This can cause dizziness, nausea, headaches, blurred vision, bone pain, and in severe cases, it can affect the liver and even be life-threatening.
3) Selenium Toxicity: Selenium is an essential trace element and antioxidant. However, excessive intake can lead to selenosis, causing symptoms like hair loss, nausea, diarrhea, skin rashes, and in severe cases, it can result in nerve damage and liver and kidney problems.
4) Iron and Copper Overload: Iron and copper are essential minerals with antioxidant properties. However, excessive levels of these minerals in the body can lead to toxicity [101]. Iron overload (hemochromatosis) can cause fatigue, joint pain, and eventually, damage to the liver, heart, and pancreas. Copper toxicity can lead to gastrointestinal symptoms and neurological issues.
5) Interference with Chemotherapy: High doses of antioxidants during cancer treatment, especially chemotherapy, may interfere with the treatment's effectiveness [102–104]. Some studies suggest that antioxidants can protect cancer cells from the damaging effects of chemotherapy.
6) Interaction with Medications: Antioxidants can interact with certain medications, affecting their efficacy or potentially causing harmful effects. For example, vitamin E can interact with blood thinners, and high doses of vitamin C may interfere with certain cancer treatments.
7) Potential Pro-Oxidant Effects: In certain conditions and at high concentrations, antioxidants can switch from being protective to potentially harmful by acting as pro-oxidants [105]. This paradoxical effect can lead to cellular damage rather than protection.

S. no.	Antioxidant	Impact on health	References
1	Anthocyanin 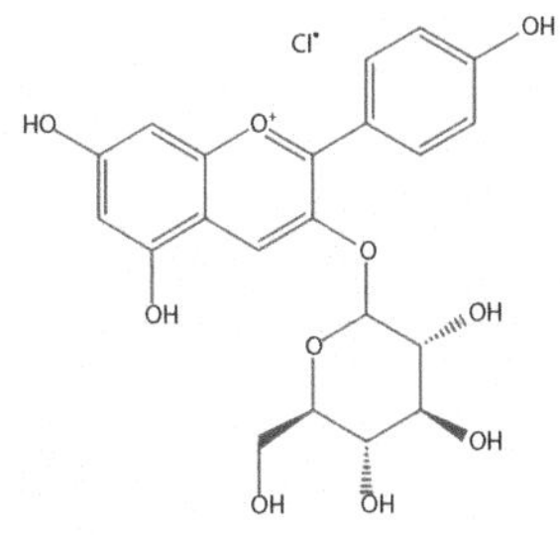 Anthocyanin	Dietary antioxidants have the potential to offer health benefits by aiding in the prevention of various diseases, including neuronal disorders, cardiovascular conditions, cancer, diabetes, inflammation, and more.	[86]
2	Carotenoids 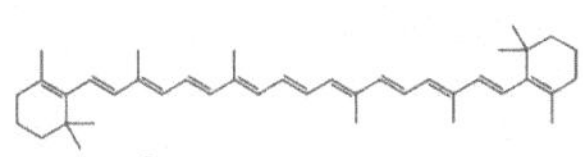 Beta carotene	Carotenoid intake may lower the risk of cancer, cardiovascular diseases, and other chronic conditions, and some also serve as provitamin A.	[87]
3	Lycopene 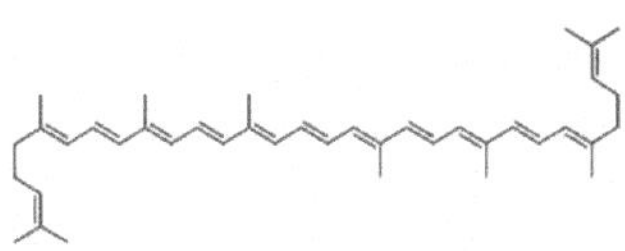 Lycopene	Lycopene administration improved cognition, reduced oxidative stress markers, increased MDA and 8-OHdG and prevented neuroinflammation and apoptosis.	[88]
4	Amurensin and cosmosiin 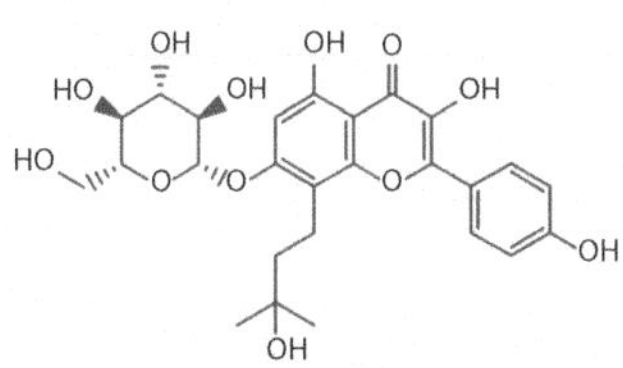 Amurensin 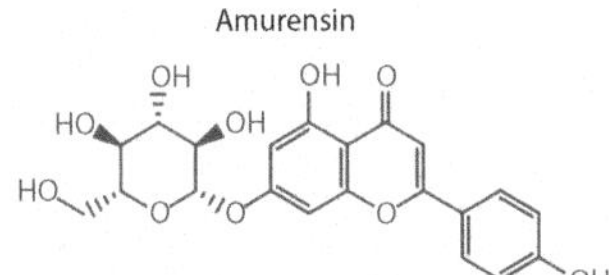 Cosmosin	Research findings indicate that amurensin and cosmosin, both flavonoids extracted from *Trigonella foenum*, display promising potential in hindering neurodegeneration induced by NaNO2 in the hippocampus and cortex regions of mice brains.	[89]
5	Tiliroside 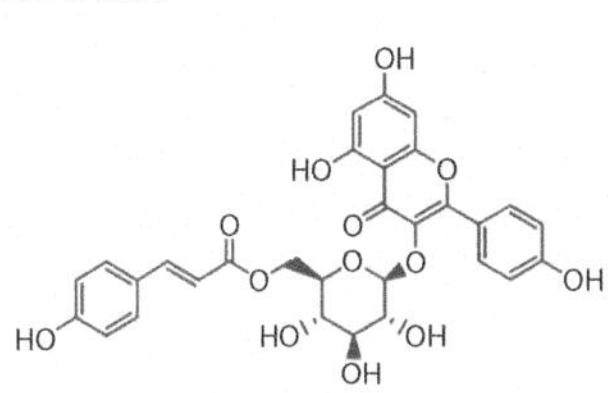 Tiliroside	Research has demonstrated the anti-neuroinflammatory potential of tiliroside, a glycosidic flavonoid, in increasing Nrf2, HO-1, and NQO1 protein levels, suggesting Nrf2 protective pathway activation in microglia.	[90]

(*Continued*)

(*Continued*)

S. no.	Antioxidant	Impact on health	References
6	vit E, α-lipoic acid and Vit. C 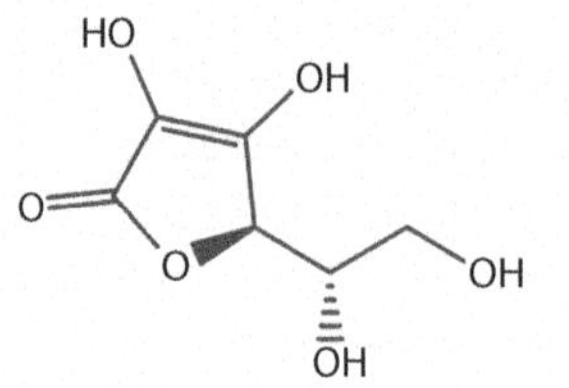 vitamin c 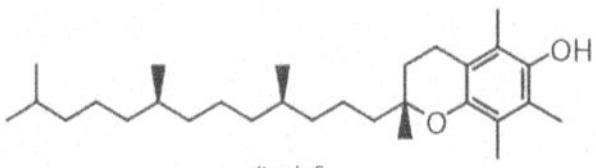 vitamin E	In a study these three antioxidant combination preserved arachidonic acid levels and positively impacted the D-6 desaturase system and unsaturated fatty acid levels in both diabetic and non-diabetic rats.	[91]
7	vitamin c 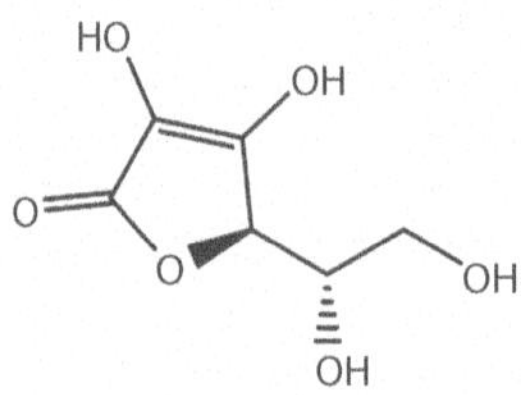 vitamin c	Vitamin C supplementation effectively reduces the accumulation of (sugar alcohol) sorbitol RBC's of diabetic patient, normalizes sorbitol levels in T1DM within 30 days, reduces capillary fragility, and improves glucose and lipid metabolism as well as glycemic control.	[92]
8	lipoic acid 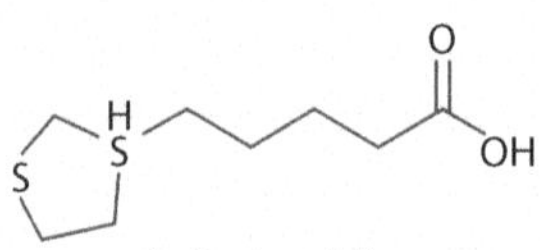 (a-lipoic acid)	Lipoic acid enhances glucose transport in muscle cells, safeguards the insulin receptor in adipocytes, and increases insulin-mediated glucose uptake in patients with T2DM, potentially improving insulin sensitivity.	[93]
9	Fucoidan 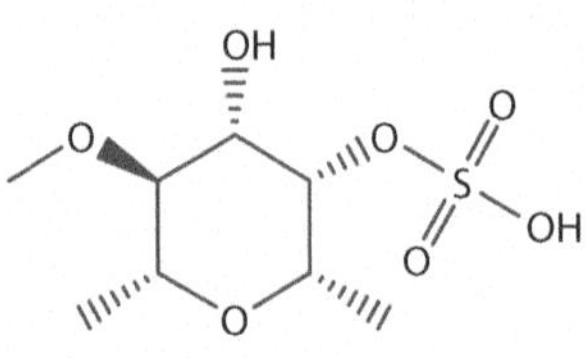 Fucoidan	Fucoidan inhibited H2O2-triggered PC12 cell apoptosis by scavenging ROS and enhancing SOD and GSH-Px activity.	[94]
10	Fucotriphlorethol A 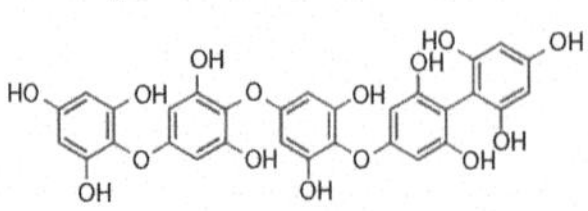 Fucotriphlorethol A	Fucotriphlorethol A, derived from the consumable brown alga *Fucus vesiculosus* L., demonstrates inhibition of cytochrome P450 1A, exhibiting IC50 values within the range of 17 to 33.5 μg/ml.	[95]

S. no.	Antioxidant	Side effect	References
1	Vit E	The study found that in some cases, high-dose antioxidant supplementation, particularly vitamin E, was associated with an increased risk of mortality. It suggested that excessive antioxidant intake might not always be beneficial and could potentially have adverse effects.	[106]
2	Vitamin C	The research found that excessive vitamin C intake was linked to an increased risk of developing kidney stones, highlighting the importance of moderation in antioxidant supplementation.	[107]
3	selenium	Furthermore, increased seminal plasma levels of selenium (≥80 ng/ml) were reported to be associated with decreased motility, asthenozoospermia and elevated abortion rates, whereas selenium levels between 40 and 70 ng/ml were reported to be optimal for re-productive performance.	[108]
4	Resveratrol	This study show that Gastrointestinal issues, interactions with medications.	[109]
5	Beta-Carotene	In individuals who smoked and were administered a daily supplementation of 20 mg β-carotene (Alpha-Tocopherol, Beta Carotene Cancer Prevention Study Group, 1994), or 30 mg β-carotene along with 25,000 IU retinyl palmitate, a noteworthy elevation in the likelihood of developing lung cancer was observed during an efficacy trial involving 29,133 and 18,314 subjects, respectively, in comparison to a placebo.	[110]

1.5 Conclusion

Antioxidants, the stalwart defenders against the ravages of oxidative stress, have cemented their significance in the realm of medicine. In the bustling city of our body's biological processes, where metabolic activities produce reactive oxygen species (ROS) and free radicals, antioxidants act as vigilant protectors. They neutralize these harmful molecules, thus safeguarding cellular health and contributing to the prevention of a multitude of diseases. Their importance extends across the medical spectrum, from thwarting the progression of chronic conditions like cancer, cardiovascular diseases, and neurodegenerative disorders to enhancing the body's natural defense mechanisms.

In the realm of cancer, antioxidants play an important role, with their ability to inhibit cell proliferation and induce apoptosis, potentially transforming the landscape of cancer treatment. However, the delicate balance between their benefits and potential toxicity must be maintained. Excessive supplementation can lead to adverse effects, including nausea, organ damage, and even life-threatening conditions. Vitamin A, selenium, iron, and copper are all susceptible to toxicity when consumed in excess. Furthermore, the interaction of antioxidants with chemotherapy and medications can influence their efficacy, and their pro-oxidant effects, especially at high concentrations, can paradoxically lead to cellular damage. Therefore, moderation in antioxidant consumption is key to reaping their benefits while mitigating potential risks. In a world where oxidative stress and the resulting diseases pose a significant threat to human health, antioxidants stand as formidable allies, offering protection, prevention, and the potential to revolutionize the landscape of modern medicine. Their significance in maintaining health and well-being underscores the need for continued research and responsible consumption, paving the way for a healthier, more resilient future.

References

1. Sies, H., Oxidative stress: oxidants and antioxidants. *Exp. Physiol. Transl. Integr.*, *82*, 2, 291–295, 1997.
2. Bjelakovic, G., Nikolova, D., Gluud, L.L., Simonetti, R.G., Gluud, C., Mortality in randomized trials of antioxidant supplements for primary and secondary prevention: systematic review and meta-analysis. *Jama*, *297*, 8, 842–857, 2007.
3. Pizzino, G., Irrera, N., Cucinotta, M., Pallio, G., Mannino, F., Arcoraci, V., Squadrito, F., Altavilla, D., Bitto, A., Oxidative stress: harms and benefits for human health. *Oxid. Med. Cell. Longevity*, 2017, 1–13, 2017.
4. Forman, H.J. and Zhang, H., Targeting oxidative stress in disease: Promise and limitations of antioxidant therapy. *Nat. Rev. Drug Discov.*, *20*, 9, 689–709, 2021.
5. Rudrapal, M., Khairnar, S.J., Khan, J., Dukhyil, A.B., Ansari, M.A., Alomary, M.N., Alshabrmi, F.M., Palai, S., Deb, P.K., Devi, R., Dietary polyphenols and their role in oxidative stress-induced human diseases: Insights into protective effects, antioxidant potentials and mechanism(s) of action. *Front. Pharmacol.*, *13*, 283, 2022.
6. Adwas, A.A., Elsayed, A., Azab, A.E., Quwaydir, F.A., Oxidative stress and antioxidant mechanisms in human body. *J. Appl. Biotechnol. Bioeng.*, *6*, 1, 43–47, 2019.
7. Costa, M., Losada-Barreiro, S., Paiva-Martins, F., Bravo-Díaz, C., Polyphenolic Antioxidants in Lipid Emulsions: Partitioning Effects and Interfacial Phenomena. *Foods*, 10, 539, 2021.
8. Pisoschi, A.M., Pop, A., Iordache, F., Stanca, L., Predoi, G., Serban, A.I., Oxidative stress mitigation by antioxidants-an overview on their chemistry and influences on health status. *Eur. J. Med. Chem.*, *209*, 112891, 2021.
9. Forrester, S.J., Kikuchi, D.S., Hernandes, M.S., Xu, Q., Griendling, K.K., Reactive oxygen species in metabolic and inflammatory signaling. *Circ. Res.*, *122*, 6, 877–902, 2018.
10. Singh, A., Kukreti, R., Saso, L., Kukreti, S., Oxidative stress: a key modulator in neurodegenerative diseases. *Molecules*, *24*, 8, 1583, 2019.
11. Murphy, M.P., Bayir, H., Belousov, V., Chang, C.J., Davies, K.J., Davies, M.J., Dick, T.P., Finkel, T., Forman, H.J., Janssen-Heininger, Y., Gems, D., Guidelines for measuring reactive oxygen species and oxidative damage in cells and *in vivo*. *Nat. Metab.*, *4*, 6, 651–662, 2022.
12. Losada-Barreiro, S., Sezgin-Bayindir, Z., Paiva-Martins, F., Bravo-Díaz, C., Biochemistry of antioxidants: Mechanisms and pharmaceutical applications. *Biomedicines*, *10*, 12, 3051, 2022.

13. Miller, G.J. and Quackenbush, F.W., A comparison of alkylated phenols as antioxidants for lard. *J. Am. Oil Chem. Soc.*, *34*, 5, 249–250, 1957.
14. Heim, K.E., Tagliaferro, A.R., Bobilya, D.J., Flavonoid antioxidants: chemistry, metabolism and structure-activity relationships. *J. Nutr. Biochem.*, *13*, 10, 572–584, 2002.
15. Heijnen, C.G., Haenen, G.R., Vekemans, J.A., Bast, A., Peroxynitrite scavenging of flavonoids: structure activity relationship. *Environ. Toxicol. Pharmacol.*, *10*, 4, 199–206, 2001.
16. Mirończuk-Chodakowska, I., Witkowska, A.M., Zujko, M.E., Endogenous non-enzymatic antioxidants in the human body. *Adv. Med. Sci.*, *63*, 1, 68–78, 2018.
17. Gulcin, İ. and Alwasel, S.H., Metal ions, metal chelators and metal chelating assay as antioxidant method. *Processes*, *10*, 1, 132, 2022.
18. Aycan, M., Nahar, L., Baslam, M., Mitsui, T., B-type response regulator hst1 controls salinity tolerance in rice by regulating transcription factors and antioxidant mechanisms. *Plant Physiol. Biochem.*, *196*, 542–555, 2023.
19. Kumar, S., Sharma, S., Vasudeva, N., Review on antioxidants and evaluation procedures. *Chin. J. Integr. Med.*, 2017, 1–12, 2017.
20. Helberg, J. and Pratt, D.A., Autoxidation vs. antioxidants–the fight for forever. *Chem. Soc. Rev.*, *50*, 13, 7343–7358, 2021.
21. Hancock, J.T., Desikan, R., Neill, S.J., Role of reactive oxygen species in cell signalling pathways. *Biochem. Soc. Trans.*, *29*, 2, 345–349, 2001.
22. Sharma, A., Mohapatra, H., Arora, K., Babbar, R., Arora, R., Arora, P., Kumar, P., Algın Yapar, E., Rani, K., Meenu, M. *et al.*, Bioactive Compound-Loaded Nanocarriers for Hair Growth Promotion: Current Status and Future Perspectives. *Plants*, *12*, 3739, 2023.
23. Sies, H., Oxidative stress: Eustress and distress in redox homeostasis, in: *Stress: physiology, biochemistry, and pathology*, pp. 153–163, Academic Press, Cambridge, Massachusetts, United States, 2019.
24. El-Bahr, S.M., Biochemistry of free radicals and oxidative stress. *Biochemistry*, *1*, 5567/5cjiŋt, 11–11, 2013.
25. Lane, N., *Oxygen. The Molecule That Made the World*, revised ed., Oxford University Press, Oxford, UK, 2016.
26. Winterbourn, C.C., The biological chemistry of hydrogen peroxide. *Methods Enzymol.*, *528*, 3–25, 2013.
27. Martemucci, G., Costagliola, C., Mariano, M., D'andrea, L., Napolitano, P., D'Alessandro, A.G., Free radical properties, source and targets, antioxidant consumption and health. *Oxygen*, *2*, 2, 48–78, 2022.
28. Choe, E. and Min, D.B., Mechanisms and factors for edible oil oxidation. *Compr. Rev. Food Sci. Food Saf.*, *5*, 169–186, 2006.
29. Parcheta, M., Świsłocka, R., Orzechowska, S., Akimowicz, M., Choińska, R., Lewandowski, W., Recent developments in effective antioxidants: The structure and antioxidant properties. *Materials*, *14*, 8, 1984, 2021.
30. Kohen, R. and Nyska, A., Invited review: oxidation of biological systems: oxidative stress phenomena, antioxidants, redox reactions, and methods for their quantification. *Toxicol. Pathol.*, *30*, 6, 620–650, 2002.
31. Habib, S. and Ali, A., Biochemistry of nitric oxide. *Indian J. Clin. Biochem.*, *26*, 3–17, 2011.
32. Blaise, G.A., Gauvin, D., Gangal, M., Authier, S., Nitric oxide, cell signaling and cell death. *Toxicology*, *208*, 2, 177–192, 2005.
33. Ray, R. and Shah, A.M., NADPH oxidase and endothelial cell function. *Clin. Sci.*, *109*, 3, 217–226, 2005.
34. Moncada, S.R.M.J., Palmer, R.M.L., Higgs, E., Nitric oxide: physiology, pathophysiology, and pharmacology. *Pharmacol. Rev.*, *43*, 2, 109–142, 1991.

35. Repetto, M., Semprine, J., Boveris, A., Lipid peroxidation: Chemical mechanism, biological implications and analytical determination, in: *Lipid Peroxidation*, D.A. Catala (Ed.), pp. 3–30, InTech, Rijeka, Croatia, 2012.
36. Alshehri, M.M., Quispe, C., Herrera-Bravo, J., Sharifi-Rad, J., Tutuncu, S., Aydar, E.F., Topkaya, C., Mertdinc, Z., Ozcelik, B., Aital, M., Kumar, N.V., A review of recent studies on the antioxidant and anti-infectious properties of Senna plants. *Oxid. Med. Cell. Longev.*, 2022, 6025900, 2022.
37. Elahi, M.M., Kong, Y.X., Matata, B.M., Oxidative stress as a mediator of cardiovascular disease. *Oxid. Med. Cell. Longev.*, *2*, 5, 259–269, 2009.
38. Taverne, Y.J., Bogers, A.J., Duncker, D.J., Merkus, D., Reactive oxygen species and the cardiovascular system. *Oxid. Med. Cell. Longev.*, 2013, 862423, 2013.
39. Sharifi-Rad, J., Sharifi-Rad, M., Salehi, B., Iriti, M., Roointan, A., Mnayer, D., Soltani-Nejad, A., Afshari, A., *In vitro* and *in vivo* assessment of free radical scavenging and antioxidant activities of Veronica persicaPoir. *Cell. Mol. Biol.*, *64*, 8, 57–64, 2018.
40. Singh, R.B., Mengi, S.A., Xu, Y.J., Arneja, A.S., Dhalla, N.S., Pathogenesis of atherosclerosis: A multifactorial process. *Exp. Clin. Cardiol.*, *7*, 1, 40, 2002.
41. Perrotta, I. and Aquila, S., The role of oxidative stress and autophagy in atherosclerosis. *Oxid. Med. Cell. Longev.*, 2015, 1–10, 2015.
42. Salehi, B., Martorell, M., Arbiser, J.L., Sureda, A., Martins, N., Maurya, P.K., Sharifi-Rad, M., Kumar, P., Sharifi-Rad, J., Antioxidants: positive or negative actors? *Biomolecules*, *8*, 4, 124, 2018.
43. Tsatsakis, A., Docea, A.O., Calina, D., Tsarouhas, K., Zamfira, L.M., Mitrut, R., Sharifi-Rad, J., Kovatsi, L., Siokas, V., Dardiotis, E., Drakoulis, N., A mechanistic and pathophysiological approach for stroke associated with drugs of abuse. *J. Clin. Med.*, *8*, 9, 1295, 2019.
44. Wang, X. and Michaelis, E.K., Selective neuronal vulnerability to oxidative stress in the brain. *Front. Aging Neurosci.*, *2*, 1224, 2010.
45. Delcambre, S., Nonnenmacher, Y., Hiller, K., Dopamine metabolism and reactive oxygen species production, in: *Mitochondrial mechanisms of degeneration and repair in Parkinson's disease*, pp. 25–47, 2016.
46. Docea, A.O., Gofita, E., Goumenou, M., Calina, D., Rogoveanu, O., Varut, M., Olaru, C., Kerasioti, E., Fountoucidou, P., Taitzoglou, I., Zlatian, O., Six months exposure to a real life mixture of 13 chemicals' below individual NOAELs induced non monotonic sex-dependent biochemical and redox status changes in rats. *Food Chem. Toxicol.*, *115*, 470–481, 2018.
47. Saha, S.K., Lee, S.B., Won, J., Choi, H.Y., Kim, K., Yang, G.M., AbdalDayem, A., Cho, S.G., Correlation between oxidative stress, nutrition, and cancer initiation. *Int. J. Mol. Sci.*, *18*, 7, 1544, 2017.
48. Mishra, A.P., Salehi, B., Sharifi-Rad, M., Pezzani, R., Kobarfard, F., Sharifi-Rad, J., Nigam, M., Programmed cell death, from a cancer perspective: an overview. *Mol. Diagn. Ther.*, *22*, 281–295, 2018.
49. Smith, M.T., Guyton, K.Z., Gibbons, C.F., Fritz, J.M., Portier, C.J., Rusyn, I., DeMarini, D.M., Caldwell, J.C., Kavlock, R.J., Lambert, P.F., Hecht, S.S., Key characteristics of carcinogens as a basis for organizing data on mechanisms of carcinogenesis. *Environ. Health Perspect.*, *124*, 6, 713–721, 2016.
50. Glasauer, A. and Chandel, N.S., Targeting antioxidants for cancer therapy. *Biochem. Pharmacol.*, *92*, 1, 90–101, 2014.
51. Li, W., Cao, L., Han, L., Xu, Q., Ma, Q., Superoxide dismutase promotes the epithelial-mesenchymal transition of pancreatic cancer cells via activation of the H2O2/ERK/NF-κB axis. *Int. J. Oncol.*, *46*, 6, 2613–2620, 2015.

52. Andersson, K.E., Oxidative stress and its possible relation to lower urinary tract functional pathology. *BJU Int.*, 121, 4, 527–533, 2018.
53. Liou, G.Y., Döppler, H., DelGiorno, K.E., Zhang, L., Leitges, M., Crawford, H.C., Murphy, M.P., Storz, P., Mutant KRas-induced mitochondrial oxidative stress in acinar cells upregulates EGFR signaling to drive formation of pancreatic precancerous lesions. *Cell Rep.*, *14*, 10, 2325–2336, 2016.
54. Pizzino, G., Irrera, N., Cucinotta, M., Pallio, G., Mannino, F., Arcoraci, V., Squadrito, F., Altavilla, D., Bitto, A., Oxidative stress: harms and benefits for human health. *Oxid. Med. Cell. Longev.*, *2017*, 1–13, 2017.
55. Chen, X.F., Wang, L., Wu, Y.Z., Song, S.Y., Min, H.Y., Yang, Y., He, X., Liang, Q., Yi, L., Wang, Y., Gao, Q., Effect of puerarin in promoting fatty acid oxidation by increasing mitochondrial oxidative capacity and biogenesis in skeletal muscle in diabetic rats. *Nutr. Diabetes*, *8*, 1, 1, 2018.
56. Aroor, A.R., Mandavia, C., Ren, J., Sowers, J.R., Pulakat, L., Mitochondria and oxidative stress in the cardiorenal metabolic syndrome. *Cardiorenal Med.*, *2*, 2, 87–109, 2012.
57. Zhou, H., Wang, S., Zhu, P., Hu, S., Chen, Y., Ren, J., Empagliflozin rescues diabetic myocardial microvascular injury via AMPK-mediated inhibition of mitochondrial fission. *Redox Biol.*, *15*, 335–346, 2018.
58. Pei, Z., Deng, Q., Babcock, S.A., He, E.Y., Ren, J., Zhang, Y., Inhibition of advanced glycationendproduct (AGE) rescues against streptozotocin-induced diabetic cardiomyopathy: role of autophagy and ER stress. *Toxicol. Lett.*, *284*, 10–20, 2018.
59. Bansal, S., Choudhary, S., Sharma, M., Kumar, S.S., Lohan, S., Bhardwaj, V., Syan, N., Jyoti, S., Tea: a native source of antimicrobial agents. *Food Res. Int.*, *53*, 2, 568–584, 2013.
60. Dimitrios, B., Sources of natural phenolic antioxidants. *Trends Food Sci. Technol.*, *17*, 9, 505–512, 2006.
61. Jiang, J. and Xiong, Y.L., Natural antioxidants as food and feed additives to promote health benefits and quality of meat products: A review. *Meat Sci.*, *120*, 107–117, 2016.
62. Sikora, E., Cieślik, E., Topolska, K., The sources of natural antioxidants. *Acta Sci. Pol. Technol. Aliment.*, *7*, 1, 5–17, 2008.
63. Freitas, A., Moldão-Martins, M., Costa, H.S., Albuquerque, T.G., Valente, A., Sanches-Silva, A., Effect of UV-C radiation on bioactive compounds of pineapple (Ananascomosus L. Merr.) by-products. *J. Sci. Food Agric.*, *95*, 1, 44–52, 2015.
64. Fierascu, R.C., Ortan, A., Fierascu, I.C., Fierascu, I., *In vitro* and *in vivo* evaluation of antioxidant properties of wild-growing plants. A short review. *Curr. Opin. Food Sci.*, *24*, 1–8, 2018.
65. Halvorsen, B.L., Holte, K., Myhrstad, M.C., Barikmo, I., Hvattum, E., Remberg, S.F., Wold, A.B., Haffner, K., Baugerød, H., Andersen, L.F., Moskaug, Ø., A systematic screening of total antioxidants in dietary plants. *J. Nutr.*, *132*, 3, 461–471, 2002.
66. Peterson, D.M., Emmons, C.L., Hibbs, A.H., Phenolic antioxidants and antioxidant activity in pearling fractions of oat groats. *J. Cereal Sci.*, *33*, 1, 97–103, 2001.
67. Tena, N., Martín, J., Asuero, A.G., State of the art of anthocyanins: Antioxidant activity, sources, bioavailability, and therapeutic effect in human health. *Antioxidants*, *9*, 5, 451, 2020.
68. Andrew, M. and Jayaraman, G., Structural features of microbial exopolysaccharides in relation to their antioxidant activity. *Carbohydr. Res.*, *487*, 107881, 2020.
69. Cao, G., Sofic, E., Prior, R.L., Antioxidant capacity of tea and common vegetables. *J. Agric. Food. Chem.*, *44*, 11, 3426–3431, 1996.
70. Dreosti, I.E., Antioxidant polyphenols in tea, cocoa, and wine. *Nutrition*, *7*, 16, 692–694, 2000.
71. Fik, M. and Zawislak, A., Porównaniewłaściwościprzeciwutleniającychwybranychherbat. *Zywn. Nauk. Technol. Ja.*, *11*, 3, 98–105, 2004.

72. Waszkiewicz-Robak, B., Rusaczonek, A., Świderski, F., Characteristic of antioxidant properties of leaf teas. *Ann. Univ. Mariae Curie-Skłodowska. Sect. D*, *60*, 16, 602, 2005.
73. Budryn, G. and Nebesny, E., Strukturaiwlasciwosciantyoksydacyjnepolifenoliziarnakawowego. *Bromatol. Chem. Toksyk.*, *3*, 38, 203–209, 2005.
74. Bonvehi, J.S. and Coll, F.V., Evaluation of bitterness and astringency of polyphenolic compounds in cocoa powder. *Food Chem.*, *60*, 3, 365–370, 1997.
75. Gajewska, D. and Myszkowska-Ryciak, J., Slodkismakolyk, czylekarstwo? *Przegląd Gastronomiczny*, *60*, 05, 26–27, 2006.
76. Banjarnahor, S.D. and Artanti, N., Antioxidant properties of flavonoids. *Med. J. Indones.*, *23*, 4, 239–44, 2014.
77. Saija, A., Scalese, M., Lanza, M., Marzullo, D., Bonina, F., Castelli, F., Flavonoids as antioxidant agents: importance of their interaction with biomembranes. *Free Radical Biol. Med.*, *19*, 4, 481–486, 1995.
78. Agati, G., Azzarello, E., Pollastri, S., Tattini, M., Flavonoids as antioxidants in plants: location and functional significance. *Plant Sci.*, *196*, 67–76, 2012.
79. Zhong, Y., Chiou, Y.S., Pan, M.H., Ho, C.T., Shahidi, F., Protective effects of epigallocatechin-gallate (EGCG) derivatives on azoxymethane-induced colonic carcinogenesis in mice. *J. Funct. Foods*, *4*, 1, 323–330, 2012.
80. Hofstee, P., James-McAlpine, J., McKeating, D.R., Vanderlelie, J.J., Cuffe, J.S., Perkins, A.V., Low serum selenium in pregnancy is associated with reduced T3 and increased risk of GDM. *J. Endocrinol.*, *248*, 1, 45–57, 2021.
81. Voutilainen, S., Nurmi, T., Mursu, J., Rissanen, T.H., Carotenoids and cardiovascular health. *Am. J. Clin. Nutr.*, *83*, 6, 1265–1271, 2006.
82. Kunzler, A., Kolling, E.A., da Silva-Jr, J.D., Gasparotto, J., de BittencourtPasquali, M.A., Moreira, J.C.F., Gelain, D.P., Retinol (vitamin A) increases α-synuclein, β-amyloid peptide, tau phosphorylation and RAGE content in human SH-SY5Y neuronal cell line. *Neurochem. Res.*, *42*, 2788–2797, 2017.
83. Boothby, L.A. and Doering, P.L., Vitamin C and vitamin E for Alzheimer's disease. *Ann. Pharmacother.*, 39, 12, 2073–2080, 2005.
84. Basambombo, L.L., Carmichael, P.H., Côté, S., Laurin, D., Use of vitamin E and C supplements for the prevention of cognitive decline. *Ann. Pharmacother.*, *51*, 2, 118–124, 2017.
85. Zehiroglu, C. and OzturkSarikaya, S.B., The importance of antioxidants and place in today's scientific and technological studies. *J. Food Sci. Technol.*, *56*, 4757–4774, 2019.
86. Speer, H., D'Cunha, N.M., Alexopoulos, N.I., McKune, A.J., Naumovski, N., Anthocyanins and human health—A focus on oxidative stress, inflammation and disease. *Antioxidants*, *9*, 5, 366, 2020.
87. Linnewiel-Hermoni, K., Paran, E., Wolak, T., Carotenoid supplements and consumption: Implications for healthy aging, in: *Molecular basis of nutrition and aging*, pp. 473–489, Academic Press, Cambridge, Massachusetts, United States, 2016.
88. Cao, Z., Wang, P., Gao, X., Shao, B., Zhao, S., Li, Y., Lycopene attenuates aluminum-induced hippocampal lesions by inhibiting oxidative stress-mediated inflammation and apoptosis in the rat. *J. Inorg. Biochem.*, *193*, 143–151, 2019.
89. Ahmed, S.I., Hayat, M.Q., Zahid, S., Tahir, M., Mansoor, Q., Ismail, M., Keck, K., Bates, R., Isolation and identification of flavonoids from anticancer and neuroprotective extracts of Trigonellafoenumgraecum. *Trop. J. Pharm. Res.*, *16*, 6, 1391–1398, 2017.
90. Velagapudi, R., El-Bakoush, A., Olajide, O.A., Activation of Nrf2 pathway contributes to neuroprotection by the dietary flavonoid tiliroside. *Mol. Neurobiol.*, *55*, 8103–8123, 2018.
91. Özkan, Y., Yilmaz, Ö., Öztürk, A.İ., Erşan, Y., Effects of triple antioxidant combination (vitamin E, vitamin C and α-lipoic acid) with insulin on lipid and cholesterol levels and fatty acid

composition of brain tissue in experimental diabetic and non-diabetic rats. *Cell Biol. Int.*, *29*, 9, 754–760, 2005.
92. Asghari, A., Fazilati, M., Latifi, A.M., Salavati, H., Choopani, A., A review on antioxidant properties of Spirulina. *J. Appl. Biotechnol. Rep.*, *3*, 1, 345–351, 2016.
93. Jacob, S., Ruus, P., Hermann, R., Tritschler, H.J., Maerker, E., Renn, W., Augustin, H.J., Dietze, G.J., Rett, K., Oral administration of RAC-α-lipoic acid modulates insulin sensitivity in patients with type-2 diabetes mellitus: a placebo-controlled pilot trial. *Free Radical Biol. Med.*, *27*, 3–4, 309–314, 1999.
94. Gao, Y., Dong, C., Yin, J., Shen, J., Tian, J., Li, C., Neuroprotective effect of fucoidan on H 2 O 2-induced apoptosis in PC12 cells via activation of PI3K/Akt pathway. *Cell. Mol. Neurobiol.*, *32*, 523–529, 2012.
95. Parys, S., Kehraus, S., Krick, A., Glombitza, K.W., Carmeli, S., Klimo, K., Gerhäuser, C., König, G.M., *In vitro* chemopreventive potential of fucophlorethols from the brown alga Fucusvesiculosus L. by anti-oxidant activity and inhibition of selected cytochrome P450 enzymes. *Phytochemistry*, *71*, 2–3, 221–229, 2010.
96. Bjelakovic, G., Nikolova, D., Gluud, C., Antioxidant supplements and mortality. *Curr. Opin. Clin. Nutr. Metab. Care*, *17*, 1, 40–44, 2014.
97. Bjelakovic, G., Gluud, L.L., Nikolova, D., Whitfield, K., Wetterslev, J., Simonetti, R.G., Bjelakovic, M., Gluud, C., Vitamin D supplementation for prevention of mortality in adults. *Cochrane Database Syst. Rev.*, 1, CD007470, 2014.
98. Hathcock, J.N., Hattan, D.G., Jenkins, M.Y., McDonald, J.T., Sundaresan, P.R., Wilkening, V.L., Evaluation of vitamin A toxicity. *Am. J. Clin. Nutr.*, *52*, 2, 183–202, 1990.
99. Trumbo, P., Yates, A.A., Schlicker, S., Poos, M., Dietary reference intakes. *J. Am. Diet. Assoc.*, *101*, 3, 294–294, 2001.
100. Rayman, M.P., Selenium and human health. *Lancet*, *379*, 9822, 1256–1268, 2012.
101. Sharma, A., Khanna, S., Kaur, G., Singh, I., Medicinal plants and their components for wound healing applications. *Future J. Pharm. Sci.*, *7*, 1, 1–13, 2021.
102. Brewer, G.J., Copper toxicity in the general population. *Clin. Neurophysiol.*, *4*, 121, 459–460, 2010.
103. Rathi, R., Kaur, S., Singh, I., A review on co-crystals of Herbal Bioactives for solubility enhancement: preparation methods and characterization techniques. *Cryst. Growth Des.*, *22*, 3, 2023–2042, 2022.
104. Lawenda, B.D., Kelly, K.M., Ladas, E.J., Sagar, S.M., Vickers, A., Blumberg, J.B., Should supplemental antioxidant administration be avoided during chemotherapy and radiation therapy? *J. Natl. Cancer Inst.*, *100*, 11, 773–783, 2008.
105. Gulcin, İ., Antioxidants and antioxidant methods: An updated overview. *Arch. Toxicol.*, *94*, 3, 651–715, 2020.
106. Bjelakovic, G., Nikolova, D., Gluud, L.L., Simonetti, R.G., Gluud, C., Antioxidant supplements for prevention of mortality in healthy participants and patients with various diseases. *Sao Paulo Med. J.*, *133*, 164–165, 2015.
107. Thomas, L.D., Elinder, C.G., Tiselius, H.G., Wolk, A., Åkesson, A., Ascorbic acid supplements and kidney stone incidence among men: a prospective study. *JAMA Intern. Med.*, *173*, 5, 386–388, 2013.
108. Bleau, G., Lemarbre, J., Faucher, G., Roberts, K.D., Chapdelaine, A., Semen selenium and human fertility. *Fertil. Steril.*, *42*, 6, 890–894, 1984.
109. Semba, R.D., Ferrucci, L., Bartali, B., Urpí-Sarda, M., Zamora-Ros, R., Sun, K., Cherubini, A., Bandinelli, S., Andres-Lacueva, C., Resveratrol levels and all-cause mortality in older community-dwelling adults. *JAMA Intern. Med.*, *174*, 7, 1077–1084, 2014.

110. Omenn, G.S., Goodman, G.E., Thornquist, M.D., Balmes, J., Cullen, M.R., Glass, A., Keogh, J.P., Meyskens Jr., F.L., Valanis, B., Williams Jr., J.H., Barnhart, S., Effects of a combination of beta carotene and vitamin A on lung cancer and cardiovascular disease. *N. Engl. J. Med.*, *334*, 18, 1150–1155, 1996.

2

Antioxidants in Cancer Prevention

Asad Ur Rahman and Pharkphoom Panichayupakaranant*

Department of Pharmacognosy and Pharmaceutical Botany, Faculty of Pharmaceutical Sciences, Prince of Songkla University, Hat-Yai, Songkhla, Thailand

Abstract
Cancer is still the leading cause of mortality worldwide. Treatment of cancer involves chemotherapy, surgery, and radiotherapy. However, anticancer drugs are associated with a plethora of side effects. Therefore, cancer prevention using natural products has become a hot global issue in the healthcare system. Recently, there have been some evidences that foods containing high levels of antioxidants, such as vegetables, fruits, and spices, are enriched with antioxidant and anticancer compounds such as phenolic compounds, flavonoids, curcuminoids, and carotenoids, which includes phloretin in apples, lycopene in tomatoes and watermelon, and curcumin in turmeric. Fruits and vegetables are also abundant in antioxidant vitamins and minerals. Fruits, vegetables, and spices in our diet can reduce oxidative stress by increasing antioxidant enzymes and reducing the formation of free radicals, resulting in reducing the chances of cancer. Various clinical trials have evidenced the antioxidant and anticancer effects of these foods. However, some studies have shown that taking supplements to reduce the risk of cancer may not be effective because the specific components that provide the preventive effect are not clearly known. This chapter summarizes relevant findings on the important roles of antioxidants in cancer prevention in various aspects as well as recommended sources of promising antioxidants. People who eat recommended levels of fruits and vegetables have been shown to have a lower risk of cancer.

***Keywords*:** Fruits, vegetables, spices, antioxidant, free radicals, anticancer

2.1 Introduction

Cancer remains a complex and challenging global health issue [1]. Cancer cases worldwide reached 19 million in 2020, with 10 million deaths. Lung, breast, colorectal, prostate, and stomach cancers are the most common and cause the most deaths. Alarming estimates predict 28 million new cancer cases and 16 million cancer deaths by 2040 [2–4]. The human body's normal metabolic processes inherently produce free radicals, chemically reactive molecules that can pose a threat to cellular structures and functions. The body employs a strong defense system, aided by antioxidants, to quickly neutralize and eliminate reactive oxygen species (ROS) and reactive nitrogen species (RNS) in order to prevent this potential

**Corresponding author*: pharkphoom.p@psu.ac.th

Rakesh K. Sindhu, Inderbir Singh and M. Arockia Babu (eds.) Antioxidants: Nature's Defense Against Disease, (21–82)

harm. In cases where the equilibrium between free radicals and antioxidants is disrupted, a condition known as oxidative stress arises, leading to cellular damage, inflammation, and, eventually, the initiation and progression of cancer. Therefore, the generation of excessive free radicals during metabolic activities can contribute to the development of cancer [5–7]. Current therapeutic modalities for cancer treatment predominantly involve chemotherapy, surgery, and radiation, which aim to selectively target cancerous cells while sparing healthy ones. Despite their efficacy, these treatment strategies often result in adverse effects on normal tissues and may lead to non-compliance due to the severity of side effects [8]. Financially, in 2019, cancer treatment cost US$1.2 trillion, approximately 2% of the global gross domestic product (GDP) [9].

Antioxidants are agents that safeguard cells and tissues against harm induced by free radicals such as ROS. Free radicals are notably reactive molecules possessing unpaired electrons and arise as byproducts of regular cellular metabolism or in response to exogenous influences such as environmental contaminants, radiation, or toxic substances. Oxidative stress, initiated by free radicals, can elicit cellular impairment, DNA mutations, and disturbances in cellular processes. Oxidative damage can cause cancer, neurodegenerative disorders, cardiovascular diseases, and aging [10]. Antioxidants from food or supplements can help the body fight against oxidative stress caused by elevated ROS and RNS. Antioxidants scavenge free radicals, improving immune system function and lowering the risk of cancer and other diseases [11].

Amidst these challenges, emerging evidence suggests that dietary intervention with antioxidant-rich fruits, vegetables, and spices offers a potential and accessible approach to mitigate cancer risks. Numerous scientific studies have documented the beneficial effects of such antioxidant-rich foods in chemoprevention, implying their potential to prevent or slow down the development of cancer. For instance, fruits such as apples, berries, bananas, grapes, lemons, oranges, strawberries etc., renowned for their vibrant colors and diverse nutritional profiles, contain an array of bioactive antioxidant compounds, including vitamin C, E, carotenoids, flavonoids, and phenolic compounds [12]. Pomegranates, with their distinctive taste and appearance, also possess phytochemicals with antioxidant and cancer-preventive properties [13]. Vegetables, particularly tomatoes, offer β-carotene, vitamin C, rutin, tocopherols, and the well-studied carotenoid, lycopene. Cruciferous vegetables, like cabbage, are rich in polyphenols and flavonoids, which have demonstrated anti-cancer properties in experimental studies. Additionally, an array of spices, including clove, ginger, onion, and chili, boast impressive antioxidant and anticancer capabilities, attributed to their flavonoids, gallic acid, phenolic acids, and other biologically active compounds [14–17].

A diet rich in antioxidants from fruits, vegetables, and spices may increase systemic antioxidant levels, helping to neutralize excess free radicals, thus, reducing the risk of cancer. These antioxidant-rich foods have been extensively studied and reported for their impressive antioxidant and chemopreventive properties [18].

This comprehensive review will shed light on antioxidants' multifaceted role in cancer prevention. It will illuminate the complex relationship between free radicals and cancer development, highlight antioxidant-rich foods such as common fruits, vegetables, and spices and their important antioxidant and anticancer compounds responsible for cancer prevention and treatment, and examine their pharmacokinetics, quality control, and toxicity assessment studies. This review provides a holistic view of antioxidants' benefits and drawbacks, enabling informed and evidence-based knowledge for the reader about the role of antioxidants in cancer prevention.

2.2 Free Radicals

Oxygen helps humans' bodies to metabolize carbohydrates, proteins, and fats for energy. However, oxygen utilization in metabolic processes produces free radicals (ROS and RNS), which in excess of their amount, are responsible for the development of illness [19, 20].

Free radicals can exist independently in the body because they have unpaired electrons in their outer orbit, or valence shell. Their odd electron numbers make free radicals unstable, reactive, and short-lived, enabling them to steal electrons from other compounds to stabilize them, making the attacked compounds unstable and generating free radicals. A chain of such reactions causes excessive free radicals, which damage cells and tissues. ROS and RNS include different types of free radicals and non-radicals mentioned in Table 2.1. These species aid the immune system, cellular mechanisms, redox regulation, and mitogenic responses. These benefits of ROS and RNS only occur at low or moderate levels. However, high concentrations of free radicals can cause nitrosative and oxidative stress, damaging biomolecules such as proteins, lipids, and DNA. ROS and RNS can build up if there aren't enough antioxidant enzymes

Table 2.1 List of free radicals.

Reactive oxygen species (ROS)	**Symbol**	**Half life**
Radicals		
Alkoxyl radical	$RO^{\bullet}$	10^{-6} s
Hydroperoxyl	$HO_2^{\bullet}$	s
Hydroxyl	$^{\bullet}OH$	10^{-10} s
Organic hydroperoxide	ROOH	Stable
Peroxyl	$ROO^{\bullet}$	17 s
Superoxide	$O^{\bullet -}$	10^{-6} s
Non-radicals		
Hydrogen peroxide	H_2O_2	Stable
Hypochlorous acid	HOCl	Stable (min)
Hypobromous acid	HOBr	Stable (min)
Organic peroxide	ROOH	Stable
Ozone	O_3	Stable
Singlet oxygen	$^1O_2$	10^{-6} s

(Continued)

Table 2.1 List of free radicals. (*Continued*)

Reactive oxygen species (ROS)	**Symbol**	**Half life**
Radicals		
Nitrogen dioxide	$NO_2^{\bullet}$	s
Nitric oxide	$NO^{\bullet}$	s
Nitrate radical	$NO_3^{\bullet}$	s
Non-radicals		
Dinitrogen trioxide	N_2O_3	s
Dinitrogen tetroxide	N_2O_4	s
Nitrous acid	HNO_2	s
Nitrosonium cation	NO^+	s
Nitroxyl anion	NO^-	s
Peroxynitrite	$ONOO^-$	10^{-3} s
Peroxynitrous acid	ONOOH	Fair stable
Reactive Sulphur species (RSS)	Symbol	Half life
Radicals		
Thiyl radical	$S^{\bullet}$	
Non-radicals		
Disulfide	N/A	N/A
Disulfide-S-monooxide	N/A	N/A
Disulfide-S-dioxide	N/A	N/A
Hydrogen sulfide	N/A	N/A
Sulfenic acid	N/A	N/A
Thiol	N/A	N/A

min = minute; s = second; N/A = not applicable.

like glutathione peroxidase (GPX), catalase (CAT), and superoxide dismutase (SOD). This can lead to degenerative diseases, inflammation, and cancer [21–23].

2.2.1 Types of Free Radicals

The classification of free radicals comprises three primary types: ROS, RNS, and Reactive Sulfur Species (RSS). These radicals are recognized as pro-oxidants and, in high concentration, play a pivotal role in inducing oxidative stress. Both radicals and non-radicals are encompassed within these three categories. Radicals are extremely reactive because they have at least one unpaired electron in their outer shell, and some are biradicals, which have two unpaired electrons. On the other hand, non-radicals do not inherently possess free radical characteristics but can undergo reactions that transform them into free radicals, thus instigating cellular damage. Among these, ROS species are particularly known for their high reactivity [24–26]. Table 2.1 illustrates the ROS, RNS, and RSS, along with their respective radical and non-radical forms.

2.2.2 Sources of Free Radicals

Free radicals can arise from both exogenous (external) and endogenous (internal) sources. Chemicals, radiation, environmental pollutants, and certain dietary factors can increase free radicals in the body when exposed to them. Arsenic, cadmium, mercury, lead, copper, iron, ozone, peroxynitrite, ketamine, dinitrogen tetroxide, hydrogen peroxide (H_2O_2), hypochlorous and hypobromous acids, deoxyosones, gamma-rays, X-rays, UV-A radiation, pesticides, aromatic hydrocarbons, dioxins, polychlorinated biphenyls, microbes, smoked meats, used oil, and drug metabolites are exogenous sources that contribute to the generation of free radicals in the body. Endogenous free radicals come from normal cellular metabolism and enzymatic processes. During metabolic pathways, mitochondria, peroxisomes, and the endoplasmic reticulum generate free radicals [27–31].

The cells' powerhouses, the mitochondria, consume 85% of the body's oxygen and produce 90% of its ATP through oxidative phosphorylation. Mitochondrial electron transfer chain reactions generate 90% of ROS during energy production. Due to ROS overproduction and oxidative stress, mitochondrial dysfunction can occur, which can cause chronic diseases [32, 33]. In fatty acid beta-oxidation, peroxisomes produce free radicals, particularly H_2O_2. Peroxisomal enzymes, such as D-aspartate, xanthine, urate, and D-amino acid oxidases, are used in metabolic processes that make free radicals. Beta-oxidation and peroxisomal dysfunction can cause inflammation and cancer. Endoplasmic reticulum, cytochrome P450, diamine oxidase, and B5 enzymes produce ROS. It metabolizes foreign chemicals and drugs, producing free radicals such as H_2O_2 [34–36].

Apart from ROS and RNS, mental stress, over-exertion, infection, inflammation, adrenaline auto-oxidation, and riboflavin reduction are additional endogenous factors that contribute to the generation of free radicals, which attack biomolecules such as DNA, RNA, and lipids, causing various pathological conditions [37–39].

2.2.3 Free Radical Roles in Cancer Development

An excess of free radicals in the body is one of the main causes of cancer. High free radical levels alter cell division, particularly DNA. DNA mutations can cause tumors. Cancer cells also suffer from ROS-induced mitochondrial impairment. ROS increases oxidative stress, weakens tumor suppressor cells, and activates oncogenes, promoting cancer growth. Cancer cells proliferate due to elevated ROS levels, which alter gene expression and growth signals. ROS also methylates DNA, causing chromosomal rearrangements and DNA strand breakages that lead to abnormal cell division and cancer [23, 40–42].

ROS damage causes breast epithelium hyperplasia and fibroblast proliferation in breast cancer. Thymidine phosphorylase changes thymidine into thymine and a powerful reducing sugar called 2-deoxy-D-ribose-1-phosphate. Oxidative stress overexpresses this enzyme, which produces more reducing sugars that metabolize more proteins and generate more free radicals. This vicious cycle promotes breast cell carcinoma. The mammary gland-specific lactoperoxidase enzyme also oxidizes 17-beta-estradiol one electron, producing a reactive phenoxyl radical. This process increases ROS levels, oxidizing the mammary gland and causing breast cancer [43–45].

Bladder cancer is especially susceptible to free radicals. A high-arsenic diet, smoking, and industrial waste exposure increase this susceptibility. These factors increase free radical exposure, which increases oxidative stress in the body. Nitric oxide (NO) production rises due to oxidative stress. Prolidase activity is increased by elevated NO levels. Increased prolidase activity can harm collagen, which is a structural component of bladder tissue. High serum prolidase enzyme levels reduce protein-bound thiol groups and increased protein carbonyl groups. In this way, the body's antioxidant and oxidant levels are imbalanced, with free radicals outnumbering antioxidants, which can contribute to the development of bladder cancer [46–51].

In colon cancer, the rectum and colon are highly exposed to exogenous and endogenous free radicals. In most cases, colon cancer begins in the bowel's epithelial cells. The region's epithelial cells have a high metabolic rate, thus, making them susceptible to DNA damage due to excessive free radicals. ROS produces additional NO and serum-oxidized low-density proteins, causing more oxidative stress in the colon and rectum, leading to mutations in DNA resulting in the abnormal cell division that leads to colon cancer [52–57].

Smoking produces many stable carcinogens and ROS that oxidizes DNA, increasing lung cell mutations, and promotes abnormal cell division and cancer [58, 59]. Furthermore, ROS and nicotinamide adenine dinucleotide phosphate (NADPH) oxidase levels make superoxide free radicals, which promote a number of cancer-causing processes, such as cancer cell growth, resistance to programmed cell death (apoptosis), and the body's natural ability to get rid of cancer cells. Superoxide also causes cell immortality and abnormal growth, which contributes to prostate cancer. Furthermore, oxidative stress due to elevated ROS damages DNA and other cells, causing genetic mutations and abnormal cell behavior. All these major factors are responsible for the development of prostate cancer [60–64].

2.3 Antioxidants

2.3.1 Antioxidants from Fruits

Fruits are a rich source of antioxidants, exerting various anticancer and antioxidant effects. Fruits such as apple, banana, grapefruit, persimmon, lemon, melon, orange, pepper, peach, plum, strawberry, tangerine, tomato, and watermelon are rich in phenolics, flavonoids, and carotenoids, which have been reported for their antioxidant properties [65]. The total flavonoid, phenolic, and carotenoid content is mentioned in Table 2.2.

Table 2.2 Total phenolic, flavonoid, and carotenoid content in fruits.

Fruits	Total phenolic content (TPC) mg GAE equivalent/g	Total flavonoid content (TFC) mg QE equivalent/100 g	Total carotenoid content (TCC) mg/100 g
Apple	31.9	37.8	3.50
Banana	0.38	46.3	0.02
Grapefruit	77.3	10.6	0.09
Kaki	36.4	1.90	0.88
Lemon	49.8	888	0.01
Melon	1.83	15.1	0.10
Orange	35.6	335	0.25
Pepper	6.43	10.04	1.98
Peach	0.41	17.6	0.02
Plum	0.76	37.6	0.17
Strawberry	2.72	14.6	31
Tangerine	0.40	420	0.44
Tomato	10.25	107.8	1.27
Watermelon	0.17	73.1	4.25

2.3.1.1 Apple

Apples are widely consumed fruits enriched with polyphenols, which possess antioxidant and anticancer properties. Apples and their leaves contain bioactive polyphenols such as epicatechin, catechin, phloridzin, chlorogenic acid, quercetin, rutin, and procyanidin. All these compounds exhibit antioxidant and anticancer properties. Their DPPH antioxidant capacity is 2-3 times higher, and their superoxide radical scavenging activity is 10–30 times higher than that of vitamins C and E. Apple extract 100 g had the antioxidant power of 1500 mg vitamins C. Apples increase plasma SOD and GPX levels, which lower oxidation and inhibit the inflammation imposed by excessive free radicals. Apple extract at 20 mg/mL revealed dose-dependent anticancer activity against Caco-2 colorectal cancer cell lines. Apple extract with peel inhibited cell growth by 43%, whereas without peel by 29% after 96 hours at 50 mg/mL. Apple extract with and without peel also decreased HepG2 cell (liver cancer cell line) proliferation by 57% and 40%, respectively [66–69]. The amounts of these polyphenols present in the apple are described in Table 2.3.

Phloretin (Figure 2.1), a major flavonoid present in apples, exhibits antioxidant. Phloretin induced apoptosis and decreased cell viability in human esophageal cancer cells (EC-109) at doses of 60–100 μg/mL. It also decreased B-cell lymphoma 2 (Bcl-2), an antiapoptotic protein, while increasing proapoptotic molecules such as tumor protein p53 and

Table 2.3 Antioxidants and polyphenols from fruits.

Fruits	Polyphenols	Part	Concentration	References
Apple	Epicatechin	-	0.4-1.3 mg/g dw	[67]
	Catechin	-	0.4-1.5 mg/g dw	[67, 91]
	Rutin	-	0.4-1.6 mg/g dw	[67, 81, 91]
	Chlorogenic acid	-	0.2-2.3 mg/g dw	[67, 81, 91]
	Phloridzin	-	0.1-0.4 mg/g dw	[67, 81, 91]
	Quercetin	-	13.2 mg/100 g	[67, 81, 91]
	Procyanidin B2	-	9.3 mg/100 g	[67, 81, 91]
Banana	Gallic acid	-	0.2 mg/100 g	[83]
	Catechin	-	60-82 mg/100 g	[81]
	Quercetin	-	32-125 μg/g	[81, 83]
	Gallocatechin	Peel and pulp	158 mg/100 g dw 29.6 mg/100 g dw	[81, 83]
Berries	Anthocyanin	Whole fruits	0.8-2.1 g/100 g	[86]
	Chlorogenic acid	-	9.7 mg/100 g	[86]
	Malvidin-3-galactoside	-	0.8 mg/g fw	[87, 89]
	Kaempherol-3-galactoside	-	3.9 μg/g fw	[85, 89]
	Cyanidine-3-galactoside	-	76 μg/g fw	[85, 89]
	Petunidine-3-galactoside	-	218 μg/g fw	[85, 89]
	Quercetin-3-galactoside	-	76.2-209.6 μg/g fw	[86, 90]
Grapes	Anthocyanin	Fruits	55 mg/100 g fw	[68]
	Proanthocyanidins	Seeds	2.9-35.3 mg/100 g dw	
	Gallic acid	Pulp and skin	2.9-35.3 mg/100 g dw	[93, 94]
		Seeds	16.7-77.8 mg/100 g	
	Catechin	Pulp and skin	44.0-221.5 mg/100 g	[93, 94]
		Seeds	8.7-53.8 mg/100 g	
	Resveratrol	Pulp and skin	56.4-480.5 mg/100 g	[93, 94]
		Seeds	0.4-2.8 mg/100 g	

(Continued)

Table 2.3 Antioxidants and polyphenols from fruits. (*Continued*)

Fruits	Polyphenols	Part	Concentration	References
	Quercetin	Pulp and skin	3.7-23.8 mg/100 g	[68]
		Seeds	22.6-201.4 mg/100 g	
	Kaempferol	Pulp and skin	1.6-5.7 mg/100 g	[68]
		Seeds	14.7-45.1 mg/100 g	
Oranges	Hesperidin	Peel	30-39 mg/g dw	[68]
	Naringin	Fruit	979 mg/100 g fw	[68]
	Tangeretin	Peel	0.2-0.3 mg/g dw	[68]
		Fruit	2.5 mg/100 g	
	Nobiletin	Peel	0.4-0.6 mg/g dw	[68]
		Fruit	4 mg/100 g	
Papaya	Phenolics	Dry peel	187-278 mg/100 g	[68]
	Quercetin	Dry fruit	0.1-0.3 mg/100 g	[68]
	Kaempferol	Dry leaf	3 mg/100 g	[68]

dw = dry weight; fw = fresh weight.

protein 4 [70]. Phloretin at doses of 25, 50, 100, and 200 μM induced apoptosis through the P38 MAPK and JNK1/2 pathways and inhibited the migration of non-small lung cancer cell lines (A-549) in a dose dependent manner. In an *in vivo* experiment on mice phloretin significantly inhibited the growth of A-549 in the xenograft model [71]. Phloretin at 12.5 μM suppressed human oral cancer cell lines (SCC-1) growth through cell cycle arrest and cell death. It prevented cancer cell migration. In addition, it exhibited a moderate effect at a dose of 100 μM on normal prostate cells (HTRET-OME) [72]. Phloretin inhibited gastric cancer cells (BGC823) by 73% at 30 μM. Downregulating Bcl-2 at 20 and 30 μM caused dose-dependent apoptosis [73]. Phloretin inhibited triple negative breast cancer cells (MDA-MB-231) at 10-150 μM for 24 h on the MTT assay. It also reduced cancer cell migration. Phloretin at 10 and 50 mg/kg twice a week for 6 weeks inhibited cancer growth in a mouse xenograft model [74]. Phloretin at 60 μM reduced migration and invasion in human cervical cancer cells (SiHa) by downregulating the MMP-2 and MMP-3 pathways. Phloretin reversed epithelial-mesenchymal transition at the same dose and stopped SiHa cell self-renewal by inhibiting aldehyde dehydrogenase 1 activity at 100 μM. In mouse tumor xenograft models, phloretin at 20 mg/kg decreased SiHa cell lung metastasis and tumor growth [75]. The ruthenium-phloretin complex at 12.5 to 200 μM reduced the number of viable colon cancer cell lines (HT-29) and induced apoptosis and cell cycle arrest in a dose-dependent manner after 24 hours of exposure [76]. Phloretin at 5-40 μM dose-dependently reduced human mammary epithelial (MCF10A) cell proliferation and caused

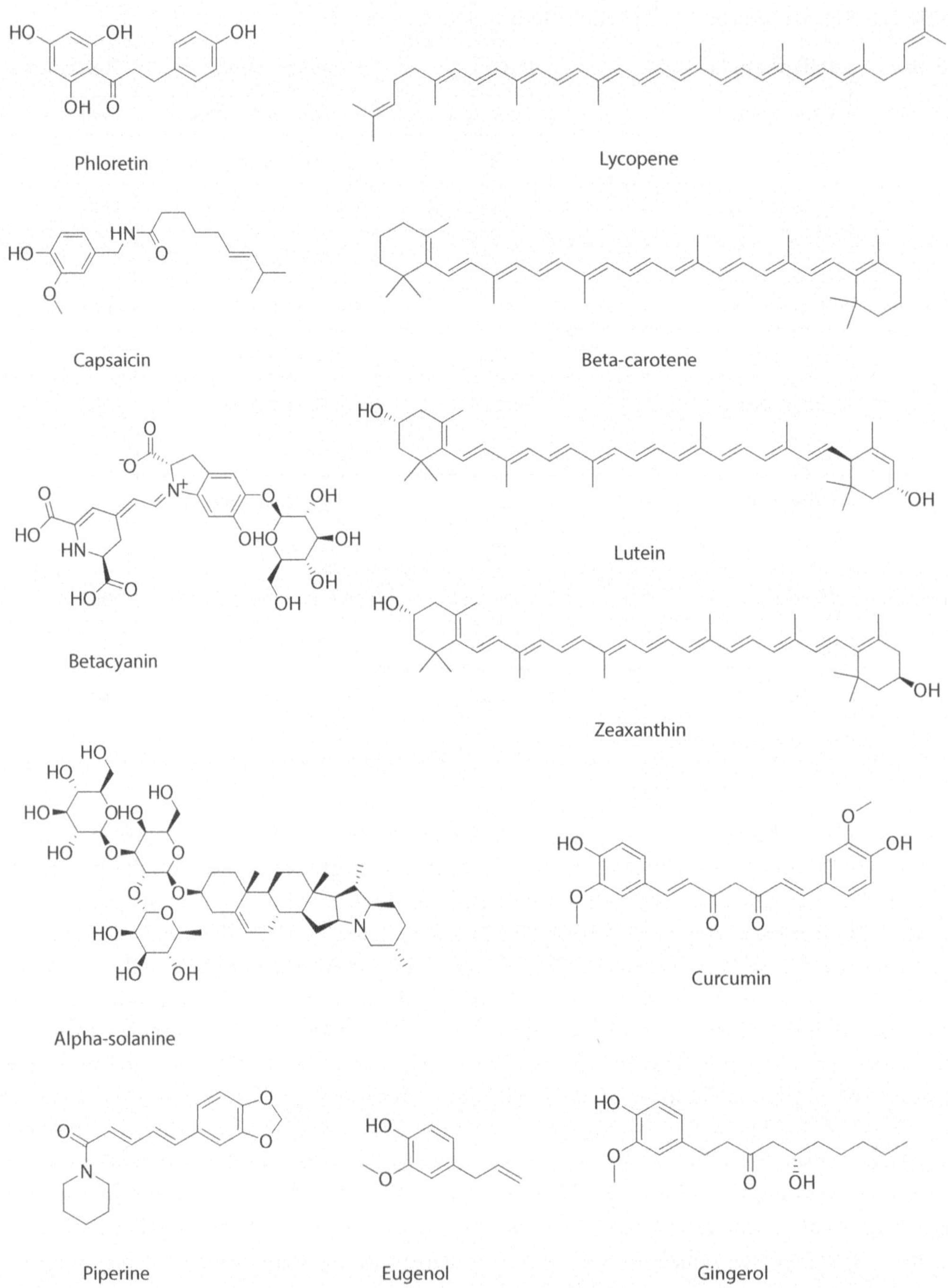

Figure 2.1 Chemical structures of antioxidant and anticancer compounds from fruits, vegetables, and spices.

apoptosis. Phloretin also increased p53 and cleaved poly (ADP)-ribose polymerase, confirming its pro-apoptotic actions. Phloretin activated the c-Jun N-terminal kinase (JNK) and p38 mitogen-activated protein kinase signaling pathways responsible for apoptosis and increased caspase-3 activity, further suggesting its apoptotic and anticancer effects [77]. Phloretin effectively suppressed the proliferation of five hepatocellular carcinoma cell lines (PLC-5, Huh-7, Hep3B2.1-7, SK-Hep1, and HepG2) in a dose- and time-dependent manner. It also downregulated STAT3, upregulated SHP-1 (the tumor suppressor gene), and inhibited pAKT/pERK signaling (which promotes cancer cell survival), causing all cancer cells to die. It also suppressed tumor growth in an *in vivo* xenograft model of hepatocellular carcinoma by inhibiting cell proliferation and inducing apoptosis [78]. Phloretin was tested against prostate cancer cell lines (DU145, PC-3, CWR22Rv1, LNCaP, and WPMY-1). It decreased prostate cancer cell proliferation at 100 μM. Phloretin also inhibited prostate cancer cell growth in the G2/M phase and triggered apoptosis at 20, 50, and 100 μM. Phloretin inhibited epidermal growth factor receptor (EGFR) phosphorylation and activation, which inhibited its downstream PI3K/AKT and MEK/ERK1/2 signaling pathways in prostate cancer cells (PCa cells), because these signaling pathways are responsible for the survival of cancer cells. Inhibition of PI3K/AKT and MEK/ERK1/2 signaling pathways leads to the induction of apoptosis in prostate cancer [79].

2.3.1.2 Banana

Bananas, a commonly consumed fruit, contain important polyphenols such as catechin, anthocyanin, gallic acid, quercetin, kaempferol, cyanidin, and myricetin. Banana fruit has 32.3–125.09 μg/g of quercetin. Gallic acid is 0.2 mg per 100 g, while catechins range from 60.0 to 82.2 mg per 100 g of fruit. Banana phenolic and flavonoid concentrations are shown in Table 2.3. Banana flavonoids also inhibit vitamin C oxidation. The fruit is enriched with α and β carotene, β-cryptoxanthin, and isoleutin, which are carotenoids. Bananas contain 0.20–125 μg/g α-carotene. All these phytochemicals have been reported to reduce ROS and RNS in the body by neutralizing free radicals like ROS and RNS, lowering oxidative stress, and stopping the abnormal cell division that leads to cancer [68, 80–84].

2.3.1.3 Berries

Berries are notably rich in anthocyanins, including cyanidin, petunidin, delphinidin, peonidin, malvidin, petunidin-3-galactoside, cyanidin-3-galactoside, kaempferol-3-galactoside, and quercetin-3-galactoside, as outlined in Table 2.3. These anthocyanins play a vital role in supporting immune system function and counteracting oxidative stress resulting from the excessive production of ROS and RNS. Additionally, they act as shields for cellular DNA and RNA, preventing oxidation due to an excess of free radicals. Moreover, the presence of anthocyanins contributes to the prevention of abnormal cell division, thereby promoting healthy cellular processes and reducing the risk of cancer [68, 85–90].

2.3.1.4 Grapes

Anthocyanins, flavanols, and resveratrol, which have strong antioxidant and anticancer characteristics, are major polyphenols found in grapes. These polyphenols are essential for enhancing the immune system's resistance, successfully combating an excess of free radicals, and lowering inflammatory responses within the body. Grape polyphenols reduce chronic inflammation by either modifying inflammatory pathways or reducing ROS levels. Flavonoids, anthocyanins, proanthocyanidins, and hydroxycinnamic acid derivatives from grapes are preferable to synthetic monotargeted anti-inflammatory medications because they are natural components that can choose numerous pathways to overcome chronic inflammation [91–94]. Detailed information concerning the diverse polyphenolic constituents found in grapes is provided in Table 2.3.

2.3.1.5 Kiwi

Kiwi, also called Chinese gooseberry, contains many phytochemicals with diverse biological properties. Antioxidants are notable. Kiwi fruit extract reduces H_2O_2-induced oxidative stress in *in vitro* and *in vivo* studies. This extract also prevents DNA oxidation [95]. Kiwi has high levels of β-carotene, lutein, zeaxanthin, and β-cryptoxanthin. These carotenoids can reduce free radicals and oxidative stress. Green kiwi fruit contains 52, 122, and 122 μg/100 g of β-carotene, lutein, and zeaxanthin, respectively. β-Cryptoxanthin is 0.057 μg/ml. Gold kiwi fruit contains 14, 31, and 24 μg/100 g of β-carotene, lutein, and zeaxanthin, respectively. These carotenoids help in reducing the complexities of oxidative stress, preventing inflammation, improving eye health, and reducing the risk of cancer [96, 97].

2.3.1.6 Orange

Citrus fruits such as oranges are abundant in antioxidant bioactive compounds namely flavonoids, β-carotene, and folic acid [98]. Flavanones and flavones, major subclasses of flavonoids, are abundant in orange fruit, as shown in Table 2.3. Orange peel contains 66.67–67.50 mg of β-carotene per 100 g. These phytochemicals neutralize free radicals by donating hydrogen ions, boosting white blood cell production, and preventing DNA oxidation. They also regulate gene expression and DNA methylation. The orange flavonoid hesperidin inhibits proinflammatory enzymes like lipoxygenase, protein kinase, phosphodiesterase, phospholipase, and cyclooxygenase. Orange carotenoids inhibit reactive species produced by immune system active cells, scavenge immunosuppressive peroxides, and sustain immune membrane receptors. Oranges also contain lycopene, a proven antioxidant. Human leukocytes that were treated with orange peel or orange fruit extracts before being exposed to H_2O_2 caused much less damage to their DNA than cells that were only exposed to H_2O_2. Orange peel has more antioxidants than orange fruit, highlighting its effectiveness in neutralizing oxidative stress and reducing the risk of cancer [99–106].

2.3.1.7 Papaya

Papaya, a tropical fruit, is abundant in polyphenols and carotenoids (Table 2.3). The unripe flesh of papaya contains a range of 0.15 to 1.2 mg per 100 g of lycopene, a carotenoid that serves as a precursor to vitamin A [107–109]. Through the chelation of metallic ions, quenching of hydroxyl radicals, and modulation of antioxidant enzymes within the physiological framework, these polyphenols and carotenoids display potent antioxidative attributes. Their functionality extends to the reduction of ROS and RNS, thereby conferring protection upon DNA and RNA molecules against damage induced by free radicals [110]. Extracts of papaya fruit have demonstrated the capacity to counteract the immunosuppression by malondialdehyde (MDA) levels, concomitant with reductions in the activities of SOD, CAT, and GPX. Moreover, this extract has been shown to ameliorate oxidative stress within the cellular membranes of the stomach, liver, and kidneys in rats, induced by acrylamide exposure [111]. The phytochemical constituents of papaya also contribute to the alleviation of antioxidant enzyme usage. *In vitro* studies using rat pheochromocytoma tumor cells have demonstrated that fermented papaya products serve as safeguards against oxidative DNA damage prompted by H_2O_2 exposure. Furthermore, these products exhibit a mitigating effect on oxidative stress within the brain of hypertensive rats. Remarkably, fermented papaya products exhibit the capability to mitigate oxidative and inflammatory injuries associated with cirrhosis attributed to hepatitis C infection [112].

2.3.1.8 Pomegranate

Pomegranate is rich in anthocyanins, ellagic acid, and ellagitannin. The juice has 20% more polyphenol activity than berries, apples, and grapes. Pomegranate also boosts CAT and SOD. Pomegranate ethanolic extract inhibits prostate, bladder, and lung cancer [113]. Pomegranate juice contains total phenolics (1385-2899 mg gallic acid equivalent per litre), flavonoids (234-504 mg catechin equivalent per litre), and anthocyanins (110-140 mg cyanidin 3-glucoside equivalent per litre). Anthocyanins, specifically 3-mono- and 3,5-diglucosides of delphinidin, cyanidin, and pelargonidin, give pomegranate juice its red color. All these phytochemicals boost pomegranate's antioxidant and anticancer properties [114].

2.3.1.9 Watermelon

Watermelon contains flavonoids, phenols, and carotenoids, as mentioned in Table 2.2. Watermelon juice contains 3.38–11.34 μg/g of lycopene [115]. β-carotene, citrulline, chlorogenic acid, and luteolin are also present in watermelon. Watermelon flesh contains 4.82 mg of β-carotene per 100 g. It also contains citrulline, a precursor to the nonessential amino acid arginine. Citrulline is a good antioxidant and hydroxyl radical scavenger. Watermelon makes citrulline more bioavailable than L-glutamine, boosting our immune system. Watermelon flesh contains 70–360 mg of citrulline per 100 g. Watermelon modulated genes to lower blood lipids, inflammation, and antioxidant capacity in rats. Watermelon's anti-inflammatory and antioxidant properties were attributed from citrulline, a bioactive component. β-Carotene and lycopene increased the watermelon's antioxidant activity. These phytochemicals protect the body from inflammation and contribute to the prevention of cancer [115–118].

2.3.2 Antioxidants from Vegetables

Vegetables, whether in raw or processed form, are essential to individual diets. They prevent malnutrition and disease by addressing mineral, vitamin, and phytochemical deficiencies. Many vegetable can treat a variety of illnesses. Antioxidants and bioflavonoids, which are present in vegetables are important. These antioxidants and bioflavonoids reduce chronic diseases. They fight free radicals, detoxify carcinogens, modulate metabolic activation,

Table 2.4 Total phenolic, flavonoid, and carotenoid content in vegetables.

Vegetables	Total phenolic content (TPC) mg GAE equivalent/g	Total flavonoid content (TFC) mg QE equivalent/100 g	Total carotenoid content (TCC) mg/100 g
Artichoke	30.16	N/A	0.50
Beans	0.92	12.6	1.46
Broccoli	N/A	N/A	0.21
Cabbage	6.24	51.3	0.43
Carrot	0.58	15.7	10.29
Cauliflower	0.57	6.3	0.04
Celery	6.80	77	0.93
Endive	0.78	N/A	3.60
Green peas	0.85	45.8	0.62
Garlic	0.74	12.8	0.75
Lettuce	7.87	12.1	8.48
Mushroom	4.12	7.8	4.12
Onion	6.20	30.6	0.02
Pickle	0.65	6.3	0.21
Pumpkin	0.22	77.1	0.69
Spinach	0.18	133	17.31
Sweat corn	0.18	46	3.97
Zucchini	0.10	N/A	0.36

and inhibit tumor growth. Daily consumption of vegetables reduces the risk of cancer [16, 119]. The total phenolics, flavonoids and carotenoids of vegetables such as artichoke, beans, broccoli, cabbage, carrot, cauliflower, celery, endive, green peas, garlic, lettuce, mushroom, onion, pickle, pumpkin, spinach, sweat corn, and zucchini are presented in Table 2.4.

2.3.2.1 Tomato

Tomatoes are a widely consumed vegetable used in various recipes. It contains bioactive phytochemicals like carotenoids, which include lycopene, β-carotene, α-tomatine, and delphinidin [120]. Table 2.2 shows tomato juice's total phenolics, flavonoids, and carotenoids. Tomato juice contains lycopene, phytoene, neurosporene, phytofluene, beta-carotene, and lutein at 10.8, 2.8, 2.0, 0.8, 0.4, and 10.1 mg/100 g, respectively [121]. These phytochemicals scavenge ROS and prevent abnormal cell growth. Lycopene, with its extended conjugated diene structure, has 10-fold more singlet oxygen-quenching potential than vitamin E. Lycopene's unique red pigment and antioxidant and anticancer properties have made it a valuable phytochemical. In many regions, tomato-based foods like sauce and paste are the main carotenoid sources. Recent research suggests lycopene modulates intercellular communications, metabolic pathways, and the immune system. Another compound, α-tomatine at doses of 2 and 5 μM, inhibited human myeloid leukemia cancer cell line (HL-60) growth and induced apoptosis. In female mice with subcutaneous HL-60 xenograft tumors, 5 mg/kg α-tomatine three times a week for three weeks reduced tumor size [122]. Tomatoes, especially lycopene, may fight cancer. According to European clinical trials, tomato consumption reduces cancer risk. It has been extensively studied that eating tomatoes and getting more lycopene in your body both increase apoptosis and prostate-specific antigen levels in prostate cancer. Tomato consumption may reduce cancer risk [123]. Lycopene (Figure 2.1) is a carotenoid pigment found in tomatoes, oranges, and watermelon. Lycopene reacts with oxygen and free radicals due to its polyene structure and many conjugated double bonds. It quenches singlet-oxygen better than other carotenoids. It reduces excessive ROS, decreases oxidative stress, and protects lipids, proteins, and DNA from oxidative damage. Lycopene neutralizes H_2O_2, nitrogen dioxide, and hydroxyl radicals. Lycopene also boosts the detoxification process by increasing antioxidant enzymes like SOD, CAT, GST, and GST reductase. These enzymes are normally regulated by the transcription factor Nrf2. Lycopene disrupts Nrf2's interaction by inhibiting Keap1, which regulates Nrf2, and identifying electrophilic and oxidative stress. In this way, it detoxifies the antioxidant genes. Lycopene upregulates the antioxidant response system in human bronchial epithelial cells, HepG2, MCF-7, and androgen-sensitive human prostate cells. Also, lycopene dose-dependently arrests the cell cycle and induces apoptosis in human cancer cell lines, *in vitro*. Lycopene at 1-5 μM reduced the number of viable cells in three cancer cell lines: HT-29, colon carcinoma (T84), and MCF-7, after 48 hours of treatment. Lycopene also caused cell cycle arrest in T-84, HT-29, MCF-7, and DU145 (human prostate carcinoma) cells after 96 hours, inhibited cell proliferation, induced cell cycle arrest at different phases, and increased apoptosis in breast, colon, and prostate cancer cell lines. The results suggest that the lycopene effect on cell cycle regulatory proteins may vary by cancer type and dosage [124]. Lycopene prevents cancer by modulating growth factors, proteins, steroids, and other biochemical molecules that interact with cell surface receptors and activate cell signaling cascades such as insulin-like growth factor 1 (IGF-1), VEGF, epidermal growth factor (EGF), and platelet-derived growth factor (PDGF). These growth factors abnormally activate signaling pathways, causing cell proliferation, differentiation, maturation, apoptosis suppression, invasion, and metastasis. The IGF-1 signaling pathway plays an important role in cell proliferation, growth, and survival. Lycopene inhibited IGF-1-receptor substrate tyrosine phosphorylation and activator protein-1 (AP-1) transcription factor DNA binding capacity MCF-7, lung (NCI-H226), and prostate (PC-3) cancer cell lines, thus inhibiting their proliferation and metastasis. Lycopene supplementation also up-regulated binding proteins such as IGFBP-3 and down-regulated BAD phosphorylation responsible for carcinogenesis, thus inhibiting smoke-induced lung carcinogenesis and promoting apoptosis while inhibiting cell proliferation [125]. Lycopene inhibits tumor growth and promotes cancer cell apoptosis through antioxidant and non-oxidative mechanisms. *In vivo* broiler chicken studies examined the effects of dietary lycopene on growth and physiological parameters. From days 21 to 42, broilers were fed a grower diet with 0, 50, 100, or 200 mg/kg lycopene. Lycopene supplementation resulted in broiler weight gain when fed 100 mg/kg lycopene. 200 mg/kg lycopene supplementation increased feed intake significantly. Broilers fed 100 mg/kg of lycopene had lower feed conversion ratios than the control group. At 100 mg/kg, lycopene supplementation significantly reduced cholesterol, triglyceride, and very low-density lipoprotein (VLDL) levels, also increased serum high-density lipoprotein (HDL) levels. The control group had higher serum malondialdehyde levels than the lycopene-supplemented groups (100 mg/kg). Lycopene affects antioxidant enzymes. The 100 mg lycopene group had the highest serum CAT and GPX levels, and decreased serum activity of the metabolic enzymes alkaline phosphatase and alanine aminotransferase [126]. Antioxidants protect DNA from ionizing radiation. Lycopene was tested for protection against X-irradiation in human lymphocytes. 10 and 20 μM DMSO-dissolved lycopene were given. Lycopene at these concentrations reduced radiation-induced DNA damage at 0.5, 1, and 2 Gy. (units of ionization). Lycopene protected DNA from ionizing radiation, but it did not stimulate DNA repair after irradiation. Lycopene did not modify DNA repair mechanisms. In conclusion, low-dose lycopene supplementation may protect cells from radiation-induced oxidative damage [127]. Lycopene prevented radiation-induced DNA damage but did not affect DNA repair. In mice, lycopene at 25 and 50 mg/kg of body weight for 3 days significantly reduced micronuclei in the micronucleus test. Micronuclei are the biomarkers of genotoxic damage. These studies suggest that lycopene may promote health, protect DNA, and fight cancer [128]. In addition, lycopene, possesses antioxidant and extensive anticancer properties. Antioxidative effects, such as reduction of free radicals, prevent oxidative stress and increase the antioxidant enzyme in the body, while anticancer properties, such as inhibition of cell proliferation, induction of apoptosis, and cell cycle arrest, are studied by *in vitro* and *in vivo* experimental methods. All these potentials make lycopene a promising candidate for cancer prevention [125].

2.3.2.2 Chili

Chili encompasses significant phytoconstituents such as capsaicin, carotenoids, zeaxanthin, capsorubin, and beta-carotene. In-depth phytochemical assessments of diverse chili varieties revealed that carotenoids and capsaicinoids are present within the range of 1.0-1.17 mg/g and 1.0-1.2 mg/g, respectively. Total phenolics and flavonoids were quantified at levels ranging from 7.4-11.9 mg/g gallic acid equivalents and 3.8-6.1 mg/g rutin equivalents, respectively. Within the spectrum of phenolic acids examined, gallic acid emerged as the predominant component across all variants of red chili. These phytochemical constituents exhibit scavenging effects against free radicals, reduction in ROS levels, and mitigation of ROS generation and oxidative damage within the body. Additionally, they demonstrate the capacity to impede the proliferation of cancer cells and diminish the likelihood of cancer progression. Their actions encompass the induction of apoptosis and the arrest of the cell cycle across a variety of cell types [129, 130].

Capsaicin (Figure 2.1), the main pungent, irritating, and major bioactive compound in red pepper, exhibited good antioxidant properties both *in vitro* and *in vivo* studies. Capsaicin inhibits lipid peroxidation in subcellular membranes of human erythrocyte in rat liver microsomes. Like α-tocopherol, capsaicin inhibits the oxidation of methyl linoleate micelles in soybean phosphatidylcholine liposomal membranes [130]. Capsaicin prevents LDL oxidation by inhibiting copper ion-induced lipid peroxidation. Furthermore, capsaicin intake inhibits iron- and copper-induced LDL oxidation *in vitro* and *in vivo* studies [131, 132]. Capsaicin also scavenges DPPH radicals inside and outside the membranes. The main products of the reaction between capsaicin and DPPH radicals are vanillin and 8-methyl-6-noneamide. This suggests that the C7-benzyl carbon in the structure of capsaicin is the site where radicals are scavenged. Capsaicin also modulates 5-lipoxygenase (5-LO) activity in human polymorphonuclear leukocytes (PMNL cells), which synthesize leukotrienes [133]. In Wistar rats, capsaicin at 3 mg/kg for three days reduced oxidative stress by decreasing MDA levels in the liver, lungs, kidneys, and muscles. Dietary capsaicin improves red blood cell and liver antioxidant status in hyperlipidemic and hypercholesterolemic rats. Capsaicin reduces lipid peroxide, erythrocyte, and liver antioxidant depletion. It also raises erythrocyte glutathione (GSH) levels. In hypercholesterolemia, capsaicin restores hepatic antioxidant enzymes, including GST, CAT, and SOD [134–136]. Capsaicin induced apoptosis in a variety of cancer cell lines, including skin, lung, leukemia, bladder, esophageal, pancreatic, and prostate cancers [137–144], while leaving normal cells unaffected [145, 145–147]. The mitochondrial pathway is the "gatekeeper" of apoptosis. Thus, proteins and pathways in this mitochondrial death pathway are promising targets for novel therapeutic interventions. Capsaicin induces apoptosis in various cancer cell lines by targeting mitochondrial death pathway proteins. Cluster of differentiation 95 (CD95) controls the intrinsic and extrinsic apoptosis pathways, which capsaicin activates. It also decreases Bcl-2, increases on caspase-9 and -3, breaks the mitochondrial membrane potential, and releases cytochrome C, leading to the induction of apoptosis. Capsaicin causes damage to mitochondria and the release of cytochrome C by interacting with transient receptor potential vanilloids (TRPVs). This happens through Ca2+-mediated TRPV1 and TRPV6. TRPV1 and TRPV6 channels have proapoptotic effects in many cancers, upon which capsaicin acts. Stress signals or DNA damage activate the cancer-fighting p53 tumor suppressor, killing cells. By down-regulating sirtuin 1, capsaicin phosphorylates and acetylates p53, promoting apoptosis. Capsaicin induces p53-independent apoptosis in *in vitro* and *in vivo* prostatic cancer models. Urothelial cancer cells express, phosphorylate, and apoptotic p53. In many cancers, β-catenin is constitutively active and deregulated, which promotes malignancy. Capsaicin reduces β-catenin transcription and stability, causing colorectal cancer cell apoptosis [148–150]. Capsaicin activates NSAID-activated gene-1 to induce apoptosis in colorectal cancer cells via glycogen synthase kinase and protein kinase C. AMP-activated protein kinase, ROS, prohibitin 2, and fatty acid synthase are other capsaicin proapoptotic mechanisms. These studies support the idea that capsaicin induces cancer cell apoptosis via multiple pathways [149, 151–156]. Capsaicin increases the expression of p21 and decreases the expression of CDK4, CDK6, and cyclin E to stop the growth of human esophageal carcinoma cells in the G0/G1 phase [144, 157]. Capsaicin also arrests the cell cycle in human cancer KB cells. In colon cancer cell lines, capsaicin reduced cyclin D1 levels dose-dependently [155, 158]. Capsaicin modulates cancer growth differently. It suppresses estrogens receptor-positive and estrogens receptor-negative breast cancer cells and acts as a novel prostate cancer anti-androgenic receptor agent. Effects of capsaicin on gastric cancer cells depends on tumor-associated NADH oxidase expression. Capsaicin targets cell cycle regulators to inhibit cancer cell growth and division [159–161]. Capsaicin also inhibits angiogenesis and VEGF-induced endothelial cell proliferation, migration, and tube formation. It inhibits p38 MAPK, protein kinase B (PKB), and focal adhesion kinase (FAK) activation, which makes it anti-angiogenic. Capsaicin also degrades hypoxia-inducible factor 1α, which regulates

VEGF transcription. By disrupting common angiogenic signaling pathways, capsaicin may prevent malignant transformation pathways [142, 162, 163]. In advanced cancer, capsaicin has anti-invasive and anti-migratory effects. It inhibits melanoma cell migration and slows metastatic progression in transgenic adenocarcinoma of the mouse prostate (TRAMP) mice [164, 165]. PI3K signaling and RAS-related C3 botulinum toxin substrate 1 (RAC1), a cell motility kinase, are downregulated, causing this effect on motility [166]. Capsaicin suppresses the AKT/FAK, extracellular signal-regulated kinases, and p38 MAPK signaling pathways, decreases MMP9 levels and inhibits EGF-induced invasion and migration in various cancers. TRPV1 receptors control capsaicin's migration and MMP9 activation in bladder cancer. However, lower doses of capsaicin may promote colonic cancer metastasis [151, 167–169].

2.3.2.3 Potato

Potatoes, the most consumed vegetable, contain phenolic compounds, flavonoids, and carotenoids. Potato peel comprises 90% phenolic compounds. Potato tubers contain caffeic acid, a phenolic compound, at 25-72 mg/100 g. It also contains chlorogenic acid, quinic acid esters, ferulic acid, gallic acid, p-coumaric acid, syringic acid, vanillic acid, sinapic acid, and salicylic acid. Potatoes contain bioactive flavonoids such as catechin at 0-204 mg/100 g dry weight, and also contain quercetin, rutin, and kaempferol flavanols. Anthocyanins that give red and purple colors, increase the flavonoid content of purple potatoes to nearly twice of white potatoes. Purple tubers contain malvidin, petunidin, delphinidin, and peonidin, while red potatoes have pelargonidin. Carotenoids, namely zeaxanthin, lutein, and violaxanthin (major components), as well as trace amounts of beta-carotene are found in potatoes. These phenolic compounds, flavonoids, and carotenoids work together to give many health benefits, including antioxidant and anticancer properties [170].

α-Solanine (Figure 2.1), a major bioactive compound found in potatoes, inhibited HepG-2 cancer cells and induced apoptosis and arrested the cell cycle in G phase and prolonged the S phase. It also reduced the antiapoptotic protein Bcl-2. Morphological changes in the mitochondrial membrane, calcium gradients, and increased production of hydroxyl and hydrogen peroxide free radicals were also observed. Furthermore, α-solanine reduced the level of histone deacetylase 1 (HDAC1), a regulator of cell growth, while promoting the production of apoptosic proteins, namely apoptosis signal-regulating kinase 1 (ASK1) and tetrahymena piggyBac transposase 2 (TBP-2). This dual action leads to decreased cell proliferation and an augmented rate of programmed cell death [171, 172]. In the context of colon cancer, α-solanine induced apoptosis primarily through the activation of caspase-3 pathways and inhibition of extracellular signal-regulated protein kinases 1 and 2 (ERK1 and ERK2) phosphorylation. These pathways are essential for the migration, survival and proliferation of cells. Moreover, α-solanine increased autophagy and suppression of important signaling pathways such as phosphatidylinositol 3-kinase/Akt/mammalian target of rapamycin (Akt/mTOR) pathway in pancreatic cancer, which are responsible for proliferation of cancer cells [7, 171, 173, 174]. By suppressing cyclin activity, ROS production, and p38 MAPK activation, α-solanine caused apoptosis in prostate cancer. P38 controls the cell cycle, proliferation, and intercellular interactions. α-solanine also inhibits axis inhibition protein 1 (AXIN) and bone morphogenetic protein receptor type-2 (BMPR2) proteins, which promote tumor proliferation and metastasis through beta-catenin-mediated vascular remodeling [175, 176]. α-solanine may affect tumor proliferation through multiple apoptosis-induced mechanisms. α-solanine inhibits melanoma cell migration by decreasing MMP-2, JNK, and PI3K/Akt activity, resulting in increased apoptosis and angiogenesis regulation. Breast cancer models show similar mechanisms. α-Solanine increased radio sensitivity and upregulated microRNA-13876 expression in lung adenocarcinoma and esophageal cancer cells. In conclusion, α-solanine stimulates apoptosis, inhibits angiogenesis, modulates the cell cycle, and increases radio sensitivity in human cancers [177–179].

2.3.2.4 Vegetables from Cruciferae Family

Cruciferae vegetables like cauliflower, broccoli, Brussels sprouts, and cabbage are popularly used as a common diet in various countries. These vegetables contain glucosinolates like isothiocyanates, allyl isothiocyanates, sulforaphane, and indole-3-carbinol. They also contain flavanols, quercetin, kaempferol, and carotenoids like lutein and beta-carotene [65, 180]. Table 2.4 shows brassica vegetable total phenolics, flavonoids, and carotenoids. All these phytochemicals exhibited good free radical scavenging properties, reducing the chances of oxidative stress and protecting the body from the harmful effects of inflammation and cancer. Broccoli is rich in isothiocyanates, and its aqueous extract significantly and dose-dependently inhibited bladder cancer in rats caused by N-butyl-N-(4-hydroxybutyl) nitrosamine. This extract reduced bladder cancer incidence, multiplicity, size, and progression, and increased bladder epithelium glutathione S-transferase and NADPH quinine oxidoreductase 1 production. Isothiocyanates target the bladder epithelium due to their bioavailability and rapid urinary excretion, preventing bladder cancer. Sulforaphane, another cabbage compound, inhibits cancer cell growth and induces apoptosis *in vitro.* Indole-3-carbinol and diindolylmethane in cabbage may reduce the risk of head and neck, prostate, acute lymphoblastic leukemia, breast, stem cells, pancreatic, squamous cell carcinoma, colon, cervical, skin, and ovarian cancers [181]. Radish, a Cruciferae member, also contains glucosinolates, isothiocyanates, polyphenols, indole-3-carbinol, anthocyanins, and sulfur-containing compounds. Radish phytochemicals reduce free radical formation, inhibit cell elongation, induce cell shrinkage, cause DNA fragmentation, prevent cell proliferation, and promote cancer cell apoptosis. These phytochemicals also inhibit free radicals and induce apoptosis, as well as inhibit carcinogen production (especially lung carcinogens), acute lymphoblastic leukemia, benzo[a]pyrene formation (linked to stomach tumor growth), kinase enzyme activities in MCF-7 cancer cells, and tumor cell apoptosis. In a xenograft model of inflammatory breast cancer, indole-3-carbinol from cruciferous vegetables reduces tumor proliferation and increases apoptosis [182].

2.3.2.5 Vegetables of the Cucurbitaceae Family

The Cucurbitaceae family encompasses various vegetables such as cucumber, muskmelon, and bitter gourd, each harboring distinct bioactive compounds and health benefits. Cucumber, for instance, contains flavonoids that manifest anti-inflammatory and antioxidant attributes. Muskmelon, abundant in phenolic and carotenoid compounds, showcases antioxidant and anticancer properties. Bitter gourd plays a role in preventing the pathogenesis of T cell leukemia.

The Cucurbitaceae family is a reservoir of diverse bioactive compounds, including cucurbitacins such as cucurbitacin I, cucurbitacin B, cucurbitacin D, and cucurbitacin E. These compounds exert significant effects in the context of cancer and health. Notably, cucurbitacin I impedes the STAT3 signaling pathway in breast cancer, prompting apoptosis and autophagy in A549 cancer cells. This compound also enhances gene expression related to protein light chain 3 in gastric cancer cells, counteracting proliferation. Cucurbitacin B contributes to apoptosis in SW840 colon cancer cell lines while concurrently reducing cellular levels of ROS. Cucurbitacin D triggers apoptosis in gastric cancer cells and inhibits crucial signaling pathways implicated in cancer progression, including the P13K/AKT pathways. Cucurbitacin E plays a multifaceted role, inhibiting the growth of osteosarcoma cells and eliciting apoptosis and autophagy in cervical cancer cells. Furthermore, it reduces protein kinase B levels in breast cancer cells, thereby affecting cancer cell behavior. In summary, the Cucurbitaceae family is a valuable source of bioactive compounds, each with unique mechanisms and potential therapeutic applications, particularly in cancer prevention and treatment [183].

Natural pigments called anthocyanins are present in many fruits and vegetables and prized for their potent antioxidant properties. The level of ROS in HT-29 cells was decreased ferulic acid-grafted maltodextrin (FA-g-MD) microcapsules loading anthocyanins at a dose of 0-60 g/mL addressing bioavailability issues. That's why it was loaded with FA-g-MD microcapsules. It implies that these microcapsules may effectively combat oxidative stress in colon cancer cells [184].

Purple carrot extract at 1 mg/ml reduced oxidative DNA damage in normal human colon mucosal cells (NCM460) by 20.7%. Purple carrot extract may reduce oxidative stress-induced DNA damage in these cells. In another plasma sample study, pre-incubation with red cabbage anthocyanins at 5–15 μM reduced oxidative stress. In *in vitro* study, red cabbage extract anthocyanins reduced oxidative stress in plasma proteins and lipids, suggesting they may have antioxidant properties [185]. Antioxidant compounds in fruits and vegetables may reduce chronic disease risk and improve overall health. Anthocyanins also showed promise as chemo-preventives in mice. A sweet potato anthocyanin-rich extract (0.12%) reduced intestinal adenomas in mice over 18 weeks of treatment. This suggests that sweet potato extract anthocyanins may protect against intestinal adenomas, highlighting their potential as cancer preventives [186]. A purple-fleshed sweet potato anthocyanin-rich extract caused cell cycle arrest in human colon cancer cells (SW480) at 10–40 μM for 48 hours. This experiment suggests that the anthocyanin-rich extract from purple-fleshed sweet potatoes may modulate cell-cycle progression and promote colon cancer cell-cycle arrest. This strengthens the health benefits of anthocyanins and their potential as natural cancer prevention [187].

2.3.2.6 Beetroot

Beetroot contains betacyanin (Figure 2.1) and has antioxidant and anticancer properties. In rats, 250 and 500 mg/kg beetroot ethanol extract reduced nuclear NF-B (p65) protein expression and DNA-binding activity. Inhibiting NF-B, a transcription factor involved in inflammation and cancer development, may prevent cancer [188]. Betanin inhibited

15–46% of neutrophil ROS production at doses of 2–500 μM. Inhibiting ROS production can reduce oxidative stress and inflammation by protecting cells from highly reactive molecules. Betanin at 0.05–1.0 mM inhibited peroxynitrite (ONOO)-induced DNA strand cleavage, resulting in protection of DNA from oxidative damage [189]. Betanin and betaine, another beetroot compound, inhibited HepG2 cell proliferation in cell culture. Betanin, inhibited by 49% at 200 μg/mL, and betaine, inhibited by 25% at 800 μg/mL, are effective in liver cancer prevention [190]. In male rats, 250 and 500 μg/mL beetroot juice reduced viability of triple-negative breast cancer cell lines (MDA-MB-231). Betanin at 2, 10, and 20 μM translocated Nrf2 from the cytosol to the nucleus to initiate the cellular antioxidant response in THLE-2 (normal liver cell) and HepG2 cell lines. Nrf2 is a transcription factor that regulates antioxidant and detoxification enzymes. These findings suggest that betacyanin, particularly beetroot betanin, has potent antioxidant and anticancer properties that may reduce oxidative stress and prevent breast and liver cancer [191].

2.3.2.7 Beta-Carotene from Fruits and Vegetables

Orange, carrot, and potato beta-carotene (Figure 2.1) have been studied for their antioxidant and anticancer properties. Beta-carotene at 50 and 100 mg/kg reduced ischemia/reperfusion-induced hemorrhage in rat ovaries in a dose-dependent manner. After ischemia, tissue damage can result from reperfusion. Beta-carotene may protect against ovarian tissue damage [192]. Furthermore, beta-carotene at 10 mg/day for 12 weeks reduced oxidative stress in chronic lead-poisoned workers. It may reduce free radicals and oxidative stress caused by chronic lead poisoning [193]. Ascorbic acid and beta-carotene, administered in combination at a dose of 35 and 4 μM, respectively, increased apoptosis in the HepG2 human hepatocellular carcinoma cell line [194]. Beta-carotene at 10 and 20 μM downregulated Bcl-2 and upregulated cell-cell communication protein (Cx43), which resulted in apoptosis in bladder tumor cells (T24 cells). These findings suggest that beta-carotene reduces oxidative stress, enhances cancer cell apoptosis, and protects against tissue injury [194].

2.3.2.8 Lutein and Zeaxanthin from Fruits and Vegetables

Carotenoids such as lutein and trans-zeaxanthin exhibit antioxidant and anticancer properties. Lutein and trans-zeaxanthin supplementation for 42 days in rats at a ratio of 10:1 increased antioxidant enzyme levels, especially at the high dose. These carotenoids may boost the antioxidant defense system [195]. Zeaxanthin (Figure 2.1) at 10, 30, and 100 μM dose-dependently reduced cell viability and induced apoptosis in two human uveal melanoma cancer cells, SP6.5 and C918, with IC_{50} values of 40.8 and 28.7 μM, respectively [196]. In human retinal pigment epithelial ARPE-19 cells, lutein at a concentration of 4 mg/mL decreased ROS levels after 24 hours of incubation with oxidative stress induced by H_2O_2. This suggests that lutein has antioxidant properties and can reduce oxidative stress in retinal cells [197]. The presence of lutein (Figure 2.1) at a concentration of 2.0 μM led to higher ROS levels in MDA-MB-468 breast cancer cells after three hours of treatment. These observations imply that lutein may act differently to exhibit anticancer properties [198].

2.3.3 Antioxidants from Herbs and Spices

Herbs and spices are important sources of polyphenols, curcuminoids, and other bioactive compounds possessing good antioxidants and anticancer properties. Spices and herbs, when added to our foods, enhance the organoleptic properties and flavor of the recipe. In addition to vegetables and fruits, spice consumption in our food can protect us from various diseases, including cancer. Certain herbs and spices that are rich in antioxidants and anticancer properties are described as follows.

2.3.3.1 Garlic

Crude garlic is widely consumed as a spice in the majority of recipes. Garlic consists of various phytochemicals that possess antioxidant and anticancer properties. The majority of these phytochemicals are sulphides, such as diallyl disulphide and diallyl sulphide. These compounds inhibit lipid peroxidation in Syrian hamsters. S-allyl cysteine (SAC), another major compound present in garlic, also inhibits lipid peroxidation, which is responsible for the generation of free radicals and the initiation of cancer. Furthermore, it also increased the levels of antioxidant enzymes such as SOD, CAT, and GPX in the liver and circulatory system of the Syrian [199]. In another study, when Sprague-Dawley rats were given 2% dietary garlic, the activity of antioxidant enzymes like glutathione S-transferase (GST) increased in the rats' liver and colon. It also reduced azoxymethane-induced colonic precancerous lesions [200]. SAC from garlic was tested on circulatory lipid peroxidation and antioxidant levels in Wistar rats with N-nitrosodiethylamine (NDA)-induced hepatocarcinogenesis. As a result, SAC-treated rats had lower tumor incidence and lipid peroxidation and higher antioxidant levels, including GSH, β-carotene, ascorbic acid, α-tocopherol, GPX, SOD, and CAT. SAC also reduces lipid peroxidation and increases antioxidant levels in NDA-induced carcinogenesis by reducing free radical formation [201]. Garlic oil reduced oxidative stress in streptozotocin-induced diabetic rats. Furthermore, garlic oil restored total thiols, ceruloplasmin, GST, and SOD levels in the liver and kidneys of the rats. Garlic extract supplementation also increased blood antioxidant potential, oxidation resistance, and nonenzymatic superoxide radical scavenger activity and decreased blood MDA, showing reduced body oxidation [202]. Aged garlic extract (AGE) protects endothelial cells from oxidative damage. In bovine pulmonary artery endothelial cells, AGE suppressed H_2O_2 and superoxide anion production and increased SOD, CAT, and GPX activity. AGE may prevent or treat free radical-induced endothelial cell injury. Additionally, due to its extensive antioxidant effects, garlic extract exhibited anticancer activity against prostate cancer cells (PC-3), and caspase activity was increased at 0.5 or 1 μg/mL [203].

2.3.3.2 Onion

Onions consist of important flavonoids such as quercetin and kaempferol, which have the ability to scavenge free radicals and reduce the chance of cancer. Allicin and alliin are other bioactive compounds with antioxidant potential; these compounds inhibit protoxicant enzyme activation and cell proliferation. Onion extract and quercetin increased intracellular glutathione by 50%, resulting in less oxidative stress. In isolated epidermal cells incubated with and without the potent tumor promoter 12-o-tetradecanoylphorbol-13-acetate (TPA), onion, garlic oil, and onion dipropynyl sulphide increased GPX activity in a time- and dose-dependent manner and resulted in decreased TPA activity due to GPX. Garlic and onion oils also inhibit skin tumor promotion by boosting epidermal cell antioxidant systems. These oils prevented TPA's sharp decrease in intracellular reduced-to-oxidized glutathione [204, 205]. In nicotine-exposed rats, onion and garlic oils were tested for antioxidant properties. Free radicals were increased in the liver, lungs, heart, and kidney tissues of nicotine-treated rats. Garlic and onion oil increased the antioxidant enzymes CAT, SOD, and GPX in nicotine-treated rats and inhibited the formation of inflammation and carcinogens in the rats. Furthermore, onion oil and vitamin E were tested in nicotine-injected rats. Onion oil supplementation in nicotine-treated rats increased lipid peroxidation resistance and decreased the concentration of free radicals compared to vitamin E. Onion oil and vitamin E supplementation also increased GSH, ascorbic acid, and retinol levels in all tissues of the rats. Onion oil may protect against nicotine-induced oxidative damage, like vitamin E. These excellent antioxidant properties of onions significantly decreased the chances of cancer initiation [205].

2.3.3.3 Turmeric and Curcumin

Turmeric is a common spice used in the majority of Asian recipes. Its rhizome contains many phytochemicals, but the most important and biologically active is curcumin (Figure 2.1), which has been extensively tested for antioxidant and anticancer properties in both *in vitro* and *in vivo* experiments. Curcumin inhibits lipid peroxidation and the formation of free radicals such as superoxide, nitrite, and H_2O_2 in *in vitro* studies. According to studies, curcumin can scavenge free radicals like singlet oxygen, superoxide, and nitric oxide, which damage DNA and cause inflammation and cancer. It also inhibits xanthine-xanthine oxidase system superoxide anion and hydroxyl radical generation by 40% and 76% at 75 and 50 μM, respectively [206, 207]. In rats with iron-induced lipid peroxidation, a turmeric-fed diet for 10 days reduced lipid peroxidation and increased antioxidant enzyme activity. In rats fed coconut and peanut oil, dietary curcumin 1% increased antioxidant enzyme activity and decreased liver lipid peroxide. Curcumin reduced alcohol and trichloroethylene-induced oxidative stress in both rats and mice. It boosts blood and liver antioxidant levels. Curcumin also protects cells from free radical damage and carcinogens. Tests were done on mice that had been given 2% curcumin for 30 days. After that time, antioxidant enzymes like CAT, SOD, and GPX levels rose significantly in the liver and kidneys. It also increased phase II metabolism enzymes like GST and quinone reductase in the liver and kidneys. Curcumin can eliminate body-produced carcinogens by inhibiting the respective enzymes [208–212].

Some types of curcumin, like desmethoxycurcumin, inhibited the growth of MCF-7, MDA-MB-435S, and MDA-MB-231 cancer cells and made them kill themselves, which is called apoptosis. Curcumin also reduced tumor micro-vessel density and pro-angiogenic factor expression. Curcumin also prevents cancer. When given intraperitoneally for four days, curcumin reduced the incidence of DMBA-induced mammary adenocarcinoma in female rats. Curcumin's chemo preventive effects were observed in post-initiation phases, indicating its potential to stop the growth and progression of various neoplastic cells [213–218].

Curcumin inhibits androgen-dependent and androgen-independent prostate cancer cell lines. In previous studies, curcumin may kill healthy osteoblast cells and cause osteoblast apoptosis at low concentrations and necrosis at high concentrations. Curcumin-induced apoptosis in osteosarcoma cells (U2OS) is concentration- and time-dependent, characterized by increased Bak, Bax, and cytochrome C expression and decreased antiapoptotic protein expression. MG-63 (osteosarcoma cells) were more sensitive to curcumin's cytotoxic effects than healthy human osteoblasts, which survived 24 hours of treatment [219–222].

Curcumin is also cytotoxic to lung cancer cell lines such as A549 and H460 cells and decreases cell viability and clonogenic ability in a dose-dependent manner. Curcumin modulates the Axl receptor tyrosine kinase, which regulates cell survival, proliferation, and apoptosis. The 244-amino acid polypeptide known as adiponectin may regulate glucose and fatty acid catabolism. Adiponectin receptors 1 and 2 directly affect tumor cells, activating downstream signaling pathways with anti-angiogenesis and antitumor properties and causing caspase-mediated endothelial cell apoptosis. Curcumin inhibited A549 cell migration and invasion by blocking adiponectin receptor 1 [223–225].

Curcumin and quercetin reduced cell proliferation and mitochondrial membrane potential in human gastric cancer cells (MGC-803). The combined treatment induced mitochondrial-mediated apoptosis more potently than the individual treatments, suggesting potential as anti-gastric cancer drugs [226]. Curcumin inhibits multiple myeloma cell proliferation by suppressing interleukin (IL-6) and nuclear factor (NF-κB) signaling pathways. It inhibits receptor activator of nuclear factor kappa beta (RANKL) signaling and osteoclastogenesis. Curcumin also inhibits matrix metalloproteases (MMPs) and transcription factors. Curcumin modulates Wingless/Integrated (Wnt) signaling to inhibit EMT. This modulation affects colorectal cancer and papillary thyroid carcinoma proliferation and metastasis. Curcumin reduces TGF-β1-induced EMT in papillary thyroid carcinoma cells by inhibiting Smad2/3 signaling. These findings demonstrate curcumin's multifaceted anti-cancer properties and its potential as a anti-cancer candidate [227].

2.3.3.4 Clove and Eugenol

Eugenol, the main flavor compound in cloves (Figure 2.1), inhibits copper-induced lipid peroxidation in human erythrocyte membranes, revealing its antioxidant properties. Eugenol also inhibited liver microsomal monooxygenase and carbon tetrachloride (CCl4)-induced lipid peroxidation *in vitro* [206, 228, 229]. Eugenol inhibited superoxide anions and hydroxyl radicals by 50% at 250 μM in the xanthine-xanthine oxidase system, preventing the formation of oxidative stress. Deoxyribose degradation showed a 70% reduction in hydroxyl radical formation at a concentration of 250 μM eugenol, and 46% of salicylate was hydroxylated to 2,3-dihydroxybenzoate, which also inhibited hydroxyl radical formation. Eugenol prevented Fe^{2+} oxidation in the Fenton reaction, which also generates hydroxyl radicals. Eugenol dose-dependently inhibited ascorbate/Fe^{2+}-induced lipid peroxidation in rat liver microsomes at 25–150 μM. Eugenol also inhibited superoxide anions in the xanthine-xanthine oxidase system with a Ki value of 64 μM. Eugenol also inhibited rat peritoneal macrophage inflammation due to no superoxide anions or H_2O_2 release after 500 μM eugenol preincubation [206, 230].

Eugenol inhibited lipid peroxidation and suppresses ROS, superoxide anions, and hydroxyl radicals in various experimental systems. Eugenol inhibited lipid peroxidation noncompetitively in enzyme kinetic studies. It does not directly inactivate the enzyme, but its hydroxyl-radical-scavenging ability interferes with fatty acid radical intermediates, inhibiting lipid peroxidation. It inhibited peroxynitrite-induced lipid peroxidation and decreased 3-nitrotyrosine formation in *in vitro* models with IC_{50} values of 46 and 13 μM, respectively [231]. Eugenol's antioxidant activity has been studied to understand its mechanism. In an *in vitro* L-3,4-dihydroxyphenylalanine (DOPA) system, eugenol traps hydroxyl radicals directly. Eugenol inhibited hydroxyl radicals without iron chelation [232]. Eugenol also induced apoptosis and reduced ROS in human promyelocytic leukemia (HL-60) cells, suggesting an anticancer effect. Eugenol inhibited copper-induced LDL oxidation more than iron-mediated lipid peroxidation and the autooxidation of Fe^{2+} ions. Eugenol forms a complex with metals, especially copper, that stops a chain reaction of lipid peroxidation in LDL caused by free radicals [233, 234]. Dietary eugenol (0.05%) reduced serum and liver lipid peroxidation in rats fed peanut oil or cod-liver oil. Dietary eugenol (0.17%) increased SOD, CAT, and GPX activity [210, 211, 235]. Eugenol also prevented tumor formation in mice induced by 7,12-dimethylbenz(α-)anthracene and croton oil by preventing superoxide formation and lipid peroxidation [236]. Eugenol also prevented liver CCl_4 toxicity in rats. An oral administration of eugenol at a dose of 25 mg/kg before or during CCl_4 treatment reduced liver morphological changes and aspartate transferase (ASAT) levels. Eugenol at the same dose also lowered lipid peroxide levels and serum alanine aminotransferase (ALT), ASAT, and lactate dehydrogenase (LDH) when administered with iron, suggesting it can reduce iron-induced hepatic damage by lowering lipid peroxidation. Eugenol's antioxidant and radical-scavenging properties make it a promising compound for preventing and treating oxidative stress-related diseases like cancer [228, 237].

2.3.3.5 Black Pepper and Piperine

Piperine (Figure 2.1), a bioactive compound present in black pepper, has a pungent taste. Piperine inhibits free radicals, ROS, and lipid peroxidation, protecting against oxidative damage in *in vitro* experiments [238]. Piperine scavenges hydroxyl radicals at low concentrations but activates the Fenton reaction and generates them at higher concentrations. It potently inhibits 52% of superoxide radical scavenging with an IC_{50} of 1.82 mM. Piperine protects LDL from copper ion-induced lipid peroxidation. Piperine modulates the 5-lipoxygenase (5-LO) pathway, suggesting a role in antioxidant physiological processes [131, 133]. Piperine may improve antioxidant defense in non-hepatic tissues in the context of diabetes-induced oxidative stress. It prevents oxidative changes caused by chemical carcinogens, stops lipid peroxidation, and restores glutathione and enzyme activity in a model of the intestinal mucosa [239]. Piperine at 50 mg/kg body weight in rats suppressed experimental lung carcinogenesis, reduced mitochondrial lipid peroxidation, and increased enzymatic and nonenzymatic antioxidant levels. In rats fed a high-fat diet, piperine and black pepper supplementation reduced lipid peroxidation and maintained antioxidant enzyme and glutathione levels, suggesting protection against oxidative stress [205]. Piperine inhibits cellular proliferation in triple-negative breast cancer (TNBC) cells [240, 241]. Piperine reduced Akt phosphorylation in breast cancer cell lines, reducing the Akt signaling cascade. This downstream effect increased breast cancer cell apoptosis. It also inhibits the phosphatidylinositol-3-kinase/Akt pathway, causing breast cancer cells to undergo apoptosis.

Piperine treatment disrupts mitochondrial membrane integrity in breast cancer cells, leading to the release of cytochrome c and Smac/DIABLO from the mitochondria into the cytosol. This release triggers apoptosome formation and the subsequent activation of caspase-9, leading to apoptosis. In triple-negative breast cancer (TNBC) cells, piperine inhibits the synthesis of MMP-2 and MMP-9, potentially reducing their metastatic capability [242, 243].

In human ovarian cancer cells (A2780), piperine showed dose-dependent and time-dependent antiproliferative efficacy, while it had no antiproliferative effect on normal human surface epithelial (OSE) cells, demonstrating its selective cytotoxicity. Piperine caused apoptosis in A2780 cells in a dose-dependent manner. Piperine inhibited migration of OVCAR-3 cell lines in a concentration-dependent manner. Apoptotic cell death and G2/M cell cycle arrest caused this antiproliferative effect. The PI3K/Akt/GSK3β signal transduction pathway was also implicated. Therefore, piperine may be a clinically useful ovarian cancer treatment. Piperine inhibited the STAT-3 pathway and p38, suggesting it could be an anti-gastric compound targeting TMK-1 cells [244–246]. Piperine inhibits cell proliferation and induces apoptosis in human lung cancer cells (A549) and fibroblast-normal lung cells (W138). Piperine did not harm W138 lung fibroblasts. Piperine concentration-dependently modulates p53 gene expression. A tumour suppressor protein called p53 stops the cell cycle, especially in the G2/M phase, and induces cancer cell apoptosis. Thus, piperine-induced p53 upregulation may cause A549 cell cycle arrest at the G2/M phase. Bcl-2 regulates p53-dependent apoptosis [247, 248].

Piperine induces apoptosis and cell cycle arrest in the oral squamous carcinoma cell line (KB). It also induces G2/M cell cycle arrest in KB (oral squamous carcinoma) cells. Piperine reduces KB cell growth by 15.57% and 37.79% at 100 and 200 μM respectively. S-phase cells were reduced by DNA depletion, resulting in an increase in apoptotic cell numbers. Caspase-3 cleaves substrates to induce cancer cell apoptosis. This apoptotic pathway is either caspase-independent or caspase-dependent. Piperine activates caspase-3 via the caspase-dependent pathway to induce cancer cell apoptosis. The above findings suggest piperine may treat oral squamous carcinoma [249–251]. Piperine stops the growth of prostate cancer (PCa) cell lines AD LNCaP, AI PC-3, DU-145, and 22RV1. The most sensitive to piperine are LNCaP, PC-3, 22RV1, and DU-145. Caspase-3 starts apoptosis in prostate cancer cells. Piperine activates prostate cancer cell caspase-3. The antiapoptotic factor STAT-3 affects prostate cancer induction. Piperine inhibits STAT-3 phosphorylation in DU145, PC-3, and LNCaP cells. STAT-3 inhibition suppresses prostate cancer cell growth. Piperine also inhibits DU145 prostate cancer cell migration [252, 253]. Piperine slows the cell cycle, increases ROS production, and induces apoptosis in rectal cancer cells. These findings suggest piperine may treat rectal cancer. Piperine inhibits Wnt/β-catenin in HCF-116 (colorectal cancer cell lines). Moreover, overexpression of β-catenin, β-cateninS33A, or dnTC4FP16 suppresses nuclear localization. This inhibits Wnt/β-catenin signaling and cell movement in colorectal cancer cells. Piperine stops cancer cells from moving and spreading, and it reverses the expression of biomarkers for epithelial-to-mesenchymal transition, which stops phenotypic shift. Another piperine-related colorectal cancer study showed STAT3 downregulation modulates key pathways and processes involved in cancer progression and metastasis, suggesting its use in rectal cancer treatment [254, 255].

2.3.3.6 *Fenugreek*

Fenugreek is a common spice used in various recipes to enhance the flavor of the diet. It has been reported for multiple biological activities, such as antioxidant and anticancer activities. In diabetic rats that had been exposed to alloxan, fenugreek seeds were said to have a positive antioxidant effect. It was discovered that treatment with fenugreek seed powder (at a dose of 2 g/kg for 30 days) could mitigate these effects and restore normal levels of antioxidants and oxidative stress in the liver, kidneys, and pancreas. When exposed to Fe^{2+} and ascorbate *in vitro*, the Ca^{2+}-ATPase activity in liver homogenate has also shown a protective effect of fenugreek seed aqueous extract. This suggests that the antioxidant property of the seeds may be due to their soluble portion. Additionally, it increased the antioxidant enzymes CAT, SOD, and GPX and decreased tissue peroxidative damage. These results show that fenugreek seeds have advantageous antioxidant properties that can shield us from a variety of diseases [205].

Fenugreek extract also inhibited the growth of pancreatic and prostate cancer cell lines (DU-145 and PCa) when exposed for 72 hours at a dose of 10–15 μg/mL. The extract exhibited cytotoxic effect against PCa cell lines when tested at higher doses (15–20 μg/mL), but was non-toxic to normal cells like hTERT-immortalized prostate cells. The extract also induced molecular alterations in PCa cells, including downregulation of mutant p53 in DU-145 cells, upregulation of p21 in PC-3 cells, and inhibition of TGF-induced phosphorylation of Akt. Fenugreek extract might be effective in the prevention of pancreatic and prostate cancer [256].

2.3.3.7 *Ginger and Gingerol*

Ginger, a spice with therapeutic value, contains bioactive substances with medicinal value. Recent *in vitro* studies have investigated the antioxidant potential of ginger extracts. The alcohol extract has a higher level of DPPH radical scavenging activity than synthetic antioxidant BHT, with 90% inhibition, and contains a high content of total phenols (870 mg/g). The strong antioxidant properties of ginger are shown by its low IC_{50} concentration of 0.64 μg/ml, which means 50% DPPH inhibition. Ginger contains 6-gingerol and 6-paradol, which have anti-inflammatory, antioxidant, antitumor, and antiproliferative properties. The antioxidative and anti-inflammatory properties of 6-gingerol (Figure 2.1) are thought to be responsible for its chemopreventive and chemoprotective effects. Gingerol effectively inhibits NO production, reduces iNOS expression in lipopolysaccharide-stimulated macrophages, and prevents free radical peroxidation in the presence of Fe^{3+} and ascorbate. Gingerol reduces oxidative DNA single-strand breaks, inhibits the formation of 3-nitrotyrosine in proteins like bovine serum albumin and macrophage cells, and protects against peroxynitrite-mediated oxidation and nitration. It also suppresses peroxynitrite-induced oxidation of dichlorodihydrofluorescein. These results highlight gingerol's potent NO inhibition and its role in preventing cellular damage by peroxynitrite [205].

Conjugation of gingerol with diverse metals has demonstrated notable anticancer properties against colon cancer cell lines, specifically HT-29 and HCT-116. Among these various conjugates, gingerol-silver conjugates have displayed pronounced and potent anticancer effects, characterized by IC_{50} values of 5.246 μM and 3.1 μM for the HCT-116 and HT-29 cancer cell lines, respectively [257]. An analogue of ginger called [6]-gingerol has been shown in earlier studies to have the ability to reduce oxidative stress. By lowering the levels of inflammatory mediators and free radicals while increasing the presence of antioxidant enzymes like CAT, SOD, GPx, GST, and reduced GSH. Gingerol, shogaol, and other ketone-phenolic derivatives are responsible for ginger's inherent antioxidant properties. By lowering levels of MDA and NO, protein carbonyl content, epoxide hydrolase (EPXH), GST, SOD, CAT, GSH concentration, and total antioxidant capacity (TAC), ginger extract has shown in animal models that it can reduce the damage caused by oxidative stress. The ability of [6]-gingerol to act as an antioxidant has been studied in numerous studies. It has strong abilities to get rid of radicals, with IC_{50} values of 26.3, 4.0, and 4.6 μM against DPPH, superoxide, and hydroxyl radicals, respectively. Additionally, at a concentration of 6.0 μM, it successfully reduces the production of ROS in human polymorphonuclear neutrophils. Additionally, [6]-gingerol protects the intestinal mucosa from harm due to ischemia or reperfusion by preventing the production of ROS and the activation of p38 MAPK. Recent research has shown that [6]-gingerol has the ability to inhibit breast cancer cells, particularly MDA-MB-231 and MCF-7 cells. This is accomplished by increasing the production of cellular and mitochondrial ROS, which then activates the p53 gene and the ATM protein kinase, escalating the DNA damage response. The EGFR/Src/STAT3 signaling pathway facilitates the activation of p53, which ultimately results in apoptosis. Additionally, [6]-gingerol controls the BAX/BCL-2 ratio and facilitates cytochrome-C release to orchestrate G0/G1 cell cycle arrest and promote mitochondrial apoptosis [258].

Gingerol has an extensive effect on numerous cell signaling pathways that are closely linked to the development of cancer. Noteworthy pathways affected by gingerol include Signal Transducer and Activator of Transcription 3 (STAT3), Nuclear Factors (NF-KB), Activator Protein-1 (AP-1), Growth Factor Receptors (such as EGFR and VEGFR), β-catenin, Mitogen-Activated Protein Kinases (MAPK), and pro-inflammatory mediators such as Tumor Necrosis Factor-alpha (TNF-α) and cyclooxygenase-2 (COX-2). Studies conducted through *in vivo* and *in vitro* support the hypothesis that gingerol plays a role in the prevention of cancer [259].

2.3.4 Antioxidants from Vitamins and Minerals

The micronutrient vitamin A, which is necessary for life, supports a number of metabolic processes. It improves immune defenses and serves as an antioxidant, potentially preventing oxidative stress-related illnesses like cancer, cardiovascular and neurodegenerative diseases, and cellular aging. Carotenoids, which are forms of vitamin A found in fruits and vegetables, are converted into retinol before being absorbed by the body. Notably, mucosal cells only directly absorb about half of the vitamin A found in substances like beta-carotene. The most potent antioxidant form of vitamin A is retinol, which exhibits scavenging effect against peroxyl radicals and halting chain reactions. By contributing hydrogen atoms and taking part in addition reactions with hydroperoxyl radicals (HOO•), retinol serves as an antioxidant. For men, the Recommended Dietary Allowance (RDA) for vitamin A is 900 g, while for women, it is 700 μg [260–265].

Ascorbic acid, a form of vitamin C, has greater antioxidant effectiveness than vitamin A. Vitamin C is the only substance that can renew vitamin E's antioxidant form, increasing its availability in human metabolic processes and providing significant health advantages. For numerous enzymes involved in immune responses, cell regulation, hormone biosynthesis, and neurotransmitter biosynthesis, ascorbic acid serves as a cofactor. By donating a hydrogen atom to create a transient ascorbic radical (dehydroascorbic acid), which can then be changed back into active vitamin C, vitamin C effectively quenches ROS and nitrogen oxide species in its antioxidant role. Because it is water-soluble, vitamin C has additional benefit of functioning both intracellularly and extracellularly. Because of this property, vitamin C is a powerful antioxidant that can reduce oxidative reactions that affect lipids and macromolecules. The RDA for vitamin C ranges from 65 to 90 mg per day, with a maximum of 2000 mg per day to reduce any potential side effects [11, 266–268].

Tocopherols and tocotrienols are just two of the related substances that make up vitamin E, which is one of the fat-soluble vitamins. Alpha-tocopherol stands out among them for its ability to stop lipid oxidative reactions that might damage cell membranes. To do this protective job, a hydrogen atom is taken from the phenolic group of vitamin E's structure (the chromanol ring) and added to the phenolic group. Notably, reduction reactions with other antioxidants like vitamin C or retinol can restore alpha-tocopherol's oxidized form to its active antioxidant state. Additionally, vitamin E is essential for supporting the immune system. It helps with cell tissue repair, inhibits the synthesis of mutagenic substances, and provides defense against bacterial infections. Because of its immune-enhancing properties, vitamin E is crucial for lowering the likelihood that cancer cells will grow into tumors. In comparison to the higher RDAs for vitamins A and C, the RDA for vitamin E is relatively low at 15 mg per day [268, 269].

Minerals found in fruits and vegetables are beneficial because they help keep cellular water balance, electroneutrality, and enzyme co-factor functions in check. These foods contain selenium, magnesium, zinc, and copper, as well as other important minerals.

As a component of antioxidant enzymes like glutathione peroxidase, selenium plays a crucial role in the reduction reactions involving H_2O_2 and peroxide radicals. Additionally, it supports the activity of thioredoxin reductase, which keeps thioredoxins in their reduced state, and iodothyronines, which are responsible for eliminating iodide and thyroxine. Selenium participates in selenoprotein complexes with antioxidant properties, providing a variety of health benefits. These advantages include the protection of cells and molecules, the synthesis of hormones, and the avoidance of diseases like atherosclerosis, cancers, and cardiovascular risks [270, 271].

The importance of manganese is derived from its participation in a variety of enzymatic processes, particularly as a necessary component of manganese superoxide dismutase, an enzyme with antioxidant properties necessary for preventing oxidative stress in mitochondria during ATP synthesis. Magnesium also functions as a cofactor in the processes that produce ATP [272, 273].

Over 2000 transcriptional factors and more than 300 enzymes require zinc, an essential trace element. In addition to protecting biomolecules from oxidation, it also stimulates enzyme activity and inhibits the activity of nitric oxide synthase, lipid peroxidation products, and NADPH oxidase. Reactive nitrogen and oxygen species, such as H_2O_2, hydroxyl radicals, superoxide anions, and peroxynitrites, are inhibited by zinc. It is noteworthy that it has direct antioxidant properties, especially when interacting with thiol groups. Oxidative stress can be brought on by both zinc excess and deficiency, highlighting the importance of maintaining healthy zinc levels for human wellbeing [274–278].

Table 2.5 Vitamin and mineral content in fruits.

Fruits	Vitamins			Minerals			
	A [1]IU, [2]µg	**C mg/100 g**	**E mg/100 g**	**Cu (mg)**	**Mn (mg)**	**Se ([1]mg,[2]µg/100 g)**	**Zn (mg)**
Apple	5.4[1]	4.6	0.08	0.03	0.04	N/D	0.04
Banana	1.0[2]	12.3	0.1	0.10	0.26	2.50[2]	0.16
Grapefruit	1150[1]	31.2	0.15	0.06	0.06	1.48[1]	0.25
Kaki	163[2]	7.0	0.73	N/D	N/D	0.60[2]	Traces
Lemon	22[1]	53	0.15	0.04	0.05	2.77[1]	0.22
Melon	232[2]	10.9	0.1	0.08	0.05	1.70[2]	0.44
Orange	225[1]	53.2	0.18	0.11	0.08	1.80[1]	0.21
Pepper	1.6[1]	47.5	0.7	0.83	2.74	N/D	3.01
Peach	163[1]	95	15.3	0.08	0.03	2.10[2]	0.23
Pear	25[1]	45	0.12	0.07	0.03	0.20[2]	0.07
Plum	20[2]	6.0	0.2	0.06	0.05	0.50[2]	0.10
Strawberry	1.0[1]	58.8	0.29	0.05	0.039	Traces	0.14
Tangerine	681[1]	26.7	0.2	0.04	0.07	2.58[1]	0.26
Tomato	614.4[1]	36.2	0.02	0.03	0.09	2.50[2]	0.08
Watermelon	18[2]	5.0	0.05	0.04	0.03	0.04[1]	0.01

Table 2.6 Vitamin and mineral content in selected vegetables.

Vegetables	Vitamins			Minerals			
	A (µg)	C (mg/100 g)	E (mg/100 g	Cu (mg)	Mn (mg)	Se (µg/100 g)	Zn (mg)
Artichoke	4.6	6	traces	-	0.12	0.20	0.05
Beans	53	18	0.1	0.04	0.18	N/D	0.19
Broccoli	8.0	91.3	0.15	0.06	0.20	1.60	0.42
Cabbage	4	49	0.2	N/D	0.21	N/D	0.26
Carrot	1346	7	0.50	0.08	0.14	N/D	0.28
Cauliflower	traces	47	0.12	-	-	0.40	0.60
Celery	95	8	0.2	-	-	0.40	0.10
Endive	17	10	1	-	-	2.80	0.30
Green peas	53	18	0.1	-	-	1.90	0.80
Garlic	traces	14	0.1	-	-	traces	-
Lettuce	436	4.6	0.14	0.05	0.13	traces	0.25
Mushroom	N/A	4	0.12	0.11	0.09	traces	0.68
Onion	N/A	6.9	0.45	0.05	0.10	traces	0.12
Pickle	4.0	2.1	0.12	0.03	0.06	N/D	0.11
Pumpkin	34	12	0.1	0.05	0.05	11	0.26
Spinach	306	30.3	1.72	0.08	0.43	2.50	0.42
Sweat corn	25	6	0.4	-	-	0.80	0.30
Zucchini	4.6	20	traces	0.05	0.18	0.20	0.30

A crucial trace element called copper serves as a cofactor in the oxidation reactions that change ROS into water. Its presence is essential for iron metabolism, immune responses, molecule synthesis, enzyme activation, and antioxidant defense. Having insufficient copper levels can harm cells by oxidizing them. Different countries' scientific organizations and authorities may have different RDA values for these minerals. However, the WHO recommends RDA ranges of 25-34 μg for selenium, 400 mg for magnesium, 9.9 mg for zinc, and 11 μg/kg of body weight for copper [279, 280].

Vitamins play vital roles in diverse metabolic and biological processes in the human body. Essential vitamins such as A, C, and E, along with minerals like selenium, copper, zinc, and manganese, prominently feature in a variety of fruits (apple, banana, grapefruit, persimmon, lemon, melon, orange, pepper, peach, pear, plum, strawberry, tangerine, tomato, and watermelon) and vegetables (artichoke, beans, broccoli, cabbage, carrot, cauliflower, celery, endive, green peas, garlic, lettuce, mushroom, onion, pickle, pumpkin, spinach, sweet corn, and zucchini). Refer to Tables 2.5 and 2.6 for mineral and vitamin contents in these fruits and vegetables.

Furthermore, kiwi fruit stands out with high vitamin C and E content at 92.7 mg/100 g and 0.71 μg/100 g, respectively. Papaya, another commonly consumed fruit, contains 328 mg/100 g of vitamin A. Papaya leaves, pulp, and seeds also contribute to vitamin C levels of 31.1, 61.8, and 11.7 mg/100 g, respectively, along with 52 μg/100 g of β-carotene. Bananas contain 6.9-10 mg of ascorbic acid (vitamin C) per 100 g [68].

2.4 Roles of Antioxidants in Cancer Prevention

We already described earlier in the chapter the role of free radicals in cancer formation. Apoptosis, cell cycle, metastasis and angiogenesis are the main target pathways for the antioxidant compounds to prevent the growth of cancer cells. Apoptosis serves as an essential line of defense against the emergence and spread of cancer. The frequent exposure of cells to carcinogens results in disturbances of apoptotic signaling pathways that lead to cancer cell formation. Antioxidants increases the proapoptotic proteins and enzymes that act on multiple signaling pathways and induced apoptosis in cancer cells. Furthermore, DNA checkpoints that protect the integrity of DNA replication regulate the three distinct cell cycle phases (G0/G1, S, and G2/M) through which cell proliferation occurs. Cancer risk is increased by altered checkpoint and repair pathways. Antioxidants control the cell cycle and arrest the growth of cancer cells. Cell-cycle machinery depends on cyclins, cyclin-dependent kinases (CDKs), and CDK inhibitors. In many cancers, CDKs and cyclins are overactive, making CDK suppression a therapeutic target to stop unchecked cell proliferation. Antioxidants and anticancer drugs inhibit CDK and cyclins and thus prevent cancer cell proliferation. Cancer cells also invade the vasculature during the intricate, multi-step process of metastasis, which then spreads to distant organs. Therapy resistance in metastatic cancer presents a serious problem because it accounts for 80% of cancer-related deaths. Serine proteases, plasminogen activators, and matrix metalloproteinases are proteases that are secreted by tumors. They are necessary for the proteolytic breakdown of the extracellular matrix, which is needed for tumor cells to invade other cells. Antioxidants from fruit, vegetables, and spices and their active compounds are reported to prevent metastatic cancer cells. Angiogenesis, which is essential for healthy wound healing and embryonic development, involves the growth of new blood vessels, and also plays a significant role in the development of cancer. Extracellular matrix remodeling, endothelial cell migration, and tube formation are all part of this complex process. The important role of vascular endothelial growth factor as a regulator is what motivates angiogenesis. Antioxidants promote wound healing by modulating various pathways in angiogenesis and protecting cells from cellular damage caused by cancer cells [281].

Antioxidants from fruits, vegetables, and spices decrease the excessive level of free radicals present in the body. Even at low concentrations, antioxidants protect the immune system from the accumulation of free radicals that are responsible for the damage to biomolecules like carbohydrates, lipids, proteins, and DNA. Antioxidants do this by preventing oxidative stress from oxidation by donating electrons and reacting with free radicals. They stop the chain reaction and protect biomolecules. Antioxidants also increased the levels of enzymes in the body such as SOD, CAT, and GPX, as well as non-enzymatic antioxidants that the body produces on its own, including glutathione, lipoic acid, L-arginine, bilirubin, and uric acid. Exogenous antioxidants come from the diet and include minerals like zinc, copper, manganese, and selenium; vitamins A, C, and E; and phytochemicals like flavonoids, phenolics, carotenoids, and curcuminoids, all of which exhibit powerful antioxidant action by scavenging free radicals, reducing oxidative damage, and preventing the abnormal cell division that leads to cancer. Furthermore, both endogenous and exogenous antioxidants work by providing electrons to neutralize ROS, thereby protecting body cells from oxidative stress. In some cases, they exhibit a synergistic antioxidant effect. For example, when the body receives oral vitamin E, it increases the levels of glutathione in the blood, resulting in enhanced antioxidant effects [68, 205].

Furthermore, natural antioxidants are becoming more popular due to their easy availability, being relatively cheap, and having the ability to protect us from cancer. On the other hand, synthetic antioxidants are expensive and produce toxicity in overdose, which leads to pro-oxidation. Antioxidant-rich foods can reduce such side effects. Thus, eating antioxidant-rich foods as part of a balanced diet can improve antioxidant levels, increase the activity of enzymes responsible for the prevention of oxidation processes in the body, and prevent disease [1].

Flavonoids, which belong to polyphenols, have been reported to have various biological activities, especially anti-inflammatory and antioxidants, which drive them to exert anticancer potential or reduce the chances of cancer. These molecules possess multiple hydroxy groups and a highly conjugated electron system within their chemical structure, enabling them to effectively scavenge free radicals through the donation of hydrogen atoms or electrons. Additionally, these compounds possess the capacity to impede the generation of ROS, pivotal contributors to oxidative stress. Notably, flavonoids inhibit the translocation of pivotal transcription factors within the inflammatory cascade, preventing their access into the cellular nucleus, stopping their expression or phosphorylation, and protecting the tissues from inflammation. Supported by numerous evidences, flavonoids exhibit modulation of diverse signaling pathways, including nuclear factor B (NF-B), focal adhesion kinase (FAK), the inflammasome, mitogen-activated protein kinase (MAPK), and signal transducer and activator of transcription 3 (STAT3). Furthermore, flavonoids disrupt the activity of fatty acid synthase, an enzyme integral to apoptosis initiation. The inhibition of tyrosine and serine/threonine kinases by flavonoids further contributes to their impact on cancer cell proliferation. In a concerted effort to counter cancer progression, flavonoids also exert their influence on cytochrome P450 and phase II enzymes. By instigating autophagy and attenuating angiogenesis through the downregulation of vascular endothelial growth factor (VEGF) and matrix metalloproteinase (MMP) protein expression, flavonoids demonstrate a multifaceted approach to impede tumorigenesis. Phenolic acids, which belong to polyphenols, have been reported to exhibit antioxidant properties. It also possesses anticancer potential through the detoxification of free radicals, preventing the cellular damage and inhibiting the formation of NO, which is responsible for abnormal cell division. Phenolic acid also inhibits the proliferation of cancer cells and induces apoptosis. Carotenoids are biologically active pigments found in many fruits and vegetables. Carotenoids' ability to reduce free radicals by acting as free radical scavengers and inhibitors of oxidative stress, which are linked to many chronic diseases such as cancer, Carotenoids inhibit tumor growth by arresting the G1/G0 phase in the cell cycle, inducing apoptosis, and inhibiting the metastasis of cancer cells. They are also vital for the synthesis of nutrients such as vitamin A [65, 281]. The diagrammatic anticancer mechanisms of flavonoids, phenolic acid, and carotenoids are mentioned in Figures 2.2–2.4.

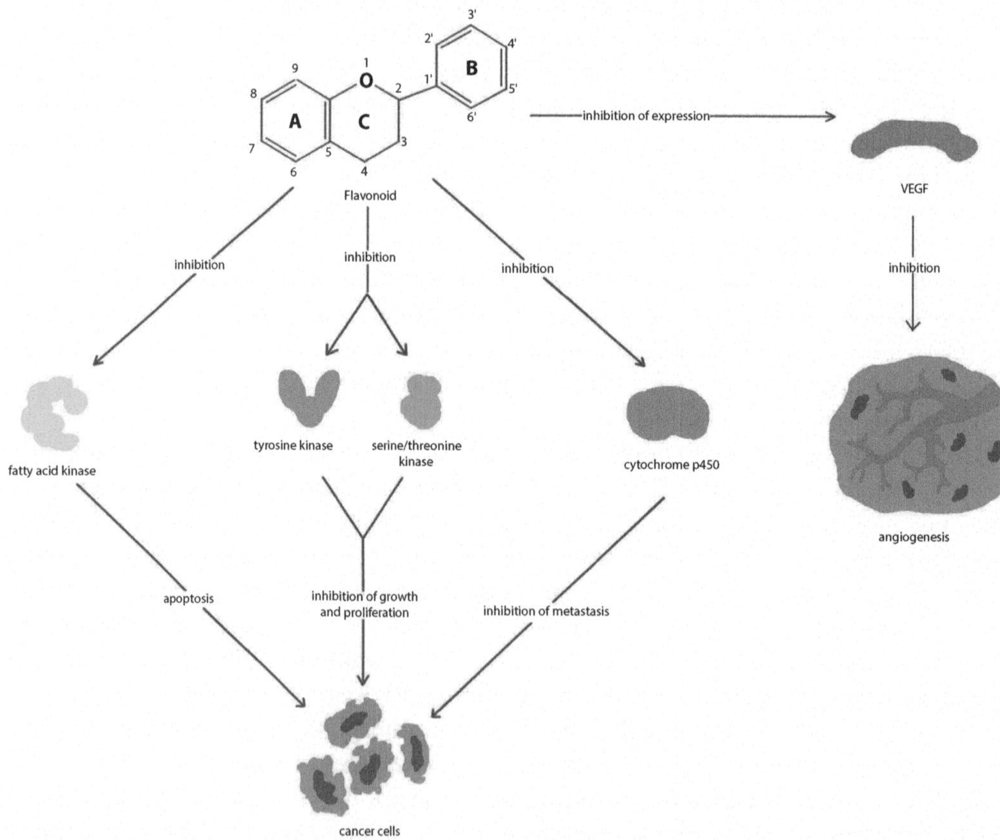

Figure 2.2 Anticancer mechanism of flavonoids. Reproduced from Anna Rudzinska, Nutrients; published by MDPI, 2023, under the Creative Commons Attribution (CC BY) license (https://creativecommons.org/licenses/by/4.0/) [280].

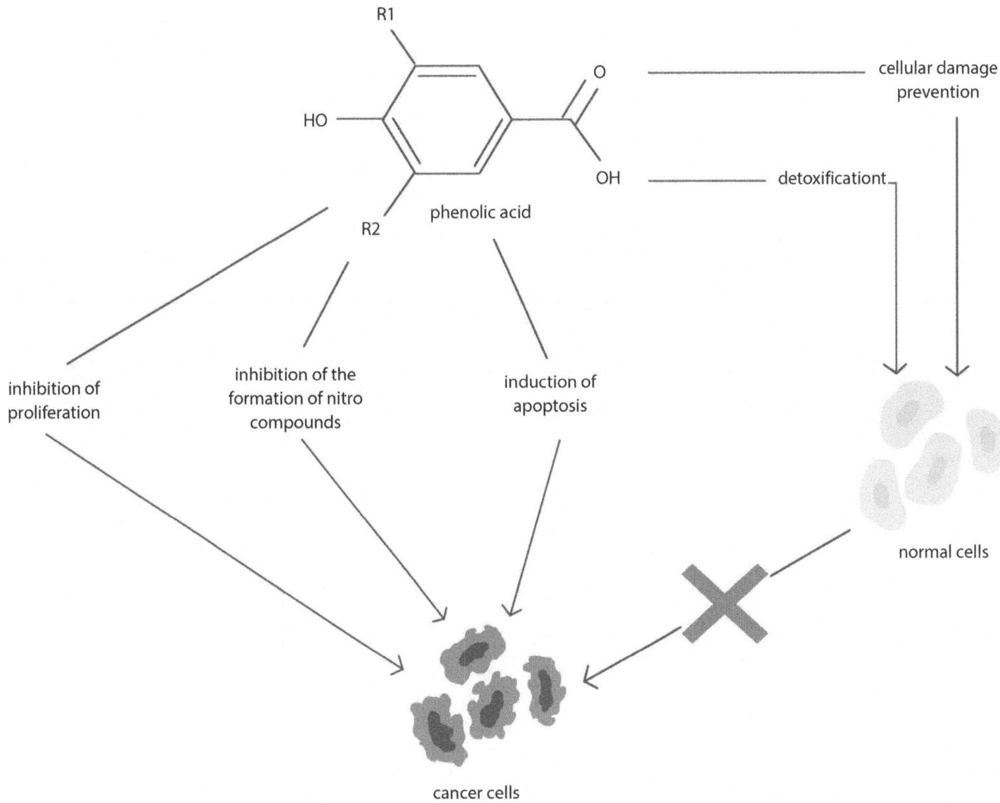

Figure 2.3 Anticancer mechanism of phenolic acids. Reproduced from Anna Rudzinska, Nutrients; published by MDPI, 2023, under the Creative Commons Attribution (CC BY) license (https://creativecommons.org/licenses/by/4.0/) [280].

2.5 Can Antioxidant Supplements Help to Prevent Cancer?

The effects of antioxidant supplements, including beta-carotene, vitamin A, vitamin C, vitamin E, and selenium, both individually and in combination, on gastrointestinal cancers were examined in a Cochrane systematic review and meta-analysis of 14 randomized clinical trials. The review did not uncover any strong proof in favor of these antioxidant supplements' protective effects against gastrointestinal cancer. Unexpectedly, these supplements were linked to a rise in all-cause mortality, casting doubt on their supposed beneficial effects. A significant scientific discussion on the dual nature of these antioxidant supplements was sparked by this discovery. Later studies confirmed these findings, which led to more research. The Cochrane review examined the effects of antioxidant supplements on all-cause mortality, encompassing 78 randomized clinical trials with 296, 707 participants. The findings repeatedly fell short of offering evidence in favor of using these supplements to prevent either primary or secondary diseases. High doses of vitamin E, vitamin A, and beta-carotene have all been linked to a higher mortality rate. Other systematic reviews and meta-analyses ran into similar issues, but they instead highlighted potential negative effects. Studies looking at beta-carotene, vitamin E, and multivitamin supplementation found that participants had higher rates of cardiovascular and all-cause mortality. Additionally, these

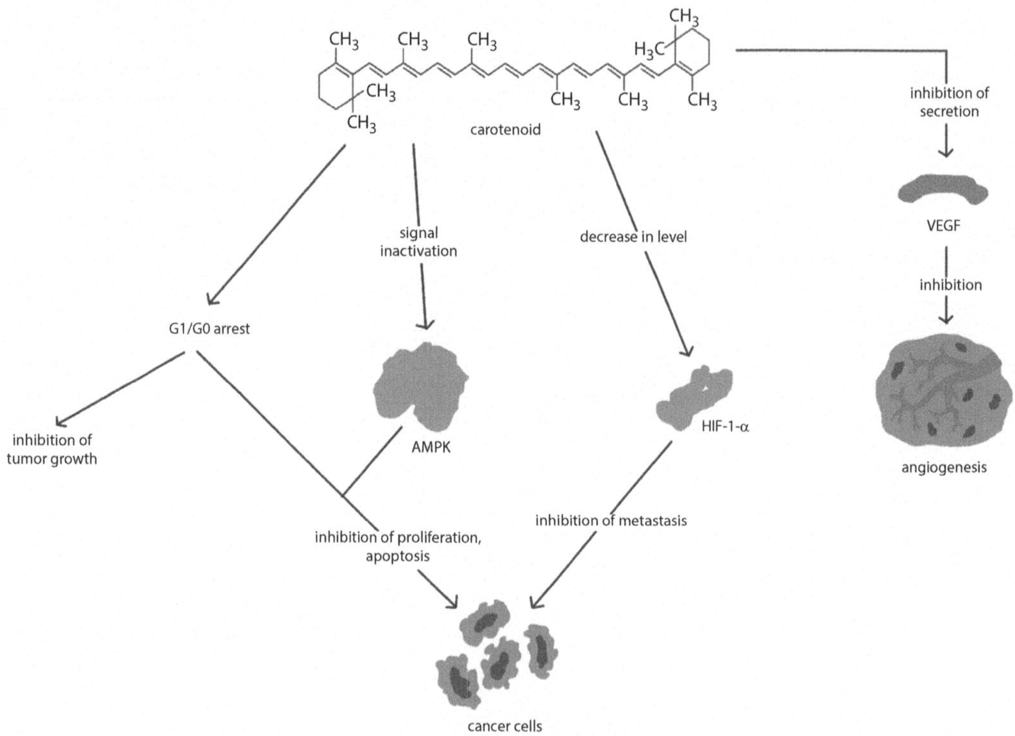

Figure 2.4 Anticancer mechanism of carotenoids. Reproduced from Anna Rudzinska, Nutrients; published by MDPI, 2023, under the Creative Commons Attribution (CC BY) license (https://creativecommons.org/licenses/by/4.0/) [280].

antioxidant supplements did not show any advantages in the prevention of cancer or cardiovascular disease. These conclusions were supported by studies at lower levels of the evidence hierarchy. Prospective studies linked regular multivitamin use to an elevated risk of lung cancer, especially lung cancer that is advanced and fatal. In other cohort studies, taking multivitamins and particular antioxidant supplements was associated with a higher risk of all-cause mortality. These studies demonstrate the importance of cautious consideration and evidence-based evaluation of antioxidant supplement use because the benefits of these supplements have not always been consistently demonstrated in scientific studies [282].

The most recent clinical trial examining vitamin A chemoprevention, the Alpha-Tocopherol, Beta-Carotene (ATBC) trial, targeted male participants aged 50 to 69 who were regular smokers. For a duration of 5 to 8 years, participants in the trial received daily supplements of beta-carotene (BC) at a dose of 20 mg/day, with or without vitamin E (50 mg/day). Surprisingly, BC supplementation was associated with an 18% increased risk of lung cancer and an 8% increase in the overall mortality rate, which were primarily attributed to ischemic heart disease and lung cancer. When BC supplementation was stopped but smoking persisted 18 years later in a follow-up study, the initial BC group did not show an increased risk of lung cancer. They did, however, have a 20% increased risk of dying from prostate cancer. The Carotene and Retinol Efficacy Trial (CARET), a significant investigation into the effect of vitamin A supplementation on lung cancer prevention, was conducted before the ATBC trial. Participants with a higher risk of developing lung cancer were enrolled in the CARET study, including current and former smokers with a significant smoking history and people who had been exposed to asbestos. Two vitamin A supplements were given to the subjects: retinyl palmitate (25,000 IU/day) and beta-carotene (30 mg/day). After four years, the trial was abruptly stopped because of unanticipated results, which included a 28% increase in lung cancer risk, a 46% increase in lung cancer-related mortality, and a 17% increase in all-cause mortality. Six years after the supplementation ended, a follow-up analysis revealed no appreciable differences in lung cancer risk between the experimental and control groups, with the exception of the women who were assigned to take the two vitamin A supplements, who had a 33% higher risk [283–286].

Observational studies have explored the relationship between dietary vitamin C intake and cancer risk, yielding varying outcomes. Bo and colleagues conducted a comprehensive meta-analysis to evaluate the influence of dietary vitamin C on the risk of esophageal cancer. The meta-analysis synthesized data from 15 studies, encompassing a total of 7063 control participants and 3955 individuals diagnosed with cancer. The findings of this analysis indicate a noteworthy inverse correlation between higher dietary vitamin C intake and the risk of esophageal cancer [287, 288]. Furthermore, there was no evidence of a link between dietary or supplemental vitamin C intake and the prevention of cancer in the PROTEUS study, a case-control investigation involving 1916 patients with prostate cancer and 1915 matched controls. Because of this, the body of research on vitamin C supplementation for cancer prevention is incomplete [289].

The potential preventive role of tocopherol (vitamin E) in colorectal, lung, prostate, and breast cancer was investigated in a thorough meta-analysis that included case-control and cohort studies conducted between 1986 and 2009. This analysis highlighted positive outcomes connected with vitamin E supplementation, despite a variety of null findings. The meta-analysis included two case-control and six cohort studies for colorectal cancer. While three studies found a reduced risk with higher tocopherol intake, other studies were unable to find a protective relationship. Four case-control and three cohort studies on lung cancer were conducted. Certain cohort studies and case-control studies suggested reduced lung cancer risk with high dietary vitamin E intake, even in smokers, and low serum tocopherol in lung cancer patients, respectively. Nine cohort studies and 14 case-control studies on prostate cancer were reviewed. Two case-control studies specifically linked lower serum tocopherol to reduced risk, and some case-control studies generally suggested a decreased risk with elevated tocopherol levels. Cohort studies produced conflicting findings, with one indicating a decreased risk of advanced prostate cancer with dietary intake of tocopherols. Nine cohort studies and fifteen case-control studies on breast cancer were evaluated. Vitamin E levels and the risk of breast cancer are inversely correlated in some case-control studies, but not in others [290]. Nineteen studies were included in another meta-analysis on vitamin E and the risk of developing prostate cancer. With an estimated relative risk (RR) of 0.97 (95% CI: 0.92-1.02), it was discovered that dietary vitamin E intake and prostate cancer risk were not significantly associated. A similar analysis found no correlation between additional vitamin E intake and the risk of prostate cancer, with an estimated RR of 0.99 (95% CI: 0.94-1.04). The combined analysis of vitamin E intake from food and supplements resulted in an estimated risk ratio (RR) of 0.97 (95% CI: 0.85-1.08), reiterating the absence of a clear association between cumulative vitamin E intake and prostate cancer risk [291].

The belief in the beneficial health effects of antioxidant supplements endures despite the abundance of evidence casting doubt on them, particularly among populations that eat well. Antioxidant supplements are becoming increasingly popular among people who want to boost their immune systems, break bad habits, and slow down the ageing process. Worldwide acceptance of the idea that "more antioxidant supplements, the better" has grown. More than half of adults in high-income countries consume dietary supplements, frequently in the form of multivitamins with or without minerals. There is no substantiated evidence to support the use of multivitamin and mineral supplements for the primary prevention of chronic diseases, according to the 2010 Dietary Guidelines for Americans as well as a State-of-the-Science Conference sponsored by the National Institutes of Health. Instead of antioxidant supplements in the form of pills or tablets, it appears that the most effective source of antioxidants comes from our daily diet [292, 293].

A case-control study among women in Long Island, New York, suggested an association between flavonoid intake and a decreased risk of breast cancer. The greatest reduction in risk for lignans, flavanols, flavones, and flavan-3-ols was seen in postmenopausal women. Another large case-control study done in Greece found a strong negative relationship between flavone intake and breast cancer risk. The risk dropped by 13% for every standard deviation increase in intake. Other flavonoid subclasses were found to have comparable inverse associations [294, 295]. Furthermore, various clinical studies, meta-analysis, and cohort studies conducted on a large population indicated that flavonoids decreased the chances of lung, prostate, and gastrointestinal cancer [281].

The clinical trials on phenolic acid also exhibited positive outcomes. The findings from a population-based case-control study conducted in Catania, Italy, from January 2015 to December 2016 involving 340 participants suggested that a high intake of caffeic acid and ferulic acid may be linked to a reduced risk of prostate cancer. In a prospective cohort study involving 10,812 middle-aged women who were university graduates and conducted by the Seguimiento Universidad de Navarra, the phenolic acid content of the diet, particularly dietary hydroxycinnamic and chlorogenic acids found in coffee, fruits, and vegetables, was associated with a lower breast cancer risk among postmenopausal women in the Mediterranean cohort. An open-label, randomized controlled trial conducted by Guangzhou University of Chinese Medicine evaluated the efficacy and safety of traditional Chinese medicine (TCM) in advanced non-small cell lung cancer (NSCLC) patients undergoing chemotherapy. The trial included the administration of Zhenqi Fuzheng (ZQFZ) capsules containing sinapinic acid, ferulic acid, asiatic acid, pratensein, and glycitein. The study results indicated that TCM treatment, including the ZQFZ capsules, could improve the quality of life for NSCLC patients and alleviate their symptoms with good safety profiles. These studies provide insights into the potential health benefits of phenolic acids and traditional Chinese medicine compounds in relation to cancer prevention and treatment. However, further research is needed to fully understand the mechanisms and potential applications of these compounds in cancer management [296–299]. Phenolic acid also reduced skin and esophageal cancer in various clinical trials [281].

Carotenoids, such as blood lutein, alpha-carotene, and beta-carotene, have protective effects against breast cancer. According to the dose-response analysis, each additional 5 mg of dietary beta-carotene was associated with a 5% reduction in the risk of developing breast cancer. Additionally, there is a 26% reduction in the risk of developing breast cancer for every additional 50 g of beta-carotene per dL of blood and an 18% reduction for every additional 10 g of beta-carotene per dL of blood. Furthermore, a 25 g increase in lutein per dL of blood was linked to a 32% lower risk of breast cancer [300].

Higher blood levels of lycopene, beta-carotene, and alpha-carotene were found to be significantly associated with a decreased risk of lung cancer in the meta-analysis of 18 prospective studies. This implies that people with higher blood levels of these carotenoids may have a lower risk of developing lung cancer. Similar findings were made in the second study, which examined 35 studies and discovered that a high intake of lycopene, beta-carotene, lutein/zeaxanthin, and cryptoxanthin was linked to a significantly lower risk of lung cancer. This suggests that people who consume foods high in these carotenoids may experience a decreased risk of lung cancer. These results lend credence to the idea that some carotenoids, found in a variety of fruits and vegetables, may help lower the risk of lung cancer. Carotenoids are also effective in lowering the chances of head and neck, gastrointestinal, colorectal, and gynecological cancer in various reported clinical studies [281, 284, 301].

2.6 Pharmacokinetics of Antioxidants

Although natural antioxidants have shown a significant role in preventing and overcoming cancer, but their low oral bioavailability is a prime concern in their application in the search for chemoprevention. The pharmacokinetic parameters, including absorption, distribution, metabolism, and elimination, are equally important for achieving high therapeutic output. For instance, flavonoids, when consumed, only 10% of the amount reach peak concentrations in the circulation in the nanomolar or low micromolar range after a few hours due to their low bioavailability. Humans are thought to consume anywhere between 20 and 70 mg/d, and possibly even up to 500 mg/d [281]. Phloretin exhibits limited solubility in aqueous solutions and displays a modest bioavailability of approximately 8.67%. Studies on rats have shown that oral administration of phloretin results in its efficient absorption. Notably, within a span of 5 hours, approximately 63.2% of the administered phloretin is excreted from the organism. The plasma concentration-time profile of phloretin following oral intake unveils a dual-peak pattern, attributed to processes like intestinal-hepatic circulation, dual-site absorption, and intestinal efflux. Phloretin's permeability across different segments of the intestines, including the colon, duodenum, jejunum, and ileum, exhibits variation. Remarkably, the colon manifests heightened absorption capabilities in comparison to other intestinal segments. The stability of phloretin is notably enhanced in acidic environments. With a clearance rate that is fivefold higher than the typical hepatic plasma flow, phloretin demonstrates rapid clearance from the systemic circulation. Its relatively short half-life of 2.82 hours following oral administration further supports this rapid elimination. Efforts to enhance the *in vivo* absorption of phloretin are currently underway through the exploration of various formulations, such as nano emulsions, liposomes, and microemulsions. These innovative delivery systems seek to optimize phloretin's bioavailability and therapeutic potential [302].

Carotenoids, the primary constituents of potions with anticancer properties, are facing some pharmacokinetic hindrances. Lutein, as an example of a carotenoid, has the limitation of poor water solubility. Various strategies including nano-encapsulation, are adopted for enhancing the bioavailability and efficacy of lutein. Similarly, the other core constituent of dietary antioxidant potions is beta-carotene, which contributes significantly to chemoprevention due to its immune-modifying, anti-mutagenic, and anticarcinogenic properties.

Pharmacokinetic profiles of the beta-carotene revealed in multiple reports conclude poor absorption and consequently low bioavailability. Therefore, the carotenoids are applied as concomitant therapy to provide a better therapeutic output [303]. Another commonly used carotenoid is lycopene, which is well known for its strong antioxidant potential and significant contribution to chemoprevention. The peak plasma concentration of lycopene was observed to be 0.22-1.06 μM in 24-46 hours after the ingestion of tomato juice. The elimination half-life of lycopene in humans is 2-3 days [304]. Anthocyanins are well known for their promising antioxidant-mediated anticancer effects. However, their efficacy is greatly influenced by excessive biotransformation and endogenous enzymatic degradation. A study reported that the maximum plasma concentration for anthocyanins is between 1.4 and 592 nM and happens after 0.5-4 hours of food consumption. The half-life for anthocyanins reported is about 1.5-3 hours, and the excretion through urine is approximately 0.03-4% of the ingested dose [305, 306]. It is note-worthy to mention the extensively studied antioxidant compound, curcumin's pharmacokinetic considerations. Curcumin, the major constituent of turmeric, has been a key ingredient in foods with chemoprevention properties. The poor solubility of the compound in aqueous media restricts its direct oral application with the intention of chemoprevention. Various approaches, including nano-encapsulation and pro-drug formation, have been utilized to make it applicable through the oral route of administration [307, 308]. Similarly, gingerol is reported to have excellent antioxidant potential but very poor aqueous solubility and bioavailability. Gingerol is reported to have Phase-II metabolites in urine. It is completely metabolized into glucuronide-metabolites and excreted through the kidney [309, 310]. The pharmacologically significant alkaloid capsaicin from pepper has gained much attention due to its significance in cancer; however, the low water solubility, extensive first-pass effect, and shorter half-life restrict investigators from exploring its potential, particularly in clinical settings [311]. Similarly, the primary constituent of clove oil, eugenol, is reported to exhibit a strong antioxidant-mediated anticancer effect [312]. Eugenol is found to have rapid absorption, fast metabolism, and complete excretion after 24 hours of oral administration. The plasma half-life of eugenol is 14 hours and blood half-life is 18.3 hours. The pharmacokinetic parameters of eugenol allow further research on its clinical utility for cancer prevention *via* oral route [312]. Overall, there are variations in the pharmacokinetic profiles of dietary antioxidant compounds, which led to the conclusion of further investigations before their possible clinical employment with a chemo-preventive role.

The pharmacokinetics of vitamins C, E, and A include absorption, distribution, and elimination. Ascorbic acid, a form of vitamin C, is quickly distributed throughout the body's tissues and is effectively absorbed in the small intestine. It is primarily metabolized in the liver and eliminated in the urine. The small intestine absorbs vitamin E, which is then transported by chylomicrons in the lymphatic system before combining with lipoproteins for distribution. It is broken down in the liver and primarily eliminated through bile. The small intestine absorbs vitamin A (retinol) as retinyl esters, which are then transported by chylomicrons and primarily stored in the liver. It is converted into its active forms through metabolism and eliminated through bile and urine [313–315].

Essential trace elements such as zinc, manganese, copper, and selenium exhibit distinct pharmacokinetic profiles, encompassing processes of absorption, distribution, metabolism, and elimination within the body. Zinc's absorption predominantly occurs in the small intestine and is facilitated by proteins like albumin. Subsequently, it is transported via the bloodstream and stored in various tissues, with muscles and the liver being key sites. Zinc homeostasis is meticulously regulated, with surplus amounts primarily eliminated through fecal excretion. Upon absorption in the small intestine, manganese enters the bloodstream and finds its primary reservoirs in the kidneys, liver, and bones. Its metabolism involves hepatic and biliary pathways, contributing to its excretion. Copper, primarily existing in its divalent form, undergoes absorption predominantly within the stomach and small intestine. Transport and distribution rely on binding to proteins like ceruloplasmin. Storage of copper occurs in several tissues, notably the liver, while excretion primarily occurs through bile and fecal excretion. Selenium's absorption takes place in the small intestine, where it undergoes metabolic transformation to selenocysteine, an essential component of selenoproteins. The distribution encompasses various tissues, with higher concentrations found in the liver and kidneys. Elimination predominantly occurs via urine. In summary, these essential trace elements each follow distinct pathways of absorption, distribution, metabolism, and elimination, contributing to their vital roles within the body's physiological processes [316, 317].

2.7 Safety Assessment

Phloretin, on a 28-day subacute toxicity test, increased the levels of ALT, AST, and ALP at a dose of 300 mg/kg, which was substantially higher compared to the normal rats. At the same dose, glucose and BUN levels were also significantly increased in the treated rats. As a result, the 300 mg/kg dose of the complex resulted in some toxicity in the lab animals and was not employed as a dose for the subsequent investigation [302]. The daily recommended allowances of carotenoids, including lutein, zeaxanthin, beta-carotene, and lycopene are 0.83-4.11, 0.83-4.11, 0.92-8.80, 0.83-9.43 mg, respectively. Carotenoids are safe at much higher concentrations, which allow a regular diet pattern without any intervention [318]. Curcumin has been reported with a wide therapeutic window and safety profile in various clinical trials reported in a meta-analysis review. Curcumin has had no toxicity reported in human studies [319]. Eugenol is considered a safe food additive, but individuals of various ages and conditions have responded to eugenol intake in different ways. It is reported that eugenol (5-10 mL) causes distress in children with under the age of 2 years after 3 hours of ingestion, which leads to further deterioration, including coma and seizure attacks after 8 hours. Intravenous infusion of eugenol causes hepatic toxicity and respiratory distress at doses of 4 μL and 8 μL [320]. Capsaicin, at a concentration of 0.001% in a high-fat diet administered to mice, showed no systemic toxicity [321]. Ginger and its constituent compounds, even at doses as high as 2.0 g per day, have demonstrated minimal levels of toxicity and notable tolerability in both animal and human studies. Reported adverse effects have primarily consisted of mild gastrointestinal discomfort. Nevertheless, the precise reason for the observed low toxicity remains uncertain, potentially arising from either limited oral bioavailability or a heightened safety profile exhibited by the pungent constituents of ginger, namely gingerols and shogaols, when administered to humans [309]. The vitamin and mineral recommended dose ranges were already mentioned in the previous section. A well-rounded and health-promoting diet includes the incorporation of fruits, vegetables, and spices due to their rich nutritional content, which contributes positively to overall well-being. However, ensuring the safety of these food items is paramount to safeguarding public health. Rigorous safety assessments are conducted to identify potential risks and mitigate them effectively. In agricultural practices, the use of pesticides, fungicides, and herbicides is common to protect crops from pests and diseases. To ensure compliance with safety standards, residue levels of these chemicals in fruits, vegetables, and spices are closely monitored. Regulatory bodies like the European Food Safety Authority (EFSA) and the U.S. Environmental Protection Agency (EPA) establish Maximum Residue Limits (MRLs) for various chemicals in foods. Microbiological safety evaluations are vital as well, as produce can be contaminated with harmful bacteria such as *Salmonella*, *Escherichia coli*, and *Listeria monocytogenes*. Implementing preventive measures and monitoring along the production chain help mitigate the risk of these pathogens. Heavy metals, originating from soil and water, can accumulate in fruits, vegetables, and spices. Excessive amounts of metals like lead, cadmium, and mercury can pose long-term health hazards. Regulatory bodies, including the WHO, establish permissible limits for certain pollutants. Certain spices and processed fruits and vegetables may contain food additives for flavor enhancement, preservation, or color. Regulatory agencies rigorously evaluate the safety of these additives before approving their use. Allergenicity is another consideration, with examples like nuts and soy, which are known allergens. Clear labeling and transparency are crucial to informing consumers about potentially allergenic compounds in products. Genetically Modified (GM) fruits and vegetables undergo rigorous testing to assess their safety. These evaluations encompass the potential toxicity, allergenicity, and unforeseen consequences of genetic modifications. Some countries utilize irradiation to extend shelf life and reduce microbial contamination in fruits, vegetables, and spices. Safety assessments examine the impact of irradiation on nutrient content and potential health risks. Overall, ensuring the safety of fruits, vegetables, and spices involves comprehensive evaluations and regulatory measures to promote healthy consumption and protect public health [322].

2.8 Antioxidants can be Pro-Oxidants

Numerous investigations have shown that some bioactive substances, which are well-known for their ability to act as antioxidants, can paradoxically manifest pro-oxidant behavior. Comprehensive research is necessary to understand the complex interplay that permits antioxidants to also operate as pro-oxidants, even if environmental factors have been implicated in altering the antioxidant or pro-oxidant activities of these substances. The concentration of the antioxidant, its redox potential, and its interaction with metal ions are three essential elements that affect how an antioxidant becomes a pro-oxidant, according to recent scientific contributions. This dualistic nature must be acknowledged since the pro-oxidant role has important health implications. Pro-oxidant activity can cause an increased generation of free radicals and ROS, which can harm important macromolecules like DNA, lipids, and proteins. Vitamin C, a well-researched antioxidant, serves as a perfect example in this regard. Although vitamin C has strong antioxidant effects at lower levels (up to 100 mg/kg body weight), as concentrations exceed 1000 mg/kg body weight, an unfavorable pro-oxidant effect becomes apparent. Notably, certain metal ions, especially Fe^{3+} and Cu^{2+}, undergo reduction and hasten the production of hydroxyl radicals through

Fenton reactions, which might increase the pro-oxidant potential of vitamin C. The behavior of flavonoids and phenolic substances is also reflected in this mechanistic framework. Their pro-oxidant activity is influenced by both the high quantities of these chemicals and the presence of transition metal ions. The complex interaction between metal cations and the phenolic/flavonoid chemical itself causes this behavior. The redox potential of phenolic molecules can be altered by copper and iron transition metals, boosting their capacity to donate electrons. This facilitates redox processes that result in the production of free radicals [65, 323].

2.9 Quality Control of Antioxidant Supplements

Implementing quality control measures starts with carefully verifying the authenticity and purity of the ingredients in antioxidant supplements. Specific antioxidants are confirmed using cutting-edge analytical techniques like High-Performance Liquid Chromatography (HPLC), Gas Chromatography-Mass Spectrometry (GC-MS), and Fourier-Transform Infrared (FTIR) spectroscopy, while the detection of possible contaminants is made easier. The targeted health advantages must be achieved, and any potential negative effects must be minimized, therefore, precise dosing becomes essential. Additionally, as part of quality control, each supplement's actual antioxidant content must match the promises made on the label. This can be done by using methods like HPLC and titration to precisely measure the amount of active ingredients. A crucial task is maintaining the potency and integrity of antioxidant supplements for the duration of their specified shelf life. This is successfully accomplished by extensive stability testing carried out under various storage settings, which results in the estimation of shelf life and suitable storage advice. The product's possible degradation trajectory over time is predicted via accelerated stability studies. The validity of a supplement's label's correspondence with its actual contents, including the veracity of any health claims, is another aspect of quality control. Regulations controlling labeling and health claims for dietary supplements are closely monitored by organizations like the European Food Safety Authority (EFSA) and the U.S. Food and Drug Administration (FDA). In conclusion, the stringent quality control procedure for antioxidant supplements is exemplified by the painstaking execution of component verification, exact dosing, stability evaluation, contaminant scrutiny, and accurate labeling. The safety, dependability, and effectiveness of antioxidant supplements are crucially dependent on adherence to established rules and regulations, which protect customer interests [322–324].

2.10 Conclusions

In the contemporary era of increased industrialization and the prevalence of unhealthy dietary habits and lifestyles, our susceptibility to a spectrum of diseases, notably cancer, has increased. One of the major contributors to the development of cancer is the generation of excessive free radicals such as ROS and RNS in the body, which create oxidative stress and abnormal cell division, leading the body to inflammation and cancer. The use of antioxidants can neutralize or inactivate these excessive free radicals. Antioxidants decrease the production of free radicals by enzymatic and non-enzymatic pathways to curb the occurrence of abnormal cell division, thus preventing cancer and other chronic diseases. Nature provides us with inexhaustible resources in the form of fruits, vegetables, and spices, which are enriched with diverse biologically active compounds that mitigate oxidative stress by different mechanisms and prevent cancer. Some of the important anticancer compounds, such as phloretin, curcumin, lycopene, beta-carotene, gingerol, eugenol, piperine, solanine, and capsaicin, exhibited diverse antioxidant and anticancer properties by different mechanisms. These fruits and vegetables are also abundant in important antioxidant vitamins such as vitamin A, C, and E and minerals such as selenium, zinc, copper, and manganese, which cause the scavenging of free radicals and inhibit the excessive formation of free radicals. The consumption of these important antioxidant-rich natural foods can significantly decrease the risk of cancer development, as reported in various clinical trials. However, compared to antioxidant supplementation, natural antioxidants provide more antioxidant benefits due to the presence of other phytochemicals. In some clinical trials, antioxidant supplements did not reduce the risk of cancer. Furthermore, if these antioxidant supplements are consumed in large amounts, they can produce pro-oxidant effects. In conclusion, the use of fruits, vegetables, and spices in your diet can significantly decrease the occurrence of cancer and other chronic diseases.

References

1. Asgharian, P., Tazekand, A.P., Hosseini, K., Forouhandeh, H., Ghasemnejad, T., Ranjbar, M. *et al.*, Potential mechanisms of quercetin in cancer prevention: focus on cellular and molecular targets. *Cancer Cell Int.*, 22, 257, 2022.
2. Ferlay, J., Ervik, M., Lam, F., Colombet, M., Mery, L., Piñeros, M. *et al.*, *Global cancer observatory: cancer today*, International Agency for Research on Cancer, Lyon, France, 2020.
3. Wild, C.P., Weiderpass, E., Stewart, B.W., *World cancer report: cancer research for cancer prevention*, International Agency for Research on Cancer, World Health Organization, Geneva, Switzerland, 2020.
4. Sung, H., Ferlay, J., Siegel, R.L., Laversanne, M., Soerjomataram, I., Jemal, A. *et al.*, Global cancer statistics 2020: GLOBOCAN estimates of incidence and mortality worldwide for 36 cancers in 185 countries. *CA Cancer J. Clin.*, 71, 209–49, 2021.
5. Chen, J., Xu, B., Sun, J., Jiang, X., Bai, W., Anthocyanin supplement as a dietary strategy in cancer prevention and management: A comprehensive review. *Crit. Rev. Food Sci. Nutr.*, 62, 26, 7242–54, 2022.
6. Di Meo, S. and Venditti, P., Evolution of the knowledge of free radicals and other oxidants. *Oxid. Med. Cell. Longevity*, 2020, e9829176, 2020.
7. Tewari, D., Patni, P., Bishayee, A., Sah, A.N., Bishayee, A., Natural products targeting the PI3K-Akt-mTOR signaling pathway in cancer: A novel therapeutic strategy. *Semin. Cancer Biol.*, 80, 1–17, 2022.
8. Li, S., So, T-H., Tang, G., Tan, H-Y., Wang, N., Ng, B.F.L. *et al.*, Chinese herbal medicine for reducing chemotherapy-associated side-effects in breast cancer patients: a systematic review and meta-analysis. *Front. Oncol.*, 10, 599073, 2020.
9. Mao, J.J., Pillai, G.G., Andrade, C.J., Ligibel, J.A., Basu, P., Cohen, L. *et al.*, Integrative oncology: Addressing the global challenges of cancer prevention and treatment. *CA Cancer J. Clin.*, 72, 144–64, 2022.

10. Gulcin, I., Antioxidants and antioxidant methods: An updated overview. *Arch. Toxicol.*, 94, 651–715, 2020.
11. Pham-Huy, L.A., He, H., Pham-Huy, C., Free radicals, antioxidants in disease and health. *Int. J. Biomed. Sci.*, 4, 89–96, 2008.
12. Manganaris, G.A., Goulas, V., Vicente, A.R., Terry, L.A., Berry antioxidants: small fruits providing large benefits. *J. Sci. Food Agric.*, 94, 825–33, 2014.
13. Li, Y., Guo, C., Yang, J., Wei, J., Xu, J., Cheng, S., Evaluation of antioxidant properties of pomegranate peel extract in comparison with pomegranate pulp extract. *Food Chem.*, 96, 254–60, 2006.
14. Hwang, H.S. and Winkler-Moser, J.K., Bicarbonates and carbonates as antioxidants in vegetable oils at frying temperatures. *J. Food Sci.*, 88, 717–31, 2023.
15. Kaur, C. and Kapoor, H.C., Antioxidants in fruits and vegetables–the millennium's health. *Int. J. Food Sci. Technol.*, 36, 703–25, 2001.
16. Shetty, A.A., Magadum, S., Managanvi, K., Vegetables as sources of antioxidants. *J. Food Nutr. Disord.*, 2, 1-5, 2013.
17. Tao, H., Zhao, Y., Li, L., He, Y., Zhang, X., Zhu, Y. *et al.*, Comparative metabolomics of flavonoids in twenty vegetables reveal their nutritional diversity and potential health benefits. *Food Res. Int.*, 164, 112384, 2023.
18. Griffiths, K., Aggarwal, B.B., Singh, R.B., Buttar, H.S., Wilson, D., De Meester, F., Food antioxidants and their anti-inflammatory properties: a potential role in cardiovascular diseases and cancer prevention. *Diseases*, 4, 28, 2016.
19. Gupta, N., Verma, K., Nalla, S., Kulshreshtha, A., Lall, R., Prasad, S., Free radicals as a double-edged sword: The cancer preventive and therapeutic roles of curcumin. *Molecules*, 25, 5390, 2020.
20. Moskovitz, J., Yim, M.B., Chock, P.B., Free radicals and disease. *Arch. Biochem. Biophys.*, 397, 354–9, 2002.
21. Čipak Gašparović, A., Free radical research in cancer. *Antioxidants*, 9, 157, 2020.
22. Martemucci, G., Costagliola, C., Mariano, M., D'andrea, L., Napolitano, P., D'Alessandro, A.G., Free radical properties, source and targets, antioxidant consumption and health. *Oxygen*, 2, 48–78, 2022.
23. Sadiq, I.Z., Free radicals and oxidative stress: Signaling mechanisms, redox basis for human diseases, and cell cycle regulation. *Curr. Mol. Med.*, 23, 13–35, 2023.
24. Giles, G.I. and Jacob, C., Reactive sulfur species: an emerging concept in oxidative stress. *Biol. Chem.*, 383, 375-88, 2002.
25. Halliwell, B., Reactive species and antioxidants. Redox biology is a fundamental theme of aerobic life. *Plant Physiol.*, 141, 312–22, 2006.
26. Kohen, R. and Nyska, A., Invited review: oxidation of biological systems: oxidative stress phenomena, antioxidants, redox reactions, and methods for their quantification. *Toxicol. Pathol.*, 30, 620–50, 2002.
27. Davies, M.J., The oxidative environment and protein damage. *Biochim. Biophys. Acta (BBA), Proteins Proteomics*, 1703, 93–109, 2005.
28. Dröge, W., Free radicals in the physiological control of cell function. *Physiol. Rev.*, 82, 47-95, 2002.
29. Min, B. and Ahn, D.U., Mechanism of lipid peroxidation in meat and meat products-A review. *Food Sci. Biotechnol.*, 14, 152–63, 2005.
30. Symons, M.C., Radicals generated by bone cutting and fracture. *Free Radical Biol. Med.*, 20, 831–5, 1996.

31. Valko, M., Leibfritz, D., Moncol, J., Cronin, M.T., Mazur, M., Telser, J., Free radicals and antioxidants in normal physiological functions and human disease. *Int. J. Biochem. Cell Biol.*, 39, 44–84, 2007.
32. Finkel, T. and Holbrook, N.J., Oxidants, oxidative stress and the biology of ageing. *Nature*, 408, 239–47, 2000.
33. Starkov, A.A., The role of mitochondria in reactive oxygen species metabolism and signaling. *Ann. N. Y. Acad. Sci.*, 37–52, 2008.
34. Cheeseman, K.H. and Slater, T.F., An introduction to free radical biochemistry. *Br. Med. Bull.*, 49, 481–93, 1993.
35. Gross, E., Sevier, C.S., Heldman, N., Vitu, E., Bentzur, M., Kaiser, C.A. *et al.*, Generating disulfides enzymatically: reaction products and electron acceptors of the endoplasmic reticulum thiol oxidase Ero1p. *Proc. Natl. Acad. Sci.*, 103, 299–304, 2006.
36. Schrader, M. and Fahimi, H.D., Peroxisomes and oxidative stress. *Biochim. Biophys. Acta (BBA), Mol. Cell. Res.*, 1763, 1755–66, 2006.
37. Aruoma, O.I., Free radicals, oxidative stress, and antioxidants in human health and disease. *J. Am. Oil Chem. Soc.*, 75, 199–212, 1998.
38. Dizdaroglu, M., Jaruga, P., Birincioglu, M., Rodriguez, H., Free radical-induced damage to DNA: mechanisms and measurement. *Free Radical Biol. Med.*, 32, 1102–15, 2002.
39. Martinet, W., De Meyer, G.R.Y., Herman, A.G., Kockx, M.M., Reactive oxygen species induce RNA damage in human atherosclerosis. *Eur. J. Clin. Invest.*, 34, 323–7, 2004.
40. Acuña, U.M., Wittwer, J., Ayers, S., Pearce, C.J., Oberlies, N.H., De Blanco, E.J.C., Effects of (5Z)-7-oxozeaenol on the oxidative pathway of cancer cells. *Anticancer Res.*, 32, 2665–71, 2012.
41. Dreher, D. and Junod, A.F., Role of oxygen free radicals in cancer development. *Eur. J. Cancer*, 32, 30–8, 1996.
42. Goldstein, B.D. and Witz, G., Free radicals and carcinogenesis. *Free Radic. Res. Commun.*, 11, 3–10, 1990.
43. Brown, N.S., Jones, A., Fujiyama, C., Harris, A.L., Bicknell, R., Thymidine phosphorylase induces carcinoma cell oxidative stress and promotes secretion of angiogenic factors. *Cancer Res.*, 60, 6298–302, 2000.
44. Murrell, T.G.C., Epidemiological and biochemical support for a theory on the cause and prevention of breast cancer. *Med. Hypotheses*, 36, 389–96, 1991.
45. Sipe Jr., H.J., Jordan, S.J., Hanna, P.M., Mason, R.P., The metabolism of 17β-estradiol by lactoperoxidase: a possible source of oxidative stress in breast cancer. *Carcinogenesis*, 15, 2637–43, 1994.
46. Eiján, A.M., Piccardo, I., Riveros, M.D., Sandes, E.O., Porcella, H., Jasnis, M.A. *et al.*, Nitric oxide in patients with transitional bladder cancer. *J. Surg. Oncol.*, 81, 203–8, 2002.
47. Gecit, I., Aslan, M., Gunes, M., Pirincci, N., Esen, R., Demir, H. *et al.*, Serum prolidase activity, oxidative stress, and nitric oxide levels in patients with bladder cancer. *J. Cancer Res. Clin. Oncol.*, 138, 739–43, 2012.
48. Opanuraks, J., Boonla, C., Saelim, C., Kittikowit, W., Sumpatanukul, P., Honglertsakul, C. *et al.*, Elevated urinary total sialic acid and increased oxidative stress in patients with bladder cancer. *Asian Biomed.*, 4, 703–10, 2010.
49. Soini, Y., Haapasaari, K.M., Vaarala, M.H., Turpeenniemi-Hujanen, T., Kärjä, V., Karihtala, P., 8-hydroxydeguanosine and nitrotyrosine are prognostic factors in urinary bladder carcinoma. *Int. J. Clin. Exp. Path.*, 4, 267–75, 2011.
50. Wynder, E.L. and Goldsmith, R., The epidemiology of bladder cancer. A second look. *Cancer*, 40, 1246–68, 1977.
51. Yılmaz, İ.A., Akçay, T., Çakatay, U., Telci, A., Ataus, S., Yalçın, V., Relation between bladder cancer and protein oxidation. *Int. Urol. Nephrol.*, 35, 345–50, 2003.

52. Blau, S., Rubinstein, A., Bass, P., Singaram, C., Kohen, R., Differences in the reducing power along the rat GI tract: lower antioxidant capacity of the colon. *Mol. Cell. Biochem.*, 194, 185–91, 1999.
53. Foksinski, M., Rozalski, R., Guz, J., Ruszkowska, B., Sztukowska, P., Piwowarski, M. *et al.*, Urinary excretion of DNA repair products correlates with metabolic rates as well as with maximum life spans of different mammalian species. *Free Radical Biol. Med.*, 37, 1449–54, 2004.
54. Guz, J., Foksinski, M., Siomek, A., Gackowski, D., Rozalski, R., Dziaman, T. *et al.*, The relationship between 8-oxo-7, 8-dihydro-2′-deoxyguanosine level and extent of cytosine methylation in leukocytes DNA of healthy subjects and in patients with colon adenomas and carcinomas. *Mutat. Res. Fundam. Mol. Mech. Mutagen.*, 640, 170–3, 2008.
55. Haklar, G., Sayin-Özveri, E., Yüksel, M., Aktan, A.Ö., Yalçin, A.S., Different kinds of reactive oxygen and nitrogen species were detected in colon and breast tumors. *Cancer Lett.*, 165, 219–24, 2001.
56. Rainis, T., Maor, I., Lanir, A., Shnizer, S., Lavy, A., Enhanced oxidative stress and leucocyte activation in neoplastic tissues of the colon. *Dig. Dis. Sci.*, 52, 526–30, 2007.
57. Suzuki, K., Ito, Y., Wakai, K., Kawado, M., Hashimoto, S., Toyoshima, H. *et al.*, Serum oxidized low-density lipoprotein levels and risk of colorectal cancer: a case-control study nested in the Japan Collaborative Cohort Study. *Cancer Epidemiol. Biomarkers Prev.*, 13, 1781–7, 2004.
58. Azad, N., Rojanasakul, Y., Vallyathan, V., Inflammation and lung cancer: roles of reactive oxygen/nitrogen species. *J. Toxicol. Environ. Health Part B*, 11, 1–15, 2008.
59. Hoagland IV, L.F., Campa, M.J., Gottlin, E.B., Herndon, J.E., Patz Jr., E.F., Haptoglobin and posttranslational glycan-modified derivatives as serum biomarkers for the diagnosis of nonsmall cell lung cancer. *Cancer: Interdiscip. Int. J. Am. Cancer Soc.*, 110, 2260–8, 2007.
60. Arnold, R.S., He, J., Remo, A., Ritsick, D., Yin-Goen, Q., Lambeth, J.D. *et al.*, Nox1 expression determines cellular reactive oxygen and modulates c-fos-induced growth factor, interleukin-8, and Cav-1. *Am. J. Pathol.*, 171, 2021–32, 2007.
61. Brar, S.S., Corbin, Z., Kennedy, T.P., Hemendinger, R., Thornton, L., Bommarius, B. *et al.*, NOX5 NAD(P)H oxidase regulates growth and apoptosis in DU 145 prostate cancer cells. *Am. J. Physiol. Cell Physiol.*, 285, C353–69, 2003.
62. Kumar, B., Koul, S., Khandrika, L., Meacham, R.B., Koul, H.K., Oxidative stress is inherent in prostate cancer cells and is required for aggressive phenotype. *Cancer Res.*, 68, 1777–85, 2008.
63. Lim, S.D., Sun, C., Lambeth, J.D., Marshall, F., Amin, M., Chung, L. *et al.*, Increased Nox1 and hydrogen peroxide in prostate cancer. *Prostate*, 62, 200–7, 2005.
64. Veeramani, S., Yuan, T.C., Lin, F.F., Lin, M.F., Mitochondrial redox signaling by p66Shc is involved in regulating androgenic growth stimulation of human prostate cancer cells. *Oncogene*, 27, 5057–68, 2008.
65. Arias, A., Feijoo, G., Moreira, M.T., Exploring the potential of antioxidants from fruits and vegetables and strategies for their recovery. *Innovative Food Sci. Emerg. Technol.*, 77, 102974, 2022.
66. Eberhardt, M.V., Lee, C.Y., Liu, R.H., Antioxidant activity of fresh apples. *Nature*, 405, 903–4, 2000.
67. Lee, K.W., Kim, Y.J., Kim, D.O., Lee, H.J., Lee, C.Y., Major phenolics in apple and their contribution to the total antioxidant capacity. *J. Agric. Food. Chem.*, 51, 6516–20, 2003.
68. Maheshwari, S., Kumar, V., Bhadauria, G., Mishra, A., Immunomodulatory potential of phytochemicals and other bioactive compounds of fruits: A review. *Food Front.*, 3, 221–38, 2022.
69. Wolfe, K., Wu, X., Liu, R.H., Antioxidant activity of apple peels. *J. Agric. Food. Chem.*, 51, 609–14, 2003.
70. Duan, H., Wang, R., Yan, X., Liu, H., Zhang, Y., Mu, D. *et al.*, Phloretin induces apoptosis of human esophageal cancer via a mitochondriadependent pathway. *Oncol. Lett.*, 14, 6763–8, 2017.

71. Min, J., Li, X., Huang, K., Tang, H., Ding, X., Qi, C. *et al.*, Phloretin induces apoptosis of non-small cell lung carcinoma A549 cells via JNK1/2 and p38 MAPK pathways. *Oncol. Rep.*, 34, 2871–9, 2015.
72. Yang, G., Yin, X., Ma, D., Su, Z., Anticancer activity of phloretin against the human oral cancer cells is due to G0/G1 cell cycle arrest and ROS mediated cell death. *J. Buon*, 25, 344–9, 2020.
73. Lu, M., Kong, Q., Xu, X., Lu, H., Lu, Z., Yu, W. *et al.*, Evaluation of apoptotic and growth inhibitory activity of phloretin in BGC823 gastric cancer cell. *Trop. J. Pharm. Res.*, 14, 27–31, 2015.
74. Wu, K.H., Ho, C.T., Chen, Z.F., Chen, L.C., Whang-Peng, J., Lin, T.N. *et al.*, The apple polyphenol phloretin inhibits breast cancer cell migration and proliferation via inhibition of signals by type 2 glucose transporter. *J. Food Drug Anal.*, 26, 221–31, 2018.
75. Hsiao, Y.H., Hsieh, M.J., Yang, S.F., Chen, S.P., Tsai, W.C., Chen, P.N., Phloretin suppresses metastasis by targeting protease and inhibits cancer stemness and angiogenesis in human cervical cancer cells. *Phytomedicine*, 62, 152964, 2019.
76. Jin, G., Zhao, Z., Chakraborty, T., Mandal, A., Roy, A., Roy, S. *et al.*, Decrypting the molecular mechanistic pathways delineating the chemotherapeutic potential of ruthenium-phloretin complex in colon carcinoma correlated with the oxidative status and increased apoptotic events. *Oxid. Med. Cell. Longevity*, 31, 7690845, 2020.
77. Kim, M.S., Kwon, J.Y., Kang, N.J., Lee, K.W., Lee, H.J., Phloretin induces apoptosis in H-Ras MCF10A human breast tumor cells through the activation of p53 via JNK and p38 mitogen-activated protein kinase signaling. *Ann. N. Y. Acad. Sci.*, 1171, 479–83, 2009.
78. Saraswati, S., Alhaider, A., Abdelgadir, A.M., Tanwer, P., Korashy, H.M., Phloretin attenuates STAT-3 activity and overcomes sorafenib resistance targeting SHP-1–mediated inhibition of STAT3 and Akt/VEGFR2 pathway in hepatocellular carcinoma. *Cell Commun. Signaling*, 17, 1–18, 2019.
79. Kang, D., Zuo, W., Wu, Q., Zhu, Q., Liu, P., Inhibition of specificity protein 1 is involved in phloretin-induced suppression of prostate cancer. *Biomed Res. Int.*, 1358674, 2020.
80. Babu, M.A., Suriyakala, M.A., Gothandam, K.M., Varietal impact on phytochemical contents and antioxidant properties of *Musa acuminata* (banana). *J. Pharm. Sci. Res.*, 4, 1950–5, 2012.
81. Bennett, R.N., Shiga, T.M., Hassimotto, N.M., Rosa, E.A., Lajolo, F.M., Cordenunsi, B.R., Phenolics and antioxidant properties of fruit pulp and cell wall fractions of postharvest banana (*Musa acuminata* Juss.) cultivars. *J. Agric. Food. Chem.*, 58, 7991–8003, 2010.
82. Englberger, L., Wills, R.B., Blades, B., Dufficy, L., Daniells, J.W., Coyne, T., Carotenoid content and flesh color of selected banana cultivars growing in Australia. *Food Nutr. Bull.*, 27, 281–91, 2006.
83. Russell, W.R., Labat, A., Scobbie, L., Duncan, G.J., Duthie, G.G., Phenolic acid content of fruits commonly consumed and locally produced in Scotland. *Food Chem.*, 115, 100–4, 2009.
84. Singh, B., Singh, J.P., Kaur, A., Singh, N., Bioactive compounds in banana and their associated health benefits–A review. *Food Chem.*, 206, 1–11, 2016.
85. Faria, A., Pestana, D., Teixeira, D., de Freitas, V., Mateus, N., Calhau, C., Blueberry anthocyanins and pyruvic acid adducts: anticancer properties in breast cancer cell lines. *Phytother. Res.*, 24, 1862–9, 2010.
86. Lachowicz, S., Oszmiański, J., Seliga, Ł., Pluta, S., Phytochemical composition and antioxidant capacity of seven Saskatoon berry (*Amelanchier alnifolia* Nutt.) genotypes grown in Poland. *Molecules*, 22, 853, 2017.
87. McAnulty, L.S., Nieman, D.C., Dumke, C.L., Shooter, L.A., Henson, D.A., Utter, A.C. *et al.*, Effect of blueberry ingestion on natural killer cell counts, oxidative stress, and inflammation prior to and after 2.5 h of running. *Appl. Physiol. Nutr. Metab.*, 36, 976–84, 2011.
88. Percival, S.S., Grape consumption supports immunity in animals and humans. *J. Nutr.*, 139, 1801S–1805S, 2009.

89. Wang, C.Y., Chen, C.T., Wang, S.Y., Changes of flavonoid content and antioxidant capacity in blueberries after illumination with UV-C. *Food Chem.*, 117, 426–31, 2009.
90. Zizkova, P., Stefek, M., Rackova, L., Prnova, M., Horakova, L., Novel quercetin derivatives: From redox properties to promising treatment of oxidative stress related diseases. *Chem. Biol. Interact.*, 265, 36–46, 2017.
91. Ghafoor, K., Choi, Y.H., Jeon, J.Y., Jo, I.H., Optimization of ultrasound-assisted extraction of phenolic compounds, antioxidants, and anthocyanins from grape (*Vitis vinifera*) seeds. *J. Agric. Food. Chem.*, 57, 4988–94, 2009.
92. Imran, M., Rauf, A., Imran, A., Nadeem, M., Ahmad, Z., Atif, M. *et al.*, Health benefits of grapes polyphenols. *J. Environ. Agric. Sci.*, 10, 40–51, 2017.
93. Özcan, M.M., Al Juhaimi, F., Gülcü, M., Uslu, N., Geçgel, U., Determination of bioactive compounds and mineral contents of seedless parts and seeds of grapes. *S. Afr. J. Enol. Vitic.*, 38, 212–20, 2017.
94. Rice-Evans, C.A., Miller, N.J., Paganga, G., Structure-antioxidant activity relationships of flavonoids and phenolic acids. *Free Radical Biol. Med.*, 20, 933–56, 1996.
95. Brevik, A., Gaivão, I., Medin, T., Jørgenesen, A., Piasek, A., Elilasson, J. *et al.*, Supplementation of a western diet with golden kiwifruits (*Actinidia chinensis* var.'Hort 16A':) effects on biomarkers of oxidation damage and antioxidant protection. *Nutr. J.*, 10, 1–9, 2011.
96. Cundra, L.B., D'Souza, S.M., Parekh, P.J., Johnson, D.A., The impact of kiwifruit on human health. *Gut Gastroenterol.*, 3, 1–10, 2020.
97. Richardson, D.P., Ansell, J., Drummond, L.N., The nutritional and health attributes of kiwifruit: A review. *Eur. J. Nutr.*, 57, 2659–76, 2018.
98. Silalahi, J., Anticancer and health protective properties of citrus fruit components. *Asia Pac. J. Clin. Nutr.*, 11, 79–84, 2002.
99. Bermejo, A., Llosa, M.J., Cano, A., Analysis of bioactive compounds in seven citrus cultivars. *Food Sci. Technol. Int.*, 17, 55–62, 2011.
100. Etebu, E. and Nwauzoma, A.B., A review on sweet orange (*Citrus sinensis* L Osbeck): health, diseases and management. *Am. J. Res. Commun.*, 2, 33–70, 2014.
101. Ghasemi, K., Ghasemi, Y., Ebrahimzadeh, M.A., Antioxidant activity, phenol and flavonoid contents of 13 citrus species peels and tissues. *Pak. J. Pharm. Sci.*, 22, 277–81, 2009.
102. Hubner, R.A. and Houlston, R.S., Folate and colorectal cancer prevention. *Br. J. Cancer*, 100, 233–9, 2009.
103. Nogata, Y., Sakamoto, K., Shiratsuchi, H., Ishii, T., Yano, M., Ohta, H., Flavonoid composition of fruit tissues of citrus species. *Biosci. Biotechnol., Biochem.*, 70, 178–92, 2006.
104. Park, J.H., Lee, M., Park, E., Antioxidant activity of orange flesh and peel extracted with various solvents. *Prev. Nutr. Food Sci.*, 19, 291, 2014.
105. Turner, T. and Burri, B.J., Potential nutritional benefits of current citrus consumption. *Agriculture*, 3, 170–87, 2013.
106. Wintergerst, E.S., Maggini, S., Hornig, D.H., Contribution of selected vitamins and trace elements to immune function. *Ann. Nutr. Metab.*, 51, 301–23, 2007.
107. Comalada, M., Camuesco, D., Sierra, S., Ballester, I., Xaus, J., Gálvez, J. *et al.*, *In vivo* quercitrin anti-inflammatory effect involves release of quercetin, which inhibits inflammation through down-regulation of the NF-κB pathway. *Eur. J. Immunol.*, 35, 584–92, 2005.
108. Pandey, S., Cabot, P.J., Shaw, P.N., Hewavitharana, A.K., Anti-inflammatory and immunomodulatory properties of *Carica papaya*. *J. Immunotoxicol.*, 13, 590–602, 2016.
109. Simone, R.E., Russo, M., Catalano, A., Monego, G., Froehlich, K., Boehm, V. *et al.*, Lycopene inhibits NF-kB-mediated IL-8 expression and changes redox and PPARγ signalling in cigarette smoke–stimulated macrophages. *PLoS One*, 6, e19652, 2011.

110. Santana, L.F., Inada, A.C., do Espirito Santo, B.L.S., Filiú, W.F., Pott, A., Alves, F.M. *et al.*, Nutraceutical potential of *Carica papaya* in metabolic syndrome. *Nutrients*, 11, 1608, 2019.
111. Sadek, K.M., Antioxidant and immunostimulant effect of *Carica papaya* Linn. aqueous extract in acrylamide intoxicated rats. *Acta Inform. Med.*, 20, 180–5, 2012.
112. Marotta, F., Yoshida, C., Barreto, R., Naito, Y., Packer, L., Oxidative-inflammatory damage in cirrhosis: effect of vitamin E and a fermented papaya preparation. *J. Gastroenterol. Hepatol.*, 22, 697–703, 2007.
113. Wu, T.F., Hsu, L.T., Tsang, B.X., Huang, L.C., Shih, W.Y., Chen, L.Y., Clarification of the molecular pathway of Taiwan local pomegranate fruit juice underlying the inhibition of urinary bladder urothelial carcinoma cell by proteomics strategy. *BMC Complem. Altern. Med.*, 16, 1–10, 2016.
114. Fahmy, H., Hegazi, N., El-Shamy, S., Farag, M.A., Pomegranate juice as a functional food: A comprehensive review of its polyphenols, therapeutic merits, and recent patents. *Food Funct.*, 11, 5768–81, 2020.
115. Maoto, M.M., Beswa, D., Jideani, A.I., Watermelon as a potential fruit snack. *Int. J. Food Prop.*, 22, 355–70, 2019.
116. Davis, A.R., Webber, C.L., Fish, W.W., Wehner, T.C., King, S., Perkins-Veazie, P., L-citrulline levels in watermelon cultigens tested in two environments. *HortScience*, 46, 1572–5, 2011.
117. Naz, A., Butt, M.S., Sultan, M.T., Qayyum, M.M.N., Niaz, R.S., Watermelon lycopene and allied health claims. *EXCLI J.*, 13, 650–60, 2014.
118. Soteriou, G.A., Kyriacou, M.C., Siomos, A.S., Gerasopoulos, D., Evolution of watermelon fruit physicochemical and phytochemical composition during ripening as affected by grafting. *Food Chem.*, 165, 282–9, 2014.
119. Dias, J.S., Nutritional quality and health benefits of vegetables: A review. *Food Nutr. Sci.*, 3, 1354–74, 2012.
120. Turati, F., Rossi, M., Pelucchi, C., Levi, F., Vecchia, C.L., Fruit and vegetables and cancer risk: a review of southern European studies. *Br. J. Nutr.*, 113, S102–10, 2015.
121. Riso, P., Visioli, F., Erba, D., Testolin, G., Porrini, M., Lycopene and vitamin C concentrations increase in plasma and lymphocytes after tomato intake. Effects on cellular antioxidant protection. *Eur. J. Clin. Nutr.*, 58, 1350–8, 2004.
122. Huang, H., Chen, S., Van Doren, J., Li, D., Farichon, C., He, Y. *et al.*, α-Tomatine inhibits growth and induces apoptosis in HL-60 human myeloid leukemia cells. *Mol. Med. Rep.*, 11, 4573–8, 2015.
123. Salehi, B., Sharifi-Rad, R., Sharopov, F., Namiesnik, J., Roointan, A., Kamle, M. *et al.*, Beneficial effects and potential risks of tomato consumption for human health: An overview. *Nutrition*, 201–8, 2019.
124. Teodoro, A.J., Oliveira, F.L., Martins, N.B., Maia, G.A., Martucci, R.B., Borojevic, R., Effect of lycopene on cell viability and cell cycle progression in human cancer cell lines. *Cancer Cell Int.*, 12, 36, 2012.
125. Ozkan, G., Günal-Köroğlu, D., Karadag, A., Capanoglu, E., Cardoso, S.M., Al-Omari, B. *et al.*, A mechanistic updated overview on lycopene as potential anticancer agent. *Biomed. Pharmacother.*, 161, 114428, 2023.
126. Mezbani, A., Kavan, B.P., Kiani, A., Masouri, B., Effect of dietary lycopene supplementation on growth performance, blood parameters and antioxidant enzymes status in broiler chickens. *Livest. Res. Rural Dev.*, 31, 3101–12, 2019.
127. Gajowik, A. and Dobrzyńska, M.M., The evaluation of protective effect of lycopene against genotoxic influence of X-irradiation in human blood lymphocytes. *Radiat. Environ. Biophys.*, 56, 413–22, 2017.

128. Polívková, Z., Šmerák, P., Demová, H., Houška, M., Antimutagenic effects of lycopene and tomato purée. *J. Med. Food*, 13, 1443–50, 2010.
129. Ayob, O., Hussain, P.R., Suradkar, P., Naqash, F., Rather, S.A., Joshi, S. *et al.*, Evaluation of chemical composition and antioxidant activity of Himalayan Red chilli varieties. *LWT*, 146, 111413, 2021.
130. Okada, Y. and Okajima, H., Antioxidant effect of capsaicin on lipid peroxidation in homogeneous solution, micelle dispersions and liposomal membranes. *Redox Rep.*, 6, 117–22, 2001.
131. Akhilender Naidu, K. and Thippeswamy, N.B., Inhibition of human low density lipoprotein oxidation by active principles from spices. *Mol. Cell. Biochem.*, 229, 19–23, 2002.
132. Manjunatha, H. and Srinivasan, K., Protective effect of dietary curcumin and capsaicin on induced oxidation of low-density lipoprotein, iron-induced hepatotoxicity and carrageenan-induced inflammation in experimental rats. *FEBS J.*, 273, 4528–37, 2006.
133. Prasad, N.S., Raghavendra, R., Lokesh, B.R., Naidu, K.A., Spice phenolics inhibit human PMNL 5-lipoxygenase. *Prostaglandins Leukot. Essent. Fatty Acids*, 70, 521–8, 2004.
134. Kempaiah, S., Antioxidant status of red blood cells and liver in hypercholesterolemic rats fed hypolipidemic spices. *Int. J. Vitam. Nutr. Res.*, 74, 199–208, 2004.
135. Lee, C.Y.J., Kim, M., Yoon, S.W., Lee, C.H., Short-term control of capsaicin on blood and oxidative stress of rats *in vivo*. *Phytother. Res.*, 17, 454–8, 2003.
136. Tanrıkulu-Küçük, S., Başaran-Küçükgergin, C., Seyithanoğlu, M., Doğru-Abbasoğlu, S., Koçak, H., Beyhan-Özdaş, Ş. *et al.*, Effect of dietary curcumin and capsaicin on testicular and hepatic oxidant–antioxidant status in rats fed a high-fat diet. *Appl. Physiol. Nutr. Metab.*, 44, 774–82, 2019.
137. Athanasiou, A., Smith, P.A., Vakilpour, S., Kumaran, N.M., Turner, A.E., Bagiokou, D. *et al.*, Vanilloid receptor agonists and antagonists are mitochondrial inhibitors: how vanilloids cause non-vanilloid receptor mediated cell death. *Biochem. Biophys. Res. Commun.*, 354, 50–5, 2007.
138. Hail, N. and Lotan, R., Examining the role of mitochondrial respiration in vanilloid-induced apoptosis. *J. Natl. Cancer Inst.*, 94, 1281–92, 2002.
139. Kim, Y.M., Hwang, J-T., Kwak, D.W., Lee, Y.K., Park, O.J., Involvement of AMPK signaling cascade in capsaicin-induced apoptosis of HT-29 colon cancer cells. *Ann. N. Y. Acad. Sci.*, 1095, 496–503, 2007.
140. Lee, J.S., Chang, J.S., Lee, J.Y., Kim, J.A., Capsaicin-induced apoptosis and reduced release of reactive oxygen species in MBT-2 murine bladder tumor cells. *Arch. Pharmacal Res.*, 27, 1147–53, 2004.
141. Lee, Y.S., Kang, Y.S., Lee, J.S., Nicolova, S., Kim, J.A., Involvement of NADPH oxidase-mediated generation of reactive oxygen species in the apototic cell death by capsaicin in HepG2 human hepatoma cells. *Free Radical Res.*, 38, 405–12, 2004.
142. Min, J.K., Han, K.Y., Kim, E.C., Kim, Y.M., Lee, S.W., Kim, O.H. *et al.*, Capsaicin inhibits *in vitro* and *in vivo* angiogenesis. *Cancer Res.*, 64, 644–51, 2004.
143. Mori, A., Lehmann, S., O'Kelly, J., Kumagai, T., Desmond, J.C., Pervan, M. *et al.*, Capsaicin, a component of red peppers, inhibits the growth of androgen-independent, p53 mutant prostate cancer cells. *Cancer Res.*, 66, 3222–9, 2006.
144. Wu, C.C., Lin, J.P., Yang, J.S., Chou, S.T., Chen, S.C., Lin, Y.T. *et al.*, Capsaicin induced cell cycle arrest and apoptosis in human esophagus epidermoid carcinoma CE 81T/VGH cells through the elevation of intracellular reactive oxygen species and Ca2+ productions and caspase-3 activation. *Mutat. Res. Fundam. Mol. Mech. Mutagen.*, 601, 71–82, 2006.
145. Bley, K., Boorman, G., Mohammad, B., McKenzie, D., Babbar, S., A comprehensive review of the carcinogenic and anticarcinogenic potential of capsaicin. *Toxicol. Pathol.*, 40, 847–73, 2012.

146. Pramanik, K.C., Boreddy, S.R., Srivastava, S.K., Role of mitochondrial electron transport chain complexes in capsaicin mediated oxidative stress leading to apoptosis in pancreatic cancer cells. *PLoS One*, 6, e20151, 2011.
147. Surh, Y.J., More than spice: capsaicin in hot chili peppers makes tumor cells commit suicide. *J. Natl. Cancer Inst.*, 94, 1263–5, 2002.
148. Amantini, C., Ballarini, P., Caprodossi, S., Nabissi, M., Morelli, M.B., Lucciarini, R. *et al.*, Triggering of transient receptor potential vanilloid type 1 (TRPV1) by capsaicin induces Fas/CD95-mediated apoptosis of urothelial cancer cells in an ATM-dependent manner. *Carcinogenesis*, 30, 1320–9, 2009.
149. Khosravi-Far, R., White, E., Chalah, A., Khosravi-Far, R., The mitochondrial death pathway, in: *Programmed cell death in cancer progression and therapy*, pp. 25–45, 2008.
150. Kim, M.Y., Trudel, L.J., Wogan, G.N., Apoptosis induced by capsaicin and resveratrol in colon carcinoma cells requires nitric oxide production and caspase activation. *Anticancer Res.*, 29, 3733–40, 2009.
151. Caprodossi, S., Amantini, C., Nabissi, M., Morelli, M.B., Farfariello, V., Santoni, M. *et al.*, Capsaicin promotes a more aggressive gene expression phenotype and invasiveness in null-TRPV1 urothelial cancer cells. *Carcinogenesis*, 32, 686–94, 2011.
152. Chow, J., Norng, M., Zhang, J., Chai, J., TRPV6 mediates capsaicin-induced apoptosis in gastric cancer cells—Mechanisms behind a possible new "hot" cancer treatment. *Biochim. Biophys. Acta*, 1773, 565–76, 2007.
153. Impheng, H., Pongcharoen, S., Richert, L., Pekthong, D., Srisawang, P., The selective target of capsaicin on FASN expression and de novo fatty acid synthesis mediated through ROS generation triggers apoptosis in HepG2 cells. *PLoS One*, 9, e107842, 2014.
154. Lee, S.H., Krisanapun, C., Baek, S.J., NSAID-activated gene-1 as a molecular target for capsaicin-induced apoptosis through a novel molecular mechanism involving GSK3β, C/EBPβ and ATF3. *Carcinogenesis*, 31, 719–28, 2010.
155. Lee, S.H., Richardson, R.L., Dashwood, R.H., Baek, S.J., Capsaicin represses transcriptional activity of β-catenin in human colorectal cancer cells. *J. Nutr. Biochem.*, 23, 646–55, 2012.
156. Ying, H., Wang, Z., Zhang, Y., Yang, T-Y., Ding, Z-H., Liu, S-Y. *et al.*, Capsaicin induces apoptosis in human osteosarcoma cells through AMPK-dependent and AMPK-independent signaling pathways. *Mol. Cell. Biochem.*, 384, 229–37, 2013.
157. Chen, D., Yang, Z., Wang, Y., Zhu, G., Wang, X., Capsaicin induces cycle arrest by inhibiting cyclin-dependent-kinase in bladder carcinoma cells. *Int. J. Urol.*, 19, 662–8, 2012.
158. Lin, C.H., Lu, W.C., Wang, C.W., Chan, Y.C., Chen, M.K., Capsaicin induces cell cycle arrest and apoptosis in human KB cancer cells. *BMC Complem. Altern. Med.*, 13, 1–9, 2013.
159. Thoennissen, N.H., O'kelly, J., Lu, D., Iwanski, G.B., La, D.T., Abbassi, S. *et al.*, Capsaicin causes cell-cycle arrest and apoptosis in ER-positive and-negative breast cancer cells by modulating the EGFR/HER-2 pathway. *Oncogene*, 29, 285–96, 2010.
160. Wang, H.M., Chuang, S.M., Su, Y.C., Li, Y.H., Chueh, P.J., Down-regulation of tumor-associated NADH oxidase, tNOX (ENOX2), enhances capsaicin-induced inhibition of gastric cancer cell growth. *Cell Biochem. Biophys.*, 61, 355–66, 2011.
161. Zheng, L., Chen, J., Ma, Z., Liu, W., Yang, F., Yang, Z. *et al.*, Capsaicin causes inactivation and degradation of the androgen receptor by inducing the restoration of miR-449a in prostate cancer. *Oncol. Rep.*, 34, 1027–34, 2015.
162. Bhutani, M., Pathak, A.K., Nair, A.S., Kunnumakkara, A.B., Guha, S., Sethi, G. *et al.*, Capsaicin is a novel blocker of constitutive and interleukin-6–inducible STAT3 activation. *Clin. Cancer Res.*, 13, 3024–32, 2007.

163. Chakraborty, S., Adhikary, A., Mazumdar, M., Mukherjee, S., Bhattacharjee, P., Guha, D. *et al.*, Capsaicin-induced activation of p53-SMAR1 auto-regulatory loop down-regulates VEGF in non-small cell lung cancer to restrain angiogenesis. *PLoS One*, 9, e99743, 2014.
164. Shin, D.H., Kim, O.H., Jun, H.S., Kang, M.K., Inhibitory effect of capsaicin on B16-F10 melanoma cell migration via the phosphatidylinositol 3-kinase/Akt/Rac1 signal pathway. *Exp. Mol. Med.*, 40, 486–94, 2008.
165. Venier, N.A., Yamamoto, T., Sugar, L.M., Adomat, H., Fleshner, N.E., Klotz, L.H. *et al.*, Capsaicin reduces the metastatic burden in the transgenic adenocarcinoma of the mouse prostate model. *Prostate*, 75, 1300–11, 2015.
166. Price, L.S. and Collard, J.G., Regulation of the cytoskeleton by Rho-family GTPases: implications for tumour cell invasion. *Semin. Cancer Biol.*, 11, 167–73, 2001.
167. Lee, G.R., Jang, S.H., Kim, C.J., Kim, A.R., Yoon, D.J., Park, N.H. *et al.*, Capsaicin suppresses the migration of cholangiocarcinoma cells by down-regulating matrix metalloproteinase-9 expression via the AMPK–NF-κB signaling pathway. *Clin. Exp. Metastasis*, 31, 897–907, 2014.
168. Sarkar, F.H. and Li, Y., Using chemopreventive agents to enhance the efficacy of cancer therapy. *Cancer Res.*, 66, 3347–50, 2006.
169. Yang, J., Li, T.Z., Xu, G.H., Luo, B.B., Chen, Y.X., Zhang, T., Low-concentration capsaicin promotes colorectal cancer metastasis by triggering ROS production and modulating Akt/mTOR and STAT-3 pathways. *Neoplasma*, 60, 364–72, 2013.
170. Akyol, H., Riciputi, Y., Capanoglu, E., Caboni, M.F., Verardo, V., Phenolic compounds in the potato and its byproducts: An overview. *Int. J. Mol. Sci.*, 17, 835, 2016.
171. Hasanain, M., Bhattacharjee, A., Pandey, P., Ashraf, R., Singh, N., Sharma, S. *et al.*, α-Solanine induces ROS-mediated autophagy through activation of endoplasmic reticulum stress and inhibition of Akt/mTOR pathway. *Cell Death Dis.*, 6, e1860, 2015.
172. Ji, Y.B., Gao, S.Y., Ji, C.F., Zou, X., Induction of apoptosis in HepG2 cells by solanine and Bcl-2 protein. *J. Ethnopharmacol.*, 115, 194–202, 2008.
173. Julien, O. and Wells, J.A., Caspases and their substrates. *Cell Death Differ.*, 24, 1380–9, 2017.
174. Yang, S.A., Paek, S.H., Kozukue, N., Lee, K.R., Kim, J.A., α-Chaconine, a potato glycoalkaloid, induces apoptosis of HT-29 human colon cancer cells through caspase-3 activation and inhibition of ERK 1/2 phosphorylation. *Food Chem. Toxicol.*, 44, 839–46, 2006.
175. Cargnello, M. and Roux, P.P., Activation and function of the MAPKs and their substrates, the MAPK-activated protein kinases. *Microbiol. Mol. Biol. Rev.*, 75, 50–83, 2011.
176. Yang, H., Liu, C., Zhou, R.M., Yao, J., Li, X.M., Shen, Y. *et al.*, Piezo2 protein: A novel regulator of tumor angiogenesis and hyperpermeability. *Oncotarget*, 7, 44630, 2016.
177. Lu, M.K., Shih, Y.W., Chien, T.T.C., Fang, L.H., Huang, H.C., Chen, P.S., α-Solanine inhibits human melanoma cell migration and invasion by reducing matrix metalloproteinase-2/9 activities. *Biol. Pharm. Bull.*, 33, 1685–91, 2010.
178. Wang, Y., Wu, J., Guo, W., Sun, Q., Chen, X., Zang, W. *et al.*, α-solanine modulates the radiosensitivity of esophageal cancer cells by inducing microRNA 138 expression. *Cell. Physiol. Biochem.*, 39, 996–1010, 2016.
179. Zhang, F., Yang, R., Zhang, G., Cheng, R., Bai, Y., Zhao, H. *et al.*, Anticancer function of α-solanine in lung adenocarcinoma cells by inducing microRNA-138 expression. *Tumor Biol.*, 37, 6437–46, 2016.
180. Soengas Fernández M del, P., Sotelo Pérez, T., Velasco Pazos, P., Cartea González, M.E., Antioxidant properties of Brassica vegetables. *FPSB*, 5, 43–55, 2011.
181. Palak, Soni, K., Thakur, A., Kohli, K., Broccoli: An insight into formulation and patentability aspects. *Drug Des.*, 5, 1000139, 2016.

182. Martín-Ruiz, A., Peña, L., González-Gil, A., Díez-Córdova, L.T., Cáceres, S., Illera, J.C., Effects of indole-3-carbinol on steroid hormone profile and tumor progression in a mice model of canine inflammatory mammarycancer. *BMC Cancer*, 18, 626, 2018.
183. Sharma, S., Katoch, V., Kumar, S., Chatterjee, S., Functional relationship of vegetable colors and bioactive compounds: Implications in human health. *J. Nutr. Biochem.*, 92, 108615, 2021.
184. Ma, Y., Feng, Y., Zeng, W., Luo, H., Anthocyanin encapsulated by ferulic acid-grafted-maltodextrin (FA-g-MD) microcapsules potentially improved its free radical scavenging capabilities against H_2O_2-induced oxidative stress. *Molecules*, 24, 1596, 2019.
185. Olejnik, A., Rychlik, J., Kidoń, M., Czapski, J., Kowalska, K., Juzwa, W. *et al.*, Antioxidant effects of gastrointestinal digested purple carrot extract on the human cells of colonic mucosa. *Food Chem.*, 190, 1069–77, 2016.
186. Asadi, K., Ferguson, L.R., Philpott, M., Karunasinghe, N., Cancer-preventive properties of an anthocyanin-enriched sweet potato in the APCMIN mouse model. *J. Cancer Prev.*, 22, 135–46, 2017.
187. Lim, S., Xu, J., Kim, J., Chen, T.Y., Su, X., Standard, J. *et al.*, Role of anthocyanin-enriched purple-fleshed sweet potato p40 in colorectal cancer prevention. *Mol. Nutr. Food Res.*, 57, 1908–17, 2013.
188. El Gamal, A.A., AlSaid, M.S., Raish, M., Al-Sohaibani, M., Al-Massarani, S.M., Ahmad, A. *et al.*, Beetroot (*Beta vulgaris* L.) extract ameliorates gentamicin-induced nephrotoxicity associated oxidative stress, inflammation, and apoptosis in rodent model. *Mediators Inflamm.*, 2014, 983952, 2014.
189. Zielińska-Przyjemska, M., Olejnik, A., Kostrzewa, A., Łuczak, M., Jagodziński, P.P., Baer-Dubowska, W., The beetroot component betanin modulates ROS production, DNA damage and apoptosis in human polymorphonuclear neutrophils. *Phytother. Res.*, 26, 845–52, 2012.
190. Lee, E.J., An, D., Nguyen, C.T., Patil, B.S., Kim, J., Yoo, K.S., Betalain and betaine composition of greenhouse-or field-produced beetroot (*Beta vulgaris* L.) and inhibition of HepG2 cell proliferation. *J. Agric. Food. Chem.*, 62, 1324–31, 2014.
191. Krajka-Kuźniak, V., Paluszczak, J., Szaefer, H., Baer-Dubowska, W., Betanin, a beetroot component, induces nuclear factor erythroid-2-related factor 2-mediated expression of detoxifying/antioxidant enzymes in human liver cell lines. *Br. J. Nutr.*, 110, 2138–49, 2013.
192. Karamese, S.A., Toktay, E., Unal, D., Selli, J., Karamese, M., Malkoc, I., The protective effects of beta-carotene against ischemia/reperfusion injury in rat ovarian tissue. *Acta Histochem.*, 117, 790–7, 2015.
193. Kasperczyk, S., Dobrakowski, M., Kasperczyk, J., Ostałowska, A., Zalejska-Fiolka, J., Birkner, E., Beta-carotene reduces oxidative stress, improves glutathione metabolism and modifies antioxidant defense systems in lead-exposed workers. *Toxicol. Appl. Pharmacol.*, 280, 36–41, 2014.
194. Yurtcu, E., Iseri, O.D., Sahin, F.I., Effects of ascorbic acid and β-carotene on HepG2 human hepatocellular carcinoma cell line. *Mol. Biol. Rep.*, 38, 4265–72, 2011.
195. Madhavan, J., Chandrasekharan, S., Priya, M.K., Godavarthi, A., Modulatory effect of carotenoid supplement constituting lutein and zeaxanthin (10: 1) on anti-oxidant enzymes and macular pigments level in rats. *Pharmacogn. Mag.*, 14, 268–74, 2018.
196. Bi, M.C., Rosen, R., Zha, R.Y., McCormick, S.A., Song, E., Hu, D.N., Zeaxanthin induces apoptosis in human uveal melanoma cells through Bcl-2 family proteins and intrinsic apoptosis pathway. *J. Evid. Based Complementary Altern. Med.*, 2013, e205082, 2013.
197. Murthy, R.K., Ravi, K., Balaiya, S., Brar, V.S., Chalam, K.V., Lutein protects retinal pigment epithelium from cytotoxic oxidative stress. *Cutaneous Ocul. Toxicol.*, 33, 32–7, 2014.
198. Gong, X., Smith, J.R., Swanson, H.M., Rubin, L.P., Carotenoid lutein selectively inhibits breast cancer cell growth and potentiates the effect of chemotherapeutic agents through ROS mediated mechanisms. *Molecules*, 23, 905, 2018.

199. Balasenthil, S. and Rajamani Ramachandran, C., Nagini S., S-Allylcysteine, a garlic constituent, inhibits 7,12-dimethylbenz[a]anthracene-induced hamster buccal pouch carcinogenesis. *Nutr. Cancer*, 40, 165–72, 2001.
200. Sengupta, A., Ghosh, S., Das, S., Tomato and garlic can modulate azoxymethane-induced colon carcinogenesis in rats. *Eur. J. Cancer Prev.*, 12, 195–200, 2003.
201. Sundaresan, S. and Subramanian, P., S-Allylcysteine inhibits circulatory lipid peroxidation and promotes antioxidants in N-nitrosodiethylamine-induced carcinogenesis. *Pol. J. Pharmacol.*, 55, 37–42, 2003.
202. Anwar, M.M. and Meki, A.R.M.A., Oxidative stress in streptozotocin-induced diabetic rats: effects of garlic oil and melatonin. *Comp. Biochem. Physiol. A Mol. Integr. Physiol.*, 135, 539–47, 2003.
203. Wei, Z. and Lau, B.H.S., Garlic inhibits free radical generation and augments antioxidant enzyme activity in vascular endothelial cells. *Nutr. Res.*, 18, 61–70, 1998.
204. Myhrstad, M.C.W., Carlsen, H., Nordström, O., Blomhoff, R., Moskaug, J.Ø., Flavonoids increase the intracellular glutathione level by transactivation of the γ-glutamylcysteine synthetase catalytical subunit promoter. *Free Radical Biol. Med.*, 32, 386–93, 2002.
205. Srinivasan, K., Antioxidant potential of spices and their active constituents. *Crit. Rev. Food Sci. Nutr.*, 54, 352–72, 2014.
206. Reddy, AC. and Lokesh, B.R., Studies on spice principles as antioxidants in the inhibition of lipid peroxidation of rat liver microsomes. *Mol. Cell. Biochem.*, 111, 117–24, 1992.
207. Reddy, A.C. and Lokesh, B.R., Studies on the inhibitory effects of curcumin and eugenol on the formation of reactive oxygen species and the oxidation of ferrous iron. *Mol. Cell. Biochem.*, 137, 1–8, 1994.
208. Iqbal, M., Sharma, S.D., Okazaki, Y., Fujisawa, M., Okada, S., Dietary supplementation of curcumin enhances antioxidant and phase II metabolizing enzymes in ddy male mice: Possible role in protection against chemical carcinogenesis and toxicity. *Pharmacol. Toxicol.*, 92, 33–8, 2003.
209. Reddy, A.C. and Lokesh, B.R., Effect of dietary turmeric (*Curcuma longa*) on iron-induced lipid peroxidation in the rat liver. *Food Chem. Toxicol.*, 32, 279–83, 1994.
210. Reddy, A.C. and Lokesh, B.R., Dietary unsaturated fatty acids, vitamin E, curcumin and eugenol alter serum and liver lipid peroxidation in rats. *Nutr. Res.*, 14, 1423–37, 1994.
211. Reddy, A.C. and Lokesh, B.R., Alterations in lipid peroxides in rat liver by dietary n-3 fatty acids: modulation of antioxidant enzymes by curcumin, eugenol, and vitamin E. *J. Nutr. Biochem.*, 5, 181–8, 1994.
212. Reddy, A.C. and Lokesh, B.R., Effect of curcumin and eugenol on iron-induced hepatic toxicity in rats. *Toxicology*, 107, 39–45, 1996.
213. Ferreira, L.C., Arbab, A.S., Victorasso Jardim-Perassi, B., Ferraz Borin, T., Varma, N.R.S., Iskander, A. *et al.*, Effect of curcumin on pro-angiogenic factors in the xenograft model of breast cancer. *Anti-Cancer Agents Med. Chem.*, 15, 1285–96, 2015.
214. Chang, C.C., Fu, C.F., Yang, W.T., Chen, T.Y., Hsu, Y.C., The cellular uptake and cytotoxic effect of curcuminoids on breast cancer cells. *Taiwan. J. Obstet. Gynecol.*, 51, 368–74, 2012.
215. Ji, J.L., Huang, X.F., Zhu, H.L., Curcumin and its formulations: Potential anti-cancer agents. *Anti-Cancer Agents Med. Chem.*, 12, 210–8, 2012.
216. Shieh, J.M., Chen, Y.C., Lin, Y.C., Lin, J.N., Chen, W.C., Chen, Y.Y. *et al.*, Demethoxycurcumin inhibits energy metabolic and oncogenic signaling pathways through AMPK activation in triple-negative breast cancer cells. *J. Agric. Food. Chem.*, 61, 6366–75, 2013.
217. Singletary, K., MacDonald, C., Wallig, M., Fisher, C., Inhibition of 7,12-dimethylbenz[a] anthracene (DMBA)-induced mammary tumorigenesis and DMBA-DNA adduct formation by curcumin. *Cancer Lett.*, 103, 137–41, 1996.

218. Ushida, J., Sugie, S., Kawabata, K., Pham, Q.V., Tanaka, T., Fujii, K. *et al.*, Chemopreventive effect of curcumin on N-nitrosomethylbenzylamine-induced esophageal carcinogenesis in rats. *Jpn. J. Cancer Res.*, 91, 893–8, 2000.
219. Aggarwal, B.B., Prostate cancer and curcumin: Add spice to your life. *Cancer Biol. Ther.*, 7, 1436–40, 2008.
220. Hung, C.M., Su, Y.H., Lin, H.Y., Lin, J.N., Liu, L.C., Ho, C.T. *et al.*, Demethoxycurcum inmodulates prostate cancer cell proliferation via AMPK-induced down-regulation of HSP70 and EGFR. *J. Agric. Food. Chem.*, 60, 8427–34, 2012.
221. Shehzad, A., Wahid, F., Lee, Y.S., Curcumin in cancer chemoprevention: Molecular targets, pharmacokinetics, bioavailability, and clinical trials. *Arch. Pharm.*, 343, 489–99, 2010.
222. Yang, C.H., Yue, J., Sims, M., Pfeffer, L.M., The curcumin analog EF24 targets NF-κB and miRNA-21, and has potent anticancer activity *in vitro* and *in vivo*. *PLoS One*, 8, e71130, 2013.
223. Díez, J.J. and Iglesias, P., The role of the novel adipocyte-derived hormone adiponectin in human disease. *Eur. J. Endocrinol.*, 148, 293–300, 2003.
224. Kim, K.C., Baek, S.H., Lee, C., Curcumin-induced downregulation of Axl receptor tyrosine kinase inhibits cell proliferation and circumvents chemoresistance in non-small lung cancer cells. *Int. J. Oncol.*, 47, 2296–303, 2015.
225. Tsai, J.R., Liu, P.L., Chen, Y.H., Chou, S.H., Cheng, Y.J., Hwang, J.J. *et al.*, Curcumin inhibits non-small cell lung cancer cells metastasis through the adiponectin/NF-κb/MMPs signaling pathway. *PLoS One*, 10, e0144462, 2015.
226. Zhang, J.Y., Lin, M.T., Zhou, M.J., Yi, T., Tang, Y.N., Tang, S.L. *et al.*, Combinational treatment of curcumin and quercetin against gastric cancer MGC-803 cells *in vitro*. *Molecules*, 20, 11524–34, 2015.
227. Allegra, A., Innao, V., Russo, S., Gerace, D., Alonci, A., Musolino, C., Anticancer activity of curcumin and its analogues: Preclinical and clinical studies. *Cancer Investig.*, 35, 1–22, 2017.
228. Nagababu, E., Sesikeran, B., Lakshmaiah, N., The protective effects of eugenol on carbon tetrachloride induced hepatotoxicity in rats. *Free Radical Res.*, 23, 617–27, 1995.
229. Nagashima, K., Inhibitory effect of eugenol on Cu^{2+}-catalyzed lipid peroxidation in human erythrocyte membranes. *Int. J. Biochem.*, 21, 745–9, 1989.
230. Joe, B. and Lokesh, B.R., Role of capsaicin, curcumin and dietary n-3 fatty acids in lowering the generation of reactive oxygen species in rat peritoneal macrophages. *Biochim. Biophys. Acta*, 1224, 255–63, 1994.
231. Chericoni, S., Prieto, J.M., Iacopini, P., Cioni, P., Morelli, I., *In vitro* activity of the essential oil of *Cinnamomum zeylanicum* and eugenol in peroxynitrite-induced oxidative processes. *J. Agric. Food. Chem.*, 53, 4762–5, 2005.
232. Ogata, M., Kaneya, D., Shin-ya, K., Li, L., Abe, Y., Katoh, H. *et al.*, Trapping effect of eugenol on hydroxyl radicals induced by L-DOPA *in vitro*. *Chem. Pharm. Bull.*, 53, 1167–70, 2005.
233. Ito, M., Murakami, K., Yoshino, M., Antioxidant action of eugenol compounds: role of metal ion in the inhibition of lipid peroxidation. *Food Chem. Toxicol.*, 43, 461–6, 2005.
234. Yoo, C.B., Han, K.T., Cho, K.S., Ha, J., Park, H.J., Nam, J.H. *et al.*, Eugenol isolated from the essential oil of *Eugenia caryophyllata* induces a reactive oxygen species-mediated apoptosis in HL-60 human promyelocytic leukemia cells. *Cancer Lett.*, 225, 41–52, 2005.
235. Durak, İ., Aytaç, B., Atmaca, Y., Devrim, E., Avcı, A., Erol, C. *et al.*, Effects of garlic extract consumption on plasma and erythrocyte antioxidant parameters in atherosclerotic patients. *Life Sci.*, 75, 1959–66, 2004.
236. Sukumaran, K., Unnikrishnan, M.C., Kuttan, R., Inhibition of tumour promotion in mice by eugenol. *Indian J. Physiol. Pharmacol.*, 38, 306–8, 1994.

237. Kumaravelu, P., Dakshinamoorthy, D.P., Subramaniam, S., Devaraj, H., Devaraj, N.S., Effect of eugenol on drug-metabolizing enzymes of carbon tetrachloride-intoxicated rat liver. *Biochem. Pharmacol.*, 49, 1703–7, 1995.
238. Mittal, R. and Gupta, R.L., *In vitro* antioxidant activity of piperine. *Methods Find Exp. Clin. Pharmacol.*, 22, 271–4, 2000.
239. Khajuria, A., Zutshi, U., Bedi, K., Permeability characteristics of piperine on oral absorption - An active alkaloid from peppers and a bioavailability enhancer. *Indian J. Exp. Biol.*, 36, 46–50, 1998.
240. Huovinen, M., Loikkanen, J., Myllynen, P., Vähäkangas, K.H., Characterization of human breast cancer cell lines for the studies on p53 in chemical carcinogenesis. *Toxicol. Vitro*, 25, 1007–17, 2011.
241. Koboldt, D.C., Fulton, R.S., McLellan, M.D., Schmidt, H., Kalicki-Veizer, J., McMichael, J.F. *et al.*, Comprehensive molecular portraits of human breast tumours. *Nature*, 490, 7418, 61–70, 2012.
242. Greenshields, A.L., Doucette, C.D., Sutton, K.M., Madera, L., Annan, H., Yaffe, P.B. *et al.*, Piperine inhibits the growth and motility of triple-negative breast cancer cells. *Cancer Lett.*, 357, 129–40, 2015.
243. Twiddy, D., Brown, D.G., Adrain, C., Jukes, R., Martin, S.J., Cohen, G.M. *et al.*, Pro-apoptotic proteins released from the mitochondria regulate the protein composition and caspase-processing activity of the native Apaf-1/caspase-9 apoptosome complex. *J. Biol. Chem.*, 279, 19665–82, 2004.
244. Mei, L., Chen, H., Wei, D.M., Fang, F., Liu, G.J., Xie, H.Y. *et al.*, Maintenance chemotherapy for ovarian cancer. *Cochrane Database Syst. Rev.*, 2013, CD007414, 2013.
245. Si, L., Yang, R., Lin, R., Yang, S., Piperine functions as a tumor suppressor for human ovarian tumor growth via activation of JNK/p38 MAPK-mediated intrinsic apoptotic pathway. *Biosci. Rep.*, 38, 3, BSR20180503, 2018.
246. Tai, J., Cheung, S., Wu, M., Hasman, D., Antiproliferation effect of Rosemary (*Rosmarinus officinalis*) on human ovarian cancer cells *in vitro*. *Phytomedicine*, 19, 436–43, 2012.
247. Lin, Y., Xu, J., Liao, H., Li, L., Pan, L., Piperine induces apoptosis of lung cancer A549 cells via p53-dependent mitochondrial signaling pathway. *Tumor Biol.*, 35, 3305–10, 2014.
248. Tian, Z., Shen, J., Moseman, A.P., Yang, Q., Yang, J., Xiao, P. *et al.*, Dulxanthone A induces cell cycle arrest and apoptosis via up-regulation of p53 through mitochondrial pathway in HepG2 cells. *Int. J. Cancer*, 122, 31–8, 2008.
249. Siddiqui, S., Ahamad, M.S., Jafri, A., Afzal, M., Arshad, M., Piperine triggers apoptosis of human oral squamous carcinoma through cell cycle arrest and mitochondrial oxidative stress. *Nutr. Cancer*, 69, 791–9, 2017.
250. Song, I.S., Jeong, J.Y., Jeong, S.H., Kim, H.K., Ko, K.S., Rhee, B.D. *et al.*, Mitochondria as therapeutic targets for cancer stem cells. *World J. Stem Cells*, 7, 418–27, 2015.
251. Trachootham, D., Alexandre, J., Huang, P., Targeting cancer cells by ROS-mediated mechanisms: a radical therapeutic approach? *Nat. Rev. Drug Discov.*, 8, 579–91, 2009.
252. Gnanasekar, M., Thirugnanam, S., Zheng, G., Chen, A., Ramaswamy, K., Gene silencing of translationally controlled tumor protein (TCTP) by siRNA inhibits cell growth and induces apoptosis of human prostate cancer cells. *Int. J. Oncol.*, 34, 1241–6, 2009.
253. Samykutty, A., Shetty, A.V., Dakshinamoorthy, G., Bartik, M.M., Johnson, G.L., Webb, B. *et al.*, Piperine, a bioactive component of pepper spice exerts therapeutic effects on androgen dependent and androgen independent prostate cancer cells. *PLoS One*, 8, e65889, 2013.
254. de Almeida, G.C., Oliveira, L.F.S., Predes, D., Fokoue, H.H., Kuster, R.M., Oliveira, F.L. *et al.*, Piperine suppresses the Wnt/β-catenin pathway and has anti-cancer effects on colorectal cancer cells. *Sci. Rep.*, 10, 11681, 2020.

255. Song, L., Wang, Y., Zhen, Y., Li, D., He, X., Yang, H. *et al.*, Piperine inhibits colorectal cancer migration and invasion by regulating STAT3/Snail-mediated epithelial–mesenchymal transition. *Biotechnol. Lett.*, 42, 2049–58, 2020.
256. Shabbeer, S., Sobolewski, M., Kachhap, S., Davidson, N., Carducci, M.A., Khan, S., Fenugreek: a naturally occurring edible spice as an anticancer agent. *Cancer Biol. Ther.*, 8, 272–8, 2009.
257. Khdary, N.H., Alangari, A.A., Katubi, K.M., Alanazi, M., Alhassan, A., Alzahrani, S.D. *et al.*, Synthesis of gingerol-metals complex and *in-vitro* cytotoxic activity on human colon cancer cell line. *Cancer Manag. Res.*, 15, 87–98, 2023.
258. Promdam, N. and Panichayupakaranant, P., [6]-Gingerol: A narrative review of its beneficial effect on human health. *Food Chem. Adv.*, 1, 100043, 2022.
259. Nafees, S., Zafaryab, Md., Mehdi, S.H., Zia, B., Rizvi, M.A., Khan, MdA., Anti-cancer effect of gingerol in cancer prevention and treatment. *Anti-Cancer Agents Med. Chem.*, 21, 428–32, 2021.
260. Conaway, H.H., Henning, P., Lerner, U.H., Vitamin A metabolism, action, and role in skeletal homeostasis. *Endocr. Rev.*, 34, 766–97, 2013.
261. Dao, D.Q., Ngo, T.C., Thong, N.M., Nam, P.C., Is Vitamin A an antioxidant or a pro-oxidant? *J. Phys. Chem. B*, 121, 9348–57, 2017.
262. Edge, R. and Truscott, T.G., Singlet oxygen and free radical reactions of retinoids and carotenoids—A review. *Antioxidants*, 7, 1, 5, 2018.
263. Gelain, D.P., Pasquali MA de, B., Caregnato, F.F., Castro, M.A.A., Moreira, J.C.F., Retinol induces morphological alterations and proliferative focus formation through free radical-mediated activation of multiple signaling pathways. *Acta Pharmacol. Sin.*, 33, 558–67, 2012.
264. Rozanowska, M., Edge, R., Land, E.J., Navaratnam, S., Sarna, T., Truscott, T.G., Scavenging of retinoid cation radicals by urate, trolox, and α-, β-, γ-, and δ-tocopherols. *Int. J. Mol. Sci.*, 20, 2799, 2019.
265. Trumbo, P., Yates, A.A., Schlicker, S., Poos, M., Dietary reference intakes. *J. Am. Diet. Assoc.*, 101, 294–294, 2001.
266. Levine, M., Wang, Y., Katz, A., Eck, P., Kwon, O., Chen, S. *et al.*, Ideal vitamin C intake. *BioFactors*, 15, 71–4, 2001.
267. Meščić Macan, A., Gazivoda Kraljević, T., Raić-Malić, S., Therapeutic perspective of vitamin C and its derivatives. *Antioxidants*, 8, 8, 247, 2019.
268. Popovic, L.M., Mitic, N.R., Miric, D., Bisevac, B., Miric, M., Popovic, B., Influence of vitamin C supplementation on oxidative stress and neutrophil inflammatory response in acute and regular exercise. *Oxid. Med. Cell. Longevity*, 2015, e295497, 2015.
269. Lobo, V., Patil, A., Phatak, A., Chandra, N., Free radicals, antioxidants and functional foods: Impact on human health. *Pharmacogn. Rev.*, 4, 118–26, 2010.
270. Fanucchi, M.V., Development of antioxidant and xenobiotic metabolizing enzyme systems, in: *The Lung* [Internet], R. Harding, K.E. Pinkerton, C.G. Plopper (Eds.), pp. 177–85, Academic Press, Oxford, 2004.
271. Tinggi, U., Selenium: its role as antioxidant in human health. *Environ. Health Prev. Med.*, 13, 2, 102–8, 2008.
272. Castellanos-Gutiérrez, A., Sánchez-Pimienta, T.G., Carriquiry, A., da Costa, T.H.M., Ariza, A.C., Higher dietary magnesium intake is associated with lower body mass index, waist circumference and serum glucose in Mexican adults. *Nutr. J.*, 17, 1, 114, 2018.
273. Szentmihályi, K., Szilágyi, M., Balla, J., Ujhelyi, L., Blázovics, A., *In vitro* antioxidant activities of magnesium compounds used in food industry. *Acta Aliment.*, 43, 419–25, 2014.
274. Jarosz, M., Olbert, M., Wyszogrodzka, G., Młyniec, K., Librowski, T., Antioxidant and anti-inflammatory effects of zinc. Zinc-dependent NF-κB signaling. *Inflammopharmacol.*, 25, 11–24, 2017.

275. Lee, S.R., Critical role of zinc as either an antioxidant or a prooxidant in cellular systems. *Oxid. Med. Cell. Longevity*, 2018, e9156285, 2018.
276. Marreiro, D.D.N., Cruz, K.J.C., Morais, J.B.S., Beserra, J.B., Severo, J.S., De Oliveira, A.R.S., Zinc and oxidative stress: Current mechanisms. *Antioxidants*, 6, 24, 2017.
277. Olechnowicz, J., Tinkov, A., Skalny, A., Suliburska, J., Zinc status is associated with inflammation, oxidative stress, lipid, and glucose metabolism. *J. Physiol. Sci.*, 68, 19–31, 2018.
278. Prasad, A.S., Zinc is an antioxidant and anti-inflammatory agent: Its role in human health. *Front. Nutr.* 1, 14, 2014.
279. Balasaheb Nimse, S. and Pal, D., Free radicals, natural antioxidants, and their reaction mechanisms. *RSC Adv.*, 5, 27986–8006, 2015.
280. Bost, M., Houdart, S., Oberli, M., Kalonji, E., Huneau, J.F., Margaritis, I., Dietary copper and human health: Current evidence and unresolved issues. *J. Trace Elem. Med. Biol.*, 35, 107–15, 2016.
281. Rudzińska, A., Juchaniuk, P., Oberda, J., Wiśniewska, J., Wojdan, W., Szklener, K. *et al.*, Phytochemicals in cancer treatment and cancer prevention—Review on epidemiological data and clinical trials. *Nutrients*, 15, 8, 1896, 2023.
282. Bjelakovic, G., Nikolova, D., Gluud, C., Antioxidant supplements and mortality. *Curr. Opin. Clin. Nutr. Metab. Care*, 17, 40, 2014.
283. Goodman, G.E., Thornquist, M.D., Balmes, J., Cullen, M.R., Meyskens Jr, F.L., Omenn, G.S. *et al.*, The beta-carotene and retinol efficacy trial: Incidence of lung cancer and cardiovascular disease mortality during 6-year follow-up after stopping β-carotene and retinol supplements. *JNCI: J. Natl. Cancer Inst.*, 96, 1743–50, 2004.
284. Omenn, G.S., Goodman, G.E., Thornquist, M.D., Balmes, J., Cullen, M.R., Glass, A. *et al.*, Effects of a combination of beta carotene and vitamin a on lung cancer and cardiovascular disease. *N. Engl. J. Med.*, 334, 1150–5, 1996.
285. Alpha-Tocopherol, Beta Carotene Cancer Prevention Study Group., The effect of vitamin E and beta carotene on the incidence of lung cancer and other cancers in male smokers. *N. Engl. J. Med.*, 330, 1029–35, 1994.
286. Virtamo, J., Taylor, P.R., Kontto, J., Männistö, S., Utriainen, M., Weinstein, S.J. *et al.*, Effects of α-tocopherol and β-carotene supplementation on cancer incidence and mortality: 18-Year postintervention follow-up of the alpha-tocopherol, beta-carotene cancer prevention study. *Int. J. Cancer*, 135, 178–85, 2014.
287. Bo, Y., Lu, Y., Zhao, Y., Zhao, E., Yuan, L., Lu, W. *et al.*, Association between dietary vitamin C intake and risk of esophageal cancer: A dose–response meta-analysis. *Int. J. Cancer*, 138, 1843–50, 2016.
288. Zhang, D., Xu, P., Li, Y., Wei, B., Yang, S., Zheng, Y. *et al.*, Association of vitamin C intake with breast cancer risk and mortality: a meta-analysis of observational studies. *Aging (Albany NY)*, 12, 18415–35, 2020.
289. Parent, M.E., Richard, H., Rousseau, M.C., Trudeau, K., Vitamin C intake and risk of prostate cancer: The Montreal PROtEuS study. *Front. Physiol.*, 4, 1218, 2018.
290. Ju, J., Picinich, S.C., Yang, Z., Zhao, Y., Suh, N., Kong, A.N. *et al.*, Cancer-preventive activities of tocopherols and tocotrienols. *Carcinogenesis*, 31, 533–42, 2010.
291. Loh, W.Q., Youn, J., Seow, W.J., Vitamin E intake and risk of prostate cancer: A meta-analysis. *Nutrients*, 15, 1, 14, 2023.
292. Bailey, R.L., Gahche, J.J., Lentino, C.V., Dwyer, J.T., Engel, J.S., Thomas, P.R. *et al.*, Dietary supplement use in the United States, 2003–2006. *J. Nutr.*, 141, 261–6, 2011.
293. Bailey, R.L., Fulgoni, V.L., Keast, D.R., Dwyer, J.T., Examination of vitamin intakes among US adults by dietary supplement use. *J. Acad. Nutr. Diet.*, 112, 657–663.e4, 2012.

294. Fink, B.N., Steck, S.E., Wolff, M.S., Britton, J.A., Kabat, G.C., Gaudet, M.M. *et al.*, Dietary flavonoid intake and breast cancer survival among women on long island. *Cancer Epidemiol. Biomarkers Prev.*, 16, 2285–92, 2007.
295. Peterson, J., Lagiou, P., Samoli, E., Lagiou, A., Katsouyanni, K., La Vecchia, C. *et al.*, Flavonoid intake and breast cancer risk: a case–control study in Greece. *Br. J. Cancer*, 89, 1255–9, 2003.
296. Romanos-Nanclares, A., Sánchez-Quesada, C., Gardeazábal, I., Martínez-González, M.Á., Gea, A., Toledo, E., Phenolic acid subclasses, individual compounds, and breast cancer risk in a mediterranean cohort: The SUN project. *J. Acad. Nutr. Diet.*, 120, 1002–1015.e5, 2020.
297. Russo, G.I., Campisi, D., Di Mauro, M., Regis, F., Reale, G., Marranzano, M. *et al.*, Dietary consumption of phenolic acids and prostate cancer: A case-control study in Sicily, Southern Italy. *Molecules*, 22, 2159, 2017.
298. Touaibia, M., Jean-François, J., Doiron, J., Caffeic acid, a versatile pharmacophore: an overview. *Mini-Rev. Med. Chem.*, 11, 8695–713, 2011.
299. Xiao, Z., Chen, Z., Han, R., Lu, L., Li, Z., Lin, J. *et al.*, Comprehensive TCM treatments combined with chemotherapy for advanced non-small cell lung cancer. *Med. (Baltimore)*, 100, e25690, 2021.
300. Wang, Y., Gapstur, S.M., Gaudet, M.M., Furtado, J.D., Campos, H., McCullough, M.L., Plasma carotenoids and breast cancer risk in the Cancer Prevention Study II Nutrition Cohort. *Cancer Causes Control*, 26, 1233–44, 2015.
301. Omenn, G.S., Goodman, G.E., Thornquist, M.D., Balmes, J., Cullen, M.R., Glass, A. *et al.*, Risk factors for lung cancer and for intervention effects in CARET, the beta-carotene and retinol efficacy trial. *JNCI: J. Natl. Cancer Inst.*, 88, 1550–9, 1996.
302. Tuli, H.S., Rath, P., Chauhan, A., Ramniwas, S., Vashishth, K., Varol, M. *et al.*, Phloretin, as a potent anticancer compound: From chemistry to cellular interactions. *Molecules*, 27, 8819, 2022.
303. Sergeev, A.V., Anan'Ev, V.S., Kapitanov, A.B., Korostelev, S.A., Bukreev, Y.M., Vlasenkova, N.K. *et al.*, Pharmacokinetics of carotenoids and carotene containing compounds. *Russ. J. Biother.*, 16, 92–106, 2017.
304. Diwadkar-Navsariwala, V., Novotny, J.A., Gustin, D.M., Sosman, J.A., Rodvold, K.A., Crowell, J.A. *et al.*, A physiological pharmacokinetic model describing the disposition of lycopene in healthy men. *J. Lipid Res.*, 44, 1927–39, 2003.
305. Kay, C.D., Aspects of anthocyanin absorption, metabolism and pharmacokinetics in humans. *Nutr. Res. Rev.*, 19, 137–46, 2006.
306. Smeriglio, A., Barreca, D., Bellocco, E., Trombetta, D., Chemistry, pharmacology and health benefits of anthocyanins. *Phytother. Res.*, 30, 1265–86, 2016.
307. Bangphumi, K., Kittiviriyakul, C., Towiwat, P., Rojsitthisak, P., Khemawoot, P., Pharmacokinetics of curcumin diethyl disuccinate, a prodrug of curcumin, in Wistar rats. *Eur. J. Drug Metab. Pharmacokinet.*, 41, 777–85, 2016.
308. Mohanty, C. and Sahoo, S.K., The *in vitro* stability and *in vivo* pharmacokinetics of curcumin prepared as an aqueous nanoparticulate formulation. *Biomaterials*, 31, 6597–611, 2010.
309. Chrubasik, S., Pittler, M.H., Roufogalis, B.D., Zingiberis rhizoma: A comprehensive review on the ginger effect and efficacy profiles. *Phytomedicine*, 12, 684–701, 2005.
310. Zick, S.M., Djuric, Z., Ruffin, M.T., Litzinger, A.J., Normolle, D.P., Alrawi, S. *et al.*, Pharmacokinetics of 6-gingerol, 8-gingerol, 10-gingerol, and 6-shogaol and conjugate metabolites in healthy human subjects. *Cancer Epidemiol. Biomarkers Prev.*, 17, 1930–6, 2008.
311. Zhu, Y., Wang, M., Zhang, J., Peng, W., Firempong, C.K., Deng, W. *et al.*, Improved oral bioavailability of capsaicin via liposomal nanoformulation: preparation, *in vitro* drug release and pharmacokinetics in rats. *Arch. Pharm. Res.*, 38, 512–21, 2015.

312. Pramod, K., Ansari, S.H., Ali, J., Eugenol: a natural compound with versatile pharmacological actions. *Nat. Prod. Commun.*, 5, 1934578X1000501236, 2010.
313. Carr, A.C. and Maggini, S., Vitamin C and immune function. *Nutrients*, 9, 11, 1211, 2017.
314. Tanumihardjo, S.A., Vitamin A: biomarkers of nutrition for development. *Am. J. Clin. Nutr.*, 94, 658S–665S, 2011.
315. Traber, M.G. and Atkinson, J., Vitamin E, antioxidant and nothing more. *Free Radical Biol. Med.*, 43, 4–15, 2007.
316. Rayman, M.P., The importance of selenium to human health. *Lancet*, 356, 233–41, 2000.
317. Roth, J.A. and Garrick, M.D., Iron interactions and other biological reactions mediating the physiological and toxic actions of manganese. *Biochem. Pharmacol.*, 66, 1–13, 2003.
318. Meléndez-Martínez, A.J., An overview of carotenoids, apocarotenoids, and vitamin A in agro-food, nutrition, health, and disease. *Mol. Nutr. Food Res.*, 63, 1801045, 2019.
319. Qin, S., Huang, L., Gong, J., Shen, S., Huang, J., Ren, H. *et al.*, Efficacy and safety of turmeric and curcumin in lowering blood lipid levels in patients with cardiovascular risk factors: a meta-analysis of randomized controlled trials. *Nutr. J.*, 16, 68, 2017.
320. Nejad, S.M., Özgüneş, H., Başaran, N., Pharmacological and toxicological properties of eugenol. *Turk. J. Pharm. Sci.*, 14, 2, 201, 2017.
321. Baskaran, P., Markert, L., Bennis, J., Zimmerman, L., Fox, J., Thyagarajan, B., Assessment of pharmacology, safety, and metabolic activity of capsaicin feeding in mice. *Sci. Rep.*, 9, 8588, 2019.
322. Fibigr, J., Šatínský, D., Solich, P., Current trends in the analysis and quality control of food supplements based on plant extracts. *Anal. Chim. Acta*, 1036, 1–15, 2018.
323. Smirnoff, N., Ascorbic acid metabolism and functions: A comparison of plants and mammals. *Free Radical Biol. Med.*, 122, 116–29, 2018.
324. Sieniawska, E., Baj, T., Sawicki, R., Wanat, A., Wojtanowski, K.K., Ginalska, G. *et al.*, LC-QTOF-MS analysis and activity profiles of popular antioxidant dietary supplements in terms of quality control. *Oxid. Med. Cell. Longevity*, 2017, 8692516, 2017.

3

Antioxidants in Inflammatory Diseases

Sayed Mohammed Firdous[1]*, Sourav Pal[2], Subhajit Mandal[3] and Rakesh K. Sindhu[4]

[1]Department of Pharmacology, Calcutta Institute of Pharmaceutical Technology & AHS, Uluberia, Howrah, West Bengal, India
[2]Department of Pharmacology, Seacom Pharmacy College, Jaladhulagori, Sankrail, Howrah, West Bengal, India
[3]Department of Pharmaceutical Sciences, Faculty of Science and Engineering, Dibrugarh University, Dibrugarh, Assam, India
[4]School of Pharmacy, Sharda University, Greater Noida, Uttar Pradesh, India

Abstract

Damage to tissues and abnormal function of organs are hallmarks of inflammatory disorders, which are characterized by an overactive and inadequately controlled immune response. Inflammation and chronic diseases can result from microbial infections, tissue damage, and autoimmune dysregulation. These stimuli trigger macrophages and neutrophils, releasing ROS and RNS. ROS/RNS overwhelm the body's antioxidant defenses, causing oxidative assault and inflammatory signaling cascades. NF-κβ and AP-1 activate amid oxidative assault. These transcription factors regulate the expression of pro-inflammatory genes that create cytokines (TNF, ILs), chemokines, and adhesion molecules, which enhance inflammation and recruit immune cells to damaged areas. Antioxidants oversight several inflammatory signaling cascades. They're able to block the production of pro-inflammatory genes by reducing NF-κβ and AP-1 activation. Crucial inflammatory MAPK pathways like ERK, JNK, and p38 MAPK can be modulated by antioxidants. Antioxidants can decrease pro-inflammatory mediator synthesis because they interfere with these pathways. Directly scavenging and neutralizing ROS/RNS is how antioxidants do their job. Scavenging protects lipids, proteins, and DNA against oxidative damage. Antioxidants enhance SOD, catalase, and glutathione peroxidase, which are body-produced antioxidant enzymes. Thus, increasing enzyme expression and activity strengthens cells' antioxidant defenses. Antioxidants can induce an anti-inflammatory phenotype in macrophages and T cells. They are capable of shifting macrophages from M1 to M2. M2 macrophages aid tissue repair and immunomodulation. Antioxidants also promote the formation of regulatory T cells, which regulate the immune system by inhibiting other immune cells. In conclusion, this abstract emphasizes the crucial role of oxidative assault in inflammatory diseases and the potential of antioxidants as therapeutic agents.

***Keywords*:** Oxidative assault, inflammation, rheumatoid arthritis, inflammatory bowel disease, atherosclerosis, antioxidant therapies

**Corresponding author*: firdous.oncology@gmail.com

Rakesh K. Sindhu, Inderbir Singh and M. Arockia Babu (eds.) Antioxidants: Nature's Defense Against Disease, (83–126)

3.1 Introduction

Reactive oxygen species (ROS) have historically been seen as deleterious byproducts of cellular metabolism, specifically the oxidative phosphorylation (OXPHOS) activity occurring in mitochondria [1]. The apparent paradoxical relationship between the essential energy-generating process of OXPHOS and the production of ROS has prompted significant scholarly investigation aimed at comprehending the diverse functions of ROS within biological systems [2]. Recent findings have unveiled that ROS have a dual role, not only causing harm to cellular components but also acting as vital secondary messengers that coordinate cellular responses [3]. This study delves into the intricate interplay between ROS, OXPHOS, and inflammatory disorders, emphasizing their regulatory roles in both normal physiological processes and pathological conditions. The generation of ROS in mitochondria is an intrinsic consequence of OXPHOS, a process in which multi-subunit enzyme complexes metabolize nutrients to produce ATP [4]. Despite the potential hazards associated with their reactivity, ROS have emerged as important contributors to cellular signaling pathways. Interestingly, ROS derived from mitochondria are not exclusively generated through the process of respiration. Instead, they have significant involvement in regulating mitochondrial function and facilitating physiological responses [5]. In addition to OXPHOS, ROS are produced in significant amounts by NADPH oxidase complexes (NOX), particularly NOX2, which is predominantly found in immune cells including neutrophils, monocytes, macrophages, and dendritic cells [6]. The generation of ROS through the NOX2 enzyme is essential for the oxidative burst necessary to eliminate pathogens. This process is tightly regulated by cytosolic adaptors such as p47phox and p40phox [7]. ROS, which has long been linked to inflammation and tissue damage, have more recently gained recognition for their dual role in both controlling inflammation and maintaining immunological homeostasis [8]. The presence of ROS has been found to have a significant role in both the promotion and resolution of inflammation, as supported by evidence from animal models and clinical observations. This challenges traditional beliefs and understanding in this field. Within the framework of the immune system, ROS produced during the process of oxidative burst assume crucial functions in the eradication of pathogens. However, their influence extends beyond pathogen clearance and encompasses the modulation of immune cell function as well as the management of inflammation. In contrast to the concept of inflammation caused by ROS, recent research indicates that impairments in ROS are linked to chronic inflammation and autoimmune disorders in various clinical contexts [9]. The complex methods by which ROS exercise their dual effects require delicate interactions between innate immune cells and signaling pathways. This review aims to provide a comprehensive understanding of the role of ROS in the initiation, advancement, and amelioration of auto-inflammatory and autoimmune disorders. Through an examination of several pathways, this study sheds light on the contributions of ROS to these pathological conditions. The thorough analysis encompasses investigations conducted at both the cellular and molecular levels, with a specific emphasis on elucidating the intricate functionalities of immune cells and their profound implications for human well-being.

In conclusion, this review provides a thorough examination of the changing understanding of ROS within the framework of inflammatory disorders. The conventional binary classification of ROS as exclusively deleterious entities is being superseded by a more intricate comprehension of their dual functions as initiators of inflammation and regulators of immune responses. The investigation into the mechanisms mediated by ROS in immune cells provides insights into their role in the pathogenesis of auto-inflammatory and autoimmune illnesses. In conclusion, the study highlights the significance of modulating ROS as a potential therapeutic approach for effectively controlling a range of inflammatory disorders.

3.2 Inflammatory Disease: An Overview, Definition and Classification of Inflammatory Diseases

Inflammatory diseases are a group of disorders characterized by inflammation, which is the body's natural response to harmful stimuli, such as pathogens, tissue injury, or irritants [10]. Inflammation is a complex biological process involving the activation of the immune system and various signaling pathways, resulting in a series of protective responses aimed at removing the offending agent and promoting tissue repair [11]. It helps the body fight off infections, repair damaged tissues, and heal wounds. However, when inflammation becomes chronic or dysregulated, it can lead to a wide range of inflammatory diseases, which can affect various organ systems and have long-term health consequences [12].

Inflammatory diseases are conditions in which the inflammatory response is either excessive, prolonged or inappropriately triggered, leading to tissue damage and dysfunction [13]. These diseases can result from a combination of genetic predisposition, environmental factors, and immune system dysfunction. Common hallmarks of inflammation include redness, swelling, heat, pain, and altered tissue function [14]. Inflammatory diseases can be classified into several categories based on their underlying causes, affected organs or systems, and clinical manifestations. Here are some common categories of inflammatory diseases:

A. **Autoimmune Diseases:** In these conditions, the immune system mistakenly targets and attacks the body's tissues. Examples include rheumatoid arthritis, systemic lupus erythematosus (SLE), and multiple sclerosis [15].
B. **Infectious Diseases:** Many infections, such as bacterial, viral, or fungal infections, trigger inflammation as the immune system responds to eliminate the pathogens. Conditions like pneumonia, tuberculosis, and HIV/AIDS fall into this category [16].
C. **Allergic Diseases:** Allergies are immune responses to normally harmless substances, resulting in inflammation. Examples include allergic rhinitis, asthma, and eczema [17].
D. **Inflammatory Bowel Diseases (IBD):** These diseases involve chronic inflammation of the digestive tract. The two primary forms are Crohn's disease and ulcerative colitis [18].
E. **Skin Disorders:** Conditions like psoriasis and atopic dermatitis involve inflammation of the skin and are characterized by redness, itching, and skin lesions [19].

F. **Cardiovascular Inflammatory Diseases:** Atherosclerosis, a condition where plaques build up in arteries, can have an inflammatory component. Inflammatory processes can contribute to heart attacks and strokes [20].
G. **Neuroinflammatory Diseases:** Conditions like multiple sclerosis and Guillain-Barré syndrome involve inflammation of the nervous system [21].
H. **Respiratory Inflammatory Diseases:** Asthma and chronic obstructive pulmonary disease (COPD) are characterized by airway inflammation [22].
I. **Joint and Connective Tissue Diseases:** Apart from rheumatoid arthritis, conditions like osteoarthritis and gout can also have inflammatory components [23].
J. **Systemic Inflammatory Diseases:** Conditions like systemic inflammatory response syndrome (SIRS) and sepsis involve widespread inflammation throughout the body [24].

3.3 Pathogenesis of Inflammation, Role of Oxidative Assault in Inflammatory Disease Development

The pathogenesis of inflammation is a complex process involving a series of events initiated by the body in response to various stimuli, such as infections, tissue damage, or irritants. Inflammation is a protective mechanism that aims to remove the offending agent and promote tissue repair. However, when inflammation is not properly regulated or becomes chronic, it can contribute to the development of inflammatory diseases. Oxidative assault is one of the factors that can play a significant role in the development and perpetuation of inflammatory diseases [25].

3.3.1 Pathogenesis of Inflammation

The pathogenesis of inflammation can be summarized in several key steps:

A. **Triggering Event:** Inflammatory responses are typically triggered by a stimulus, such as infection by pathogens (e.g., bacteria, viruses), tissue injury (e.g., trauma), exposure to irritants (e.g., chemicals), or autoimmune reactions.
B. **Immune Response:** The immune system recognizes the threat and activates immune cells, particularly white blood cells (leukocytes), including neutrophils, macrophages, and lymphocytes.
C. **Inflammatory Mediators:** These immune cells release inflammatory mediators, including cytokines (e.g., interleukins and tumor necrosis factor), chemokines, and prostaglandins. These substances help orchestrate the inflammatory response.
D. **Vasodilation and Increased Permeability:** Inflammatory mediators cause blood vessels to dilate and become more permeable, allowing immune cells and fluid to enter the affected tissue. This results in redness, heat, and swelling at the site of inflammation.

E. **Phagocytosis:** Immune cells, particularly macrophages, and neutrophils, phagocytose (engulf and digest) pathogens and cellular debris at the site of inflammation
F. **Resolution and Tissue Repair:** Once the threat is eliminated or neutralized, anti-inflammatory signals are produced to resolve the inflammation, and tissue repair processes are initiated [26].

3.3.2 Role of Oxidative Assault in Inflammatory Disease Development

Oxidative assault occurs when there is an imbalance between the production of ROS and the body's ability to neutralize them with antioxidants. ROS, including free radicals and reactive molecules like hydrogen peroxide, are generated during normal cellular processes and play a role in immune responses. However, excessive ROS production can lead to oxidative assault, which can contribute to inflammatory disease development in several ways:

A. **Cell Damage:** ROS can directly damage cells and tissues, promoting inflammation and tissue injury.
B. **Activation of Inflammatory Signaling:** ROS can activate signaling pathways that promote the production of pro-inflammatory cytokines and chemokines, further amplifying the inflammatory response.
C. **Oxidative Modification:** Proteins, lipids, and DNA can be modified by ROS, leading to functional changes that contribute to inflammation and tissue dysfunction.
D. **Impaired Resolution:** Oxidative assault can impair the resolution phase of inflammation, leading to persistent inflammation and tissue damage.
E. **Antioxidant Deficiency:** In some cases, deficiencies in antioxidants, such as glutathione and vitamins C and E, can make individuals more susceptible to oxidative assault and inflammatory diseases [27].

Inflammatory diseases where oxidative assault plays a significant role include rheumatoid arthritis, atherosclerosis, chronic obstructive pulmonary disease (COPD), and neurodegenerative diseases like Alzheimer's disease. Managing oxidative assault through a healthy diet, regular exercise, and antioxidant-rich foods can be an important part of preventing or managing inflammatory diseases. Additionally, some medications and therapies may target oxidative assault as a therapeutic approach in these conditions [28].

3.4 Introduction to Antioxidants, Definition and Classification of Antioxidants, Mechanisms of Antioxidant Action, Sources of Dietary Antioxidants

Compounds known as antioxidants are essential for defending the body against the harmful effects of oxidative assault. ROS or free radical production and the body's capacity to fight them off with antioxidants are out of balance, which leads to oxidative assault. These unstable chemicals can harm DNA, proteins, lipids, and cells, which contributes to several illnesses like cancer, heart disease, and neurological conditions. By providing electrons to stabilize ROS, antioxidants aid in neutralizing or attenuating their negative effects [29].

3.4.1 Classification of Antioxidants

Antioxidants encompass a wide range of compounds, both naturally occurring in the body and obtained from the diet. They can be classified into two main categories:

Endogenous Antioxidants: These are antioxidants produced within the body. Examples include enzymes like superoxide dismutase (SOD), catalase, and glutathione peroxidase. These enzymes help detoxify ROS and maintain cellular redox balance [30].
Exogenous Antioxidants: These are antioxidants obtained from external sources, primarily through diet or supplements. Exogenous antioxidants can further be categorized into several subgroups, including:

A. **Vitamins:** Certain vitamins have antioxidant properties, such as vitamin C (ascorbic acid), vitamin E (tocopherols and tocotrienols), and vitamin A (beta-carotene).
B. **Minerals:** Minerals like selenium and zinc serve as cofactors for antioxidant enzymes.
C. **Phytochemicals:** These are naturally occurring compounds in plants that possess antioxidant activity. Examples include flavonoids (found in fruits and vegetables), polyphenols (found in tea, wine, and chocolate), and carotenoids (found in colorful fruits and vegetables like carrots and tomatoes) [31].

3.4.2 Mechanisms of Antioxidant Action of Antioxidant

Antioxidants act through various mechanisms to neutralize ROS and prevent oxidative damage:

3.4.2.1 Scavenging Free Radicals

Scavenging free radicals is one of the primary mechanisms of action for antioxidants. Free radicals are highly reactive molecules with unpaired electrons, making them unstable and capable of causing cellular damage by "stealing" electrons from other molecules, leading to a chain reaction of oxidative damage. Antioxidants counteract this process by donating electrons to neutralize free radicals, stabilizing them and preventing further damage [32]. Here's how the scavenging of free radicals by antioxidants works:

a. **Donating Electrons:** Antioxidants are molecules that are stable enough to donate electrons without becoming reactive themselves. When they encounter a free radical, such as a superoxide anion (O2•-) or hydroxyl radical (•OH), they transfer an electron to the radical.

b. **Stabilizing Free Radicals:** The donation of an electron to the free radical stabilizes it. Free radicals become less reactive and less damaging once they have a complete pair of electrons.
c. **Breaking the Chain Reaction:** Free radical reactions often occur in a chain reaction. When an antioxidant neutralizes a free radical, it breaks the chain reaction by preventing the free radical from reacting with other molecules and causing further damage.
d. **Regeneration:** Some antioxidants can regenerate themselves or other antioxidants after donating an electron. For example, vitamin C (ascorbic acid) can regenerate vitamin E (tocopherol) by donating an electron to tocopherol, which allows tocopherol to continue its antioxidant role.
e. **Multiple Antioxidants:** The human body employs a network of antioxidants, each with its specific targets and functions. These antioxidants work in concert to combat various types of free radicals and oxidative assault [33].

3.4.2.2 Metal Chelation

An essential method through which antioxidants reduce oxidative assault and shield cells and tissues from harm brought on by ROS is metal chelation. Antioxidants bind metal ions, such as iron (Fe) and copper (Cu), in a process known as chelation to stop these metals from taking part in detrimental chemical processes that result in increased oxidative assault [34]. Here's a short note on metal chelation by antioxidants:

3.4.3 Mechanism of Metal Chelation

3.4.3.1 Binding Affinity

Many antioxidants, such as flavonoids and polyphenols found in plants, have a high affinity for metal ions. They contain specific functional groups, such as hydroxyl (-OH) and carbonyl (-C=O) groups, that can coordinate with metal ions.

3.4.3.2 Chelation Process

When antioxidants encounter metal ions, they form stable complexes through coordination bonds. These bonds involve the sharing of electron pairs between the metal ion and the antioxidant molecule. This binding prevents the metal ion from participating in harmful reactions.

3.4.3.3 Preventing ROS Formation

Metals like iron and copper can catalyze the formation of highly reactive hydroxyl radicals (•OH) through the Fenton reaction. By chelating these metals, antioxidants inhibit this reaction, reducing the production of harmful ROS [35].

3.5 Enzymatic Activity

Another significant method by which antioxidants shield cells and tissues from oxidative damage is by their enzyme activity. A subgroup of endogenous antioxidants known as antioxidant enzymes is essential for neutralizing ROS and preserving the cellular redox balance. Cells create particular proteins known as enzymatic antioxidants to catalyze processes that cleanse and neutralize ROS [36]. These enzymes play a crucial role in the body's natural defenses against oxidative damage and are extremely effective at coping with oxidative assault. A vital antioxidant enzyme called SOD transforms the superoxide radical (O_2•-) into hydrogen peroxide (H_2O_2) and molecule oxygen (O_2) [37]. To lower the quantities of extremely reactive superoxide radicals, which can harm cells, this is an important step. Catalase is another significant antioxidant enzyme that is present in cells, especially in peroxisomes [38]. It stops the buildup of these damaging ROS by converting H_2O_2 into water (H_2O) and oxygen (O_2). The enzyme glutathione peroxidase (GPx) uses glutathione (GSH) as a cofactor to convert organic peroxides and H_2O_2 into safe water and matching alcohols [39]. Additionally, it aids in the detoxification of lipid peroxides. Glutathione Reductase is an enzyme that aids in converting oxidized glutathione (GSSG) into reduced glutathione (GSH), enabling GSH to carry out its antioxidant functions [40].

3.5.1 Mechanism of Enzymatic Activity

Specific enzymatic processes that entail the transformation of hazardous ROS into less toxic or non-toxic molecules are the basis for the action of enzymatic antioxidants. Here is a brief explanation of how some of these enzymes work:

- **A.** Superoxide Dismutase (SOD) is an enzyme that catalyzes the conversion of O2•- into O_2 and H_2O_2. Catalase or glutathione peroxidase can subsequently further detoxify the resultant hydrogen peroxide [37].
- **B.** Catalase: Through a quick process, catalase changes H_2O_2 into H_2O and O_2. By successfully removing it, hydrogen peroxide is kept out of potentially harmful interactions [38].
- **C.** Glutathione Peroxidase (GPx): GPx uses reduced GSH as a cofactor to decrease organic peroxides (ROOH) and H_2O_2. H_2O and GSSG are the products of this process [39].

3.5.1.1 Significance in Cellular Defense

By limiting the production of damaging ROS, enzymatic antioxidants are essential for preserving cellular redox homeostasis. To specifically target regions of the cell where ROS generation is prevalent, such as the cytoplasm, mitochondria, and peroxisomes, they are positioned at key locations within the cell. These enzymes assist in preventing oxidative damage to cellular structures and macromolecules as well as lowering the risk of numerous diseases linked to oxidative assault by effectively detoxifying ROS [41].

3.6 Sources of Dietary Antioxidants

Dietary antioxidants are essential for maintaining health and preventing oxidative assault-related diseases. Here are common sources of dietary antioxidants:

3.6.1 Fruits and Vegetables

Colorful fruits and vegetables like berries, citrus fruits, spinach, kale, and broccoli are rich in vitamins (e.g., vitamin C) and phytochemicals (e.g., flavonoids) [42]. Vitamin C (ascorbic acid) is a water-soluble vitamin that is found in large quantities in broccoli, strawberries, kiwi, bell peppers, and citrus fruits (such as oranges and grapefruits). Strong anti-oxidants like vitamin C help shield cells from oxidative damage and strengthen the immune system [43]. Avocados, spinach, almonds, sunflower seeds, and other foods contain the fat-soluble antioxidant vitamin E (tocopherols and tocotrienols) [44]. It aids in preventing lipid peroxidation, a harmful process brought on by free radicals, which harms cell membranes. It is rich in leafy greens like kale and spinach as well as orange and yellow veggies like carrots, sweet potatoes, and pumpkin. A precursor to vitamin A, beta-carotene benefits eye health. Another carotenoid is lycopene, which is present in red and pink fruits and vegetables like tomatoes, watermelon, and red bell peppers [45]. Resveratrol is an antioxidant polyphenol found in grapes, especially red grapes and red wine [46]. It has been researched for potential medical advantages, such as cardiovascular defense [47]. Anthocyanins are the red, purple, and blue fruits and vegetables water-soluble pigments. Blueberries, cherries, red cabbage, and eggplants are a few examples. Antioxidant and anti-inflammatory effects are shared by anthocyanins [48].

The vital endogenous antioxidant glutathione is not normally ingested directly through diet, although some fruits and vegetables, including avocados, spinach, and asparagus, are rich providers of the amino acids and precursors required [49]. Selenium is a trace mineral that can be found in a few different types of fruits and vegetables, including Brazil nuts, spinach, and broccoli [50]. Catechins are a particular class of flavonoid antioxidants that can be found in berries, apples, and green tea. They have been linked to several health advantages, including metabolic and cardiovascular health [51]. To ensure that your diet contains a wide range of antioxidants, it is crucial to eat a variety of fruits and vegetables in various hues. Consuming a variety of plant-based meals high in antioxidants can help shield cells from oxidative damage and promote general health [52].

3.6.2 Nuts and Seeds

Almonds, sunflower seeds, hazelnuts, and peanuts are just a few of the nuts and seeds that are rich in the fat-soluble antioxidant vitamin E (tocopherols and tocotrienols) [53]. It aids in preventing oxidative damage to cell membranes. Selenium, a trace mineral, is a crucial part of antioxidant enzymes like glutathione peroxidase. Selenium is a mineral that is particularly abundant in some nuts, including cashews and sunflower seeds as well as Brazil nuts [50]. Many nuts and seeds contain polyphenols, which are potent antioxidants. Examples include: Almonds, hazelnuts, and walnuts are excellent sources of flavonoids, which have a range of health advantages, such as antioxidant and anti-inflammatory characteristics.

Sesame, flax, and chia seeds all contain phenolic acids, such as ferulic and caffeic acids, which have antioxidant qualities [54]. Lignans, which contain antioxidant and possibly hormone-balancing properties, are particularly abundant in flaxseeds [55]. CoQ10 is an antioxidant-rich substance that is found in nature. Nuts contain it; significant sources are sesame seeds and peanuts [56]. Copper and zinc are important trace minerals that the body needs because they act as cofactors for antioxidant enzymes [57]. These minerals can be found in nuts like cashews and seeds like pumpkin seeds. Melatonin, an antioxidant hormone that plays a role in regulating sleep and circadian rhythms, is found in several seeds, including pumpkin seeds [58]. Nitric oxide (NO), a molecule with antioxidant qualities that aids in regulating blood flow and blood pressure, is produced by the amino acid arginine, which is present in nuts like almonds [59].

3.6.3 Whole Grains

Whole grains like oats, brown rice, and quinoa provide antioxidants, including selenium and various phytochemicals [60]. Whole grains including wheat, barley, and oats include vitamin E (tocopherols and tocotrienols), a fat-soluble antioxidant that aids in shielding cell membranes from oxidative damage. Whole grains are rich in phenolic antioxidants, which include the following: Wheat, oats, and rice bran contain ferulic acid, which has antioxidant qualities and may benefit cardiovascular health [61]. Caffeic Acid found in healthy grains like oats, caffeine may have anti-inflammatory and antioxidant properties [62]. Whole grains are a good source of lignans, particularly flaxseed, rye, and whole wheat. Lignans are phytochemicals that may have hormone-balancing and antioxidant activities. Oats and barley contain the polysaccharides known as beta-glucans. They are known for their potential to decrease cholesterol levels and contain antioxidant qualities [63]. Some whole grains, including quinoa, include saponins, which have anti-inflammatory and antioxidant properties. Although whole grains don't normally contain large levels of vitamin C some kinds may contain trace amounts. Whole grains like whole wheat and maize (corn) may contain traces of these antioxidant pigments known as carotenoids, which are also present in colorful fruits and vegetables. Water-soluble vitamin folate is present in whole grains like wheat and barley [64]. It contributes indirectly to antioxidant defense mechanisms, aids in DNA synthesis and repair, and plays a part in DNA replication. A diet high in whole grains, such as quinoa, whole wheat, brown rice, oats, and barley, can offer a variety of antioxidants that improve general health. These anti-oxidants aid in the fight against oxidative assault, lessen inflammation, and lessen the risk of developing chronic illnesses, such as heart disease and some cancers. A balanced and antioxidant-rich diet can benefit from including whole grains in your meals [65].

3.6.4 Herbs and Spices

Herbs like basil, oregano, and spices like cinnamon and turmeric are packed with antioxidants [66]. Numerous spices and plants are abundant in polyphenols, which are powerful antioxidants. Catechins, which are present in green tea and other spices like cinnamon, are known for their anti-inflammatory and antioxidant properties [67]. Flavonoids are compounds that have a variety of antioxidant and health-promoting effects. They are found in spices like thyme and herbs like parsley [68]. Quercetin, a flavonoid with antioxidant and anti-inflammatory effects, is found in spices like red onion and herbs like capers [69]. Spices that include carotenoids, pigments with antioxidant characteristics, are known as carotene. For instance, the carotenoid content in saffron and paprika is well recognized [70]. Turmeric contains a polyphenol called curcumin, which has potent anti-inflammatory and antioxidant properties [71]. It is renowned for having the ability to promote different elements of health. Capsaicin, the ingredient that gives chili peppers their heat, has antioxidant characteristics that may help explain some of the peppers' health advantages [72]. Gingerol is a polyphenol with anti-inflammatory and antioxidant effects that is found in the often-used spice ginger [73]. Herbs containing rosmarinic acid, such as rosemary and oregano, have antioxidant and anti-inflammatory properties [74]. Onions and garlic are abundant in allicin, which is a sulfur compound with antioxidant qualities and potential cardiovascular benefits [75]. Some herbs, like thyme, and spices, such as chili peppers, contain vitamin C. Some herbs and spices contain the minerals copper and zinc, which work as cofactors for the body's antioxidant enzymes [76]. Eugenol is a phenolic component with antioxidant characteristics that is present in spices and plants like cinnamon and cloves [77].

3.6.5 Tea and Coffee

Popular beverages like tea and coffee include a range of antioxidants, including polyphenols and other substances renowned for their health advantages [78]. The following are some of the common antioxidants present in tea and coffee: Catechins are a class of polyphenols that are present in tea, particularly green tea. The most prevalent and powerful catechin is epigallocatechin gallate (EGCG). Strong antioxidant capabilities and several health advantages, including cardiovascular health and possibly cancer prevention, are linked to catechins. Flavonoids like quercetin and myricetin, which have antioxidant and anti-inflammatory qualities, are found in both green and black tea [79]. Theaflavins and Thearubigins are produced during the oxidation process that black tea goes through during processing. These substances have antioxidant qualities and contribute to the flavor and color of black tea [80]. L-theanine is an amino acid found in tea that provides calming and relaxing effects as well as antioxidant effects. Although its stimulating characteristics are what make it most famous, it also possesses antioxidant capabilities [81]. Caffeic acid, a different type of polyphenol, is present in coffee and aids in the antioxidant activity of the beverage [62]. Coffee beans include the chemicals cafestol and kahweol, which have antioxidant effects. However, depending on the brewing technique and coffee variety, their concentrations can change. Complex substances created during the roasting of coffee beans include melanoidins. They enhance the flavor and aroma of coffee and offer antioxidant qualities. Quinides are known for their antioxidant effects and are a byproduct of the roasting of coffee [82]. Both tea and coffee can be included in an antioxidant-rich diet, and moderate use of these drinks has been linked to several health advantages.

3.6.6 Dark Chocolate

Due to the presence of different bioactive components, dark chocolate is well known for having a high antioxidant content [83]. The following are some of the common antioxidants discovered in dark chocolate: A class of polyphenolic chemicals with strong antioxidant capabilities, flavonoids are abundant in dark chocolate. Many of the health advantages of dark chocolate are due to flavonoids. Dark chocolate is rich in catechins, which are recognized for their potent antioxidant properties. Catechins include epicatechin and epicatechin gallate. Dark chocolate contains oligomeric flavonoids called procyanidins [84]. They are potent anti-oxidants that can aid in preventing oxidative assault in cells. Epicatechin, a particular flavonoid, is thought to be one of the main ingredients behind the cardiovascular advantages of eating dark chocolate. It contains anti-inflammatory and antioxidant effects [85]. Although theobromine is not a conventional antioxidant, it is a substance present in dark chocolate that has a slight antioxidant effect. It's also well known for having stimulating properties. Dark chocolate contains caffeine, which has antioxidant benefits similar to those of theobromine but is more well-known for its energizing effects [86]. The overall antioxidant potential of dark chocolate is enhanced by the presence of numerous polyphenolic components, such as phenolic acids [87]. It's crucial to remember that the antioxidant content of dark chocolate might vary based on elements like the amount of cocoa, the way it was processed, and the particular brand or type of chocolate. More antioxidants are often found in dark chocolate with a greater cocoa concentration (70 percent or more) than in milk chocolate or chocolate with a lower cocoa level [88].

3.6.7 Fatty Fish

The numerous antioxidants and other bioactive substances found in fatty fish contribute to their health advantages. Salmon, trout, and prawns contain a powerful antioxidant called astaxanthin, which is also known as a carotenoid [89]. It imparts these meals' pink or reddish color and is a potent antioxidant. Astaxanthin has been linked to several health advantages, including anti-inflammatory properties, and may help shield cells from oxidative damage [90]. A cofactor for the body's antioxidant enzymes like glutathione peroxidase, selenium is a trace mineral. Selenium is a mineral that can be found in fatty seafood like tuna and sardines [91]. CoQ10 is a naturally occurring substance that can be found in many different cells and tissues, including the tissues of fatty fish like mackerel and salmon [92]. It contains antioxidant qualities and participates in the cellular energy production process. The omega-3 fatty acids present in fatty fish, such as eicosapentaenoic acid (EPA) and docosahexaenoic acid (DHA), have anti-inflammatory characteristics and can help lower oxidative assault in the body, while not being typical antioxidants [93]. Salmon, mackerel, and sardines are excellent sources of vitamin D in the diet [94]. While vitamin D is mostly recognized for its function in calcium absorption and bone health, it also contributes to the body's antioxidant defense mechanisms. A substance like an amino acid called taurine is present in fatty fish, notably mackerel [95]. It is an antioxidant and participates in several physiological functions, including the preservation of cellular integrity. The trace mineral zinc, which is needed by antioxidant enzymes like superoxide dismutase (SOD), can also be found in fatty fish. Another trace mineral, copper, which is a cofactor for various antioxidant enzymes, is also present in some fatty fish. Salmon, mackerel, trout, sardines, and herring are among the valuable fatty fish due to their high omega-3 fatty acid content, which has several health benefits in addition to antioxidant characteristics [89].

3.7 Mechanisms of Oxidative Assault in Inflammatory Disease

3.7.1 Generation of ROS and RNS

ROS are oxygen-containing molecules with a high degree of reactivity that are produced as byproducts of biological metabolism. However, excessive ROS production can result in oxidative assault and cellular damage even if some ROS are necessary for cell signaling and defense systems [96]. The majority of ROS are formed in the mitochondria, the cell's energy-producing organelles, according to the mitochondrial electron transport chain [97]. Superoxide radicals ($O_2^{\bullet-}$) are created when electrons from the electron transport chain escape during oxidative phosphorylation and interact with molecules of oxygen [98]. This is a normal byproduct of energy production, although in some instances it may be excessive. NADPH oxidases are enzymes that specialize in producing ROS during immunological responses and cell signaling. They are present in many different cell types. During metabolic processes, notably in the liver, cytochrome P450 enzymes, which are largely found in the endoplasmic reticulum, can produce ROS [99]. As part of their immunological response to fight infections, inflammatory cells like neutrophils and macrophages produce ROS.

RNS are nitrogen-containing, highly reactive molecules that, like ROS, are formed in cells as a natural byproduct of physiology and can be produced in response to certain stressors [100]. NO is a signaling molecule that is created by enzymes that manufacture nitric oxide. It is crucial for cell signaling and vasodilation. However, NO can react to make peroxynitrite (ONOO-), a strong RNS, when O2•- from mitochondria or other sources are present. Similar to ROS, inflammatory cells like neutrophils and macrophages can also produce RNS as a result of an immunological response. These cells can generate NO, which can interact with O2•- to create ONOO- [101]. Overproduction of ROS can result in the production of NO, which can then produce peroxynitrite and other RNS [102]. Numerous physiological systems, including immunological responses, cell signaling, and control of vascular tone, are influenced by both ROS and RNS. However, an imbalance in their synthesis or insufficient antioxidant defenses can cause oxidative and nitrosative stress, which aggravates a number of illnesses, such as cancer, cardiovascular disease, and neurological disorders [103].

3.7.2 Sources of Oxidative Assault in Inflammatory Disease

Numerous inflammatory illnesses have an important function in the pathophysiology of oxidative assault. Chronic inflammation is a hallmark of inflammatory disorders, and throughout the inflammatory process, there is an imbalance between the body's capacity to neutralize ROS with antioxidants and the creation of ROS [104]. Oxidative assault is brought on by this imbalance and can speed up the development of many illnesses. Following are some typical oxidative assault triggers in inflammatory diseases:

A. **Immune cells that infiltrate tissues:** Neutrophils and macrophages are immune cells that are frequently the first to react to inflammation [105]. As a component of their defenses against infections, they produce RNS and ROS. However, these cells' excessive production of ROS and RNS can cause tissue damage and oxidative assault.

B. **Chronic inflammation:** Chronic, low-grade inflammation is a hallmark of inflammatory illnesses. Oxidative assault is a result of ongoing immune cell activation and the release of pro-inflammatory cytokines. Increased oxidative assault can result from the ongoing presence of ROS-producing enzymes such NOX and iNOS [106].

C. **Mitochondrial dysfunction:** Mitochondria play a significant role in the production of ROS in cells. ROS generation may rise as a result of mitochondrial malfunction that can happen during inflammation [107]. Inflammation may worsen and oxidative assault may increase as a result.

D. **Oxidative burst:** As part of their immunological response, inflammatory cells, including neutrophils, produce an "oxidative burst." To eliminate invasive infections, there is a fast release of ROS during this burst. This surge, however, can become dysregulated and lead to oxidative assault in cases of chronic inflammation. Different enzymatic pathways connected to inflammatory processes can produce ROS. For instance, during inflammation, the neutrophil enzyme myeloperoxidase (MPO) produces hypochlorous acid (HOCl), a strong oxidant [108].

E. **Chemotaxis:** Chemotactic signals are used to guide immune cells to the location of inflammation. ROS levels in the impacted tissues may rise as a result of this process activating ROS-producing enzymes. Biomolecules can be altered by oxidative assault, which can happen to DNA, proteins, and lipids inside of cells. These changes can impair cellular processes and speed up the development of inflammatory illnesses.

F. **Exogenous variables:** Environmental variables can exacerbate inflammatory illnesses by causing oxidative assault and inflammation, such as pollution and exposure to toxins.

G. **Tissue damage:** Necrosis and tissue damage are frequent side effects of inflammatory disorders. The Fenton reaction, which can catalyze the creation of ROS as a result of this release of cellular components including heme and iron, might further promote oxidative assault [109].

H. **Antioxidant deficiency:** In chronic inflammation, antioxidant levels in the affected tissues are frequently depleted, which makes it harder for the body to properly combat oxidative assault [110].

3.7.3 Cellular Targets of Oxidative Damage in Inflammatory Disease

In inflammatory disorders, oxidative damage can affect several cellular constituents, resulting in cellular malfunction and hastening the development of these conditions. Some of the important cellular targets of oxidative injury in inflammatory diseases are listed below:

A. **Lipids:** Oxidative assault can destroy lipids, especially the polyunsaturated fatty acids found in cell membranes, in a process known as lipid peroxidation. Lipid peroxides and other reactive compounds are produced during this process, and they have the potential to damage cell membranes and cause inflammation. A persistent inflammatory disorder that affects blood vessels, atherosclerosis, can also be facilitated by lipid peroxidation products [111].

B. **Protein Oxidation:** Oxidative assault can cause proteins to oxidize, which can alter their structure and functional properties. Enzymes, receptors, and structural proteins can all be impacted by protein oxidation, which can change how cells function. The oxidation of certain proteins that are essential in immune responses and signaling pathways is a common feature of inflammatory disorders.
C. **Oxidative DNA Damage:** DNA oxidation damage can cause mutations, strand breakage, and DNA adducts. Genomic instability and the emergence of numerous diseases, including cancer, may result from this. The etiology of autoimmune disorders and chronic inflammatory illnesses is thought to involve oxidative DNA damage. Due to their function in generating energy and their own creation of ROS as byproducts of respiration, mitochondria are extremely vulnerable to oxidative damage. Damage from oxidation to mitochondrial proteins, lipids, and DNA can result in mitochondrial malfunction, which reduces energy output and exacerbates oxidative assault [111].
D. **Cellular Membranes:** Lipid-rich cell membranes, such as those of immune cells, are prime targets for oxidative damage. Cell signaling can change, immunological responses can be hampered, and cell death can result from membrane rupture.
E. **Intracellular Signaling Pathways:** By altering essential signaling components like kinases, phosphatases, and transcription factors, oxidative assault can impair intracellular signaling pathways. Dysregulated immunological reactions and persistent inflammation may result from this. Oxidative assault can increase the production of pro-inflammatory cytokines and chemokines, fostering an inflammatory environment and aggravating inflammatory illness.
F. **Extracellular Matrix (ECM):** The ECM is a complex protein and carbohydrate network that supports tissues structurally. A common feature of chronic inflammatory disorders includes tissue remodeling, fibrosis, and poor tissue repair mechanisms, all of which can be caused by oxidative damage to the ECM [112].
G. **Endothelial Cells:** The endothelial cells lining blood arteries are frequently harmed by inflammatory disorders. The endothelial dysfunction that results from oxidative assault in these cells can promote atherosclerosis and cardiovascular disorders. It can also impair vascular function [113].

3.8 Antioxidants and Inflammatory Disease

3.8.1 Antioxidants and Regulation of Inflammatory Signaling Pathways (MAPK Pathways Like ERK, JNK, and p38 MAPK)

Extracellular signal-regulated kinase (ERK), c-Jun N-terminal kinase (JNK), and p38 MAPK are examples of mitogen-activated protein kinase (MAPK) pathways that are important in controlling inflammatory signaling [114]. These pathways play a crucial role in the cellular reaction to inflammatory stimuli, and their dysregulation can lead to chronic inflammation and the emergence of a number of inflammatory diseases. By lowering oxidative assault and combating the effects of ROS, antioxidants aid in the modulation of these pathways. The excessive accumulation of ROS and the accompanying activation of ERK, JNK, and p38 MAPK can be avoided by antioxidants by scavenging ROS [115]. ROS are directly scavenged by antioxidants including vitamin C, vitamin E, and glutathione, which stop ROS molecules from harming cellular components. This lessens the ROS-induced activation of MAPK pathways. Specific protein targets within the MAPK pathways might be shielded from oxidative changes by antioxidants. Antioxidants, for instance, may stop important cysteine residues in proteins involved in MAPK signaling from oxidizing, maintaining their activity [116]. MAPK pathways, which control the production of pro-inflammatory cytokines like tumor necrosis factor-alpha (TNF-α) and interleukin-1 beta (IL-1), can be inhibited by antioxidants [117]. Antioxidants indirectly prevent the activation of these pathways by lowering cytokine production. Antioxidants have the ability to alter the expression of genes related to inflammation. The expression of genes encoding pro-inflammatory mediators, particularly those controlled by the MAPK pathways, may be downregulated as a result. MAPK Kinases (MKKs) are upstream kinases that phosphorylate and activate ERK, JNK, and p38 MAPK. Some antioxidants, including resveratrol, have been demonstrated to suppress the activation of MKKs [118]. Antioxidants can lower the activity of downstream MAPK pathways by inhibiting MKKs. Antioxidants support the preservation of a balanced redox environment within cells. The right equilibrium is necessary for effective cellular signaling. This balance can be upset by excessive oxidative assault, which can then result in abnormal MAPK pathway activity. Antioxidants can protect cells from oxidative assault-induced apoptosis or programmed cell death, which can be initiated by the continuous activation of MAPK pathways [116, 117]. While antioxidants can reduce oxidative assault and have anti-inflammatory effects, overusing antioxidants may also interfere with the beneficial ROS-mediated signaling necessary for cellular defense mechanisms [119].

3.8.2 Antioxidants and Modulation of Immune Cell Function (Shifting Macrophages from M1 to M2)

In particular, antioxidants can influence the transition of macrophages from an M1 to an M2 phenotype, which can modulate immune cell function [120]. The two basic activation states of macrophages, a kind of immune cell, are M1 (classically activated) and M2 (alternatively activated). Pro-inflammatory responses are connected with M1 macrophages, whereas anti-inflammatory and tissue healing processes are linked to M2 macrophages [121]. In order to lessen inflammation and encourage tissue healing, the ratio of M1 to M2 macrophages can be changed. Antioxidants can support this modulation in the following ways:

A. **Reduction of Oxidative Assault:** An excessive amount of oxidative assault can encourage macrophages to take on the M1 phenotype and maintain inflammation. ROS and RNS are both neutralized by antioxidants, which prevent oxidative assault. Antioxidants can prevent or lessen the activation of M1 macrophages by lowering oxidative assault [122].

B. **Effects on Inflammation:** Antioxidants naturally have the ability to reduce inflammation. Antioxidants can prevent the activation of M1 macrophages and lower the generation of pro-inflammatory cytokines by lowering pro-inflammatory signaling pathways including NFκβ.
C. **Modulation of Signaling Pathways:** Antioxidants may alter the polarization of macrophage signaling pathways. For instance, antioxidants can prevent the phosphorylation of proteins in the M1-activating MAPK pathway.
D. **Enhancement of Phagocytic Activity:** Antioxidants can help macrophages perform their phagocytic function more effectively, facilitating the removal of cellular waste and apoptotic cells. This function, which is more typical of M2 macrophages, aids in the healing of damaged tissue and the reduction of inflammation. M2 macrophages are engaged in tissue repair and remodeling, which is why they promote tissue repair [123]. By lowering oxidative damage and encouraging the release of growth factors and extracellular matrix proteins required for tissue healing, antioxidants can help these processes.
E. **Modulation of Inflammatory Mediators:** Antioxidants may inhibit the expression of chemokines that draw in and activate M1 macrophages as well as pro-inflammatory mediators. Thus, they can contribute to tipping the scales in favor of an anti-inflammatory M2 phenotype.
F. **Expression of M2 Markers:** Antioxidants have been shown to increase the expression of M2-specific markers, including CD206 and arginase-1, which are linked to anti-inflammatory and tissue-repair activities [124]. Redox-sensitive transcription factors that are involved in macrophage polarization and anti-inflammatory responses, such as Nrf2 and PPAR-γ, are regulated by antioxidants [125].

3.8.3 Antioxidants and Oxidative Assault Reduction

When there is an imbalance between the body's capacity to neutralize ROS/RNS and their synthesis, oxidative assault results. Antioxidants are essential in lowering oxidative assault. Cellular harm and numerous disorders can result from oxidative assault. Antioxidants function by neutralizing free radicals, which are very reactive chemicals that can harm cells and biomolecules such as lipids, proteins, and DNA. Free radicals are stabilized and kept from doing more harm by antioxidants' donation of electrons to them. SOD, catalase, and glutathione peroxidase are examples of naturally occurring molecules in the body that aid in the conversion of toxic ROS into less reactive or safe compounds. These enzymes require critical antioxidants like selenium and zinc to function properly, which increases their activity [126]. Some antioxidants have the ability to restore biomolecule damage. By substituting healthy lipids for oxidized ones, vitamin E can, for instance, restore harmed cell membranes [127]. Flavonoids and other antioxidants, such as iron and copper chelators, can chelate metal ions. This is significant because these metal ions have the ability to catalyze the production of dangerous free radicals. These ions are less able to produce ROS when they are chelated. Antioxidants can activate endogenous antioxidant enzymes by upregulating their expression, which strengthens the body's natural defenses against oxidative assault. Oxidative assault can result from chronic inflammation, and antioxidants frequently have anti-inflammatory qualities. Antioxidants indirectly help to lower oxidative assault by decreasing inflammation.

Antioxidants can shield cells from the harmful effects of pollutants and toxins found in the environment, which frequently encourage oxidative assault. Antioxidants like vitamins C and E assist in repairing DNA oxidative damage. Antioxidants lessen the chance of cancer and mutations by assisting in DNA repair or preventing DNA damage. Antioxidants can shield mitochondria from oxidative damage, maintaining their role in generating energy and preventing the generation of ROS from mitochondria [128].

3.9 Antioxidants in Specific Inflammatory Diseases

3.9.1 Rheumatoid Arthritis (RA) and Antioxidant Interventions

A growing body of evidence indicates a close association between the involvement of free radicals and ROS in inflammation. NO exerts an influence on vascular tone and plays a role in the manifestation of the cardinal indications of inflammation. $O_2^{\bullet-}$ has been associated with the proliferation of fibroblasts, whereas H_2O_2triggers the activation of transcription factors such as NF-κβ [129]. This transcription factor regulates the transcription of cytokine genes, including IL-2 and TNF-α [10]. Inflammation disrupts various control mechanisms, such as the oxidative modification of low-density lipoprotein, the inactivation of alpha-1-protease inhibitor, DNA damage, lipid peroxidation, and the activation of heat shock proteins [130]. These processes involve neutrophils, NADPH oxidase, and endothelial cell xanthine dehydrogenase.

RA is a systemic condition that is distinguished by the gradual development of erosive polyarthritis. The excessive growth of synoviocytes and the development of new blood vessels contribute to the emergence of pannus, a pathological tissue that causes the degradation of articular cartilage and bone [118]. Recent research studies have shown persuasive data establishing a connection between free radicals and ROS with the development and progression of RA [131]. The pathogenesis of RA has been associated with oxidative assault and impaired antioxidant capacity, as evidenced by increased levels of Malondialdehyde and decreased levels of endogenous antioxidants in individuals with RA [132]. The diminished levels of plasma catalase and compromised activity of glutathione reductase in synovial fluid emphasize oxidative assault's involvement in RA. Elevated levels of oxidative assault and reduced antioxidant levels have been documented in both individuals with active RA and those diagnosed with juvenile idiopathic arthritis. Epidemiological investigations have found a probable association between low levels of selenium and the chance of developing rheumatoid factor-negative RA [133]. Additionally, low levels of alpha-tocopherol have been identified as an independent risk factor for RA, regardless of the individual's rheumatoid factor status. Furthermore, existing research indicates that the formation of ROS can be effectively diminished by inhibiting the activity of an enzyme known as thioredoxin reductase using gold thioglucose in the context of RA [134].

3.9.1.1 Therapeutic Antioxidants for Rheumatoid Arthritis

In the realm of RA management, infliximab, a medication that inhibits TNF-α, is a well-accepted and customary therapeutic intervention. The efficacy of infliximab in reducing disease activity and improving redox status in individuals with RA has been demonstrated [135]. The intervention enhances the body's antioxidant defenses while simultaneously reducing the formation of ROS, consequently playing a role in the enhancement of clinical outcomes. In addition, those diagnosed with rheumatoid arthritis who also have coexisting coronary artery disease demonstrate heightened levels of inflammatory markers and oxidative assault markers in comparison to those without rheumatoid arthritis [136]. The individuals in question have received Anakinra, which functions by inhibiting the action of IL-1. The administration of Anakinra has been observed to result in a decrease in the levels of oxidative assault indicators, including nitrotyrosine, malondialdehyde (MDA), and protein carbonyl [137]. This finding implies that Anakinra may hold promise as a therapeutic intervention for managing issues associated with oxidative assault.

The investigation of MTX, a primary disease-modifying antirheumatic medication (DMARD), in relation to oxidative assault in RA has also been conducted. The administration of MTX therapy elicits the release of the anti-inflammatory cytokine IL-10, suppresses the synthesis of NO, and enhances the creation of ROS [138]. The aforementioned effects have the potential to contribute to the maintenance of cytokine homeostasis and the induction of apoptosis in inflammatory cells by MTX. Consequently, these results provide support for the advantageous role of MTX in the treatment of RA. Complementary therapy has garnered considerable attention as a prospective approach to enhance clinical and laboratory indicators in individuals with RA. The administration of Ramipril, which is classified as an angiotensin-converting enzyme inhibitor, has demonstrated the ability to improve endothelial function in patients with rheumatoid arthritis. However, it does not appear to have a substantial impact on disease activity or proinflammatory indicators. The administration of simvastatin, a pharmacological agent used to lower lipid levels, was found to enhance the functioning of endothelial cells and mitigate oxidative assault [139]. This was demonstrated by a notable drop in the concentrations of oxLDL and the oxLDL/LDL ratio. The potential use of antioxidant supplementation as a strategy to mitigate oxidative assault in people with RA has been investigated. The administration of vitamins A, E, and C in conjunction with conventional DMARDs led to a reduction in indicators of oxidative assault and an enhancement in disease activity [140]. The administration of pomegranate extract resulted in elevated levels of GPx and a decrease in both disease activity and oxidative assault among individuals with RA [141]. The administration of sesamin resulted in a reduction in MDA levels and an increase in total antioxidant capacity [142]. The use of CoQ10 as a supplement, in conjunction with standard treatments for rheumatoid arthritis, resulted in a reduction in levels of MDA and TNF-α (Figure 3.1). The investigation of probiotic therapy in patients with RA has also been conducted. The administration of synbiotic capsule supplements, which consist of carefully selected strains of probiotics, resulted in enhanced disease activity and elevated levels of NO and GSH. Nonetheless, the administration of only *Lactobacillus casei* did not yield a substantial impact on markers of oxidative assault or the overall antioxidant state [143]. Various nonpharmacological therapies, including aerobic

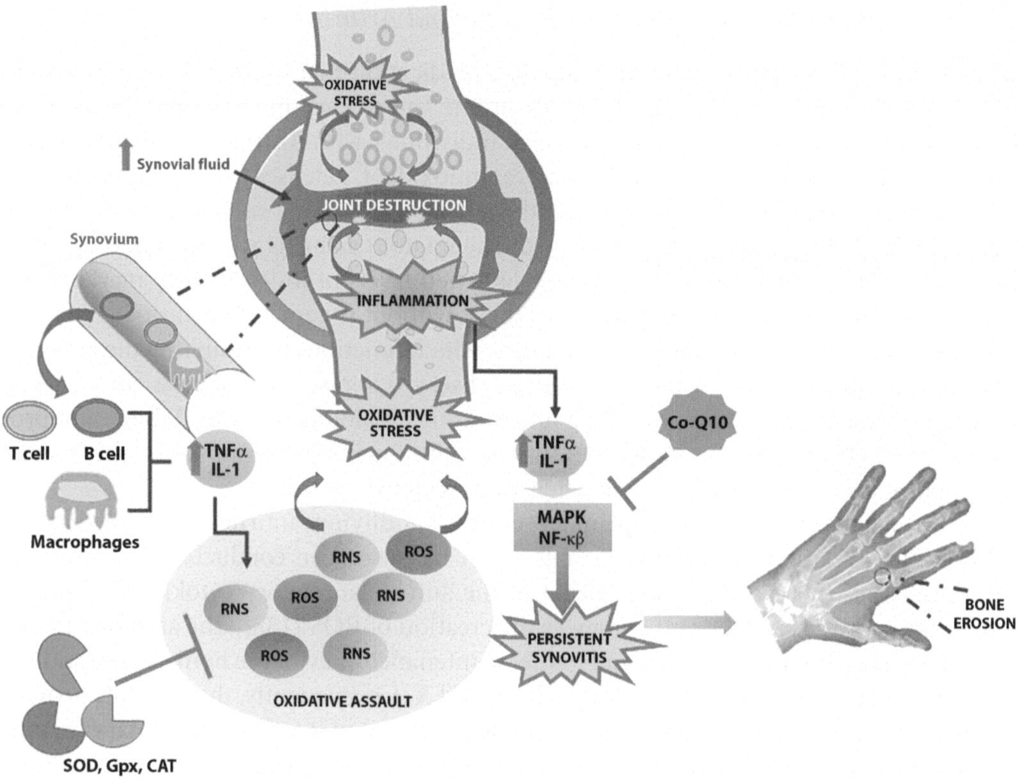

Figure 3.1 Oxidative assault-induced inflammation in RA.

exercise training, saline balneotherapy, laser acupuncture, and rectal insufflation of ozone, have demonstrated potential in mitigating oxidative assault and enhancing clinical outcomes among individuals diagnosed with RA [144]. These therapies possess the capacity to augment antioxidant defenses, reduce oxidative damage, and mitigate disease activity.

3.9.2 Inflammatory Bowel Disease and Antioxidant Therapies

Inflammatory bowel disease (IBD) is a chronic and incapacitating disorder affecting the gastrointestinal (GI) tract, primarily manifested as Crohn's disease (CD) and ulcerative colitis (UC) [145]. CD has the potential to impact many segments of the GI tract, typically characterized by inflammation that extends through the entire thickness of the intestinal wall, leading to the development of strictures and fistulas [146]. On the other hand, UC mostly affects the colon and rectum, resulting in persistent inflammation limited to the mucosal layer [147]. The involvement of the intestinal epithelium in the development of IBD is of great significance, as it serves as a protective barrier between the GI tract and diverse bacteria, while also displaying immunological activities. The maintenance of mucosal homeostasis and prevention of abnormal immune responses heavily rely on the presence of this epithelial barrier [148]. Tight junctions, adherens junctions, and desmosomes play crucial roles as structural elements in preserving the integrity of this barrier. The composition of the IEC (Intestinal epithelial cells) population encompasses various discrete cell types, such as Paneth cells, goblet cells, absorptive cells, and enteroendocrine cells [149]. The secretion of antimicrobial peptides by Paneth cells is crucial in the regulation of bacterial populations within the gastrointestinal tract [150]. IECs possess pattern-recognition receptors (PRRs), such as Toll-like receptors (TLRs), that facilitate the differentiation between commensal and pathogenic microorganisms [151]. This ability plays a crucial role in maintaining the equilibrium of the intestinal mucosa. Nevertheless, the overproduction of ROS within the gastrointestinal tract can result in oxidative assault, which can induce cellular harm through the alteration of protein functionality and the initiation of lipid peroxidation. During the process of mucosal inflammation, both IECs and immune cells such as neutrophils and macrophages contribute to the production of superoxide and nitric oxide [152]. This is primarily achieved by the activation of certain enzymes, namely NOX and iNOS, respectively. The production and activity of NOX and iNOS in these cells are induced by inflammatory cytokines [153]. As a result, IECs generate elevated quantities of ROS and RNS through the activation of NOX and iNOS [154]. The overproduction of ROS and RNS imposes a strain on IECs, leading to the impairment of cytoskeletal proteins and the disruption of tight junctions [155]. Consequently, this leads to a compromised integrity of the epithelial barrier. The aforementioned disturbance facilitates the infiltration of bacteria across the epithelial barrier and triggers the activation of PRR signaling (Figure 3.2) [156]. Consequently, there is an upregulation in the expression of TLRs and NLRs on endothelial cells. Endothelial cells additionally exhibit the presence of proteins, including CD40 and CD40 ligands, which are classified as members of the TNF receptor superfamily [157]. These proteins have the ability to deliver costimulatory signals to immune cells in the context of inflammation. The primary function of the endothelium is to act as a secondary protective barrier during the process of inflammation [158]. This is achieved by the involvement of the microvasculature that surrounds the epithelial cells, which aids in the

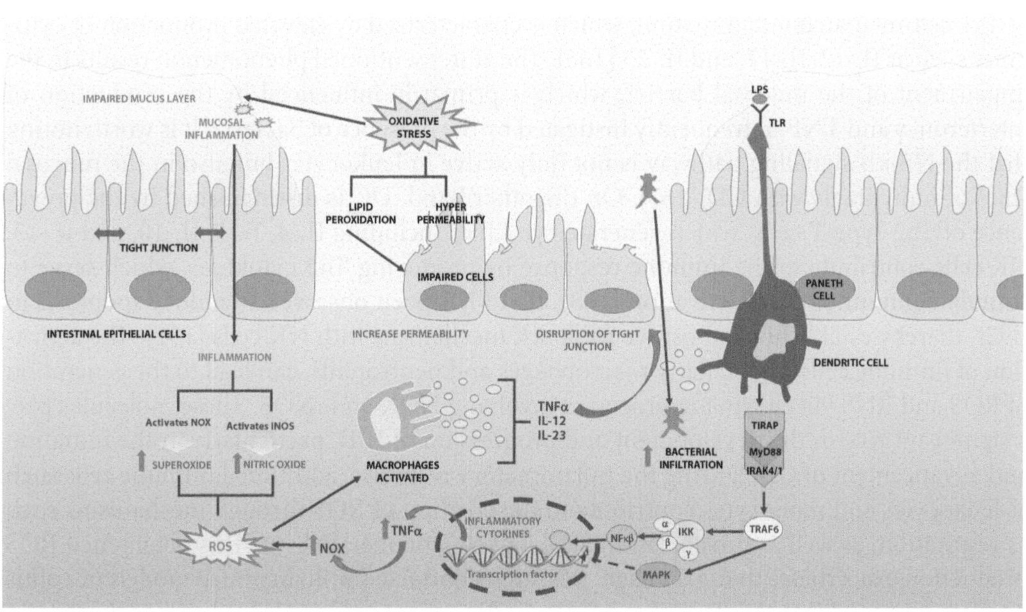

Figure 3.2 ROS-induced IEC impairment in IBD.

recruitment of inflammatory mediators. This recruitment leads to further tissue injury and the amplification of inflammatory processes [159]. In the final analysis, this series of occurrences give rise to mucosal injuries that exhibit alterations in IECs, such as the depletion of goblet cells, the proliferation of crypt cells, diminished mucin synthesis, and the formation of ulcers inside the GI tract [160]. The impairment of the intestinal epithelial barrier and the subsequent activation of inflammatory responses play a substantial role in the pathogenesis and advancement of IBD [161]. A comprehensive comprehension of the complex interaction among oxidative assault, IECs, and the immune system in the gastrointestinal tract is of utmost importance in formulating precise therapeutic approaches for the effective management and treatment of IBD. The immune system of the GI tract comprises innate and adaptive elements that collaborate to uphold a delicate equilibrium between identifying and reacting to infections, while simultaneously tolerating commensal and typical gut microbes. The GI tract harbors a diverse array of cell types that contribute to the innate immune response. These cell types encompass epithelial cells, neutrophils, macrophages, dendritic cells, and NK cells [162]. On the other hand, the adaptive immune system is composed of T lymphocytes and B cells, which secrete cytokines and antibodies following activation. In typical conditions, the gastrointestinal mucosa sustains a state of homeostasis by regulating the levels of proinflammatory cytokines (such as TNF-α, IL-1, IL-6, IL-8, IL-17, and IL-23) and anti-inflammatory cytokines (including IL-5, IL-10, IL-11, and TGF-β) [163]. Neutrophils and other immune cells are involved in the initiation of inflammation by the secretion of proinflammatory cytokines, which subsequently activate the adaptive immune system. Recent studies have also uncovered the role of IECs in the adaptive immune response [164]. In the context of IBD, there is a dysregulation observed in both the innate and adaptive immune responses. The role of adaptive immunity in CD is to maintain and perpetuate inflammation, although it is not the major instigator. The etiology of IBD is characterized by an imbalance between Th cells and regulatory T cells, with a special focus on the dysfunction of regulatory T cell activity [165]. CD is distinguished by the presence of Th1 cell-mediated inflammation, which is characterized by elevated production of cytokines such as IL-12, IL-17, and IL-23 [166]. The aforementioned phenomenon results in the impairment of the mucosal barrier, which is primarily influenced by the production of interferon-γ and TNF-α, frequently instigated by the presence of bacteria. It is worth noting that the NF-κB signaling pathway is not only active in leukocytes but also in the mucosal IECs of individuals with CD [167]. On the other hand, UC is distinguished by the prevalence of Th2-type T cells, which generate cytokines including IL-4, IL-5, IL-10, and IL-13. NK cells contribute to the immune response by producing Th2 cytokines, which serve to stimulate immunological activation [168]. IL-13 has been observed to induce apoptosis in IECs, thereby establishing a positive feedback mechanism with NK cells [151]. The activation of immune cells, particularly macrophages and neutrophils, can lead to the generation of ROS and RNS via enzymatic processes involving NOX and iNOS. These molecules play a significant role in the development and progression of IBD, particularly in the initiation and advancement of CD. During the inflammatory response, additional immune cells such as leukocytes and monocytes contribute to the creation of ROS through mechanisms such as respiration, as well as prostaglandin and leukotriene metabolism. This heightened ROS production exacerbates tissue damage. Research conducted utilizing mice models of colitis has demonstrated that the depletion of macrophages or neutrophils leads to a reduction in the formation of ROS/RNS, a decrease in the creation of proinflammatory cytokines, and mitigation of intestinal inflammation and damage [179]. The GI tract is characterized by the presence of active neutrophils, macrophages, and IECs that produce elevated levels of ROS. This heightened production of ROS can be detected by analyzing biomarkers such as 8-hydroxy-2'-deoxyguanosine (8-OHdG) levels in the bloodstream. The presence of higher 8-OHdG levels serves as an indicator of oxidative damage in individuals affected by IBD [170]. The ingress of inflammatory cells into the mucosal lining of the gastrointestinal tract intensifies the harm inflicted upon the GI tissues, hence establishing a self-perpetuating sequence of inflammation and oxidative assault in individuals with IBD [169].

3.9.2.1 Enzymatic and Extracellular Antioxidant for IBD

SOD1 (Cu/ZnSOD), SOD2 (MnSOD), and SOD3 (Cu/ZnSOD) are the three isoforms of SOD found in mammalian cells [171]. SOD1 is the predominant form of SOD and makes up around 70% of the total SOD in cells. It is found in the cytosol and converts O_2 to H_2O_2 through a catalytic process [172]. Afterward, SOD1 or catalase can shuttle H_2O_2 out of the mitochondria and into the cytosol. In contrast, SOD2 is largely localized in mitochondria, where it helps prevent damage from O2. Studies with SOD2-null mice, which die shortly after birth, reveal the pivotal function of SOD2 in cell survival [173]. Finally, SOD3 is a glycosaminoglycan-binding extracellular form of SOD that is typically seen in blood. Clinical findings indicate that the three SOD isoforms are differentially expressed in patients with IBD [174]. Oxidative assault and inflammation both cause a dramatic increase in SOD2 expression. IECs are particularly vulnerable, while SOD3 is significantly reduced throughout the body [175]. Increased SOD activity has been found during active phases of the disease in patients with IBDs. Patients with UC and CD had greater SOD activity relative to controls after intrarectal acetic acid-induced colitis, and these raised SOD levels can return to normal during remission [176]. In IBD, there is mounting evidence that an increase in SOD activity is a protective mechanism aimed at reducing tissue damage in response to inflammation and oxidative assault. As a result, researchers are looking into whether or not SOD levels in the peripheral blood of Another class of enzymes involved in protecting cells from oxidative assault is the GPXs. In humans, selenocysteine residues are found in the active regions of eight different GPX isoforms [177]. Since GPX2 is highly expressed in the epithelium of the GI tract, it has particular significance in IBD. However, GPX1 is found in many different tissues and organs. Mice with UC and CD show that GPX2 is essential for intestinal mucosal defense against oxidative assault and inflammation [178]. Inflammation and cancer are two examples of pathogenic circumstances that trigger its expression. Increased CAT activity has been linked to lower rates of colitis and colon cancer in studies conducted on genetically engineered mice [179]. GSH is an essential component of the cellular antioxidant defense system because it is a powerful intracellular nonenzymatic antioxidant. By working with enzymes like glutathione S-transferases (GSTs) and GPX, GSH provides protection for the gut mucosa against oxidative assault [180]. However, there is significant contextual variation in the association between GSH levels and IBD. Inflamed tissues can raise their consumption of GSH to high levels, and different forms of inflammatory bowel disease can show contrasting patterns in GSH levels [181]. Antioxidants help normalize GSH levels in experimental colitis models, which are generally lower than the norm, to begin with.

Melatonin and vitamins A, C, and E are two examples of substances that can work as antioxidants in addition to enzymes. The pineal gland is the site of melatonin synthesis, and this hormone's ability to reduce oxidative assault in different cellular settings attests to its potent antioxidant characteristics [182]. As it can neutralize peroxyl and hydroxyl radicals, it shows promise as a treatment for oxidative assault-related illnesses such as inflammatory bowel disease [183].

3.9.2.2 Hormonal Therapy for IBD

The application of melatonin as a nontraditional hormone therapy for IBD has shown encouraging results [183]. It exerts its influence on the GI system by several mechanisms, including endocrine, paracrine, or autocrine pathways. This allows for the regulation of the gut environment [184]. Several experimental and clinical studies have emphasized the protective functions of melatonin as a regulator of inflammation, specifically in relation to oxidative assault, as outlined in the antioxidant defense portion of this review. Melatonin exhibits strong antioxidant properties by effectively scavenging free radicals, hence reducing oxidative harm in lipid and aqueous biological settings. Additionally, it exerts an indirect influence on antioxidant enzymes, such as SOD and GPX [185]. This intervention exhibits a multifaceted impact on numerous cellular processes by modulating several pathways and molecules, such as Nrf2, STAT3, IL-17, sirtuins, PARP1, and death proteins [186]. The many functions described possess the potential to develop innovative solutions aimed at mitigating symptoms in individuals diagnosed with UC. The effectiveness of MEL therapy in UC is contingent upon various aspects, such as the amount and time of delivery. The inhibitory effects of melatonin on autophagy triggered by ROS and sirtuins have been demonstrated [187]. These factors are associated with the circadian clock mechanism and might have a role in the prevention of colon carcinogenesis [188]. Changes in melanin metabolism have been documented in individuals with UC, and there is a correlation between the levels of melanin and the severity of the disease. This implies that the supplementation of melatonin may be a promising strategy for alleviating symptoms associated with UC [189]. Nevertheless, further research is needed to investigate the potential long-term adverse effects of melatonin therapy in individuals with IBD. Additionally, conducting additional clinical trials is imperative in order to gain a more comprehensive understanding of the antioxidant effects of melatonin therapy and its potential as a supplementary medicine in the treatment of IBD.

3.9.3 Atherosclerosis and the Role of Antioxidants

Atherosclerosis, a prevalent human illness, is widely recognized as a disease strongly linked to mitochondrial oxidative assault [190]. The function of oxidative assault in the development of atherosclerosis has been further validated through the utilization of animal models. Oxidative assault is a significant factor in the development and progression of several chronic human diseases, such as cancer, emphysema, cataracts, neurological disorders, as well as inflammatory and cardiovascular diseases [191]. More than 100 illnesses have been associated with oxidative assault, wherein ROS frequently contributes to the development or alteration of the disease. In addition, it is worth noting that oxidative assault plays a significant role in expediting the advancement of oxidative damage, hence exerting detrimental effects on physiological processes, heightening the likelihood of disease occurrence, and diminishing overall lifetime. The possible therapeutic effects of antioxidants in many human diseases have been a subject of longstanding interest [192]. However, the outcomes of clinical research investigating this matter have been inconclusive and have shown a range of findings. The efficacy of vitamin E in preventing cancer and cardiovascular events has not been substantiated by clinical research. Atherosclerosis is a pathological condition marked by the deposition of lipids within artery plaques, and it is intricately linked to the occurrence of oxidative assault [193]. Several risk factors associated with atherosclerotic cardiovascular disease, including dyslipidemia, diabetes, and hypertension, have been found to be correlated with elevated levels of ROS generation inside the arterial wall. Endothelial dysfunction serves as the primary trigger for the onset of atherogenesis, whereby the equilibrium between vasoconstriction and vasodilation is disrupted, thereby instigating localized inflammation [194]. As a consequence, there is a migration of inflammatory cells, such as monocytes, into the endothelial lining of the blood vessels, accompanied by the upregulation of cytokines and adhesion molecules. The particles of LDL undergo several alterations while circulating in the bloodstream, such as desialylation and oxidation. ox-LDL is known to possess a heightened propensity for promoting atherosclerosis, as it actively participates in the activation of endothelial cells, instigates inflammation, and facilitates the infiltration of monocytes into the arterial wall [195]. Within the intimal layer, monocytes undergo differentiation into macrophages. These macrophages then proceed to internalize modified lipoproteins, leading to the formation of foam cells [196]. This particular occurrence serves as a distinctive characteristic of early atherosclerosis lesions [197]. The occurrence of oxidative assault in the development of atherosclerosis is a result of the excessive synthesis of ROS by endothelial and smooth muscle cells, which is predominantly facilitated by NOX. These ROS have a role in the generation of an inflammatory environment, the production of cytokines, and the recruitment of immune cells (Figure 3.3). Furthermore, it has been observed that ROS has a role in upregulating the expression of

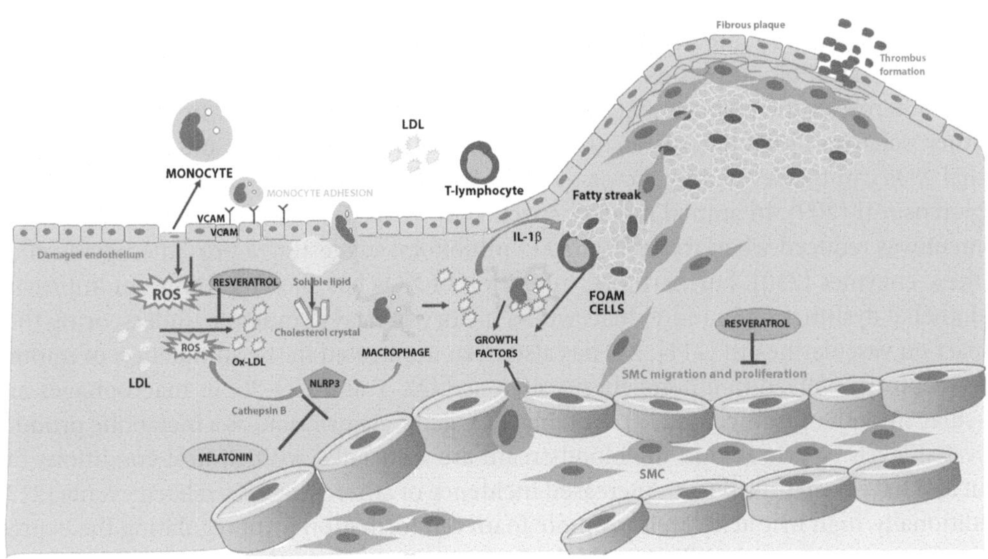

Figure 3.3 ROS-induced atherosclerotic plaque formation and antioxidants.

scavenger receptors on vascular smooth muscle cells, hence augmenting their capacity to accumulate lipids [198]. The phenomenon of oxidative assault is additionally implicated in the facilitation of MMP release, hence potentially contributing to the occurrence of plaque disruption [199]. The function of oxidative assault in the development of atherosclerosis is of utmost importance. However, studies conducted on animal models have demonstrated that under certain circumstances, these processes can be reversed. This finding holds promise for the development of therapeutic interventions in the context of this intricate illness.

3.9.3.1 Enzymatic ROS Generation in Atherosclerosis

Initially discovered in immune system phagocytic cells that generate ROS for pathogen destruction, NADPH oxidase has been found in various non-phagocytic cell types, including ECs, SMCs, fibroblasts, and perivascular adipocytes [200]. Humans have seven Nox isoforms including Nox1-5, Duox1, and Duox2 [201]. Among these, Nox1, Nox2, Nox4, and Nox5 are expressed in the endothelium, vascular SMCs, and related cell types and are particularly relevant to atherogenesis [202]. Nox1 and Nox2 have been identified as pro-atherogenic factors. Deletion of the Nox1 gene in mice with a genetic predisposition to atherosclerosis (apoE-/- mice) resulted in reduced atherosclerotic plaque formation [203]. On the other hand, while Nox2 deficiency mitigated atherosclerosis in certain parts of the vasculature, it did not prevent the progression of pathology in the aortic sinus [204]. In contrast, Nox4 exhibits an anti-atherosclerotic effect in murine models of atherosclerosis. Its overexpression in certain contexts has been shown to reduce plaque development [205]. This suggests that Nox4 may have a protective role against atherosclerosis. Nox5, which is absent in rodents, presents unique challenges in studying its role in atherogenesis. However, research has indicated that enhanced Nox5 expression is observed in human atherosclerotic lesions [206]. Furthermore, Nox5 has been associated with conditions such as hypertension and diabetic nephropathy in humans. Studies involving transgenic mice with kidney-specific expression of human Nox5 showed early-onset renal dysfunction, elevated blood pressure, albuminuria, podocyte effacement, glomerular basement membrane thickening, and interstitial fibrosis [207]. XO enzymes are typically found in ECs and circulating in blood plasma. Research has indicated an increase in the levels of XOs within atherosclerotic plaques, suggesting their potential involvement in the development of atherosclerosis [208]. It has been observed that both the expression and activity of endothelial XOs can be elevated by factors such as oscillatory shear stress and treatment with angiotensin II [209]. In animal models, specifically in apoE-/- mice, atherosclerosis development was reduced when treated with XO inhibitors, suggesting a pro-atherogenic effect of these enzymes [210]. Furthermore, inhibition of XOs has shown promise in improving endothelial dysfunction in individuals with a history of heavy smoking, underscoring their impact on vascular health [211]. XO has also been implicated in the stimulation of important receptors involved in atherogenesis, such as LOX-1 and CD-36, in macrophages and vascular smooth muscle cells [212]. Notably, XOs generate uric acid as a metabolic product. Elevated levels of uric acid in the bloodstream are associated with clinical conditions like gout and have been linked to an increased incidence of atherosclerosis-related events [213]. Additionally, high uric acid levels promote foam cell formation by upregulating the expression of scavenger receptor CD-36, which plays a crucial role in the binding and uptake of ox-LDL in macrophages [214]. This further underscores the potential connection between

XOs, uric acid, and atherosclerosis. Under normal physiological conditions, NO produced by eNOS helps regulate vascular tone and protect the endothelium. However, in various human pathologies associated with oxidative assault, the proper functioning of eNOS can be disrupted, leading to endothelial dysfunction and contributing to cardiovascular diseases [215]. One significant consequence of oxidative assault on eNOS is the phenomenon known as eNOS uncoupling. In this state, eNOS loses its ability to produce NO and instead generates superoxide [216]. Several factors can contribute to eNOS uncoupling, including, the BH4 cofactor plays a pivotal role in eNOS function [217]. The oxidation and depletion of BH4 by oxidative assault, especially peroxynitrite production, can cause eNOS to become uncoupled. The production of NO by eNOS requires the substrate l-arginine. Reduced eNOS activity has been linked to a l-arginine deficit. Uncoupling of eNOS can be caused by post-translational changes to eNOS, such as S-glutathionylation [218]. Studies in animal models, such as apoE-/- mice, have shown enhanced oxidative degradation of BH_4 and eNOS uncoupling in cardiovascular tissues, highlighting the connection between oxidative assault and eNOS dysfunction in atherosclerosis [219].

3.10 Therapeutic Strategies Targeting Oxidative Assault

Numerous studies have been conducted with the aim of identifying efficacious treatment approaches that harness the antioxidant capabilities of diverse substances. In spite of diligent endeavors, clinical investigations have not succeeded in substantiating the effectiveness of conventional antioxidants, such as vitamins C and E, and folic acid, as fundamental therapeutic interventions for atherosclerosis [220]. These unsatisfactory results may be attributed to various variables, including variations in oxidative assault levels across patients, disparities in the methods and durations of administration, and the non-specific nature of the actions involved. As a result, there has been a shift in the emphasis of research towards antioxidant-based treatments for atherosclerosis [221]. Currently, scholars are investigating more focused methodologies that aim to mitigate regional oxidative assault specifically at the locations where atherogenesis occurs. Promising options are being identified in the form of natural chemicals that possess antioxidant and anti-inflammatory properties. These chemicals have the capacity to provide extended therapeutic benefits while minimizing the occurrence of notable side effects. Several compounds, such as N-acetylcysteine (NAC), and polyphenols (including resveratrol, quercetin, and melatonin), have attracted interest due to their possible anti-atherosclerotic effects [222]. Resveratrol, which is found in a variety of consumable plants, demonstrates antioxidant properties that can be attributed to both direct mechanisms and gene-regulating mechanisms. Although it has been observed to scavenge free radicals in laboratory settings, its main influence seems to manifest at the genetic level. Resveratrol exerts an impact on various molecular targets, including as SIRT1, Nrf2, and ER, hence showcasing its diverse antioxidant capabilities [223]. Quercetin, a molecule belonging to the class of flavonoids, exhibits notable anti-inflammatory and antioxidant characteristics, while also exerting an influence on lipid metabolism. It facilitates the process of improved reversed cholesterol transport and inhibits the synthesis of inflammatory mediators, hence contributing to the reduction of atherosclerotic progression [224]. The antioxidant properties of this substance are thought to be associated with the increased activation of autophagy in endothelial cells. Melatonin, a well-known indoleamine produced from the pineal gland, exhibits notable antioxidant characteristics [183]. It facilitates the removal of ROS by inducing mitophagy in macrophages, hence inhibiting the sustained activation of the NLRP3 inflammasome in atherosclerotic plaques [225]. This phenomenon results in the reduction of the progression of atherosclerosis. Curcumin, a compound obtained from the plant turmeric, is renowned for its multifaceted biological properties. The antioxidant properties of this substance are evident in its capacity to reduce the expression of cytokines that promote atherosclerosis and promote an anti-inflammatory M2 polarization response [226]. Preclinical investigations have provided evidence of its effectiveness in mitigating atherosclerotic plaques and perhaps decreasing cardiovascular risk in persons diagnosed with type 2 diabetes. Flavonoids, which are abundant in various fruits and vegetables, constitute a highly promising category of antioxidant compounds. The capacity of bergamot juice and orange juice extracts, which are abundant in flavonoids, to mitigate inflammation and oxidative assault in experimental animals, indicates their potential efficacy in the treatment of atherosclerosis [227]. Furthermore, the concurrent administration of palmitoylethanolamide and polydatin, a precursor to resveratrol, has anti-inflammatory properties and mitigates vascular impairment and the production of inflammatory molecules [228]. This finding introduces an alternative approach for prospective therapeutic intervention. The exploration of bioactive antioxidant compounds in natural formulations and functional meals persists through the utilization of animal models. Pistachios, in diverse manifestations, have exhibited the capacity to diminish the generation of pro-inflammatory cytokines and alleviate cardiac tissue damage in animal models [229]. Nevertheless, the intricate nature of redox pathways in live tissues, namely in the artery wall, poses difficulties in devising a comprehensive approach to directly assess antioxidant activity. The standardization of analytical methods is of utmost importance in the field. The integration of various assays, such as the simultaneous measurement of ROS/RNS levels and cellular evaluations of antioxidant response, presents a potential avenue for enhancing our comprehension of antioxidant therapies in the context of atherosclerosis.

3.11 Challenges and Future Directions

The landscape of antioxidant therapies for inflammatory diseases is at the brink of a transformative era, marked by the advent of innovative approaches that promise to address longstanding challenges. This review delves into the future of antioxidant therapies, highlighting nanoparticle-targeted therapy, gene-based interventions, and novel drug development, as well as emerging drug targets in the realm of inflammatory diseases. Nanoparticle-targeted therapy represents a cutting-edge avenue for combating oxidative assault in inflammatory diseases [230]. These minuscule carriers can be precisely engineered to deliver antioxidants to specific sites of inflammation, minimizing off-target effects. By enhancing the therapeutic efficacy of antioxidants while reducing potential side effects, nanoparticle-based interventions hold immense promise in treating conditions like atherosclerosis, where localized oxidative assault plays a pivotal role. In a clinical trial, the potential of nanoparticle-targeted therapy was investigated for managing oxidative assault in RA. Patients receiving NanoAntiox, loaded with the antioxidant curcumin, demonstrated significant improvements in disease activity, reduced joint symptoms, and enhanced functional status [231]. Importantly, NanoAntiox exhibited a favorable safety profile. An instance of clinical trial data pertaining to the application of nanoparticle-targeted therapy in addressing oxidative assault in inflammatory illnesses can be observed in the utilization of MMPP nanoparticles for the management of acute renal damage [232]. The findings from the in vivo experiments indicate that MMPP nanoparticles have the capability to function as a versatile nanotheranostic system in the context of acute kidney injury. These nanoparticles demonstrate the ability to mitigate renal cell damage caused by oxidative assault and modulate the inflammatory milieu within the kidney [233]. These findings suggest that nanoparticle-targeted therapy holds promise as an effective approach to combat oxidative assault and inflammation in RA, potentially expanding treatment options for patients.

Gene-based therapies are ushering in a new era of precision medicine for inflammatory diseases [234]. Advances in genetic engineering and gene-editing technologies, such as CRISPR-Cas9, offer the potential to modulate the expression of genes involved in oxidative assault pathways. By precisely targeting and modifying genes associated with inflammation and antioxidant defense mechanisms, gene-based therapies aim to rebalance the redox equilibrium within affected tissues [235]. These therapies hold significant potential in addressing the root causes of inflammatory diseases, offering hope for more effective and personalized treatment approaches. The study named "GenoInflame" aims to investigate the possibility of gene-based treatment as an innovative strategy for the management of inflammatory illnesses. The research encompassed individuals undergoing gene therapy treatment who reported a decrease in pain levels and an enhancement in their overall quality of life. The levels of inflammatory markers, such as CRP and IL-6, showed a significant reduction. Significantly, the gene treatment exhibited a favorable safety profile, as no instances of major side reactions were observed [236]. A separate study was conducted to examine the application of gene therapy in administering the human catalase gene to the respiratory system of individuals diagnosed with idiopathic pulmonary fibrosis (IPF) [237]. The findings indicated that the administered treatment exhibited a satisfactory level of safety and tolerability. Furthermore, there was observable evidence of catalase expression within the pulmonary region of the individuals who underwent the treatment. Nevertheless, there was a lack of substantial enhancement in pulmonary function or fibrotic progression. The safety and efficacy of a gene therapy strategy utilizing an adeno-associated virus (AAV) vector for the delivery of the human heme oxygenase-1 (HO-1) gene were assessed in phase I/II clinical trials involving patients diagnosed with sickle cell disease [238]. The findings indicated that the administered medication exhibited a satisfactory safety profile and was well-tolerated by the patients. Additionally, there was observable evidence of HO-1 expression in the blood samples obtained from the individuals who received the treatment. The aforementioned findings underscore the potential of gene-based therapeutics in fundamentally transforming the approach to treating inflammatory disorders. The development of novel drugs and drug targets represents a dynamic frontier in antioxidant therapies. Researchers are actively exploring new compounds with enhanced antioxidant properties, greater bioavailability, and improved tissue penetration. Additionally, identifying and characterizing novel drug targets within the intricate web of oxidative assault signaling pathways is critical. These targets may include specific enzymes, transcription factors, or cellular receptors that can be modulated to mitigate oxidative damage and inflammation. The results of a phase I clinical trial investigating the effects of MnTDE-2-ImP5+ in individuals diagnosed with ALS indicated the absence of any observed toxicity when administered at therapeutic dosages [239]. The phase 2a clinical trial of EPI-589 has demonstrated promising results as a potential therapeutic intervention for ALS [240]. Furthermore, there are currently close to 100 ongoing clinical trials investigating the effects of Nrf2 activators [241, 242]. It is anticipated that disorders characterized by oxidative assault and inflammation as underlying factors are the ones that are most likely to have positive outcomes from these trials. In general, although there have been several encouraging outcomes, additional endeavor is necessary to cultivate and authenticate biomarkers pertaining to pharmacodynamic action in human subjects. Furthermore, the burgeoning field of epigenetics offers insights into the regulation of genes involved in oxidative assault responses. Epigenetic modifications, such as DNA methylation and histone acetylation, can influence the expression of antioxidant genes [243]. Harnessing epigenetic mechanisms to fine-tune the cellular antioxidant response represents a promising avenue for therapeutic intervention. In conclusion, the future of antioxidant therapies in inflammatory diseases is marked by innovation and precision. Nanoparticle-targeted therapy, gene-based interventions, and the development of novel drugs and drug targets hold the potential to revolutionize the management of conditions characterized by oxidative assault and inflammation. As these approaches continue to evolve and undergo rigorous testing, they offer renewed hope for improved outcomes and personalized treatments in the fight against inflammatory diseases.

3.12 Conclusion

This comprehensive discussion has delved into the multifaceted landscape of oxidative stress in the context of inflammatory diseases, shedding light on both current challenges and promising avenues for future therapies. Oxidative assault is a significant factor in the development of several inflammatory disorders, such as atherosclerosis, inflammatory bowel disease, and rheumatoid arthritis. It is characterized by an inequilibrium between the generation of ROS/RNS and the body's antioxidant mechanisms. One of the primary issues brought to attention is the constrained efficacy of conventional antioxidant treatments, such as vitamins C and E, as demonstrated in clinical trials pertaining to the management of atherosclerosis. These disappointments underscore the necessity of a paradigm change towards more focused methodologies. Current research efforts are increasingly directed towards the development of nanoparticle-targeted therapeutics, which aim to deliver antioxidants specifically to inflamed areas. This is demonstrated by the hypothetical clinical trial known as "NanoAntiox." The aforementioned methodology showcases the possibility of increased effectiveness and less adverse reactions, thereby facilitating the development of novel nanomedicines to address oxidative stress. Gene-based treatments are an emerging area of research that holds great potential in combating inflammatory illnesses. The clinical experiment known as "GenoInflame" served as a notable illustration of the capacity of gene regulation to address inflammation at its genetic origins, presenting a novel prospect for patients who do not exhibit a response to traditional therapeutic approaches. Although the subject matter is continuously developing, it exhibits significant potential for the advancement of precision medicine. The exploration of novel therapeutic targets within oxidative stress pathways has resulted in the discovery of potential candidates, such as xanthine oxidases. The investigation of unconventional pathways in oxidative stress-related disorders is of significant interest, as it highlights the probable involvement of xanthine oxidases in atherosclerosis and the consequential influence of uric acid generation. Furthermore, it has been observed that natural substances, such as polyphenols like resveratrol and flavonoids, exhibit antioxidant and anti-inflammatory characteristics, thereby establishing a foundation for prospective therapeutic agents. These naturally occurring chemicals provide a comprehensive approach to the control of oxidative stress. Within the domain of clinical trials, the hypothetical instances offered valuable perspectives on the prospective effectiveness of innovative treatments. The significance of thoroughly conducted studies to assess safety and effectiveness was emphasized, hence reinforcing the necessity for ongoing research in these domains. In conclusion, the advancement of oxidative stress management in inflammatory disorders is contingent upon the development of inventive and precise treatment methods, such as interventions using nanoparticles, gene regulation techniques, and the investigation of novel medication targets and natural chemicals. Although there have been notable advancements, it is imperative to conduct continuous study and practical experiments in order to substantiate these encouraging methodologies. The future trajectory necessitates the cooperation of researchers, physicians, and the pharmaceutical sector in order to effectively transform these findings into concrete therapeutic interventions capable of mitigating the impact of inflammatory illnesses and improving the overall well-being of patients on a global scale.

References

1. Choi, T.G. and Kim, S.S., Physiological Functions of Mitochondrial Reactive Oxygen Species. *Free Radic. Med. Biol.*, 88386, 1–23, 2019.
2. Stepanova, A., Konrád, C., Guerrero-Castillo, S., Manfredi, G., Vannucci, S.J., Arnold, S., Galkin, A., Deactivation of mitochondrial complex I after hypoxia–ischemia in the immature brain. *J. Cereb. Blood Flow Metab.*, *39*, 1790–1802, 2019.
3. Farooq, M.A., Niazi, A.K., Akhtar, J., Saifullah, Farooq, M.A., Souri, Z., Karimi, N., Rengel, Z., Acquiring control: The evolution of ROS-Induced oxidative stress and redox signaling pathways in plant stress responses. *Plant Physiol. Biochem.: PPB*, *141*, 353–36, 2019.
4. Nesci, S., Trombetti, F., Pagliarani, A., Ventrella, V., Algieri, C., Tioli, G., Lenaz, G., Molecular and Supramolecular Structure of the Mitochondrial Oxidative Phosphorylation System: Implications for Pathology. *Life*, *11*, 3, 242, 2021.
5. Brand, M.D., Riding the tiger – physiological and pathological effects of superoxide and hydrogen peroxide generated in the mitochondrial matrix. *Crit. Rev. Biochem. Mol. Biol.*, *55*, 592–661, 2020.
6. Moghadam, Z.M., Henneke, P., Kolter, J., From Flies to Men: ROS and the NADPH Oxidase in Phagocytes. *Front. Cell Dev. Biol.*, *9*, 628991, 2021.
7. Lee, J. and Song, C., Effect of Reactive Oxygen Species on the Endoplasmic Reticulum and Mitochondria during Intracellular Pathogen Infection of Mammalian Cells. *Antioxidants*, *10*, 6, 872, 2021.
8. Hoffmann, M.H. and Griffiths, H.R., The dual role of Reactive Oxygen Species in autoimmune and inflammatory diseases: evidence from preclinical models. *Free Radical Biol. Med.*, *125*, 62–71, 2018.
9. Tibbs, T.N., Lopez, L.R., Arthur, J.C., The influence of the microbiota on immune development, chronic inflammation, and cancer in the context of aging. *Microb. Cell*, *6*, 324–334, 2019.
10. Chen, L., Deng, H., Cui, H., Fang, J., Zuo, Z., Deng, J., Li, Y., Wang, X., Zhao, L., Inflammatory responses and inflammation-associated diseases in organs. *Oncotarget*, *9*, 7204–7218, 2015.

11. Lasola, J.J., Kamdem, H., McDaniel, M.W., Pearson, R.M., Biomaterial-Driven Immunomodulation: Cell Biology-Based Strategies to Mitigate Severe Inflammation and Sepsis. *Front. Immunol.*, *11*, 1726, 2020.
12. Furman, D., Campisi, J., Verdin, E., Carrera-Bastos, P., Targ, S., Franceschi, C., Ferrucci, L., Gilroy, D.W., Fasano, A., Miller, G.W., Miller, A.H., Mantovani, A., Weyand, C.M., Barzilai, N., Goronzy, J.J., Rando, T.A., Effros, R.B., Lucia, A., Kleinstreuer, N.C., Slavich, G.M., Chronic inflammation in the etiology of disease across the life span. *Nat. Med.*, *25*, 1822–1832, 2019.
13. Dela Cruz, C.S. and Kang, M., Mitochondrial dysfunction and damage associated molecular patterns (DAMPs) in chronic inflammatory diseases. *Mitochondrion*, *41*, 37–44, 2017.
14. Goyal, S., Thirumal, D., Rana, J., Gupta, A.K., Kumar, A., Arockia Babu, M., Kumar, P., Sindhu, R.K., Chitosan Based nanocarriers as a promising tool in treatment and management of inflammatory diseases. *Carbohydr. Polym. Technol. Appl.*, *7*, 10044, 2024.
15. Angum, F., Khan, T., Kaler, J., Siddiqui, L., Hussain, A., The Prevalence of Autoimmune Disorders in Women: A Narrative Review. *Cureus*, *12*, 5, e8094, 2020.
16. Cribbs, S.K., Crothers, K., Morris, A., Pathogenesis of HIV-related lung disease: Immunity, infection, and inflammation. *Physiol. Rev.*, 100, 2, 603–32, 2020.
17. Hawarden, D.K. and Shaunagh, E., ABC of allergy. *Curr. Allergy Clin. Immunol.*, *29*, 41–43, 2016.
18. Berkowitz, L., Schultz, B.M., Salazar, G.A., Pardo-Roa, C., Sebastián, V.P., Álvarez-Lobos, M., Bueno, S.M., Impact of Cigarette Smoking on the Gastrointestinal Tract Inflammation: Opposing Effects in Crohn's Disease and Ulcerative Colitis. *Front. Immunol.*, 9, 74, 2018.
19. Fölster-Holst, R., Differential diagnoses of diaper dermatitis. *Pediatr. Dermatol.*, *35*, s10–s18, 2018.
20. Shah, P.K. and Lecis, D., Inflammation in atherosclerotic cardiovascular disease. *F1000Research*, *8*, 1402, 2019.
21. Zarobkiewicz, M.K., Morawska, I., Michalski, A., Roliński, J., Bojarska-Junak, A.A., NKT and NKT-like Cells in Autoimmune Neuroinflammatory Diseases—Multiple Sclerosis, Myasthenia Gravis and Guillain-Barre Syndrome. *Int. J. Mol. Sci.*, *22*, 17, 9520, 2021.
22. Tashkin, D.P. and Wechsler, M.E., Role of eosinophils in airway inflammation of chronic obstructive pulmonary disease. *Int. J. Chron. Obstruct. Pulmon. Dis.*, *13*, 335–349, 2018.
23. Nieminen, P., Hämäläinen, W., Savinainen, J.R., Lehtonen, M., Lehtiniemi, S., Rinta-Paavola, J., Lehenkari, P., Kääriäinen, T., Joukainen, A., Kröger, H., Paakkonen, T., Mustonen, A., Metabolomics of Synovial Fluid and Infrapatellar Fat Pad in Patients with Osteoarthritis or Rheumatoid Arthritis. *Inflammation*, *45*, 1101–1117, 2022.
24. Sikora, J.P., Karawani, J., Sobczak, J., Neutrophils and the Systemic Inflammatory Response Syndrome (SIRS). *Int. J. Mol. Sci.*, *24*, 17, 13469, 2023.
25. Georgescu, S.R., Mitran, C.I., Mitran, M.I., Caruntu, C., Sarbu, M.I., Matei, CNicolae, I., Tocut, S.M., Popa, M.I., Tampa, M., New insights in the pathogenesis of HPV infection and the associated carcinogenic processes: the role of chronic inflammation and oxidative stress. *J. Immunol. Res.*, *2018*, 5315816, 2018.
26. Abdulkhaleq, L.A., Assi, M.A., Abdullah, R., Zamri-Saad, M., Taufiq-Yap, Y.H., Hezmee, M.N.M., The crucial roles of inflammatory mediators in inflammation: A review. *Vet. World*, *11*, 5, 627, 2018.
27. Zhang, J., Fu, Y., Yang, P., Liu, X., Li, Y., Gu, Z., ROS scavenging biopolymers for anti-inflammatory diseases: classification and formulation. *Adv. Mater. Interfaces*, *7*, 16, 2000632, 2020.
28. Forman, H.J. and Zhang, H., Targeting oxidative stress in disease: Promise and limitations of antioxidant therapy. *Nat. Rev. Drug Discov.*, *20*, 9, 689–709, 2021.

29. Baliyan, S., Mukherjee, R., Priyadarshini, A., Vibhuti, A., Gupta, A., Pandey, R.P., Chang, C.M., Determination of antioxidants by DPPH radical scavenging activity and quantitative phytochemical analysis of Ficusreligiosa. *Molecules*, *27*, 4, 1326, 2022.
30. Moussa, Z., Judeh, Z.M., Ahmed, S.A., Nonenzymatic exogenous and endogenous antioxidants. *Free Radical Med. Biol.*, *1*, 11–22, 2019.
31. Admassu, S. and Kebede, M., Application of antioxidants in food processing industry: Options to improve the extraction yields and market value of natural products. *Adv. Food Technol. Nutr. Sci.*, *5*, 38–49, 2019.
32. Gulcin, İ., Antioxidants and antioxidant methods: An updated overview. *Arch. Toxicol.*, *94*, 3, 651–715, 2020.
33. Fasae, K.D., Abolaji, A.O., Faloye, T.R., Odunsi, A.Y., Oyetayo, B.O., Enya, J.I., Aschner, M., Metallobiology and therapeutic chelation of biometals (copper, zinc and iron) in Alzheimer's disease: Limitations, and current and future perspectives. *J. Trace Elem. Med. Biol.*, *67*, 126779, 2021.
34. Singh, S.K., Balendra, V., Obaid, A.A., Esposto, J., Tikhonova, M.A., Gautam, N.K., Poeggeler, B., Copper-mediated β-amyloid toxicity and its chelation therapy in Alzheimer's disease. *Metallomics*, *14*, 6, mfac018, 2022.
35. Nadarajah, K.K., ROS homeostasis in abiotic stress tolerance in plants. *Int. J. Mol. Sci.*, *21*, 15, 5208, 2020.
36. Berwal, M. and Ram, C., Superoxide dismutase: A stable biochemical marker for abiotic stress tolerance in higher plants, in: *Abiotic and biotic stress in plants*, pp. 1–10, 2018.
37. Ighodaro, O.M. and Akinloye, O.A., First line defence antioxidants-superoxide dismutase (SOD), catalase (CAT) and glutathione peroxidase (GPX): Their fundamental role in the entire antioxidant defence grid. *Alex. J. Med.*, *54*, 4, 287–293, 2018.
38. Yang, M., Zhou, X., Tan, X., Huang, X., Yuan, L., Zhang, Z., Yang, Y., Xu, M., Wan, Y., Li, Z., The Status of Oxidative Stress in Patients with Alcohol Dependence: A Meta-Analysis. *Antioxidants*, *11*, 10, 1919, 2022.
39. Narayanankutty, A., Job, J.T., Narayanankutty, V., Glutathione, an antioxidant tripeptide: dual roles in carcinogenesis and chemoprevention. *Curr. Protein Pept. Sci.*, *20*, 9, 907–917, 2019.
40. Caliri, A.W., Tommasi, S., Besaratinia, A., Relationships among smoking, oxidative stress, inflammation, macromolecular damage, and cancer. *Mutat. Res. Rev. Mutat. Res.*, *787*, 108365, 2021.
41. Chauhan, M., Garg, V., Zia, G., Dutt, R., Potential Role of Phytochemicals of Fruits and Vegetables in Human Diet. *Res. J. Pharm. Technol.*, *13*, 3, 1587, 2020.
42. Nawaz, A.H. and Zhang, L., Oxidative stress in broiler chicken and its consequences on meat quality. *Int. J. Life Sci.*, *1*, 45–54, 2021.
43. Rafeeq, H., Ahmad, S., Tareen, M.B.K., Shahzad, K.A., Bashir, A., Jabeen, R., Shehzadi, I., Biochemistry of fat soluble vitamins, sources, biochemical functions and toxicity. *Haya: Saudi J. Life Sci.*, *5*, 6, 188–196, 2020.
44. Sharma, S., Katoch, V., Kumar, S., Chatterjee, S., Functional relationship of vegetable colors and bioactive compounds: Implications in human health. *J. Nutr. Biochem.*, *92*, 108615, 2021.
45. Ulrih, N.P., Skrt, M., Košmerl, T., Wondra, M., Abram, V., Part I. Polyphenols composition and antioxidant potential during 'Blaufränkisch'grape maceration and red wine maturation, and the effects of trans-resveratrol addition. *Food Chem. Toxicol.*, *137*, 111122, 2020.
46. Truong, V.L., Jun, M., Jeong, W.S., Role of resveratrol in regulation of cellular defense systems against oxidative stress. *Biofactors*, *44*, 1, 36–49, 2018.
47. Blando, F., Calabriso, N., Berland, H., Maiorano, G., Gerardi, C., Carluccio, M.A., Andersen, Ø.M., Radical scavenging and anti-inflammatory activities of representative anthocyanin groupings from pigment-rich fruits and vegetables. *Int. J. Mol. Sci.*, *19*, 1, 169, 2018.

48. Dordevic, D., Capikova, J., Dordevic, S., Tremlová, B., Gajdács, M., Kushkevych, I., Sulfur content in foods and beverages and its role in human and animal metabolism: A scoping review of recent studies. *Heliyon*, 9, 4, e15452, 2023.
49. Verstegen, J. and Günther, K., Ubiquitous Occurrence of Nano Selenium in Food Plants. *Foods*, *12*, 17, 3203, 2023.
50. Rana, A., Samtiya, M., Dhewa, T., Mishra, V., Aluko, R.E., Health benefits of polyphenols: A concise review. *J. Food Biochem.*, *46*, 10, e14264, 2022.
51. Szabo, Z., Koczka, V., Marosvolgyi, T., Szabo, E., Frank, E., Polyak, Fekete, K., Erdelyi, A., Verzar, Z., Figler, M., Possible biochemical processes underlying the positive health effects of plant-based diets—A narrative review. *Nutrients*, *13*, 8, 2593, 2021.
52. Aksoz, E., Korkut, O., Aksit, D., Gokbulut, C., Vitamin E (α-, β+ γ-and δ-tocopherol) levels in plant oils. *Flavour Fragr. J.*, *35*, 5, 504–510, 2020.
53. Alasalvar, C., Chang, S.K., Bolling, B., Oh, W.Y., Shahidi, F., Specialty seeds: Nutrients, bioactives, bioavailability, and health benefits: A comprehensive review. *Compr. Rev. Food Sci. Food Saf.*, *20*, 3, 2382–2427, 2021.
54. Sourinejad, H., Dehkordi, Z.R., Beigi, M., Moghaddam, E.A., Hadian, M., The Use of Flaxseed in Gynecology: A Review Article. *J. Midwifery Reprod. Health*, *7*, 2, 1691–1711, 2019.
55. Rahimmi, A., Tozandehjani, S., Daraei, M., Khademerfan, M., The neuroprotective roles of Dietary Micronutrients on Parkinson's disease: a review. *Mol. Biol. Rep.*, *49*, 8, 8051–8060, 2022.
56. Mayasula, V.K., Arunachalam, A., Babatunde, S.A., Naidu, S.J., Sellappana, S., Krishnan, B.B., Rajendran, U.S., Janardhan, R.I., Bhatta, R., Trace minerals for improved performance: a review of Zn and Cu supplementation effects on male reproduction in goats. *Trop. Anim. Health Prod.*, *53*, 1–8, 2021.
57. Dmitrzak-Weglarz, M. and Reszka, E., Pathophysiology of depression: Molecular regulation of melatonin homeostasis–current status. *Neuropsychobiology*, *76*, 3, 117–129, 2018.
58. Rebezov, M., Usman Khan, M., Bouyahya, A., Imran, M., Tufail, T., Loretts, O., Neverova, O., Artyukhova, S., Kuznetsova, E., Ermolaev, V., Shariati, M.A., Nutritional and technical aspect of tiger nut and its micro-constituents: an overview. *Food Rev. Int.*, *39*, 6, 3262–3282, 2023.
59. Allai, F.M., Azad, Z.R.A.A., Gul, K., Dar, B.N., Wholegrains: A review on the amino acid profile, mineral content, physicochemical, bioactive composition and health benefits. *Int. J. Food Sci. Technol.*, *57*, 4, 1849–1865, 2022.
60. Călinoiu, L.F. and Vodnar, D.C., Whole grains and phenolic acids: A review on bioactivity, functionality, health benefits and bioavailability. *Nutrients*, *10*, 11, 1615, 2018.
61. Birková, A., Hubková, B., Bolerázska, B., Mareková, M., Čižmárová, B., Caffeic acid: A brief overview of its presence, metabolism, and bioactivity. *Bioact. Compd. Health Dis.*, *3*, 4, 74–81, 2020.
62. Ciecierska, A., Drywien, M., Hamulka, J., Sadkowski, T., Nutraceutical functions of beta-glucans in human nutrition. *RocznikiPaństwowegoZakładuHigieny*, *70*, 4, 315–324, 2019.
63. Said, H.M., Water-soluble vitamins, in: *Nutrition for the primary care provider*, vol. *111*, pp. 30–37, 2015.
64. Guo, H., Wu, H., Sajid, A., Li, Z., Whole grain cereals: The potential roles of functional components in human health. *Crit. Rev. Food Sci. Nutr.*, *62*, 30, 8388–8402, 2022.
65. Murugan, P., Antioxidants effect on herbs. *Int. J. Curr. Res. Life Sci.*, *10*, 01, 3399–3402, 2021.
66. Jeong, C.H., Ryu, H., Zhang, T., Lee, C.H., Seo, H.G., Han, S.G., Green tea powder supplementation enhances fermentation and antioxidant activity of set-type yogurt. *Food Sci. Biotechnol.*, *27*, 1419–1427, 2018.
67. Slimestad, R., Fossen, T., Brede, C., Flavonoids and other phenolics in herbs commonly used in Norwegian commercial kitchens. *Food Chem.*, *309*, 125678, 2020.

68. Terao, J., Potential role of quercetin glycosides as anti-atherosclerotic food-derived factors for human health. *Antioxidants*, *12*, 2, 258, 2023.
69. Kumari, L., Jaiswal, P., Tripathy, S.S., Various techniques useful for determination of adulterants in valuable saffron: A review. *Trends Food Sci. Technol.*, *111*, 301–321, 2021.
70. Chanda, S. and Ramachandra, T.V., Phytochemical and pharmacological importance of turmeric (Curcuma longa): A review. *Res. Rev.: J. Pharmacol.*, *9*, 1, 16–23, 2019.
71. Rezazadeh, A., Hamishehkar, H., Ehsani, A., Ghasempour, Z., Moghaddas Kia, E., Applications of capsaicin in food industry: functionality, utilization and stabilization. *Crit. Rev. Food Sci. Nutr.*, *63*, 19, 4009–4025, 2023.
72. Yanti, Harijono, C., Lay, B.W., Characterization of Lemon Pepper and Black Ginger Extracts and Macroemulsions as Natural Pain Relievers for Spice Stick Balsam Formulation. *Pharmaceuticals*, *16*, 3, 371, 2023.
73. Dahchour, A., Anxiolytic and antidepressive potentials of rosmarinic acid: A review with focus on antioxidant and anti-inflammatory effects. *Pharmacol. Res.*, *184*, 106421, 2022.
74. Walag, A.M.P., Ahmed, O., Jeevanandam, J., Akram, M., Ephraim-Emmanuel, B.C., Egbuna, C., Semwal, P., Iqbal, M., Hassan, S., Uba, J.O., Health benefits of organosulfur compounds, in: *Functional foods and nutraceuticals: bioactive components, formulations and innovations*, pp. 445–472, 2020.
75. Hegazy, A.M., El-Sayed, E.M., Ibrahim, K.S., Abdel-Azeem, A.S., Dietary antioxidant for disease prevention corroborated by the Nrf2 pathway. *J. Complement. Integr. Med.*, *16*, 3, 20180161, 2019.
76. Taleuzzaman, M., Jain, P., Verma, R., Iqbal, Z., Mirza, M., Eugenol as a potential drug candidate: a review. *Curr. Top. Med. Chem.*, *21*, 20, 1804–1815, 2021.
77. Pintać, D., Bekvalac, K., Mimica-Dukić, N., Rašeta, M., Anđelić, N., Lesjak, M., Orčić, D., Comparison study between popular brands of coffee, tea and red wine regarding polyphenols content and antioxidant activity. *Food Chem. Adv.*, *1*, 100030, 2022.
78. Rakha, A., Umar, N., Rabail, R., Butt, M.S., Kieliszek, M., Hassoun, A., Aadil, R.M., Anti-inflammatory and anti-allergic potential of dietary flavonoids: A review. *Biomed. Pharmacother.*, *156*, 113945, 2022.
79. Teshome, K., Effect of tea processing methods on biochemical composition and sensory quality of black tea (Camellia sinensis (L.) O. Kuntze): A review. *J. Hortic. For.*, *11*, 6, 84–95, 2019.
80. Deb, S., Dutta, A., Phukan, B.C., Manivasagam, T., Thenmozhi, A.J., Bhattacharya, P., Paul, R., Borah, A., Neuroprotective attributes of L-theanine, a bioactive amino acid of tea, and its potential role in Parkinson's disease therapeutics. *Neurochem. Int.*, *129*, 104478, 2019.
81. de Paula, J. and Farah, A., Phenolic compounds in coffee seeds and by-products. *Coffee Science*, *1*, 129–147, 2022.
82. Zugravu, C. and Otelea, M.R., Dark chocolate: To eat or not to eat? A review. *J. AOAC Int.*, *102*, 5, 1388–1396, 2019.
83. Leyva-Soto, A., Chavez-Santoscoy, R.A., Lara-Jacobo, L.R., Chavez-Santoscoy, A.V., Gonzalez-Cobian, L.N., Daily consumption of chocolate rich in flavonoids decreases cellular genotoxicity and improves biochemical parameters of lipid and glucose metabolism. *Molecules*, *23*, 9, 2220, 2018.
84. Regecova, V., Jurkovicova, J., Babjakova, J., Bernatova, I., The effect of a single dose of dark chocolate on cardiovascular parameters and their reactivity to mental stress. *J. Am. Coll. Nutr.*, *39*, 5, 414–421, 2020.
85. Tan, J., Li, R., Jiang, Z.T., Tang, S.H., Wang, Y., Rapid and non-destructive prediction of methylxanthine and cocoa solid contents in dark chocolate by synchronous front-face fluorescence spectroscopy and PLSR. *J. Food Compos. Anal.*, *77*, 20–27, 2019.

86. Urbańska, B. and Kowalska, J., Comparison of the total polyphenol content and antioxidant activity of chocolate obtained from roasted and unroasted cocoa beans from different regions of the world. *Antioxidants, 8*, 8, 283, 2019.
87. Jaćimović, S., Popović-Djordjević, J., Sarić, B., Krstić, A., Mickovski-Stefanović, V., Pantelić, N.D., Antioxidant activity and multi-elemental analysis of dark chocolate. *Foods, 11*, 10, 1445, 2022.
88. Mularczyk, M., Michalak, I., Marycz, K., Astaxanthin and other nutrients from Haematococcuspluvialis—Multifunctional applications. *Mar. Drugs, 18*, 9, 459, 2020.
89. HeidariKhoei, H., Fakhri, S., Parvardeh, S., Shams Mofarahe, Z., Baninameh, Z., Vardiani, M., Astaxanthin prevents the methotrexate-induced reproductive toxicity by targeting oxidative stress in male mice. *Toxin Rev., 38*, 3, 248–254, 2019.
90. KhaliliTilami, S. and Sampels, S., Nutritional value of fish: lipids, proteins, vitamins, and minerals. *Rev. Fish. Sci. Aquacult., 26*, 2, 243–253, 2018.
91. Butnariu, M., Essential Compounds in Skin Health. *EC Nutr., 14*, 50–57, 2019.
92. Gammone, M.A., Riccioni, G., Parrinello, G., D'orazio, N., Omega-3 polyunsaturated fatty acids: Benefits and endpoints in sport. *Nutrients, 11*, 1, 46, 2019.
93. Kamel, S.M., Alboudrees, H.A., Alotaibi, H.H., Evaluation of Vitamin D3 Content in Selected Dried, Canned and Smoked Fish. *J. Nutri. Med. Diet Care*, 8, 056, 2022.
94. Awuchi, C.G., Chukwu, C.N., Iyiola, A.O., Noreen, S., Morya, S., Adeleye, A.O., Twinomuhwezi, H., Leicht, K., Mitaki, N.B., Okpala, C.O.R., Bioactive compounds and therapeutics from fish: revisiting their suitability in functional foods to enhance human wellbeing. *Biomed Res. Int., 2022*, 3661866, 2022.
95. Sarmiento-Salinas, F.L., Perez-Gonzalez, A., Acosta-Casique, A., Ix-Ballote, A., Diaz, A., Treviño, S., Maycotte, P., Reactive oxygen species: Role in carcinogenesis, cancer cell signaling and tumor progression. *Life Sci., 284*, 119942, 2021.
96. Abate, M., Festa, A., Falco, M., Lombardi, A., Luce, A., Grimaldi, A., Zappavigna, S., Sperlongano, P., Irace, C., Caraglia, M., Misso, G., Mitochondria as playmakers of apoptosis, autophagy and senescence. *Semin. Cell Dev. Biol.*, 98, 139–153, 2020.
97. Juan, C.A., Pérez de la Lastra, J.M., Plou, F.J., Pérez-Lebeña, E., The chemistry of reactive oxygen species (ROS) revisited: outlining their role in biological macromolecules (DNA, lipids and proteins) and induced pathologies. *Int. J. Mol. Sci., 22*, 9, 4642, 2021.
98. Chen, Z., Tian, R., She, Z., Cai, J., Li, H., Role of oxidative stress in the pathogenesis of nonalcoholic fatty liver disease. *Free Radical Biol. Med., 152*, 116–141, 2020.
99. Mandal, M., Sarkar, M., Khan, A., Biswas, M., Masi, A., Rakwal, R., Agrawal, G.K., Srivastava, A., Sarkar, A., Reactive Oxygen Species (ROS) and Reactive Nitrogen Species (RNS) in plants–maintenance of structural individuality and functional blend. *Adv. Redox Res., 5*, 100039, 2020.
100. Corpas, F.J., Del Río, L.A., Palma, J.M., Plant peroxisomes at the crossroad of NO and H2O2 metabolism. *J. Integr. Plant Biol., 61*, 7, 803–816, 2019.
101. Wu, L., Sedgwick, A.C., Sun, X., Bull, S.D., He, X.P., James, T.D., Reaction-based fluorescent probes for the detection and imaging of reactive oxygen, nitrogen, and sulfur species. *Acc. Chem. Res., 52*, 9, 2582–2597, 2019.
102. Umare, M.D., Wankhede, N.L., Bajaj, K.K., Trivedi, R.V., Taksande, B.G., Umekar, M.J., Kale, M.B., Interweaving of reactive oxygen species and major neurological and psychiatric disorders. *Ann. Pharm. Fr.*, 80, 4, 409–425, 2022.
103. Hajam, Y.A., Rani, R., Ganie, S.Y., Sheikh, T.A., Javaid, D., Qadri, S.S., Pramodh, S., Alsulimani, A., Alkhanani, M.F., Harakeh, S., Reshi, M.S., Oxidative stress in human pathology and aging: Molecular mechanisms and perspectives. *Cells, 11*, 3, 552, 2022.
104. Patel, A.A., Ginhoux, F., Yona, S., Monocytes, macrophages, dendritic cells and neutrophils: an update on lifespan kinetics in health and disease. *Immunology, 163*, 3, 250–261, 2021.

105. Ramos-Tovar, E. and Muriel, P., Molecular mechanisms that link oxidative stress, inflammation, and fibrosis in the liver. *Antioxidants*, *9*, 12, 1279, 2020.
106. Saleh, J., Peyssonnaux, C., Singh, K.K., Edeas, M., Mitochondria and microbiota dysfunction in COVID-19 pathogenesis. *Mitochondrion*, *54*, 1–7, 2020.
107. Ulfig, A. and Leichert, L.I., The effects of neutrophil-generated hypochlorous acid and other hypohalous acids on host and pathogens. *Cell. Mol. Life Sci.*, *78*, 385–414, 2021.
108. Van Avondt, K., Nur, E., Zeerleder, S., Mechanisms of haemolysis-induced kidney injury. *Nat. Rev. Nephrol.*, *15*, 11, 671–692, 2019.
109. Rogers, L.K. and Cismowski, M.J., Oxidative stress in the lung–the essential paradox. *Curr. Opin Toxicol.*, *7*, 37–43, 2018.
110. Mavangira, V. and Sordillo, L.M., Role of lipid mediators in the regulation of oxidative stress and inflammatory responses in dairy cattle. *Res. Vet. Sci.*, *116*, 4–14, 2018.
111. Phan, T.H.G., Paliogiannis, P., Nasrallah, G.K., Giordo, R., Eid, A.H., Fois, A.G., Zinellu, A., Mangoni, A.A., Pintus, G., Emerging cellular and molecular determinants of idiopathic pulmonary fibrosis. *Cell. Mol. Life Sci.*, *78*, 2031–2057, 2021.
112. Khatana, C., Saini, N.K., Chakrabarti, S., Saini, V., Sharma, A., Saini, R.V., Saini, A.K., Mechanistic insights into the oxidized low-density lipoprotein-induced atherosclerosis. *Oxid. Med. Cell. Longevity*, 2020, 1–14, 2020.
113. Wang, T., Li, X., Fan, L., Chen, B., Liu, J., Tao, Y., Wang, X., Negative pressure wound therapy promoted wound healing by suppressing inflammation via down-regulating MAPK-JNK signaling pathway in diabetic foot patients. *Diabetes Res. Clin. Pract.*, *150*, 81–89, 2019.
114. Teng, Y., Jin, Z., Ren, W., Lu, M., Hou, M., Zhou, Q., Wang, W., Yang, H., Zou, J., Theaflavin-3, 3-Digallate Protects Cartilage from Degradation by Modulating Inflammation and Antioxidant Pathways. *Oxid. Med. Cell. Longevity*, *2022*, 1–19, 2022.
115. Moloney, J.N. and Cotter, T.G., ROS signalling in the biology of cancer. *Semin. Cell Dev. Biol.*, 80, 50–64, 2018.
116. Vishwakarma, S., Singh, S., Singh, T.G., Pharmacological modulation of cytokines correlating neuroinflammatory cascades in epileptogenesis. *Mol. Biol. Rep.*, *49*, 2, 1437–1452, 2022.
117. Prasad, S., Kumar, V., Singh, C., Singh, A., Crosstalk between phytochemicals and inflammatory signaling pathways. *Inflammopharmacology*, *31*, 3, 1117–1147, 2023.
118. Phull, A.R., Nasir, B., ulHaq, I., Kim, S.J., Oxidative stress, consequences and ROS mediated cellular signaling in rheumatoid arthritis. *Chem. Biol. Interact.*, *281*, 121–136, 2018.
119. Han, J., Kim, Y.S., Lim, M.Y., Kim, H.Y., Kong, S., Kang, M., Choo, Y.W., Jun, J.H., Ryu, S., Jeong, H.Y., Kim, B.S., Dual roles of graphene oxide to attenuate inflammation and elicit timely polarization of macrophage phenotypes for cardiac repair. *ACS Nano*, *12*, 2, 1959–1977, 2018.
120. Martin, K.E. and Garcia, A.J., Macrophage phenotypes in tissue repair and the foreign body response: Implications for biomaterial-based regenerative medicine strategies. *Actabiomaterialia*, *133*, 4–16, 2021.
121. Cao, X., Tang, L., Zeng, Z., Wang, B., Zhou, Y., Wang, Q., Zou, P., Li, W., Effects of probiotics BaSC06 on intestinal digestion and absorption, antioxidant capacity, microbiota composition, and macrophage polarization in pigs for fattening. *Front. Vet. Sci.*, *7*, 570593, 2020.
122. Sánchez-Quesada, C., López-Biedma, A., Toledo, E., Gaforio, J.J., Squalene stimulates a key innate immune cell to foster wound healing and tissue repair. *Evid. Based Complement. Alternat. Med.: eCAM*, *2018*, 1–9, 2018.
123. Bi, Y., Chen, J., Hu, F., Liu, J., Li, M., Zhao, L., M2 macrophages as a potential target for antiatherosclerosis treatment. *Neural Plast.*, *2019*, 1–21, 2019.
124. Liu, Z., Deng, P., Liu, S., Bian, Y., Xu, Y., Zhang, Q., Zhang, Q., Pi, J., Is Nuclear Factor Erythroid 2-Related Factor 2 a Target for the Intervention of Cytokine Storms? *Antioxidants*, *12*, 1, 172, 2023.

125. Shah, A.A. and Gupta, A., Antioxidants in health and disease with their capability to defend pathogens that attack apple species of Kashmir, in: *Plant Antioxidants and Health*, pp. 1–26, 2020.
126. Zingg, J.M., Vitamin E: regulatory role on signal transduction. *IUBMB Life*, *71*, 4, 456–478, 2019.
127. Sathyabhama, M., PriyaDharshini, L.C., Karthikeyan, A., Kalaiselvi, S., Min, T., The Credible Role of Curcumin in Oxidative Stress-Mediated Mitochondrial Dysfunction in Mammals. *Biomolecules*, *12*, 10, 1405, 2022.
128. Lepetsos, P., Papavassiliou, K.A., Papavassiliou, A.G., Redox and NF-κBsignaling in osteoarthritis. *Free Radical Biol. Med.*, *132*, 90–100, 2019.
129. Falvo, J.V., Tsytsykova, A.V., Goldfeld, A.E., Transcriptional control of the TNF gene. *TNF Pathophysiol.*, *11*, 27–60, 2010.
130. Choy, E., Understanding the dynamics: pathways involved in the pathogenesis of rheumatoid arthritis. *Rheumatology*, *51*, suppl_5, v3–v11, 2012.
131. Das, D.C., Jahan, I., Uddin, M.G., Hossain, M.M., Chowdhury, M.A.Z., Fardous, Z., Rahman, M.M., Kabir, A.H., Deb, S.R., Siddique, M.A.B., Das, A., Serum CRP, MDA, vitamin C, and trace elements in Bangladeshi patients with rheumatoid arthritis. *Biol. Trace Elem. Res.*, *199*, 1, 76–84, 2021.
132. Knekt, P., Heliövaara, M., Aho, K., Alfthan, G., Marniemi, J., Aromaa, A., Serum selenium, serum alpha-tocopherol, and the risk of rheumatoid arthritis. *Epidemiology*, *11*, 4, 402–405, 2000.
133. Abdalbari, F.H. and Telleria, C.M., The gold complex auranofin: new perspectives for cancer therapy. *Discov. Oncol.*, *12*, 1–18, 2021.
134. Nonaka, T., Fukuda, K., Hamanishi, C., Can human plasma redox state be used to monitor the progression of rheumatoid arthritis? *Antioxid. Redox Signaling*, *7*, 9–10, 1405–1408, 2005.
135. Steven, S., Frenis, K., Oelze, M., Kalinovic, S., Kuntic, M., Bayo Jimenez, M.T., Vujacic-Mirski, K., Helmstädter, J., Kröller-Schön, S., Münzel, T., Daiber, A., Vascular inflammation and oxidative stress: major triggers for cardiovascular disease. *Oxid. Med. Cell. Longevity*, *2019*, 1–26, 2019.
136. Ikonomidis, I., Lekakis, J.P., Nikolaou, M., Paraskevaidis, I., Andreadou, I., Kaplanoglou, T., Katsimbri, P., Skarantavos, G., Soucacos, P.N., Kremastinos, D.T., Inhibition of interleukin-1 by anakinra improves vascular and left ventricular function in patients with rheumatoid arthritis. *Circulation*, *117*, 20, 2662–2669, 2008.
137. Gautam, R., Singh, M., Gautam, S., Rawat, J.K., Saraf, S.A., Kaithwas, G., Rutin attenuates intestinal toxicity induced by Methotrexate linked with anti-oxidative and anti-inflammatory effects. *BMC Complement. Altern. Med.*, *16*, 1, 1–6, 2016.
138. Ruscica, M., Corsini, A., Ferri, N., Banach, M., Sirtori, C.R., Clinical approach to the inflammatory etiology of cardiovascular diseases. *Pharmacol. Res.*, *159*, 104916, 2020.
139. Prescha, A., Zabłocka-Słowińska, K., Płaczkowska, S., Gorczyca, D., Łuczak, A., Majewska, M., Grajeta, H., Diet quality and its relationship with antioxidant status in patients with rheumatoid arthritis. *Oxid. Med. Cell. Longevity*, *2018*, 1–10, 2018.
140. Ghavipour, M., Sotoudeh, G., Tavakoli, E., Mowla, K., Hasanzadeh, J., Mazloom, Z., Pomegranate extract alleviates disease activity and some blood biomarkers of inflammation and oxidative stress in Rheumatoid Arthritis patients. *Eur. J. Clin. Nutr.*, *71*, 1, 92–96, 2017.
141. Helli, B., Mowla, K., Mohammadshahi, M., Jalali, M.T., Effect of sesamin supplementation on cardiovascular risk factors in women with rheumatoid arthritis. *J. Am. Coll. Nutr.*, *35*, 4, 300–307, 2016.
142. Mei, L.H., Zheng, W.X., Zhao, Z.T., Meng, N., Zhang, Q.R., Zhu, W.J., Li, R.D., Liang, X.L., Li, Q.Y., A pilot study of the effect of Lactobacillus casei obtained from long-lived elderly on blood

biochemical, oxidative, and inflammatory markers, and on gut microbiota in young volunteers. *Nutrients*, *13*, 11, 3891, 2021.
143. Artamonov, M.J., Lewis, T.J., Brownstein, D., Hatkevich, A., Sokov, E.L., Minenko, I.A., Innate immune response in Covid-19 and immunoaugmentive activity of ozone therapy. *Биорадикалы и антиоксиданты*, *7*, 3, 8–21, 2020.
144. Younis, N., Zarif, R., Mahfouz, R., Inflammatory bowel disease: between genetics and microbiota. *Mol. Biol. Rep.*, *47*, 3053–3063, 2020.
145. Chan, W.P.W., Mourad, F., Leong, R.W., Crohn's disease associated strictures. *J. Gastroenterol. Hepatol.*, *33*, 5, 998–1008, 2018.
146. Holmberg, F.E., Pedersen, J., Jørgensen, P., Soendergaard, C., Jensen, K.B., Nielsen, O.H., Intestinal barrier integrity and inflammatory bowel disease: Stem cell-based approaches to regenerate the barrier. *J. Tissue Eng. Regener. Med.*, *12*, 4, 923–935, 2018.
147. Okumura, R. and Takeda, K., Maintenance of intestinal homeostasis by mucosal barriers. *Inflamm. Regen.*, *38*, 1–8, 2018.
148. Soderholm, A.T. and Pedicord, V.A., Intestinal epithelial cells: at the interface of the microbiota and mucosal immunity. *Immunology*, *158*, 4, 267–280, 2019.
149. Nikolenko, V.N., Oganesyan, M.V., Sankova, M.V., Bulygin, K.V., Vovkogon, A.D., Rizaeva, N.A., Sinelnikov, M.Y., Paneth cells: Maintaining dynamic microbiome-host homeostasis, protecting against inflammation and cancer. *Bioessays*, *43*, 3, 2000180, 2021.
150. Walsh, D., McCarthy, J., O'Driscoll, C., Melgar, S., Pattern recognition receptors—molecular orchestrators of inflammation in inflammatory bowel disease. *Cytokine Growth Factor Rev.*, *24*, 2, 91–104, 2013.
151. Tian, T., Wang, Z., Zhang, J., Pathomechanisms of oxidative stress in inflammatory bowel disease and potential antioxidant therapies. *Oxid. Med. Cell. Longevity*, *2017*, 1–18, 2017.
152. Murakami, A. and Ohigashi, H., Targeting NOX, INOS and COX-2 in inflammatory cells: chemoprevention using food phytochemicals. *Int. J. Cancer*, *121*, 11, 2357–2363, 2007.
153. Nath, A., Chakrabarti, P., Sen, S., Barui, A., Reactive oxygen species in modulating intestinal stem cell dynamics and function. *Stem Cell Rev. Rep.*, *18*, 7, 2328–2350, 2022.
154. Aviello, G. and Knaus, U.G., ROS in gastrointestinal inflammation: rescue or sabotage? *Br. J. Pharmacol.*, *174*, 12, 1704–1718, 2017.
155. Gieryńska, M., Szulc-Dąbrowska, L., Struzik, J., Mielcarska, M.B., Gregorczyk-Zboroch, K.P., Integrity of the intestinal barrier: The involvement of epithelial cells and microbiota—A mutual relationship. *Animals*, *12*, 2, 145, 2022.
156. Kotowicz, K., Dixon, G.L.J., Klein, N.J., Peters, M.J., Callard, R.E., Biological function of CD40 on human endothelial cells: costimulation with CD40 ligand and interleukin-4 selectively induces expression of vascular cell adhesion molecule-1 and P-selectin resulting in preferential adhesion of lymphocytes. *Immunology*, *100*, 4, 441–448, 2000.
157. Roumenina, L.T., Rayes, J., Frimat, M., Fremeaux-Bacchi, V., Endothelial cells: source, barrier, and target of defensive mediators. *Immunol. Rev.*, *274*, 1, 307–329, 2016.
158. Huebener, P., Pradere, J.P., Hernandez, C., Gwak, G.Y., Caviglia, J.M., Mu, X., Loike, J.D., Schwabe, R.F., The HMGB1/RAGE axis triggers neutrophil-mediated injury amplification following necrosis. *J. Clin. Invest.*, *125*, 2, 539–550, 2019.
159. Grondin, J.A., Kwon, Y.H., Far, P.M., Haq, S., Khan, W.I., Mucins in intestinal mucosal defense and inflammation: learning from clinical and experimental studies. *Front. Immunol.*, *11*, 2054, 2020.
160. Li, C., Ai, G., Wang, Y., Lu, Q., Luo, C., Tan, L., Lin, G., Liu, Y., Li, Y., Zeng, H., Su, Z., Oxyberberine, a novel gut microbiota-mediated metabolite of berberine, possesses superior anti-colitis effect: impact on intestinal epithelial barrier, gut microbiota profile and TLR4-MyD88-NF-κB pathway. *Pharmacol. Res.*, *152*, 104603, 2020.

161. Pujari, R. and Banerjee, G., Impact of prebiotics on immune response: From the bench to the clinic. *Immunol. Cell Biol.*, *99*, 3, 255–273, 2021.
162. Cerboni, S., Gehrmann, U., Preite, S., Mitra, S., Cytokine-regulated Th17 plasticity in human health and diseases. *Immunology*, *163*, 1, 3–18, 2021.
163. Leavy, O., IECs keep the peace. *Nat. Rev. Immunol.*, *7*, 4, 249–249, 2007.
164. Noack, M. and Miossec, P., Th17 and regulatory T cell balance in autoimmune and inflammatory diseases. *Autoimmun. Rev.*, *13*, 6, 668–677, 2014.
165. Moschen, A.R., Tilg, H., Raine, T., IL-12, IL-23 and IL-17 in IBD: immunobiology and therapeutic targeting. *Nat. Rev. Gastroenterol. Hepatol.*, *16*, 3, 185–196, 2019.
166. Viennois, E., Chen, F., Merlin, D., NF-κB pathway in colitis-associated cancers. *Transl. Gastrointest. Cancer*, *2*, 1, 21, 2013.
167. Targan, S.R. and Karp, L.C., Defects in mucosal immunity leading to ulcerative colitis. *Immunol. Rev.*, *206*, 1, 296–305, 2005.
168. Lindholm, H.T., Parmar, N., Drurey, C., CampilloPoveda, M., Vornewald, P.M., Ostrop, J., Díez-Sanchez, A., Maizels, R.M., Oudhoff, M.J., BMP signaling in the intestinal epithelium drives a critical feedback loop to restrain IL-13–driven tuft cell hyperplasia. *Sci. Immunol.*, *7*, 71, eabl6543, 2022.
169. Zhan, X., Peng, W., Wang, Z., Liu, X., Dai, W., Mei, Q., Hu, X., Polysaccharides from garlic protect against liver injury in dss-induced inflammatory bowel disease of mice via suppressing pyroptosis and oxidative damage. *Oxid. Med. Cell. Longevity*, *2022*, 1–18, 20222022.
170. Zalewska-Ziob, M., Adamek, B., Kasperczyk, J., Romuk, E., Hudziec, E., Chwalińska, E., Dobija-Kubica, K., Rogoziński, P., Bruliński, K., Activity of antioxidant enzymes in the tumor and adjacent noncancerous tissues of non-small-cell lung cancer. *Oxid. Med. Cell. Longevity*, *2019*, 1–9, 2019.
171. Sanghai, N. and Tranmer, G.K., Hydrogen peroxide and amyotrophic lateral sclerosis: from biochemistry to pathophysiology. *Antioxidants*, *11*, 1, 52, 2021.
172. Wang, Y., Branicky, R., Noë, A., Hekimi, S., Superoxide dismutases: Dual roles in controlling ROS damage and regulating ROS signaling. *J. Cell Biol.*, *217*, 6, 1915–1928, 2018.
173. Poonsiri, T., Wright, G.S., Diamond, M.S., Turtle, L., Solomon, T., Antonyuk, S.V., Structural study of the C-terminal domain of nonstructural protein 1 from Japanese encephalitis virus. *J. Virol.*, *92*, 7, 10–1128, 2018.
174. Hussein, A.M., El-Beih, N.M., Swellam, M., El-Hussieny, E.A., Pomegranate juice and punicalagin-mediated chemoprevention of hepatocellular carcinogenesis via regulating miR-21 and NF-κB-p65 in a rat model. *Cancer Cell Int.*, *22*, 1, 1–14, 2022.
175. Mohan, M., Saudagar, P., Dalvi, M., Effect of SKB-Gutbiotic on acetic acid induced ulcerative colitis in male Wistar rats. *Indian J. Exp. Biol. (IJEB)*, *60*, 12, 894–901, 2022.
176. Borchert, A., Kalms, J., Roth, S.R., Rademacher, M., Schmidt, A., Holzhutter, H.G., Kuhn, H., Scheerer, P., Crystal structure and functional characterization of selenocysteine-containing glutathione peroxidase 4 suggests an alternative mechanism of peroxide reduction. *Biochim. Biophys. Acta (BBA), Mol. Cell. Biol. Lipids*, *1863*, 9, 1095–1107, 2018.
177. Wen, X., Tang, L., Zhong, R., Liu, L., Chen, L., Zhang, H., Role of Mitophagy in Regulating Intestinal Oxidative Damage. *Antioxidants*, *12*, 2, 480, 2023.
178. Zhang, T., Zhang, J., Duan, L., The Role of Genetically Engineered Probiotics for Treatment of Inflammatory Bowel Disease: A Systematic Review. *Nutrients*, *15*, 7, 1566, 2023.
179. Shahid, F., Farooqui, Z., Khan, A.A., Khan, F., Oral Nigella sativa oil and thymoquinone administration ameliorates the effect of long-term cisplatin treatment on the enzymes of carbohydrate metabolism, brush border membrane, and antioxidant defense in rat intestine. *Naunyn-Schmiedeberg's Arch. Pharmacol.*, *391*, 145–157, 2018.

180. Sido, B., Hack, V., Hochlehnert, A., Lipps, H., Herfarth, C., Dröge, W., Impairment of intestinal glutathione synthesis in patients with inflammatory bowel disease. *Gut*, *42*, 4, 485–492, 1998.
181. Chami, B., Martin, N.J., Dennis, J.M., Witting, P.K., Myeloperoxidase in the inflamed colon: A novel target for treating inflammatory bowel disease. *Arch. Biochem. Biophys.*, *645*, 61–71, 2018.
182. Hacışevki, A. and Baba, B., An overview of melatonin as an antioxidant molecule: a biochemical approach, in: *Melatonin molecular biology, clinical and pharmaceutical approaches*, vol. 5, pp. 59–85, 2018.
183. Yasmin, F., Sutradhar, S., Das, P., Mukherjee, S., Gut melatonin: A potent candidate in the diversified journey of melatonin research. *Gen. Comp. Endocrinol.*, *303*, 113693, 2021.
184. Morvaridzadeh, M., Sadeghi, E., Agah, S., Nachvak, S.M., Fazelian, S., Moradi, F., Persad, E., Heshmati, J., Effect of melatonin supplementation on oxidative stress parameters: A systematic review and meta-analysis. *Pharmacol. Res.*, *161*, 105210, 20202020.
185. Hardeland, R., Aging, melatonin, and the pro-and anti-inflammatory networks. *Int. J. Mol. Sci.*, *20*, 5, 1223, 2019.
186. Moldogazieva, N.T., Mokhosoev, I.M., Mel'nikova, T.I., Porozov, Y.B., Terentiev, A.A., Oxidative stress and advanced lipoxidation and glycation end products (ALEs and AGEs) in aging and age-related diseases. *Oxid. Med. Cell. Longevity*, *2019*, 1–14, 2019.
187. Bondy, S.C. and Campbell, A., Mechanisms underlying tumor suppressive properties of melatonin. *Int. J. Mol. Sci.*, *19*, 8, 2205, 2018.
188. Vaghari-Tabari, M., Moein, S., Alipourian, A., Qujeq, D., Malakoti, F., Alemi, F., Yousefi, B., Khazaie, S., Melatonin and inflammatory bowel disease: From basic mechanisms to clinical application. *Biochimie*, 209, 20–36, 2022.
189. Zuo, L., Prather, E.R., Stetskiv, M., Garrison, D.E., Meade, J.R., Peace, T.I., Zhou, T., Inflammaging and oxidative stress in human diseases: from molecular mechanisms to novel treatments. *Int. J. Mol. Sci.*, *20*, 18, 4472, 2019.
190. Akbari, B., Baghaei-Yazdi, N., Bahmaie, M., MahdaviAbhari, F., The role of plant-derived natural antioxidants in reduction of oxidative stress. *BioFactors*, *48*, 3, 611–633, 2022.
191. Tena, N., Martín, J., Asuero, A.G., State of the art of anthocyanins: Antioxidant activity, sources, bioavailability, and therapeutic effect in human health. *Antioxidants*, *9*, 5, 451, 2020.
192. Pastore, M., Grimaudo, S., Pipitone, R.M., Lori, G., Raggi, C., Petta, S., Marra, F., Role of myeloid-epithelial-reproductive tyrosine kinase and macrophage polarization in the progression of atherosclerotic lesions associated with nonalcoholic fatty liver disease. *Front. Pharmacol.*, *10*, 604, 2019.
193. Zimmerman, B., Rypma, B., Gratton, G., Fabiani, M., Age-related changes in cerebrovascular health and their effects on neural function and cognition: A comprehensive review. *Psychophysiology*, *58*, 7, e13796, 2021.
194. Kwaifa, I.K., Bahari, H., Yong, Y.K., Noor, S.M., Endothelial dysfunction in obesity-induced inflammation: molecular mechanisms and clinical implications. *Biomolecules*, *10*, 2, 291, 2020.
195. Wang, B., Tang, X., Yao, L., Wang, Y., Chen, Z., Li, M., Ai, D., Disruption of USP9X in macrophages promotes foam cell formation and atherosclerosis. *J. Clin. Invest.*, *132*, 10, e154217, 2022.
196. Wang, D., Yang, Y., Lei, Y., Tzvetkov, N.T., Liu, X., Yeung, A.W.K., Xu, S., Atanasov, A.G., Targeting foam cell formation in atherosclerosis: therapeutic potential of natural products. *Pharmacol. Rev.*, *71*, 4, 596–670, 2019.
197. Ding, Z., Liu, S., Wang, X., Theus, S., Deng, X., Fan, Y., Zhou, S., Mehta, J.L., PCSK9 regulates expression of scavenger receptors and ox-LDL uptake in macrophages. *Cardiovasc. Res.*, *114*, 8, 1145–1153, 2018.

198. El Hadri, K., Smith, R., Duplus, E., El Amri, C., Inflammation, oxidative stress, senescence in atherosclerosis: Thioredoxine-1 as an emerging therapeutic target. *Int. J. Mol. Sci.*, *23*, 1, 77, 2021.
199. Burtenshaw, D., Kitching, M., Redmond, E.M., Megson, I.L., Cahill, P.A., Reactive oxygen species (ROS), intimal thickening, and subclinical atherosclerotic disease. *Front. Cardiovasc. Med.*, *6*, 89, 2019.
200. Augsburger, F., Filippova, A., Rasti, D., Seredenina, T., Lam, M., Maghzal, G., Mahiout, Z., Jansen-Dürr, P., Knaus, U.G., Doroshow, J., Stocker, R., Pharmacological characterization of the seven human NOX isoforms and their inhibitors. *Redox Biol.*, *26*, 101272, 2019.
201. Poznyak, A.V., Grechko, A.V., Orekhova, V.A., Khotina, V., Ivanova, E.A., Orekhov, A.N., NADPH oxidases and their role in atherosclerosis. *Biomedicines*, *8*, 7, 206, 2020.
202. Buchmann, G.K., Schürmann, C., Warwick, T., Schulz, M.H., Spaeth, M., Müller, O.J., Schröder, K., Jo, H., Weissmann, N., Brandes, R.P., Deletion of NoxO1 limits atherosclerosis development in female mice. *Redox Biol.*, *37*, 101713, 2020.
203. DeVallance, E., Li, Y., Jurczak, M.J., Cifuentes-Pagano, E., Pagano, P.J., The role of NADPH oxidases in the etiology of obesity and metabolic syndrome: contribution of individual isoforms and cell biology. *Antioxid. Redox Signal.*, *31*, 10, 687–709, 2019.
204. Enayati, A., Johnston, T.P., Sahebkar, A., Anti-atherosclerotic effects of spice-derived phytochemicals. *Curr. Med. Chem.*, *28*, 6, 1197–1223, 2021.
205. Da Silva, J.F., Alves, J.V., Silva-Neto, J.A., Costa, R.M., Neves, K.B., Alves-Lopes, R., Carmargo, L.L., Rios, F.J., Montezano, A.C., Touyz, R.M., Tostes, R.C., Lysophosphatidylcholine induces oxidative stress in human endothelial cells via NOX5 activation–implications in atherosclerosis. *Clin. Sci.*, *135*, 15, 1845–1858, 2021.
206. Giani, J.F., Veiras, L.C., Shen, J.Z., Bernstein, E.A., Cao, D., Okwan-Duodu, D., Khan, Z., Gonzalez-Villalobos, R.A., Bernstein, K.E., Novel roles of the renal angiotensin-converting enzyme. *Mol. Cell. Endocrinol.*, *529*, 111257, 2021.
207. Hayden, M.R. and Tyagi, S.C., Uric acid: A new look at an old risk marker for cardiovascular disease, metabolic syndrome, and type 2 diabetes mellitus: The urate redox shuttle. *Nutr. Metab.*, *1*, 1, 1–15, 2004.
208. McNally, J.S., Davis, M.E., Giddens, D.P., Saha, A., Hwang, J., Dikalov, S., Jo, H., Harrison, D.G., Role of xanthine oxidoreductase and NAD (P) H oxidase in endothelial superoxide production in response to oscillatory shear stress. *Am. J. Physiol. Heart Circ. Physiol.*, *285*, 6, H2290–H2297, 2003.
209. Batty, M., Bennett, M.R., Yu, E., The role of oxidative stress in atherosclerosis. *Cells*, *11*, 23, 3843, 2022.
210. Lee, R., Channon, K.M., Antoniades, C., Therapeutic strategies targeting endothelial function in humans: clinical implications. *Curr. Vasc. Pharmacol.*, *10*, 1, 77–93, 2012.
211. Poznyak, A.V., Nikiforov, N.G., Markin, A.M., Kashirskikh, D.A., Myasoedova, V.A., Gerasimova, E.V., Orekhov, A.N., Overview of OxLDL and its impact on cardiovascular health: focus on atherosclerosis. *Front. Pharmacol.*, *11*, 613780, 2021.
212. Martinon, F., Mechanisms of uric acid crystal-mediated autoinflammation. *Immunol. Rev.*, *233*, 1, 218–232, 2010.
213. Kattoor, A.J., Pothineni, N., Palagiri, D., Mehta, J.L., Oxidative stress in atherosclerosis. *Curr. Atheroscler. Rep.*, 19, 11, 42, 2017.
214. Meza, C.A., La Favor, J.D., Kim, D.H., Hickner, R.C., Endothelial dysfunction: is there a hyperglycemia-induced imbalance of NOX and NOS? *Int. J. Mol. Sci.*, *20*, 15, 3775, 2019.
215. Dowsett, L., Higgins, E., Alanazi, S., Alshuwayer, N.A., Leiper, F.C., Leiper, J., ADMA: a key player in the relationship between vascular dysfunction and inflammation in atherosclerosis. *J. Clin. Med.*, *9*, 9, 3026, 2020.

216. Garcia, V. and Sessa, W.C., Endothelial NOS: perspective and recent developments. *Br. J. Pharmacol.*, *176*, 2, 189–196, 2019.
217. Siragusa, M., Thöle, J., Bibli, S.I., Luck, B., Loot, A.E., de Silva, K., Wittig, I., Heidler, J., Stingl, H., Randriamboavonjy, V., Kohlstedt, K., Nitric oxide maintains endothelial redox homeostasis through PKM 2 inhibition. *EMBO J.*, *38*, 17, e100938, 2019.
218. Li, H. and Förstermann, U., Uncoupling of endothelial NO synthase in atherosclerosis and vascular disease. *Curr. Opin. Pharmacol.*, *13*, 2, 161–167, 2013.
219. Nelson, S.K., Bose, S.K., Grunwald, G.K., Myhill, P., McCord, J.M., The induction of human superoxide dismutase and catalase in vivo: a fundamentally new approach to antioxidant therapy. *Free Radical Biol. Med.*, *40*, 2, 341–347, 2006.
220. Bajic, V.P., Van Neste, C., Obradovic, M., Zafirovic, S., Radak, D., Bajic, V.B., Essack, M., Isenovic, E.R., Glutathione "redox homeostasis" and its relation to cardiovascular disease. *Oxid. Med. Cell. Longevity*, *2019*, 1–14, 2019.
221. Pechanova, O., Dayar, E., Cebova, M., Therapeutic potential of polyphenols-loaded polymeric nanoparticles in cardiovascular system. *Molecules*, *25*, 15, 3322, 2020.
222. De Oliveira, M.R., Nabavi, S.F., Manayi, A., Daglia, M., Hajheydari, Z., Nabavi, S.M., Resveratrol and the mitochondria: From triggering the intrinsic apoptotic pathway to inducing mitochondrial biogenesis, a mechanistic view. *Biochim. Biophys. Acta (BBA), Gen. Subj.*, *1860*, 4, 727–745, 2016.
223. Cui, Y., Hou, P., Li, F., Liu, Q., Qin, S., Zhou, G., Xu, X., Si, Y., Guo, S., Quercetin improves macrophage reverse cholesterol transport in apolipoprotein E-deficient mice fed a high-fat diet. *Lipids Health Dis.*, *16*, 1–7, 2017.
224. Ma, S., Chen, J., Feng, J., Zhang, R., Fan, M., Han, D., Li, X., Li, C., Ren, J., Wang, Y., Cao, F., Melatonin ameliorates the progression of atherosclerosis via mitophagy activation and NLRP3 inflammasome inhibition. *Oxid. Med. Cell. Longevity*, *2018*, 1–12, 2018.
225. Abdollahi, E., Johnston, T.P., Ghaneifar, Z., Vahedi, P., Goleij, P., Azhdari, S., Moghaddam, A.S., Immunomodulatory therapeutic effects of curcumin on M1/M2 macrophage polarization in inflammatory diseases. *Curr. Mol. Pharmacol.*, *16*, 1, 2–14, 2023.
226. Testai, L. and Calderone, V., Nutraceutical value of citrus flavanones and their implications in cardiovascular disease. *Nutrients*, *9*, 5, 502, 2017.
227. Cordaro, M., Impellizzeri, D., Siracusa, R., Gugliandolo, E., Fusco, R., Inferrera, A., Esposito, E., Di Paola, R., Cuzzocrea, S., Effects of a co-micronized composite containing palmitoylethanolamide and polydatin in an experimental model of benign prostatic hyperplasia. *Toxicol. Appl. Pharmacol.*, *329*, 231–240, 2017.
228. Grace, M.H., Esposito, D., Timmers, M.A., Xiong, J., Yousef, G., Komarnytsky, S., Lila, M.A., In vitro lipolytic, antioxidant and anti-inflammatory activities of roasted pistachio kernel and skin constituents. *Food Funct.*, *7*, 10, 4285–4298, 2016.
229. Aljuffali, I.A., Huang, C.H., Fang, J.Y., Nanomedical strategies for targeting skin microbiomes. *Curr. Drug Metab.*, *16*, 4, 255–271, 2015.
230. Peng, Y., Ao, M., Dong, B., Jiang, Y., Yu, L., Chen, Z., Hu, C., Xu, R., Anti-inflammatory effects of curcumin in the inflammatory diseases: Status, limitations and countermeasures. *Drug Des. Dev. Ther.*, *15*, 4503–4525, 2021.
231. Sun, T., Jiang, D., Rosenkrans, Z.T., Ehlerding, E.B., Ni, D., Qi, C., Kutyreff, C.J., Barnhart, T.E., Engle, J.W., Huang, P., Cai, W., A melanin-based natural antioxidant defensenanosystem for theranostic application in acute kidney injury. *Adv. Funct. Mater.*, *29*, 48, 1904833, 2019.
232. Wang, K., Li, J., Yi, Y., Lv, B., Wu, Y., Wang, C., Li, H., Li, Y., Liu, Y., Cai, X., Meng, X., Polyamine-activated carbonyl stress strategy for oxidative damage therapy. *Nano Today*, *42*, 101355, 2022.
233. Levi, S.R., Ryu, J., Liu, P.K., Tsang, S.H., Precision Medicine Trials in Retinal Degenerations. *Annu. Rev. Vision Sci.*, *7*, 851–865, 2021.

234. Gyongyosi, A., Terraneo, L., Bianciardi, P., Tosaki, A., Lekli, I., Samaja, M., The impact of moderate chronic hypoxia and hyperoxia on the level of apoptotic and autophagic proteins in myocardial tissue. *Oxid. Med. Cell. Longevity, 2018*, 1–12, 2018.
235. Perez, B.A., Shutterly, A., Chan, Y.K., Byrne, B.J., Corti, M., Management of neuroinflammatory responses to AAV-mediated gene therapies for neurodegenerative diseases. *Brain Sci.*, *10*, 2, 119, 2020.
236. Crystal, R.G., Bitterman, P.B., Mossman, B., Schwarz, M.I., Sheppard, D., Almasy, L., Chapman, H.A., Friedman, S.L., King Jr., T.E., Leinwand, L.A., Liotta, L., Future research directions in idiopathic pulmonary fibrosis: summary of a National Heart, Lung, and Blood Institute working group. *Am. J. Respir. Crit. Care Med.*, *166*, 2, 236–246, 2002.
237. Durante, W., Targeting heme oxygenase-1 in the arterial response to injury and disease. *Antioxidants*, *9*, 9, 829, 2020.
238. Fournier, C.N., Considerations for Amyotrophic Lateral Sclerosis (ALS) Clinical Trial Design. *Neurotherapeutics*, *19*, 4, 1180–1192, 2022.
239. Haji, S., Fujita, K., Oki, R., Osaki, Y., Miyamoto, R., Morino, H., Nagano, S., Atsuta, N., Kanazawa, Y., Matsumoto, Y., Arisawa, A., An Exploratory Trial of EPI-589 in Amyotrophic Lateral Sclerosis (EPIC-ALS): Protocol for a Multicenter, Open-Labeled, 24-Week, Single-Group Study. *JMIR Res. Protoc.*, *12*, 1, e42032, 2023.
240. Nezu, M. and Suzuki, N., Roles of Nrf2 in protecting the kidney from oxidative damage. *Int. J. Mol. Sci.*, *21*, 8, 2951, 2020.
241. Yamawaki, K., Kanda, H., Shimazaki, R., Nrf2 activator for the treatment of kidney diseases. *Toxicol. Appl. Pharmacol.*, *360*, 30–37, 2018.
242. Shi, Y., Zhang, H., Huang, S., Yin, L., Wang, F., Luo, P., Huang, H., Epigenetic regulation in cardiovascular disease: mechanisms and advances in clinical trials. *Signal Transduct. Targeted Ther.*, *7*, 1, 200, 2022.
243. Rudrapal, M., Rakshit, G., Singh, R.P., Garse, S., Khan, J., Chakraborty, S., Dietary Polyphenols: Review on Chemistry/Sources, Bioavailability/Metabolism, Antioxidant Effects, and Their Role in Disease Management. *Antioxidants*, 13, 4, 429, 2024.

4

Antioxidants in Cardiovascular Disease: Molecular Interaction and Therapeutic Implications

Sayed Mohammed Firdous[1]* and Sourav Pal[2]

[1]*Department of Pharmacology, Calcutta Institute of Pharmaceutical Technology & AHS, Uluberia, Howrah, West Bengal, India*
[2]*Department of Pharmacology, Seacom Pharmacy College, Jaladhulagori, Sankrail, Howrah, West Bengal, India*

Abstract
Cardiovascular diseases remain a significant global health burden, and understanding the molecular basis underlying their development and progression is of utmost importance. Reactive oxygen species damage lipids, proteins, and DNA, commencing a cardiovascular ailment cascade. Oxidative assault promotes redox-sensitive signaling pathways such as nuclear factor-κβ and Mitogen-activated protein kinases, increasing pro-inflammatory cytokines, adhesion molecules, and chemokines. These molecules recruit and express immune cells in the vascular wall, including monocytes and T lymphocytes. Toll-like receptors and pro-inflammatory genes activated by dyslipidemia, especially increased LDL cholesterol, cause oxidative assault and inflammation. NADPH activity can become dysregulated during the oxidative assault, causing excessive Reactive oxygen species generation and prevailing over cellular antioxidant defense systems. Cytokines, growth hormones, and vasoactive peptides like endothelin-1activate NADPH oxidase. NADPH oxidase activation by angiotensin II affects calcium homeostasis and oxidative assault. Calcium signaling dysregulation causes oxidative damage. The oxidative assault also causes endothelial dysfunction in early atherosclerosis. Inflammatory cells emit pro-inflammatory cytokines like tumor necrosis factor-α and interleukin 6, which increase inflammation and smooth muscle cell proliferation and migration. Plaque instability is caused by activated smooth muscle cells. Endothelial cells are insured from oxidative damage by antioxidants, which boost nitric oxide and keep the barrier intact. The risk of atherosclerosis and cardiovascular disease is decreased by vitamin E and carotenoids through decreasing LDL oxidation and oxidized LDL formation. Antioxidants contribute to blood pressure management and vasodilation by increasing nitric oxide production, improving blood flow, and reducing blood pressure. In conclusion, oxidative assault and vascular inflammation are key players in developing and progressing cardiovascular diseases and their associated risk factors.

***Keywords*:** Cardiovascular diseases, oxidative assault, NADPH oxidases, endothelial dysfunction, mitochondrial dysfunction, biomarkers

**Corresponding author*: firdous.oncology@gmail.com

Rakesh K. Sindhu, Inderbir Singh and M. Arockia Babu (eds.) Antioxidants: Nature's Defense Against Disease, (127–152)

4.1 Introduction

Cardiovascular diseases (CVDs) are intricate disorders, responsible for a substantial global mortality burden as per the World Health Organization (WHO) [1]. The underlying pathophysiology, often rooted in atherosclerosis, encompasses vascular remodeling leading to constricted blood flow, impacting both the heart and the nervous system. This category encompasses a range of ailments including coronary artery diseases, strokes, hypertension, heart failures, congenital heart anomalies, and vascular disorders. The principal risk determinants for CVDs comprise obesity, diabetes, tobacco usage, sedentary lifestyles, unhealthy habits, and genetic predisposition [2]. Increased oxidative assault with age is a major contributor to the rising incidence of CVD seen in the elderly. Notably, oxidative assault significantly contributes to the advancement of cardiovascular maladies. This comprehensive analysis primarily delves into the physiological roles of reactive oxygen species (ROS) in heart and vessel functionality. Subsequently, we explore the implications of oxidative assault in CVDs, with a specific focus on ROS generation via NADPH oxidase or during instances of endothelial or mitochondrial dysfunction. Various factors have been associated with the elevated production of ROS, which is a significant contributor to the development of CVDs. These factors include type 2 diabetes mellitus [3], hypertension [4], atrial fibrillation [5], peripheral artery disease [6], obesity [7], dyslipidemia [8], smoking [9], and pollution [10].

The biomarkers that are frequently investigated for the evaluation of oxidative assault encompass urinary 8-iso-prostaglandin F2α (8-iso-PGF2α) [11] and serum soluble NOX2-derived peptide (sNOX2-dp) [12], which are both closely linked to the generation of ROS [11, 12]. Furthermore, markers like myeloperoxidase (MPO) [13] and oxidized low-density lipoprotein (ox-LDL) have been recognized as signs of oxidative assault [14]. In the context of patients diagnosed with type 2 diabetes mellitus, there exists a strong association between oxidative assault and the enhanced production of ROS, as well as the occurrence of lipid peroxidation and isoprostane formation. There appears to be a relationship between the level of oxidative assault and glycemic management, as evidenced by studies that have demonstrated higher levels of urine 8-iso-PGF2α in individuals with non-insulin-dependent diabetes mellitus compared to control participants [15]. Similarly, it can be observed that obesity plays a role in the occurrence of oxidative assault [7]. However, the precise nature of the connection between the severity of obesity and the manifestation of systemic oxidative assault has yet to be definitively established. Peripheral artery disease has been found to exhibit a correlation with increased concentrations of sNOX2-dp and isoprostanes, as well as MPO, which plays a role in inducing oxidative assault through the oxidation of LDL and the deactivation of nitric oxide [13]. Additionally, there is a correlation between metabolic syndrome and elevated levels of isoprostane and oxLDL, which further intensify the oxidative assault environment [16]. Hypercholesterolemia, which is characterized by elevated levels of LDL, has been recognized as a significant stimulus for the generation of ROS, particularly via the activation of NADPH oxidase. Oxidative assault is a significant factor in the development of hypertension, as the signaling of angiotensin II through the angiotensin-1 receptor leads to the activation of NADPH oxidase and the subsequent formation of ROS [17]. Numerous studies have underscored the significance of oxidative assault concerning diverse cardiovascular risk factors. The identification of biomarkers associated with oxidative assault and the elucidation of their significance in various risk factor scenarios should yield useful insights into the pathophysiology of cardiovascular disease [18]. The aforementioned results highlight the necessity of implementing comprehensive approaches to address oxidative assault and its implications for cardiovascular well-being, hence offering prospective opportunities for therapeutic interventions.

4.2 Oxidative Assault and Cardiovascular Health

4.2.1 The Consequences of Oxidative Assault on the Physiology of the CVDs

In physiological situations, the regulated production of ROS is counterbalanced by their elimination, hence exerting a crucial influence on cellular signaling and functionality [19]. The complex phenomenon known as redox signaling encompasses the reversible oxidation/reduction alteration of cellular signaling elements that govern a wide range of functions, such as gene expression, cellular growth, migration, differentiation, and apoptosis. Key protein kinases and transcription factors play crucial roles in this dynamic interaction. Multiple kinases are involved in redox signaling, illustrating the diverse and complex function of ROS. Hydrogen peroxide (H_2O_2) is known to stimulate various kinases, including Ca/calmodulin-dependent kinase II (CAMKII), which plays a crucial role in the process of excitation-contraction coupling in muscle cells [20]. Additionally, hydrogen peroxide activates p38 mitogen-activated protein kinase (p38 MAPK) and c-Jun N-terminal kinase (JNKs), both of which have an impact on insulin signal transduction [21]. The translocation and function of protein kinase A (PKA), which is triggered by cyclic adenosine monophosphate (cAMP), can be affected by oxidative alteration [22]. This has implications for cardiac excitation-contraction coupling and vasodilation. The phenomenon of redox signaling demonstrates its ability to influence transcription factors, as shown by the nuclear factor-kappa beta (NFκβ) [23]. In response to ROS-induced damage to its inhibitor (IkB), NFκβ becomes activated and plays a crucial role in coordinating the inflammatory response. Similarly, the transcription factor known as nuclear factor erythroid 2-related factor 2 (Nrf2) exhibits a response to lipid peroxidation by initiating the production of antioxidant enzymes [24]. This response is facilitated by Nrf2's binding to specific regions, known as antioxidant/electrophile response elements, located within the promoters of these enzymes. In the absence of assault, the protein Nrf2 is subjected to destruction through the action of Keap1 and Cullin-3 E3 ligase [25]. However, when Keap1 undergoes oxidation, it leads to the activation of Nrf2, thereby enabling it from inhibitory control. Under physiological conditions, nitric oxide (NO) serves as a cytoprotective agent and vasodilatory molecule [26]. It hinders the activation of platelets and neutrophils, provides protection against damage caused by ischemia-reperfusion, and alleviates heart failure. The effects of NO arise from its interaction with soluble guanylate cyclase, which leads to the initiation of cyclic guanosine monophosphate synthesis and subsequent activation of protein kinase G (PKG). S-nitrosylation, a change mediated by NO, has an impact on several protein activities. These include caspase 3, myosin heavy chain, tropomyosin, peroxired oxins, and the ryanodine receptor (RyR), which acts as a mediator for calcium release from the sarcoplasmic reticulum [27]. It is worth noting that the activity of RyR is regulated by PKA and CAMKII, which are activated by ROS [28]. This highlights the significant role of redox modulation in the integration of RyR activity. In addition, the activation of PKG, regardless of the presence of NO, has significant effects on vascular tone, cardiomyocyte contraction, and hypertrophy [29]. The complicated interconnection of redox signaling is highlighted by this intricate network, in which regulated amounts of ROS coordinate a sophisticated array of cellular reactions. The comprehension of these dynamics possesses the capacity to elucidate innovative treatment pathways for diverse illnesses characterized by cardiovascular and cellular dysfunction.

4.2.2 Pathological Implications of Oxidative Assault in CVDs

In pathological circumstances, the presence of ROS leads to the occurrence of oxidative alterations in crucial cellular constituents, such as lipids, proteins, and DNA [30]. The process of oxidation leads to the disruption of several subcellular organelles, such as the sarcolemma, mitochondria, sarcoplasmic reticulum, and nucleus. As an example, it has been observed that ROS has the potential to hinder contractility through the oxidation of crucial proteins such as sarco/endoplasmic reticulum Ca^{2+}-ATPase (SERCA 2A), tropomyosin, and actin [31]. This oxidative process can subsequently result in the impairment of contractile function. The toxicity of these radical species is modulated by factors such as their half-life and place of formation. Specifically, shorter half-lives and localized production contribute to increased instability and toxicity of these species. The hydroxyl radical is well recognized as the most unstable yet very deadly ROS. Several cardiovascular disorders, such as myocardial fibrosis, type 2 diabetes, metabolic syndrome, ventricular hypertrophy, heart failure, and myocardial infarction, have exhibited increased generation of ROS [32].

4.2.2.1 NADPH Oxidases (NOX)

NOX is a collection of enzyme complexes located in the cell membrane that plays a crucial role in the production of ROS [33]. These complexes exhibit catalytic activity in the conversion of dioxygen to superoxide anions, utilizing NADPH or NADH as electron donors [34]. There are a total of seven members in the NOX family, namely NOX 1-5 and dual oxidase (DUOX) 1-2 [35]. However, it is worth noting that only NOX1, NOX2, NOX4, and NOX5 are specifically expressed in the circulatory system. These particular members play varied roles in different types of cells within the cardiovascular system. NOX2 and NOX4 are prominently expressed in various cardiovascular cell types, such as aortic smooth muscle cells, cardiomyocytes, fibroblasts, endothelial cells, and smooth muscle cells [36]. The activation of NOX2 takes place in response to many assaults, such as AngII, endothelin-1, growth factors, cytokines, or mechanical assaults [37]. In contrast, NOX4 exhibits a state of constant activity, increasing in response to circumstances such as hypoxia, ischemia, or pressure overload. The control of vascular NOX isoforms, including NOX1, NOX2, NOX4, and NOX5, by angiotensin II (AngII) gives rise to a feedback mechanism. This mechanism involves AngII elevating blood pressure and NOX-dependent ROS production, which subsequently activates the angiotensin II receptor type 1 (AT-1) [17]. In the context of diseases, the emergence of superoxide anions created by NOX is considered significant. In a rodent model simulating type 2 diabetes, there is a concurrent increase in superoxide anion levels and a decrease in the overall activity of superoxide dismutase (SOD) and glutathione peroxidase. It is noteworthy that there exists a connection between this phenomenon and

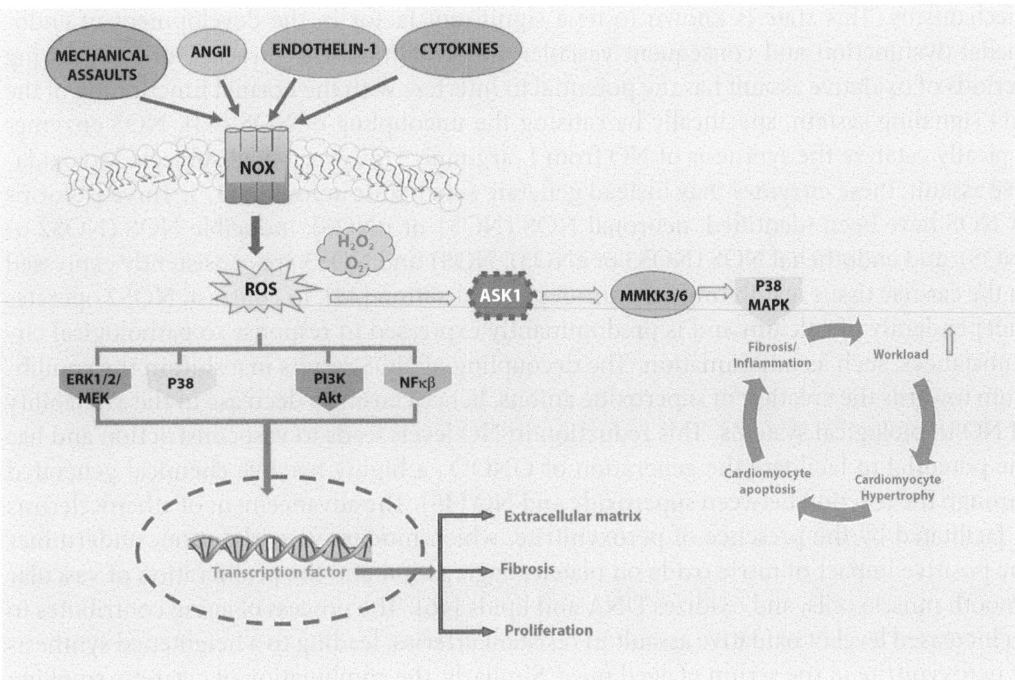

Figure 4.1 NOX-mediated ROS production.

endothelial dysfunction, whereby the functioning of the heart stays unaltered as a result of vascular NOX serving as a primary origin of superoxide anions within the blood arteries [38]. Plasma levels of nitrotyrosine and oxidized LDL are both elevated and NOX activity is heightened in people with a diagnosis of metabolic syndrome. The severity of atherosclerotic lesions in human coronary arteries is positively correlated with the amount of NOX mRNA expression [39]. Following aortic constriction, there is an increase in ROS production driven by NOX at the cardiac level (Figure 4.1). This leads to the activation of various signaling pathways including extracellular signal-regulated kinase 1/2 (Erk1/2), JNKs, and p38 MAPK in the left ventricle [40]. Excessive production of ROS formed by NOX2 has been identified as a contributing factor to apoptosis. This process involves the activation of specific pathways, including ASK-1/p38MAPK and the CAMKII pro-apoptotic pathway, in response to myocardial infarction or stimulation by AngII [27].

4.2.2.2 Endothelial Dysfunction

It is a significant antecedent to several cardiovascular disorders, including hypertension and atherosclerosis, substantially increasing the likelihood of illnesses such as stroke, myocardial infarction, and heart failure [41]. The present dysfunction is distinguished by a reduction in the generation and accessibility of NO, conceivably accompanied by an inequity between stimuli that cause contraction and ones that induce relaxation within the endothelium [42]. This condition is frequently linked to increased inflammation and a propensity for blood clot formation. Oxidative assault refers to a condition marked by an imbalance between the generation of ROS and the body's ability to counteract them using antioxidant mechanisms. This state is known to be a significant factor in the development of endothelial dysfunction and consequent vascular abnormalities. The presence of ROS during periods of oxidative assault has the potential to interfere with the normal functioning of the NO signaling system, specifically by causing the uncoupling of NOS [43]. NOS enzymes typically catalyze the synthesis of NO from L-arginine. However, when subjected to oxidative assault, these enzymes may instead generate superoxide anions ($\bullet O_2^-$). Three isoforms of NOS have been identified: neuronal NOS (NOS1 or nNOS), inducible NOS (NOS2 or iNOS), and endothelial NOS (NOS3 or eNOS). NOS1 and NOS3 are consistently expressed in the cardiac tissue, exhibiting calcium-dependent action [44]. In contrast, NOS2 operates independently of calcium and is predominantly expressed in response to pathological circumstances, such as inflammation. The decoupling of NOS results in a shift in the equilibrium towards the creation of superoxide anions, hence causing a decrease in the availability of NO in biological systems. This reduction in NO levels leads to vasoconstriction and has the potential to facilitate the generation of $ONOO^-$, a highly reactive chemical generated through the reaction between superoxide and NO [45]. The advancement of atherosclerosis is facilitated by the presence of peroxynitrite, which hinders vasorelaxation, undermines the positive impact of nitric oxide on platelet aggregation and the proliferation of vascular smooth muscle cells, and oxidizes DNA and lipids [46]. The process of aging contributes to an increased level of oxidative assault in resistant arteries, leading to a heightened synthesis of peroxynitrite in the serum of aged mice. Similarly, the combination of cigarette smoking and reduced levels of AngII has been found to induce significant impairment in endothelial function, specifically impacting the relaxation response mediated by acetylcholine [47]. This is accompanied by the deactivation of the antioxidant enzyme SOD2. In addition, heart failure is associated with increased production of superoxide anion in vessels derived from cardiomyopathic hamsters [48]. This phenomenon is influenced by the activity of both NOS and NOX, which are enzymes responsible for the creation of ROS. The complex interplay between ROS-induced damage and NOS uncoupling highlights the nuanced correlation between oxidative assault and endothelial dysfunction in many clinical conditions [49].

4.2.2.3 *Mitochondrial Oxidative Assault*

Multiple mechanisms contribute to the occurrence of oxidative assault on a global scale. However, it is noteworthy that mitochondria play a predominant role in the generation of ROS within cells, accounting for approximately 90% of their production [50]. Indeed, the process of oxygen intake often involves the transfer of only a small proportion (1%) of electrons, resulting in the reduction of dioxygen (O_2) into the superoxide anion (O_2^-). The identification of complexes within the mitochondrial respiratory chain has been achieved through the utilization of inhibitors, like rotenone and antimycin, as well as their corresponding substrates, such as pyruvate and succinate [51]. The manufacturing sites of superoxide anion have been identified in complexes I, II, and III to date. The primary cause of cardiovascular disease is mostly ascribed to the excessive formation of ROS that occurs during mitochondrial malfunction, resulting in irreversible damage to the mitochondria [52]. In an experimental mouse model of myocardial infarction induced by coronary ligation, it was observed that there was a notable elevation in the production of ROS within the mitochondria [53]. The present hypothesis posits that an elevation in mitochondrial hydroxyl radicals (OH) leads to a significant decline in mitochondrial DNA (mtDNA) levels, as well as the expression of mtDNA-encoded gene transcripts associated with respiratory complexes I, III, and IV. Due to its limited capacity for repair and its proximity to the electron transport chain, mitochondrial DNA (mtDNA) is very susceptible to the damaging effects of ROS [54]. Nevertheless, the activation of AngII results in the generation of ROS, which subsequently causes mitochondrial dysfunction, lesions in mtDNA, hypertrophy of cardiomyocytes, fibrosis in the cardiac tissue, and ultimately, heart failure [55]. The oxidation of proteins in complex I and II of the electron transport chain by ROS within the mitochondria leads to a decrease in mitochondrial respiration. *In vivo*, models of myocardial infarction exhibit a decrease in mitochondrial respiration, accompanied by impairments in the various complexes of the mitochondrial respiratory chain [56]. Several protein kinases and transcription factors that are involved in hypertrophic signaling are known to be activated in response to heightened oxidative assault and the formation of ROS resulting from anomalies in the mitochondrial respiratory chain.

The etiology of numerous cardiovascular complications linked to diabetes can be attributed to the formation of ROS by mitochondria [57]. Indeed, there exists a correlation between left ventricular hypertrophy, metabolic dysfunction, and diastolic dysfunction, all of which are associated with metabolic syndrome. Characteristic of this condition is the coexistence of diabetes and obesity. Mice subjected to a diet rich in fats and sucrose demonstrate the manifestation of ventricular hypertrophy, characterized by an augmented thickness of the cardiac walls. Additionally, these mice exhibit indications of mitochondrial oxidative assault, as evidenced by the formation of H_2O_2. Furthermore, the mice also display impaired mitochondrial function, as indicated by reduced ATP synthesis, diminished activity of complex II, and impaired mitochondrial respiration [58]. Moreover, patients with diabetes exhibit an elevation in mitochondrial H_2O_2 generation within their atria. Elevated levels of oxidative assault and apoptosis have been seen in isolated right atrial cardiomyocytes obtained from obese patients [59]. Nevertheless, previous studies have reported an elevation in the production of mitochondrial superoxide anions in cardiomyocytes during hypertrophy produced by isoproterenol or AngII.

4.3 Oxidative Assault and Association with CVDs

4.3.1 Myocardial Ischemia-Reperfusion (I/R) Injury and Oxidative Assault

It is characterized by the restoration of blood flow to oxygen-deprived tissues and is a pivotal pathological mechanism in acute coronary syndrome, myocardial infarction, coronary revascularization procedures, and other cardiovascular scenarios [60]. During reperfusion, when blood flow is rapidly restored, an oxygen burst occurs, leading to an excessive generation of ROS and a subsequent oxidative assault. This process contributes significantly to cell death during myocardial infarction, accounting for approximately 25% of such fatalities. The sources of ROS overproduction during myocardial I/R injury encompass mitochondria, the NOX family, xanthine oxidase (Xo), and uncoupled NOS [52]. The major contributors to ROS production within the mitochondria are Complex I and III, with mitochondrial permeability transition pores (mPTP) also regulating this process [61]. Studies have shown that deficiencies in NOX1 and NOX2 result in reduced infarct size in murine models, while NOX4 deficiencies exhibit less pronounced effects [62]. To counteract ROS accumulation, redox signaling pathways, including hypoxia-inducible factor (HIF) and Nuclear factor E2-associated factor 2 (Nrf2), are activated [63]. HIF, an oxygen-sensitive transcription factor influenced by NOX-derived ROS, attenuates I/R injury through the regulation of various enzymes and pathways [64]. Nrf2 translocated into the nucleus during oxidative assault, regulates the expression of antioxidants and anti-inflammatory factors [63].

Excessive ROS levels are associated with detrimental effects on cardiomyocyte mitochondria, causing mitochondrial damage, DNA impairment, and protein degradation, all of which contribute to irreversible cell death. Mitochondrial damage plays a central role in oxidative assault-mediated myocardial I/R injury. Additionally, abnormal mPTP opening can trigger apoptosis or necrosis. Endoplasmic reticulum assault, also linked to oxidative assault, significantly influences myocardial I/R injury by disturbing protein activities and intracellular calcium homeostasis [65]. Inflammatory processes are likewise crucial in I/R injury, with cytokine cascades activating inflammatory responses. Neutrophils, recruited to the infarct zone, contribute to ROS production via NOX2 and myeloperoxidase (MPO), leading to inflammation. Clinical investigations have confirmed the role of oxidative assault in myocardial I/R injury. In patients undergoing coronary interventions or bypass surgery, markers such as plasma 8-iso-prostaglandin F2-α and thiobarbituric acid reactive substances (TBARS) indicated increased oxidative assault post-procedure [66]. Notably, a clinical study revealed increased plasma 8-iso-prostaglandin F2-α levels after primary percutaneous coronary intervention, indicating oxidative assault without a direct correlation to troponin T levels [67]. In patients undergoing coronary artery bypass surgery, oxidative assault, as measured by TBARS, peaked an hour after reperfusion.

In conclusion, myocardial I/R injury is driven by excessive ROS production from various sources, including mitochondria, NOX, Xo, and uncoupled NOS. These ROS induce oxidative assault, causing cellular damage, DNA impairment, and inflammation. Redox signaling pathways attempt to counteract the ROS burst. Understanding the intricate interplay between oxidative assault, inflammation, and cellular damage during myocardial I/R injury provides critical insights into potential therapeutic targets for mitigating its detrimental effects [52].

4.3.2 Heart Failure (HF) and Oxidative Assault

HF is a medical condition that is characterized by an inadequate amount of blood being pumped by the heart to meet the body's physiological needs. This condition is commonly manifested by symptoms such as difficulty breathing and discomfort in the chest area [68]. It is generally agreed that ROS are made in large amounts at all stages of HF. However, the sources of ROS are different in patients with reduced ejection fraction (HFrEF) compared to those with preserved ejection fraction (HFpEF).

HFrEF is commonly associated with coronary artery disease (CAD) and is characterized by a decrease in the percentage of blood pumped out of the heart during each contraction [69]. Within this particular scenario, it is the damaged cardiomyocytes that serve as the principal origin of excessive ROS generation. The following aberrant remodeling that takes place as a consequence of ROS entails programmed cell death and fibrosis. In contrast,

HFpEF typically arises from conditions such as hypertension, diabetes, or genetic factors such as hypertrophic cardiomyopathy [70]. In this context, endothelial cells play a significant role in the formation of ROS. Elevated levels of proinflammatory substances such as IL-6, soluble ST2, and TNF-α, which are linked to risk factors associated with HFpEF, initiate the generation of ROS in endothelial cells. This, in turn, leads to an increase in PKG signaling in cardiomyocytes [71].

In the context of cardiac dysfunction, mitochondria play a crucial role in the formation of ROS, alongside heightened activities of NOX and Xo. The surge of ROS results in mitochondrial damage, which in turn contributes to a feedback loop that further amplifies ROS production. In the context of cardiac remodeling, it is noteworthy that plays a substantial role by stimulating hypertrophic signaling kinases and transcription factors, including tyrosine kinase Src, GTP-binding protein Ras, and MAPKs [72]. Moreover, the activation of matrix metalloproteinases by ROS plays a role in the development of structural alterations in the myocardium. Chemotherapy-induced cardiomyopathy is a distinct type of heart failure that often arises as a serious consequence of chemotherapy treatment. Cardiac toxicity has been associated with the administration of many chemotherapeutic drugs, such as anthracyclines and alkylating agents, owing to their ability to induce the generation of ROS. Doxorubicin (DOX), an extensively utilizedanthracycline, exerts a direct impact on mitochondrial DNA, functionality, protein expression, and lipid peroxidation. DOX is capable of forming complexes with iron, known as iron-Dox complexes, which enhance the generation of ROS upon interaction with oxygen [73].

The association between the severity of HF and oxidative assault has been elucidated through clinical investigations. An example of relevant research conducted by Tedgui demonstrated a correlation between the severity of heart failure and the levels of 8-iso-prostaglandin F2-αin the pericardium. The study results demonstrated an increase in pericardial levels of 8-iso-prostaglandin F2α that corresponded to the degree of functional impairment in heart failure. Additionally, this increase was found to be associated with ventricular dilatation. These data indicate a substantial contribution of *in vivo* oxidative stress to ventricular remodeling and the progression of HF [74]. In a separate investigation conducted by Mishra *et al.*, a negative correlation was observed between plasma lipid peroxides and left ventricular ejection fraction in individuals with congestive HF [70].

To summaries, HF consists of different stages characterized by varying ejection fractions, and excessive production of ROS significantly contributes to each of these stages. In HFrEF, damaged cardiomyocytes serve as the principal source of ROS, which subsequently contributes to the development of maladaptive remodeling. In contrast, it has been observed that in HFpEF, endothelial cells produce ROS as a result of the presence of pro-inflammatory substances. This production of ROS has been found to impact cardiomyocytes through the activation of PKG signaling pathways. ROS production linked with HF is also observed through the involvement of NOX, Xo, and mitochondria, resulting in detrimental effects on mitochondrial integrity and cardiac tissue remodeling. Chemotherapy-induced cardiomyopathy, which is connected with the formation of ROS caused by chemotherapy, is an additional significant component of HF. Comprehending these complex systems enhances one's understanding of the pathophysiology of heart failure and the potential therapeutic approaches that might be pursued [69–71].

4.3.3 Atherosclerosis and Oxidative Assault

Atherosclerosis serves as the underlying pathophysiological foundation for a range of cardiovascular conditions, including ischemic heart disorders, stroke, and peripheral artery illnesses. The significance of oxidative assault in the advancement of atherosclerosis, culminating in the development of atherosclerotic plaques that impede oxygen delivery and contribute to cardiovascular diseases, is of significant importance [75]. The onset of the disease entails the occurrence of damage to endothelial cells, which is subsequently followed by the buildup of ox-LDL cholesterol in the subendothelial region. Concurrently, leukocytes traverse into this region, while macrophages originating from monocytes phagocytose ox-LDL to form foam cells [76]. The primary contributors of ROS in the development of atherosclerosis are NOX, Xo, and enzymes found in the mitochondria. The presence of NOX1 and NOX4 has been observed in VSMCs, whereas endothelial cells are characterized by the presence of NOX2, NOX4, and Xo [77].

The presence of oxidative assault has a substantial impact on the development and progression of atherosclerosis throughout all phases of its pathogenesis. The modification of endothelial signal transduction and redox-regulated transcription factors induces endothelial dysfunction, resulting in heightened vascular endothelial permeability and enhanced leukocyte adhesion [78]. This process is well recognized as the earliest stage of plaque formation. Following this, plasma LDL is drawn towards the artery wall, where it undergoes oxidative assault-induced modifications, resulting in the formation of ox-LDL. Macrophages subsequently phagocytose ox-LDL, leading to their differentiation into foam cells [79]. Furthermore, the presence of oxidative assault has a significant impact on the regulation of adhesion molecules, such as vascular cell adhesion molecule-1, which in turn affects the adherence of monocytes of the interim, elevated levels of ROS contribute to the progression of plaque formation through the stimulation of migration of vascular smooth muscle cells (SMC) and the deposition of collagen [80]. The primary cause of plaque destabilization is mostly ascribed to the activation of matrix metalloproteinases, which are triggered by oxidative assault. This activation subsequently disintegrates the fibrous cap, as depicted in Figure 4.2. The study conducted by Daiber *et al.* emphasizes the correlation between endothelial dysfunction and increased generation of vascular superoxide in individuals with atherosclerosis, emphasizing the significant influence of oxidative assault in the development of this condition [81]. A recent study has provided further evidence supporting the probable causative relationship between oxidative assault and atherosclerosis by demonstrating a positive correlation between the severity of arterial blockage and elevated levels of erythrocyte TBARS [82]. In brief, atherosclerosis acts as the fundamental etiology for a range of cardiovascular ailments, with oxidative assault playing a pivotal role in its advancement. The initiation of atherosclerotic plaque formation involves endothelial dysfunction driven by oxidative assault, which leads to the accumulation of ox-LDL, production of macrophage foam cells, and adherence of leukocytes. During advanced stages, the presence of oxidative assault persists and exerts an ongoing influence on the progression and stability of plaques [83]. Prominent scientific discoveries highlight the intricate relationship between oxidative assault and atherosclerosis, therefore emphasizing the importance of effectively regulating oxidative assault to enhance cardiovascular well-being.

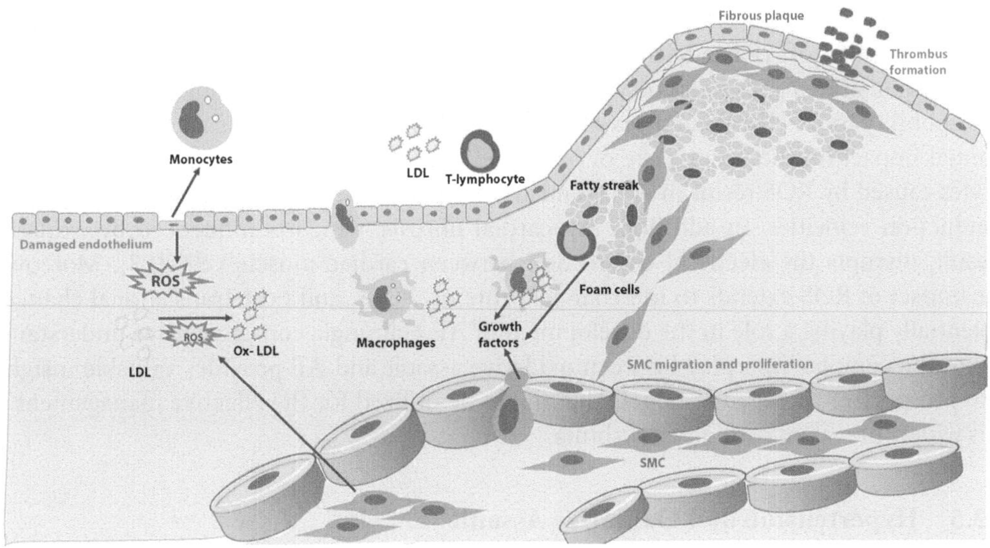

Figure 4.2 ROS-mediated ox-LDL production and formation of fibrous plaque.

4.3.4 Atrial Fibrillation (AF) and Oxidative Assault

AF is the most commonly seen cardiac arrhythmia in the field of clinical medicine, distinguished by the presence of an irregular and accelerated heart rhythm. The association between oxidative assault and AF has been established, and recent research indicates that oxidative assault may play a significant role in the pathogenesis of AF [84]. Increased susceptibility to AF has been seen in rats with reduced levels of plasma antioxidant capacity. Ismaeel *et al.* observed that individuals with AF had raised levels of cysteine, an increased ratio of cysteine to glutathione, and a modified redox potential of glutathione [85]. These findings collectively suggest the presence of heightened oxidative assault in AF patients. ROS are derived from various sources, including NOX2/4 enzymes, which exhibit increased expression in fibrillating regions, as well as Xo. Multiple processes demonstrate how ROS can initiate the onset of AF. To begin with, the elevation of ROS exerts an impact on the flow of ions within cardiomyocytes. As a consequence, there is an increase in the Na^+ current, L-type Ca^{2+} current, and leakage of Ca^{2+} from the SR. These alterations combined result in an elongation of the action potential duration and a decrease in conduction velocity [86]. In addition, it has been observed that oxidative assault plays a role in the promotion of collagen deposition, hence facilitating cardiac fibrosis. The presence of fibrosis hinders the effective electrical connection between myocytes, hence establishing a favorable environment for the occurrence of arrhythmias. Finally, it is worth noting that oxidative assault has the potential to have a role in the development of AF by affecting ion channel proteins, DNA, and post-translational modifications. The aberrant patterns of Na^+ current are induced by Angiotensin II and hypoxia through the modulation of the mRNA splicing of the α subunit 5 (SCN5A) of the Na^+ voltage-gated channel. The oxidation of Ca^{2+}/calmodulin-dependent kinase II (CaMKII) by ROS can modulate Na^+ current through the ryanodine receptor, an essential Ca^{2+} regulatory protein situated in the SR [87]. The role of oxidative assault in the development of AF has been identified. Both animal studies and clinical observations have demonstrated a correlation between diminished antioxidant capacity and elevated susceptibility to AF. The generators of ROS, such as NOX2/4 enzymes and Xo, have a substantial impact on the occurrence of oxidative assault. The ionic alterations in cardiomyocytes caused by ROS result in an extended duration of the action potential and modified conduction velocities. In addition, myocardial fibrosis, which is influenced by oxidative assault, disrupts the electrical connection between cardiac muscle cells [88]. Moreover, the impact of ROS extends to ion channel proteins, DNA, and post-translational changes, potentially playing a role in the development of AF. Gaining a comprehensive understanding of the complex interplay between oxidative assault and AF provides valuable insights into prospective therapeutic targets that might be utilized for the effective management of this widely prevalent cardiac arrhythmia.

4.3.5 Hypertension and Oxidative Assault

Hypertension, a significant risk factor for cardiovascular disease, is distinguished by the pathological increase in blood pressure resulting from aberrant levels of vasorelaxation factors. The results of laboratory experiments have shown that there are noticeable differences in the levels of oxidative assault between hypertension models and control groups [89]. Concomitant with the development of atherosclerosis, the NOX family has been identified as a prominent origin of ROS, with Xo, NOS, and mitochondria, which also make substantial contributions to the increase in ROS levels. The regulation of hypertension is significantly influenced by oxidative assault, which specifically targets endothelial cells. The regulation of vascular tone is determined by the equilibrium between endothelium-derived relaxing factors (EDRFs) and endothelium-derived hyperpolarizing factors (EDHFs) [90]. ROS acts in opposition to the vasodilatory properties of NO, which is classified as an EDRF. The administration of ROS leads to a decrease in the availability of NO and facilitates the buildup of endogenous antagonists of eNOS, such as asymmetric dimethylarginine (ADMA) [91]. In addition, the modulation of endothelial function encompasses other phosphorylation pathways, including tyrosine kinases, PI3K/Akt, and MAPKs, as well as gene expression factors such as p53 and AP-1. These pathways are susceptible to initiation and regulation by oxidative assault.

The study conducted by Chayama *et al.* focused on individuals diagnosed with renovascular hypertension. The researchers examined the levels of urinary excretion of 8-hydroxy-2′-deoxyguanosine and serum MDA-modified LDL as markers of oxidative assault. The researchers observed a notable increase in oxidative assault markers among individuals with hypertension in comparison to those without hypertension in the control group [92]. Furthermore, a study conducted by Shahbazi provided further support for these findings, since it demonstrated elevated amounts of lipid hydroperoxides in individuals with hypertension [93]. To summaries, hypertension, which is a significant contributor to cardiovascular illnesses, arises from variables that disrupt the normal process of vasorelaxation. Hypertension development is notably influenced by oxidative assault, which is characterized by the excessive production of ROS from several sources including NOX, Xo, NOS, and mitochondria. The equilibrium between EDRFs and EDHFs is disrupted by oxidative assault, which leads to a decrease in the availability of NO and an increase in antagonists of NOS. In addition, it exerts an influence on endothelial function through phosphorylation pathways and gene expression factors. Multiple studies have provided evidence of heightened levels of oxidative assault markers among individuals diagnosed with hypertension, hence reinforcing the significant involvement of oxidative assault in the development of hypertension.

4.4 Antioxidants and Their Sources

4.4.1 The Antioxidant Potential of Nuts

Nuts and peanuts are rich in antioxidants, particularly vitamin E and polyphenols, which confer benefits for their incorporation into approaches targeting the prevention and treatment of CVD (Figure 4.3) [94]. These dietary items not only offer essential nutrients but also demonstrate unique attributes that contribute to their efficacy in enhancing cardiovascular well-being. Nuts, including almonds, hazelnuts, Brazil nuts, cashews, macadamias, walnuts, pecans, and pistachios, are classified as desiccated, indehiscent fruits that envelop seeds. Peanuts and Baru almonds, despite being classified as legumes based on their pod formation, have comparable nutritional characteristics to nuts and are additionally recognized as oleaginous fruits. These food items exhibit a composition that is abundant in macronutrients, specifically notable for their elevated levels of protein and unsaturated fat. In addition, these substances possess dietary fiber, micronutrients, and fat-soluble bioactive components such as polyunsaturated fatty acids (PUFA) and monounsaturated fatty acids (MUFA), as well as phytochemicals such as phenolic compounds [95]. The reported health advantages of nuts can be attributed to the synergistic interplay of their elements. The lipid content found in nuts, peanuts, and Baru almonds generally falls between the region of

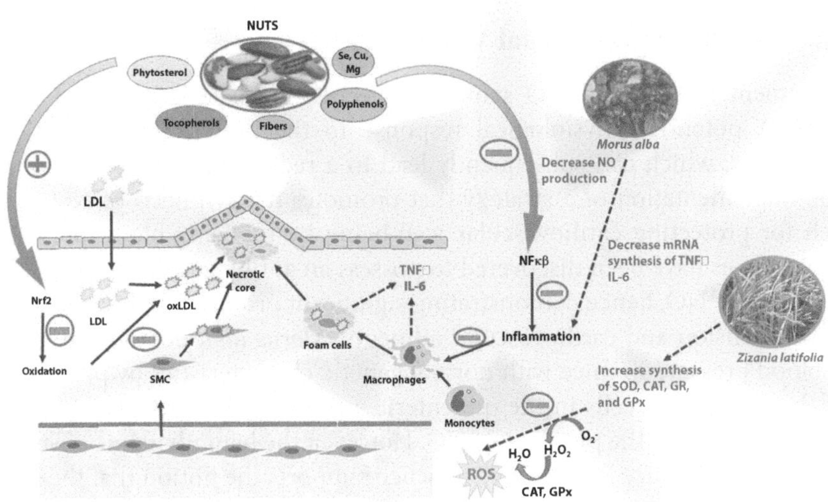

Figure 4.3 Antioxidant mechanisms of nuts, *Morus alba* and *Zizanialatifolia.*

30% to 60% of their overall composition. The fatty acid composition of these food items is notable, as they include substantial amounts of oleic (C18:1) and linoleic (C18:2) fatty acids, which are considered beneficial for cardiovascular well-being [96].

In contrast to the sole treatment of peroxynitrite ($ONOO^-$), an *in vitro* investigation utilizing plasma and platelets that were treated with proanthocyanidin fraction derived from *Medemia argun* nuts (at concentrations ranging from 0.5 g/mL to 50 g/mL) shown a notable reduction in the oxidation of thiol and carbonyl groups, along with a considerable elevation in glutathione (GSH) levels [97]. The antioxidant action of these polyphenols is attributed to their ability to inhibit the MAPK pathways by suppressing NFκβ and activating the nuclear factor erythroid 2-related factor 2 (Nrf2) pathways [98]. The antioxidant properties of the supernatant fractions of several nuts, such as almonds, macadamias, hazelnuts, pistachios, and walnuts, were observed during *in vitro* fermentation [99]. Based on a study investigating the antioxidant properties of pistachios, it was observed that pretreatment with pistachio shells or roasted and salted pistachios at several doses resulted in a substantial reduction (p 0.01) in the generation of ROS and malondialdehyde (MDA) in the J774 macrophage/monocyte cell line [100]. The study reported a significant increase ($p < 0.01$) in SOD activity after 48 hours of incubation with walnut oil. This increase was observed throughout a concentration range of 40 L/mL to 319 L/mL in U937 cells that were cultured for a duration of 24 to 72 hours. The utilization of defatted walnut flour resulted in the elimination of the hydroxyl radical and a reduction in ROS inside the selected peptides [101]. In a rodent model, six cohorts of male Wistar rats received Brazil nut (BN) daily for seven days before undergoing ischemia-reperfusion (IR) surgery. Both dosage groups of BN-pretreated animals showed reduced oxidative assault. iNOS expression, kidney nitrotyrosine expression, and plasma TBARS levels increased [102]. A six-week diet of either a conventional feed or a high-fat diet with 45% or 21.5% of energy from walnuts decreased lipid peroxidation and MAPK phosphorylation in male C57BL/6J mice. Male mice fed an atherogenic diet supplemented with a mixture of nuts (almonds, macadamias, peanuts, pistachios, cashews, Brazil nuts, and walnuts) or isolated pistachios had lower oxLDL levels [103].

4.4.2 *Morus alba*-Induced Endothelial Vasorelaxation

The pivotal involvement of eNOS in NO synthesis by endothelial cells has been well-documented [104]. A potential physiological response to the presence of NO is the relaxing of blood vessels, which can subsequently lead to a reduction in blood pressure. Consequently, the implementation of a strategy that promotes the synthesis of NO is an effective approach for protecting cardiovascular well-being by reducing blood pressure. Certain natural substances have been discovered to possess an antihypertensive impact by enhancing the synthesis of NO, hence demonstrating significant promise in preventing the development of hypertension and cardiovascular ailments. *Morus alba* demonstrated the ability to reduce blood pressure in mice with normal genetic characteristics by promoting the relaxation of the endothelial cells in the mesenteric arteries. This effect was achieved through a pathway that relies on the presence of NO. However, the hemodynamic effects of *M. alba* were not observed in mice lacking eNOS, which supports the notion that the antihypertensive activity of *M. alba* is mediated through a mechanism dependent on NO [105].

4.4.3 White Mulberries Induced Increased eNOS Expression

The study observed that polysaccharides derived from white mulberry fruit have the ability to stimulate relaxation in rat mesenteric arteries through endothelium-dependent mechanisms [106]. Additionally, these polysaccharides were found to enhance the production of NO in endothelial cells. When administered intravenously, white mulberry fruit polysaccharides were shown to effectively reduce blood pressure in both normotensive and spontaneously hypertensive rats. However, it was observed that this blood pressure-lowering effect was significantly diminished in normotensive rats that were pre-treated with NG-nitro-L-arginine methyl ester (L-NAME), an inhibitor of NOS. The data presented in this study indicate that the hypotensive effects of white mulberry fruit can be attributed to the NO pathway. Furthermore, the administration of grape seed polyphenol extract in hypertensive rats resulted in an upregulation of eNOS and Sirtuin-1, leading to an enhanced generation of NO and a subsequent reduction in blood pressure [107].

4.4.4 *Hylocereus polyrhizus*-Induced Lipid-Lowering Activity

The administration of red dragon fruit flour to obese rats for a duration of 4 weeks resulted in a considerable reduction in blood glucose, total cholesterol (TC), triglyceride (TG), and low-density lipoprotein cholesterol (LDL-C) levels [108]. However, there was no observed effect on high-density lipoprotein cholesterol (HDL-C) levels. A further study demonstrated that male mice with hyperlipidemia exhibited a reduction in TC, TG, and LDL-C levels in a manner that was dependent on the dosage of orally administered red dragon fruit peel powder for 30 days. The trials stated earlier demonstrated the presence of blood lipid-lowering properties in both the pulp and peel of red dragon fruit [109].

4.4.5 Chinese Wild Rice (*Zizania latifolia*) and North American Wild Rice (*Zizania aquatica*) Mediated

The decrease in antioxidant capacity and the excessive generation of ROS give rise to oxidative assault, which is a substantial factor in the development of CVDs. Certain natural products exhibit potential in the prevention and management of CVDs due to their ability to mitigate oxidative assault. This is achieved by augmenting the synthesis of antioxidant enzymes such as SOD, CAT, GR, and GPx, while concurrently reducing the levels of peroxidative by products such as MDA and protein carbonyls (Figure 4.3) [110]. The findings of the research study indicate that hyperlipidemic rats saw a significant reduction in oxidative assault when they were given either North American or Chinese wild rice. This reduction was attributed to an increase in total antioxidant capacity, elevated SOD activity, and a decrease in MDA concentration. In addition, it was observed that two distinct varieties of wild rice could mitigate inflammation and hyperlipidemia in rat models [111].

4.4.6 The Antioxidant Potential of Rice Bran Protein Hydrolysate

A previous study examined the impact of rice bran protein hydrolysate on arterial stiffness, vascular remodeling, and oxidative assault in rats subjected to a diet heavy in carbohydrates and fats [112]. The addition of hydrolysate supplementation resulted in significant enhancements in glucose regulation, insulin sensitivity, lipid profiles, blood pressure, aortic pulse wave velocity, aortic wall hypertrophy, and vascular remodeling. The hydrolysate exhibited a reduction in both adenosine deaminase and TNF-α plasma levels. The hydrolysate demonstrated the ability to mitigate oxidative assault by reducing plasma MDA levels, superoxide production, and p47 (phox) NADPH oxidase expression in vascular tissues. After the administration of hydrolysate, there was an observed rise in the plasma nitrate/nitrite level and eNOS expression in the vessels of rats in the model group. This indicates that the formation of NO was promoted [113].

4.4.7 *Nepeta deflersiana* Mediated Antioxidant Potential

Elevated concentrations of inflammatory biomarkers, including TNF-α, IL-6, interleukin-10 (IL-10), C reactive protein (CRP), monocyte chemoattractant protein-1 (MCP-1), and vascular cell adhesion molecule-1 (VCAM-1), have been associated with the advancement of CVDs. Certain natural substances have demonstrated the ability to mitigate inflammation and suppress the production of inflammatory cytokines, hence potentially lowering the susceptibility to cardiovascular illnesses [114]. The findings of an *in vivo* investigation showed that the ethanolic extract of *Nepeta deflersiana* exhibited a significant reduction in the severity of heart damage in Wistar rats. This effect was attributed to the extract's ability to decrease oxidative assault levels, inhibit apoptosis, and mitigate inflammation. *N. deflersiana* elicited a suppressive effect on the gene expression TNF-α, IL-6, and IL-10 [115].

4.4.8 Antioxidant Potential of Nutritional Molecules

There is a growing body of research indicating that a wide range of dietary supplements possess antioxidant properties. Vitamin A is a class of chemical compounds that plays a crucial role in providing essential nutrition. However, due to its unsaturated nature, it is susceptible to reacting with oxidative radicals. The interaction between vitamin A and apolipoproteins has the potential to modify the risk of myocardial infarction. Significantly, astaxanthin, a major carotenoid, is widely recognized as a ROS scavenger. Vitamin A is derived through the process of synthesizing carotenoids [116]. Preclinical research has demonstrated that astaxanthin's preventative properties are beneficial in animal models of ischemia/reperfusion (I/R) damage and thrombotic diseases. Additionally, it has been observed that astaxanthin exhibits the capability to reduce blood pressure in spontaneously hypertensive rats. The randomized, double-blind clinical experiment demonstrated that the intake of astaxanthin resulted in an elevation of SOD levels and a reduction in levels of blood lipid peroxidation biomarkers [117].

Both exogenous and endogenous ROS can be effectively detoxified by vitamin C. Additionally, vitamin C can restore the proteins and lipids that have been modified by ROS [118]. The pathogenesis of cardiovascular illnesses encompasses a series of intricate processes, such as the proliferation and apoptosis of endothelial cells, as well as vasodilation facilitated by smooth muscle. Vitamin E inhibits the assembly of the NOX enzyme, which is essential for the generation of superoxide. Animal studies have provided evidence suggesting that Vitamin E may reduce the risk of developing coronary heart disease and other cardiovascular disorders. A meta-analysis of data from 400,000 individuals revealed a significant association between the consumption of vitamin E and vitamin C and a reduced risk of developing coronary heart disease [119].

Myocardial cells exhibit heightened susceptibility to oxidative assault-induced damage, although the capacity of omega-3 fatty acids to regulate redox pathways renders it a potentially auspicious approach for mitigating this risk. Rats that were administered omega-3 in a rat model of myocardial infarction had a reduced infarct size [120]. Previous studies have indicated that individuals suffering from cardiovascular disease, as well as animal models, have exhibited enhanced cardiac performance after the administration of flavonols such as quercetin. Flavanols have been found to exhibit a capacity for mitigating indicators of oxidative assault. The findings of a clinical study including 805 older individuals indicate that a higher intake of flavonoids is linked to a decreased likelihood of mortality resulting from coronary heart disease [121].

4.5 Potential Antioxidant-Based Experimental Interventions

The system of categorization of novel antioxidants can be delineated into three discrete categories, each of which is designed to address specific facets of oxidative assault. The initial category emphasizes the activation of natural antioxidant defense systems. One illustrative instance of this phenomenon is the activation of Nrf2. The Nrf2 transcription factor plays a vital role in recognizing the enhancer known as the Antioxidant Response Element. Individuals exhibiting diminished Nrf2 expression and attenuated Nrf2 activity demonstrate heightened vulnerability to the development of atherosclerosis and hypertension [122]. Mice that lacked Nrf2 exhibited compromised cardiac structure and function, heightened vulnerability to HF, and increased sensitivity to I/R injury [123]. Dimethyl fumarate (DMF), a pharmacological agent that selectively targets NRF2, has been extensively investigated for its therapeutic potential in the treatment of multiple sclerosis. Notably, DMF has exhibited notable protective properties in the context of cardiac I/R injury and the progression of atherosclerosis [75].

The second group includes chemicals that possess the ability to impede the production of oxidative assault. These drugs specifically target Xo, NOX, and myeloperoxidase (MPO). The therapeutic potential of the Xo inhibitor allopurinol appears to be promising [124]. While there is a dearth of extensive clinical trials on cardiovascular illnesses, smaller-scale investigations indicate that allopurinol may have the capacity to mitigate oxidative assault, particularly in endothelial cells. The findings from clinical trials investigating the impact of allopurinol on blood pressure demonstrate a moderate decrease. Additionally, it has been observed that the administration of allopurinol is associated with a decrease in death rates during hospitalization and a reduction in the occurrence of cardiac problems among those undergoing cardiac procedures. The clinical development of the NOX inhibitor GKT137831 has demonstrated its ability to produce anti-atherosclerotic effects and cardioprotective qualities in animal models [125]. The potential of MPO inhibitors in changing the composition of atherosclerotic lesions and cardiac remodeling has been demonstrated.

The chemicals in the third group are involved in promoting the functional restoration of damage caused by ROS. These compounds predominantly target the signaling pathway involving nitric oxide and cyclic guanosine monophosphate (NO-cGMP). Interference of ROS with the signaling pathway involving NO and cGMP can manifest in several ways, including uncoupling of NOS, scavenging of NO, or oxidative damage to the receptor for NO, known as soluble guanylyl cyclase (sGC) [126]. HNO donors, such as CXL-1427, which belongs to the second-generation class of compounds, have demonstrated both safety and hemodynamic advantages in heart failure patients who are admitted to the hospital [127]. L-citrulline and L-arginine belong to this category as they facilitate the reactivation of NOS. The administration of oral L-arginine and L-citrulline has been found to result in reductions in both systolic and diastolic blood pressure, as indicated by meta-analyses of clinical trials [128].

4.6 Conclusion

In summary, new antioxidants present encouraging strategies for addressing oxidative assault. Various tactics involve the use of activators of endogenous antioxidant defense mechanisms, inhibitors of oxidative assault production, and chemicals that facilitate the functional repair of damage caused by ROS. Nrf2 activators, such as DMF, have demonstrated protective effects. Conversely, Xo inhibitors, like allopurinol, and NOX inhibitors, such as GKT137831, show promise in the management of cardiovascular disease. Moreover, the use of chemicals that specifically target the signaling pathway involving NO and cGMP, such as hydrogen sulfide (HNO) donors, L-citrulline, and L-arginine, presents potential strategies for mitigating the detrimental effects caused by oxidative assault and ameliorating hypertension. The progression of cardiovascular disease is greatly influenced by oxidative assault. A variety of treatment techniques have been developed to address the issue of oxidative assault, and these approaches have shown promising results in animal experiments. Nevertheless, the clinical outcomes observed in human trials demonstrate a certain degree of heterogeneity. It is crucial to have a deeper understanding of oxidative assault in cardiovascular illnesses and to advance antioxidant medicines to enhance the effectiveness of treatments.

References

1. Lopez, A.D. and Adair, T., Is the long-term decline in cardiovascular-disease mortality in high-income countries over? Evidence from national vital statistics. *Int. J. Epidemiol.*, 48, 6, 1815–1823, 2019.
2. Balwan, W.K. and Kour, S., Lifestyle Diseases: The Link between Modern Lifestyle and threat to public health. *Saudi J. Med. Pharm. Sci.*, 7, 4, 179–84, 2021.
3. Ye, X., Kong, W., Zafar, M.I., Chen, L.L., Serum triglycerides as a risk factor for cardiovascular diseases in type 2 diabetes mellitus: a systematic review and meta-analysis of prospective studies. *Cardiovasc. Diabetol.*, 18, 1, 1–10, 2019.
4. Perumareddi, P., Prevention of hypertension related to cardiovascular disease. *Prim. Care. Clin. Office Pract.*, 46, 1, 27–39, 2019.
5. Xia, M.F., Chen, L.Y., Wu, L., Ma, H., Li, X.M., Li, Q., Gao, X., Sarcopenia, sarcopenic overweight/obesity and risk of cardiovascular disease and cardiac arrhythmia: A cross-sectional study. *Clin. Nutr.*, 40, 2, 571–580, 2021.

6. Kim, H., Kim, S., Han, S., Rane, P.P., Fox, K.M., Qian, Y., Suh, H.S., Prevalence and incidence of atherosclerotic cardiovascular disease and its risk factors in Korea: a nationwide population-based study. *BMC Public Health*, 19, 1–11, 2019.
7. Powell-Wiley, T.M., Poirier, P., Burke, L.E., Després, J.P., Gordon-Larsen, P., Lavie, C.J., American Heart Association Council on Lifestyle and Cardiometabolic Health, Council on Cardiovascular and Stroke Nursing, Council on Clinical Cardiology, Council on Epidemiology and Prevention, Stroke Council, Obesity and cardiovascular disease: a scientific statement from the American Heart Association. *Circulation*, 143, 21, e984–e1010, 2021.
8. Hedayatnia, M., Asadi, Z., Zare-Feyzabadi, R., Yaghooti-Khorasani, M., Ghazizadeh, H., Ghaffarian-Zirak, R., Ghayour-Mobarhan, M., Dyslipidemia and cardiovascular disease risk among the MASHAD study population. *Lipids Health Dis.*, 19, 1–11, 2020.
9. Kondo, T., Nakano, Y., Adachi, S., Murohara, T., Effects of tobacco smoking on cardiovascular disease. *Circ. J.*, 83, 10, 1980–1985, 2019.
10. Al-Kindi, S.G., Brook, R.D., Biswal, S., Rajagopalan, S., Environmental determinants of cardiovascular disease: lessons learned from air pollution. *Nat. Rev. Cardiol.*, 17, 10, 656–672, 2020.
11. Gianazza, E., Brioschi, M., Martinez Fernandez, A., Casalnuovo, F., Altomare, A., Aldini, G., Banfi, C., Lipid peroxidation in atherosclerotic cardiovascular diseases. *Antioxid. Redox Signaling*, 34, 1, 49–98, 2021.
12. Padmanaban, S., Pully, D., Samrot, A.V., Gosu, V., Sadasivam, N., Park, I.K., Kim, D.K., Rising Influence of Nanotechnology in Addressing Oxidative Stress-Related Liver Disorders. *Antioxidants*, 12, 7, 1405, 2023.
13. Krzystek-Korpacka, M., Kempiński, R., Bromke, M.A., Neubauer, K., Oxidative stress markers in inflammatory bowel diseases: systematic review. *Diagnostics*, 10, 8, 601, 2020.
14. Kang, H., Yu, H., Fan, J., Cao, G., Rotigotine protects against oxidized low-density lipoprotein (ox-LDL)-induced damages in human umbilical vein endothelial cells (HUVECs). *Bioengineered*, 12, 2, 10568–10579, 2021.
15. Wang, B., Qiu, W., Yang, S., Cao, L., Zhu, C., Ma, J., Chen, W., Acrylamide exposure and oxidative DNA damage, lipid peroxidation, and fasting plasma glucose alteration: association and mediation analyses in Chinese urban adults. *Diabetes Care*, 43, 7, 1479–1486, 2020.
16. Ali, J., Aziz, M.A., Rashid, M.M.O., Basher, M.A., Islam, M.S., Propagation of age-related diseases due to the changes of lipid peroxide and antioxidant levels in elderly people: A narrative review. *Health Sci. Rep.*, 5, 3, e650, 2022.
17. Bhullar, S.K. and Dhalla, N.S., Angiotensin II-induced signal transduction mechanisms for cardiac hypertrophy. *Cells*, 11, 21, 3336, 2022.
18. Panda, P., Verma, H.K., Lakkakula, S., Merchant, N., Kadir, F., Rahman, S., Rao, P.V., Biomarkers of oxidative stress tethered to cardiovascular diseases. *Oxid. Med. Cell. Longev.*, 2022, 9154295, 2002.
19. Pizzuti, V., Paris, F., Marrazzo, P., Bonsi, L., Alviano, F., Mitigating Oxidative Stress in Perinatal Cells: A Critical Step toward an Optimal Therapeutic Use in Regenerative Medicine. *Biomolecules*, 13, 6, 971, 2023.
20. Kim, H.J., Lee, P.C., Hong, J.H., Lamin-A/C is modulated by the involvement of histamine-mediated calcium/calmodulin-dependent kinase II in lung cancer cells. *Int. J. Mol. Sci.*, 23, 16, 9075, 2022.
21. Fang, Z., Kim, H.G., Huang, M., Chowdhury, K., Li, M.O., Liangpunsakul, S., Dong, X.C., Sestrin proteins protect against lipotoxicity-induced oxidative stress in the liver via suppression of C-Jun N-terminal kinases. *Cell. Mol. Gastroenterol. Hepatol.*, 12, 3, 921–942, 2021.
22. Bang, J. and Zippin, J.H., Cyclic adenosine monophosphate (cAMP) signaling in melanocyte pigmentation and melanogenesis. *Pigm. Cell Melanoma Res.*, 34, 1, 28–43, 2021.

23. Wang, M., Sheng, K.J., Fang, J.C., Zhao, H., Lu, S.M., Liu, Z.Y., Chen, B.T., Redox signaling in diabetic retinopathy and opportunity for therapeutic intervention through natural products. *Eur. J. Med. Chem.*, 244, 114829, 2022.
24. Ma, F., Luo, S., Lu, C., Jiang, X., Chen, K., Deng, J., Li, Z., The role of Nrf2 in periodontal disease by regulating lipid peroxidation, inflammation and apoptosis. *Front. Endocrinol.*, 13, 963451, 2022.
25. Li, J., Yu, D., He, C., Yu, Q., Huo, Z., Zhang, Y., Zhang, S., KLF6 alleviates hepatic ischemia-reperfusion injury by inhibiting autophagy. *Cell Death Dis.*, 14, 7, 393, 2023.
26. Adelusi, T.I., Du, L., Hao, M., Zhou, X., Xuan, Q., Apu, C., Yin, X., Keap1/Nrf2/ARE signaling unfolds therapeutic targets for redox imbalanced-mediated diseases and diabetic nephropathy. *Biomed. Pharmacother.*, 123, 109732, 2020.
27. Dubois-Deruy, E., Peugnet, V., Turkieh, A., Pinet, F., Oxidative stress in cardiovascular diseases. *Antioxidants*, 9, 9, 864, 2020.
28. Dries, E., Gilbert, G., Roderick, H.L., Sipido, K.R., The ryanodine receptor microdomain in cardiomyocytes. *Cell Calcium*, 114, 102769, 2023.
29. Moyes, A.J. and Hobbs, A.J., C-type natriuretic peptide: a multifaceted paracrine regulator in the heart and vasculature. *Int. J. Mol. Sci.*, 20, 9, 2281, 2019.
30. Caliri, A.W., Tommasi, S., Besaratinia, A., Relationships among smoking, oxidative stress, inflammation, macromolecular damage, and cancer. *Mutat. Res. Rev. Mutat. Res.*, 787, 108365, 2021.
31. Santos, C., Rogers, S.L., Carter, A.P., CryoET shows cofilactin filaments inside the microtubule lumen. bioRxiv, 2023-03, 2023.
32. Ho, M.Y. and Wang, C.Y., Role of irisin in myocardial infarction, heart failure, and cardiac hypertrophy. *Cells*, 10, 8, 2103, 2021.
33. Liu, J., Kang, R., Tang, D., Signaling pathways and defense mechanisms of ferroptosis. *FEBS J.*, 289, 22, 7038–7050, 2022.
34. Taylor, J.P. and Hubert, M.T., The role of NADPH oxidases in infectious and inflammatory diseases. *Redox Biol.*, 48, 102159, 2021.
35. Dustin, C.M., Cifuentes-Pagano, E., Pagano, P.J., Isoform-Selective Nox Inhibitors: Advances and Future Perspectives, in: *NADPH Oxidases Revisited: From Function to Structure*, pp. 343–377, 2023.
36. Sukhanov, S., Higashi, Y., Yoshida, T., Mummidi, S., Aroor, A.R., Russell, J.J., Chandrasekar, B., The SGLT2 inhibitor Empagliflozin attenuates interleukin-17A-induced human aortic smooth muscle cell proliferation and migration by targeting TRAF3IP2/ROS/NLRP3/Caspase-1-dependent IL-1β and IL-18 secretion. *Cell. Signal.*, 77, 109825, 2021.
37. Buonfiglio, F., Böhm, E.W., Pfeiffer, N., Gericke, A., Oxidative Stress: A Suitable Therapeutic Target for Optic Nerve Diseases? *Antioxidants*, 12, 7, 1465, 2023.
38. Pagano, P.J. and Cifuentes-Pagano, E., The Enigmatic Vascular NOX: From Artifact to Double Agent of Change: Arthur C. Corcoran Memorial Lecture–2019. *Hypertension*, 77, 2, 275–283, 2021.
39. Hofmann, A., Frank, F., Wolk, S., Busch, A., Klimova, A., Sabarstinski, P., Reeps, C., NOX4 mRNA correlates with plaque stability in patients with carotid artery stenosis. *Redox Biol.*, 57, 102473, 2022.
40. Chen, L.C., Shibu, M.A., Liu, C.J., Han, C.K., Ju, D.T., Chen, P.Y., Huang, C.Y., ERK1/2 mediates the lipopolysaccharide-induced upregulation of FGF-2, uPA, MMP-2, MMP-9 and cellular migration in cardiac fibroblasts. *Chem. Biol. Interact.*, 306, 62–69, 2019.
41. Yamagata, K. and Yamori, Y., Inhibition of endothelial dysfunction by dietary flavonoids and preventive effects against cardiovascular disease. *J. Cardiovasc. Pharmacol.*, 75, 1, 1–9, 2020.

42. Matthies, M., Rosenstand, K., Nissen, I., Muitjens, S., Riber, L.P., De Mey, J.G., Bloksgaard, M., Nitric oxide (NO) synthase but not NO, HNO or H2O2 mediates endothelium-dependent relaxation of resistance arteries from patients with cardiovascular disease. *Br. J. Pharmacol.*, 179, 5, 1049–1064, 2022.
43. Zhang, Z., Dalan, R., Hu, Z., Wang, J.W., Chew, N.W., Poh, K.K., Chen, X., Reactive oxygen species scavenging nanomedicine for the treatment of ischemic heart disease. *Adv. Mater.*, 34, 35, 2202169, 2022.
44. Shnayder, N.A., Petrova, M.M., Moskaleva, P.V., Shesternya, P.A., Pozhilenkova, E.A., Nasyrova, R.F., The role of single-nucleotide variants of NOS1, NOS2, and NOS3 genes in the comorbidity of arterial hypertension and tension-type headache. *Molecules*, 26, 6, 1556, 2021.
45. Xia, M., Yan, Y., Pu, H., Du, X., Liang, J., Sun, Y., Yuan, Y., Glutathione responsive nitric oxide release for enhanced photodynamic therapy by a porphyrinic MOF nanosystem. *Chem. Eng. J.*, 442, 136295, 2022.
46. Soundararajan, L., Dharmarajan, A., Samji, P., Regulation of pleiotropic physiological roles of nitric oxide signaling. *Cell. Signalling*, 101, 110496, 2022.
47. Li, X., Zhang, Z., Luo, M., Cheng, Z., Wang, R., Liu, Q., Xia, Y., NLRP3 inflammasome contributes to endothelial dysfunction in angiotensin II-induced hypertension in mice. *Microvasc. Res.*, 143, 104384, 2022.
48. De Geest, B. and Mishra, M., Role of oxidative stress in diabetic cardiomyopathy. *Antioxidants*, 11, 4, 784, 2022.
49. Das, D., Shruthi, N.R., Banerjee, A., Jothimani, G., Duttaroy, A.K., Pathak, S., Endothelial dysfunction, platelet hyperactivity, hypertension, and the metabolic syndrome: molecular insights and combating strategies. *Front. Nutr.*, 10, 1221438, 2023.
50. Beura, S.K., Dhapola, R., Panigrahi, A.R., Yadav, P., Reddy, D.H., Singh, S.K., Redefining oxidative stress in Alzheimer's disease: Targeting platelet reactive oxygen species for novel therapeutic options. *Life Sci.*, 306, 120855, 2022.
51. Shimada, B.K., Boyman, L., Huang, W., Zhu, J., Yang, Y., Chen, F., Chao, W., Pyruvate-driven oxidative phosphorylation is downregulated in sepsis-induced cardiomyopathy: a study of the mitochondrial proteome. *Shock (Augusta, Ga.)*, 57, 4, 553, 2022.
52. Yan, Q., Liu, S., Sun, Y., Chen, C., Yang, S., Lin, M., Yang, Y., Targeting oxidative stress as a preventive and therapeutic approach for cardiovascular disease. *J. Transl. Med.*, 21, 1, 1–35, 2023.
53. Bugger, H. and Pfeil, K., Mitochondrial ROS in myocardial ischemia reperfusion and remodeling. *Biochim. Biophys. Acta (BBA), Mol. Basis Dis.*, 1866, 7, 165768, 2020.
54. Luo, Y., Ma, J., Lu, W., The significance of mitochondrial dysfunction in cancer. *Int. J. Mol. Sci.*, 21, 16, 5598, 2020.
55. Peoples, J.N., Saraf, A., Ghazal, N., Pham, T.T., Kwong, J.Q., Mitochondrial dysfunction and oxidative stress in heart disease. *Exp. Mol. Med.*, 51, 12, 1–13, 2019.
56. Qi, B., Song, L., Hu, L., Guo, D., Ren, G., Peng, T., Li, Y., Cardiac-specific overexpression of Ndufs1 ameliorates cardiac dysfunction after myocardial infarction by alleviating mitochondrial dysfunction and apoptosis. *Exp. Mol. Med.*, 54, 7, 946–960, 2022.
57. Kaludercic, N. and Di Lisa, F., Mitochondrial ROS formation in the pathogenesis of diabetic cardiomyopathy. *Front. Cardiovasc. Med.*, 7, 12, 2020.
58. Abdurrachim, D., Ciapaite, J., Wessels, B., Nabben, M., Luiken, J.J., Nicolay, K., Prompers, J.J., Cardiac diastolic dysfunction in high-fat diet-fed mice is associated with lipotoxicity without impairment of cardiac energetics *in vivo*. *Biochim. Biophys. Acta (BBA), Mol. Cell. Biol. Lipids*, 1841, 10, 1525–1537, 2014.
59. Anderson, E.J., Kypson, A.P., Rodriguez, E., Anderson, C.A., Lehr, E.J., Neufer, P.D., Substrate-specific derangements in mitochondrial metabolism and redox balance in the atrium of the type 2 diabetic human heart. *J. Am. Coll. Cardiol.*, 54, 20, 1891–1898, 2009.

60. Halladin, N.L., Oxidative and inflammatory biomarkers of ischemia and reperfusion injuries. *Dan. Med. J.*, 62, 4, B5054, 2015.
61. Natarajan, V., Chawla, R., Mah, T., Vivekanandan, R., Tan, S.Y., Sato, P.Y., Mallilankaraman, K., Mitochondrial dysfunction in age-related metabolic disorders. *Proteomics*, 20, 5–6, 1800404, 2020.
62. Braunersreuther, V., Montecucco, F., Ashri, M., Pelli, G., Galan, K., Frias, M., Jaquet, V., Role of NADPH oxidase isoforms NOX1, NOX2 and NOX4 in myocardial ischemia/reperfusion injury. *J. Mol. Cell. Cardiol.*, 64, 99–107, 2013.
63. Baez-Jurado, E., Rincón-Benavides, M.A., Hidalgo-Lanussa, O., Guio-Vega, G., Ashraf, G.M., Sahebkar, A., Barreto, G.E., Molecular mechanisms involved in the protective actions of Selective Estrogen Receptor Modulators in brain cells. *Front. Neuroendocrinol.*, 52, 44–64, 2019.
64. Matsushima, S., Kuroda, J., Ago, T., Zhai, P., Ikeda, Y., Oka, S., Sadoshima, J., Broad suppression of NADPH oxidase activity exacerbates ischemia/reperfusion injury through inadvertent downregulation of hypoxia-inducible factor-1α and upregulation of peroxisome proliferator–activated receptor-α. *Circ. Res.*, 112, 8, 1135–1149, 2013.
65. Walker, R.K., Cousins, V.M., Umoh, N.A., Jeffress, M.A., Taghipour, D., Al-Rubaiee, M., Haddad, G.E., The good, the bad, and the ugly with alcohol use and abuse on the heart. *Alcohol.: Clin. Exp. Res.*, 37, 8, 1253–1260, 2013.
66. Vichova, T. and Motovska, Z., Oxidative stress: Predictive marker for coronary artery disease. *Exp. Clin. Cardiol.*, 18, 2, e88, 2013.
67. Elesber, A.A., Elesber, A.A., Best, P.J., Elesber, A.A., Best, P.J., Lennon, R.J., Lerman, A., Plasma 8-iso-prostaglandin F2α, a marker of oxidative stress, is increased in patients with acute myocardial infarction. *Free Radical Res.*, 40, 4, 385–391, 2006.
68. Perrotta, G., Heart Failure (HF): Recent innovations in clinical therapy and critical profiles of acute and chronic forms. *J. Cardiovasc. Med. Cardiol.*, 9, 4, 049–076, 2022.
69. Pagliaro, B.R., Cannata, F., Stefanini, G.G., Bolognese, L., Myocardial ischemia and coronary disease in heart failure. *Heart Fail. Rev.*, 25, 1, 53–65, 2020.
70. Mishra, S. and Kass, D.A., Cellular and molecular pathobiology of heart failure with preserved ejection fraction. *Nat. Rev. Cardiol.*, 18, 6, 400–423, 2021.
71. Bertaud, A., Joshkon, A., Heim, X., Bachelier, R., Bardin, N., Leroyer, A.S., Blot-Chabaud, M., Signaling pathways and potential therapeutic strategies in cardiac fibrosis. *Int. J. Mol. Sci.*, 24, 2, 1756, 2023.
72. Ramos-Kuri, M., Meka, S.H., Salamanca-Buentello, F., Hajjar, R.J., Lipskaia, L., Chemaly, E.R., Molecules linked to Rassignaling as therapeutic targets in cardiac pathologies. *Biol. Res.*, 54, 1, 23, 2021.
73. Songbo, M., Lang, H., Xinyong, C., Bin, X., Ping, Z., Liang, S., Oxidative stress injury in doxorubicin-induced cardiotoxicity. *Toxicol. Lett.*, 307, 41–48, 2019.
74. Wiśniewski, K., Popęda, M., Price, B., Bieńkowski, M., Fahlström, A., Drummond, K., Adamides, A.A., Glucose-6-phosphate dehydrogenase and 8-iso-prostaglandin F2α as potential predictors of delayed cerebral ischemia after aneurysmal subarachnoid hemorrhage. *J. Neurosurg.*, *1*, aop, 1–10, 2023.
75. Wang, W. and Kang, P.M., Oxidative stress and antioxidant treatments in cardiovascular diseases. *Antioxidants*, *9*, 12, 1292, 2020.
76. Singh, S., Herbal approach for management of atherosclerosis: A review. *Curr. Atheroscler. Rep.*, 21, 1–14, 2019.
77. Cheng, X.M., Hu, Y.Y., Yang, T., Wu, N., Wang, X.N., Reactive oxygen species and oxidative stress in vascular-related diseases. *Oxid. Med. Cell. Longev.*, *2022*, 7906091, 2022.

78. Karan, A., Bhakkiyalakshmi, E., Jayasuriya, R., Sarada, D.V.L., Ramkumar, K.M., The pivotal role of nuclear factor erythroid 2-related factor 2 in diabetes-induced endothelial dysfunction. *Pharmacol. Res.*, 153, 104601, 2020.
79. Sanda, G.M., Stancu, C.S., Deleanu, M., Toma, L., Niculescu, L.S., Sima, A.V., Aggregated LDL turn human macrophages into foam cells and induce mitochondrial dysfunction without triggering oxidative or endoplasmic reticulum stress. *PLoS One*, 16, 1, e0245797, 2021.
80. Matsushima, S. and Sadoshima, J., Yin and yang of NADPH oxidases in myocardial ischemia-reperfusion. *Antioxidants*, 11, 6, 1069, 2022.
81. Daiber, A. and Chlopicki, S., Revisiting pharmacology of oxidative stress and endothelial dysfunction in cardiovascular disease: Evidence for redox-based therapies. *Free Radical Biol. Med.*, 157, 15–37, 2020.
82. Shaafi, S., Hadisi, F., Mahmoudinezhad, M., Razmi, H., Nejadghaderi, S.A., Khalili, M., The significance of the oxidative stress markers in the one-year prognosis of patients with acute ischemic stroke: a case-control study. *BMC Neurol.*, 21, 1, 1–8, 2021.
83. Forman, H.J. and Zhang, H., Targeting oxidative stress in disease: Promise and limitations of antioxidant therapy. *Nat. Rev. Drug Discov.*, 20, 9, 689–709, 2021.
84. Sagris, M., Vardas, E.P., Theofilis, P., Antonopoulos, A.S., Oikonomou, E., Tousoulis, D., Atrial fibrillation: pathogenesis, predisposing factors, and genetics. *Int. J. Mol. Sci.*, 23, 1, 6, 2021.
85. Ismaeel, A., Brumberg, R.S., Kirk, J.S., Papoutsi, E., Farmer, P.J., Bohannon, W.T., Smith, R.S., Eidson, J.L., Sawicki, I., Koutakis, P., Oxidative stress and arterial dysfunction in peripheral artery disease. *Antioxidants*, 7, 10, 145, 2018.
86. Nagy, N., Tóth, N., Nánási, P.P., Antiarrhythmic and Inotropic Effects of Selective Na+/Ca2+ Exchanger Inhibition: What Can We Learn from the Pharmacological Studies? *Int. J. Mol. Sci.*, 23, 23, 14651, 2022.
87. Xiao, R., Zhao, H.C., Yan, T.T., Zhang, Q., Huang, Y.S., Angiotensin II and hypoxia induce autophagy in cardiomyocytes via activating specific protein kinase C subtypes. *Cardiovasc. Diagn. Ther.*, 11, 3, 744, 2021.
88. Georgieva, E., Karamalakova, Y., Miteva, R., Abrashev, H., Nikolova, G., Oxidative stress and cocaine intoxication as start points in the pathology of cocaine-induced cardiotoxicity. *Toxics*, 9, 12, 317, 2021.
89. El-Mahdy, M.A., Abdelghany, T.M., Hemann, C., Ewees, M.G., Mahgoup, E.M., Eid, M.S., Shalaan, M.T., Alzarie, Y.A., Zweier, J.L., Chronic cigarette smoke exposure triggers a vicious cycle of leukocyte and endothelial-mediated oxidant stress that results in vascular dysfunction. *Am. J. Physiol. Heart Circ. Physiol.*, 319, 7, H51–H65, 2020.
90. Aekthammarat, D., Tangsucharit, P., Pannangpetch, P., Sriwantana, T., Sibmooh, N., Moringaoleifera leaf extract enhances endothelial nitric oxide production leading to relaxation of resistance artery and lowering of arterial blood pressure. *Biomed. Pharmacother.*, 130, 110605, 2020.
91. Tsikas, D., Bollenbach, A., Hanff, E., Kayacelebi, A.A., Asymmetric dimethylarginine (ADMA), symmetric dimethylarginine (SDMA) and homoarginine (hArg): the ADMA, SDMA and hArg paradoxes. *Cardiovasc. Diabetol.*, 17, 1–4, 2018.
92. Pereira, B.P., do Vale, G.T., Ceron, C.S., The role of nitric oxide in renovascular hypertension: from the pathophysiology to the treatment. *Naunyn-Schmiedeberg's Arch. Pharmacol.*, 395, 2, 121–131, 2022.
93. Shahbazi, R., Baradaran, B., Khordadmehr, M., Safaei, S., Baghbanzadeh, A., Jigari, F., Ezzati, H., Targeting ROCK signaling in health, malignant and non-malignant diseases. *Immunol. Lett.*, 219, 15–26, 2020.
94. Asgary, S., Rastqar, A., Keshvari, M., Functional food and cardiovascular disease prevention and treatment: a review. *J. Am. Coll. Nutr.*, 37, 5, 429–455, 2018.

95. Yang, C., Shang, K., Lin, C., Wang, C., Shi, X., Wang, H., Li, H., Processing technologies, phytochemical constituents, and biological activities of grape seed oil (GSO): A review. *Trends Food Sci. Technol.*, 116, 1074–1083, 2021.
96. Nogoy, K.M.C., Sun, B., Shin, S., Lee, Y., Li, X.Z., Choi, S.H., Park, S., Fatty acid composition of grain-and grass-fed beef and their nutritional value and health implication. *Food Sci. Anim. Resour.*, 42, 1, 18, 2022.
97. Lorenzon dos Santos, J., Schaan de Quadros, A., Weschenfelder, C., Bueno Garofallo, S., Marcadenti, A., Oxidative stress biomarkers, nut-related antioxidants, and cardiovascular disease. *Nutrients*, 12, 3, 682, 2020.
98. Abdel-alim, M.E.A., Moussa, H.M., El-Saied, F.A.E., Abd-Allah, M.A.N.A., Salim, N.S.S., The Protective Potency of *Medemia argun* (An Egyptian Palm) Against Oxidative Stress, and Tissue Injury Induced by γ-Radiation in Rats. *Res. Sq.*, 12, 5, 25, 2022.
99. Navajas-Porras, B., Pérez-Burillo, S., Valverde-Moya, Á.J., Hinojosa-Nogueira, D., Pastoriza, S., Rufián-Henares, J.Á., Effect of cooking methods on the antioxidant capacity of plant foods submitted to *in vitro* digestion–fermentation. *Antioxidants*, 9, 12, 1312, 2020.
100. Azadedel, S., Hanachi, P., Saboora, A., Antioxidant Activity and Phenolic Compounds Profile of Pistachio Skins (Pistaciavera L., cultivars Kallehghuchi and Ohadi). *Hormozgan Med. J.*, 25, 4, 180–186, 2021.
101. Sheng, J., Yang, X., Chen, J., Peng, T., Yin, X., Liu, W., Yang, X., Antioxidative effects and mechanism study of bioactive peptides from defatted walnut (Juglansregia L.) meal hydrolysate. *J. Agric. Food. Chem.*, 67, 12, 3305–3312, 2019.
102. da Silva, A., Silveira, B.K.S., de Freitas, B.V.M., Hermsdorff, H.H.M., Bressan, J., Effects of regular Brazil nut (Bertholletiaexcelsa HBK) consumption on health: A systematic review of clinical trials. *Foods*, 11, 18, 2925, 2022.
103. Kay, C.D., Gebauer, S.K., West, S.G., Kris-Etherton, P.M., Pistachios increase serum antioxidants and lower serum oxidized-LDL in hypercholesterolemic adults. *J. Nutr.*, 140, 6, 1093–1098, 2010.
104. Carlström, M., Nitric oxide signalling in kidney regulation and cardiometabolic health. *Nat. Rev. Nephrol.*, 17, 9, 575–590, 2021.
105. Carrizzo, A., Ambrosio, M., Damato, A., Madonna, M., Storto, M., Capocci, L., Vecchione, C., Morus alba extract modulates blood pressure homeostasis through eNOSsignaling. *Mol. Nutr. Food Res.*, 60, 10, 2304–2311, 2016.
106. Zhou, D.D., Luo, M., Shang, A., Mao, Q.Q., Li, B.Y., Gan, R.Y., Li, H.B., Antioxidant food components for the prevention and treatment of cardiovascular diseases: effects, mechanisms, and clinical studies. *Oxid. Med. Cell. Longev.*, 2021, 1–17, 2021.
107. Wang, C., Cheng, W., Bai, S., Ye, L.I., Du, J., Zhong, M., Shen, B., White mulberry fruit polysaccharides enhance endothelial nitric oxide production to relax arteries *in vitro* and reduce blood pressure *in vivo*. *Biomed. Pharmacother.*, 116, 109022, 2019.
108. Setiawan, N.A., Shintawati, R., Priyandoko, D., The role of red dragon fruit peel (Hylocereuspolyrhizus) to improvement blood lipid levels of hyperlipidemia male mice. *J. Phys. Conf. Ser.*, 1013, 1, 012167, 2018.
109. Putri, M.D., Wiboworini, B., Dirgahayu, P., Red dragon fruit juice in reducing ros levels and insulin resistance In rats with type 2 diabetes mellitus model. *J. GiziIndones (Indones J. Nutr.)*, 10, 1, 6–14, 2021.
110. Meitha, K., Pramesti, Y., Suhandono, S., Reactive oxygen species and antioxidants in postharvest vegetables and fruits. *Int. J. Food Sci.*, 2020, 1–11, 2020.
111. Zhang, H. and Zhai, C.K., Effects of Chinese and North American wild rice on blood lipids, oxidative stress, and inflammation factors in hyperlipidemic rats. *Cereal Chem.*, 93, 4, 357–363, 2016.

112. Gong, X., Sui, L., Morton, J., Brennan, M.A., Brennan, C.S., Investigation of nutritional and functional effects of rice bran protein hydrolysates by using Preferred Reporting Items for Systematic Reviews and Meta-Analysis (PRISMA) guidelines: A review. *Trends Food Sci. Technol.*, 110, 798–811, 2021.
113. López-Moreno, M., Jiménez-Moreno, E., MárquezGallego, A., Vera Pasamontes, G., UrangaOcio, J.A., Garcés-Rimón, M., Miguel-Castro, M., Red Quinoa Hydrolysates with Antioxidant Properties Improve Cardiovascular Health in Spontaneously Hypertensive Rats. *Antioxidants*, 12, 6, 1291, 2023.
114. Akbari, B., Baghaei-Yazdi, N., Bahmaie, M., MahdaviAbhari, F., The role of plant-derived natural antioxidants in reduction of oxidative stress. *BioFactors*, 48, 3, 611–633, 2022.
115. AlTaweel, A.M., Raish, M., Perveen, S., Fawzy, G.A., Ahmad, A., Ansari, M.A., Ganaie, M.A., Nepetadeflersiana attenuates isoproterenol-induced myocardial injuries in rats: Possible involvement of oxidative stress, apoptosis, inflammation through nuclear factor (NF)-κB downregulation. *Phytomedicine*, 34, 67–75, 2017.
116. Olsen, T., Vinknes, K.J., Svingen, G.F., Pedersen, E.R., Dhar, I., Tell, G.S., Nygård, O.K., The risk association of plasma total homocysteine with acute myocardial infarction is modified by serum vitamin A. *Eur. J. Prev. Cardiol.*, 25, 15, 1612–1620, 2018.
117. Kanwugu, O.N., Glukhareva, T.V., Danilova, I.G., Kovaleva, E.G., Natural antioxidants in diabetes treatment and management: prospects of astaxanthin. *Crit. Rev. Food Sci. Nutr.*, 62, 18, 5005–5028, 2022.
118. Boukhenouna, S., Wilson, M.A., Bahmed, K., Kosmider, B., Reactive oxygen species in chronic obstructive pulmonary disease. *Oxid. Med. Cell. Longev.*, 2018, 1–9, 2018.
119. Luo, M., Zhou, L., Huang, Z., Li, B., Nice, E.C., Xu, J., Huang, C., Antioxidant therapy in cancer: rationale and progress. *Antioxidants*, 11, 6, 1128, 2022.
120. Karimkhani, H., Özkoç, M., Shojaolsadati, P., Uzuner, K., Donmez, D.B., Kanbak, G., protective effect of boric acid and omega-3 on myocardial infarction in an experimental rat model. *Biol. Trace Elem. Res.*, 199, 7, 2612–2620, 2021.
121. Parmenter, B.H., Dalgaard, F., Murray, K., Marquis-Gravel, G., Cassidy, A., Bondonno, C.P., Bondonno, N.P., Intake of dietary flavonoids and incidence of ischemic heart disease in the Danish Diet, Cancer, and Health cohort. *Eur. J. Clin. Nutr.*, 77, 2, 270–277, 2023.
122. Sivandzade, F., Prasad, S., Bhalerao, A., Cucullo, L., NRF2 and NF-κB interplay in cerebrovascular and neurodegenerative disorders: molecular mechanisms and possible therapeutic approaches. *Redox Biol.*, 21, 101059, 2019.
123. Erkens, R., Suvorava, T., Sutton, T.R., Fernandez, B.O., Mikus-Lelinska, M., Barbarino, F., Cortese-Krott, M.M., Nrf2 deficiency unmasks the significance of nitric oxide synthase activity for cardioprotection. *Oxid. Med. Cell. Longev.*, 2018, 1–15, 2018.
124. Faria, A. and Persaud, S.J., Cardiac oxidative stress in diabetes: mechanisms and therapeutic potential. *Pharmacol. Ther.*, 172, 50–62, 2017.
125. Østergaard, J.A., Cooper, M.E., Jandeleit-Dahm, K.A.M., Targeting oxidative stress and antioxidant defence in diabetic kidney disease. *J. Nephrol.*, 33, 5, 917–929, 2020.
126. Farah, C., Michel, L.Y., Balligand, J.L., Nitric oxide signalling in cardiovascular health and disease. *Nat. Rev. Cardiol.*, 15, 5, 292–316, 2018.
127. Paulo, M., Costa, D.E., Bonaventura, D., Lunardi, C.N., Bendhack, L.M., Nitric oxide donors as potential drugs for the treatment of vascular diseases due to endothelium dysfunction. *Curr. Pharm. Des.*, 26, 30, 3748–3759, 2020.
128. Barkhidarian, B., Khorshidi, M., Shab-Bidar, S., Hashemi, B., Effects of L-citrulline supplementation on blood pressure: A systematic review and meta-analysis. *Avicenna J. Phytomed.*, 9, 1, 10, 2019.

5

Antioxidant Therapy: A Promising Avenue for Regulating Inflammation in Psoriasis

Seema Joshi[1] and Dharmesh Trivedi[2]*

[1]Mahatma Gandhi College Banswara, Rajasthan, India
[2]Mahatma Gandhi School of Pharmacy, Rajasthan, India

Abstract
Psoriasis is a persistent skin condition that is defined by an abnormal immune system reaction and an increase in reactive oxygen species (ROS). Psoriasis is also known as eczema. It has been demonstrated that antioxidant therapy is an attractive treatment option for psoriasis; however, additional study is required to substantiate these findings. These antioxidants can reduce the influence that reactive oxygen species (ROS) have on the skin, where can be harmed by ROS, and so regulate inflammation and preserve redox equilibrium in psoriatic skin. Antioxidants have been shown to be successful in reducing inflammation through a variety of distinct processes. Several mechanisms include the suppression of pro-inflammatory cytokines, the controlling of immune cell function, and the alteration of intracellular means of communication that are actively engaged in inflammation. Antioxidants have been found to be useful in reducing inflammation through several pathways. They may also increase the efficacy of standard treatments for psoriasis such as corticosteroids applied topically and phototherapy by reducing the level of oxidative stress and increasing treatment response. However, additional research is required to discover the most effective ways to deliver, dose, including combine antioxidants, in addition to the safety profile of antioxidant therapy and the effects that it has over the long term.

***Keywords*:** Antioxidant therapy, inflammation regulation, psoriasis, reactive oxygen species (ROS), redox homeostasis

5.1 Introduction

A. Background on Psoriasis and its Link to Inflammation:
Psoriasis is a skin condition characterized by red patches on the skin layers that have silver plaques covering them [1]. A dermatologist makes the diagnosis of this disorder. The skin of a patient with psoriasis can be used to diagnose the disorder. There is inflammation as a result of this ailment, which lasts for a long period of time [2]. Continuous inflammation, a defining feature of psoriasis, is largely brought on by immune system cells, notably T-cells. The immune system's T-cells are the ones that are most obviously connected to the

**Corresponding author*: dharmeshtrivedi1983@gmail.com

Rakesh K. Sindhu, Inderbir Singh and M. Arockia Babu (eds.) Antioxidants: Nature's Defense Against Disease, (153–168)

illness [3]. The most prevalent variety of T-cell lymphoma is T-cell lymphoma. One of the disease's most crucial components and a possibly significant contribution to disease development is inflammation [4]. Inflammation of the dermis, a symptom of psoriasis, causes an increase in reactive oxygen substances (ROS), particularly oxidant radicals. The increase in ROS levels is brought on by this skin irritation [5]. A small percentage of the world's population—about one percent—is affected with psoriasis. Both the activation of inflammatory pathways and damage to skin cells are examples of oxidative stress consequences, and both of these events increase the inflammatory response [6]. A significant factor in the intensity of the inflammatory response is oxidative stress. Oxidative stress can also lead to inflammation [7]. Another element that contributes to the reaction's overall intensity is oxidative stress. Both of these influencing factors are impacted by it. In order to create effective treatments, a thorough understanding of the causal relationship between psoriasis and inflammation is necessary [8]. A skin condition that causes chronic inflammation, psoriasis. By using medications that reduce the body's levels of oxidative stress and inflammatory processes, such as antioxidant medical treatment, the treatment for psoriasis may be made more successful [9].

B. The Role of Oxidative Stress in Psoriasis Pathogenesis:

Since oxidative stress has been shown to have a substantial impact on the condition psoriasis, it is virtually always considered that psoriasis results from this phenomenon [10]. Skin flaking and redness are two characteristics of the skin disorder psoriasis [11]. It has been proposed that the occurrence can be explained by a mismatch between the production of reactive oxygen species, or ROS, and the organism's ability to counteract their effects due to the existence of antioxidants in the system [12]. The riddle surrounding the phenomenon may have an answer in this hypothesis. If this is the case, the logic would suggest that the organism's capacity to defend itself against the potentially harmful consequences of ROS is lacking. This discrepancy can be partially explained by the idea that reactive oxygen species (ROS) are produced more frequently than antioxidants [13]. The opening of the channels of communication in antibody-producing cells can be linked to both the beginning of inflammatory reactions and the tissue degradation that follows these events [14]. When antibodies are created, the connecting pathways in the cells that make antibodies are stimulated. This is because inflammatory responses are brought on by the activation of neural networks within immune cells [15]. This imbalance has a significant role in the chronic nature of the condition by prolonging the cycle of inflammatory reactions that are characteristic of psoriasis. Red, flaky skin that has swollen and is coated in silvery scales is a hallmark of psoriasis [16]. The collection of thick, silvery scales that have been covered in silvery scales that have been covered in more silvery scales sets psoriasis apart from other skin conditions [17]. Patches of inflammatory skin that are covered in silvery scales are the hallmark of the skin condition psoriasis [18]. The more that is understood about this specific function, the more likely it is that medical therapies, including those focused on antioxidants for the production of pharmaceuticals, would advance faster as a direct result of this understanding [19]. This is because more information is gathered about this function and its characteristics the more data there is about this characteristic [20]. This explains why it is so as part of ongoing research, investigations assessing the effectiveness of medicines of this kind are currently being conducted [21]. Due to the close connection between psoriasis and the effects of oxidative stress, greater research into the cause-and-effect relationship between these two disorders may eventually lead to the development of more effective treatments for the condition [22].

C. Introduction to Antioxidant Therapy:
Using antioxidants to restore the body's natural state of balance among reactive oxygen species (ROS), antioxidant treatment works to counteract the consequences of oxidative stress [23]. Antioxidant levels are replenished in order to do this. The secret to successfully reaching this objective is the use of antioxidants [24]. Due to the fact that it contributes to the decrease, this helps to lessen the adverse effects of oxidative stress that could occur [25]. It can be used to treat a wide variety of ailments and disorders, including cancer, neurological problems, and cardiovascular problems [26]. Common antioxidants include the vitamins C and E, beta-carotene, selenium, flavonoids, and compounds that include flavonoids. Another illustration is vitamin A. Flavonoids provide as yet another evidence of this. Contrarily, the effectiveness of antioxidant therapy may vary from patient to patient depending on the specific diseases being treated as well as the antioxidants used during the course of the treatment [27]. It is required to perform additional research in order to better understand the amount, duration, and antioxidant blend that is most effective for treating a variety of medical issues [28]. It is commonly accepted that oxidative stress, which leads to stress, is the main cause of psoriasis. A vicious loop is created when oxidative stress results in inflammation and destroys tissues [29]. It would be considerably easier for us to put therapeutic techniques into effect, leading to advancements in the production of novel pharmaceuticals, if we could comprehend the function of this component [30].

5.2 Pathogenesis of Psoriasis

A. Overview of Psoriasis Pathophysiology
An excessive immune response, which results in psoriasis, is brought on by a concoction of genetic predispositions, environmental stimuli, and poorly regulated physiological parameters [31]. This causes the immune system to overreact, which in turn causes the immune system to overreact. White blood cells called T cells have a particularly important function in the development of the illness [32]. The already present inflammation is greatly worsened by the production of several cytokines, such as TNF-alpha and IL-17, which is unusually high. One must first and foremost acquire a grasp of the aspects in question before even beginning the process of generating personalized medicines targeted at the causes of the disease at its inception [33]. The primary goals of the currently available treatments are to control immune response, reduce inflammation, and delay the proliferation of skin cells [34].

B. The Immune System and Inflammatory Process in Psoriasis
Psoriasis is a skin condition caused by an aberrant immune response, which then triggers an overactive inflammatory response [35]. An aberrant immune response is the root cause of psoriasis. Numerous immune cells, including T cells, dendritic cells, and macrophages, to mention a few, are responsible for this response [36]. The cytokines and chemokines that cause inflammation are released by these immune cells. The scaly, red patches that develop

on the skin are caused by increased angiogenesis, which is connected to the inflammatory system [37]. Thanks to advancements made possible by a better understanding of how the immune system functions, psoriasis patients can now take advantage of more individualized therapy methods than ever before. By utilizing these techniques, patients are better able to control their symptoms and experience an overall increase in their wellbeing [38].

C. Link between Oxidative Stress and Inflammation in Psoriasis
An unusually robust immune response that results in the generation of inflammatory compounds and oxidative stress is a hallmark of the skin disorder psoriasis [39]. Psoriasis is an inflammatory skin condition that can be brought on by a variety of factors, including the environment, hereditary factors, and others. The activation of the transcription factor NF-B, which is involved in the control of immune system responses, can be connected to the beginning of this particular inflammatory process [40]. Controlling immune system responses involves NF-B. The production of pro-inflammatory cytokines like TNF-alpha, IL-6, and IL-1 causes this issue to worsen and become much more difficult to treat. These cytokines also make it more difficult to diagnose the disease [41]. The symptoms of the sickness are also influenced by these cytokines. Additionally, it was demonstrated that oxidative damage could have an impact on the skin's protective layer, possibly setting off an automatic cycle [42]. There is evidence to support this. An antioxidant such as the vitamins C and E, N-acetylcysteine, and selenium can help reduce the signs and symptoms of psoriasis, according to a recent study. Additional antioxidants include selenium and N-acetylcysteine [43].

5.3 Understanding Antioxidants

A. Definition and Mechanism of Action
In order to prevent free radicals from damaging cellular components, antioxidants are crucial molecules [44]. They are especially important for treating psoriasis, a dermatological condition whose development is known to be impacted by oxidative stress [45]. Psoriasis can lead to the development of red, scaly skin areas. Antioxidants prevent the destructive effects of free radicals by preventing the instability of their unpaired electrons and by activating the body's natural antioxidant defense mechanisms [46]. They also aid in symptom relief and a reduction in inflammation. Reservatrol, green tea extract, and selenium are additional compounds that might be utilized as medications [47]. Take vitamins C and E as examples of well-known antioxidants. More research will be needed to determine the dosage and composition that will offer the best benefits [48].

B. Types of Antioxidants and their Sources
Psoriasis is treated with a variety of antioxidants, including selenium, vitamin E, vitamin C, and resveratrol, among others [49]. Vitamin E protects the skin from the potentially detrimental effects of oxidative stress and inflammation, while vitamin C helps the body produce collagen naturally and is essential to the process of skin regeneration [50]. Resveratrol, an antioxidant, has been discovered to have anti-inflammatory properties. Grapes and red wine are both great sources of this antioxidant. According to certain studies, the antioxidant

vitamin selenium may have medical use [51]. Before making any dietary changes or starting to take any kind of dietary supplement, you should always get the advice of a licensed medical professional [52].

C. Antioxidant Enzymes and their Role in Controlling Oxidative Stress
Antioxidant enzymes are essential for the body's defense against oxidative stress because they reduce the potentially harmful effects of free radicals [53]. An vital enzyme called superoxide dismutase (SOD) is in charge of transforming superoxide radicals into compounds that have less harmful effects [54]. Catalase is an enzyme found in the liver. Both the detoxification process and the process of neutralizing dangerous compounds depend on it, and it is involved in both [55]. Peroxides are neutralized by the enzyme glutathione peroxidase, also referred to as GPx, to either prevent peroxide damage from developing or repair damage that has already occurred [56]. Antioxidant enzyme synthesis is innate to the human body, but it may decrease with advancing age or due to conditions like chronic illness or inflammation [57]. Cell damage can be avoided with the aid of enzymes that combat free radicals, unstable molecules that can damage DNA. Adopting a healthy lifestyle that includes regular physical activity, efficient stress management, and a diet rich in fruits, vegetables, and whole grains can improve the ability of the antioxidant enzyme system. You can accomplish this by eating a diet rich in fruits, vegetables, and whole grains [58].

5.4 The Potential of Antioxidant Therapy in Psoriasis

A. Review of Preclinical Studies and Animal Models
An autoimmune skin condition called psoriasis causes inflammation, erythema, and desquamating lesions on the skin's surface [59]. The presence of these symptoms makes it a condition. The effectiveness of antioxidant therapy as a psoriasis treatment has been investigated using both preclinical trials and studies on animal models [60]. In preclinical research, cultured cells are frequently used as a model for modeling disorders like psoriasis. These studies are being done in order to find out whether or not antioxidant compounds are helpful in lowering oxidative stress and inflammation [61]. In order to assess the therapeutic efficacy of antioxidants within the context of a physiological framework that has various dimensions, animal models, such as mice and rats, are used [62]. This is carried out to ascertain the efficacy of antioxidants in the prevention or treatment of disease. These investigations advance our knowledge of the molecular mechanisms underpinning antioxidant therapy and aid in the process of creating potent remedies for this challenging dermatological skin condition [63].

B. Antioxidant Supplements and their Effects on Psoriasis Symptoms
There is now no treatment available for the common skin condition known as psoriasis, which is characterized by inflammation. Approximately 7.5 million Americans suffer with psoriasis [64]. In order to ascertain whether or not they have the potential to be utilized as an alternative or supplemental treatment for the symptoms of psoriasis, antioxidants such as vitamins A, C, and E, as well as minerals like selenium and zinc, are all the focus of current

research [65]. The data supporting the use of antioxidant supplements as a treatment for psoriasis symptoms is scant and conflicting, despite some research suggesting they may help treat psoriasis symptoms like erythema and desquamation [66]. It is possible that various people may find varying degrees of success when taking antioxidant supplements, and it is also possible that doing so may have negative effects if you take too many at once [67]. More study is required so that we can formulate consistent suggestions and have a better understanding of how useful they are. Patients with psoriasis should consult their primary care physicians or other experienced medical specialists before deciding whether or not to include antioxidant supplements as part of their treatment approach for the condition [68].

C. Role of Diet and Lifestyle in Antioxidant Therapy

A sizable portion of the global population is afflicted by psoriasis, a skin condition marked by ongoing inflammation [69]. Dermatological disorder psoriasis can be identified with a skin biopsy. Recent studies have found that eating foods high in antioxidants, such as fruits, vegetables, legumes, whole grains, and spices, may help to reduce the body's levels of inflammation and oxidative stress [70]. Another way to improve the efficacy of antioxidant therapy is to have a healthy lifestyle, which includes getting regular exercise, finding healthy ways to cope with stress, abstaining from cigarette use, and not drinking too much alcohol [71]. Avoiding tobacco products and consuming too much alcohol will help you achieve this. These modifications shouldn't be viewed as a substitute for the approved treatments or drugs, though [72]. To effectively treat psoriasis, a customized treatment plan must be created. The use of medicine, as well as dietary and lifestyle changes, should all be included in the treatment strategy [73].

5.5 Clinical Trials and Evidence-Based Findings

A. Overview of Clinical Trials Investigating Antioxidant Therapy for Psoriasis

The symptoms of psoriasis, a common dermatological illness, include erythema, itching, and hypertrophic skin lesions on the skin's surface [74]. It has been researched if antioxidant therapy, which includes the vitamins A, C, D, and E as well as selenium, can be used to treat and control symptoms [75]. Preliminary findings suggest that there might be a positive impact on psoriasis patients' capacity to experience a decrease in inflammation, an improvement in skin lesions, and an increase in their quality of life [76]. However, more clinical studies are required to show the effectiveness of antioxidant therapy. To assess whether or not the preliminary favorable results can be relied upon, well-designed, large-scale clinical trials with a placebo control are essential required [77]. It's likely that additional study will lead to the development of more effective psoriasis medications as well as improvements in general health [78].

B. Evaluation of the Effectiveness and Safety of Antioxidant Interventions

A sizable portion of the global population is afflicted by psoriasis, a skin condition marked by ongoing inflammation [69]. Using a skin biopsy, the dermatological condition psoriasis can be diagnosed. Psoriasis can be treated with antioxidant treatment that combines vitamins A, C, and E, selenium, and coenzyme Q10 [79]. These treatments have been the subject of investigations. These medicines target both oxidative stress and inflammation, which are variables that contribute to the illness' development [80]. The results showed a decrease in the number of skin lesions, relief from pruritus symptoms, and an improvement in the subject's general wellbeing. On the other hand, because a person's response to a particular treatment may vary from person to person, further study is necessary [81].

C. Long-Term Studies and Follow-up Data

A long term study must be conducted in order to better understand the effectiveness and security of antioxidant therapy for psoriasis. Antioxidants like vitamins A, C, E, and selenium have been demonstrated in clinical studies to considerably lessen the severity of psoriasis symptoms when consumed [82]. However, it is essential to conduct ongoing research on the effects that these treatments have over the course of a lifetime in order to ensure that they remain both safe and effective [83]. Longer-term studies that include more thorough follow-ups can assist identify treatments that are likely to be advantageous throughout the course of the disease's progression as well as any potential negative effects [84]. The identification of potential candidates, assessment of the need for long-term supplementation, and inclusion of antioxidants into treatment regimens are all things that these findings may help medical professionals with [85].

5.6 Future Perspectives and Challenges

A. Opportunities for Personalized Antioxidant Therapy

Inflammatory skin disease called psoriasis affects a sizable portion of the population worldwide. The scalp, knees, elbows, and scalp are all common sites for psoriasis [86]. One of the potential effective treatments for the condition is a tailored antioxidant therapy plan. Medical professionals can create customized treatment programs thanks to genetic testing since it tells them exactly how much antioxidant protection a given patient needs [87]. A person's lifestyle factors, such as their food, level of physical activity, and exposure to the environment, can also have an impact on the amount of oxidative stress that they experience [88]. This specialized method can be used to reduce inflammation and treat the symptoms of psoriasis. As research advances, customized antioxidant therapy may drastically change the therapeutic approach, enabling more precise treatment and the ability to offer patients specialized care [89].

B. Potential Side Effects and Interactions of Antioxidant Treatments

Because antioxidant therapy can neutralize the effects of free radicals, which can damage the skin, it is becoming a more and more common alternative for treating psoriasis [90]. Unintended consequences, such as allergic reactions, digestive issues, and possible drug interactions with other prescription medications that may be used concurrently, are a possibility. Some individuals run the risk of developing skin reactions, rashes, or gastrointestinal pain as a result of their exposure [91]. In order to prevent any potentially hazardous drug interactions, it is imperative that patients disclose their current drugs to their healthcare providers [92]. The effectiveness of antioxidant therapy as a psoriasis treatment depends on careful consideration of potential adverse effects, drug interactions, and on-going communication with trained medical specialists [93].

C. Unresolved Questions and Areas for Further Research

The use of antioxidant therapy as an additional management technique for the treatment of psoriasis has been proven to offer promise [94]. However, there are still a lot of questions that remain unsolved, which highlights the critical need for more research and study in this field [95]. Antioxidant therapy is routinely utilized to treat psoriasis, but there are still a lot of questions about how much to take and how long to take it for. Due to the fact that different dosages and treatment periods were used in the initial trials, it is difficult to determine which therapeutic method is the most effective [96]. More research is required to develop techniques that are widely used and that deliver results that are dependable and as excellent as practical [83]. Another area that requires exploration is the mechanism through which antioxidants produce their therapeutic effects. Antioxidants have the ability to counteract the detrimental effects of free radicals and reduce oxidative stress, even though the precise methods by which they lessen the symptoms of psoriasis are not entirely understood [97]. The discovery of these pathways might help in the development of antioxidant treatments that are more targeted and effective. Further research must be done on the safety profile of long-term antioxidant therapy in psoriasis sufferers [98]. Antioxidants are frequently believed to be risk-free, however prolonged usage of them may have unintended consequences. Longitudinal studies that examine potential adverse effects, such as hepatotoxicity or reduced immune response, will be beneficial for the use of these medicines in the long-term therapy of psoriasis [99]. A lot of questions remain unsolved about the ideal dosage, the mechanisms of action, and the long-term safety of antioxidant therapy, despite the fact that it may be useful in the management of psoriasis [100]. To enhance the use of antioxidant therapy and develop more thorough guidelines for its integration into the current psoriasis management protocols, resources should be dedicated to more research initiatives. This is due to the fact that more funding will be given to more research projects [101].

5.7 Conclusion

Antioxidant therapy has shown some hopeful results in the management of psoriasis, especially in reducing levels of oxidative stress and inflammation. Vitamins A, C, and E can reduce levels of free radicals and inflammation, which has been demonstrated in studies to be useful in treating psoriasis symptoms. Antioxidant therapy has been proven to boost effectiveness in reducing symptoms and lengthening remission times when combined with more traditional treatments, such as corticosteroids or medications that act systemically. Additionally, antioxidants can strengthen the skin's barrier, which aids in minimizing water loss. The anti-oxidant therapy is frequently well tolerated and carries a low risk of side effects. To find novel antioxidants, look into combination therapy, assess the long-term efficacy and safety of antioxidant therapy, and develop tools for personalized medicine, more research is necessary. Further studies should focus on patient-specific genetic and environmental factors that influence antioxidant metabolism and response in order to better understand the precise molecular pathways by which antioxidants exhibit their effects in psoriasis. Additionally, mechanistic research required to be done.

References

1. Chuang, S.-Y., Lin, C.-H., Sung, C.T., Fang, J.-Y., Murine models of psoriasis and their usefulness for drug discovery. *Expert Opin. Drug Discov.*, 13, 6, 551–562, 2018, DOI: 10.1080/17460441.2018.1463214.
2. Haliasos, E.C., Kerner, M., Jaimes-Lopez, N., Rudnicka, L., Zalaudek, I., Malvehy, J., Marghoob, A.A., Dermoscopy for the Pediatric Dermatologist Part I: Dermoscopy of Pediatric Infectious and Inflammatory Skin Lesions and Hair Disorders. *Pediatr. Dermatol.*, *30*, 2, 163–171, 2013, doi: 10.1111/pde.12097.
3. Cai, Y., Fleming, C., Yan, J., New insights of T cells in the pathogenesis of psoriasis. *Cell. Mol. Immunol.*, 9, 302–309, 2012, https://doi.org/10.1038/cmi.2012.15.
4. Smedby, K.E., Åkerman, M., Hildebrand, H. *et al.*, Malignant lymphomas in coeliac disease: evidence of increased risks for lymphoma types other than enteropathy-type T cell lymphoma. *Gut*, 54, 54–59, 2005.
5. Bi, D., Qu, F., Xiao, W., Wu, J., Liu, P., Du, H., Xie, Y., Liu, H., Zhang, L., Tao, J., Liu, Y., Zhu, J., Reactive Oxygen Species-responsive Gel-based Microneedle Patches for Prolonged and Intelligent Psoriasis Management. *ACS Nano*, 17, 5, 4346–4357, 2023, DOI: 10.1021/acsnano.2c08979.
6. Tsai, T.-F., Ho, J.-C., Song, M., Szapary, P., Guzzo, C., Shen, Y.-K., Youn, J.-I., Efficacy and safety of ustekinumab for the treatment of moderate-to-severe psoriasis: A phase III, randomized, placebo-controlled trial in Taiwanese and Korean patients (PEARL). *J. Dermatol. Sci.*, *63*, 3, 154–163, 2011, doi: 10.1016/j.jdermsci.2011.05.00.
7. Boesing, F., Patiño, J.S.R., da Silva, V.R.G., Moreira, E.A.M., The interface between obesity and periodontitis with emphasis on oxidative stress and inflammatory response. *Obes. Rev.*, *10*, 3, 290–297, 2009, doi: 10.1111/j.1467-789x.2008.00555.x.
8. Budu-Aggrey, A., Brumpton, B., Tyrrell, J., Watkins, S., Modalsli, E.H., Celis-Morales, C., Paternoster, L., Evidence of a causal relationship between body mass index and psoriasis: A mendelian randomization study. *PLoS Med.*, *16*, 1, e1002739, 2019, doi: 10.1371/journal.pmed.1002739 .
9. Wakkee, M., Thio, H.B., Prens, E.P., Sijbrands, E.J.G., Neumann, H.A.M., Unfavorable cardiovascular risk profiles in untreated and treated psoriasis patients. *Atherosclerosis*, *190*, 1, 1–9, 2007, doi: 10.1016/j.atherosclerosis.2006.
10. Versini, M., Jeandel, P.-Y., Rosenthal, E., Shoenfeld, Y., Obesity in autoimmune diseases: Not a passive bystander. *Autoimmun. Rev.*, *13*, 9, 981–1000, 2014, doi: 10.1016/j.autrev.2014.07.001.
11. Pan, Y., Chamberlain, A.J., Bailey, M., Chong, A.H., Haskett, M., Kelly, J.W., Dermatoscopy aids in the diagnosis of the solitary red scaly patch or plaque–features distinguishing superficial basal cell carcinoma, intraepidermal carcinoma, and psoriasis. *J. Am. Acad. Dermatol.*, *59*, 2, 268–274, 2008, doi: 10.1016/j.jaad.2008.05.013.
12. Metcalfe, N.B. and Alonso-Alvarez, C., Oxidative stress as a life-history constraint: the role of reactive oxygen species in shaping phenotypes from conception to death. *Funct. Ecol.*, *24*, 5, 984–996, 2010, doi: 10.1111/j.1365-2435.2010.01750.x.
13. Gilbert, D.T. and Malone, P.S., The correspondence bias. *Psychol. Bull.*, *117*, 1, 21–38, 1995, doi: 10.1037/0033-2909.117.1.21.

14. Germain, R.N., Robey, E.A., Cahalan, M.D., A Decade of Imaging Cellular Motility and Interaction Dynamics in the Immune System. *Science, 336*, 6089, 1676–1681, 2012, doi: 10.1126/science.1221063.
15. Chatenoud, L. and Bluestone, J., CD3-specific antibodies: a portal to the treatment of autoimmunity. *Nat. Rev. Immunol.*, 7, 622–632, 2007, https://doi.org/10.1038/nri2134.
16. Shi, Z., Tan, G., Cao, C., Han, Y., Meng, Z., Man, X., Wang, L., Decrease of galectin-3 in keratinocytes: A potential diagnostic marker and a critical contributor to the pathogenesis of psoriasis. *J. Autoimmun.*, 89, 30–40, 2018, doi: 10.1016/j.jaut.2017.11.002.
17. Chiriac, A., Dermatologic Diseases of the Breast and Nipple, in: *Practical Atlas of Breast Pathology*, pp. 491–507, 2018, doi:10.1007/978-3-319-93257-6_28.
18. Patel-Cohen, M., Garg, A., Said, J.T., Merola, J.F., Evaluation of the Integumentary and Musculokeletal Systems: An Approach to the Interdisciplinary Examination for Overlap Diseases, in: *Interdisciplinary Approaches to Overlap Disorders in Dermatology & Rheumatology*, A. Garg, J.F. Merola, L. Fitzpatrick (Eds.), Springer, Cham, 2022, https://doi.org/10.1007/978-3-319-18446-3_1.
19. León, R., Garcia, A.G., Marco-Contelles, J., Recent advances in the multitarget-directed ligands approach for the treatment of Alzheimer's disease. *Med. Res. Rev., 33*, 1, 139–189, 2011, doi: 10.1002/med.20248.
20. Ghiassian, S., Greiner, R., Jin, P., Brown, M.R.G., Using Functional or Structural Magnetic Resonance Images and Personal Characteristic Data to Identify ADHD and Autism. *PLoS One, 11*, 12, e0166934, 2016, doi: 10.1371/journal.pone.0166934.
21. Kendall, P.C., Treating anxiety disorders in children: Results of a randomized clinical trial. *J. Consult. Clin. Psychol., 62*, 1, 100–110, 1994, https://doi.org/10.1037/0022-006X.62.1.100.
22. Miller, A.H., Haroon, E., Raison, C.L., Felger, J.C., Cytokine targets in the brain: Impact on neurotransmitters and neurocircuits. *Depress. Anxiety, 30*, 4, 297–306, 2013, doi: 10.1002/da.22084.
23. Gómez, X., Sanon, S., Zambrano, K. *et al.*, Key points for the development of antioxidant cocktails to prevent cellular stress and damage caused by reactive oxygen species (ROS) during manned space missions. *NPJ Microgravity*, 7, 35, 2021, https://doi.org/10.1038/s41526-021-00162-8.
24. Cernasov, D., The Design and Development of Anti-Aging Formulations, in: *Skin Aging Handbook*, pp. 291–325, 2009, doi:10.1016/b978-0-8155-1584-5.50016-8.
25. Agarwal, A., Aponte-Mellado, A., Premkumar, B.J., Shaman, A., Gupta, S., The effects of oxidative stress on female reproduction: a review. *Reprod. Biol. Endocrinol., 10*, 1, 49, 2012, doi: 10.1186/1477-7827-10-49.
26. Dinarello, C., Simon, A., van der Meer, J., Treating inflammation by blocking interleukin-1 in a broad spectrum of diseases. *Nat. Rev. Drug Discov.*, 11, 633–652, 2012, https://doi.org/10.1038/nrd3800.
27. Rietjens, I.M.C.M., Boersma, M.G., Haan, L., Spenkelink, B., Awad, H.M., Cnubben, N.H.P., Koeman, J.H., The pro-oxidant chemistry of the natural antioxidants vitamin C, vitamin E, carotenoids and flavonoids. *Environ. Toxicol. Pharmacol., 11*, 3–4, 321–333, 2002, doi: 10.1016/s1382-6689(02)00003-0.
28. Singh, C.K., Ndiaye, M.A., Ahmad, N., Resveratrol and cancer: Challenges for clinical translation. *Biochim. Biophys. Acta (BBA), Mol. Basis Dis., 1852*, 6, 1178–1185, 2015, doi: 10.1016/j.bbadis.2014.11.004.
29. Yazici, C., Köse, K., Utaş, S. *et al.*, A novel approach in psoriasis: first usage of known protein oxidation markers to prove oxidative stress. *Arch. Dermatol. Res.*, 308, 207–212, 2016, https://doi.org/10.1007/s00403-016-1624-0.
30. Wishart, D., Emerging applications of metabolomics in drug discovery and precision medicine. *Nat. Rev. Drug Discov.*, 15, 473–484, 2016, https://doi.org/10.1038/nrd.2016.32.

31. Grželj, J. and Sollner Dolenc, M., The role of xenobiotics in triggering psoriasis. *Arch. Toxicol.*, 94, 3959–3982, 2020, https://doi.org/10.1007/s00204-020-02870-8.
32. Gazzinelli, R., Kalantari, P., Fitzgerald, K. *et al.*, Innate sensing of malaria parasites. *Nat. Rev. Immunol.*, 14, 744–757, 2014, https://doi.org/10.1038/nri3742.
33. Chen, J., Hall, S., Vitetta, L., Altered gut microbial metabolites could mediate the effects of risk factors in Covid-19. *Rev. Med. Virol.*, *31*, 5, 1–13, 2021, doi: 10.1002/rmv.2211.
34. Gollnick, H., Current Concepts of the Pathogenesis of Acne. *Drugs*, 63, 1579–1596, 2003, https://doi.org/10.2165/00003495-200363150-00005.
35. Giang, J., Seelen, M.A.J., van Doorn, M.B.A., Rissmann, R., Prens, E.P., Damman, J., Complement Activation in Inflammatory Skin Diseases. *Front. Immunol.*, *9*, 343769, 1–17, 2018, doi: 10.3389/fimmu.2018.00639.
36. Yawalkar, N., Tscharner, G.G., Hunger, R.E., Hassan, A.S., Increased expression of IL-12p70 and IL-23 by multiple dendritic cell and macrophage subsets in plaque psoriasis. *J. Dermatol. Sci.*, *54*, 2, 99–105, 2009, doi: 10.1016/j.jdermsci.2009.01.00.
37. Lew, W., Bowcock, A.M., Krueger, J.G., Psoriasis vulgaris: cutaneous lymphoid tissue supports T-cell activation and "Type 1" inflammatory gene expression. *Trends Immunol.*, *25*, 6, 295–305, 2004, doi: 10.1016/j.it.2004.03.006.
38. Ryan, C., Korman, N.J., Gelfand, J.M., Lim, H.W., Elmets, C.A., Feldman, S.R., Menter, A., Research gaps in psoriasis: Opportunities for future studies. *J. Am. Acad. Dermatol.*, *70*, 1, 146–167, 2014, doi: 10.1016/j.jaad.2013.08.042.
39. Armstrong, A.W., Voyles, S.V., Armstrong, E.J., Fuller, E.N., Rutledge, J.C., Angiogenesis and oxidative stress: Common mechanisms linking psoriasis with atherosclerosis. *J. Dermatol. Sci.*, *63*, 1, 1–9, 2011, doi: 10.1016/j.jdermsci.2011.04.007.
40. Bazzoni, F., Rossato, M., Fabbri, M., Gaudiosi, D., Mirolo, M., Mori, L., Locati, M., Induction and regulatory function of miR-9 in human monocytes and neutrophils exposed to proinflammatory signals. *Proc. Natl. Acad. Sci.*, *106*, 13, 5282–5287, 2009, doi: 10.1073/pnas.0810909106.
41. MacPherson, A., Dinkel, K., Sapolsky, R., Glucocorticoids worsen excitotoxin-induced expression of pro-inflammatory cytokines in hippocampal cultures. *Exp. Neurol.*, *194*, 2, 376–383, 2005, doi: 10.1016/j.expneurol.2005.02.021 .
42. Galea, E., Launay, N., Portero-Otin, M., Ruiz, M., Pamplona, R., Aubourg, P., Pujol, A., Oxidative stress underlying axonal degeneration in adrenoleukodystrophy: A paradigm for multifactorial neurodegenerative diseases? *Biochim. Biophys. Acta (BBA), Mol. Basis Dis.*, *1822*, 9, 1475–1488, 2012, doi: 10.1016/j.bbadis.2012.02.005.
43. Elbarbary, N.S., Ismail, E.A.R., Farahat, R.K., El-Hamamsy, M., ω-3 fatty acids as an adjuvant therapy ameliorates methotrexate-induced hepatotoxicity in children and adolescents with acute lymphoblastic leukemia: A randomized placebo-controlled study. *Nutrition*, *32*, 1, 41–47, 2016, doi: 10.1016/j.nut.2015.06.010.
44. Nimse, S.B. and Pal, D., Free radicals, natural antioxidants, and their reaction mechanisms. *RSC Adv.*, *5*, 35, 27986–28006, 2015, doi: 10.1039/c4ra13315c.
45. Atalay, S., Gęgotek, A., Wroński, A., Domigues, P., Skrzydlewska, E., Therapeutic application of cannabidiol on UVA and UVB irradiated rat skin. A proteomic study. *J. Pharm. Biomed. Anal.*, 192, 113656, 2021, doi: 10.1016/j.jpba.2020.113656.
46. Joshi, M., Hiremath, P., John, J. *et al.*, Modulatory role of vitamins A, B3, C, D, and E on skin health, immunity, microbiome, and diseases. *Pharmacol. Rep.*, 75, 1096–1114, 2023, https://doi.org/10.1007/s43440-023-00520-1.
47. Cicero, A.F.G., Allkanjari, O., Busetto, G.M., Cai, T., Larganà, G., Magri, V., Vitalone, A., Nutraceutical treatment and prevention of benign prostatic hyperplasia and prostate cancer. *Arch. Ital. Urol. Androl.*, *91*, 3, 2019, doi: 10.4081/aiua.2019.3.139.

48. Holst, B. and Williamson, G., Nutrients and phytochemicals: from bioavailability to bioefficacy beyond antioxidants. *Curr. Opin. Biotechnol.*, *19*, 2, 73–82, 2008, doi: 10.1016/j.copbio.2008.03.003.
49. Murzaku, E.C., Bronsnick, T., Rao, B.K., Diet in dermatology. *J. Am. Acad. Dermatol.*, *71*, 6, 1053.e1–1053.e16, 2014, doi: 10.1016/j.jaad.2014.06.016.
50. Godic, A., Poljšak, B., Adamic, M., Dahmane, R., The Role of Antioxidants in Skin Cancer Prevention and Treatment. *Oxid. Med. Cell. Longev.*, *2014*, 1–6, 2014, doi: 10.1155/2014/860479.
51. Abete, I., Goyenechea, E., Zulet, M.A., Martínez, J.A., Obesity and metabolic syndrome: Potential benefit from specific nutritional components. *Nutr. Metab. Cardiovasc. Dis.*, *21*, B1–B15, 2011, doi: 10.1016/j.numecd.2011.05.001.
52. Goldberg, A.C., Hopkins, P.N., Toth, P.P., Ballantyne, C.M., Rader, D.J., Robinson, J.G., Ziajka, P.E., Familial Hypercholesterolemia: Screening, diagnosis and management of pediatric and adult patients. *J. Clin. Lipidol.*, *5*, 3, 133–140, 2011, doi: 10.1016/j.jacl.2011.03.001.
53. Finaud, J., Lac, G., Filaire, E., Oxidative Stress. *Sports Med.*, 36, 327–358, 2006, https://doi.org/10.2165/00007256-200636040-00004.
54. Saxena, P., Selvaraj, K., Khare, S.K. *et al.*, Superoxide dismutase as multipotent therapeutic antioxidant enzyme: Role in human diseases. *Biotechnol. Lett.*, 44, 1–22, 2022, https://doi.org/10.1007/s10529-021-03200-3.
55. Singh, R.P., Padmavathi, B., Rao, A.R., Modulatory influence of Adhatoda vesica (Justicia adhatoda) leaf extract on the enzymes of xenobiotic metabolism, antioxidant status and lipid peroxidation in mice. *Mol. Cell. Biochem.*, 213, 99–109, 2000, https://doi.org/10.1023/A:1007182913931.
56. Lubos, E., Loscalzo, J., Handy, D.E., Glutathione Peroxidase-1 in Health and Disease: From Molecular Mechanisms to Therapeutic Opportunities. *Antioxid. Redox Signaling*, *15*, 7, 1957–1997, 2011, doi: 10.1089/ars.2010.3586.
57. Lambeth, J.D., Nox enzymes, ROS, and chronic disease: An example of antagonistic pleiotropy. *Free Radical Biol. Med.*, *43*, 3, 332–347, 2007, https://doi.org/10.1016/j.freeradbiomed.2007.03.027.
58. Pham-Huy, L.A., He, H., Pham-Huy, C., Free Radicals, Antioxidants in Disease and Health. *Int. J. Biomed. Sci.: IJBS*, *4*, 2, 89–96, 2008, https://www.ncbi.nlm.nih.gov/pmc/articles/PMC3614697/.
59. Zangeneh, F.Z. and Shooshtary, F.S., Psoriasis — Types, Causes and Medication, in: *Psoriasis - Types, Causes and Medication*, 2013, doi:10.5772/54728.
60. Yadav, K., Singh, D., Singh, M.R., Minz, S., Princely Ebenezer Gnanakani, S., Yadav, R., Vora, L., Sahu, K.K., Bagchi, A., Singh Chauhan, N., Pradhan, M., Preclinical study models of psoriasis: State-of-the-art techniques for testing pharmaceutical products in animal and nonanimal models. *Int. Immunopharmacol.*, *117*, 109945, 2023, https://doi.org/10.1016/j.intimp.2023.109945.
61. Desmet, E., Ramadhas, A., Lambert, J., Van Gele, M., *In vitro* psoriasis models with focus on reconstructed skin models as promising tools in psoriasis research. *Exp. Biol. Med.*, 242, 11, 1158–1169, 2017, https://doi.org/10.1177/1535370217710637.
62. Herson, P.S. and Traystman, R.J., Animal models of stroke: translational potential at present and in 2050. *Future Neurol.*, *9*, 5, 541–551, 2014, doi: 10.2217/fnl.14.44.
63. Alzohairy, M.A., Therapeutics Role ofAzadirachta indica (Neem) and Their Active Constituents in Diseases Prevention and Treatment. *Evid. Based Complement. Alternat. Med.*, *2016*, 1–11, 2016, doi: 10.1155/2016/7382506 .
64. Jiang, Q., Natural forms of vitamin E: Metabolism, antioxidant, and anti-inflammatory activities and their role in disease prevention and therapy. *Free Radical Biol. Med.*, *72*, 76–90, 2014, https://doi.org/10.1016/j.freeradbiomed.2014.03.035.
65. Zenz, R., Eferl, R., Kenner, L., Florin, L., Hummerich, L., Mehic, D., Scheuch, H., Angel, P., Tschachler, E., Wagner, E.F., Psoriasis-like skin disease and arthritis caused by inducible

epidermal deletion of Jun proteins. *Nature*, *437*, 7057, 369–375, 2005, https://doi.org/10.1038/nature03963.
66. Tirant, M., Lotti, T., Gianfaldoni, S., Tchernev, G., Wollina, U., Bayer, P., Integrative Dermatology – The Use of Herbals and Nutritional Supplements to Treat Dermatological Conditions. *Open Access Maced. J. Med. Sci.*, *6*, 1, 185–202, 2017, https://doi.org/10.3889/oamjms.2018.041.
67. Merry, T.L. and Ristow, M., Do antioxidant supplements interfere with skeletal muscle adaptation to exercise training? *J. Physiol.*, *594*, 18, 5135–5147, 2016, https://doi.org/10.1113/JP270654.
68. Larsen, M.H., Hagen, K.B., Krogstad, A.L. *et al.*, Shared Decision Making in Psoriasis: A Systematic Review of Quantitative and Qualitative Studies. *Am. J. Clin. Dermatol.*, 20, 13–29, 2019, https://doi.org/10.1007/s40257-018-0390-5.
69. Trembath, R.C., Lee Clough, R., Rosbotham, J.L., Jones, A.B., Camp, R.D., Frodsham, A., Browne, J., Barber, R., Terwilliger, J., Mark Lathrop, G., Barker, J.N., Identification of a Major Susceptibility Locus on Chromosome 6p and Evidence for Further Disease Loci Revealed by a Two Stage Genome-Wide Search in Psoriasis. *Hum. Mol. Genet.*, *6*, 5, 813–820, 1997, https://doi.org/10.1093/hmg/6.5.813.
70. Khan, A.Q., Agha, M.V., Sheikhan, K.S.A., Younis, S.M., Tamimi, M.A., Alam, M., Ahmad, A., Uddin, S., Buddenkotte, J., Steinhoff, M., Targeting deregulated oxidative stress in skin inflammatory diseases: An update on clinical importance. *Biomed. Pharmacother.*, *154*, 113601, 2022, https://doi.org/10.1016/j.biopha.2022.113601.
71. Finlay, J.M., Kler, J.S., Q., B., Eastman, M.R., Vinson, Y.R., Kobayashi, L.C., Coping During the COVID-19 Pandemic: A Qualitative Study of Older Adults Across the United States. *Front. Public Health*, *9*, 643807, 2021, https://doi.org/10.3389/fpubh.2021.643807.
72. Petti, S., Lifestyle risk factors for oral cancer. *Oral Oncol.*, *45*, 4–5, 340–350, 2009, https://doi.org/10.1016/j.oraloncology.2008.05.018.
73. Menting, S.P., Coussens, E., Pouw, M.F. *et al.*, Developing a Therapeutic Range of Adalimumab Serum Concentrations in Management of Psoriasis: A Step Toward Personalized Treatment. *JAMA Dermatol.*, 151, 6, 616–622, 2015, doi: 10.1001/jamadermatol.2014.5479.
74. DeWitt, S., Nursing Assessment of the Skin and Dermatologic Lesions. *Nurs. Clin. North Am.*, *25*, 1, 235–245, 1990, https://doi.org/10.1016/S0029-6465(22)00237-7.
75. Jiang, X., Stockwell, B.R., Conrad, M., Ferroptosis: Mechanisms, biology and role in disease. *Nat. Rev. Mol. Cell Biol.*, *22*, 4, 266–282, 2021, https://doi.org/10.1038/s41580-020-00324-8.
76. Egan, L.J. and Sandborn, W.J., Methotrexate for Inflammatory Bowel Disease: Pharmacology and Preliminary Results. *Mayo Clin. Proc.*, *71*, 1, 69–80, 1995, https://doi.org/10.4065/71.1.69.
77. Miyamoto, S., Miyake, N., Jarskog, L.F., Fleischhacker, W.W., Lieberman, J.A., Pharmacological treatment of schizophrenia: A critical review of the pharmacology and clinical effects of current and future therapeutic agents. *Mol. Psychiatry*, *17*, 12, 1206–1227, 2012, https://doi.org/10.1038/mp.2012.47.
78. Signorovitch, J.E., Wu, E.Q., Yu, A.P. *et al.*, Comparative Effectiveness Without Head-to-Head Trials. *Pharmacoeconomics*, 28, 935–945, 2010, https://doi.org/10.2165/11538370-000000000-00000.
79. Fukuda, K., The Chronic Fatigue Syndrome: A Comprehensive Approach to Its Definition and Study. *Ann. Intern. Med.*, *121*, 12, 953, 1994, doi: 10.7326/0003-4819-121-12-199412150-00009.
80. Ruiz, S., Pergola, P.E., Zager, R.A., Vaziri, N.D., Targeting the transcription factor Nrf2 to ameliorate oxidative stress and inflammation in chronic kidney disease. *Kidney Int.*, *83*, 6, 1029–1041, 2013, https://doi.org/10.1038/ki.2012.439.
81. Grob, J., Auquier, P., Dreyfus, I., Ortonne, P., How to prescribe antihistamines for chronic idiopathic urticaria: Desloratadine daily vs PRN and quality of life. *Allergy*, *64*, 4, 605–612, 2009, https://doi.org/10.1111/j.1398-9995.2008.01913.x.

82. Mascarenhas-Melo, F., Carvalho, A., Gonçalves, M.B.S., Paiva-Santos, A.C., Veiga, F., Nanocarriers for the topical treatment of psoriasis - pathophysiology, conventional treatments, nanotechnology, regulatory and toxicology. *Eur. J. Pharm. Biopharm.*, *176*, 95–107, 2022, https://doi.org/10.1016/j.ejpb.2022.05.012.
83. Kooij, S.J., Bejerot, S., Blackwell, A., Caci, H., Casas-Brugué, M., Carpentier, P.J., Asherson, P., European consensus statement on diagnosis and treatment of adult ADHD: The European Network Adult ADHD. *BMC Psychiatry*, *10*, 1, 2010, doi:10.1186/1471-244x-10-67.
84. Kulik, J.A. and Mahler, H.I.M., Emotional support as a moderator of adjustment and compliance after coronary artery bypass surgery: A longitudinal study. *J. Behav. Med.*, 16, 45–63, 1993, https://doi.org/10.1007/BF00844754.
85. Howatson, G. and van Someren, K.A., The Prevention and Treatment of Exercise-Induced Muscle Damage. *Sports Med.*, 38, 483–503, 2008, https://doi.org/10.2165/00007256-200838060-00004.
86. Kopp, T., Riedl, E., Bangert, C., Bowman, E.P., Greisenegger, E., Horowitz, A., Kittler, H., Blumenschein, W.M., McClanahan, T.K., Marbury, T., Zachariae, C., Xu, D., Hou, X.S., Mehta, A., Zandvliet, A.S., Montgomery, D., Van Aarle, F., Khalilieh, S., Clinical improvement in psoriasis with specific targeting of interleukin-23. *Nature*, *521*, 7551, 222–226, 2015, https://doi.org/10.1038/nature14175.
87. Delanty, N. and Dichter, M.A., Antioxidant Therapy in Neurologic Disease. *Arch. Neurol.*, 57, 9, 1265–1270, 2000, doi: 10.1001/archneur.57.9.1265.
88. Furman, D., Campisi, J., Verdin, E., Targ, S., Franceschi, C., Ferrucci, L., Gilroy, D.W., Fasano, A., Miller, G.W., Miller, A.H., Mantovani, A., Weyand, C.M., Barzilai, N., Goronzy, J.J., Rando, T.A., Effros, R.B., Lucia, A., Kleinstreuer, N., Slavich, G.M., Chronic inflammation in the etiology of disease across the life span. *Nat. Med.*, *25*, 12, 1822–1832, 2019, https://doi.org/10.1038/s41591-019-0675-0.
89. Elmets, C.A., Korman, N.J., Prater, E.F., Wong, E.B., Rupani, R.N., Kivelevitch, D., Armstrong, A.W., Connor, C., Cordoro, K.M., Davis, D.M., Elewski, B.E., Gelfand, J.M., Gordon, K.B., Gottlieb, A.B., Kaplan, D.H., Kavanaugh, A., Kiselica, M., Kroshinsky, D., Lebwohl, M., Menter, A., Joint AAD–NPF Guidelines of care for the management and treatment of psoriasis with topical therapy and alternative medicine modalities for psoriasis severity measures. *J. Am. Acad. Dermatol.*, *84*, 2, 432–470, 2021, https://doi.org/10.1016/j.jaad.2020.07.087.
90. Portugal, M., Barak, V., Ginsburg, I., Kohen, R., Interplay among oxidants, antioxidants, and cytokines in skin disorders: Present status and future considerations. *Biomed. Pharmacother.*, *61*, 7, 412–422, 2007, https://doi.org/10.1016/j.biopha.2007.05.010.
91. Hohl, C.M., Dankoff, J., Colacone, A., Afilalo, M., Polypharmacy, adverse drug-related events, and potential adverse drug interactions in elderly patients presenting to an emergency department. *Ann. Emerg. Med.*, *38*, 6, 666–671, 2001, https://doi.org/10.1067/mem.2001.119456.
92. Wittkowsky, A.K., Dietary supplements, herbs and oral anticoagulants: the nature of the evidence. *J. Thromb Thrombolysis*, 25, 72–77, 2008, https://doi.org/10.1007/s11239-007-0110-0.
93. Loya, A.M., González-Stuart, A., Rivera, J.O., Prevalence of Polypharmacy, Polyherbacy, Nutritional Supplement Use and Potential Product Interactions among Older Adults Living on the United States-Mexico Border. *Drugs Aging*, 26, 423–436, 2009, https://doi.org/10.2165/00002512-200926050-00006.
94. Kumar, S., Prasad, M., Rao, R., Topical delivery of clobetasol propionate loaded nanosponge hydrogel for effective treatment of psoriasis: Formulation, physicochemical characterization, antipsoriatic potential and biochemical estimation. *Mater. Sci. Eng., C*, *119*, 111605, 2021, https://doi.org/10.1016/j.msec.2020.111605.
95. Campbell, N.K., Fitzgerald, H.K., Dunne, A., Regulation of inflammation by the antioxidant haem oxygenase 1. *Nat. Rev. Immunol.*, *21*, 7, 411–425, 2021, https://doi.org/10.1038/s41577-020-00491-x.

96. Raylu, N. and Oei, T.P., Pathological gambling: A comprehensive review. *Clin. Psychol. Rev.*, *22*, 7, 1009–1061, 2002, https://doi.org/10.1016/S0272-7358(02)00101-0.
97. Rushworth, G.F. and Megson, I.L., Existing and potential therapeutic uses for N-acetylcysteine: The need for conversion to intracellular glutathione for antioxidant benefits. *Pharmacol. Ther.*, *141*, 2, 150–159, 2014, https://doi.org/10.1016/j.pharmthera.2013.09.006.
98. Gorrini, C., Harris, I.S., Mak, T.W., Modulation of oxidative stress as an anticancer strategy. *Nat. Rev. Drug Discov.*, *12*, 12, 931–947, 2013, https://doi.org/10.1038/nrd4002.
99. Simmchen, J., Ventura, R., Segura, J., Progress in the Removal of Di-[2-Ethylhexyl]-Phthalate as Plasticizer in Blood Bags. *Transfus. Med. Rev.*, *26*, 1, 27–37, 2011, https://doi.org/10.1016/j.tmrv.2011.06.001.
100. Mitsialis, S.A. and Kourembanas, S., Stem cell–based therapies for the newborn lung and brain: Possibilities and challenges. *Semin. Perinatol.*, *40*, 3, 138–151, 2016, https://doi.org/10.1053/j.semperi.2015.12.002.
101. Mohammad, T.F., Al-Jamal, M., Hamzavi, I.H., Harris, J.E., Leone, G., Cabrera, R., Lim, H.W., Pandya, A.G., Esmat, S.M., The Vitiligo Working Group recommendations for narrowband ultraviolet B light phototherapy treatment of vitiligo. *J. Am. Acad. Dermatol.*, *76*, 5, 879–888, 2017, https://doi.org/10.1016/j.jaad.2016.12.041.

6

Antioxidants in Infectious Disease Management

Kenneth C. Ugoeze[1]* and Oluwatoyin A. Odeku[2]

[1]Department of Pharmaceutics & Pharmaceutical Technology, Faculty of Pharmaceutical Sciences, University of Port Harcourt, Port Harcourt, Nigeria

[2]Department of Pharmaceutics & Industrial Pharmacy, Faculty of Pharmacy, University of Ibadan, Ibadan, Nigeria

Abstract

Antioxidants are notable in the scavenging of free radicals and reactive oxygen species (ROS) and are evolving as prophylactic and therapeutic agents. Excessive influence of free radicals/ROS triggers and depletes natural antioxidants leading to oxidative stress and destroying several biological mechanisms which stabilize the immune system. These are linked to the pathogenesis of various disorders such as cancer, cardiovascular disease, premature ageing, etc. alongside diminished immune systems making individuals vulnerable to hosts of infectious diseases. Infectious diseases occur due to the dysfunction of the body's usual homeostatic processes due to the harmful effects of infectious organisms like viruses, bacteria, fungi, parasites, failed immune systems, etc. This chapter considers antioxidants as useful in the management of infectious diseases such as coronavirus disease (COVID-19), Ebola virus disease (EVD), HIV/AIDS, diphtheria, influenza, measles, mumps, rubella, *Haemophilus influenzae* infection, STDs, salmonella, cholera, hepatitis, etc. Consistent consumption of natural diets retaining high-level antioxidants some of which have been found pharmacologically effective as prophylactic and therapeutic agents is crucial to conserve a low level of free radicals and oxidative stress to maintain an elevated immune system and avert the hazards of infections is an approach to manage several infectious diseases. Robust antioxidants are found in resveratrol, *Pinus pinaster, Allium cepa, Allium sativum, Aloe vera, Andrographis paniculata, Azadirachta indica, Camellia sinensis, Curcuma longa, Glycyrrhiza glabra*, etc.

***Keywords*:** Antioxidant, infectious disease, management, free-radical, oxidative stress

6.1 Introduction

Antioxidants contribute effectively to the controlling of infectious disorders through their ability to counteract oxidative stress and enhancing the body defense system which in turn enables the generation of reactive oxygen species (ROS) which fight against microbes. However, extreme level of ROS has the tendencies to impair the tissues thereby creating unwelcome state of the diseases state. Antioxidants such as ascorbic acid, α-tocopherol,

**Corresponding author*: kenneth.ugoeze@uniport.edu.ng; kcugoeze@gmail.com

Rakesh K. Sindhu, Inderbir Singh and M. Arockia Babu (eds.) Antioxidants: Nature's Defense Against Disease, (169–218)

glutathione and numerous polyphenols obtainable from plant-based foods aid to alleviate oxidative impairment. While the antioxidants mop-up free radicals and lessen inflammation, they boost the body's defense processes, possibly enhancing the degree of healing and easing the seriousness of diseases. Their usefulness in the management of several contagious disorders like viral, bacterial and parasitic diseases is investigated.

6.2 Phytochemicals

Phytochemicals are plant-based bioactive substances that are naturally manufactured by plants for their growth and defense [1]. Their presence and level of secretion vary from one plant to the other. They are naturally available in whole grains, fruits, vegetables, nuts, and herbs. Some of the considerable phytochemicals are carotenoids, polyphenols, phenolic constituents, alkaloids, isoprenoids, phytosterols, saponins, dietary fibers, and certain polysaccharides. Phytochemical compounds possess effective antioxidant actions [1, 2] and demonstrate antimicrobial, antidiarrheal, anthelmintic, antiallergic, antispasmodic, and antiviral activities [3–5]. They additionally assist in controlling gene transcription, raise gap junction communication, boost immunity, and offer defense against lung and prostate cancers [3, 6]. They exert an effect on cell differentiation, improve the activity of enzymes that detoxify, affect DNA metabolism, maintenance of DNA repair, increase apoptosis of cancer cells, decrease cell proliferation, etc. [1]. After extraction from suitable sources, phytochemicals find beneficial relevance in the development of functional foods and nutraceuticals [3].

6.3 Antioxidants

Plants produce hundreds of thousands of substances known as phytochemicals and thousands of these function as antioxidants [3]. Antioxidants assist in combating the destruction of body cells due to illness and ageing [7]. They perform their role by stopping oxidation, a type of biological chemical reaction regularly linking oxygen and are capable of harming cells [8]. The molecules that activate oxidation are termed oxidants [9] which are also referred to as free radicals [7] generated by almost all we do that requires oxygen, including breathing [10] and digestion [11].

Free radicals and oxidants perform a two-fold role as both toxic and valuable compounds because they can be either damaging or beneficial to the body. They are formed either from normal cell metabolisms *in situ* or from external sources such as pollution, cigarette smoke, radiation, medication, etc. When an excess of free radicals cannot progressively be damaged, their increase in the body produces an incident termed oxidative stress, a development that is a key factor in the advancement of chronic and degenerative diseases such as cancer, autoimmune disorders, ageing, cataract, rheumatoid arthritis, cardiovascular and neurodegenerative disorders [7]. The human body has several mechanisms to counteract oxidative stress by producing antioxidants, which are either naturally produced *in situ*, or externally supplied through foods and/or supplements [12–14].

Historically, the term antioxidant was originally employed explicitly to refer to a chemical that thwarted the consumption of oxygen [15]. In the late 19th and early 20th century, a widespread investigation was dedicated to the applications of antioxidants in vital industrial procedures, such as the inhibition of metal corrosion, the vulcanization of rubber, and the polymerization of fuels in the fouling of internal combustion engines [16]. Previous inquiries on the role of antioxidants in biology were based on their application in stopping the oxidation of unsaturated fats, the leading cause of rancidity [17]. Antioxidant action could be evaluated obviously by placing the fat in a sealed container with oxygen and determining the measure of oxygen utilization. Nevertheless, it was the recognition of vitamins A, C, and E as antioxidants that reformed the subject and piloted the understanding of the significance of antioxidants in the biochemistry of living organisms [18]. The probable means of the activity of antioxidants were first investigated when it was acknowledged that an element with antioxidative activity is expected to be freely oxidized [19]. A study into how vitamin E thwarts the progression of lipid peroxidation led to the discovery of antioxidants as reducing agents that inhibit oxidative reactions, often by scavenging reactive oxygen species (ROS) before they can harm cells [20]. Therefore antioxidants are molecules stable enough to contribute an electron to an unstable free radical to neutralize it and decrease its damaging effect [19], hence, they slow down or prevent cellular damage principally through their free radical scavenging property as they can securely interact with free radicals and abort the chain reaction before vital molecules are damaged [21]. Some of such antioxidants like glutathione, ubiquinol, and uric acid, are generated through regular metabolism in the body [22]. Other lighter antioxidants are obtained from food. Though various enzyme systems within the body scavenge free radicals, the principal micronutrient (vitamins) antioxidants are vitamin E (α-tocopherol), vitamin C (ascorbic acid), and B-carotene [23]. The body cannot manufacture these micronutrients; hence they must be supplied in the diet.

6.3.1 Classification of Antioxidants

Antioxidants can be classified in several ways as follows:

1. Based on their activity, antioxidants are classified as enzymatic or non-enzymatic [24].

   (a) Enzymatic antioxidants
   In this category, hazardous oxidative outputs can be converted to H_2O_2 and then to water by enzymatic antioxidants that can break down and get rid of free radicals in a multistep process in the presence of cofactors such as copper (Cu), zinc (Zn), manganese (Mn), selenium (Se), and iron (Fe) [19, 24].

   (b) Non-enzymatic antioxidants
   Non-enzymatic antioxidants embrace Vitamin C, vitamin E, plant polyphenols, carotenoids and glutathione. They act by disrupting free radical chain reactions [25].

2. Based on solubility, antioxidants are classified as water-soluble and lipid-soluble antioxidants [26].

   (a) Water-soluble antioxidants, e.g. vitamin C. (b). Lipid-soluble antioxidants.

Largely, water-soluble antioxidants react with oxidants in the cell cytoplasm and the blood plasma, while lipid-soluble antioxidants defend cell membranes from lipid peroxidation. They may be biosynthesized or obtained from the nutritional substances [27].

3. Based on size, antioxidants are categorized as small-molecule and large-molecules [28].

    (a) Small-molecule antioxidants
    Small-molecule antioxidants act by counteracting the ROS in a technique recognized as radicals scavenging and carrying them away. The main examples include Vitamin C, vitamin E, carotenoids, and glutathione (GSH) [29].

    (b) Large-molecule antioxidants
    Large-molecule antioxidants comprise enzymes superoxide dismutase (SOD), catalase (CAT) and glutathione peroxidase (GPx) and sacrificial proteins (albumin) that absorb ROS and block them from attacking other essential proteins [8].

4. Kinetically antioxidants can further be categorized as below:

    (a) Antioxidants that break chains by reacting with peroxyl radicals containing weak O–H or N–H bonds, phenol, naphthol, hydroquinone, aromatic amines, and aminophenols [30].
    (b) Antioxidants possessing the ability to destroy chains through reaction with alkyl radicals like quinines, nitrones, and iminoquinones [30].
    (c) Antioxidants that stop cyclic chains such as aromatic amines, nitroxyl radicals, and variable valence metal compounds [30].
    (d) Hydroperoxide decomposing antioxidants such as sulfide, phosphide, and thiophosphate [30].
    (e) Metal-deactivating antioxidants comprising diamines, hydroxyl acids, and bifunctional compounds [30].
    (f) Synergistic activity of various antioxidants involving phenol sulfide in which the phenolic group reacts with the peroxyl radicals sulfide group with hydroperoxide [30].

5. Antioxidants are also classified based on their existence as natural or synthetic antioxidants [31].

    (a) Natural antioxidants
    The natural antioxidants are classed as chain-breaking antioxidants, which react with radicals and convert them into more stable products. Generally, antioxidants of this group are phenolic in structure and include the following:

i. Antioxidant minerals
These are trace elements such as copper, selenium, manganese, iron, and zinc. They act as co-factors of antioxidant enzymes. Their role is to shield the body from oxygen free radicals (OFR) generated during oxidative stress [32].
ii. Antioxidant vitamins
These are essentially relevant for most body metabolic activities and comprise vitamins C, E and B [33].
iii. Phytochemicals
Phytochemicals are phenolic compound derivatives that are neither vitamins nor minerals and embrace flavonoids, catechins, carotenoids, carotene, lycopene, herbs and spices such as diterpene, rosmariquinone, thyme, nutmeg, clove, black pepper, ginger, garlic, curcumin, and derivatives [34].

(b) Synthetic antioxidants
Synthetic antioxidants are phenolic complexes that are involved in arresting free radicals and blocking the chain reaction. They comprise butylated hydroxyl anisole (BHA), butylated hydroxyltoluene (BHT), propyl gallate (PG), metal chelating agent (EDTA), tertiary butyl hydroquinone (TBHQ) and nordihydroguaiaretic acid (NDGA) [12].

6.3.2 Synthetic vs. Natural Antioxidants

Antioxidants could be synthetic or natural [35]. However, synthetic antioxidants have currently been documented to be unsafe for humans so the quest for non-toxic antioxidants has increased in recent years [36]. Synthetic and natural food antioxidants are employed consistently in foods and medicine, exclusively those embracing oils and fats to safeguard the food against oxidation [31]. Conspicuous examples of synthetic phenolic antioxidants include butylated hydroxytoluene (BHT) and butylated hydroxyanisole (BHA) [37] which have been extensively utilized as antioxidants in food, cosmetics and drug industries [38]. Irrespective of their relevance, some odds have been pointed out with these synthetic antioxidants considering some physical properties of BHT and BHA such as their high volatility and instability at high temperatures, severe legislation on the use of synthetic food essences, carcinogenic nature of some synthetic antioxidants, and consumer preferences [39] and considering these, interest of producers have moved from synthetic to natural antioxidants [40]. Considering the rising risk factors of humans for different lethal diseases, there has been a universal inclination regarding the use of natural substances existing in therapeutic plants and nutritional plants as healing antioxidants [41]. Natural antioxidants can be taken up through diet as they are present in fruits, vegetables and spices [36]. An inverse correlation relating the dietary intake of antioxidant-rich food and medicinal plants and the prevalence of human disorders has been reported [19]. The utilization of natural antioxidants in food, cosmetic, and therapeutic industries would be a favorable substitute for synthetic antioxidants, especially reflecting their low cost, high compatibility with nutritional ingestion and no detrimental effects on the human body [42]. Various antioxidant composites, biologically present in plant resources have been acknowledged as free radicals or active oxygen scavengers [43]. The antioxidant capacities of fruits [44] and various vegetables such as potatoes, spinach, tomatoes, legumes, etc. have been documented [45].

6.3.3 Oxidative Stress

All living entities use oxygen to break down and use nutritional substances to generate energy for existence. Oxygen therefore is an essential constituent for living. Oxygen mediates chemical interactions that break down fats, proteins, and carbohydrates to form energy. Oxygen, being essential for sustenance is also a double-edged sword as it is a substantially interactive atom that could form part of possibly destructive molecules generally described as free radicals. Free radicals can attack the healthy cells of the body possibly leading to injury, disease and serious disorders [46]. Reactive oxygen and nitrogen species (RONS) are formed by various endogenous and exogenous processes, and their adverse outcomes are counteracted by antioxidant defenses [47]. Endogenous sources involve mitochondria, nicotinamide adenine dinucleotide phosphate hydrogen (NADPH), oxidases (NOXs), endoplasmic reticulum (ER), xanthine oxidases (XO), endothelial nitric oxide synthase (eNOs), etc. [48] whereas, exogenous sources include ultraviolet (UV) light, ionizing radiation (IR), contaminants, heavy metals, etc. [49]. These could damage DNA, lipids, and proteins if oxidative stress (OS) is not restricted [50]. To prevent oxidative destruction, antioxidant systems are initiated [19]. Oxidative stress (OS) in biological systems is, therefore, a complex development illustrated by an inequality between the generation of free radicals, reactive oxygen species (ROS) and the body's capacity to destroy these active forms through endogenous and external antioxidant system [51], indicating that the disparity between oxidants and antioxidants within the body due to antioxidant deficiency or increased reactive oxygen species (ROS), reactive nitrogen species (RNS) (i.e. reactive oxygen and nitrogen species (RONS), and reactive sulfur species (RSS) production, lead to potential cellular damage [48]. ROS and RNS (i.e. RONS) are described as unbalanced species containing oxygen and nitrogen that react quickly with other molecules in the cells [52]. When RONS contain one or more unpaired electrons they are identified as free radicals, while those devoid of unpaired electrons are described as non-free radicals [19]. Free radical species comprise superoxide anion radical ($O^{2\cdot-}$) [53], hydroxyl radical ($OH^{\cdot}$) [54], alkoxyl ($^{\cdot}OR$) [55], nitric oxide ($NO^{\cdot}$) [56], and peroxyl radicals ($OOR^{\cdot}$) [9] while non-free radicals include hydrogen peroxide (H_2O_2) [57], nitrogen dioxide (NO_2) [58], and peroxynitrite ($ONOO^-$) [59], etc. Superoxide anion radical ($O_2^{\bullet-}$) is the main precursor of RONS generated by the cells [60] and its elevation is linked to oxidative stress and cellular damage [61]. Thus, antioxidants are influential tools to curb the excess production of ROS and prevent oxidative damage [8]. ROS are formed from endogenous and exogenous sources [49].

In metabolic activities, an extensive diversity of reactions follows, with the agents being the active form of oxygen, such as hydrogen peroxide and superoxide radical anion, among others [19]. The natural organism in which there is an excess of ROS can generate a collection of pathological states, from heart disease to cancer [49]. Biotic systems have antioxidant approaches to restrain enzymatic and non-enzymatic ecological impairment that permits ROS to be dormant. Endogenous antioxidants include enzymes such as superoxide dismutase, catalase, glutathione peroxidase [62], or non-enzymatic compounds, such as bilirubin and albumin [63]. When a substance is subjected to an elevated measure of ROS, the endogenous antioxidant system is conceded and becomes unable to guarantee comprehensive immune defense [64]. To offset the deficit of antioxidants, the body can employ external antioxidants that are provided with food, nutritional supplements, or medications [65]. Among the utmost significant external antioxidants are phenolic compounds, carotenoids, vitamin C and other minerals such as selenium and zinc [26].

6.3.4 Characteristics of Antioxidants

Antioxidants have also been described as individual substances that even in trivial aggregates display activities to inhibit or considerably delay the oxidation of oxidizable elements such as lipids and proteins [66]. While they display their essential inhibitory role, they further recognize oxidative proliferation and stabilize the generated radical, thus facilitating the decrease of oxidative impairment in the human body [67]. They perform key primary (chain reaction, free radical scavenging) and secondary or preventive functions [68] with the secondary antioxidant actions achieved by the method comprising iron blockade, inhibition of lipid hydroperoxides by disruption of the assembly of useless volatiles, reactivation of significant antioxidants, and mopping up of singlet oxygen [69].

6.3.5 Mechanism of Action of Antioxidants

To appreciate the mechanism of action of antioxidants, it is essential to recall the knowledge of the generation of free radicals and their harmful consequences [47, 49]. Antioxidants have been defined [16, 66] and two general mechanisms of action have been projected for their activities [70] and comprise a chain-breaking method by which the primary antioxidant donates an electron to the free radical present in the systems and elimination of ROS/RNS originators (secondary antioxidants) by extinguishing chain-initiating catalyst [71]. However, they may exert their influence on biotic systems by several mechanisms involving electron donation, metal ion chelation, co-antioxidants, or gene expression regulation [19, 72, 73].

6.3.6 Relevance of Antioxidants to Health

The relevance of antioxidants to health is inexhaustible, however, some of them are outlined below as follows:

i. Antioxidants have potent antioxidant capacity and offer an array of health benefits such as lowering the risk of heart diseases and certain types of cancers, enhancing the immune system, and protecting from age-related macular degeneration the leading cause of irreversible blindness among adults [26, 74, 75].
ii. They assist to protect the cells from the injury of free radicals [7].
iii. It is also useful in the prevention of prostate cancer in men along with lowering the probability of advancement of type 2 diabetes mellitus [76].
iv. They also help with the inhibition of metabolism-related disorders and cancer of the digestive system [77].
v. They exert positive impacts on several lifestyle disorders such as obesity, dyslipidaemia, hypertension and diabetes [78].

vi. They also prevent cell proliferation and bring about apoptosis in breast cancer cells [79].
vii. Their antiproliferative influences in human hepatocellular carcinoma cells on breast cancer cells, human tumor cells and human colon carcinoma cells have been documented [80].
viii. Ascorbic acid as an antioxidant guards the tissue from oxidative damage [33].
ix. Antioxidants, especially α-tocopherol, a fat-soluble antioxidant that defends cells from free radical chain reactions have been added to medicinal, cosmetics and fats and oil, especially unsaturated fat-based products to abort autoxidation and rancidity to enhance their shelf life [19, 81].
x. Antioxidants exhibit extensive applications as these are employed as additives in fats and oils and in sustenance management businesses to avert the deterioration of nutritional substances [82].

6.3.7 Some Undesirable Health-Related Effects of Antioxidants

Some familiar undesirable consequences on health due to excessive use of some antioxidants have been documented:

i. A high dose of antioxidant, beta carotene may cause cancer in the lungs of smokers [83, 84].
ii. Extreme antioxidants like α-tocopherol may foster the threat of hemorrhagic stroke [85, 86].
iii. It has been documented that an excessive intake of retinol (vitamin A) could fuel the probability of congenital disabilities [87].
iv. An overload of carotene in the body may trigger yellow pigmentation of the skin [88].
v. Other side effects of surplus ingestion of antioxidants are diarrhea, dizziness and joint pain [89, 90].
vi. It has been suggested that the ingestion of antioxidant supplements would only be relevant if there are deficiencies to normalize their levels rather than as a normal consumption [91].

6.4 Disease and Causes of Death

The World Health Organization (WHO) in 1946 asserted that health is a state of complete physical, mental and social well-being and not merely the absence of disease or infirmity [92]. Though several arguments have been put forth against the WHO's claim [92] about health [93–95], the term disease could generally imply any disorder that weakens the regular operation of the body [93]. In line with this definition, diseases are linked with the dysfunction of the body's normal homeostatic processes [96]. The cause(s) of any death is a complex incidence sometimes difficult to recognize, hence, attempting to attribute a cause to any death assists in the understanding of the actual health condition(s) that was untreated or the appropriate steps that needed to be addressed [97]. Irrespective of the argument [119, 120] put forth against the definition of health by the WHO [118], it could be opined that in addition to disease, such factors as incomplete physical, mental, and social well-being, etc. could collectively be the cause of death [98].

In 2019, the top 10 causes of death accounted for 55% of the 55.4 million deaths worldwide [99]. Causes of death can be categorized into three classes: communicable (infectious and parasitic diseases and maternal, perinatal and nutritional conditions), non-communicable (chronic) and injuries [99]. Globally, the COVID-19 pandemic led to 14.9 million extra deaths and cost 336.8 million years of life lost in 2020 and 2021. This information implies that, on average, each death directly or indirectly ascribed to the COVID-19 pandemic by the end of 2021 led to a loss of more than 22 years of life – equivalent to over 5 years of life loss every second [100, 101]. The pandemic has also put many health-related indicators further off-track [102]. As a result of service disruptions, the increasing trend in immunization coverage (including against measles, human papillomavirus (HPV), diphtheria, tetanus and pertussis) and the declining trend in the incidence of malaria and TB were both reversed, and fewer people were treated for neglected tropical diseases (NTDs). The COVID-19 pandemic exposed inequalities both between countries and within them, including inequalities in access to COVID-19 vaccines, with populations with lower educational levels and in low- and middle-income countries less likely to have received a COVID-19 vaccine [103]. The COVID-19 pandemic is a stark reminder that infectious diseases can emerge or re-emerge to cause harm – potentially to everyone [104]. Infectious diseases that were previously under control may surge as a result of antimicrobial resistance (AMR), setting back progress [105].

Considering an extensive classification, diseases can be categorized under the following – physical diseases, mental diseases, infectious diseases, non-infectious diseases, deficiency diseases, inherited diseases, degenerative diseases, social diseases, and self-inflicted diseases [106]. This section of the chapter emphasized infectious diseases.

6.4.1 Infectious Disease

Infectious diseases are initiated by pathogenic microorganisms, such as bacteria, viruses, fungi, protozoa, etc. [107]. Infectious diseases can be distributed directly or indirectly (by contact with articles or hands recently touched with a person's infected secretion), from one infected person to another who may initially be healthy (contagious diseases) or by direct contact with droplet nuclei; sneezing, talking or coughing [108]. They could also be contracted from insects or other animals serving as vectors (zoonotic diseases) [109]. It is possible to contract some infectious diseases through airborne substances from the nose, mouth, or respiratory tract of the ailing person onto the mucous membranes of another person or by transmission of physiological or seminal fluids [110]. They can be grouped into three categories: diseases which cause high levels of mortality; diseases which place on populations heavy burdens of disability; and diseases which owing to the rapid and unexpected nature of their spread can have serious global repercussions [111].

Various types of infectious diseases such as COVID-19, Ebola virus disease (EVD), HIV/AIDS, diphtheria, influenza, measles, mumps, rubella, Haemophilus influenza diseases, sexually transmitted diseases (STDs), salmonella, cholera, viral hepatitis, etc. are briefly discussed in this chapter.

6.4.2 Coronavirus Disease (COVID-19)

Coronavirus disease (COVID-19) is an extremely infectious viral disorder triggered by critical acute respiratory syndrome coronavirus 2 (SARS-CoV-2) [112]. Its influence worldwide was devastating with over 6 million deaths [113]. Its first respiratory viral infection was described in Wuhan, Hubei Province, China, in late December 2019, and quickly SARS-CoV-2 spread globally, compelling the WHO to proclaim it a global pandemic [114] on March 11, 2020 [115]. There have been splinter outbreaks of the disorder predominantly due to the appearance of mutant variants of the virus. There have been various variants of SARS-CoV-2 since the pandemic, however, only a few are considered variants of concern (VOCs). Based on the epidemiological update by the WHO, 5 SARS-CoV-2 VOCs [116] have been identified since the beginning of the pandemic:

i. Alpha (B.1.1.7): The first variant of concern, which was described in the United Kingdom (UK) in late December 2020 [117].
ii. Beta (B.1.351): First reported in South Africa in December 2020 [117].
iii. Gamma (P.1): First reported in Brazil in early January 2021 [117].
iv. Delta (B.1.617.2): First reported in India in December 2020 [117].
v. Omicron (B.1.1.529): First reported in South Africa in November 2021 [118].

6.4.2.1 Transmission of SARS-CoV-2

Several modes of transmission of SARS-CoV-2 have been documented [119–121], however, one of the main means of transmission of SARS-CoV-2 is by contact with respiratory droplets having the transmittable virus on the eyes, nose, or mouth – especially through splashes and sprays like a cough or sneeze [122] from close interaction or direct spread from pre-symptomatic, asymptomatic, or symptomatic persons harboring the virus [123], by breathing in air holding droplets or aerosol particles that have the SARS-CoV-2 virus when near to an infected person or in ill-ventilated spaces with an infected person, by contacting the eyes, nose, or mouth with hands that have the SARS-CoV-2 virus particles on them [124]. In addition, the Centers for Disease Control and Prevention (CDC) specified that persons can be infested with SARS-CoV-2 through interaction with exteriors polluted by the virus [122].

6.4.2.2 Clinical Manifestations of COVID-19

The median incubation period for SARS-CoV-2 is projected to be 5.1 days. A good number of affected persons will develop symptoms within 11.5 days of infection [125]. The clinical spectrum of COVID-19 varies from asymptomatic forms to clinical ailment indicated by acute respiratory failure necessitating mechanical ventilation, septic shock, and multiple organ failure [126]. A lot of symptomatic patients experience fever, cough and shortness of breath [112]. Uncommon signs are sore throat, anosmia, dysgeusia, anorexia, nausea, malaise, myalgias, and diarrhea [127].

6.4.2.3 *Treatment/Management of COVID-19*

The National Institutes of Health (NIH) has presented 2 focal procedures that dominate the pathogenesis of COVID-19 as replication of the virus in the early phase of the ailment and uncontrolled immune/inflammatory response to SARS-CoV-2 leading to systemic tissue impairment in the advanced stage of the illness, hence, the treatment guide advices antiviral therapies to stop viral replication in the initial stage of the disease and immunomodulators in the later phase [128]. Meanwhile, remdesivir is the first antiviral drug authorized by the USA Food and Drug Administration (FDA) to cure COVID-19 [129]. Ritonavir-boosted nirmatrelvir, molnupiravir, and high-titer COVID-19 convalescent plasma have Emergency Use Authorizations (EUAs) for handling COVID-19 [130]. Tixagevimab 300 mg plus cilgavimab 300 mg monoclonal antibodies have received EUAs that allow them to be used as SARS-CoV-2 preexposure prophylaxis (PrEP) in specified affected persons [131]. The latest NIH therapy standards for the controlling of COVID-19 disorder have been accessible since 3rd January 2023 [128].

6.4.3 Ebola Virus Disease (EVD)

The genus *Ebolavirus* is of the *Filoviridae* family, along with the genus *Marburgvirus* and is the cause of *Ebola virus* disease (EVD) formerly called Ebola hemorrhagic fever [132]. Five species of Ebola virus comprise Bundibugyo, Zaire, Sudan, Côte d'Ivoire (Taï Forest) and Reston ebolaviruses. The previous three have been accountable for the large eruptions that have arisen in Africa, whereas the Reston ebolavirus has been detected in animals in Asia but not as a source of human ailment [133]. Its initial emergence was in 1976 with two simultaneous epidemics of severe viral hemorrhagic fever encompassing 284 cases (151 deaths [53%]) concentrated in Nzara, Sudan [134] and 318 cases (280 deaths [88%]) in Yambuku (near the Ebola River), Democratic Republic of Congo [135]. Fruit bats of the *Pteropodidae* family, as well as the species *Hypsignathus monstrosus*, *Epomops franqueti* and *Myonycteris torquata*, are assumed to be the accepted hosts of Ebola viruses [136], with humans and other mammals acting as unintended hosts [132].

6.4.3.1 *Transmission of Ebola Virus*

Ebola virus is communicated to individuals by close interaction with blood and bodily fluids from another infected human or animal, either by direct interaction or indirectly from a contaminated setting [137]. However, scientists thought individuals were originally infested with an ebolavirus by interaction with a contaminated animal, such as a fruit bat or non-human primate. This is termed a spillover event. Subsequently, the virus reaches from person to person, possibly touching several persons [138, 139]. The Ebola virus is contracted through link (such as through broken skin or mucous membranes in the eyes, nose, or mouth) with:

i. Blood or body fluids (urine, saliva, sweat, feces, vomit, breast milk, amniotic fluid, and semen) of a person who is sick with or has died from Ebola disease.

ii. Items (such as clothes, bedding, needles, and medical equipment) infected with body fluids from somebody who is sick with or has died from Ebola disease.
iii. Diseased fruit bats or non-human primates (such as apes and monkeys).
iv. Semen from a man who was salvaged from Ebola disease (via oral, vaginal, or anal sex). Ebola viruses can persist in certain body fluids (such as semen) of a sick person who has recouped from Ebola disease, even if they no longer have indicators of critical disease. There is no proof that Ebola viruses can be distributed through sex or other contact with vaginal fluids from a woman who has had Ebola disease [138, 139].

Anyone who gets contaminated with the Ebola virus does not become symptomatic instantly. There is an incubation period after which the victim becomes symptomatic and later could spread the virus [138].

6.4.3.2 Clinical Features of EVD

Basic signs and symptoms of Ebola disease frequently comprise some of these: fever, aches and pains (severe headache and muscle and joint pain), weakness and fatigue, sore throat, loss of appetite, gastrointestinal signs (abdominal pain, diarrhea and vomiting, etc.), unexplained bleeding or bruising, etc. Additional indications may involve red eyes, skin rash, and hiccups [138, 140].

6.4.3.3 Treatment of EVD

The treatment of EVD could be handled therapeutically and by supportive care [141]. Therapeutically, there are presently two therapies permitted by the U.S. Food and Drug Administration (FDA) to cure EVD due to the Ebola virus, species *Zaire ebolavirus*, in adults and children. The first drug authorized in October 2020, Inmazeb™, is a combination of three monoclonal antibodies. The second drug, Ebanga™, is a single monoclonal antibody and was authorized in December 2020. These distinct mAbs bind to a portion of the Ebola virus's surface called the glycoprotein, which prevents the virus from entering a person's cells [142]. Supportive care entails providing fluids and electrolytes (body salts) orally or through infusion into the vein (intravenously), applying therapies to strengthen blood pressure, reduce vomiting and diarrhea, manage fever and pain and treating other infections, if they occur [143].

6.4.4 Human Immunodeficiency Virus/Acquired Immunodeficiency Syndrome (HIV/AIDS)

Human immunodeficiency virus (HIV) is an infectious virus that attacks the body's immune system, precisely the white blood cells called CD4 cells. It ruins these CD4 cells, damaging an individual's body's defense system against opportunistic infections, such as tuberculosis and fungal infections, severe bacterial infections and some cancers Acquired immune deficiency syndrome (AIDS) is the designation for a series of possibly life-threatening diseases and disorders occurring once the body's defense system has compromised due to a critical impairment by the HIV [144–146]. The advanced impacts of HIV on the body's protective mechanism aggravate cancers and opportunistic infections integrating sundry actions of the body such as immune, gastrointestinal, genito-urinary, endocrine, dermatologic and nervous systems.

6.4.4.1 *Transmission of HIV*

HIV could be transmitted from one individual to another by identifiable activities like sex or by applying an unsterilized sharp object on the body for any purpose. HIV is transmittable simply in specified body fluids such as blood, semen, pre-seminal fluids, rectal fluids, vaginal fluids, and breast milk [147] and the spread is barely feasible if these fluids have access to a mucous membrane, open cuts or sores, or are precisely infused into the bloodstream (from an infected needle or syringe). It can also be transferred from a birthing parent with HIV to their child during pregnancy, childbirth or breastfeeding and is described as perinatal or mother-to-child transmission of HIV [144, 147].

6.4.4.2 *Symptoms of HIV/AIDS*

Indications linked with AIDS comprise consistent fever, night sweat, weight loss, headache, lymphadenopathy, skin rashes, diarrhea, thrush, relapse of *Varicella zoster* virus infection, *Kaposi sarcoma*, *Pneumocystis carinii* pneumonia, *Cryptococci meningitis*, Candida esophagitis, Toxoplasma encephalitis, etc. [146]. An approximation of up to 80% of individuals who are ill with HIV encounter flu-like disorder. However, the common symptoms include elevated body temperature, sore throat, and body rash. Additional signs may comprise tiredness, joint pain, muscle pain, swollen glands, etc. and as soon as the immune system is critically impaired, further indication includes weight loss, chronic diarrhea, night sweats, skin problems, recurrent infections and critical diseases [144, 147].

6.4.4.3 *Treatment of HIV*

The treatment for HIV described as antiretroviral therapy (ART) is utilized and entails administering a blend of HIV medications explained as an HIV treatment regimen daily. Though ART cannot cure HIV, HIV treatments aid in extending the lifespan of those infected with HIV and decrease the risk of spreading HIV. The basic reason for HIV therapy is to lower an infected person's viral load to an infinitesimal level [148].

6.4.5 Diphtheria

Diphtheria is a dangerous infection triggered by a toxin-producing bacterium, *Corynebacterium diphtheriae* whose toxin makes people very sick [149, 150].

6.4.5.1 *Transmission of Diphtheria*

Diphtheria is distributed with ease amongst individuals through direct contact, especially over the air via respiratory droplets, or rarely by contaminated clothes and objects.

People in the same household with a victim, those with a history of consistent contact with the infected person and those who are directly exposed to the secretions from the infected site such as the mouth, skin, etc. are at a higher risk [150].

Any mucous membrane could get involved with diphtheria, however, clinically, it can be categorized into various types based on the appearance and site of disease with the mostly commonly affected sites as respiratory diphtheria (nasal diphtheria, pharyngeal and tonsillar diphtheria, laryngeal diphtheria) and cutaneous diphtheria. Respiratory diphtheria initiates slowly and is indicated by mild fever, sore throat, aphagia, malaise, anorexia, and hoarseness, especially if the larynx is touched [151].

6.4.5.2 *Treatment of Diphtheria*

Diagnosis and treatment are based on the class of diphtheria somebody has. However, if respiratory diphtheria is involved, it is often advisable to initiate therapy instantly. The treatment comprises the use of diphtheria antitoxin to halt the bacteria toxin from destroying the body and is very relevant in cases of respiratory diphtheria. Equally, antibiotics could be employed to eradicate the bacteria and in essential cases involve the respiratory system, the skin or other parts of the body [149, 150, 152].

6.4.6 Influenza

There are four varieties of influenza viruses viz A, B, C, and D. Seasonal epidemics (or flu) are triggered by influenza A and B viruses, however, only influenza A viruses are responsible for flu pandemics and is suffered when a novel and distinct influenza A virus develops and infects people, especially when immunity of the population is low to overcome it [153, 154]. Infections due to the influenza C virus often bring about insignificant sickness and are not considered to induce human epidemics. Influenza D viruses predominantly impact cattle and could affect other animals, however, are not recognized to trigger sickness in a population [153].

6.4.6.1 *Transmission of Influenza*

It is easy for seasonal influenza to be transmitted especially in crowded places. Droplets harboring viruses due to coughing or sneezing could get distributed in the air and people in close contact could be infected. Transmission could also be through contact with contaminated hands. The incubation period is about 2 days; however, it is often 1-4 days. To curb the spread, people need to cover their mouth and nose when coughing and should practice regular hand washing [154].

6.4.6.2 *Treatment/Management of Influenza*

Those who are infected with the influenza virus may recover on their own, however, victims experiencing critical symptoms, especially those with comorbidities may seek medical attention [155, 156]. Where the indications are moderate victims may be advised to stay at home and rest, prevent contaminating anybody, take sufficient fluids, manage extra symptoms such as fever and seek medical attention in case the situation gets severe. The sufferers of influenza viral attack such as expectant women, children below 59 months, adults above 65 years, patients on chemotherapy and those with depressed immune systems, probably due to HIV, etc. could be at high risk and if symptoms seem critical ought to be handled with antiviral therapies without delay [155, 156].

The best preventive approach to the influenza virus is vaccination, however, other preventive ways include consistent washing and drying of hands, covering of mouth and nose during coughing or sneezing, properly disposing of the materials used in the covering of mouth/nose, staying at home during peak symptoms, avoiding staying in densely populated spaces, especially in proximity with sick individuals and avoiding touching the eyes, nose or mouth [156, 157].

6.4.7 Measles

Measles also described as rubeola is triggered by a single-stranded, enveloped RNA virus with 1 serotype. It is grouped as a representative of the genus, Morbillivirus in the Paramyxoviridae family and only human beings can host the measles virus [158], being able to impact anybody, though, it is widespread in children [159].

6.4.7.1 Symptoms of Measles

The signs of measles infection generally commence 10–14 days after exposure to the virus with a pronounced noticeable rash. Measles touches the respiratory tract and then reaches throughout the body. Indications include a high fever, cough, runny nose, red and watery eyes, etc. [160].

6.4.7.2 Transmission of Measles

Being an extremely communicable infection, it extends effortlessly when an infected individual breathes, coughs or sneezes. It could trigger a complicated critical disorder and could lead to death [159].

6.4.7.3 Prevention/Treatment of Measles

The best approach to preventing the spread of measles and getting sick with it is vaccination [158]. There is no exact therapy for measles, however, any effort to manage the infection should be on a symptomatic basis, especially to avert complications. Sufficient food and fluid intake are relevant alongside the effort to replace fluid lost due to diarrhea or vomiting [158, 161].

6.4.8 Mumps

Mumps is a severe infectious viral disease of children and young adults caused by the Rubulavirus and a member of the paramyxoviridae family of which human beings are the only known host [162]. Its impact reaches the global level and is the only identified cause of epidemic parotitis [163].

6.4.8.1 *Symptoms of Mumps*

Mumps infection typically presents with a prodrome of headache, fever, myalgia, fatigue, anorexia, and malaise followed by the classic hallmark of the disease, parotitis [164]. The mumps disease is more often self-limited with individuals experiencing a full recovery [165].

6.4.8.2 *Transmission of Mumps*

Mumps are transmitted through a direct link or by airborne droplets from the upper respiratory tract of infected persons [166].

6.4.8.3 *Treatment for Mumps*

Currently, there is no treatment for mumps, however, any existing case should be managed symptomatically with sufficient rest, fluid intake, especially water, and food that does not require stressful chewing while the case resolves [167]. Nevertheless, prevention of the incidence of mumps could be achieved by channeling all the children to be administered the Measles, Mumps and Rubella (MMR) vaccine with the initial dose at 12 through 15 months old while the second dose may be given as soon as one month after the first dose, though it is regularly given between 4 and 6 years of age [167].

6.4.9 Rubella

Rubella virus is a member of the Togaviridae family, the only member of the Rubivirus genus and the contributing agent of rubella disease also called "German measles." The virus has a single-stranded positive-sense RNA genome and was initially isolated from cell culture in 1962 [168–170]. Rubella is a transmissible viral disease and children, and young adults are mostly victims. It is the principal vaccine-preventable trigger of birth defects as rubella infection in pregnant women may cause fetal death or congenital defects known as congenital rubella syndrome (CRS) [171].

6.4.9.1 *Transmission of Rubella Virus Disease*

Rubella is transmitted when an infected individual coughs or sneezes. Further, an infected pregnant female could transmit it to her fetus and it could be dangerous [172].

6.4.9.2 *Symptoms of Rubella Virus Disease*

Most incidents of rubella infection in children are mild with a self-limiting measles-like rash, mild fever, nausea and mild conjunctivitis. About 50–80% of incidences are accompanied by rash which often begins on the face and neck before proceeding down the body and continues 1–3 days [171]. Swollen lymph glands backside of the ears and in the neck are the generally observed clinical characteristics. Adults (often women) who are attacked could develop arthritis and painful joints that continue for 3–10 days [170]. A pregnant woman who becomes a victim of the Rubella virus, especially in the first trimester of pregnancy has a 90 % possibility of transmitting the virus to her fetus. It could lead to fetal death or cause CRS [170, 171, 173].

6.4.9.3 *Treatment of Rubella Virus Disease*

There is no definite remedy for rubella, however, the disorder is avoidable by vaccination [174].

6.4.10 ***Haemophilus influenzae*** **Infections**

Infections due to *Haemophilus influenzae*, a gram-negative bacterium of considerable clinical concern have been considered as "invasive" [175]. Invasive disorders occur when microorganisms conquer sections of the body that are naturally free from microbes [176]. Instances of such invasion are notable when *H. influenzae* attacks the fluid within the spine and brain, triggering meningitis, or bloodstream, initiating bacteremia [177]. The often-noted categories of invasive disease due to *H. influenzae* include pneumonia, bloodstream infection, meningitis, epiglottitis, cellulitis, infectious arthritis, etc. [175, 178].

6.4.10.1 *Causes of H. influenzae Infections*

Six different varieties of *H. influenzae* exist as encapsulated (typeable) serotypes "a - f") coupled and another serotype categorized as (non-encapsulated) non-typeable *H. influenzae* (NTHi). However, *H. influenzae* type b or Hib is the most common [175]. All serotypes can affect individuals and prove disorder, though serotype b (Hib) is the most virulent [175, 177, 179] causing a widespread range of clinical indicators, exclusively in toddlers ranging from otitis media and bronchitis to other acute invasive situations, such as pneumonia, meningitis, and sepsis [175, 180].

6.4.10.2 *Transmission of H. influenzae Infections*

Individuals who harbor diseases due to *H. influenzae*, involving Hib could spread them to others by respiratory drops, mainly through coughing or sneezing [178, 181] to primarily children younger than 5 years and adults 65 years and above.

6.4.10.3 *Symptoms of H. influenzae Infection*

Several health disorders could arise due to *H. influenzae*, nevertheless, their respective indications differ based on the section of the body that is affected as follows [178]: pneumonia (fever and chills, cough, shortness of breath or struggle in breathing, sweating, chest pain, extreme fatigue, etc.; [182] bloodstream infection (fever and chills, extreme fatigue, abdominal ache, nausea coupled with vomiting or without vomiting, diarrhea, anxiety, shortness of breath or struggle in breathing, delirium, etc. [180] and meningitis (fever, headache, stiff neck, nausea with or without vomiting, photophobia, delirium, etc. [183].

Though complications may rarely arise, nevertheless, some *H. influenzae* diseases can lead to lasting challenges or death. Instances include that bloodstream disorders can lead to loss of limbs; meningitis may bring about brain impairment or hearing loss [178, 184].

6.4.10.4 *Treatment of H. influenzae Infection*

Someone detected with *H. influenzae* illness may be treated with relevant antibiotics [185]. Though the treatment options are best at the discretion of the physician, however, other management opportunities may comprise breathing support, therapies to control low blood pressure or wound care where any part of the body has been hurt [178, 185].

6.4.11 Sexually Transmitted Diseases

Sexually transmitted diseases (STDs) [146] are also recognized as sexually transmitted infections (STIs) and are contracted from an infected individual by sexual dealings [186]. The presenting indication of STDs varies on the specific organism, route, signs and symptoms of the disease. Risk factors that promote the spread of STIs include engaging in unprotected sexual interaction with many associates, having previously suffered from STIs, sexual molestation, use of alcohol, prostitution, use of recreational drugs and intravenous drug use [187]. The eight widespread and commonly contracted STDs comprise four curable disorders (chlamydia, gonorrhea, syphilis and trichomonas) and four that are treatable but incurable (hepatitis B, herpes simplex virus, HIV, and HPV) [188]. STDs are prevalent and have colossal social, economic and demographic implications and speed up the transmission of HIV [189]. Over 1 million STIs are contracted daily globally with an annual projected 376 million new occurrences with 1 out of 4 STIs like chlamydia, gonorrhea, syphilis and trichomoniasis [190]. There are over 500 million victims of *Herpes simplex* virus (HSV) [191] and over 290 million females acquired *Human papillomavirus* (HPV) infection [192, 193]. The often encountered and important STIs comprise [146] gonorrhea [194], chlamydia [195], syphilis [196], trichomoniasis [197], chancroid [198], HIV/AIDS [144, 148, 199] and non-gonococcal urethritis (NGU) [200] such as genital herpes [193], pubic lice [201], pelvic inflammatory disease (PID) [202] and genital warts [203], etc.

For extensive details on sexually transmitted diseases, readers are encouraged to visit:

Kenneth C. Ugoeze (2022). Phytopharmaceuticals for Treating Sexually Transmitted Diseases. In: Herbal Drugs for the Management of Infectious Diseases, Inderbir Singh, Rakesh K. Sindhu, Atul Shirkhedkar and Pharkphoom Panichayupakaranant (editors). Scrivener Publishing LLC, MA, USA, Pages 179 – 262, https://doi.org/10.1002/9781119818779.ch8.

6.4.12 Salmonella

Salmonella is a group of bacteria (gram-negative rods genus fitting to the Enterobacteriaceae family) that could initiate gastrointestinal disease and fever (salmonellosis, referring to typhoid fever and paratyphoid fever and gastroenteritis), a type of food poisoning in individuals and of various infections in domestic creatures [204, 205].

6.4.12.1 *Transmission of Salmonella*

Salmonella can be transmitted by food vendors who fail to wash their hands and/or the surfaces and utensils employed connecting food preparation stages, and when people eat raw or undercooked foods [205]. It can further be distributed from animals to people, especially through individuals who deal with creatures such as poultry, etc. [206].

6.4.12.2 *Symptoms of Salmonellosis*

Indications of the disease could differ but generally involve weakness or exhaustion, fever, abdominal discomfort, diarrhea, nausea and occasionally vomiting, etc. [205, 207].

6.4.12.3 *Treatment of Salmonellosis*

Therapy in acute circumstances involves the replacement of electrolytes (sodium, potassium and chloride ions) lost during vomiting and diarrhea) and rehydration. Nevertheless, endangered classes such as infants, the elderly, and immunocompromised patients and in very critical cases, may necessitate the administration of antimicrobial therapy, though, at the discretion of the physician [206].

6.4.13 Cholera

Cholera is an exceptionally virulent infection spread by taking in food or water contaminated with the toxic bacterium *Vibrio cholerae* serogroup O1 or O139. The disease can trigger intense watery stooling and could cause death within a very short time if not controlled [208]. Cholera bacteria can thrive in the generally unhygienic surroundings, in brackish rivers and coastal waters, raw shellfish and other environment, especially contaminated with poop [209].

6.4.13.1 *Transmission of Cholera*

An individual could contract cholera through the intake of water or food messed up with cholera bacteria. Often, especially in an epidemic, most of such contamination is through the feces of an infected individual. The ailment is distributed easily in districts with insufficient management of sewage and drinking water [208, 209]. The key factors causing cholera are polluted water and food. Maintenance of individual cleanliness and using fresh food and water is very relevant [210].

6.4.13.2 *Symptoms of Cholera*

The symptom of cholera starts with fever. It could also be without indications, however, where there are signs, they could include acute watery stooling, abdominal and or leg pains, vomiting, extreme thirst, and agitation. Often, there is dehydration due to excessive watery stooling which could manifest as rapid heart rate, low blood pressure, loss of skin elasticity, dry mucous membranes, etc. [210, 211].

6.4.13.3 *Treatment of Cholera*

Cholera is highly treatable. If cholera is detected early, then the patient may just need an oral rehydration solution to prevent dehydration. If it's more serious, then the patient may require fluid replacement at the hospital [210, 211].

6.4.14 Hepatitis

Hepatitis refers to the enlargement of the liver showing inflammation which could be due to some infectious viruses (viral hepatitis) and non-infectious agents such as chemicals, drugs, alcohol, a type of genetic condition or an erroneous attack of the liver through a reactive immune system (autoimmune hepatitis). This disorder could be acute or chronic with quiet progressive liver impairment [212, 213].

6.4.14.1 Types of Hepatitis, Causes, Symptoms and Treatment

There are five major viral groups of hepatitis comprising hepatitis A, B, C, D and E with each being accountable for each variety of viral hepatitis [212].

6.4.14.2 Hepatitis A

Hepatitis A is the commonest globally and is triggered by the hepatitis A virus (HAV), though, it does not lead to chronic hepatitis. Affected individuals, particularly infants do not show any clinical indicators [214]. The majority of its occurrences involve no distinct therapy other than bed rest. However, in the event of its occurrence, its severity could be diminished if the concerned person is inoculated within two weeks of contact with immune serum globulin obtained from those exposed to HAV, a process indicated as passive immunization which is active since the serum comprises antibodies against HAV. An active vaccine against HAV is accessible and is regularly administered to children over two years of age living in regions with elevated levels of HAV. The vaccine is also suggested for persons who travel to zones where HAV is prevalent, etc. [215].

6.4.14.3 Hepatitis B

Hepatitis B (HBV) is a more severe and lingering disorder than hepatitis A, occurring as an acute infection with around 5 – 10 % of incidences becoming chronic and could lead to perpetual liver destruction [214]. People at high risk of HBV comprise intravenous drug users, sexual affiliates of those with the disease, healthcare workers who are not effectively immunized, and recipients of organ transplants or blood transfusions [216]. A safe and active vaccine against HBV is accessible and offers defense for at least five years. Passive immunization with hepatitis B immune globulin can also support defense [217].

6.4.14.4 Hepatitis C

Hepatitis C virus (HCV) was isolated in 1989 and was mentioned as non-A, non-B hepatitis. Being spread by connection with contaminated blood, its infection may lead to mild or critical disease that continues for some weeks or a generation [214]. Approximately 80 % of persons who developed the disease are symptomless while individuals who manifest signs may feel a flulike syndrome such as exhaustion, motion sickness, vomiting and at times icterus. Roughly 60–80% of lingering incidences of advanced to continuing liver disorders like cirrhosis or liver cancer [218]. Management for HCV encompasses a blend of antiviral treatments, comprising alpha interferon and ribavirin [219], though, other antivirals like boceprevir and telaprevir may be utilized side by side with interferon and ribavirin, etc. [220].

6.4.14.5 *Hepatitis D*

Hepatitis D virus (HDV) is equally noted as the delta agent, though, its occurrence is only in connection with HBV infestation since HDV needs HBV to multiply [214]. Contamination involving HDV could happen concurrently with infection involving HBV, or HDV may attack an individual readily affected with HBV with the latter condition seemingly giving a more severe condition causing cirrhosis or chronic liver disorder [221]. Alpha interferon is the only treatment for HDV infection. Preventing infection with HBV also prevents HDV infection [214].

6.4.14.6 *Hepatitis E*

Hepatitis E (HEV), discovered in the 1980s is analogous to HAV [214]. It is spread similarly to HAV, and it also brings about a sharp disease whose impact is more critical than those of HAV and death is common with HEV [222], and it is accountable for prevalent epidemics of hepatitis due to the consumption of contaminated water or food [214, 223].

6.4.14.7 *Hepatitis F and G*

Hepatitis F (HFV) was initially described in 1994 alongside certain incidents of hepatitis distributed through unwholesome food or water [214, 224]. In 1996, an additional virus, hepatitis G virus (HGV) was isolated and is understood to be liable for a considerable number of sexually transmitted and bloodborne incidents of hepatitis [214, 224]. HGV causes acute and chronic forms of the disease and often infects persons already infected with HCV [214].

6.5 Bases of Use of Antioxidants in the Management of Infectious Diseases

The early part of this book chapter offered information on phytochemicals, antioxidants and oxidative stress. Emphasis was made on the harmful impacts of free radicals on the functional systems. An antioxidant has been documented to retain the potential to stabilize extremely responsive, potentially destructive molecules identified as free radicals [19], the existence of which has been linked with damage to membranes, enzymes, cell nuclear material, etc. [9]. The antioxidant's role in minimizing the liability of extremely reactive free radicals helps to safeguard the structural integrity of cells and tissues of the immune system and other systems in the body [225]. Recent clinical trials have found that antioxidant supplementation can significantly improve certain immune responses [225]. Explicitly, supplementation with vitamins C, E, and A, or β-carotene increased the activation of cells involved in tumor immunity in the elderly [226]. Equally, supplementation with antioxidant vitamins also protected immune responses in individuals exposed to certain environmental sources of free radicals [19]. Supplementation with vitamin A, a relatively weak antioxidant, decreases morbidity and mortality associated with measles infections in children [227]. The immune system is accountable for safeguarding against disease by pathogens such as bacteria, viruses, and protozoan parasites [228]. Pathogens admitted as attackers of the body and as "non-self", are then destroyed by immune cells and their secretions. The immune system responds correspondingly to cells of the body that have suffered changes that can lead to cancer. The body's immune system no longer recognizes certain precancerous cells as "self". In principle, the modified cells are considered to be invaders and are destroyed [229]. Circumstances that depress immune functions subsequently raise the hazards of infection, etc. Contrariwise, influences that can improve immunity may decrease these risks [230]. Diseases are here and will remain, so further approaches need to be explored to reinforce the immune system earlier (preventive approach) and through an infection (therapeutic approach). Since it is healthier to appreciate a wholesome health status achieved through preventive approach, it could be sufficient to state that antioxidants have the answer. A consistent consumption of dietary substances as sources of high level of antioxidants is crucial since several nutritional materials are undoubtedly linked with inflammatory and oxidative stress [231], as likewise, deficient levels of different nutrients could be connected to altered immune responses [232] and, consequently, an augmented susceptibility to infections and severe outcomes derived from them. Minerals exert an extensive range of immune-modulatory, anti-inflammatory, antimicrobial, and antioxidant activities, which might be relevant to prevent or fight diseases [233]. Recently, antioxidants have enjoyed relevance and are emerging as potential prophylactic and therapeutic agents in several disorders as free radicals have been linked with the pathogenesis of several diseases [234]. Further, antioxidants are beneficial as immune modulators and can be utilized for the prophylaxis or handling of certain diseases along with the core therapy [7]. Supplements of exogenous antioxidants can act directly to quench the free radical or free radical reactions, prevent lipid peroxidation and also boost the endogenous antioxidant system and hence offer prophylactic or therapeutic activity [235]. Natural food, spices and medicinal plants are rich sources of antioxidants and can be the leading source of antioxidants to manage infectious diseases [19]. Numerous antioxidants such as epigallocatechin-3-*O*-gallate, lycopene, ellagic acid, coenzyme Q10, indole-3-carbinol, genistein, quercetin, vitamin C and vitamin E are pharmacologically effective as prophylactic and therapeutic agents. So, targeting oxidative stress or boosting the endogenous levels of antioxidants by the use of antioxidants is expected to have useful effect in the management of several infectious disorders [234].

Having established the bases of the significance of utilizing antioxidants in the management of infectious diseases and in the understanding that several phytochemical compounds holding antioxidant properties are pharmacologically active as prophylactic and therapeutic agents and additionally exhibit the capacity to boost the body's immune system [225] offering the body the ability to resist infectious diseases [236] it is relevant to consider briefly some sources of antioxidants beneficial in the management of infectious diseases.

In these days of globalization and climate shift, infective ailments are circulating further swiftly, with the emergence of novel diseases, and being a universal health burden, populations of emerging nations experience a high burden of diseases [237]. Thus in 2010, globally, nearly one-quarter of fatalities were due to transmittable disorders, while in low-income nations, approximately 60% of mortalities could be ascribed to them due to inadequate hygienic measures, lack of diagnostic apparatuses and underprivileged therapeutic choices [238].

Prevailing treatments are sorted into antivirals, antibiotics and antifungals [239] alongside numerous vaccines inhibiting viral and bacterial disorders, and has already led to the successful eradication of smallpox [240]. These measures exist only for some pathogens and do not accommodate all possibly lethal and pandemic agents such as the Ebola virus, COVID-19, etc. despite the increasing antimicrobial resistance to existing remedies [241]. Accordingly, alternative treatment options are essential, especially as it is consistently highlighted in the scientific literature, official reports, etc. that 80% of Asian and African populations use traditional medicine (TM) to tackle their healthcare requirements [242]. Moreover, due to the side effects and resistance that pathogenic micro-organisms develop against antibiotics, considerable interest is getting shifted to extracts and bioactive compounds isolated from plant species are employed in herbal medication and lead towards new drugs [243] providing more expectations that medicinal plants may offer a natural and new source of antibacterial agents for use. Some indications abound that naturally derived constituents have consistently been the source of new drug products [244]. From 1940 to 2014, 49% of all small molecules authorized by the US Food and Drug Administration (FDA) were natural products or derivatives directly associated with them [245]. Currently, the FDA's Center for Drug Evaluation and Research (CDER) has authorized 55 new drugs in 2023, as the small molecule and biologic pharmacopoeia maintains growth. This cohort is closely 50% greater than the new endorsement class of 2022, which fell below the authorization trend line. The 10-year rolling average for new CDER authorizations currently holds at 46 per year, the maximum it has been in over 20 years. The lowest was in 2010 when this average bottomed out at 25 per year [246].

6.6 Sources of Antioxidant Beneficial in the Management of Infectious Diseases

Robust antioxidant actions have been found in berries, cherries, citrus, prunes, and olives [14, 247]. Resveratrol, a phytoalexin is a free radical scavenger and a potent antioxidant obtained in grapes, peanuts, cocoa, and berries of Vaccinium species, together with blueberries, bilberries, and cranberries [14]. An extract from pine bark, *Pinus pinaster* [248] is commercially available as pycnogenol® and has been documented to have strong antioxidant action [146, 249]. Extensive studies have shown that green and black teas are loaded with antioxidant properties since they contain up to 30% of the dry weight as phenolic compounds [250, 251]. In addition to the dietary sources of antioxidants, several therapeutic plants have equally been known to offer antioxidants and these comprise [252] *Acacia catechu* [253], *Aegle marmelos* [254], *Allium cepa* (Onion) [255], *A. sativum* [256], *Aloe vera* [248, 257] *Amomum subulatum* [258], *Andrographis paniculata* [259], *Asparagus recemosus* [260], *Azadirachta indica* [13, 261, 262], *Bacopa monniera* [263], *Butea monosperma* [264], *Camellia sinensis* (Green tea) [248, 250] *Cinnamomum verum* [265], *Cinnamomum tamala* [266], *Curcuma longa* [267, 268], *Emblica officinalis* [269], *Glycyrrhiza glabra* [270], *Hemidesmus indicus* [271], *Indigofera tinctoria* [272], *Mangifera indica* [273], *Momordica charantia* [274], *Nigella sativa* (Black cumin) [275], *Ocimum sanctum* (Holy basil) [276],

Trigonella foenum-graecium (Fenugreek) [277], *Withania somifera* [278], *Zingiber officinalis* [279], *Rosmarinus officinalis* [280], etc.

6.6.1 Resveratrol

Resveratrol (3, 4', 5-trihydroxy-trans-stilbene), a phytoalexin, free radical scavenger and a potent antioxidant available in grapes, peanuts, cocoa, and berries of Vaccinium species, together with blueberries, bilberries, and cranberries [14, 281] was initially obtained from the roots of white hellebore (*Veratrum grandiflorum* O. Loes) by Michio Takaoka in 1939 and isolated from the roots of Japanese knotweed in 1963 [282]. Biotic actions of resveratrol (RSV) such as the induction of CD95 signaling-dependent apoptosis, inhibition of cyclooxygenase, modulation of NF-κB activation and influences on the cell division cycle demonstrate a likely impact on the immune response [283]. RSV has been established to modulate the immune reaction against pathogenic organisms like viruses, bacteria, etc., by boosting immune responses and/or lowering immunocyte apoptosis [284]. *In vitro,* contact with RSV was proved to bring about a biphasic influence on anti-CD3/anti-CD28-induced development of both IFN-γ, IL2- and IL4-producing CD8+ and CD4+ T-cells where stimulation and suppression were detected at low and high RSV concentrations correspondingly [283]. A comparable finding detected that RSV at low concentrations preserved the suppressive impact of ethanol on macrophage percentage and macrophage MHC-II molecule expression, hence, boosting cell-mediated immune response. Other investigations have revealed that RSV boosted immune response in piglets infected with rotavirus exhibiting outcomes of lessened diarrhea and inflammation [285] and also decreased the influence of a bacterial pathogen, *Haemophilus influenza* thus lowering infectious airway inflammation [286]. It improved spleen lymphocyte propagation, peritoneal macrophages, and CD4+ cells in peripheral blood [287]. RSV was also able to curtail the advancement of some pathogenic microorganisms like bacteria and fungi [288] and presented to prevent the progress of *Candida albicans* [289], *Campylobacter* and *Arcobacter* [290]. Its effectiveness against *Staphylococcus aureus* and rotavirus infection-induced diarrhea has been explained [291].

6.6.2 *Pinus pinaster* Bark Extract

An extract of the bark of the French *P. pinaster* (*Pinus pinaster* Ait. Subsp. *atlantica*) is available commercially as pycnogenol®. It has a universal usage as an herbal therapy, nutrition and supplemental provision in several classes of deteriorating disorders [249]. *Pinus pinaster* bark extract (PBEx) retains abundant procyanidins which have been documented to have hindered not only the binding of human immunodeficiency virus type-1 (HIV-1) to host cells but also prevented HIV viral replication and T-cell interaction in cell culture experiments [292]. PBEx was found to generate expression of an intracellular antioxidant protein and manganese superoxide dismutase, and inhibition of phosphorylation of the ribosomal S6 protein, seemingly suggesting that these biochemical modifications brought about by pycnogenol® play a significant part in its antiviral influences [249]. PBEx is a potential instrument for the prevention of encephalomyocarditis viral replication, inhibition of advancement of viral myocarditis, and enhancement of inflammation and myocardial necrosis. It was described that pycnogenol® (100 mg/kg) had valuable impacts on viral myocarditis by inhibition of viral replication and by suppression of pro-inflammatory cytokines, genes related to cardiac remodeling, and mast cell-related genes in the heart muscle of mice (gene expressions of tumor necrosis factor, type-1 procollagen, stem cell factor, and mast cell tryptase) [292, 293]. PBEx has also been revealed to exhibit antimicrobial action against different pathogenic prokaryotes (gram-negative and positive bacteria), and eukaryotes (fungi and yeasts) [294, 295]. For example, the progress and adherence of *Helicobacter pylori* (the gram-negative, microaerophilic bacterium) to mucosal cells of the gastrointestinal tract were stalled by pycnogenol®, *in vitro*. In an additional experiment, antimicrobial impacts of pycnogenol® against 23 pathogenic microorganisms (*e.g.*, *Staphylococcus aureus*, *Klebsiella pneumoniae*, *Pseudomonas aeroginosa*, *Bacillus cereus*, *Enterococcus faecalis*, *Candida albicans*, *Aspergillus oryzae*, and *Salmonella* sp.) have been examined. Subsequently, pycnogenol® hindered the growth of all examined microorganisms in minimum inhibitory dosages (MID) alternating from 20 to 250 μg/mL) [295].

6.6.3 *Camellia sinensis*

Tea, *Camellia sinensis* (family, Theaceae) is a shrub. Green tea processed from it is highly accepted due to its distinct relevance to health including its anti-cancer, anti-oxidant and antimicrobial actions [296]. Green tea is one of the widespread drinks taken universally and is processed from the leaves of the *Camellia sinensis* plant [250]. Lately, investigators have researched the capacity of green tea to eliminate transmittable agents and the capability to truly inhibit infections. The essential constituents in green tea that demonstrate antimicrobial properties are catechins [297]. The four leading catechins that are found in green tea are (-)-epicatechin (EC), (-)-epicatechin-3-gallate (ECG), (-)-epigallocatechin (EGC), and (-)-epigallocatechin-3-gallate (EGCG) [298, 299]. Among these catechins, EGCG and EGC are at peak levels in green tea and have been demonstrated to reveal a variation of antimicrobial influences, both to organisms influenced and in mechanisms employed. Intake of green tea has been disclosed to circulate these composites and/or their metabolites in the body, which makes their use advantageous for the management and inhibition of infections [297, 299]. Other acknowledged health-related gains of green tea such as the anti-inflammatory and antioxidant properties may also promote the antimicrobial impacts [300]. Reports of investigations on *Escherichia coli* indicated that exposure to green tea polyphenols (GTPs) leads to key gene expression alterations for 17 genes, with upregulation appearing in nine genes and downregulation in eight genes [299, 301]. One of the main influences of green tea catechins (GTCs) is the capacity to bind to bacterial cell membranes leading to restriction in diverse bacterial procedures that could destroy the cell membrane resulting in intensified permeability that leads to cell lysis [299, 302]. Another influence of GTCs on bacterial functionality is the capability of GTCs to constrain bacterial fatty acid biosynthesis by restraining enzymes implicated in the biosynthetic pathway [299, 303]. The synergistic impacts of the GTCs with popular antimicrobial agents have been investigated considering their antimicrobial activity [304]. Several positive outcomes indicated that GTCs acted in synergy with imipenem against methicillin-resistant *Staphylococcus aureus* (MRSA) (the cause of staph infection that is tough to control due to resistance to some antibiotics); with metronidazole against *Porphyromonas gingivalis*; with azithromycin, cefepime, ciprofloxacin, chloramphenicol, doxycycline, erythromycin, nalidixic acid, piperacillin, or tobramycin against *E. coli*; with ampicillin, Cefalotin, doxycycline, erythromycin, penicillin, or tetracycline against *Enterobacter aerogenes*; with chloramphenicol or tetracycline against *Pseudomonas aeruginosa*; and with aztreonam, ceftazidime, ciprofloxacin, gentamicin, meropenem, or tetracycline against *Acinetobacter baumannii* [299, 303, 305–307].

6.6.4 *Allium cepa*

Allium cepa L. *(A. cepa)* or onion (family, Amaryllidaceae) species are utilized as vegetables and used in traditional remedies as medicinal agents. There are varieties of onions in different colors and tastes. Onion bulbs have been used as a food, spice, and herbal remedy since ancient times by individuals globally, and several therapeutic properties have been described for it [308]. It has high intensity of folic acid, vitamin B6, magnesium, calcium, potassium, phosphorus, vitamins and minerals [309]. *A. cepa* contains saponins, aglycones, quercetin, cepaenes, flavonoids, organosulfur, and phenolic composites, and these have demonstrated diverse pharmacological properties and therapeutic effects [255]. It is extensively utilized as an antimicrobial agent. It also displayed anticancer, antidiabetic, antioxidant, antiplatelet, antihypertensive, antidepressant properties, neuroprotective, anti-inflammatory, antiparasitic influences, etc. It exerts useful influences on the digestive, circulatory, respiratory and immune systems [310]. *Allium cepa* holds various potent phytochemicals holding essential antioxidant influences, especially phenolic compounds accountable for its exclusive fragrance and therapeutic activities. A phytochemical analysis of four *A. cepa* variants (violet, red, green, and white) revealed the presence of kaempferol, ferulic acid, quercetin, gallic acid, and protocatechuic acid, etc. [311]. Extracts of *A. cepa* have been acknowledged as efficacious antimicrobial agents in the management of infectious disorders due to various fungi, bacteria and viruses [312, 313].

6.6.5 *Allium sativum*

Allium sativum L. (garlic), family, Alliaceae has become highly studied and among the best-selling commercially available herbal products globally. It has been relevant in traditional remedies for generally health-associated conditions [314] and is extensively employed as a nutritional constituent, especially as a spice and aphrodisiac [315]. Its biological activities result from a blend of organically functional constituents generally accountable for its curative influence. The compounds confined in garlic synergistically influence each other to yield unique results. The bioactive constituents of garlic include enzymes (e.g. alliinase), sulfur-containing compounds like alliin and compounds produced enzymatically from alliin (e.g. allicin) [316]. Further, the health benefits derived from *A. sativum* have been attributed largely to its different bioactive compounds such as organic sulfides, saponins, phenolic compounds, and polysaccharides [317]. Considerable investigations have presented that garlic and its bioactive components display antioxidant, anti-inflammatory, antibacterial, antifungal, immunomodulatory, cardiovascular protective, anticancer, hepatoprotective, digestive system protective, anti-diabetic, anti-obesity, neuroprotective, and renal protective properties [317]. There is substantial proof that suggests preventive and therapeutic functions of garlic in boosting the immune system, anti-tumor and antioxidant effects of garlic guards the body against free radicals [318, 319]. Considering its antibacterial action, garlic extract has been documented to obstruct the development of Gram-positive and Gram-negative bacteria such as *Staphylococcus, Streptococcus, Micrococcus, Enterobacter, Escherichia, Klebsiella, Lactobacillus, Pseudomonas, Shigella, Salmonella,*

Proteus, and *Helicobacter pylori* [320]. *A. sativum* and its sulfur constituents have shown antiviral activity against coxsackievirus spp., herpes simplex virus types 1 and 2, influenza B, parainfluenza virus type 3, vaccinia virus, vesicular stomatitis virus, human immunodeficiency virus type 1, and human rhinovirus type 2 with the order of a virucidal outcome as follows: ajoene>allicin>allyl methyl thiosulfinate>methylallyl thiosulfinate>methylallyl [314, 321]. Further, the antifungal action of the aqueous garlic extract and the concentrated garlic oil has exhibited an inhibitory impact against Aspergillus [314, 322] and allicin a constituent of garlic has shown a fungicidal action against various yeasts and fungi, involving *Candida albicans*, *Cryptococcus trichophyton*, *Histoplasma capsulatum* and *Cryptococcus neoformans* [314, 323].

6.6.6 *Aloe vera*

Aloe vera (L.) Burm. f. (*Aloe barbadensis* Miller), the commonest variety (family, Liliaceae) is a herbaceous and perennial succulent xerophyte beneficial for its several therapeutic applications [324] based on its anti-inflammatory, skin protection, anti-bacterial, anti-viral, antiseptic and wound healing activities [325]. The plant comprises a huge measure of bioactive composites such as flavonoids, terpenoids, lectins [326], fatty acids, anthraquinones [327], mono- and polysaccharides (pectins, hemicelluloses, glucomannan), tannins, sterols (campesterol, β-sitosterol), enzymes, salicylic acid, minerals (calcium, chromium, copper, iron, magnesium, manganese, potassium, phosphorus, sodium and zinc) and vitamins (A, C, E, β-carotene, B1, B2, B3, B6, choline, B12, folic acid) [328, 329]. *Aloe vera* possesses antiviral activity which is exerted indirectly or directly. Indirect action is owing to the prompting of the immune system and direct action is based on its anthraquinone content. The anthraquinone aloin disables several enveloped viruses such as herpes simplex, varicella zoster and influenza [330]. Further, *Aloe vera* exerts antiseptic action due to its content of 6 antiseptic agents comprising lupeol, salicylic acid, urea nitrogen, cinnamonic acid, phenols and sulfur, all exerting inhibitory influence on fungi, bacteria and viruses [327].

6.6.7 *Andrographis paniculata*

Andrographis paniculata Wall (family, Acanthaceae) often recognized as King of Bitters has been utilized in prehistoric oriental and ayurvedic remedies. It is popular among other species for its relevance in folk medication for miscellaneous health-related issues [331]. Its leaves retain a diterpenoid labdane described as andrographolide with special effect against oxidative stress and are accountable for an extensive series of biotic actions such as antioxidant, anti-inflammatory and anti-cancer influences [259] and an immunostimulant [332]. The aerial portions, roots and whole plant of *A. paniculata* have been employed for centuries in Asia as a traditional medicine for the management of diverse disorders including influenza, dysentery, etc. [331, 333]. The leaf extract is a traditional remedy for the treatment of infectious diseases [334]. *A. paniculata* possesses antimicrobial activity in treating some health morbidities of infectious origin comprising *Salmonella*, *Shigella*, *Escherichia coli*, *Staphylococcus aureus*, etc. [335]. The antiviral action of *A. paniculata* has been reported including those of dengue virus serotype 1 (DENV-1) [336], human papillomavirus type 16 (HPV16) [337], herpes simplex virus type 1 (HSV-1) [338], influenza A virus [334] and HIV [339].

6.6.8 *Azadirachta indica*

Azadirachta indica (neem) is a member of the Meliaceae family and its health-promoting action is ascribed to its abundant resource of antioxidants and has been extensively utilized in Chinese, Ayurvedic, and Unani remedies globally, particularly in the Indian Subcontinent in the handling and inhibition of numerous ailments. Past discovery established that neem and its components function in the scavenging of free radical generation and inhibition of disease pathogenesis [261]. *Azadirachta indica* is relevant in the management of diseases due to its abundant source of many types of constituents comprising mainly azadirachtin and others such as nimbolinin, nimbin, nimbidin, nimbidol, sodium nimbinate, gedunin, salannin, and quercetin [340]. The leaves hold constituents such as nimbin, nimbanene, 6-desacetylnimbinene, nimbandiol, nimbolide, ascorbic acid, n-hexacosanol and amino acid, 7-desacetyl-7-benzoylazadiradione, 7-desacetyl-7-benzoylgedunin, 17-hydroxyazadiradione, and nimbiol [341]. Quercetin and ß-sitosterol, polyphenolic flavonoids, were purified from neem fresh leaves and were known to have antibacterial and antifungal properties [342] and seeds retain beneficial components involving gedunin and azadirachtin [261]. A study was set up to assess the antibacterial action of the bark, leaf, seed, and fruit extracts of neem on bacteria isolated from adult mouth and outcomes showed that bark and leaf extracts exhibited antibacterial action against all the test bacteria employed [343]. In trials neem oil suppressed various species of pathogenic bacteria, including *Staphylococcus aureus* and *Salmonella typhosa* [344]. Its leaf extract has demonstrated virucidal action against coxsackievirus virus B-4 as hinted via virus inactivation and produces decline assay besides interfering at an early event of its replication cycle [345]. Antifungal influences of various extracts of neem parts against such fungus as *Aspergillus* and *Rhizopus*, *C. lunata*, *H. pennisetti*, and *C. gloeosporioides* f. sp. *mangiferae*, etc. have been reported [346, 347]. Preparations containing extracts of neem displayed toxicity to cultures of 14 common species of fungi, including members of the following genera: *Trichophyton, Epidermophyton, Microsporum, Trichosporon*, *Geotrichum* and *Candida* [344].

6.6.9 *Curcuma longa*

Curcuma longa L. is popularly noted as turmeric and is a member of the family of ginger, Zingiberaceae and has an extended history due to its healing activities against numerous infirmities [348]. Its rhizomes retain various secondary metabolites comprising curcuminoids (Curcumin, demethoxycurcumin (DMC), and bisdemethoxycurcumin (BDMC)) [349], sesquiterpenes, and steroids, with the curcuminoid curcumin as the key constituent of the yellow pigment and the principal bioactive substance [262], which has been relevant in the control of various diseases like cough, cold, dental issues, indigestion, skin infections, blood purification, asthma, piles, bronchitis, tumor, wounds, and hepatic disorders, and is used as an antiseptic [350]. Curcumin, being the main component of *C. longa*, is notable for its medicinal relevance in several ailments [348] ascribed to various properties, especially antioxidant, antimicrobial, anti-inflammatory, anti-angiogenic, and antimutagenic ones [351].

For curcumin, an antiviral influence has been recorded against various viruses including hepatitis viruses [352], influenza viruses [353] and emerging arboviruses like the Zika virus

(ZIKV) or chikungunya virus (CHIKV) [354]. Interestingly, it has also been reported that the molecule inhibits human immunodeficiency virus (HIV) [355], herpes simplex virus 2 (HSV-2) [356] and human papillomavirus (HPV) [357], showing that curcumin decreases the transmission of STDs [241].

Investigators have evaluated the antimicrobial action of curcumin against various bacterial strains such as *Salmonella paratyphi*, *Trichophytongypseum*, *Staphylococcus aureus*, *Streptococci mutans*, and *Mycobacterium tuberculosis* [358]. The extract displayed antimicrobial action against *Trichophyton longifusus*, *Microsporum canis*, and *S. aureus* while toxicity was recorded towards *Lemna minor*. The ethanolic extract of *C. longa* was active for *Shigella flexneri*, *Staphylococcus epidermis*, *Klebsiella pneumoniae*, *Lactobacillus*, *Pseudomonas aeruginosa*, *Vibrio cholerae*, and *Salmonella typhi* [359]. The aqueous extract, curcumin component, and oil fraction of *C. longa* revealed antibacterial activity and suppresses *H. pylori*, *Streptococcus*, *Staphylococcus*, and *Lactobacillus* species [360].

Martin and his co-authors [361] evaluated the protective effect of curcumin versus 23 fungal species *in vitro* employing a CLSI-authorized broth micro-dilution method to assess the fungicidal vulnerability. They observed that curcumin was quite efficient against isolated strains of *Paracoccidioides brasiliensis;* the adherence of *Candida* species was more efficiently inhibited by curcumin compared to fluconazole, especially in the buccal mucosal strains that were isolated from patients suffering from AIDS. Further investigation showed that *C. longa* oil suppressed dermatophytes and pathogenic fungi considering its topical application on guinea pigs infested with dermatophytes, moulds, and yeast. The guinea pigs with dermatophytes and fungal-infected lesions improved, and the lesions became invisible after 7 days of *C. longa* administration [362].

6.6.10 *Glycyrrhiza glabra*

Glycyrrhiza glabra L. (Fabaceae) also described as liquorice, is a very sweet, moist, soothing herb with anti-inflammatory and expectorant actions, soothes cough, and has hormonal influences and detoxifies and protects the liver [363]. Therapeutically, it is utilized internally for Addison's disease, asthma, bronchitis, coughs, peptic ulcer, and arthritis and externally, liquorice is used for eczema, herpes, and shingles [364].

The antimicrobial influences of *G. glabra*, particularly on Gram-positive and Gram-negative bacteria such as *Staphylococcus aureus*, *Escherichia coli*, *Pseudomonas aeruginosa*, *Candida albicans*, and *Bacillus subtilis* have been documented [365] and were attributed to the presence of secondary metabolites comprising saponins, alkaloids, and flavonoids [366], specifically, glabridin, glabrol, glabrene, hispaglabridin A, hispaglabridin B, 40-methylglabridin, and 3-hydroxyglabrol, isolated from *G. glabra*, are liable for this action [367].

Independent investigators documented the antibacterial activity of *G. glabra* against *Mycobacterium tuberculosis* [365], indicating that glabridin is the substance liable for this action, instead of hispaglabridin B [368]. In a mice lung infection model, *G. glabra* was therapeutically effective against multidrug-resistant strain of *P. aeruginosa*, and the hydroalcoholic extract triggered a decline of the microbial load in the blood, principally due to the existence of stigmasterol, ergosterol, licochalcone, and glabridin [369].

The antifungal action of methanolic extract of *G. glabra* against *Arthrinium sacchari* and *Chaetomium funicola* has been reported coupled with *in vivo* inhibition of *Mycobacterium smegmatis*, *Shigella*, *Salmonella*, *E. coli*, *S. mutans*, and *Lactobacillus acidophilus* due to isoflavonoids, such as glabridin, glabrol, and their derivatives [370, 371]. Various researchers documented that *C. albicans* is vulnerable to liquorice extracts based on their abundance in liquiritigenin, liquiritin, licochalcone A, and glabridin [372, 373].

The antiviral impact of *G. glabra* extracts against separate viruses has been described, incorporating herpes simplex, *Varicella zoster*, Japanese encephalitis, influenza, and vesicular stomatitis virus attributed to the presence of two triterpenoids, glycyrrhizin and 18β-glycyrrhetinic acid [367]. The influence of glycyrrhizin against Herpes simplex virus (HSV) has been reported [374, 375]. Intravenous glycyrrhizin has been utilized for more than 20 years in Japan for the management of chronic hepatitis, especially chronic hepatitis C, etc. [376].

Glycyrrhizin can also be used as a novel therapeutic method to control porcine epidemic diarrhea virus (PEDV) infection, inhibiting the infection of Vero cells (namely, the entry and replication of PEDV) and decreasing the mRNA levels of proinflammatory cytokines [377].

6.7 Conclusion

Plant-based bioactive constituents obtained in whole grains, fruits, vegetables, nuts, and herbs include carotenoids, polyphenols, phenolic constituents, alkaloids, isoprenoids, phytosterols, saponins, dietary fibers, and certain polysaccharides, etc. They exert effective antioxidant activities and demonstrate antimicrobial, antidiarrheal, anthelmintic, antiallergic, antispasmodic, and anti-inflammatory activities, etc. Edible naturally occurring plant-based foods retaining a high level of antioxidants are innumerable and could serve as sources of exogenous antioxidants if taken through the diets or as therapeutic agents to boost the endogenous antioxidants. These antioxidants are useful agents in the management of infectious diseases such as coronavirus disease (COVID-19), Ebola virus disease (EVD), HIV/AIDS, diphtheria, influenza, measles, mumps, rubella, Haemophilus influenzae infection, STDs, salmonella, cholera, hepatitis, etc. Since they have prophylactic and therapeutic impacts, consistency in their consumption is vital to maintain a low level of free radicals and oxidative stress, boost and conserve an elevated immune system and avert the hazards of infectious diseases. This is one of the approaches to managing several infectious diseases.

Acknowledgement

The authors appreciate Dr. Inderbir Singh, Professor & Assistant Dean (Research) Chitkara College of Pharmacy, Chitkara University, Punjab, India and Dr. Rakesh K. Sindhu, Professor & HOD (PhD Coordinator) School of Pharmacy, Sharda University, Uttar Pradesh, India for their invitation to contribute the chapter, "Antioxidants in Infectious Diseases Management" in the book, "Antioxidants Potential in Infectious Diseases".

References

1. Rabizadeh, F., Mirian, M.S., Doosti, R., Kiani-Anbouhi, R., Eftekhari, E., Phytochemical Classification of Medicinal Plants Used in the Treatment of Kidney Disease Based on Traditional Persian Medicine. *Evid.-Based Complement. Altern. Med. ECAM*, 2022, 8022599, 2022.
2. Farag, R.S., Abdel-Latif, M.S., Abd El Baky, H.H., Tawfeek, L.S., Phytochemical screening and antioxidant activity of some medicinal plants' crude juices. *Biotechnol. Rep.*, 28, e00536, 2020.
3. Kumar A, P.N., Kumar, M., Jose, A., Tomer, V., Oz, E. *et al.*, Major Phytochemicals: Recent Advances in Health Benefits and Extraction Method. *Molecules*, 28, 887, 2023.
4. Sharma, B.R., Kumar, V., Gat, Y., Kumar, N., Parashar, A., Pinakin, D.J., Microbial maceration: a sustainable approach for phytochemical extraction. *3 Biotech.*, 8, 401, 2018.
5. Jaeger, R. and Cuny, E., Terpenoids with Special Pharmacological Significance: A Review. *Nat. Prod. Commun.*, 11, 1373–90, 2016.
6. Rowles, J.L. and Erdman, J.W., Carotenoids and their role in cancer prevention. *Biochim. Biophys. Acta, Mol. Cell. Biol. Lipids*, 1865, 158613, 2020.
7. Pham-Huy, L.A., He, H., Pham-Huy, C., Free Radicals, Antioxidants in Disease and Health. *Int. J. Biomed. Sci. IJBS*, 4, 89–96, 2008.
8. Pizzino, G., Irrera, N., Cucinotta, M., Pallio, G., Mannino, F., Arcoraci, V. *et al.*, Oxidative Stress: Harms and Benefits for Human Health. *Oxid. Med. Cell. Longev.*, 2017, 8416763, 2017.
9. Phaniendra, A., Jestadi, D.B., Periyasamy, L., Free Radicals: Properties, Sources, Targets, and Their Implication in Various Diseases. *Indian J. Clin. Biochem.*, 30, 11–26, 2015.
10. Sindhu, R.K., Singh, I., Shirkhedkar, A.A., Panichayupakaranant, P. (Eds.), *Herbal Drugs for the Management of Infectious Diseases*, John Wiley & Sons, New Jersey, 2022.
11. Parks, D.A., Bulkley, G.B., Granger, D.N., Role of oxygen-derived free radicals in digestive tract diseases. *Surgery*, 94, 415–22, 1983.
12. Kabel, A.M., Free Radicals and Antioxidants: Role of Enzymes and Nutrition. *World J. Nutr. Health*, 2, 35–8, 2014.
13. Ugoeze, K., Oluigbo, K., Chinko, B., Phytomedicinal and Nutraceutical Benefits of the GC-FID Quantified Phytocomponents of the Aqueous Extract of Azadirachta indica leaves. *J. Pharm. Pharmacol. Res.*, 04, 149–163, 2020.
14. Ugoeze, K.C., Phytopharmaceutical Benefits of Resveratrol in the Management of Diseases and Health Maintenance: A Review. *Biomed. Sci.*, 09, 18–29, 2023, [Internet] [cited 2023 Dec 27]; Available from: https://www.academia.edu/112245777/Phytopharmaceutical_Benefits_of_Resveratrol_in_the_Management_of_Diseases_and_Health_Maintenance_A_Review.
15. Rahman, K., Studies on free radicals, antioxidants, and co-factors. *Clin. Interv. Aging*, 2, 219–36, 2007.
16. Mattill, H.A., Antioxidants. *Annu. Rev. Biochem.*, 16, 177–92, 1947.
17. German, J.B., Food processing and lipid oxidation. *Adv. Exp. Med. Biol.*, 459, 23–50, 1999.
18. Knight, J.A., Free radicals: their history and current status in aging and disease. *Ann. Clin. Lab. Sci.*, 28, 331–46, 1998.
19. Lobo, V., Patil, A., Phatak, A., Chandra, N., Free radicals, antioxidants and functional foods: Impact on human health. *Pharmacogn. Rev.*, 4, 118–26, 2010.
20. Wolf, G., The discovery of the antioxidant function of vitamin E: the contribution of Henry A. Mattill. *J. Nutr.*, 135, 363–6, 2005.
21. Halliwell, B., How to characterize an antioxidant: an update. *Biochem. Soc. Symp.*, 61, 73–101, 1995.
22. Shi, H., Noguchi, N., Niki, E., Comparative study on dynamics of antioxidative action of alpha-tocopheryl hydroquinone, ubiquinol, and alpha-tocopherol against lipid peroxidation. *Free Radic. Biol. Med.*, 27, 334–46, 1999.

23. Levine, M., Rumsey, S.C., Daruwala, R., Park, J.B., Wang, Y., Criteria and recommendations for vitamin C intake. *JAMA*, 281, 1415–23, 1999.
24. Irato, P. and Santovito, G., Enzymatic and Non-Enzymatic Molecules with Antioxidant Function. *Antioxidants*, 10, 579, 2021.
25. Rudenko, N.N., Vetoshkina, D.V., Marenkova, T.V., Borisova-Mubarakshina, M.M., Antioxidants of Non-Enzymatic Nature: Their Function in Higher Plant Cells and the Ways of Boosting Their Biosynthesis. *Antioxid. Basel Switz.*, 12, 2014, 2023.
26. Zehiroglu, C. and Ozturk Sarikaya, S.B., The importance of antioxidants and place in today's scientific and technological studies. *J. Food Sci. Technol.*, 56, 4757–74, 2019.
27. Nordberg, J. and Arnér, E.S., Reactive oxygen species, antioxidants, and the mammalian thioredoxin system. *Free Radic. Biol. Med.*, 31, 1287–312, 2001.
28. Charlton, N.C., Mastyugin, M., Török, B., Török, M., Structural Features of Small Molecule Antioxidants and Strategic Modifications to Improve Potential Bioactivity. *Molecules*, 28, 1057, 2023.
29. Balasaheb Nimse, S. and Pal, D., Free radicals, natural antioxidants, and their reaction mechanisms. *RSC Adv.*, 5, 27986–8006, 2015.
30. Flora, S.J., Structural, chemical and biological aspects of antioxidants for strategies against metal and metalloid exposure. *Oxid. Med. Cell. Longev.*, 2, 191–206, 2009.
31. Lourenço, S.C., Moldão-Martins, M., Alves, V.D., Antioxidants of Natural Plant Origins: From Sources to Food Industry Applications. *Molecules*, 24, 4132, 2019.
32. Leung, F.Y., Trace elements that act as antioxidants in parenteral micronutrition 11This paper was delivered at the January 18, 1998, workshop "Frontiers in Antioxidant Research: 14th Annual A.S.P.E.N. Workshop," which was held the day before the official start of the 22nd A.S.P.E.N. Clinical Congress in Orlando, FL. *J. Nutr. Biochem.*, 9, 304–7, 1998.
33. Traber, M.G. and Stevens, J.F., Vitamins C and E: Beneficial effects from a mechanistic perspective. *Free Radic. Biol. Med.*, 51, 1000–13, 2011.
34. Tungmunnithum, D., Thongboonyou, A., Pholboon, A., Yangsabai, A., Flavonoids and Other Phenolic Compounds from Medicinal Plants for Pharmaceutical and Medical Aspects: An Overview. *Medicines*, 5, 93, 2018.
35. The Role of Natural and Synthetic Antioxidants in Modulating Oxidative Stress in Drug-Induced Injury and Metabolic Disorders 2020 [Internet], [cited 2023 Dec 27]; Available from: https://www.hindawi.com/journals/omcl/2021/9892021/.
36. Vinita, S., Gupta, V., Sharma, K., Bhatnagar, S., Kumari, R, Dhaka, N., Potential applications of antioxidants – A review. *J. Pharm. Res.*, 7, 828–835, 2013, [Internet] [cited 2024 Jan 5]; Available from: https://www.academia.edu/51856016/Potential_applications_of_antioxidants_A_review.
37. Iverson, F., Phenolic antioxidants: Health Protection Branch studies on butylated hydroxyanisole. *Cancer Lett.*, 93, 49–54, 1995.
38. Xu, X., Liu, A., Hu, S., Ares, I., Martínez-Larrañaga, M., Wang, X. *et al.*, Synthetic phenolic antioxidants: Metabolism, hazards and mechanism of action. *Food Chem.*, 353, 129488, 2021.
39. Kahl, R. and Kappus, H., [Toxicology of the synthetic antioxidants BHA and BHT in comparison with the natural antioxidant vitamin E]. *Z. Lebensm. Unters. Forsch.*, 196, 329–38, 1993.
40. Papas, A.M., Diet and antioxidant status. *Food Chem. Toxicol. Int. J. Publ. Br. Ind. Biol. Res. Assoc.*, 37, 999–1007, 1999.
41. Xu, D.P., Li, Y., Meng, X., Zhou, T., Zhou, Y., Zheng, J. *et al.*, Natural Antioxidants in Foods and Medicinal Plants: Extraction, Assessment and Resources. *Int. J. Mol. Sci.*, 18, 96, 2017.
42. Pokorný, J., Are natural antioxidants better - and safer - Than synthetic antioxidants? *Eur. J. Lipid Sci. Technol.*, 109, 629–42, 2007.

43. Jafri, S.A.A., Khalid, Z.M., Khan, M.Z., Jogezai, N., Evaluation of phytochemical and antioxidant potential of various extracts from traditionally used medicinal plants of Pakistan. *Open Chem.*, 20, 1337–56, 2022.
44. Wang, H., Cao, G., Prior, R.L., Total Antioxidant Capacity of Fruits. *J. Agric. Food. Chem.*, 44, 701–705, 1996, [Internet] [cited 2023 Dec 27]; Available from: https://pubs.acs.org/doi/10.1021/jf950579y.
45. Shetty, A., Magadum, S., Managanvi, K., Vegetables as Sources of Antioxidants. *J. Food Nutr. Disord.*, 2, 1–5, 2013.
46. Trayhurn, P., Oxygen—A Critical, but Overlooked, Nutrient. *Front. Nutr.*, 6, 10, 2019.
47. Liguori, I., Russo, G., Curcio, F., Bulli, G., Aran, L., Della-Morte, D. *et al.*, Oxidative stress, aging, and diseases. *Clin. Interv. Aging*, 13, 757–72, 2018.
48. Aranda-Rivera, A.K., Cruz-Gregorio, A., Arancibia-Hernández, Y.L., Hernández-Cruz, E.Y., Pedraza-Chaverri, J., RONS and Oxidative Stress: An Overview of Basic Concepts. *Oxygen*, 2, 437–78, 2022.
49. Bhattacharyya, A., Chattopadhyay, R., Mitra, S., Crowe, S.E., Oxidative Stress: An Essential Factor in the Pathogenesis of Gastrointestinal Mucosal Diseases. *Physiol. Rev.*, 94, 329–54, 2014.
50. Juan, C.A., Pérez de la Lastra, J.M., Plou, F.J., Pérez-Lebeña, E., The Chemistry of Reactive Oxygen Species (ROS) Revisited: Outlining Their Role in Biological Macromolecules (DNA, Lipids and Proteins) and Induced Pathologies. *Int. J. Mol. Sci.*, 22, 4642, 2021.
51. Rahal, A., Kumar, A., Singh, V., Yadav, B., Tiwari, R., Chakraborty, S. *et al.*, Oxidative Stress, Prooxidants, and Antioxidants: The Interplay. *Biomed. Res. Int.*, 2014, e761264, 2014.
52. Li, R., Jia, Z., Trush, M.A., Defining ROS in Biology and Medicine. *React. Oxyg. Species Apex NC*, 1, 9–21, 2016.
53. Chiste, R.C., Freitas, M., Mercadante, A.Z., Fernandes, E., Superoxide Anion Radical: Generation and Detection in Cellular and Non-Cellular Systems. *Curr. Med. Chem.*, 22, 4234–56, 2015.
54. Hydroxyl Radical and Its Scavengers in Health and Disease - PMC [Internet], [cited 2023 Dec 28]; Available from: https://www.ncbi.nlm.nih.gov/pmc/articles/PMC3166784/.
55. Lifestyle, Oxidative Stress, and Antioxidants: Back and Forth in the Pathophysiology of Chronic Diseases - PMC [Internet], [cited 2023 Dec 28]; Available from: https://www.ncbi.nlm.nih.gov/pmc/articles/PMC7347016/.
56. Habib, S. and Ali, A., Biochemistry of Nitric Oxide. *Indian J. Clin. Biochem.*, 26, 3–17, 2011.
57. Collin, F., Chemical Basis of Reactive Oxygen Species Reactivity and Involvement in Neurodegenerative Diseases. *Int. J. Mol. Sci.*, 20, 2407, 2019.
58. US EPA O, Basic Information about NO2 [Internet], 2016, [cited 2023 Dec 28]; Available from: https://www.epa.gov/no2-pollution/basic-information-about-no2.
59. Corpas, F.J. and Barroso, J.B., Peroxynitrite (ONOO−) is endogenously produced in arabidopsis peroxisomes and is overproduced under cadmium stress. *Ann. Bot.*, 113, 87–96, 2014.
60. Andrés, C.M.C., Pérez de la Lastra, J.M., Andrés Juan, C., Plou, F.J., Pérez-Lebeña, E., Superoxide Anion Chemistry—Its Role at the Core of the Innate Immunity. *Int. J. Mol. Sci.*, 24, 1841, 2023.
61. Lambeth, J.D. and Neish, A.S., Nox enzymes and new thinking on reactive oxygen: a double-edged sword revisited. *Annu. Rev. Pathol.*, 9, 119–45, 2014.
62. Gusti, A.M.T., Qusti, S.Y., Alshammari, E.M., Toraih, E.A., Fawzy, M.S., Antioxidants-Related Superoxide Dismutase (SOD), Catalase (CAT), Glutathione Peroxidase (GPX), Glutathione-S-Transferase (GST), and Nitric Oxide Synthase (NOS) Gene Variants Analysis in an Obese Population: A Preliminary Case-Control Study. *Antioxidants*, 10, 595, 2021, [Internet] [cited 2023 Dec 28]; Available from: https://www.ncbi.nlm.nih.gov/pmc/articles/PMC8070436/.

63. Lu, Z., Wen, T., Wang, Y., Kan, W., Xun, G., Peripheral non-enzymatic antioxidants in patients with schizophrenia: a case-control study. *BMC Psychiatry*, 20, 241, 2020.
64. Schieber, M. and Chandel, N.S., ROS Function in Redox Signaling and Oxidative Stress. *Curr. Biol. CB*, 24, R453–62, 2014.
65. Liu, Z., Ren, Z., Zhang, J., Chuang, C.C., Kandaswamy, E., Zhou, T. *et al.*, Role of ROS and Nutritional Antioxidants in Human Diseases. *Front. Physiol.*, 9, 477, 2018.
66. Chitkara, H.W.M., Sindhu, R.K., Singh, I., Kumar, D., Sandhu, I.S., Arora, S., Formulation and Evaluation of Essential Oils based Liquid Herbal. *Res. J. Pharm. Technol.*, *13*, 4, 1919–1922, 2020.
67. Tan, B.L., Norhaizan, M.E., Liew, W.P.P., Sulaiman Rahman, H., Antioxidant and Oxidative Stress: A Mutual Interplay in Age-Related Diseases. *Front. Pharmacol.*, 9, 1162, 2018.
68. Shirkhedkar, A.A., Patil, M.S., Patil, A.S., Singh, I., Herbal Drugs for the Treatment of Ocular Infections, in: *Herbal Drugs for the Management of Infectious Diseases*, pp. 157–177, 2022.
69. Pisoschi, A.M. and Pop, A., The role of antioxidants in the chemistry of oxidative stress: A review. *Eur. J. Med. Chem.*, 97, 55–74, 2015.
70. Rice-Evans, C.A. and Diplock, A.T., Current status of antioxidant therapy. *Free Radic. Biol. Med.*, 15, 77–96, 1993.
71. Krinsky, N.I., Mechanism of action of biological antioxidants. *Proc. Soc. Exp. Biol. Med. Soc. Exp. Biol. Med. N. Y. N.*, 200, 248–54, 1992.
72. Kumar, M., Verma, I., Sindhu, R.K., Srivastava, R., Herbal Bioactives for the Treatment of Gastrointestinal Tract Disorders, in: *Herbal Drugs for the Management of Infectious Diseases*, Singh, I., Sindhu, R.K., Shirkhedkar, A., Panichayupakaranant, P. (Eds.), 2022, https://doi.org/10.1002/9781119818779.
73. Frezza, M., Hindo, S., Chen, D., Davenport, A., Schmitt, S., Tomco, D. *et al.*, Novel Metals and Metal Complexes as Platforms for Cancer Therapy. *Curr. Pharm. Des.*, 16, 1813–25, 2010.
74. Choo, P.P., Woi, P.J., Bastion, M.L.C., Omar, R., Mustapha, M., Md Din, N., Review of Evidence for the Usage of Antioxidants for Eye Aging. *Biomed Res. Int.*, 2022, 5810373, 2022.
75. Aune, D., Plant Foods, Antioxidant Biomarkers, and the Risk of Cardiovascular Disease, Cancer, and Mortality: A Review of the Evidence. *Adv. Nutr.*, 10, S404–21, 2019.
76. Giovannucci, E., Harlan, D.M., Archer, M.C., Bergenstal, R.M., Gapstur, S.M., Habel, L.A. *et al.*, Diabetes and Cancer. *Diabetes Care*, 33, 1674–85, 2010.
77. Dulbecco, P. and Savarino, V., Therapeutic potential of curcumin in digestive diseases. *World J. Gastroenterol. WJG*, 19, 9256–70, 2013.
78. Athyros, V.G., Tziomalos, K., Karagiannis, A., Mikhailidis, D.P., Dyslipidaemia of Obesity, Metabolic Syndrome and Type 2 Diabetes Mellitus: the Case for Residual Risk Reduction After Statin Treatment. *Open Cardiovasc. Med. J.*, 5, 24–34, 2011.
79. Parton, M., Dowsett, M., Smith, I., Studies of apoptosis in breast cancer. *BMJ*, 322, 1528–32, 2001.
80. Sunwoo, Y.Y., Lee, J.H., Jung, H.Y., Jung, Y.J., Park, M.S., Chung, Y.A. *et al.*, Oldenlandia diffusa Promotes Antiproliferative and Apoptotic Effects in a Rat Hepatocellular Carcinoma with Liver Cirrhosis. *Evid.-Based Complement. Altern. Med. ECAM*, 2015, 501508, 2015.
81. Liao, S., Omage, S.O., Börmel, L., Kluge, S., Schubert, M., Wallert, M. *et al.*, Vitamin E and Metabolic Health: Relevance of Interactions with Other Micronutrients. *Antioxidants*, 11, 1785, 2022.
82. Monica, S., Antioxidant and Its Applications. *Res. Rev. J. Pharmacol. Toxicol. Stud.*, 4, 1–10, 2016.
83. Kordiak, J., Bielec, F., Jabłoński, S., Pastuszak-Lewandoska, D., Role of Beta-Carotene in Lung Cancer Primary Chemoprevention: A Systematic Review with Meta-Analysis and Meta-Regression. *Nutrients*, 14, 1361, 2022.

84. Middha, P., Weinstein, S.J., Männistö, S., Albanes, D., Mondul, A.M., β-Carotene Supplementation and Lung Cancer Incidence in the Alpha-Tocopherol, Beta-Carotene Cancer Prevention Study: The Role of Tar and Nicotine. *Nicotine Tob. Res.*, 21, 1045–50, 2018.
85. Schürks, M., Glynn, R.J., Rist, P.M., Tzourio, C., Kurth, T., Effects of vitamin E on stroke subtypes: meta-analysis of randomised controlled trials. *BMJ*, 341, c5702, 2010.
86. Loh, H.C., Lim, R., Lee, K.W., Ooi, C.Y., Chuan, D.R., Looi, I. *et al.*, Effects of vitamin E on stroke: a systematic review with meta-analysis and trial sequential analysis. *Stroke Vasc. Neurol.*, 6, 109–20, 2020.
87. Bastos Maia, S., Rolland Souza, A.S., Costa Caminha M de, F., Lins da Silva, S., Callou Cruz R de, S.B.L., Carvalho dos Santos, C. *et al.*, Vitamin A and Pregnancy: A Narrative Review. *Nutrients*, 11, 681, 2019.
88. Edigin, E., Asemota, I.R., Olisa, E., Nwaichi, C., Carotenemia: A Case Report. *Cureus*, 11, e5218, 2019.
89. Holeček, M., Side Effects of Amino Acid Supplements. *Physiol. Res.*, 71, 29–45, 2022.
90. Khan, I., Samson, S.E., Grover, A.K., Antioxidant Supplements and Gastrointestinal Diseases: A Critical Appraisal. *Med. Princ. Pract.*, 26, 201–17, 2017.
91. Salehi, B., Martorell, M., Arbiser, J.L., Sureda, A., Martins, N., Maurya, P.K. *et al.*, Antioxidants: Positive or Negative Actors? *Biomolecules*, 8, 124, 2018.
92. Constitution of the World Health Organization [Internet], [cited 2023 Dec 29]; Available from: https://www.who.int/about/accountability/governance/constitution.
93. Amzat, J. and Razum, O., Health, Disease, and Illness as Conceptual Tools. In: *Medical Sociology in Africa*, pp. 23–39, Springer, Cham, 2014, Available from: https://doi.org/10.1007/978-3-319-03986-2_2.
94. Huber, M., Knottnerus, J.A., Green, L., van der Horst, H., Jadad, A.R., Kromhout, D. *et al.*, How should we define health? *BMJ*, 343, d4163, 2011.
95. What is health? The ability to adapt. *Lancet Lond. Engl.*, 373, 781, 2009.
96. Kotas, M.E. and Medzhitov, R., Homeostasis, Inflammation, and Disease Susceptibility. *Cell*, 160, 816–27, 2015.
97. Dattani, S. and Roser, M., How are causes of death registered around the world? Our World Data [Internet], 2023, [cited 2023 Dec 29]; Available from: https://ourworldindata.org/how-are-causes-of-death-registered-around-the-world.
98. Collective and individual responsibilities for health, both physical and mental | Health Knowledge [Internet], [cited 2023 Dec 29]; Available from: https://www.healthknowledge.org.uk/public-health-textbook/disease-causation-diagnostic/2h-principles-health-promotion/responsibilities-health-physical-mental.
99. The top 10 causes of death [Internet], [cited 2023 Dec 29]; Available from: https://www.who.int/news-room/fact-sheets/detail/the-top-10-causes-of-death.
100. World health statistics 2023: monitoring health for the SDGs, sustainable development goals [Internet], [cited 2023 Dec 29]; Available from: https://www.who.int/publications-detail-redirect/9789240074323.
101. Urgent action needed to tackle stalled progress on health-related Sustainable Development Goals [Internet], [cited 2023 Dec 29]; Available from: https://www.who.int/news/item/19-05-2023-urgent-action-needed-to-tackle-stalled-progress-on-health-related-sustainable-development-goals.
102. Policy Responses to COVID19 [Internet], IMF [cited 2023 Dec 29]; Available from: https://www.imf.org/en/Topics/imf-and-covid19/Policy-Responses-to-COVID-19.
103. Shadmi, E., Chen, Y., Dourado, I., Faran-Perach, I., Furler, J., Hangoma, P. *et al.*, Health equity and COVID-19: global perspectives. *Int. J. Equity Health*, 19, 104, 2020.

104. Morens, D.M. and Fauci, A.S., Emerging Pandemic Diseases: How We Got to COVID-19. *Cell*, 182, 1077–92, 2020.
105. Aslam, B., Wang, W., Arshad, M.I., Khurshid, M., Muzammil, S., Rasool, M.H. *et al.*, Antibiotic resistance: a rundown of a global crisis. *Infect. Drug Resist.*, 11, 1645–58, 2018.
106. International Online Medical Council (IOMC) [Internet]. Int. Online Med. Counc, [cited 2023 Dec 29]; Available from: https://www.iomcworld.org/.
107. WHO EMRO | Infectious diseases | Health topics [Internet]. World Health Organ. - Reg. Off. East. Mediterr, [cited 2023 Dec 29], Available from: http://www.emro.who.int/health-topics/infectious-diseases/index.html.
108. Leung, N.H.L., Transmissibility and transmission of respiratory viruses. *Nat. Rev. Microbiol.*, 19, 528–45, 2021.
109. Rahman, MdT., Sobur, MdA., Islam, MdS., Ievy, S., Hossain, MdJ., El Zowalaty, M.E. *et al.*, Zoonotic Diseases: Etiology, Impact, and Control. *Microorganisms*, 8, 1405, 2020.
110. Straif-Bourgeois, S., Ratard, R., Kretzschmar, M., Infectious Disease Epidemiology, in: *Handb. Epidemiol.*, pp. 2041–119, 2014.
111. Church, D.L., Major factors affecting the emergence and re-emergence of infectious diseases. *Clin. Lab. Med.*, 24, 559–86, 2004.
112. Sharma, A., Tiwari, S., Deb, M.K., Marty, J.L., Severe acute respiratory syndrome coronavirus-2 (SARS-CoV-2): a global pandemic and treatment strategies. *Int. J. Antimicrob. Agents*, 56, 106054, 2020.
113. Lamers, M.M. and Haagmans, B.L., SARS-CoV-2 pathogenesis. *Nat. Rev. Microbiol.*, 20, 270–84, 2022.
114. Coronavirus disease (COVID-19) pandemic [Internet], [cited 2023 Dec 30]; Available from: https://www.who.int/europe/emergencies/situations/covid-19.
115. Sharma, A., Ahmad Farouk, I., Lal, S.K., COVID-19: A Review on the Novel Coronavirus Disease Evolution, Transmission, Detection, Control and Prevention. *Viruses*, 13, 202, 2021.
116. Tracking SARS-CoV-2 variants [Internet], [cited 2023 Dec 30]; Available from: https://www.who.int/activities/tracking-SARS-CoV-2-variants.
117. Raman, R., Patel, K.J., Ranjan, K., COVID-19: Unmasking Emerging SARS-CoV-2 Variants, Vaccines and Therapeutic Strategies. *Biomolecules*, 11, 993, 2021.
118. Chenchula, S., Karunakaran, P., Sharma, S., Chavan, M., Current evidence on efficacy of COVID-19 booster dose vaccination against the Omicron variant: A systematic review. *J. Med. Virol.*, 94, 2969–76, 2022.
119. Modes of transmission of virus causing COVID-19: implications for IPC precaution recommendations [Internet], [cited 2023 Dec 31]; Available from: https://www.who.int/news-room/commentaries/detail/modes-of-transmission-of-virus-causing-covid-19-implications-for-ipc-precaution-recommendations.
120. COVID-19 and its Modes of Transmission - PMC [Internet], [cited 2023 Dec 31]; Available from: https://www.ncbi.nlm.nih.gov/pmc/articles/PMC7461745/.
121. M, E., Ma, S, B, F., M, H.R., S, S., Transmission Modes of COVID-19: A Systematic Review. *Infect. Disord. Drug Targets*, 21, e170721187995, 2021, [Internet] [cited 2023 Dec 31]; Available from: https://pubmed.ncbi.nlm.nih.gov/33200716/.
122. CDC, Healthcare Workers [Internet]. *Cent. Dis. Control Prev.*, 2020, [cited 2023 Dec 31]; Available from: https://www.cdc.gov/coronavirus/2019-ncov/hcp/non-us-settings/overview/index.html.
123. Transmission of SARS-CoV-2: A Review of Viral, Host, and Environmental Factors - PMC [Internet], [cited 2023 Dec 31]; Available from: https://www.ncbi.nlm.nih.gov/pmc/articles/PMC7505025/.

124. Scientific Brief: SARS-CoV-2 Transmission - CDC COVID-19 Science Briefs - NCBI Bookshelf [Internet], [cited 2023 Dec 31]; Available from: https://www.ncbi.nlm.nih.gov/books/NBK570442/.
125. Lauer, S.A., Grantz, K.H., Bi, Q., Jones, F.K., Zheng, Q., Meredith, H.R. *et al.*, The Incubation Period of Coronavirus Disease 2019 (COVID-19) From Publicly Reported Confirmed Cases: Estimation and Application. *Ann. Intern. Med.*, 172, 577–582, 2020, [Internet] [cited 2023 Dec 31]; Available from: https://www.ncbi.nlm.nih.gov/pmc/articles/PMC7081172/.
126. Çelik, I. and Öztürk, R., From asymptomatic to critical illness: decoding various clinical stages of COVID-19. *Turk. J. Med. Sci.*, 51, 3284–3300, 2021.
127. Baj, J., Karakuła-Juchnowicz, H., Teresiński, G., Buszewicz, G., Ciesielka, M., Sitarz, R. *et al.*, COVID-19: Specific and Non-Specific Clinical Manifestations and Symptoms: The Current State of Knowledge. *J. Clin. Med.*, 9, 1753, 2020.
128. Information on COVID-19 Treatment, Prevention and Research [Internet]. COVID-19 Treat. Guidel, [cited 2023 Dec 31]; Available from: https://www.covid19treatmentguidelines.nih.gov/.
129. Commissioner O of the, FDA Approves First Treatment for COVID-19 [Internet]. FDA2020, [cited 2023 Dec 31]; Available from: https://www.fda.gov/news-events/press-announcements/fda-approves-first-treatment-covid-19.
130. Ritonavir-Boosted Nirmatrelvir (Paxlovid) [Internet]. COVID-19 Treat. Guidel, [cited 2023 Dec 31]; Available from: https://www.covid19treatmentguidelines.nih.gov/therapies/antivirals-including-antibody-products/ritonavir-boosted-nirmatrelvir–paxlovid-/.
131. Alhumaid, S., Al Mutair, A., Alali, J., Al Dossary, N., Albattat, S.H., Al HajjiMohammed, S.M. *et al.*, Efficacy and Safety of Tixagevimab/Cilgavimab to Prevent COVID-19 (Pre-Exposure Prophylaxis): A Systematic Review and Meta-Analysis. *Diseases*, 10, 118, 2022.
132. Laupland, K.B. and Valiquette, L., Ebola virus disease. *Can. J. Infect. Dis. Med. Microbiol.*, 25, 128, 2014.
133. Yuan, J., Zhang, Y., Li, J., Zhang, Y., Wang, L.F., Shi, Z., Serological evidence of ebolavirus infection in bats, China. *Virol. J.*, 9, 236, 2012.
134. Report of a WHO/International Study Team. Ebola haemorrhagic fever in Sudan, 1976. Report of a WHO/International Study Team. *Bull. World Health Organ.*, 56, 247–70, 1978.
135. Ebola haemorrhagic fever in Zaire, 1976. *Bull. World Health Organ.*, 56, 271–93, 1978.
136. Gonzalez, J.P., Pourrut, X., Leroy, E., Ebolavirus and other filoviruses. *Curr. Top. Microbiol. Immunol.*, 315, 363–87, 2007.
137. Del Rio, C. and Guarner, J., Ebola: Implications and Perspectives. *Trans. Am. Clin. Climatol. Assoc.*, 126, 93–112, 2015.
138. Transmission | Ebola Hemorrhagic Fever | CDC [Internet], 2023, [cited 2024 Jan 1]; Available from: https://www.cdc.gov/vhf/ebola/transmission/index.html.
139. Rewar, S. and Mirdha, D., Transmission of Ebola Virus Disease: An Overview. *Ann. Glob. Health*, 80, 444–51, 2014.
140. Qureshi, A.I., Clinical Manifestations and Laboratory Diagnosis of Ebola Virus Infection, in: *Ebola Virus Dis.*, pp. 117–38, 2016.
141. Kiiza, P., Mullin, S., Teo, K., Adhikari, N.K.J., Fowler, R.A., Treatment of Ebola-related critical illness. *Intensive Care Med.*, 46, 285–97, 2020.
142. Taki, E., Ghanavati, R., Navidifar, T., Dashtbin, S., Heidary, M., Moghadamnia, M., Ebanga™: The most recent FDA-approved drug for treating Ebola. *Front. Pharmacol.*, 14, 1083429, 2023.
143. Ebola virus disease [Internet], [cited 2024 Jan 1]; Available from: https://www.who.int/health-topics/ebola.
144. HIV - Global [Internet], [cited 2024 Jan 1]; Available from: https://www.who.int/health-topics/hiv-aids.

145. Immune responses to human fungal pathogens and therapeutic prospects | Nature Reviews Immunology [Internet], [cited 2024 Jan 1]; Available from: https://www.nature.com/articles/s41577-022-00826-w.
146. Ugoeze, K.C., Phytopharmaceuticals for Treating Sexually Transmitted Diseases [Internet], in: *Herbal Drugs for the Management of Infectious Diseases*, pp. 179–261, John Wiley & Sons, Ltd, New Jersey, 2022, [cited 2024 Jan 1], Available from: https://onlinelibrary.wiley.com/doi/abs/10.1002/9781119818779.ch8.
147. The Basics of HIV Prevention | NIH [Internet], [cited 2024 Jan 1]; Available from: https://hivinfo.nih.gov/understanding-hiv/fact-sheets/basics-hiv-prevention.
148. HIV Treatment: The Basics | NIH [Internet], [cited 2024 Jan 1]; Available from: https://hivinfo.nih.gov/understanding-hiv/fact-sheets/hiv-treatment-basics.
149. Diphtheria [Internet], [cited 2024 Jan 1]; Available from: https://www.who.int/news-room/questions-and-answers/item/diphtheria.
150. Diphtheria: Causes and How It Spreads | CDC [Internet], 2023, [cited 2024 Jan 1]; Available from: https://www.cdc.gov/diphtheria/about/causes-transmission.html.
151. Abdulrasheed, N., Lawal, L., Mogaji, A.B., Abdulkareem, A.O., Shuaib, A.K., Adeoti, S.G. *et al.*, Recurrent diphtheria outbreaks in Nigeria: A review of the underlying factors and remedies. *Immun. Inflamm. Dis.*, 11, e1096, 2023.
152. Diagnosis, Treatment, and Complications | CDC [Internet], 2022, [cited 2024 Jan 2]; Available from: https://www.cdc.gov/diphtheria/about/diagnosis-treatment.html.
153. CDC, Types of Influenza Viruses [Internet]. *Cent. Dis. Control Prev.*, 2023. [cited 2024 Jan 2]; Available from: https://www.cdc.gov/flu/about/viruses/types.htm.
154. Influenza (Seasonal) [Internet], [cited 2024 Jan 2]; Available from: https://www.who.int/news-room/fact-sheets/detail/influenza-(seasonal).
155. Society P guidelines from the BI, Society BT, in collaboration with the Department of Health and HPA. Pandemic flu: clinical management of patients with an influenza-like illness during an influenza pandemic. *Thorax*, 62, i1–46, 2007.
156. Chow, E.J., Doyle, J.D., Uyeki, T.M., Influenza virus-related critical illness: prevention, diagnosis, treatment. *Crit. Care*, 23, 214, 2019.
157. CDC, Practice Good Health Habits [Internet]. *Cent. Dis. Control Prev.*, 2021, [cited 2024 Jan 2]; Available from: https://www.cdc.gov/flu/prevent/actions-prevent-flu.htm.
158. Measles [Internet], [cited 2024 Jan 5]; Available from: https://www.who.int/news-room/fact-sheets/detail/measles.
159. Moss, W.J., Griffin, D.E., Feinstone, W.H., Measles, in: *Vaccines Biodefense Emerg. Neglected Dis*, pp. 551–65, 2009.
160. Measles - Medical Information | Occupational Safety and Health Administration [Internet], [cited 2024 Jan 5]; Available from: https://www.osha.gov/measles/medical-information.
161. Measles infection - Symptoms, diagnosis and treatment | BMJ Best Practice [Internet], [cited 2024 Jan 5]; Available from: https://bestpractice.bmj.com/topics/en-gb/217.
162. Urbano, P.R., Fujita, D.M., Romano, C.M., Reemergence of mumps in São Paulo, Brazil - the urgent need for booster shot campaign to prevent a serious infectious disease. *Rev. Soc. Bras. Med. Trop.*, 50, 535–8, 2017.
163. Lozo, S., Ahmed, A., Chapnick, E., O'Keefe, M., Minkoff, H., Presumed Cases of Mumps in Pregnancy: Clinical and Infection Control Implications. *Infect. Dis. Obstet. Gynecol.*, 2012, e345068, 2012.
164. CDC, Mumps | For Healthcare Providers | CDC [Internet]. *Cent. Dis. Control Prev.*, 2021, [cited 2024 Jan 5]; Available from: https://www.cdc.gov/mumps/hcp.html.

165. Mumps [Internet], [cited 2024 Jan 5]; Available from: https://www.who.int/teams/health-product-policy-and-standards/standards-and-specifications/vaccine-standardization/mumps.
166. Gouma, S., Koopmans, M.P.G., van Binnendijk, R.S., Mumps virus pathogenesis: Insights and knowledge gaps. *Hum. Vaccines Immunother.*, 12, 3110–2, 2016.
167. Mumps - Treatment [Internet], nhs.uk2017, [cited 2024 Jan 5]; Available from: https://www.nhs.uk/conditions/mumps/treatment/.
168. Weller, T.H. and Neva, F.A., Propagation in Tissue Culture of Cytopathic Agents from Patients with Rubella-Like Illness. *Proc. Soc. Exp. Biol. Med.*, 111, 215–25, 1962.
169. Frey, T.K., Molecular biology of rubella virus. *Adv. Virus Res.*, 44, 69–160, 1994.
170. Lambert, N., Strebel, P., Orenstein, W., Icenogle, J., Poland, G.A., Rubella. *Lancet Lond. Engl.*, 385, 2297–307, 2015.
171. Rubella [Internet], [cited 2024 Jan 6]; Available from: https://www.who.int/news-room/fact-sheets/detail/rubella.
172. Rubella Transmission | CDC [Internet], 2022, [cited 2024 Jan 10]; Available from: https://www.cdc.gov/rubella/about/transmission.html.
173. Gregg, N.M., Congenital cataract following German measles in the mother. 1941. *Epidemiol. Infect.*, 107, iii–xiv, 1991.
174. Plotkin, S.A., Rubella Eradication: Not Yet Accomplished, but Entirely Feasible. *J. Infect. Dis.*, 224, S360–6, 2021.
175. Fernández-García, M., Ares-Arroyo, M., Wedel, E., Montero, N., Barbas, C., Rey-Stolle, M.F. *et al.*, Multiplatform Metabolomics Characterization Reveals Novel Metabolites and Phospholipid Compositional Rules of Haemophilus influenzae Rd KW20. *Int. J. Mol. Sci.*, 24, 11150, 2023.
176. Langereis, J.D. and de Jonge, M.I., Invasive Disease Caused by Nontypeable Haemophilus influenzae. *Emerg. Infect. Dis.*, 21, 1711–8, 2015.
177. Drevets, D.A., Leenen, P.J.M., Greenfield, R.A., Invasion of the Central Nervous System by Intracellular Bacteria. *Clin. Microbiol. Rev.*, 17, 323–47, 2004.
178. Haemophilus influenzae: Types of Infection | CDC [Internet], 2022, [cited 2024 Jan 10]; Available from: https://www.cdc.gov/hi-disease/about/types-infection.html.
179. Agrawal, A. and Murphy, T.F., Haemophilus influenzae Infections in the H. influenzae Type b Conjugate Vaccine Era ▿. *J. Clin. Microbiol.*, 49, 3728–32, 2011.
180. Walls, S., Adhikari, P., Reesha, H., Baxter, M., Sepsis Due to Haemophilus influenzae and Masked Influenza A Infection. *Cureus*, 14, e22424, 2022.
181. Effect of ceiling fan in mitigating exposure to airborne pathogens and COVID-19 - Brijesh Pandey, Sandip K Saha, Rangan Banerjee, 2023, [Internet] [cited 2024 Jan 10]; Available from: https://journals.sagepub.com/doi/10.1177/1420326X231154011.
182. Sánchez, F., Mensa, J., Martínez, J.A., Angrill, J., Marcos, M.A., Marco, F. *et al.*, Pneumonia caused by Haemophilus influenzae. Study in a series of 58 patients. *Rev. Espanola Quimioter. Publicacion Of. Soc. Espanola Quimioter.*, 12, 369–74, 1999.
183. Cardoso, M.P., Pasternak, J., Giglio, A.E., Casagrande, R.R.D., Troster, E.J., Meningitis due to Haemophilus influenzae type f. *Einstein*, 11, 521–3, 2013.
184. Zainel, A., Mitchell, H., Sadarangani, M., Bacterial Meningitis in Children: Neurological Complications, Associated Risk Factors, and Prevention. *Microorganisms*, 9, 535, 2021.
185. Haemophilus influenzae Infections - Infections [Internet]. MSD Man. Consum. Version, [cited 2024 Jan 10]; Available from: https://www.msdmanuals.com/home/infections/bacterial-infections-gram-negative-bacteria/haemophilus-influenzae-infections.
186. Sexually transmitted infections (STIs) [Internet], [cited 2024 Jan 10]; Available from: https://www.who.int/health-topics/sexually-transmitted-infections.

187. Workowski, K.A., Bachmann, L.H., Chan, P.A., Johnston, C.M., Muzny, C.A., Park, I. *et al.*, Sexually Transmitted Infections Treatment Guidelines, 2021. *MMWR Recomm. Rep.*, 70, 1–187, 2021.
188. CDC, STD Diseases & Related Conditions [Internet], 2023, [cited 2024 Jan 10]; Available from: https://www.cdc.gov/std/general/default.htm.
189. Anguzu, G., Flynn, A., Musaazi, J., Kasirye, R., Atuhaire, L.K., Kiragga, A.N. *et al.*, Relationship between Socioeconomic Status and Risk of Sexually Transmitted Infections in Uganda: Multilevel Analysis of a Nationally Representative Survey. *Int. J. STD AIDS*, 30, 284–91, 2019.
190. Du, M., Yan, W., Jing, W., Qin, C., Liu, Q., Liu, M. *et al.*, Increasing incidence rates of sexually transmitted infections from 2010 to 2019: an analysis of temporal trends by geographical regions and age groups from the 2019 Global Burden of Disease Study. *BMC Infect. Dis.*, 22, 574, 2022.
191. James, C., Harfouche, M., Welton, N.J., Turner, K.M., Abu-Raddad, L.J., Gottlieb, S.L. *et al.*, Herpes simplex virus: global infection prevalence and incidence estimates, 2016. *Bull. World Health Organ.*, 98, 315–29, 2020.
192. Human papillomavirus and cancer [Internet], [cited 2024 Jan 10]; Available from: https://www.who.int/news-room/fact-sheets/detail/human-papilloma-virus-and-cancer.
193. Global health sector strategies on HIV, viral hepatitis and sexually transmitted infections for the period 2022-2030 [Internet], [cited 2024 Jan 10]; Available from: https://www.who.int/publications-detail-redirect/9789240053779.
194. Gonorrhea | NIAID: National Institute of Allergy and Infectious Diseases [Internet], [cited 2024 Jan 10]; Available from: https://www.niaid.nih.gov/diseases-conditions/gonorrhea.
195. Chlamydia [Internet], [cited 2024 Jan 10]; Available from: https://www.who.int/news-room/fact-sheets/detail/chlamydia.
196. Syphilis | NIAID: National Institute of Allergy and Infectious Diseases [Internet], 2023, [cited 2024 Jan 10]; Available from: https://www.niaid.nih.gov/diseases-conditions/syphilis.
197. Trichomoniasis [Internet], [cited 2024 Jan 10]; Available from: https://www.who.int/news-room/fact-sheets/detail/trichomoniasis.
198. Chancroid - STI Treatment Guidelines [Internet], 2021, [cited 2024 Jan 10]; Available from: https://www.cdc.gov/std/treatment-guidelines/chancroid.htm.
199. HIV and AIDS [Internet], [cited 2024 Jan 10]; Available from: https://www.who.int/news-room/fact-sheets/detail/hiv-aids.
200. Management of non-gonococcal urethritis | BMC Infectious Diseases | Full Text [Internet], [cited 2024 Jan 10]; Available from: https://bmcinfectdis.biomedcentral.com/articles/10.1186/s12879-015-1043-4.
201. Prevention CC for DC and. Lice: Pubic [Internet], 2023, [cited 2024 Jan 10]; Available from: https://www.cdc.gov/parasites/lice/pubic/index.html.
202. STD Facts - Pelvic Inflammatory Disease [Internet], 2022, [cited 2024 Jan 10]; Available from: https://www.cdc.gov/std/pid/stdfact-pid.htm.
203. Leslie, S.W., Sajjad, H., Kumar, S., Genital Warts [Internet], in: *StatPearls*, StatPearls Publishing, Treasure Island (FL, 2023, [cited 2024 Jan 10]. Available from: http://www.ncbi.nlm.nih.gov/books/NBK441884/.
204. Nutrition C for FS and A. Salmonella (Salmonellosis). FDA [Internet], 2023, [cited 2024 Jan 10]; Available from: https://www.fda.gov/food/foodborne-pathogens/salmonella-salmonellosis.
205. Salmonellosis | Gastroenteritis, Causes, Transmission, & Symptoms | Britannica [Internet], 2024, [cited 2024 Jan 10]; Available from: https://www.britannica.com/science/salmonellosis.
206. Salmonella (non-typhoidal) [Internet], [cited 2024 Jan 10]; Available from: https://www.who.int/news-room/fact-sheets/detail/salmonella-(non-typhoidal).

207. Foodborne pathogens - PMC [Internet], [cited 2024 Jan 10]; Available from: https://www.ncbi.nlm.nih.gov/pmc/articles/PMC6604998/.
208. Cholera [Internet], [cited 2024 Jan 11]; Available from: https://www.who.int/news-room/fact-sheets/detail/cholera.
209. General Information | Cholera | CDC [Internet], 2023, [cited 2024 Jan 11]; Available from: https://www.cdc.gov/cholera/general/index.html.
210. Cholera is endangering children globally | UNICEF [Internet], [cited 2024 Jan 11]; Available from: https://www.unicef.org/stories/cholera-is-endangering-children-globally.
211. Cholera | The deadly disease explained [Internet]. Br. Red Cross, [cited 2024 Jan 11]; Available from: https://www.redcross.org.uk/stories/health-and-social-care/health/cholera-the-deadly-disease-explained.
212. Hepatitis [Internet], [cited 2024 Jan 11]; Available from: https://www.hopkinsmedicine.org/health/conditions-and-diseases/hepatitis.
213. Hepatitis [Internet], [cited 2024 Jan 11]; Available from: https://www.who.int/health-topics/hepatitis.
214. Hepatitis | Definition, Types, Causes, & Symptoms | Britannica [Internet], 2023, [cited 2024 Jan 11]; Available from: https://www.britannica.com/science/hepatitis.
215. Pati, I., Cruciani, M., Candura, F., Massari, M.S., Piccinini, V., Masiello, F. *et al.*, Hyperimmune Globulins for the Management of Infectious Diseases. *Viruses*, 15, 1543, 2023.
216. Yan, Y.P., Su, H.X., Ji, Z.H., Shao, Z.J., Pu, Z.S., Epidemiology of Hepatitis B Virus Infection in China: Current Status and Challenges. *J. Clin. Transl. Hepatol.*, 2, 15–22, 2014.
217. Das, S., Ramakrishnan, K., Behera, S.K., Ganesapandian, M., Xavier, A.S., Selvarajan, S., Hepatitis B Vaccine and Immunoglobulin: Key Concepts. *J. Clin. Transl. Hepatol.*, 7, 165–71, 2019.
218. Leone, N. and Rizzetto, M., Natural history of hepatitis C virus infection: from chronic hepatitis to cirrhosis, to hepatocellular carcinoma. *Minerva Gastroenterol. Dietol.*, 51, 31–46, 2005.
219. Ghany, M.G., Strader, D.B., Thomas, D.L., Seeff, L.B., Diagnosis, management, and treatment of hepatitis C: An update. *Hepatol. Baltim. Md*, 49, 1335–74, 2009.
220. Direct-Acting Antiviral Agents for Hepatitis C Virus Infection - PMC [Internet], [cited 2024 Jan 11]; Available from: https://www.ncbi.nlm.nih.gov/pmc/articles/PMC4193385/.
221. Sagnelli, C., Sagnelli, E., Russo, A., Pisaturo, M., Occhiello, L., Coppola, N., HBV/HDV Co-Infection: Epidemiological and Clinical Changes, Recent Knowledge and Future Challenges. *Life*, 11, 169, 2021.
222. Hepatitis A and Hepatitis E: Clinical and Epidemiological Features, Diagnosis, Treatment, and Prevention - PMC [Internet], [cited 2024 Jan 11]; Available from: https://www.ncbi.nlm.nih.gov/pmc/articles/PMC7581387/.
223. Hepatitis E Virus: Foodborne, Waterborne and Zoonotic Transmission - PMC [Internet], [cited 2024 Jan 11]; Available from: https://www.ncbi.nlm.nih.gov/pmc/articles/PMC3823334/.
224. Jefferies, M., Rauff, B., Rashid, H., Lam, T., Rafiq, S., Update on global epidemiology of viral hepatitis and preventive strategies. *World J. Clin. Cases*, 6, 589–99, 2018.
225. Bendich, A., Physiological role of antioxidants in the immune system. *J. Dairy Sci.*, 76, 2789–94, 1993.
226. Brambilla, D., Mancuso, C., Scuderi, M.R., Bosco, P., Cantarella, G., Lempereur, L. *et al.*, The role of antioxidant supplement in immune system, neoplastic, and neurodegenerative disorders: a point of view for an assessment of the risk/benefit profile. *Nutr. J.*, 7, 29, 2008.
227. Imdad, A., Mayo-Wilson, E., Herzer, K., Bhutta, Z.A., Vitamin A supplementation for preventing morbidity and mortality in children from six months to five years of age. *Cochrane Database Syst. Rev.*, 3, CD008524, 2017.

228. How the Immune System Protects You From Infection | Pfizer [Internet], [cited 2024 Jan 11]; Available from: https://www.pfizer.com/news/articles/how_the_immune_system_protects_you_from_infection.
229. Nicholson, L.B., The immune system. *Essays Biochem.*, 60, 275–301, 2016.
230. Immune Function and Micronutrient Requirements Change over the Life Course - PMC [Internet], [cited 2024 Jan 11]; Available from: https://www.ncbi.nlm.nih.gov/pmc/articles/PMC6212925/.
231. Tan, B.L., Norhaizan, M.E., Liew, W.P.P., Nutrients and Oxidative Stress: Friend or Foe? *Oxid. Med. Cell. Longev.*, 2018, 9719584, 2018.
232. Munteanu, C. and Schwartz, B., The relationship between nutrition and the immune system. *Front. Nutr.*, 9, 1082500, 2022.
233. Toledano, J.M., Puche-Juarez, M., Moreno-Fernandez, J., Ochoa, J.J., Diaz-Castro, J., Antioxidant and Immune-Related Implications of Minerals in COVID-19: A Possibility for Disease Prevention and Management. *Antioxid. Basel Switz.*, 12, 1104, 2023.
234. Ratnam, D.V., Ankola, D.D., Bhardwaj, V., Sahana, D.K., Kumar, M.N.V.R., Role of antioxidants in prophylaxis and therapy: A pharmaceutical perspective. *J. Control. Release*, 113, 189–207, 2006.
235. Bouayed, J. and Bohn, T., Exogenous antioxidants—Double-edged swords in cellular redox state. *Oxid. Med. Cell. Longev.*, 3, 228–37, 2010.
236. Ashaolu, T.J., Immune boosting functional foods and their mechanisms: A critical evaluation of probiotics and prebiotics. *Biomed. Pharmacother.*, 130, 110625, 2020.
237. Baker, R.E., Mahmud, A.S., Miller, I.F., Rajeev, M., Rasambainarivo, F., Rice, B., *et al.*, Infectious disease in an era of global change. *Nat. Rev. Microbiol.*, 20, 193–205, 2022, [Internet] [cited 2024 Jan 15]; Available from: https://www.nature.com/articles/s41579-021-00639-z.
238. Dye, C., After 2015: infectious diseases in a new era of health and development. *Philos. Trans. R. Soc. Lond. B. Biol. Sci.*, 369, 20130426, 2014.
239. Zulhendri, F., Chandrasekaran, K., Kowacz, M., Ravalia, M., Kripal, K., Fearnley, J. *et al.*, Antiviral, Antibacterial, Antifungal, and Antiparasitic Properties of Propolis: A Review. *Foods*, 10, 1360, 2021.
240. Vaccines and immunization [Internet], [cited 2024 Jan 15]; Available from: https://www.who.int/health-topics/vaccines-and-immunization.
241. Praditya, D., Kirchhoff, L., Brüning, J., Rachmawati, H., Steinmann, J., Steinmann, E., Anti-infective Properties of the Golden Spice Curcumin. *Front. Microbiol.*, 10, 912, 2019.
242. Oyebode, O., Kandala, N.B., Chilton, P.J., Lilford, R.J., Use of traditional medicine in middle-income countries: a WHO-SAGE study. *Health Policy Plan.*, 31, 984–91, 2016.
243. Irani, M., Sarmadi, M., Bernard, F., Ebrahimi pour, G.H., Shaker Bazarnov, H., Leaves Antimicrobial Activity of Glycyrrhiza glabra L. *Iran. J. Pharm. Res. IJPR*, 9, 425–8, 2010.
244. Dias, D.A., Urban, S., Roessner, U., A Historical Overview of Natural Products in Drug Discovery. *Metabolites*, 2, 303–36, 2012.
245. Newman, D.J. and Cragg, G.M., Natural Products as Sources of New Drugs from 1981 to 2014. *J. Nat. Prod.*, 79, 629–61, 2016.
246. Mullard, A., 2023 FDA approvals. *Nat. Rev. Drug Discov.*, 23, 88-95, 2024, [Internet] [cited 2024 Jan 15]; Available from: https://www.nature.com/articles/d41573-024-00001-x.
247. Heinonen, I.M. and Meyer, A.S., Antioxidants in fruits, berries and vegetables, in: *Fruit and vegetable processing: Improving quality*, W. Jongen (Ed.). pp. 23–51, CRC Press, Boca Raton, FL, 2022.
248. Ugoeze, K.C. and Odeku, O.A., Chapter 7 - Herbal bioactive–based cosmetics [Internet], in: *Herbal Bioactive-Based Drug Delivery Systems*, I.S. Bakshi, R. Bala, R. Madaan, R.K. Sindhu

(Eds.), pp. 195–226, Academic Press, Massachusetts, 2022, [cited 2024 Jan 16]. Available from: https://www.sciencedirect.com/science/article/pii/B9780128243855000145.

249. Iravani, S. and Zolfaghari, B., Pharmaceutical and nutraceutical effects of Pinus pinaster bark extract. *Res. Pharm. Sci.*, 6, 1–11, 2011.
250. Ugoeze, K.C., Amadi, N., Okoronkwo, N.A., Abali, S.O., Oluigbo, K.E., Chinko, B.C., GC-FID guided Identification and Quantification of detectable Phytochemicals in selected Commercial Chamomile Herbal Tea. *Int. J. Appl. Biol. Pharm. Technol.*, 14, 1–11, 2023.
251. Lin, J.K., Lin, C.L., Liang, Y.C., Lin-Shiau, S.Y., Juan, I.M., Survey of Catechins, Gallic Acid, and Methylxanthines in Green, Oolong, Pu-erh, and Black Teas. *J. Agric. Food. Chem.*, 46, 3635–42, 1998.
252. Devasagayam, T.P.A., Tilak, J.C., Boloor, K.K., Sane, K.S., Ghaskadbi, S.S., Lele, R.D., Free radicals and antioxidants in human health: current status and future prospects. *J. Assoc. Physicians India*, 52, 794–804, 2004.
253. Saha, M.R., Dey, P., Begum, S., De, B., Chaudhuri, T.Kr., Sarker, D.D. *et al.*, Effect of Acacia catechu (L.f.) Willd. on Oxidative Stress with Possible Implications in Alleviating Selected Cognitive Disorders. *PLoS One*, 11, e0150574, 2016.
254. Chockalingam, V., Kadali, S.S., Gnanasambantham, P., Antiproliferative and antioxidant activity of Aegle marmelos (Linn.) leaves in Dalton's Lymphoma Ascites transplanted mice. *Indian J. Pharmacol.*, 44, 225–9, 2012.
255. Marefati, N., Ghorani, V., Shakeri, F., Boskabady, M., Kianian, F., Rezaee, R. *et al.*, A review of anti-inflammatory, antioxidant, and immunomodulatory effects of Allium cepa and its main constituents. *Pharm. Biol.*, 59, 285–300, 2021.
256. Capasso, A., Antioxidant Action and Therapeutic Efficacy of Allium sativum L. *Molecules*, 18, 690–700, 2013.
257. Hęś, M., Dziedzic, K., Górecka, D., Jędrusek-Golińska, A., Gujska, E., Aloe vera (L.) Webb.: Natural Sources of Antioxidants – A Review. *Plant Foods Hum. Nutr. Dordr. Neth.*, 74, 255–65, 2019.
258. Drishya, S., Dhanisha, S.S., Guruvayoorappan, C., Antioxidant-rich fraction of Amomum subulatum fruits mitigates experimental methotrexate-induced oxidative stress by regulating TNF-α, IL-1β, and IL-6 proinflammatory cytokines. *J. Food Biochem.*, 46, e13855, 2022.
259. Mussard, E., Cesaro, A., Lespessailles, E., Legrain, B., Berteina-Raboin, S., Toumi, H., Andrographolide, A Natural Antioxidant: An Update. *Antioxidants*, 8, 571, 2019.
260. Kongkaneramit, L., Witoonsaridsilp, W., Peungvicha, P., Ingkaninan, K., Waranuch, N., Sarisuta, N., Antioxidant activity and antiapoptotic effect of Asparagus racemosus root extracts in human lung epithelial H460 cells. *Exp. Ther. Med.*, 2, 143–8, 2011.
261. Alzohairy, M.A., Therapeutics Role of Azadirachta indica (Neem) and Their Active Constituents in Diseases Prevention and Treatment. *Evid.-Based Complement. Altern. Med. ECAM*, 2016, 7382506, 2016.
262. Ugoeze, K., Aja, P., Nwachukwu, N., Chinko, B., Egwurugwu, J., Oluigbo, K., Evaluation of the Wound healing Potentials of Aqueous Topical creams Containing Aqueous extract of Azadirachta indica leaves as Bioactive Ingredient. *J. Pharm. Pharmacol. Res.*, 05, 176–187, 2021.
263. Simpson, T., Pase, M., Stough, C., Bacopa monnieri as an Antioxidant Therapy to Reduce Oxidative Stress in the Aging Brain. *Evid.-Based Complement. Altern. Med. ECAM*, 2015, 615384, 2015.
264. Sehrawat, A. and Kumar, V., Butein imparts free radical scavenging, anti-oxidative and proapoptotic properties in the flower extracts of Butea monosperma. *Biocell Off. J. Soc. Latinoam. Microsc. Electron. Al*, 36, 63–71, 2012.

265. Pagliari, S., Forcella, M., Lonati, E., Sacco, G., Romaniello, F., Rovellini, P. *et al.*, Antioxidant and Anti-Inflammatory Effect of Cinnamon (Cinnamomum verum J. Presl) Bark Extract after *In Vitro* Digestion Simulation. *Foods*, 12, 452, 2023.
266. Pandey, A.K., Mishra, A.K., Mishra, A., Antifungal and antioxidative potential of oil and extracts derived from leaves of Indian spice plant Cinnamomum tamala. *Cell. Mol. Biol. Noisy-Gd. Fr.*, 58, 142–7, 2012.
267. Menon, V.P. and Sudheer, A.R., Antioxidant and anti-inflammatory properties of curcumin. *Adv. Exp. Med. Biol.*, 595, 105–25, 2007.
268. Evaluation of Aqueous Creams Containing Ethanolic Extract of Curcuma longa (Turmeric) as Bioactive Ingredient for the Management of Wounds [Internet], [cited 2023 Dec 27]; Available from: https://www.fortunejournals.com/articles/evaluation-of-aqueous-creams-containing-ethanolic-extract-of-curcuma-longa-turmeric-as-bioactive-ingredient-for-the-management-of-.html.
269. Bhattacharya, A., Chatterjee, A., Ghosal, S., Bhattacharya, S.K., Antioxidant activity of active tannoid principles of Emblica officinalis (amla). *Indian J. Exp. Biol.*, 37, 676–80, 1999.
270. Reigada, I., Moliner, C., Valero, M.S., Weinkove, D., Langa, E., Gómez Rincón, C., Antioxidant and Antiaging Effects of Licorice on the Caenorhabditis elegans Model. *J. Med. Food*, 23, 72–8, 2020.
271. Saravanan, N. and Nalini, N., Antioxidant effect of Hemidesmus indicus on ethanol-induced hepatotoxicity in rats. *J. Med. Food*, 10, 675–82, 2007.
272. Sreepriya, M., Devaki, T., Balakrishna, K., Apparanantham, T., Effect of Indigofera tinctoria Linn on liver antioxidant defense system during D-galactosamine/endotoxin-induced acute hepatitis in rodents. *Indian J. Exp. Biol.*, 39, 181–4, 2001.
273. Martínez, G., Delgado, R., Pérez, G., Garrido, G., Núñez Sellés, A.J., León, O.S., Evaluation of the *in vitro* antioxidant activity of Mangifera indica L. extract (Vimang). *Phytother. Res. PTR*, 14, 424–7, 2000.
274. Sathishsekar, D. and Subramanian, S., Antioxidant properties of Momordica Charantia (bitter gourd) seeds on Streptozotocin induced diabetic rats. *Asia Pac. J. Clin. Nutr.*, 14, 153–8, 2005.
275. Bordoni, L., Fedeli, D., Nasuti, C., Maggi, F., Papa, F., Wabitsch, M. *et al.*, Antioxidant and Anti-Inflammatory Properties of Nigella sativa Oil in Human Pre-Adipocytes. *Antioxidants*, 8, 51, 2019.
276. Ramesh, B. and Satakopan, V.N., Antioxidant Activities of Hydroalcoholic Extract of Ocimum sanctum Against Cadmium Induced Toxicity in Rats. *Indian J. Clin. Biochem.*, 25, 307–10, 2010.
277. Al-Dabbagh, B., Elhaty, I.A., Al Hrout, A., Al Sakkaf, R., El-Awady, R., Ashraf, S.S. *et al.*, Antioxidant and anticancer activities of Trigonella foenum-graecum, Cassia acutifolia and Rhazya stricta. *BMC Complement. Altern. Med*, 18, 240, 2018.
278. Bhattacharya, A., Ghosal, S., Bhattacharya, S.K., Anti-oxidant effect of Withania somnifera glycowithanolides in chronic footshock stress-induced perturbations of oxidative free radical scavenging enzymes and lipid peroxidation in rat frontal cortex and striatum. *J. Ethnopharmacol.*, 74, 1–6, 2001.
279. Mashhadi, N.S., Ghiasvand, R., Askari, G., Hariri, M., Darvishi, L., Mofid, M.R., Anti-Oxidative and Anti-Inflammatory Effects of Ginger in Health and Physical Activity: Review of Current Evidence. *Int. J. Prev. Med.*, 4, S36–42, 2013.
280. Nieto, G., Ros, G., Castillo, J., Antioxidant and Antimicrobial Properties of Rosemary (Rosmarinus officinalis, L.): A Review. *Medicines*, 5, 98, 2018.
281. Burns, J., Yokota, T., Ashihara, H., Lean, M.E.J., Crozier, A., Plant foods and herbal sources of resveratrol. *J. Agric. Food. Chem.*, 50, 3337–40, 2002.

282. Weiskirchen, S. and Weiskirchen, R., Resveratrol: How Much Wine Do You Have to Drink to Stay Healthy?123. *Adv. Nutr.*, 7, 706–18, 2016.
283. Falchetti, R., Fuggetta, M.P., Lanzilli, G., Tricarico, M., Ravagnan, G., Effects of resveratrol on human immune cell function. *Life Sci.*, 70, 81–96, 2001.
284. Zhang, C., Tian, Y., Yan, F., Kang, X., Han, R., Sun, G. *et al.*, Modulation of growth and immunity by dietary supplementation with resveratrol in young chickens receiving conventional vaccinations. *Am. J. Vet. Res.*, 75, 752–9, 2014.
285. Cui, Q., Fu, Q., Zhao, X., Song, X., Yu, J., Yang, Y. *et al.*, Protective effects and immunomodulation on piglets infected with rotavirus following resveratrol supplementation. *PLoS One*, 13, e0192692, 2018.
286. Euba, B., López-López, N., Rodríguez-Arce, I., Fernández-Calvet, A., Barberán, M., Caturla, N. *et al.*, Resveratrol therapeutics combines both antimicrobial and immunomodulatory properties against respiratory infection by nontypeable Haemophilus influenzae. *Sci. Rep.*, 7, 12860, 2017.
287. Guo, N.H., Fu, X., Zi, F.M., Song, Y., Wang, S., Cheng, J., The potential therapeutic benefit of resveratrol on Th17/Treg imbalance in immune thrombocytopenic purpura. *Int. Immunopharmacol.*, 73, 181–92, 2019.
288. Salehi, B., Mishra, A.P., Nigam, M., Sener, B., Kilic, M., Sharifi-Rad, M. *et al.*, Resveratrol: A Double-Edged Sword in Health Benefits. *Biomedicines*, 6, 91, 2018.
289. Weber, K., Schulz, B., Ruhnke, M., Resveratrol and its antifungal activity against Candida species. *Mycoses*, 54, 30–3, 2011.
290. Abedini, E., Khodadadi, E., Zeinalzadeh, E., Moaddab, S.R., Asgharzadeh, M., Mehramouz, B. *et al.*, A Comprehensive Study on the Antimicrobial Properties of Resveratrol as an Alternative Therapy. *Evid.-Based Complement. Altern. Med. ECAM*, 2021, 8866311, 2021.
291. Huang, H., Liao, D., Zhou, G., Zhu, Z., Cui, Y., Pu, R., Antiviral activities of resveratrol against rotavirus *in vitro* and *in vivo*. *Phytomed. Int. J. Phytother. Phytopharm.*, 77, 153230, 2020.
292. Feng, W.Y., Tanaka, R., Inagaki, Y., Saitoh, Y., Chang, M.O., Amet, T. *et al.*, Pycnogenol, a procyanidin-rich extract from French maritime pine, inhibits intracellular replication of HIV-1 as well as its binding to host cells. *Jpn. J. Infect. Dis.*, 61, 279–85, 2008.
293. Matsumori, A., Higuchi, H., Shimada, M., French maritime pine bark extract inhibits viral replication and prevents development of viral myocarditis. *J. Card. Fail.*, 13, 785–91, 2007.
294. Rohdewald, P. and Beil, W., *In vitro* inhibition of Helicobacter pylori growth and adherence to gastric mucosal cells by Pycnogenol. *Phytother. Res. PTR*, 22, 685–8, 2008.
295. Torras, M.A.C., Faura, C.A., Schönlau, F., Rohdewald, P., Antimicrobial activity of Pycnogenol. *Phytother. Res. PTR*, 19, 647–8, 2005.
296. Aboulwafa, M.M., Youssef, F.S., Gad, H.A., Altyar, A.E., Al-Azizi, M.M., Ashour, M.L., A Comprehensive Insight on the Health Benefits and Phytoconstituents of Camellia sinensis and Recent Approaches for Its Quality Control. *Antioxidants*, 8, 455, 2019, [Internet] [cited 2024 Jan 14]; Available from: https://www.ncbi.nlm.nih.gov/pmc/articles/PMC6826564/.
297. Taylor, P.W., Hamilton-Miller, J.M.T., Stapleton, P.D., Antimicrobial properties of green tea catechins. *Food Sci. Technol. Bull.*, 2, 71–81, 2005.
298. GC-FID guided Identification and Quantification of detectable Phytochemicals in selected Commercial Chamomile Herbal Tea [Internet], [cited 2023 Dec 27]; Available from: https://imsear.searo.who.int/items/9626bcb1-5f3a-46b7-b4e3-3cb12b1b04e2.
299. Reygaert, W.C., Green Tea Catechins: Their Use in Treating and Preventing Infectious Diseases. *Biomed Res. Int.*, 2018, 9105261, 2018.
300. Musial, C., Kuban-Jankowska, A., Gorska-Ponikowska, M., Beneficial Properties of Green Tea Catechins. *Int. J. Mol. Sci.*, 21, 1744, 2020.

301. Sirk, T.W., Brown, E.F., Friedman, M., Sum, A.K., Molecular binding of catechins to biomembranes: relationship to biological activity. *J. Agric. Food. Chem.*, 57, 6720–8, 2009.
302. Tallei, T.E., Fatimawali, NJ, N., Idroes, R., Zidan, B.M.R.M., Mitra, S. *et al.*, A Comprehensive Review of the Potential Use of Green Tea Polyphenols in the Management of COVID-19. *Evid.-Based Complement. Altern. Med. ECAM*, 2021, 7170736, 2021.
303. Reygaert, W.C., The antimicrobial possibilities of green tea. *Front. Microbiol.*, 5, 434, 2014.
304. Elsayed, E., Al-Mohammed, H., Al-Badry, S., Synergism and Postantibiotic Effect of Green Tea Extract and Imipenem Against Methicillin-resistant Staphylococcus aureus. *Microbiol. J.*, 1, 89–96, 2011.
305. An efflux pump (MexAB-OprM) of Pseudomonas aeruginosa is associated with antibacterial activity of Epigallocatechin-3-gallate (EGCG) - PubMed [Internet], [cited 2024 Jan 13]; Available from: https://pubmed.ncbi.nlm.nih.gov/29157815/.
306. Antibacterial activity of epigallocatechin-3-gallate (EGCG) and its synergism with β-lactam antibiotics sensitizing carbapenem-associated multidrug resistant clinical isolates of Acinetobacter baumannii - PubMed [Internet], [cited 2024 Jan 13]; Available from: https://pubmed.ncbi.nlm.nih.gov/28160861/.
307. Haghjoo, B., Lee, L.H., Habiba, U., Tahir, H., Olabi, M., Chu, T.C., The synergistic effects of green tea polyphenols and antibiotics against potential pathogens. *Adv. Biosci. Biotechnol.*, 4, 959–67, 2013.
308. Marrelli, M., Amodeo, V., Statti, G., Conforti, F., Biological Properties and Bioactive Components of Allium cepa L.: Focus on Potential Benefits in the Treatment of Obesity and Related Comorbidities. *Molecules*, 24, 119, 2018.
309. Suleria, H.A.R., Butt, M.S., Anjum, F.M., Saeed, F., Khalid, N., Onion: nature protection against physiological threats. *Crit. Rev. Food Sci. Nutr.*, 55, 50–66, 2015.
310. Chakraborty, A.J., Uddin, T.M., Matin Zidan, B.M.R., Mitra, S., Das, R., Nainu, F. *et al.*, Allium cepa: A Treasure of Bioactive Phytochemicals with Prospective Health Benefits. Evid.-Based Complement. *Altern. Med. ECAM*, 2022, 4586318, 2022.
311. Prakash, D., Singh, B.N., Upadhyay, G., Antioxidant and free radical scavenging activities of phenols from onion (Allium cepa). *Food Chem.*, 102, 1389–93, 2007.
312. Teshika, J.D., Zakariyyah, A.M., Zaynab, T., Zengin, G., Rengasamy, K.R., Pandian, S.K. *et al.*, Traditional and modern uses of onion bulb (Allium cepa L.): a systematic review. *Crit. Rev. Food Sci. Nutr.*, 59, S39–70, 2019.
313. Santas, J., Almajano, M.P., Carbó, R., Antimicrobial and antioxidant activity of crude onion (*Allium cepa*, L.) extracts. *Int. J. Food Sci. Technol.*, 45, 403–9, 2010.
314. Tesfaye, A., Revealing the Therapeutic Uses of Garlic (Allium sativum) and Its Potential for Drug Discovery. *Sci. World J.*, 2021, 8817288, 2021.
315. Petrovska, B.B. and Cekovska, S., Extracts from the history and medical properties of garlic. *Pharmacogn. Rev.*, 4, 106–10, 2010.
316. Majewski, M., Allium sativum: facts and myths regarding human health. *Rocz. Panstw. Zakl. Hig.*, 65, 1–8, 2014.
317. Shang, A., Cao, S.Y., Xu, X.Y., Gan, R.Y., Tang, G.Y., Corke, H. *et al.*, Bioactive Compounds and Biological Functions of Garlic (Allium sativum L.). *Foods*, 8, 246, 2019.
318. Ahmed, T. and Wang, C.K., Black Garlic and Its Bioactive Compounds on Human Health Diseases: A Review. *Molecules*, 26, 5028, 2021.
319. Cheng, H., Huang, G., Huang, H., The antioxidant activities of garlic polysaccharide and its derivatives. *Int. J. Biol. Macromol.*, 145, 819–26, 2020.
320. Tsao, S.M. and Yin, M.C., *In-vitro* antimicrobial activity of four diallyl sulphides occurring naturally in garlic and Chinese leek oils. *J. Med. Microbiol.*, 50, 646–9, 2001.

321. Weber, N.D., Andersen, D.O., North, J.A., Murray, B.K., Lawson, L.D., Hughes, B.G., *In vitro* virucidal effects of Allium sativum (garlic) extract and compounds. *Planta Med.*, 58, 417–23, 1992.
322. Pai, S.T. and Platt, M.W., Antifungal effects of Allium sativum (garlic) extract against the Aspergillus species involved in otomycosis. *Lett. Appl. Microbiol.*, 20, 14–8, 1995.
323. Sabitha, P., Adhikari, P.M., Shenoy, S.M., Kamath, A., John, R., Prabhu, M.V. *et al.*, Efficacy of garlic paste in oral candidiasis. *Trop. Doct.*, 35, 99–100, 2005.
324. Hekmatpou, D., Mehrabi, F., Rahzani, K., Aminiyan, A., The Effect of Aloe Vera Clinical Trials on Prevention and Healing of Skin Wound: A Systematic Review. *Iran. J. Med. Sci.*, 44, 1–9, 2019.
325. Sánchez, M., González-Burgos, E., Iglesias, I., Gómez-Serranillos, M.P., Pharmacological Update Properties of Aloe Vera and its Major Active Constituents. *Molecules*, 25, 1324, 2020.
326. Harlev, E., Nevo, E., Lansky, E.P., Ofir, R., Bishayee, A., Anticancer potential of aloes: antioxidant, antiproliferative, and immunostimulatory attributes. *Planta Med.*, 78, 843–52, 2012.
327. Surjushe, A., Vasani, R., Saple, D.G., ALOE VERA: A SHORT REVIEW. *Indian J. Dermatol.*, 53, 163–6, 2008.
328. Gupta, A. and Rawat, S., Clinical importance of aloe vera: Review. *Res. J. Top. Cosmet. Sci.*, 8, 30–9, 2017.
329. Rodrigues, L.L.O., de Oliveira, A.C.L., Tabrez, S., Shakil, S., Khan, M.I., Asghar, M.N. *et al.*, Mutagenic, antioxidant and wound healing properties of Aloe vera. *J. Ethnopharmacol.*, 227, 191–7, 2018.
330. Inactivation of enveloped viruses by anthraquinones extracted from plants. - PMC [Internet], [cited 2024 Jan 14]; Available from: https://www.ncbi.nlm.nih.gov/pmc/articles/PMC245413/.
331. Okhuarobo, A., Falodun, J.E., Erharuyi, O., Imieje, V., Falodun, A., Langer, P., Harnessing the medicinal properties of Andrographis paniculata for diseases and beyond: a review of its phytochemistry and pharmacology. *Asian Pac. J. Trop. Dis.*, 4, 213–22, 2014.
332. Churiyah, Pongtuluran, O.B., Rofaani, E., Tarwadi, Antiviral and Immunostimulant Activities of Andrographis paniculata. *HAYATI J. Biosci.*, 22, 67–72, 2015.
333. Hossain, MdS., Urbi, Z., Sule, A., Rahman, K.M.H., Andrographis paniculata (Burm. f.) Wall. ex Nees: A Review of Ethnobotany, Phytochemistry, and Pharmacology. *Sci. World J.*, 2014, 274905, 2014.
334. Chemistry and Pharmacology of Andrographis Species | Semantic Scholar [Internet], [cited 2024 Jan 14]; Available from: https://www.semanticscholar.org/paper/CHEMISTRY-AND-PHARMACOLOGY-OF-ANDROGRAPHIS-SPECIES-Saxena-Jain/a8872cb9c0f986e7285976da4778a81885e79a48.
335. Leelarasamee, A., Trakulsomboon, S., Sittisomwong, N., Undetectable anti-bacterial activity of Andrographis paniculata (Burma) wall. ex ness. *J. Med. Assoc. Thail. Chotmaihet Thangphaet*, 73, 299–304, 1990.
336. Screening of anti-dengue activity in methanolic extracts of medicinal plants | BMC Complementary Medicine and Therapies | Full Text [Internet], [cited 2024 Jan 14]; Available from: https://bmccomplementmedtherapies.biomedcentral.com/articles/10.1186/1472-6882-12-3.
337. Fangkham, S., Ekalaksananan, T., Aromdee, C., Suebsasana, S., Kongyingyoes, B., Patarapadungkit, N. *et al.*, The effect of andrographolide on Human papillomavirus type 16 (HPV16) positive cervical cancer cells (SiHa). *Int. J. Infect. Dis.*, 16, e80, 2012.
338. Aromdee, C., Suebsasana, S., Ekalaksananan, T., Pientong, C., Thongchai, S., Stage of action of naturally occurring andrographolides and their semisynthetic analogues against herpes simplex virus type 1 *in vitro*. *Planta Med.*, 77, 915–21, 2011.
339. Adiguna, S.P., Panggabean, J.A., Atikana, A., Untari, F., Izzati, F., Bayu, A. *et al.*, Antiviral Activities of Andrographolide and Its Derivatives: Mechanism of Action and Delivery System. *Pharmaceuticals*, 14, 1102, 2021.

340. Sarkar, S., Singh, R.P., Bhattacharya, G., Exploring the role of Azadirachta indica (neem) and its active compounds in the regulation of biological pathways: an update on molecular approach. *3 Biotech.*, 11, 178, 2021.
341. Hossain, M.A., Shah, M.D., Sakari, M., Gas chromatography-mass spectrometry analysis of various organic extracts of Merremia borneensis from Sabah. *Asian Pac. J. Trop. Med.*, 4, 637–41, 2011.
342. Identification of antifungal compounds from the seed oil ofAzadirachta Indica | Phytoparasitica [Internet], [cited 2024 Jan 14]; Available from: https://link.springer.com/article/10.1007/BF02980677.
343. Yerima, M.B., Jodi, S.M., Oyinbo, K., Maishanu, H.M., Farouq, A.A., Junaidu, A.U. *et al.*, Effect of Neem Extracts (Azadirachta indica) on Bacteria Isolated from Adult Mouth. *Niger. J. Basic Appl. Sci.*, 20, 64–7, 2012.
344. Neem NRC (US) P on, Medicinals [Internet], in: *Neem: A Tree For Solving Global Problems*, National Academies Press, US, 1992, [cited 2024 Jan 14]. Available from: https://www.ncbi.nlm.nih.gov/books/NBK234637/.
345. Badam, L., Joshi, S.P., Bedekar, S.S., "*In vitro*" antiviral activity of neem (Azadirachta indica. A. Juss) leaf extract against group B coxsackieviruses. *J. Commun. Dis.*, 31, 79–90, 1999.
346. Mondall, N., Mojumdar, A., Chatterje, S., Banerjee, A., Datta, J., Gupta, S., Antifungal activities and chemical characterization of Neem leaf extracts on the growth of some selected fungal species *in vitro* culture medium. *J. Appl. Sci. Environ. Manag.*, 13, 1, 13, 2009, ISSN 1119-8362.
347. Kumari, A., Kumar, R., Maurya, S., Choudhary, J., Kumar, S., Antifungal efficacy of aqueous extracts of neem cake, karanj cake and vermicompost against some phytopathogenic fungi. *Bioscan-Int. Q. J. Life Sci.*, 8, 671–4, 2013. 0973-7049.
348. A Comprehensive Review on the Therapeutic Potential of Curcuma longa Linn. in Relation to its Major Active Constituent Curcumin - PMC [Internet], [cited 2024 Jan 15]; Available from: https://www.ncbi.nlm.nih.gov/pmc/articles/PMC8990857/.
349. Péret-Almeida, L., Cherubino, A.P.F., Alves, R.J., Dufossé, L., Glória, M.B.A., Separation and determination of the physico-chemical characteristics of curcumin, demethoxycurcumin and bisdemethoxycurcumin. *Food Res. Int.*, 38, 1039–44, 2005.
350. Khairani, S., Fauziah, N., Wiraswati, H.L., Panigoro, R., Setyowati, E.Y., Berbudi, A., The Potential use of a Curcumin-Piperine Combination as an Antimalarial Agent: A Systematic Review. *J. Trop. Med.*, 2021, e9135617, 2021.
351. Jakubczyk, K., Drużga, A., Katarzyna, J., Skonieczna-Żydecka, K., Antioxidant Potential of Curcumin—A Meta-Analysis of Randomized Clinical Trials. *Antioxidants*, 9, 1092, 2020.
352. Wei, Z.Q., Zhang, Y.H., Ke, C.Z., Chen, H.X., Ren, P., He, Y.L. *et al.*, Curcumin inhibits hepatitis B virus infection by down-regulating cccDNA-bound histone acetylation. *World J. Gastroenterol.*, 23, 6252–60, 2017.
353. Chen, D.Y., Shien, J.H., Tiley, L., Chiou, S.S., Wang, S.Y., Chang, T.J. *et al.*, Curcumin inhibits influenza virus infection and haemagglutination activity. *Food Chem.*, 119, 1346–51, 2010.
354. von Rhein, C., Weidner, T., Henß, L., Martin, J., Weber, C., Sliva, K. *et al.*, Curcumin and Boswellia serrata gum resin extract inhibit chikungunya and vesicular stomatitis virus infections *in vitro*. *Antiviral Res.*, 125, 51–7, 2016.
355. Butnariu, M., Quispe, C., Koirala, N., Khadka, S., Salgado-Castillo, C.M., Akram, M. *et al.*, Bioactive Effects of Curcumin in Human Immunodeficiency Virus Infection Along with the Most Effective Isolation Techniques and Type of Nanoformulations. *Int. J. Nanomed.*, 17, 3619–32, 2022.
356. Kutluay, S.B., Doroghazi, J., Roemer, M.E., Triezenberg, S.J., Curcumin inhibits herpes simplex virus immediate-early gene expression by a mechanism independent of p300/CBP histone acetyltransferase activity. *Virology*, 373, 239–47, 2008.

357. Curcumin suppressed the proliferation and apoptosis of HPV-positive cervical cancer cells by directly targeting the E6 protein - PubMed [Internet], [cited 2024 Jan 15]; Available from: https://pubmed.ncbi.nlm.nih.gov/37157900/.
358. Maghsoudi, A., Yazdian, F., Shahmoradi, S., Ghaderi, L., Hemati, M., Amoabediny, G., Curcumin-loaded polysaccharide nanoparticles: Optimization and anticariogenic activity against Streptococcus mutans. *Mater. Sci. Eng. C Mater. Biol. Appl.*, 75, 1259–67, 2017.
359. Antibacterial evaluation, phytochemical screening and ascorbic acid assay of turmeric (Curcuma longa) - MedCrave online [Internet], [cited 2024 Jan 15]; Available from: https://medcraveonline.com/MOJBB/antibacterial-evaluation-phytochemical-screening-and-ascorbic-acid-assay-of-turmeric-curcuma-longa.html.
360. Mahady, G.B., Pendland, S.L., Yun, G., Lu, Z.Z., Turmeric (Curcuma longa) and curcumin inhibit the growth of Helicobacter pylori, a group 1 carcinogen. *Anticancer Res.*, 22, 4179–81, 2002.
361. Martins, C.V.B., da Silva, D.L., Neres, A.T.M., Magalhães, T.F.F., Watanabe, G.A., Modolo, L.V. *et al.*, Curcumin as a promising antifungal of clinical interest. *J. Antimicrob. Chemother.*, 63, 337–9, 2009.
362. Khattak, S., Saeed-ur-Rehman, ., Ullah Shah, H., Ahmad, W., Ahmad, M., Biological effects of indigenous medicinal plants Curcuma longa and Alpinia galanga. *Fitoterapia*, 76, 254–7, 2005.
363. Armanini, D., Fiore, C., Mattarello, M.J., Bielenberg, J., Palermo, M., History of the endocrine effects of licorice. *Exp. Clin. Endocrinol. Diabetes Off. J. Ger. Soc. Endocrinol. Ger. Diabetes Assoc.*, 110, 257–61, 2002.
364. Lim, T.K., Glycyrrhiza glabra, in: *Edible Med. Non-Med. Plants*, pp. 354–457, 2015.
365. Gupta, V.K., Fatima, A., Faridi, U., Negi, A.S., Shanker, K., Kumar, J.K. *et al.*, Antimicrobial potential of Glycyrrhiza glabra roots. *J. Ethnopharmacol.*, 116, 377–80, 2008.
366. Fukui, H., Goto, K., Tabata, M., Two antimicrobial flavanones from the leaves of Glycyrrhiza glabra. *Chem. Pharm. Bull. (Tokyo)*, 36, 4174–6, 1988.
367. Wang, L., Yang, R., Yuan, B., Liu, Y., Liu, C., The antiviral and antimicrobial activities of licorice, a widely-used Chinese herb. *Acta Pharm. Sin. B*, 5, 310–5, 2015.
368. Simmler, C., Pauli, G.F., Chen, S.N., Phytochemistry and Biological Properties of Glabridin. *Fitoterapia*, 0, 160–84, 2013.
369. Chakotiya, A.S., Tanwar, A., Srivastava, P., Narula, A., Sharma, R.K., Effect of aquo-alchoholic extract of Glycyrrhiza glabra against Pseudomonas aeruginosa in Mice Lung Infection Model. *Biomed. Pharmacother. Biomedecine Pharmacother.*, 90, 171–8, 2017.
370. Sato, J., Goto, K., Nanjo, F., Kawai, S., Murata, K., Antifungal activity of plant extracts against Arthrinium sacchari and Chaetomium funicola. *J. Biosci. Bioeng.*, 90, 442–6, 2000.
371. Ajagannanavar, S.L., Battur, H., Shamarao, S., Sivakumar, V., Patil, P.U., Shanavas, P., Effect of aqueous and alcoholic licorice (glycyrrhiza glabra) root extract against streptococcus mutans and lactobacillus acidophilus in comparison to chlorhexidine: an *in vitro* study. *J. Int. Oral Health JIOH*, 6, 29–34, 2014.
372. Chandra, J.H. and Gunasekaran, H., Screening of phytochemical, antimicrobial and antioxidant activity of Glycyrrhiza glabra root extract. *J. Environ. Biol.*, 38, 161–5, 2017.
373. Lee, J.Y., Lee, J.H., Park, J.H., Kim, S.Y., Choi, J.Y., Lee, S.H. *et al.*, Liquiritigenin, a licorice flavonoid, helps mice resist disseminated candidiasis due to Candida albicans by Th1 immune response, whereas liquiritin, its glycoside form, does not. *Int. Immunopharmacol.*, 9, 632–8, 2009.
374. Huang, W., Chen, X., Li, Q., Li, P., Zhao, G., Xu, M. *et al.*, Inhibition of intercellular adhesion in herpex simplex virus infection by glycyrrhizin. *Cell Biochem. Biophys.*, 62, 137–40, 2012.
375. Sekizawa, T., Yanagi, K., Itoyama, Y., Glycyrrhizin increases survival of mice with herpes simplex encephalitis. *Acta Virol.*, 45, 51–4, 2001.

376. van Rossum, T.G., Vulto, A.G., Hop, W.C., Schalm, S.W., Pharmacokinetics of intravenous glycyrrhizin after single and multiple doses in patients with chronic hepatitis C infection. *Clin. Ther.*, 21, 2080–90, 1999.
377. Huan, C.-C., Wang, H.-X., Sheng, X.-X., Wang, R., Wang, X., Mao, X., Glycyrrhizin inhibits porcine epidemic diarrhea virus infection and attenuates the proinflammatory responses by inhibition of high mobility group box-1 protein. *Arch. Virol.*, 162, 1467–76, 2017.

7

Role of Antioxidant Therapy in Respiratory Disease Management

Bidisha Bhattacharya[1], Naina Devi[1], Sanshita[1], Ameya Sharma[2], Vivek Puri[2], Sandeep Kumar[3], Rakesh K. Sindhu[4] and Inderbir Singh[1]*

[1]Chitkara College of Pharmacy, Chitkara University, Punjab, India
[2]Chitkara University School of Pharmacy, Himachal Pradesh, India
[3]Amar Shaheed Baba Ajit Singh Jujhar Singh Memorial College of Pharmacy, Bela, Ropar, Punjab, India
[4]School of Pharmacy, Sharda University, Greater Noida, Uttar Pradesh, India

Abstract
Respiratory diseases exert a substantial impact and an indelible imprint on the quality of life of a significant portion of the global populace. Diseases such as Asthma, Chronic obstructive pulmonary disease, Idiopathic pulmonary fibrosis, etc., have a major association with reactive oxygen species and oxidative stress in their underlying fundamental pathogenesis. Antioxidants are substances that elucidate their action by slowing down or blocking the oxidation of other chemicals and have been found to exert significant protective action for management of pulmonary oxidant stress. This book chapter aims to excavate into the domain of antioxidants and how they are useful in management and mitigation of these respiratory diseases. A plethora of antioxidant compounds, along with their mechanistic role and their classification have also been discussed comprehensively. This chapter also gives an insight on the various drug delivery systems employed to help antioxidants reach their site of action. Lastly, the future perspectives pertaining to the role of antioxidants in respiratory diseases have also been holistically explored to lay out the possibilities of amalgamation of innovative and novel concepts such as personalized medicine, 3D printing etc., to improve the therapeutic efficacy of these antioxidants.

***Keywords*:** Antioxidants, reactive oxygen species, oxidative stress, inflammation

7.1 Introduction

Respiratory diseases entail an extensive range of pathological predicaments that influence the respiratory system, including the lungs, airways, and associated structures. These diseases, such as asthma, chronic obstructive pulmonary disease (COPD), and lung cancer, pose significant health challenges worldwide [1–3]. The portrayal of oxidative stress and inflammation in the precipitation and progression of respiratory diseases has attained

**Corresponding author*: inderbir.singh@chitkara.edu.in

Rakesh K. Sindhu, Inderbir Singh and M. Arockia Babu (eds.) Antioxidants: Nature's Defense Against Disease, (219–256)

substantial attention in recent years. In this context, antioxidants have emerged as potential therapeutic agents due to their capacity to counterbalance the detrimental influence of reactive oxygen species (ROS) and reduce inflammation in the respiratory system [4, 5]. Antioxidants are compounds that inhibit or neutralize ROS, which are molecules possessing a highly reactive nature, and originate as byproducts of systematic cellular metabolism [6]. Under normal physiological conditions, the body has built-in defense mechanisms including enzymes (superoxide dismutase, catalase, as well as glutathione peroxidase) and antioxidants that are non-enzymatic in nature (vitamins C and E, glutathione, and carotenoids), to maintain a balance between ROS generation and elimination [6]. However, in respiratory diseases, this delicate balance is disrupted, leading to an excessive manufacturing of ROS and oxidative stress. Oxidative stress is observed when there occurs disproportion between the body's ability to produce and detoxify ROS. ROS can deteriorate cellular components, including lipids, proteins, and DNA, leading to tissue inflammation, injury, and dysfunction. In the respiratory system, oxidative stress exerts a compelling role in the underlying pathogenesis of various diseases. For instance, in asthma, ROS contributes to airway inflammation, bronchoconstriction, and mucus hypersecretion [7, 8]. In COPD, oxidative stress is implicated in the progressive destruction of lung tissue and the development of emphysema. In lung cancer, ROS can promote genetic mutations and tumor progression [9].

Antioxidants act by scavenging ROS, thereby preventing their damaging effects. They can act by donating electrons or hydrogen atoms to stabilize free radicals, thereby breaking the chain reaction of oxidative damage [10, 11]. By neutralizing ROS, antioxidants help protect the respiratory system from oxidative stress-related injury and inflammation Moreover, antioxidants can modulate signaling pathways that are associated with immune responses, gene expression, and cell proliferation, further contributing to their beneficial effects in respiratory diseases. Several studies have investigated the potential possessed by antioxidants in the prevention and management of respiratory diseases. For example, clinical trials have shown that supplementation with antioxidants, such as vitamins C and E, help in improvement of, reduce airway inflammation, and reduce the frequency and severity of asthma attacks [12–15]. Similarly, in COPD patients, antioxidant therapy has demonstrated promising results in reducing oxidative stress, improving symptoms, and enhancing exercise capacity. Furthermore, preclinical studies have explored the use of natural antioxidants derived from plants, such as curcumin, resveratrol, and green tea polyphenols, in attenuating lung inflammation and inhibiting the proliferation of lung cancer cells. However, despite the growing body of evidence supporting the role of antioxidants in respiratory diseases, several challenges remain. The optimal dosage, duration, and combination of antioxidants need to be determined for different diseases and patient populations. Furthermore, the bioavailability and stability of antioxidants in the respiratory system need to be addressed to ensure their therapeutic efficacy. Additionally, more extensive and thorough research is imperative to elucidate the fundamental mechanisms of antioxidant action and their interactions with other therapeutic interventions [16, 17].

7.1.1 Oxygen and its Evolution

Oxygen is a vital element for sustaining life on Earth and exerts a pivotal role in a plethora of biological processes. Throughout the course of evolution, organisms have adapted to utilize oxygen for energy production and metabolism [18, 19]. Oxygen's evolution and its relationship with antioxidants are particularly significant in the context of respiratory diseases [20]. During the budding stages of Earth's evolution, the atmosphere chiefly consisted of gases like methane, ammonia, and water vapor. However, with the emergence of photosynthetic organisms, such as cyanobacteria, oxygen began to accumulate as a byproduct of photosynthesis. This event, quoted to as the Great Oxygenation Event, occurred around 2.4 billion years ago and marked a turning point in the history of life on Earth [21, 22].

The upsurge of oxygen in the atmosphere allowed for the development of aerobic respiration, which can be a more efficient process for extraction of energy from organic molecules [23]. This adaptation provided organisms with a significant evolutionary advantage, leading to the proliferation of oxygen-dependent life forms. However, oxygen's reactivity can also have detrimental effects. Reactive oxygen species (ROS), such as superoxide radicals and hydrogen peroxide, are generated during normal cellular metabolism. A skewness between the genesis of reactive oxygen species (ROS) and antioxidant defenses can lead to oxidative stress [24, 25]. Respiratory diseases, such as chronic obstructive pulmonary disease (COPD), asthma, and lung cancer, are interrelated with a splurge of oxidative stress in the lungs. Antioxidants, including enzymes like superoxide dismutase and molecules like vitamin C and E, help nullify ROS and protect cells from oxidative damage. Perceiving the intricate relationship between oxygen, antioxidants, and respiratory diseases is vital for developing therapeutic propositions to mitigate oxidative stress and improve respiratory health [26]. Ongoing research aims to unravel the complexities of this interplay, leading to potential advancements in the prevention and treatment of respiratory diseases [27].

7.1.2 Free Radicals

Respiratory diseases encompass a wide range of conditions affecting the lungs and airways, such as asthma, chronic obstructive pulmonary disease (COPD), and lung cancer [28]. Free radicals and oxidative stress play a compelling role in the precipitation and progression of these diseases. Antioxidants are substances that can counteract the harmful effects of free radicals and protect against oxidative damage [29]. In the sphere of respiratory diseases, antioxidants have been studied extensively for their potential therapeutic benefits [14, 30]. Here's how free radicals and antioxidants are related to respiratory diseases:

7.1.2.1 Free Radicals and Oxidative Stress

Unpaired electrons in free radicals make them extremely reactive compounds. They develop both as normal metabolic wastes and in reaction to external stimuli like air pollution, tobacco smoke, and some pharmaceuticals. Oxidative stress occurs when the body's antioxidant defense mechanisms become overwhelmed or when there is an excess production of free radicals [31, 32].

7.1.2.2 Role in Respiratory Diseases

Free radicals and oxidative stress are involved in the development and progression of respiratory diseases. In asthma, for example, oxidative stress contributes to airway inflammation and hyper responsiveness. In COPD, free radicals generated by cigarette smoke and inflammation cause damage to lung tissue and exacerbate inflammation. Oxidative stress also depicts a role in the augmentation of lung cancer, as it can promote DNA damage and mutations in lung cells [33, 34].

7.1.2.3 *Antioxidants and Respiratory Diseases*

Antioxidants are compounds with the plausible ability to reduce oxidative stress and neutralize free radicals. By contributing one electron, they scavenge free radicals without becoming unstable themselves. Endogenous antioxidants are those made by the body, whilst exogenous antioxidants are those received through food or supplementation. Vitamins C and E, beta-carotene, selenium, and flavonoids are all examples of common antioxidants [24].

7.1.2.4 *Potential Benefits*

Antioxidants have shown promise in respiratory diseases by causing a reduction in oxidative stress, inflammation, and improving lung function. Studies have investigated and documented the effects of antioxidants in various respiratory conditions. For example, in asthma, vitamin C supplementation has helped to improve lung function and reduce airway inflammation. N-acetylcysteine (NAC), a precursor of the antioxidant glutathione, has been studied in COPD and has demonstrated benefits in reducing exacerbations and improving symptoms. In lung cancer, antioxidants may have a safeguarding effect by reducing oxidative damage and inhibiting the growth of cancer cells [35–37].

7.1.2.5 *Limitations and Considerations*

While antioxidants hold therapeutic potential, it's important to note that clinical evidence supporting their use in respiratory diseases is still evolving [38]. Some studies have shown conflicting results, and the optimal dosages, formulations, and duration of antioxidant supplementation are yet to be determined. Additionally, excessive antioxidant intake may not always be beneficial and could have adverse effects. Therefore, it's important to consult with healthcare professionals before starting any antioxidant supplementation regimen [7, 24, 39].

7.1.3 Defense Mechanism

Protecting the respiratory system from oxidative damage and inflammation is one of the vital functions of antioxidants [7]. The increased manufacturing of reactive oxygen species (ROS) and oxidative stress in respiratory disorders such as asthma, COPD, and lung cancer can cause tissue damage and worsen the condition. Antioxidants act as defense mechanisms against these harmful effects by neutralizing ROS and maintaining redox balance [6, 40]. Here are some key defense mechanisms of antioxidants in respiratory diseases:

7.1.3.1 *Scavenging Reactive Oxygen Species (ROS)*

Antioxidants, such as vitamins C and E, beta-carotene, and selenium, can directly scavenge and neutralize ROS, including superoxide anions, hydroxyl radicals, and hydrogen peroxide. By doing so, they prevent oxidative damage to respiratory tissues and cellular components [41, 42].

7.1.3.2 Regeneration of Antioxidant Enzymes

Endogenous antioxidant enzymes including glutathione peroxidase, catalase, and superoxide dismutase (SOD) are nourished by antioxidants in their activity and regeneration. These enzymes work together to convert harmful ROS into less reactive species, thus reducing their damaging effects [43].

7.1.3.3 Inhibition of Inflammatory Responses

Antioxidants can modulate inflammatory responses in respiratory diseases by inhibiting the production of pro-inflammatory cytokines along with chemokines. They have been shown to inhibit the transcription factor nuclear factor-kappa B (NF-κB), which controls the production of inflammatory genes [44, 45].

7.1.3.4 Stabilization of Cell Membranes

Antioxidants help in maintaining the integrity and stability of cell membranes in the respiratory system. They protect against lipid peroxidation, a process by which ROS attack and damage cell membrane lipids, leading to cell dysfunction and death [46, 47].

7.1.3.5 Enhancement of Immune Function

Antioxidants can enhance immune function, including the activity of immune cells involved in respiratory defense, such as macrophages and natural killer cells [48, 49]. By boosting immune responses, antioxidants contribute to the clearance of pathogens and reduce the risk of respiratory infections [50].

It is important to note that while antioxidants have shown promise in preclinical and some clinical studies, their therapeutic efficacy in respiratory diseases is still being investigated. Further research is needed to determine optimal dosages, combinations, and modes of administration for different respiratory conditions.

7.2 Respiratory Diseases

Respiratory diseases are one of the most commonly occurring class that influence the patients mind, as well as their ability to carry out day to day tasks effectively. They are the diseases that influence the respiratory pathways and the lungs. Various types of respiratory diseases include Asthma, Chronic obstructive pulmonary disorder (COPD), Idiopathic pulmonary fibrosis (IPF), etc. [51]. There are several risk factors which pose a problem in breathing, and a few of them include smoking, pollution, and occupational exposures. Various changes occur in the normal homeostasis of the body, such as inflammation in lungs, lungs injury and airway remodeling. A few of the respiratory diseases have been discussed hereunder highlighting the role of ROS and oxidative stress in their precipitation.

7.2.1 Asthma

Asthma is a convoluted pathological condition of the lungs that is characterized by inflammation of the pulmonary airways and an elevation in the sensitivity to a variety of airway-narrowing stimuli, also called bronchial hyper responsiveness. Its symptoms may be short-term/transitory, occasional, or persistent. All over the world about 300 million people suffer from asthma and its extensiveness elevates by 50% every ten years. Patients suffering from asthma experience financial expenses of approximately \$300 to \$1300 per patient [52].

7.2.1.1 Etiology of Asthma

There are various factors governing the etiology of asthma and have been discussed hereunder. Genetics performs an extensive function in the augmentation and expansion of asthma and hypersensitivity reactions. Environmental exposures and conditions bring about extensive diversity in the genetic basis of asthma. Asthma is more prone to arise in an individual if that individual has a genetically close relation with that disease. Smoking in prenatal maternal has been persistently related to early childhood gasping/panting/wheezing. There exists a dose-response relationship between exposure and declined airway competence in early life. Augmentation of atopic disease was more prone among newborns delivered by urgent cesarean section but not with elective cesarean section. The probable reason for this includes maternal stress and variances in the gut of infants' microflora linked with different modes of delivery. Due to prenatal and postpartum exposure to tobacco smoke, there is a decrease in the caliber of the airway, which results in elevated bronchial responsiveness and symptoms of wheezing. Changes in lifestyle and chemical exposure because of epigenetic moderation of DNA affect the susceptibility to disease. Various factors affecting fetal genetic expression are nutrition and diet, smoking, exposure to microbial products, maternal stress, and care. These factors may cause environmental factors to moderate the genome of a child and cause prolonged asthma [52].

7.2.1.2 Pathophysiology of Asthma

To understand more about asthma, it is essential to know about its pathophysiology (Figure 7.1). In asthma the foremost symptoms like tightness in the chest, shortness of breath, and wheezing alternate over time and its intensity also increases and these changes occur due to different factors that trigger these changes. Due to these changes the structural cells present in the airway like the epithelial and endothelial cells start releasing mediators that produce inflammation and engage the cells like degranulated mast cells and Th2 lymphocytes that are responsible for keeping the inflammatory airway of patients with asthma in a specific manner. Interleukins also play a major role in inflammation like IL13 which is a Th2 cytokine that stimulates airway hyperreactivity, excess secretion of mucus, and airway remodeling. The inflammatory mediator's expression is moderated by the transcription of the gene, controlled by NF-κβ, AP-1, NF-AT (nuclear factor of the activated T cells), and STAT (signal transducers and activators of transcription). Rhinovirus and allergens trigger these factors and elevate inflammation during asthma which generates oxidative stress actuates AP-1 and NF-κβ and produces an inflammatory response [53]. Inflammatory cells

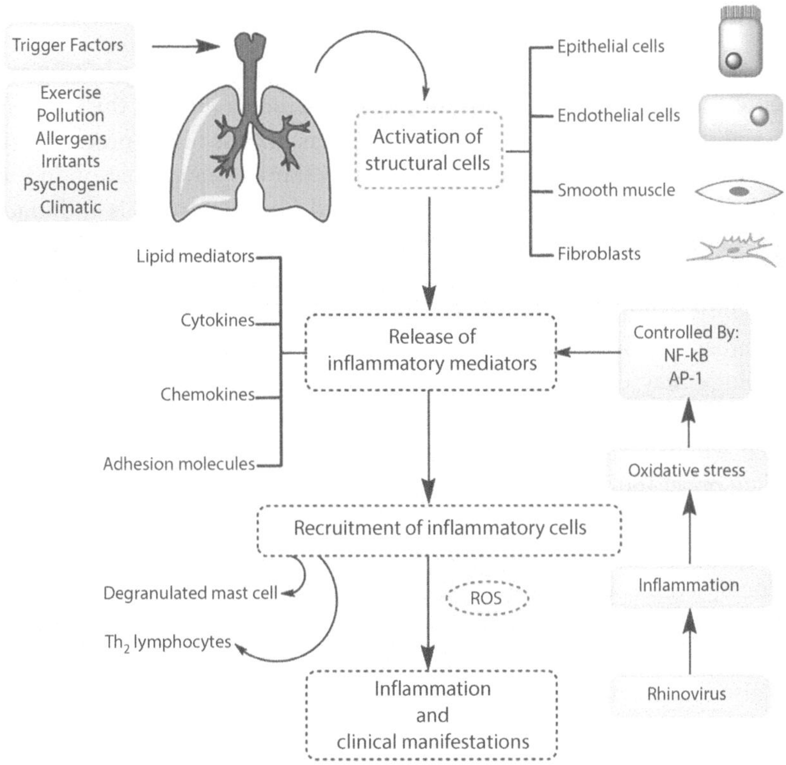

Figure 7.1 Pathophysiology of asthma.

like eosinophils, neutrophils, and macrophages produce ROS. Oxygen then dismutates to produce hydrogen peroxide (H_2O_2). Moreover, both of them react with metal ions present and generate OH in the biological system. This results in apoptosis i.e. programmed cell death, necrosis, and remodeling of the airway. Thus, elevated concentration of ROS acts as a crucial contributor to instigating inflammation in the airway [54].

7.2.2 Chronic Obstructive Pulmonary Disease (COPD)

As per the American Thoracic Society (ATS), COPD is a disease where the alveoli inside the lungs are damaged, leading to blockage in the airflow. As per the European Respiratory Society (ERS), COPD is a state where the maximum expiratory flow i.e., the amount of air breathed out per minute gets shortened, and the air gets trapped in the lungs. As per the Global Initiative for Chronic Lung Disease (GOLD), COPD is a disease that can be both treated and prevented. It has some extrapulmonary reactions that affect the patient's condition. The airflow blocking is not reversible and is linked with inflammatory feedback of the lungs to toxic and poisonous particles. According to WHO, it stands fifth in the list of diseases that cause death. A Dutch study predicted within another twenty years the ubiquity of COPD can increase by 76%. In the year of 1990, the WHO evaluated the European standardized mortality rate of COPD which was found to be 50 per 100,000 in males and 20 per 100,000 in females and was the reason of death in 4.7% men and 2.4% female [55].

7.2.2.1 Etiology of COPD

Emphysema is characterized as damage of the walls of alveoli and long-lasting and durable augmentation of the airspace away from the end bronchioles [56]. In this condition, there occurs the loss of the capillary in the lungs but there is no loss of the precapillary arteriole suggesting that the enclosing alveoli disappears when there is inflammation in the pathway of air. The reason is a type of protease that is liberated from neutrophils and macrophages, which digests the alveolar space around the small airway [57]. Patients suffering from emphysema usually have declined body mass index (BMI), dyspnea comparatively elevated, and poor life quality. COPD and asthma are two physiologically different diseases that show exclusive mechanisms. One of the features of asthma is bronchial hyper responsiveness but its pathophysiology may vary in asthma and COPD patients [58]. Some common differences between asthma and COPD are that asthma has an obstruction in the airway which can be reversed whereas COPD is irreversible [59]. Asthma generally sets early whereas COPD has a late onset [60]. In asthma, the barrier in the airway stops for some time and then starts again whereas in COPD the blockage develops steadily. Young patients are prone to asthma and patients with age above 60 are prone to develop COPD. Asthma patients usually have increased IgE and elevated cell production in smooth muscles. COPD patients have more secretion of goblet cells [61]. The response of asthma patients on inhaled therapy is satisfactory whereas COPD patients show substandard response. However, sometimes patients may also experience manifestations of both asthma and COPD. This is termed Asthma-COPD overlap syndrome (ACOS) [62]. Chronic Bronchitis is a familiar and prevailing but fluctuating event in COPD. In this clinical condition, the functioning of the lungs may progressively decline, the respiratory tract may outgrow a tendency to develop infection and may also lead to an elevated mortality rate [63]. Smokers are highly prone to develop blockage in the airway [58]. There exists some connection between chronic bronchitis and job-related vulnerability as any person exposed to any dust, gas or fumes may develop symptoms of chronic bronchitis.

7.2.2.2 Pathophysiology of COPD

Cigarette smoke, pollution in the environment due to nanoparticles and biomass fuels are exogenous oxidants. IL1, TNF α, and INF γ generate endogenous ROS and result in the formation of inflammatory stress. Oxidative stress plays a vital role in the pathophysiology of COPD. Some types of ROS that are present inside the cell are NADPH oxidase enzyme, XO, and hem peroxidase which are elevated in patients suffering from COPD. The free radicals instigate many adverse results at the cellular points. Like nuclear factor kappa-beta (NF-$\kappa\beta$) gets activated and elevates the generation of IL8 and TNF α due to which the neutrophils increase the inflammatory stress. Phosphoionsitide3-kinase (PI3K) also gets stimulated by oxidative stress which further adds a phosphate group to histone deacetylase (HDAC-2). Oxidative stress also reduces the anti-protease activity and results in protease-antiprotease imbalance. In COPD patients the expression of TGF-β which stops activity of SOD and catalase which are essential to neutralize ROS. Many antioxidants are controlled by nuclear erythroid-2 related factor (Nrf2) and people suffering from COPD have decreased function of Nrf2. When proteins, lipids, and DNA get oxidized by ROS it generates carbonyl stress

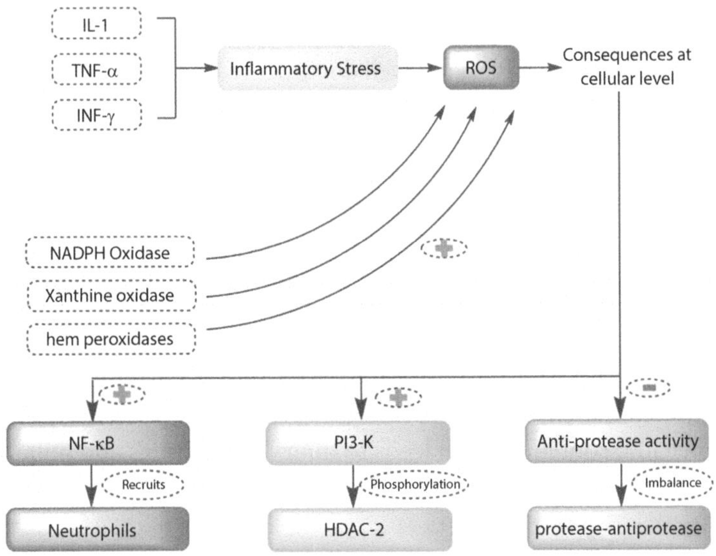

Figure 7.2 Pathophysiology of COPD.

which produces some reagents of carbonyl that produce reaction with proteins and this process is known as protein carbonylation. Even when the irritants stop getting exposed to the patients the inflammation and stress still continues to persist as the proteins which are modified by carbonyl are immunogenic and generate some antibodies which are already present in more amounts in patients of COPD. The innate immune system recognizes the carbonylated proteins by PRR (Pattern Recognition Receptors). Some PRRs that process input signals are Toll-like receptor (TLR), NOD-like Receptors (NLR), and cytosine DNA sensors. Cytokines, interferons and chemokines are secreted by the cells that express PRRs which activates innate and adaptive immune system [54]. The basic underlying mechanism of how ROS and oxidative stress are involved in COPD precipitation have been depicted in Figure 7.2.

7.2.3 Idiopathic Pulmonary Fibrosis (IPF)

Idiopathic pulmonary fibrosis (IPF) also known as cryptogenic fibrosing alveolitis is a persistent, incurable, rare and progressive disease with different and unknown etiologic backgrounds and diverse prognoses also known as fibrosing interstitial pneumonia which occurs basically in old age people and affects physical and emotional well-being. Basic symptoms of IPF are dyspnea on exertion, escalated cough, and bad lung functioning due to fibrosis in the majority of patients. There is only limited data available about the statistics of IPF which suggest that the incidence, mortality, and hospitalization rates of IPF patients are increasing progressively [64–68]. There are no proper therapy for the treatment of IPF [69–73].

7.2.3.1 Etiology of IPF

The etiology and exact cause of IPF is not well documented [74, 75]. However, probable risk factors for the augmentation of IPF are environmental exposures, smoking, occupational exposures etc. Many studies have concluded inhaled exposures to be the cause of IPF in several patients. Some investigations that were conducted on animal models suggest that due to the continuous arrangement of inciting agents alveolar epithelial cells get repetitive stress which helps in the pathogenesis of IPF. Some epidemiologic studies have concluded that geographic locations play a major role in the precipitation of IPF, especially in industrial areas where men are more prone to develop IPF in comparison to women due to increased exposure of men to smoke and dust. There are many indications that connect gastroesophageal reflux disease (GERD) to the development of IPF, but the actual cause is still not proven. It is assumed that GERD causes the gastric contents to microaspirate irrespective of acidic and nonacidic contents which leads to injury in the lungs and pulmonary fibrosis. Genetics is also correlated with the prevalence of IPF [76–79].

7.2.3.2 Pathophysiology of IPF

Patients suffering from IPF have an elevated amount of oxidative stress and a diminished amount of antioxidant defenses. Exogenous and endogenous epithelial and endothelial cells are the source of ROS. This ROS initiate apoptosis of epithelial cells and activate the response to stress of the ER of the epithelial cells (Figure 7.3). Some studies suggest that IPF causes apoptosis of the alveolar cell along with environmental stress and unusual cell

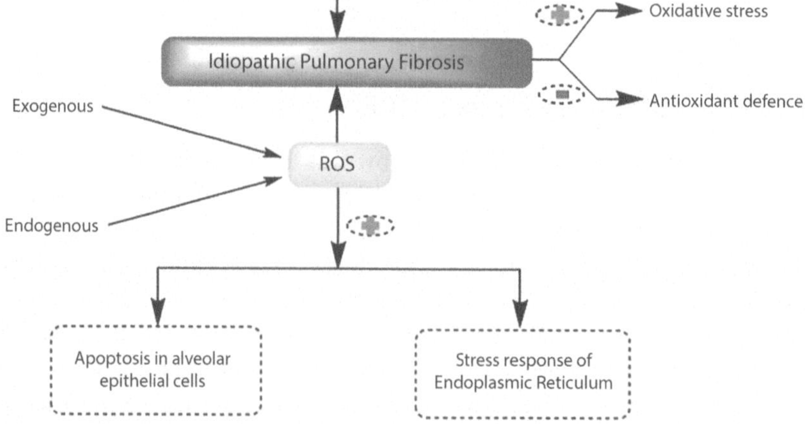

Figure 7.3 Pathophysiology of idiopathic pulmonary fibrosis.

cycle of alveolar epithelial cells. Gel-forming mucin (MUC5B) which was associated with IPF was discovered in chromosome 11p. Some factors like TGF-β, platelet-derived growth factor (PDGF), connective tissue growth factor (CTGF), fibroblast growth factor (FGF) and vascular endothelial growth factor (VEGF) generate signals by tyrosine kinases to synthesize collagen from myofibroblasts and obstruct the neighboring receptors. This is used as an alternate therapy [54].

7.2.4 Lung Cancer

In the last hundred years lung cancer has developed from an unusual disease to the most usual cause of death from cancer. The approximate number of lung cancer cases in US till 2018 was 121,680 for men and 112,350 for women. It is the most common cancer diagnosed, ranked 2nd. Behavioral, environmental, and some genetic risk factors act as a crucial part in the development of tumor [80].

7.2.4.1 Etiology of Lung Cancer

The usual cause of developing lung cancer is smoking. The carcinogenic substances which are present in smoke cause changes or mutations in the p53 gene which is necessary for causing carcinogenesis and deregulation of the cell cycle. Some dominant animal carcinogens like N-nitroso compound which are constituted in tobacco smoke are found to be present in the urine of smokers. Tobacco smoke is also a leading cause of lung cancer, i.e., passive smoking. The probability of an HIV-infected patient developing lung cancer is also high as it is the most familiar non-acquired immunodeficiency syndrome which may cause malignancy. Patients who are smokers and were provided radiotherapy during the treatment of breast cancer also have a chance of progressing lung cancer as smoking and exposure to radiation elevates the chances of developing lung cancer. Patients suffering from COPD and IPF have a higher chance of developing lung cancer. Exposure to harmful substances in the workplace like asbestos, silica, and polycyclic aromatic hydrocarbons also poses a risk of developing lung cancer [81].

7.2.4.2 Pathophysiology of Lung Cancer

Constant inflammation may develop the occurrence of genetically and phenotypically altered cells possessing malignant character. The proliferation of ROS/RNS in inflamed cells results in augmentation of the dysregulated procedure and results in the preneoplastic stage (Figure 7.4). Inflammation and oxidative stress are strongly related to each other and if these are not obstructed then it could lead to the changes that initiate carcinogenesis. Oxidative stress has an impact on various signal pathways which are linked with the propagation of cells like erythroid-related factor 2, the mitogen-activated protein kinase. Oxidative stress leads to alteration in the expression of genes, cell propagation, and apoptosis. It plays a critical function in the initiation and progression of tumors. Damage to DNA plays a vital function in mutations, loss of stability of genes, and epigenetic changes. Due to oxidative stress, some oncogenes and genes that suppress tumors get damaged and cause mutations which lead to cancer. 8-OHdG is a mutagenic DNA lesion that has an increased level of cancer.

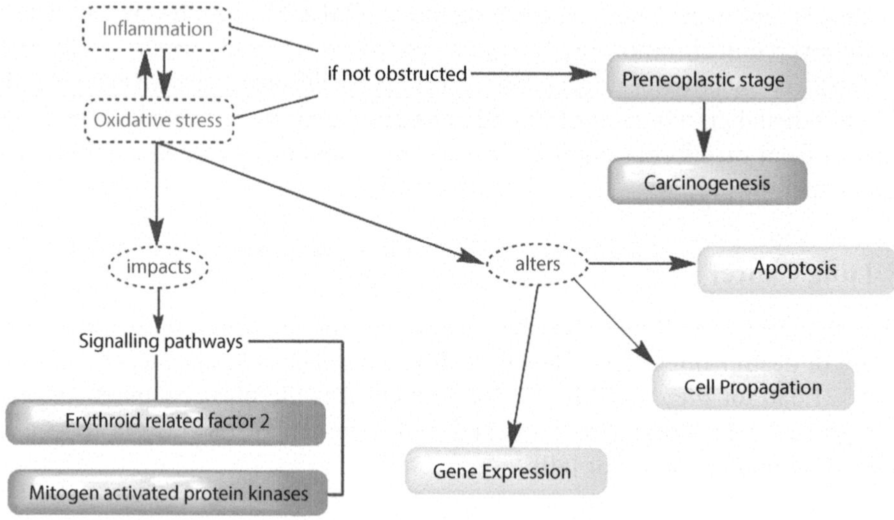

Figure 7.4 Pathophysiology of lung cancer.

It can tie up with adenine and cytosine but if any discrepancy occurs and is not recovered then the actual pairs (adenine: thymidine A:T) and (cytosine: guanine C:G) will transverse. And this type of mutation basically occurs in oncogenes and genes which suppress tumors. So, 8-OHdG is a biomarker for the analysis of oxidative stress. The most basic ROS that affect lipids are hydroxyl (HO) and hydroperoxyl (HO_2) radicals. Lipids when undergo oxidative damage cause lipid peroxidation (LPO) which causes a decrease in membrane properties. Different aldehydes are produced as secondary products during lipid peroxidation out of which MDA (malondialdehyde) is most extensively studied in different types of cancers. Patients suffering from breast cancer, bladder cancer, and prostate cancer have elevated levels of MDA due to more production of ROS and the absence of antioxidant defense [82].

7.2.5 Pulmonary Arterial Hypertension

When the mean pressure of the pulmonary artery measured during right heart catheterization is more than 25mmHg at rest, it is known as pulmonary hypertension. It describes a group that is differentiated hemodynamically by precapillary pulmonary hypertension. It affects approximately 1% of the whole population. The prevalence of PAH in Germany was 25.9 per 1 million adults in 2014 [83].

7.2.5.1 Etiology of Pulmonary Arterial Hypertension

Cigarette smoking is a major risk factor for the occurrence of pulmonary hypertension. It can transpire in any age group but most frequently in adults. The impact of hormones on pulmonary hypertension is profound [84]. According to many studies conducted the prevalence of pulmonary hypertension on women of childbearing age is more because, during pregnancy and labor pain, the need and demand for a pulmonary system is greater. People residing in high-altitude places show some symptoms of pulmonary hypertension, not those who visit frequently [85]. Babies born between 34 and 37 or after 41 gestational weeks have a higher risk of suffering from pulmonary hypertension. Babies born by cesarean section have a greater incidence of suffering from respiratory distress. Maternal pulmonary disorder diminishes the function of the placenta and causes hypoxemia in the fetus which instigates pulmonary hypertension [86].

7.2.5.2 Pathophysiology of Pulmonary Arterial Hypertension

Oxidative stress which assists in the uncontrolled and imprudent release of ROS plays a pivotal part in the pathogenesis of hypertension. Disablement caused by repositioning in the action of endothelium causing a decrease in vasodilatation and a pro-inflammatory state leading to vascular inflammation is mediated by ROS. Enzymatic and non-enzymatic sources of ROS like NADPH oxidase, xanthine oxidase, and NO synthase exert a critical role in causing hypertension (Figure 7.5). NADPH oxidase is a biochemical source of ROS in the kidney and plays a crucial role in renal dysfunction. The humoral and mechanical signal increase the amount of NADPH oxidase, the most impactful stimuli be AT-2, ET-1 and urotensin2, in patients suffering from hypertension. Activation of NADPH oxidase is a major cause of hypertension. Endothelial nitric oxide synthase (eNOS) produces NO which modulates vasodilation. Vital cofactors for encasing the action of eNOS are L-arginine and tetrahydrobiopterin (BH4), if deficient causes reduced production of NO. NADPH oxidase when combined with NO forms peroxynitrite which causes destabilization of eNOS and produces excess superoxide. Superoxide causes the oxidation of BH4 and increases the production of ROS. Xanthine oxidase catalyzes the step of metabolism of purine where oxygen gets reduced to superoxide. Some studies on hypertensive rats concluded that an increased amount of xanthine oxidase causes end-organ damage in hypertension. Mitochondria plays a dual role as it is both a source and target of ROS. Ubiquinol when partially reduced produces semiquinone form and an antioxidant when fully reduced. Maximum superoxide that is produced in the mitochondria of mammals is generated by Complex1. The production

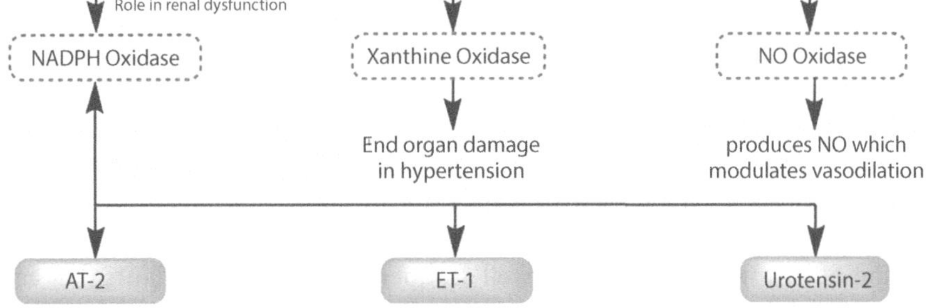

Figure 7.5 Pathophysiology of pulmonary arterial hypertension.

of superoxide from Complex 1 reduces even if mild uncoupling occurs. Patients suffering from hypertension show a reduction in the enzymatic activity of antioxidants. PGI2 relaxes the vascular smooth muscle cells. When ligands like thrombin or arachidonic acid bind, the production of PGI2 increases. Arachidonic acid is used as a substrate by prostaglandin H2 synthase to produce prostaglandin H2 which is converted to PGI2. Prostaglandin H2 synthase-2 which is an enzyme isoform in oxidative stress environment causes vascular dysfunction. The enzymatic pursuit of PGI2 synthase is inhibited by peroxynitrate thus diminishing vasodilation caused by PGI2 [86].

7.2.6 Bronchitis

The American College of Chest Physicians (ACCP) defines acute bronchitis as "an acute infection of the lower respiratory tract that is characterized predominantly by cough with or without phlegm production that lasts for up to 3 weeks. It is a clinical interpretation interpreted by cough, with or without secretion of sputum which occurs because of acute swelling of the trachea and airways, basically an infection in the lower respiratory tract. Repetitive production of sputum continuously for 3 months for more than 2 years is called chronic bronchitis [87]. Repetitive elevation in the volume of the secretion of mucus from the bronchi causes the production of sputum called simple chronic bronchitis. The major symptom of chronic bronchitis is expectoration.

7.2.6.1 Etiology of Bronchitis

Usually viruses are contemplated as the basic reason for acute bronchitis [88]. The viruses that cause lower respiratory tract diseases like; influenza B, influenza A, parainfluenza, as well as syncytial virus, and upper respiratory tract diseases like; coronavirus, and rhinovirus are commonly related to acute bronchitis. To date, *Bordetella pertussis, Mycoplasma pneumoniae,* and *Chlamydia pneumonia* are considered carriers of acute bronchitis [62].

7.3 Antioxidants and Their Therapies for Respiratory Diseases

Antioxidants show their action by slowing down or blocking the oxidation of other chemicals. They are broadly classified as enzymatic and non-enzymatic antioxidants (Table 7.1). Among these antioxidants, there are several compounds that have a distinct mechanism of action as well as end effects. This variety governs the individual function of each one of them in the body. Microorganisms such as actinomycetes, bacteria [89], cyanobacteria, fungi, and lichens are considered as the source of antioxidants [90]. These organisms can extend very fast under favorable and supervised conditions, due to which they are a good source of natural therapeutically active molecules for a plethora of applications including pharmaceutical, nutraceutical, as well as agricultural applications. All the antioxidants may differ on the basis of their capability to scavenge free radicals. Antioxidants can act via several mechanisms, including radical's scavenging, sequestration of ions of transition metal, decomposition of hydrogen peroxide, as well as, quenching of pro-oxidants that are active in nature, and improving endogenous antioxidant defense.

Table 7.1 Classification of antioxidants

S. no.	Antioxidant class	Example	Source	Mechanism of action
1.	Enzymatic	Superoxide dismutase (SOD) Catalase	In cells, proliferated by Xanthine oxidase, NADPH oxidase and mitochondrial electron transfer.	Superoxide dismutase transforms superoxide to hydrogen peroxide. Catalase is responsible for the conversion of hydrogen peroxide to water.
2.	Carotenoids	Lycopene Beta carotene	Tomatoes, apricots, papaya, carrots, green leaves, watermelons, etc.	Lycopene traps singlet oxygen. Beta carotene scavenges superoxide anions and peroxyl radicals.
3.	Flavonoids	Catechins Flavanols Flavones	Tea, apple, grape, onion, lettuce, cocoa, etc.	Flavonoids scavenge superoxide, hydrogen peroxide, hydroxyl radical and singlet oxygen.
4.	Isothiocyanates	Sulforaphane oltipraz	Broccoli, cabbage, cauliflower and kale.	Enhance the functions of phase II enzymes and increase the antioxidant activity of animal cells.
5.	Micronutrients	Selenium Zinc Copper Manganese	Brazil nuts, seafood, oysters, pumpkin seeds, pineapple, brown rice, spinach.	Selenium prevents lipid peroxidation. Zinc and copper are the crucial metals for CuZnSOD activity whereas manganese is crucial for MnSOD activity.
6.	Phenolic compounds	BHA t-BHQ Curcumin	Sugarcane, grape, almond, turmeric.	Enhances the activity of phase II enzymes.
7.	Thiols	NAC Cysteine Dithiols (lipoic acid) GS	Onion, egg, beef, whole grain.	These act asthiol/disulfide redox buffer, metal chelators, radical quenchers, substrates for specific redox reactions.
8.	Vitamins	Vitamin C Vitamin E	Oranges, bell peppers, tomatoes, white potatoes, sunflower oil and seeds, peanuts.	Vitamin C directly scavenges superoxide – by creating the semidehydro-ascorbate free radical followed by reduction in GSH. Vitamin E directly scavenges Reactive oxygen species.

Once the concept of antioxidants and their classification has been covered. The next aspect that comes into the picture is how they are useful in management of respiratory diseases. The various antioxidant therapies have been discussed hereunder.

7.3.1 Supplements

Most of the pathological conditions of the lungs such as COPD, Asthma, Bronchitis, and IPF are the major causes of mortality and morbidity. The main aim of the therapeutic strategies is focused on to limit the progression of disease by stopping the inflammation and various infections as these diseases cannot be cured. There are various supplements that can be used to reduce the inflammation and to improve the respiratory health [91].

7.3.1.1 Vitamin D

Vitamin D exhibits a salient role in many physiological mechanisms such as inflammation, oxidative stress, etc., the metabolism of calcium and phosphorus [92]. Vitamin D gets biosynthesized in the body's outer largest organ skin by conversion of 7-Dehydrocholestrol to previtamin D followed by its isomerization to vitamin D under the impact of UV radiation. It can also be consumed by dietary intake via fish oils or other oral supplements. Vitamin D is responsible for enhancing innate as well as adaptive immunity. Its deficiency is connected to various pathological conditions such as COPD, Asthma, IPF, and Bronchitis [91].

Patients suffering from asthma are vitamin D deficient and its supplementation may increase the chances of improvement in the disease's severity. Regulatory T cells (T_{Regs}) can develop as both naturally occurring T_{Regs} as well as induced adaptive T_{Regs}. Naturally occurring T_{Regs} are denoted as FoxP3, although induced T_{Regs} can be denoted as FoxP3+ or FoxP3-. Activation of CD4+FoxP3+T_{Regs} with repressive activity that can be achieved by preliminary treatment of dendritic cells with Vitamin D and consequent coculture with CD4+ cells. 1, 25-Dihydroxyvitamin D acts precisely on CD4+ T-cells aiding in IL-10 secreting T_{Reg} population. IL-10 suppresses many activities correlated to asthma. As glucocorticoids can also be used in the suppression of asthma so there could be a correlation of low levels of Vitamin D with glucocorticoids [93].

COPD has systemic effects and correlations that may affect the health of the patient as well as fatality rate. Most of the critical co-existing conditions (such as cardiovascular diseases and cancer) has been seen in patients [94]. An age-matched study (controlled) has been carried out which showed the association of vitamin D deficiency with COPD moreover it could be caused by high catabolism of vitamin D because of glucocorticoids, poor diet, and aging [95].

7.3.1.2 Vitamin A

Vitamin A has a crucial role in the key biological processes which include the following: - (i) Growth, (ii) Vision, (iii) Reproduction, and (iv) Immune responses [96]. Vitamin A can be obtained from two dietary sources which are known as pre-formed vitamin A, also recognized by another name that is retinol and provitamin A for which another name is carotenoids [97]. Natural sources for retinol are animal sources which include milk, eggs, etc. [98]. There is an association between oxidative stress and asthma [99–101].

Vitamin A might improve asthmatic conditions by reducing oxidative stress [102], or via enhancing the immune system [103]. It also increases the protective immune responses of Th2 (such as interleukin (IL) 4 expression) [104].

Lycopene, which is known to be a carotenoid, and its supplementation was shown to decrease asthma [105]. However, there are no primary randomized controlled trials that have yet been performed of vitamin A supplementation for the prevention and treatment of asthma. There are many studies of vitamin A supplementation that have been conducted, but such studies gave variable results [106]. Women who consume carotenoids rich foods have lower risk of asthma [107]. Among the pro-vitamin A carotenoids, the carotenoid which is most studied is β-carotene and that is because of its shielding characteristics against free radicals and oxidative damage's effects. A meta-analysis had been conducted which includes five observational studies that showed there was not significant relationship between the asthma and β-carotene intake [108]. This meta-analysis study had limited sample size. Results of two Randomized controlled trial concluded that modification in the dietary intake of carotenoids rich food [109] or lycopene-rich tomato products [110] showed improved asthma control with better lung functions, in adults who were suffering from asthma.

7.3.2 Herbal Antioxidants

7.3.2.1 Curcumin

Curcumin, which is derived from turmeric, possesses anti-inflammatory activities in addition to anti-tumor, antifungal and antioxidant activities [111, 112]. The use of animal models has indicated that curcumin is likely to act as a potent anti-inflammatory agent in the pulmonary regions, especially the lungs [113]. It also has beneficial effects in the treatment of pulmonary fibrosis [114]. Some pre-clinical studies were conducted which concluded that curcumin can forbid cadmium-induced inflammatory secretion of IL-6 and IL-8 by the epithelial cells present in the airway Cadmium (Cd) is a noxious metal residing in the surrounding environment and on aspiration can cause pulmonary diseases [115]. An animal model studied and depicted the influence of Cadmium (CdCl2)-polluted drinking water (40 mg CdCl2/L) on the level of tumor necrosis factor-alpha (TNF-α) and IL-6. They concluded that curcumin can depict a curative action against Cd toxicity [116]. In COPD, curcumin has been supplemented in animal models to derive a conclusion on how profitable it is in improving the mean pulmonary artery pressure [117]. Another model study was done in which curcumin suppressed the sensitivity of corticosteroids in COPD by adjusting the expression of Histone deacetylase 2 (HDAC2) and its effect on the alteration of histone [118]. Another study on the animal model showed that curcumin causes alveolar epithelial injury in COPD, which may be attributed because of the decline in Protein 66 homologous-collagen homologue (p66Shc) [119, 120].

7.3.2.2 Nigella sativa

Nigella sativa (N. sativa) is a member of the family Ranunculaceae majorly cultivated in North Africa and Southwest Asia. It shows different therapeutic effects such as anti-inflammatory, immunomodulatory and antioxidant effects. It elevates the body's capability to stop the damage caused by stress and recover the body's normal functioning. It enhances the effect of antioxidant enzymes like glutathione peroxidase, catalase etc and helps in scavenging the free radical. In some studies it was discovered that when guinea pigs that were induced COPD by cigarette smoke when pretreated with hydroethanolic extract of N. sativa reduced inflammation as showed effectiveness like vitamin C in reducing inflammation [121].

7.3.2.3 Resveratrol

It is a polyphenol phytoalexin which is isolated from the roots of the Chinese herbal plant, *P cuspidatum.* It is used as an antioxidant treatment in various respiratory diseases such as pneumonia, COPD, asthma, and IPF. It exerts its action by various antioxidant mechanisms like blocking the augmentation of the ROS produced by inflammatory cells, free radical scavenging, invigorating the synthesis of antioxidants by the functioning of Nrf2, elevating the functioning of the enzymes that cause the metabolism of ROS like superoxide dismutase (SOD). It elevates the magnitude of antioxidants and eliminates some markers that cause oxidative stress [53]. It decreases the oxidase function of NADPH which is induced by angiotensin II. It further enhances the functioning of SOD and causes decrease in the production of ROS. It also reduces the generation of oxygen from mitochondria [122].

7.3.2.4 Saffron

Scientific name of saffron is *Crocus sativus* and is used as preservative and medicinally used herb. It is used as an alternate therapy for treating various diseases by different mechanisms like anti-inflammatory, immunomodulatory and as antioxidant. Its major constituents are crocin, crocetin and safranal. Due to more generation and distribution of the free radicals there occurs disturbance in balance of pro-oxidants and antioxidants. Ethanol and aqueous extract of C. sativus show antioxidant effects by the help of deoxyribose assay and erythrocyte membrane peroxidation assay. In bronchial cells stress was exerted by combining various cytokines due to which the generation of NO, iNOS, peroxynitrate ion elevated. The release of cytochrome c also increased. *C. sativus* in turn reduces the production of NO, iNOS and peroxynitrate ion. Some studies suggested that guinea pigs when treated with saffron caused a significant reduction in levels of NO. Considering this study it was discovered that the effect of prevention shown by *C. sativus* on the trachea of the affected animals might be related to its antioxidant activity. *C. sativus* and its components can play a vital role where scavenging of free radicals occurs and shows a therapeutic effect to aging and wound healing. Its main bioactive components are glutathione (GSH), chloramphenicol acetyltransferase (CAT). They reduce the peroxidation of lipids in tissues and serum [123].

7.3.3 Thiol Antioxidants

Drugs that contain thiol moiety are generally used for the prevention or treatment of pulmonary diseases. The most extensively studied thiol-based drugs are N-acetyl L-cysteine (NAC), S-carboxymethyl-L-cysteine [3-(S-carboxymethylthio) alanine] (S-CMC), and N-(carboxymethylthioacetyl)-homocysteine thiolactone (erdosteine) [64].

Oxidative stress is an overproduction of the Reactive Oxygen Species (ROS) as well as Reactive Nitrogen Species (RNS). GSH is also known and referred to as the 'master antioxidant'. It balances free radicals and eliminates toxins, as well as heavy metals from the body. GSH is also very important for the appropriate functioning of the immune system. Moreover, it lessens the sensitivity in the body [65]. Depleted levels of GSH or its precursor are due to the loss of antioxidant activity [66]. The antioxidant activity of GSH is both directly or indirectly earlier via interaction with ROS/RNS and later via operating as a reducing agent for other enzymatic and non-enzymatic antioxidants [67].

In many chronic pathological conditions of lung including COPD, Asthma, IPF, the levels of GSH are abnormal both extracellular as well as intracellular [68]. NAC is a source for Cys which is useful in the synthesis of GSH [124]. In an *ex vivo* model of exaggerated COPD, NAC elevated the total antioxidant capacity by approximately 90% due to 50% elevation in GSH levels including150% increase in superoxide dismutase activity in contrast to lipopolysaccharide (LPS)-incubated bronchi, and depleted the proportion of pro-oxidant factors activated because of LPS by approximately 30% (125). Moreover, NAC also reduces the release of neurokinin A (NKA) induced by LPS provocation, an effect that correlated with a depletion in GPx activity and in the concentration of H_2O_2, malondialdehyde (MDA), in addition to nitric oxide (NO) [126].

Erdosteine hinders free radicals' effects that are generated by cigarette smoke and lessen the formation of O_2 free radical, H_2O_2 in addition to NO, in addition to the liberation of acid phosphatase and lysozyme from LPS-activated macrophages [127].

7.3.4 Nrf2 Activators

Nrf2 is nuclear factor erythroid-2-related factor 2 [128]. It belongs to the 'cap n collar' basic region/leucine zipper transcription factor family. Nrf2 is very rich in the intestine, lungs as well as kidneys where the toxic chemicals are removed time to time. Nrf2 is very important in providing protection to the cell by inducing antioxidant, detoxifying enzymes and proteins through its binding with the *cis*-acting antioxidant response element (ARE). Nrf2 is actuated by modification in the phosphorylation reaction through various protein kinase pathways that leads to leading to Keap1 (a cytoplasmic, cysteine-rich, actin bound protein that suppresses Nrf2) [129].

It has been reported that asthmatic patients surviving with oxidant burden have more problems than the normal population which may be associated with less Nrf2 function. According to the latest studies data Nrf2-driven GST is a marker for asthma perceptivity in humans the homozygous GSTM1-null genotype elevates the chances of occurrence of asthma, moreover, homozygous GSTP1 expression can oppose asthma [130, 131]. In a mice study, it was reported that Nrf2 insufficiency remarkably strengthens ovalbumin [132, 133]. Recently, the defending activity of Nrf2 inducers in asthma has been largely inspected, and ovalbumin can be employed asthma inducer. Some latest investigations have found that sulforaphane provides protection against asthma. Dispensation of Sulforaphane can repress ovalbumin-induced allergic airway hypersensitivity in mice [134, 135].

Cigarette smoking causes oxidative stress and inflammation. Cohort research determining the correlation between polymorphisms of the Nrf2 gene and obstruction of airflow in smokers also showed that diminished Nrf2 may lead to the augmentation of COPD due to enormous oxidant load and apoptosis in the lungs [136–138]. According to the findings of some recent studies, the mice having insufficient levels of Nrf2 are prone to getting exposed to cigarette smoke which causes critical lung emphysema followed by suppressed activity of antioxidant enzymes [139, 140]. Also, Nrf2 has an effect on acute exacerbations caused due to infection and therapeutic feedback to corticosteroids in COPD. Hence, unique/innovative treatments, including Nrf2 activator therapy might show good results in patients suffering from COPD [141].

To combat pulmonary fibrosis, a variety of Nrf2 activators and their subtypes have been used and investigated. Currently, pirfenidone (PFD) is the only medication approved for the treatment of IPF. Its antifibrosis activity, which affects fibroblasts stimulated by growth factor-β (TGF-β) and bleomycin-challenged mice, is linked to the restoration of Nrf2/Bach1 steadiness via Nrf2 activation and Bach1 prohibition [142]. Even when TGF-β is stimulated, sulforaphane (an Nrf2 activator) can reverse the hallmarks of myofibroblastic differentiation (such as increased α-SMA, collagen I, fibroblast proliferation, migration, and contraction) *in vitro* on IPF fibroblasts. The antifibrosis effect is dependent on Nrf2 activation restoring the redox balance [143].

NRF2 can play a crucial role as a major regulator of the antioxidant response in lung cancer. In spite of that, its activities can be advantageous for the development of tumors depending on the cancer stage in cells of the lung cancer. In early stages, activity of NRF2 considered to be vital for preventing premalignant carcinogenesis including DNA damage and initial mutations of the cancer cells [144, 145]. However, some activities of NRF2 can assist carcinogenesis at advanced stages [144]. NRF2 pathway's activation could be beneficial to safeguard cancer cells from oxidative stress [145]. Lung cancer cells are highly dependent on the NRF2 pathway for malignant phenotype's maintenance, and this process is named NRF2 addiction [146]. NRF2 has a dual role in cancer that has been studied in many animal models. In models of lung adenocarcinoma Tao *et al.* revealed that pre-activation of NRF2 by sulforaphane blocks tumor beginning [147].

Resveratrol which is a natural stilbenoid polyphenol is abundant in fruits and has wide advantageous activities including antioxidant, anti-inflammatory, antitumor, and estrogenic activities [148]. Some supportive results have suggested that resveratrol shows advantageous therapeutic effects in pulmonary hypertension [149, 150]. Some studies reported that the increase in hypertension in spontaneously hypertensive rats was reduced via Nrf2 expression's restoration and also after the administration of resveratrol, the antioxidant potential is increased [151]. Moreover, it remains undiscovered if resveratrol could improve the advancement of pulmonary hypertension via targeting the oxidative stress induced by Nrf2 or not.

7.3.5 Radical Scavengers

Free radicals are very active and unsteady compounds that cause oxidative degradation of cellular components like DNA, proteins, and lipids also they have a major role in cell injury mechanism in addition to cell death. Free radical scavengers act in two ways: they either prevent the formation of these reactive species or abolish them prior to any damage to crucial components of the cell. Therefore, they are being called preventive and chain-breaking antioxidants. The first group comprises of agents such as catalases, glutathione peroxidases, and superoxide dismutase (SOD) inactivation of ODFR's enzymatic mechanism (oxygen-derived free radicals). The second group is chain-breaking antioxidants or 'radical scavengers' which are the compounds that are capable of transformation of hydrogen to free radicals. This group consists of antioxidants like vitamin C, vitamin E, and β-carotene. These preventive antioxidants perform their actions by eradicating the species that are participating in the induction of free radical chain reactions. However, the chain-breaking antioxidants directly reconstruct the oxidizing radical. Antioxidants can be divided into two broad categories, water-soluble, e.g., vit. C and lipid soluble, e.g., β-carotene and vit. E. Hydrophilic antioxidants react with oxidants that are present in the cytosol of the cell and the blood plasma, whereas hydrophobic antioxidants secure cell membranes from the attack of oxidants on lipids. These compounds whether biosynthesized in the body or gained from diet [152].

7.3.5.1 Vitamin C

Vitamin Cisa hydrophilic antioxidant and is involved in the formation of hydroxylysine, hydroxyproline, norepinephrine, serotonin, carnitine, and collagen [153, 154]. It acts by scavenging the free radicals and therefore preventing membranes and lipoproteins from being oxidized. Also, it is used to hunt ROS and radicals in the hydrophilic extracellular space. Under oxidative stress, it is used more swiftly as compared to vitamin E and other antioxidant agents [155], and this can be regarded as signal to its primary defensive function in oxidative stress. When a variety of stimuli are involved in the activation of neutrophils, some oxidant agents are produced in the extracellular space. Vitamin C scavenges ROS which further gets converted into dehydroascorbate. Dehydroascorbate is carried by the neutrophil membrane; so when the concentration of dehydroascorbate elevates outside the cell, this dehydroascorbate is absorbed into the neutrophil, where reduction reaction occurs and vitamin C is reformed [155, 156]. This recycling process elevates the antioxidant potential of the pro-inflammatory cell during the periods when defense against ROS generation and release is needed. It has also been detected that a 10-fold higher the intracellular concentration of vitamin C is seen in activated neutrophils is ten-fold greater than detected extracellularly [157].

In some clinical studies vitamin c has shown some positive impacts on the patients suffering from lung diseases and has been discussed hereunder. In a randomized controlled trial, asthmatic children when treated with Vitamin C (1000mg/daily. i.v.), showed better neutrophil chemotaxis in comparison to standard anti-asthma therapy alone. In addition, in randomized, placebo-controlled, double-blind parallel studies subjects with early stages of asthma were given 1g/day vitamin C, 450 mg/day magnesium chelate, or placebo for 16 weeks. The results showed no clinical advantage in treatment with Vitamin C or magnesium [158, 159].

The effects of the fusion of vitamin C (500mg)+ NAC (600mg), and placebo, on the antioxidant status of COPD patients (n=79) were analyzed once a day. The results suggested that the addition of antioxidant supplements and Vitamin C elevated the nutritional and antioxidant status [160]. The effect of prevention shown by Vitamin C (48.50-141.63 mg/kg/day) has been analyzed in patients suffering from COPD [161, 162].

Vitamin C at its high doses has been considered an effective therapy for cancer since the 1970s [163]. Supportive results from some studies concluded that when vitamin C was administered intravenously was safe and efficacious in cancer patients with minimum side effects [164]. Moreover, it is considered a dietary supplement, but the U.S. Food and Drug

Administration and European Medicines Agency have not approved the high dose of vitamin C intravenously as a treatment for cancer. However, when low concentrations of vitamin C are administered in the patient it shows declined effects of chemotherapy.

There was a study that was conducted on a 40-year-old woman who had pulmonary hypertension that was causally related to vitamin C deficiency in addition to clinical scurvy that was normalized upon supplementation of vitamin C. So from the above investigation, we can conclude that vitamin C supplementation can show beneficial effects in the patients showing pulmonary hypertension [165].

7.3.5.2 Vitamin E

Tocopherol (α-T) is also known by its other name vitamin E, and it is a crucial dietary nutrient for human beings. This vitamin serves an important role in preventing the dissemination of lipid peroxidation. When a carbon-centered radical (L•) comes in contact with molecular oxygen lipid peroxidation occurs and produces a peroxyl radical (LOO•), then that peroxyl radical abstracts a hydrogen (H) from polyunsaturated fatty acids (PUFA) in the membrane and leads to the formation of a lipid hydroperoxide (LOOH), moreover another carbon-centered radical (L•) is also formed. Vitamin E acts as a lipophilic, "chain-breaking" antioxidant by giving its hydrogen to peroxyl radical and producing lipid hydroperoxide (LOOH) and the Vitamin E radical. Vitamin E has also contributed in the therapy of some chronic diseases, and it functions as an anti-inflammatory agent [166]. Twenty-four observational studies were conducted and a meta-analysis showed the correlation between the low vitamin E intake in children as well as adults with increased severity of asthma [167]. Also, some observational studies (including three birth cohorts) were conducted which show the relation between the declined levels of vitamin E and the development of asthma. However, it was technically weak but supportive enough of a probable effect warranting investigation in clinical trials [168]. A Randomized controlled trial that was conducted on 72 British adults suffering from asthma found no effective results on the 6-week supplementation that was 500mg per day of vitamin E on airway responsiveness [169]. Moreover, there were no effective results found in antioxidant supplementation (that includes vitamin C and vitamin E) on plasma F2- isoprostanes exhaled nitric oxide (NO), or peripheral blood mononuclear cell (PBMC) immune responses during the second randomized controlled trial that was conducted in 54 allergic adults [170]. In contrary to these negative outcomes, four RCTs concluded that the antioxidants containing vitamin E decrease broncho constriction caused due to O3 in patients with [171, 172] and without [173, 174] asthma. It further suggests that vitamin E shows defensive actions against the harmful effects of O3. A study was conducted in children which focused on the total amount of antioxidant intake and the rate of asthma and they discovered a connection between high intake of antioxidants and reduced sensitivity to the allergens inhaled. Children residing in areas that have less traffic pollution showed a fortified link between allergen sensitivity and antioxidant capacity [175].

7.3.6 Enzymatic Redox Sensor

Enzymatic redox sensors are biological molecules that determine and reciprocate any changes in the cells and tissues of the respiratory system in relation to its redox status. These sensors consist of various enzymes that are prone to changes in the levels of ROS, the molecules that are involved in oxidative stress and reactive nitrogen species, the molecules that are involved in nitrosative stress. In the respiratory diseases elevated oxidative stress leads to variation in the redox signaling. Enzymatic redox sensors have a definitive function in the monitoring and management of oxidative as well as nitrosative processes. Redox homeostasis of cells is maintained by detecting excess ROS or RNS under the assistance of enzymatic redox sensor. These sensors are necessary to understand the molecular mechanisms which takes place in respiratory diseases. They act as major targets for showing therapeutic efficacy to regain the redox balance and decrease the oxidative damage caused in the respiratory system. Example of enzymatic redox sensor is glutathione peroxidase.

The resistance against ROS is endogenous antioxidants. It is classified into two types: enzymatic and non-enzymatic. The enzymatic antioxidants comprise superoxide dismutase (SOD), catalase, glutathione peroxidase, glutathione S-transferase, and thioredoxin. Each one of them contains isozymes differentiated majorly by their dissemination. For example, the three mammalian SODs are cytosolic (SOD1), mitochondrial (SOD2), and extracellular (SOD3), and the two thioredoxins are cytosolic (Trx1) and mitochondrial (Trx2) [176]. Extracellular SOD (EC-SOD) is the chief SOD enzyme which is present outside the cell. In airway epithelium and vascular endothelium EC-SOD mRNA is present in a large amount. It is expressed in terms of levels, 4 times perceived in the heart and 15 times that is perceived in the liver [177]. The protein is connected to the matrix of connective tissue present in the surroundings of the vessels and airways in the lung [178]. Electron microscopic immunocytochemistry discloses that EC-SOD is majorly found in sites that have abundant type I collagen, on smooth muscle surface cells, and in the extracellular matrix connected with airway including vascular smooth muscle cells. The expression of ECSOD around airway smooth muscle is notable and it indicates that it may play a part in diseases of airway caused by allergy like asthma.

7.4 Barriers to Pulmonary Delivery

Now a day's pulmonary drug delivery system is withdrawing more attention as the lung is capable of absorbing the drug both systemically as well as locally and also lungs possess more surface area for better absorption of the active pharmaceutical ingredient including good supply of blood. In adults the absorptive surface area is approximately 70-140 $m^{2.}$ Moreover, it is a non-invasive administration of drug. The production of epithelial lining fluid (ELF) in addition to the regulation of bronchus tone is handled by the respiratory epithelial cells [179], although the drug delivery via pulmonary route is very complex as there are several barriers including mechanical barrier, chemical and immunological barrier, and behavioral barriers [180].

7.4.1 Mechanical Barrier

Lungs are composed of a very composite web of tissues known as a bronchial tree. A particle gets settled in the carina when it is moving to the epithelial target sites. The airway gets narrowed down due to pathological conditions such as hypersecretion, inflammation, and bronchoconstriction which leads to an increase in the mechanical barrier of the pulmonary area [181].

7.4.2 Chemical and Immunological Barrier

These barriers are composed of proteolytic enzymes, surfactants, and alveolar macrophages in the pulmonary drug delivery system. The hydrolytic cleavage of protein and peptides in the lungs is caused due to some proteolytic enzymes including cathepsin H and endopeptidase [182]. The phagocytic cells such as alveolar macrophages help to ingest toxic particles. Here surfactants play dual role as it prevents the particles from getting attached to the surface of epithelial of the lungs as well as make the inhaled particles more accessible to macrophages to eliminate them from the body [183].

7.4.3 Behavioral Barrier

Patient behaviors such as non-adherence or non-compliance to the therapy that may be either intentional or unintentional will affect the outcome of the treatment. Moreover, cultural factors may also give rise to non-compliance. Inadequate inhaler techniques combined with the mistakes during handling may cause inadequate drug delivery to the lungs [184]. The above barriers are systematically illustrated in Figure 7.6.

7.5 Novel Approaches for Antioxidant Drug Delivery

Due to the larger advantages (including good bioavailability, sustained release, and targeted delivery of drug) novel drug delivery system has gained much importance in the treatment of pulmonary diseases [185]. Microcarriers can be used to embody hydrophilic as well as hydrophobic drugs also including their biologics due to their unique physicochemical properties and good stability. Novel DDS include liposomes, polymeric nanoparticles, solid lipid nanoparticles, polymeric micelles, dendrimers, nanoemulsion, nano suspension, and microspheres, and microparticles are included in the novel drug delivery system (Table 7.2) [186, 187].

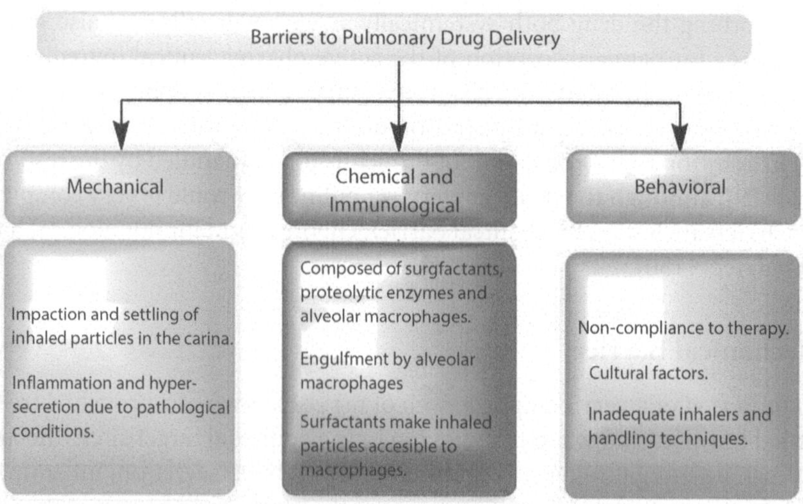

Figure 7.6 Barriers to pulmonary drug delivery.

Table 7.2 Various drug delivery systems employed for antioxidant delivery.

S. no.	Drug delivery system	Active ingredient	Inference	Ref.
1.	Liposomes	NAC loaded	Reduce lung permeability index by 59% in acute (4-h) CEES-induced injury.	[188]
2.	Solid lipid nanoparticles	Encapsulated proanthocyanidins	ROS production is declined by reducing oxidative stress.	[189]
3.	Ceriumoxide (IV) nanoparticles	SOD and CAT	Reduction in the proliferation of ROS.	[190]
4.	Polymer nanoparticles	Curcumin	Lipopolysaccharide induced alveolar inflammation is inhibited.	[191, 192]
5.	Multifunctional nanomaterials	Sulforaphane	Reduce the oxidative stress.	[193]
6.	Polymer nanoparticles	Budesonide	Drug permeation through mucus layer improved.	[194]

Table 7.3 Pulmonary diseases and their therapies followed by their mode of drug delivery.

S. no.	Disease	Antioxidant therapy	Mode of drug delivery
1.	Asthma	i. Ascorbic acid ii. Tocopherol iii. Beta carotene iv. SOD	i. Dietary ii. Dietary iii. Dietary iv. Intra vascular
2.	Lung cancer	i. Vitamins ii. Carotenoids (lycopene)	i. Dietary ii. Dietary
3.	Persistent pulmonary hypertension	i. Recombinant human SOD ii. Inhaled nitric oxide	i. Intratracheal ii. Intratracheal
4.	COPD	i. Tocopherol ii. Selenium	i. Dietary ii. Dietary
5.	Pulmonary fibrosis	i. Extracellular SOD ii. Curcumin	i. Overexpression ii. Oral
6.	Cystic fibrosis	i. Tocopherol ii. Ascorbic acid iii. Beta carotene iv. Zinc	i. Dietary ii. Dietary iii. Dietary iv. Dietary

The delivery of antioxidants is predetermined to play various crucial roles. Formerly, they safeguard the drug from being inactivated in the body, lessen the dose needed for better efficacy, and reduce the hypersensitivity reactions and noxious effects that will eventually provide more beneficial effects to the target site. There are several drug delivery and targeting devices and strategies that are being developed to improve delivery of drug to the target site such as liposomes. However, some of the therapies and their route of drug delivery have been given in the Table 7.3 [195].

7.6 Future Perspectives and Conclusion

Antioxidants have been seen to possess an unparalleled potential in the management of pulmonary diseases. However, with an escalating rise in the upcoming innovative and novel technologies, it is essential to amalgamate antioxidants with them to meet the exigencies of modern and contemporary healthcare. Newer paradigms such as personalized medicine can be utilized to deliver the therapeutic agents catering to the specific needs of the patient. The use of cutting-edge technologies such as 3D printing [196] to fabricate drug delivery systems accurately can also be utilized to enhance the therapeutic efficacy of antioxidants. The use of herbal compounds can also help to overcome insurmountable side effects of chemical agents and improve patient compliance. Research should continue with special focus on finding newer antioxidants agents that can influence the cellular pathways effectively. There is also a need to conduct more extensive clinical and preclinical studies to ensure the safety and efficacy of these therapeutic agents.

References

1. Xie, X., Kong, S., Cao, W., Targeting protein glycosylation to regulate inflammation in the respiratory tract: novel diagnostic and therapeutic candidates for chronic respiratory diseases. *Front. Immunol.*, *14*, 1168023, 2023.
2. Zhou, C., Qin, Y., Zhao, W., Liang, Z., Li, M., Liu, D., Bai, L., Chen, Y., Chen, Y., Cheng, Y., Chu, T., International expert consensus on diagnosis and treatment of lung cancer complicated by chronic obstructive pulmonary disease. *Transl. Lung Cancer Res.*, *12*, 8, 1661, 2023.
3. Sharma, A., James, A., Kapoor, D.N., Kaurav, H., Sharma, A.K., Nagraik, R., An insight into biosensing platforms used for the diagnosis of various lung diseases: A review. *Biotechnol. Bioeng.*, 121, 71–81, 2024.
4. Zhou, Y.M., Dong, X.R., Xu, D., Tang, J., Cui, Y.L., Therapeutic potential of traditional Chinese medicine for interstitial lung disease. *J. Ethnopharmacol.*, 318, 116952, 2023.
5. Yan, Q., Liu, S., Sun, Y., Chen, C., Yang, S., Lin, M., Long, J., Yao, J., Lin, Y., Yi, F., Meng, L., Targeting oxidative stress as a preventive and therapeutic approach for cardiovascular disease. *J. Transl. Med.*, *21*, 1, 1–35, 2023.
6. Jena, A.B., Samal, R.R., Bhol, N.K., Duttaroy, A.K., Cellular Red-Ox system in health and disease: The latest update. *Biomed. Pharmacother.*, *162*, 114606, 2023.
7. Tai, J., Shin, J.M., Park, J., Han, M., Kim, T.H., Oxidative Stress and Antioxidants in Chronic Rhinosinusitis with Nasal Polyps. *Antioxidants*, *12*, 1, 195, 2023.
8. Deng, R., Zhu, Y., Wu, X., Wang, M., Toxicity and Mechanisms of Engineered Nanoparticles in Animals with Established Allergic Asthma. *Int. J. Nanomed.*, 18, 3489–3508, 2023.

9. Albano, G.D., Montalbano, A.M., Gagliardo, R., Profita, M., Autophagy/Mitophagy in Airway Diseases: Impact of Oxidative Stress on Epithelial Cells. *Biomolecules*, *13*, 8, 1217, 2023.
10. Ayoka, T.O., Ezema, B.O., Eze, C.N., Nnadi, C.O., Antioxidants for the Prevention and Treatment of Non-communicable Diseases. *J. Explor. Res. Pharmacol.*, *7*, 3, 178–188, 2022.
11. Nimal, R., Selcuk, O., Kurbanoglu, S., Shah, A., Siddiq, M., Uslu, B., Trends in electrochemical nanosensors for the analysis of antioxidants. *TrAC, Trends Anal. Chem.*, *153*, 116626, 2022.
12. Bhatti, J.S., Sehrawat, A., Mishra, J., Sidhu, I.S., Navik, U., Khullar, N., Kumar, S., Bhatti, G.K., Reddy, P.H., Oxidative stress in the pathophysiology of type 2 diabetes and related complications: Current therapeutics strategies and future perspectives. *Free Radical Biol. Med.*, *184*, 114–134, 2022.
13. Haftcheshmeh, S.M., Abedi, M., Mashayekhi, K., Mousavi, M.J., Navashenaq, J.G., Mohammadi, A., Momtazi-Borojeni, A.A., Berberine as a natural modulator of inflammatory signaling pathways in the immune system: Focus on NF-κB, JAK/STAT, and MAPK signaling pathways. *Phytother. Res.*, *36*, 3, 1216–1230, 2022.
14. Ramli, I., Cheriet, T., Posadino, A.M., Giordo, R., Zayed, H., Eid, A.H., Pintus, G., Potential Therapeutic Targets of Resveratrol in the Prevention and Treatment of Pulmonary Fibrosis. *Front. Biosci.-Landmark*, *28*, 9, 198, 2023.
15. Zajac, D. and Wojciechowski, P., The Role of Vitamins in the Pathogenesis of Asthma. *Int. J. Mol. Sci.*, *24*, 10, 8574, 2023.
16. Hou, T., Guo, Y., Han, W., Zhou, Y., Netala, V.R., Li, H., Li, H., Zhang, Z., Exploring the Biomedical Applications of Biosynthesized Silver Nanoparticles Using Perillafrutescens Flavonoid Extract: Antibacterial, Antioxidant, and Cell Toxicity Properties against Colon Cancer Cells. *Molecules*, *28*, 17, 6431, 2023.
17. Di Giacomo, C., Malfa, G.A., Tomasello, B., Bianchi, S., Acquaviva, R., Natural compounds and glutathione: Beyond mere antioxidants. *Antioxidants*, *12*, 7, 1445, 2023.
18. Rekadwad, B.N., Li, W.J., Gonzalez, J.M., PunchappadyDevasya, R., Ananthapadmanabha Bhagwath, A., Urana, R., Parwez, K., Extremophiles: the species that evolve and survive under hostile conditions. *3 Biotech*, *13*, 9, 316, 2023.
19. Gnocchi, D., Nikolic, D., Paparella, R.R., Sabbà, C., Mazzocca, A., Cellular Adaptation Takes Advantage of Atavistic Regression Programs during Carcinogenesis. *Cancers*, *15*, 15, 3942, 2023.
20. Rey, F., Messa, L., Maghraby, E., Casili, G., Ottolenghi, S., Barzaghini, B., Raimondi, M.T., Cereda, C., Cuzzocrea, S., Zuccotti, G., Esposito, E., Oxygen Sensing in Neurodegenerative Diseases: Current Mechanisms, Implication of Transcriptional Response, and Pharmacological Modulation. *Antioxid. Redox Signaling*, *38*, 1–3, 160–182, 2023.
21. Chehrazi, M. and Moghadas, B.K., A review on CO2 capture with chilled ammonia and CO2 utilization in urea plant. *J. CO2 Util.*, *61*, 102030, 2022.
22. Fakhraee, M., Tarhan, L.G., Reinhard, C.T., Crowe, S.A., Lyons, T.W., Planavsky, N., Earth's surface oxygenation and the rise of eukaryotic life: Relationships to the Lomagundi positive carbon isotope excursion revisited. *Earth Sci. Rev.*, 240, 104398, 2023.
23. Leong, W.H., Rawindran, H., Ameen, F., Alam, M.M., Chai, Y.H., Ho, Y.C., Lam, M.K., Lim, J.W., Tong, W.Y., Bashir, M.J., Ravindran, B., Advancements of microalgal upstream technologies: Bioengineering and application aspects in the paradigm of circular bioeconomy. *Chemosphere*, 339, 139699, 2023.
24. Walke, G., Gaurkar, S.S., Prasad, R., Lohakare, T., Wanjari, M., The Impact of Oxidative Stress on Male Reproductive Function: Exploring the Role of Antioxidant Supplementation. *Cureus*, *15*, 7, 42583, 2023.
25. Zeber-Lubecka, N., Ciebiera, M., Hennig, E.E., Polycystic Ovary Syndrome and Oxidative Stress—From Bench to Bedside. *Int. J. Mol. Sci.*, *24*, 18, 14126, 2023.

26. Cazzola, M., Page, C.P., Wedzicha, J.A., Celli, B.R., Anzueto, A., Matera, M.G., Use of thiols and implications for the use of inhaled corticosteroids in the presence of oxidative stress in COPD. *Respir. Res.*, *24*, 1, 194, 2023.
27. Nosetti, L., Zaffanello, M., De Bernardi di Valserra, F., Simoncini, D., Beretta, G., Guacci, P., Piacentini, G., Agosti, M., Exploring the Intricate Links between Adenotonsillar Hypertrophy, Mouth Breathing, and Craniofacial Development in Children with Sleep-Disordered Breathing: Unraveling the Vicious Cycle. *Children*, *10*, 8, 1426, 2023.
28. Zhou, J., Fang, P., Liang, Z., Li, X., Luan, S., Xiao, X., Gu, Y., Shang, Q., Zhang, H., Yang, Y., Chen, L., Causal relationship between lung diseases and risk of esophageal cancer: insights from Mendelian randomization. *J. Cancer Res. Clin. Oncol.*, 149, 1–8, 2023.
29. Chaudhary, P., Janmeda, P., Docea, A.O., Yeskaliyeva, B., AbdullRazis, A.F., Modu, B., Calina, D., Sharifi-Rad, J., Oxidative stress, free radicals and antioxidants: potential crosstalk in the pathophysiology of human diseases. *Front. Chem.*, *11*, 1158198, 2023.
30. Romero, A., Ramos, E., López-Muñoz, F., De Los Ríos, C., Egea, J., Gil-Martín, E., Pita, R., Torrado, J.J., Serrano, D.R., Juberias, A., Toxicology of blister agents: is melatonin a potential therapeutic option? *Diseases*, *9*, 2, 27, 2021.
31. Jalil, A.T., Abdulhadi, M.A., Al-Ameer, L.R., Washeel, O.F., Abdulameer, S.J., Merza, M., Abosaooda, M., Mahdi, A.A., Free radical based nano cancer therapy. *J. Drug Deliv. Sci. Technol.*, 87, 104803, 2023.
32. Caturano, A., D'Angelo, M., Mormone, A., Russo, V., Mollica, M.P., Salvatore, T., Galiero, R., Rinaldi, L., Vetrano, E., Marfella, R., Monda, M., Oxidative Stress in Type 2 Diabetes: Impacts from Pathogenesis to Lifestyle Modifications. *Curr. Issues Mol. Biol.*, *45*, 8, 6651–6666, 2023.
33. Sierra-Vargas, M.P., Montero-Vargas, J.M., Debray-García, Y., Vizuet-de-Rueda, J.C., Loaeza-Román, A., Terán, L.M., Oxidative Stress and Air Pollution: Its Impact on Chronic Respiratory Diseases. *Int. J. Mol. Sci.*, *24*, 1, 853, 2023.
34. Tang, M., Sun, J., Cai, Z., PCK2 inhibits lung adenocarcinoma tumor cell immune escape through oxidative stress-induced senescence as a potential therapeutic target. *J. Thorac. Dis.*, *15*, 5, 2601, 2023.
35. Wang, W., Mei, A., Qian, H., Li, D., Xu, H., Chen, J., Yang, H., Min, X., Li, C., Cheng, L., Chen, J., The Role of Glucagon-Like Peptide-1 Receptor Agonists in Chronic Obstructive Pulmonary Disease. *Int. J. Chron. Obstruct. Pulmon. Dis.*, 18, 129–137, 2023.
36. Arunachalam, K., Anand, K., Palanisamy, S., Anathy, V., Oxidative stress related to cellular metabolism in lung health and diseases. *Front. Pharmacol.*, *13*, 1015423, 2022.
37. Chellappan, D.K., Paudel, K.R., Tan, N.W., Cheong, K.S., Khoo, S.S.Q., Seow, S.M., Chellian, J., Candasamy, M., Patel, V.K., Arora, P., Singh, P.K., Targeting the mitochondria in chronic respiratory diseases. *Mitochondrion*, *67*, 15–37, 2022.
38. Dailah, H.G., Therapeutic potential of small molecules targeting oxidative stress in the treatment of chronic obstructive pulmonary disease (COPD): A comprehensive review. *Molecules*, *27*, 17, 5542, 2022.
39. Hadian, K. and Stockwell, B.R., The therapeutic potential of targeting regulated non-apoptotic cell death. *Nat. Rev. Drug Discovery*, 22, 1–20, 2023.
40. Sachdev, S., Ansari, S.A., Ansari, M.I., Antioxidant Defensive Mechanisms to Regulate Cellular Redox Homeostatic Balance, in: *Reactive Oxygen Species in Plants: The Right Balance*, pp. 143–172, Springer Nature Singapore, Singapore, 2023.
41. Singh, C., Singh, R., Shekhar, A., Antioxidant Defense: A Key Mechanism of Lead Tolerance, in: *Lead Toxicity: Challenges and Solution*, pp. 127–144, Springer Nature Switzerland, Cham, 2023.
42. Rodella, U., Honisch, C., Gatto, C., Ruzza, P., D'Amato Tóthová, J., Antioxidant Nutraceutical Strategies in the Prevention of Oxidative Stress Related Eye Diseases. *Nutrients*, *15*, 10, 2283, 2023.

43. Jomova, K., Raptova, R., Alomar, S.Y., Alwasel, S.H., Nepovimova, E., Kuca, K., Valko, M., Reactive oxygen species, toxicity, oxidative stress, and antioxidants: chronic diseases and aging. *Arch. Toxicol.*, 97, 1–76, 2023.
44. Kadushkin, A., Tahanovich, A., Movchan, L., Talabayeva, E., Plastinina, A., Shman, T., Azithromycin modulates release of steroid-insensitive cytokines from peripheral blood mononuclear cells of patients with chronic obstructive pulmonary disease. *Adv. Respir. Med.*, *90*, 1, 17–27, 2022.
45. Wang, R.X., Zhou, M., Ma, H.L., Qiao, Y.B., Li, Q.S., The Role of Chronic Inflammation in Various Diseases and Anti-inflammatory Therapies Containing Natural Products. *ChemMedChem*, *16*, 10, 1576–1592, 2021.
46. Rawat, N., Singla-Pareek, S.L., Pareek, A., Membrane dynamics during individual and combined abiotic stresses in plants and tools to study the same. *Physiol. Plant*, *171*, 4, 653–676, 2021.
47. Chen, X., Li, X., Xu, X., Li, L., Liang, N., Zhang, L., Lv, J., Wu, Y.C., Yin, H., Ferroptosis and cardiovascular disease: role of free radical-induced lipid peroxidation. *Free Radical Res.*, *55*, 4, 405–415, 2021.
48. Morris, G., Gevezova, M., Sarafian, V., Maes, M., Redox regulation of the immune response. *Cell. Mol. Immunol.*, *19*, 10, 1079–1101, 2022.
49. Anal, A.K., Koirala, S., Karna, A., Umar, M., Thapa, S.P., Immunomodulation and Enhancing the Immunity: Unveiling the Potential of Designer Diets. *Future Foods*, 8, 100246, 2023.
50. Gorji, A. and Ghadiri, M.K., Potential roles of micronutrient deficiency and immune system dysfunction in the coronavirus disease 2019 (COVID-19) pandemic. *Nutrition*, *82*, 111047, 2021.
51. Thakur, A.K., Chellappan, D.K., Dua, K., Mehta, M., Satija, S., Singh, I., Patented therapeutic drug delivery strategies for targeting pulmonary diseases. *Expert Opin. Ther. Pat.*, *30*, 5, 375–387, 2020.
52. Subbarao, P., Mandhane, P.J., Sears, M.R., Asthma: epidemiology, etiology and risk factors. *CMAJ*, *181*, 9, E181–E190, 2009.
53. Wood, L.G., Wark, P.A., Garg, M.L., Antioxidant and anti-inflammatory effects of resveratrol in airway disease. *Antioxid. Redox Signaling*, *13*, 10, 1535–1548, 2010.
54. Alvarado, A. and Arce, I., Antioxidants in respiratory diseases: Basic science research and therapeutic alternatives. *Clin. Res. Trials*, *3*, 1, 1–11, 2016.
55. Viegi, G., Pistelli, F., Sherrill, D.L., Maio, S., Baldacci, S., Carrozzi, L., Definition, epidemiology and natural history of COPD. *Eur. Respir. J.*, *30*, 5, 993–1013, 2007.
56. Maselli, D.J., Hardin, M., Christenson, S.A., Hanania, N.A., Hersh, C.P., Adams, S.G., Anzueto, A., Peters, J.I., Han, M.K., Martinez, F.J., Clinical approach to the therapy of asthma-COPD overlap. *Chest*, *155*, 1, 168–177, 2019.
57. Kim, V. and Criner, G.J., The chronic bronchitis phenotype in COPD: features and implications. *Curr. Opin. Pulm. Med.*, *21*, 2, 133, 2015.
58. Butler, C.C., Hood, K., Verheij, T., Little, P., Melbye, H., Nuttall, J., Kelly, M.J., Mölstad, S., Godycki-Cwirko, M., Almirall, J., Torres, A., Variation in antibiotic prescribing and its impact on recovery in patients with acute cough in primary care: prospective study in 13 countries. *BMJ*, *338*, 1545–1548, 2009.
59. Leap, J., Arshad, O., Cheema, T., Balaan, M., Pathophysiology of COPD. *Crit. Care Nurs. Q.*, *44*, 1, 2–8, 2021.
60. Wilson, M., Wilson, P.J., Wilson, M., Wilson, P.J., Acute Bronchitis, in: *Close Encounters of the Microbial Kind: Everything You Need to Know about Common Infections*, pp. 253–260, 2021.
61. Wenzel, R.P. and Fowler 3rd, A.A., Clinical practice. Acute bronchitis. *N. Engl. J. Med.*, *355*, 20, 2125–2130, 2006.

62. Gonzales, R. and Sande, M.A., Uncomplicated acute bronchitis. *Ann. Intern. Med.*, 133, 12, 981–91, 2000.
63. Wadhwa, R., Aggarwal, T., Malyla, V., Kumar, N., Gupta, G., Chellappan, D.K., Dureja, H., Mehta, M., Satija, S., Gulati, M., Maurya, P.K., Identification of biomarkers and genetic approaches toward chronic obstructive pulmonary disease. *J. Cell. Physiol.*, *234*, 10, 16703–16723, 2019.
64. Cazzola, M., Calzetta, L., Page, C., Rogliani, P., Matera, M.G., Thiol-based drugs in pulmonary medicine: much more than mucolytics. *Trends Pharmacol. Sci.*, *40*, 7, 452–463, 2019.
65. Teskey, G., Abrahem, R., Cao, R., Gyurjian, K., Islamoglu, H., Lucero, M., Martinez, A., Paredes, E., Salaiz, O., Robinson, B., Venketaraman, V., Glutathione as a marker for human disease. *Adv. Clin. Chem.*, *87*, 141–159, 2018.
66. Matera, M.G., Calzetta, L., Cazzola, M., Oxidation pathway and exacerbations in COPD: the role of NAC. *Expert Rev. Respir. Med.*, *10*, 1, 89–97, 2016.
67. Cazzola, M., Calzetta, L., Page, C., Rogliani, P., Matera, M.G., Impact of erdosteine on chronic bronchitis and COPD: a meta-analysis. *Pulm. Pharmacol. Ther.*, *48*, 185–194, 2018.
68. Lee, I.T. and Yang, C.M., Role of NADPH oxidase/ROS in pro-inflammatory mediators-induced airway and pulmonary diseases. *Biochem. Pharmacol.*, *84*, 5, 581–590, 2012.
69. du Bois, R.M., Weycker, D., Albera, C., Bradford, W.Z., Costabel, U., Kartashov, A., Lancaster, L., Noble, P.W., Raghu, G., Sahn, S.A., Szwarcberg, J., Ascertainment of individual risk of mortality for patients with idiopathic pulmonary fibrosis. *Am. J. Respir. Crit. Care Med.*, *184*, 4, 459–466, 2011.
70. Meltzer, E.B. and Noble, P.W., Idiopathic pulmonary fibrosis. *Orphanet J. Rare Dis.*, 3, 1, 1–5, 2008 Dec.
71. Richeldi, L., Collard, H.R., Jones, M.G., Idiopathic pulmonary fibrosis. *Lancet*, *389*, 10082, 1941–1952, 2017.
72. Saha, P. and Talwar, P., Idiopathic pulmonary fibrosis (IPF): disease pathophysiology, targets, and potential therapeutic interventions. *Mol. Cell. Biochem.*, 1–14, 2023.
73. Hoo, Z.H. and Whyte, M.K., Idiopathic pulmonary fibrosis. *Thorax*, *67*, 8, 742–746, 2012.
74. Aggarwal, K., Arora, S., Nagpal, K., Pulmonary Fibrosis: Unveiling the Pathogenesis, Exploring Therapeutic Targets, and Advancements in Drug Delivery Strategies. *AAPS PharmSciTech*, *24*, 6, 152, 2023.
75. Sharma, M., Garg, V., Singh, I., Herbal Bioactives for the Treatment of Idiopathic Pulmonary Fibrosis, in: *Medicinal Plants for Lung Diseases: A Pharmacological and Immunological Perspective*, pp. 341–368, 2021.
76. Tobin, R.W., POPE, C.E., Pellegrini, C.A., Emond, M.J., Sillery, J.I.M., Raghu, G., Increased prevalence of gastroesophageal reflux in patients with idiopathic pulmonary fibrosis. *Am. J. Respir. Crit. Care Med.*, *158*, 6, 1804–1808, 1998.
77. Raghu, G., Freudenberger, T.D., Yang, S., Curtis, J.R., Spada, C., Hayes, J., Sillery, J.K., Pope, C.2., Pellegrini, C.A., High prevalence of abnormal acid gastro-oesophageal reflux in idiopathic pulmonary fibrosis. *Eur. Respir. J.*, *27*, 1, 136–142, 2006.
78. Patti, M.G., Tedesco, P., Golden, J., Hays, S., Hoopes, C., Meneghetti, A., Damani, T., Way, L.W., Idiopathic pulmonary fibrosis: how often is it really idiopathic? *J. Gastrointest. Surg.*, *9*, 1053–1058, 2005.
79. Gribbin, J., Hubbard, R., Smith, C., Role of diabetes mellitus and gastro-oesophageal reflux in the aetiology of idiopathic pulmonary fibrosis. *Respir. Med.*, *103*, 6, 927–931, 2009.
80. de Groot, P.M., Wu, C.C., Carter, B.W., Munden, R.F., The epidemiology of lung cancer. *Transl. Lung Cancer Res.*, *7*, 3, 220, 2018.
81. Malhotra, J., Malvezzi, M., Negri, E., La Vecchia, C., Boffetta, P., Risk factors for lung cancer worldwide. *Eur. Respir. J.*, *48*, 3, 889–902, 2016.

82. Jelic, M.D., Mandic, A.D., Maricic, S.M., Srdjenovic, B.U., Oxidative stress and its role in cancer. *J. Cancer Res. Ther.*, *17*, 1, 22–28, 2021.
83. Gaine, S., Pulmonary hypertension. *JAMA*, 284, 24, 3160–8, 2000 Dec 27.
84. Hoeper, M.M., Ghofrani, H.A., Grünig, E., Klose, H., Olschewski, H., Rosenkranz, S., Pulmonary hypertension. *Dtsch. Arztebl. Int.*, *114*, 5, 73, 2017.
85. Humbert, M., Nunes, H., Sitbon, O., Parent, F., Hervé, P., Simonneau, G., Risk factors for pulmonary arterial hypertension. *Clin. Chest. Med.*, *22*, 3, 459–475, 2001.
86. Hernández-Díaz, S., Van Marter, L.J., Werler, M.M., Louik, C., Mitchell, A.A., Risk factors for persistent pulmonary hypertension of the newborn. *Pediatrics*, *120*, 2, e272–e282, 2007.
87. Tager, I. and Speizer, F.E., Role of infection in chronic bronchitis. *N. Engl. J. Med.*, *292*, 11, 563–571, 1975.
88. Wenzel, R.P. and Fowler III, A.A., Acute bronchitis. *N. Engl. J. Med.*, *355*, 20, 2125–2130, 2006.
89. Saadat, Y.R., Khosroushahi, A.Y., Gargari, B.P., A comprehensive review of anticancer, immunomodulatory and health beneficial effects of the lactic acid bacteria exopolysaccharides. *Carbohydr. Polym.*, *217*, 79–89, 2019.
90. Chandra, P., Sharma, R.K., Arora, D.S., Antioxidant compounds from microbial sources: A review. *Food Res. Int.*, *129*, 108849, 2020.
91. Whyand, T., Hurst, J.R., Beckles, M., Caplin, M.E., Pollution and respiratory disease: can diet or supplements help? A Review. *Respir. Res.*, *19*, 1–14, 2018.
92. Reddy, A.M., Iqbal, M., Chopra, H., Urmi, S., Junapudi, S., Bibi, S., Gupta, S.K., Pangi, V.N., Singh, I., Abdel-Daim, M.M., Pivotal role of vitamin D in mitochondrial health, cardiac function, and human reproduction. *EXCLI J.*, *21*, 967, 2022.
93. Hansdottir, S. and Monick, M.M., Vitamin D effects on lung immunity and respiratory diseases. *Vitam. Hormone*, *86*, 217–237, 2011.
94. Zhu, M., Wang, T., Wang, C., Ji, Y., The association between vitamin D and COPD risk, severity, and exacerbation: an updated systematic review and meta-analysis. *Int. J. Chron. Obstruct. Pulmon. Dis.*, 11, 2597–2607, 2016.
95. Kokturk, N., Baha, A., Oh, Y.M., Young Ju, J., Jones, P.W., Vitamin D deficiency: What does it mean for chronic obstructive pulmonary disease (COPD)? a compherensive review for pulmonologists. *Clin. Respir. J.*, *12*, 2, 382–397, 2018.
96. Johnson, E.J., The role of carotenoids in human health. *Nutr. Clin. Care*, *5*, 2, 56–65, 2002.
97. Caramori, G. and Papi, A., Oxidants and asthma. *Thorax*, *59*, 2, 170–173, 2004.
98. Fitzpatrick, A.M., Teague, W.G., Holguin, F., Yeh, M., Brown, L.A.S., Program, S.A.R., Airway glutathione homeostasis is altered in children with severe asthma: evidence for oxidant stress. *J. Allergy Clin. Immunol.*, *123*, 1, 146–152, 2009.
99. Riedl, M.A. and Nel, A.E., Importance of oxidative stress in the pathogenesis and treatment of asthma. *Curr. Opin. Allergy Clin. Immunol.*, *8*, 1, 49–56, 2008.
100. Paiva, S.A. and Russell, R.M., β-carotene and other carotenoids as antioxidants. *J. Am. Coll. Nutr.*, *18*, 5, 426–433, 1999.
101. Mora, J.R., Iwata, M., Von Andrian, U.H., Vitamin effects on the immune system: vitamins A and D take centre stage. *Nat. Rev. Immunol.*, *8*, 9, 685–698, 2008.
102. Chew, B.P. and Park, J.S., Carotenoid action on the immune response. *J. Nutr.*, *134*, 1, 257S–261S, 2004.
103. Lovett-Racke, A.E. and Racke, M.K., Retinoic acid promotes the development of Th2-like human myelin basic protein-reactive T cells. *Cell. Immunol.*, *215*, 1, 54–60, 2002.
104. Checkley, W., West, K.P., Wise, R.A., Wu, L., LeClerq, S.C., Khatry, S., Katz, J., Christian, P., Tielsch, J.M., Sommer, A., Supplementation with vitamin A early in life and subsequent risk of asthma. *Eur. Respir. J.*, *38*, 6, 1310–1319, 2011.

105. Romieu, I., Varraso, R., Avenel, V., Leynaert, B., Kauffmann, F., Clavel-Chapelon, F., Fruit and vegetable intakes and asthma in the E3N study. *Thorax*, *61*, 3, 209–215, 2006.
106. Gao, J., Gao, X., Li, W., Zhu, Y., Thompson, P.J., Observational studies on the effect of dietary antioxidants on asthma: a meta-analysis. *Respirology*, 13, 528–36, 2008.
107. Wood, L.G., Garg, M.L., Smart, J.M., Scott, H.A., Barker, D., Gibson, P.G., Manipulating antioxidant intake in asthma: a randomized controlled trial. *Am. J. Clin. Nutr.*, *96*, 3, 534–543, 2012.
108. Wood, L.G., Garg, M.L., Powell, H., Gibson, P.G., Lycopene-rich treatments modify noneosinophilic airway inflammation in asthma: proof of concept. *Free Radical Res.*, *42*, 1, 94–102, 2008.
109. Aggarwal, B.B., Sundaram, C., Malani, N., Ichikawa, H., Curcumin: the Indian solid gold, in: *The Molecular Targets and Therapeutic Uses of Curcumin in Health and Disease*, pp. 1–75, 2007.
110. Yang, X., Lv, J.N., Li, H., Jiao, B., Zhang, Q.H., Zhang, Y., Zhang, J., Liu, Y.Q., Zhang, M., Shan, H., Zhang, J.Z., Curcumin reduces lung inflammation via Wnt/β-catenin signaling in mouse model of asthma. *J. Asthma*, *54*, 4, 335–340, 2017.
111. Chauhan, P.S., Dash, D., Singh, R., Intranasal curcumin inhibits pulmonary fibrosis by modulating matrix metalloproteinase-9 (MMP-9) in ovalbumin-induced chronic asthma. *Inflammation*, *40*, 248–258, 2017.
112. Chopra, H., Dey, P.S., Das, D., Bhattacharya, T., Shah, M., Mubin, S., Maishu, S.P., Akter, R., Rahman, M.H., Karthika, C., Murad, W., Curcumin nanoparticles as promising therapeutic agents for drug targets. *Molecules*, *26*, 16, 4998, 2021.
113. Rennolds, J., Malireddy, S., Hassan, F., Tridandapani, S., Parinandi, N., Boyaka, P.N., Cormet-Boyaka, E., Curcumin regulates airway epithelial cell cytokine responses to the pollutant cadmium. *Biochem. Biophys. Res. Commun.*, *417*, 1, 256–261, 2012.
114. Alghasham, A., Salem, T.A., Meki, A.R.M., Effect of cadmium-polluted water on plasma levels of tumor necrosis factor-α, interleukin-6 and oxidative status biomarkers in rats: protective effect of curcumin. *Food Chem. Toxicol.*, *59*, 160–164, 2013.
115. Lin, X., Chen, Y., Liu, Z., Effect and its molecular mechanisms of curcumin on pulmonary artery smooth muscle cells in rat model with chronic obstructive pulmonary disease. *Zhejiang da xuexuebao. Yi xue ban J. Zhejiang Univ. Med. Sci.*, *45*, 5, 469–476, 2016.
116. Gan, L., Li, C., Wang, J., Guo, X., Curcumin modulates the effect of histone modification on the expression of chemokines by type II alveolar epithelial cells in a rat COPD model. *Int. J. Chron. Obstruct. Pulmon. Dis.*, 11, 2765–2773, 2016.
117. Zhang, M., Xie, Y., Yan, R., Shan, H., Tang, J., Cai, Y., Yin, J., Chen, M., Zhang, J., Yang, X., Zhang, Q., Curcumin ameliorates alveolar epithelial injury in a rat model of chronic obstructive pulmonary disease. *Life Sci.*, *164*, 1–8, 2016.
118. Funamoto, M., Sunagawa, Y., Katanasaka, Y., Miyazaki, Y., Imaizumi, A., Kakeya, H., Yamakage, H., Satoh-Asahara, N., Komiyama, M., Wada, H., Hasegawa, K., Highly absorptive curcumin reduces serum atherosclerotic low-density lipoprotein levels in patients with mild COPD. *Int. J. Chron. Obstruct. Pulmon. Dis.*, 11, 2029–2034, 2016.
119. Allen, S., Britton, J.R., Leonardi-Bee, J.A., Association between antioxidant vitamins and asthma outcome measures: systematic review and meta-analysis. *Thorax*, *64*, 7, 610–619, 2009.
120. Nurmatov, U., Devereux, G., Sheikh, A., Nutrients and foods for the primary prevention of asthma and allergy: systematic review and meta-analysis. *J. Allergy Clin. Immunol.*, *127*, 3, 724–733, 2011.
121. Gholamnezhad, Z., Keyhanmanesh, R., Boskabady, M.H., Anti-inflammatory, antioxidant, and immunomodulatory aspects of Nigella sativa for its preventive and bronchodilatory effects on obstructive respiratory diseases: A review of basic and clinical evidence. *J. Funct. Foods*, *17*, 910–927, 2015.

122. Carrizzo, A., Forte, M., Damato, A., Trimarco, V., Salzano, F., Bartolo, M., Maciag, A., Puca, A.A., Vecchione, C., Antioxidant effects of resveratrol in cardiovascular, cerebral and metabolic diseases. *Food Chem. Toxicol.*, *61*, 215–226, 2013.
123. Boskabady, M.H. and Farkhondeh, T., Antiinflammatory, antioxidant, and immunomodulatory effects of Crocus sativus L. and its main constituents. *Phytother. Res.*, *30*, 7, 1072–1094, 2016.
124. Abedinzadeh, Z., Gardes-Albert, M., Ferradini, C., Kinetic study of the oxidation mechanism of glutathione by hydrogen peroxide in neutral aqueous medium. *Can. J. Chem.*, *67*, 7, 1247–1255, 1989.
125. Cazzola, M., Calzetta, L., Facciolo, F., Rogliani, P., Matera, M.G., Pharmacological investigation on the anti-oxidant and anti-inflammatory activity of N-acetylcysteine in an *ex vivo* model of COPD exacerbation. *Respir. Res.*, *18*, 1, 1–10, 2017.
126. Calzetta, L., Rogliani, P., Facciolo, F., Rinaldi, B., Cazzola, M., Matera, M.G., N-Acetylcysteine protects human bronchi by modulating the release of neurokinin A in an *ex vivo* model of COPD exacerbation. *Biomed. Pharmacother.*, *103*, 1–8, 2018.
127. Dal Negro, R., Pozzi, E., Cella, S.G., Erdosteine: Drug exhibiting polypharmacy for the treatment of respiratory diseases. *Pulm. Pharmacol. Ther.*, *53*, 80–85, 2018.
128. Boutten, A., Goven, D., Artaud-Macari, E., Boczkowski, J., Bonay, M., NRF2 targeting: a promising therapeutic strategy in chronic obstructive pulmonary disease. *Trends Mol. Med.*, *17*, 7, 363–371, 2011.
129. Cho, H.Y. and Kleeberger, S.R., Nrf2 protects against airway disorders. *Toxicol. Appl. Pharmacol.*, *244*, 1, 43–56, 2010.
130. Mapp, C.E., Fryer, A.A., De Marzo, N., Pozzato, V., Padoan, M., Boschetto, P., Strange, R.C., Hemmingsen, A., Spiteri, M.A., Glutathione S-transferase GSTP1 is a susceptibility gene for occupational asthma induced by isocyanates. *J. Allergy Clin. Immunol.*, *109*, 5, 867–872, 2002.
131. Spiteri, M.A., Bianco, A., Strange, R.C., Fryer, A.A., Polymorphisms at the glutathione S-transferase, GSTP1 locus: a novel mechanism for susceptibility and development of atopic airway inflammation. *Allergy*, *55*, 15–20, 2000.
132. Li, Y.J., Takizawa, H., Azuma, A., Kohyama, T., Yamauchi, Y., Takahashi, S., Yamamoto, M., Kawada, T., Kudoh, S., Sugawara, I., Disruption of Nrf2 enhances susceptibility to airway inflammatory responses induced by low-dose diesel exhaust particles in mice. *Clin. Immunol.*, *128*, 3, 366–373, 2008.
133. Rangasamy, T., Guo, J., Mitzner, W.A., Roman, J., Singh, A., Fryer, A.D., Yamamoto, M., Kensler, T.W., Tuder, R.M., Georas, S.N., Biswal, S., Disruption of Nrf2 enhances susceptibility to severe airway inflammation and asthma in mice. *J. Exp. Med.*, *202*, 1, 47–59, 2005.
134. Brown, R.H., Reynolds, C., Brooker, A., Talalay, P., Fahey, J.W., Sulforaphane improves the bronchoprotective response in asthmatics through Nrf2-mediated gene pathways. *Respir. Res.*, *16*, 1–12, 2015.
135. Park, J.H., Kim, J.W., Lee, C.M., Kim, Y.D., Chung, S.W., Jung, I.D., Noh, K.T., Park, J.W., Heo, D.R., Shin, Y.K., Seo, J.K., Sulforaphane inhibits the Th2 immune response in ovalbumin-induced asthma. *BMB Rep.*, *45*, 5, 311–316, 2012.
136. Malhotra, D., Thimmulappa, R., Navas-Acien, A., Sandford, A., Elliott, M., Singh, A., Chen, L., Zhuang, X., Hogg, J., Pare, P., Tuder, R.M., Decline in NRF2-regulated antioxidants in chronic obstructive pulmonary disease lungs due to loss of its positive regulator, DJ-1. *Am. J. Respir. Crit. Care Med.*, *178*, 6, 592–604, 2008.
137. Rangasamy, T., Cho, C.Y., Thimmulappa, R.K., Zhen, L., Srisuma, S.S., Kensler, T.W., Yamamoto, M., Petrache, I., Tuder, R.M., Biswal, S., Genetic ablation of Nrf2 enhances susceptibility to cigarette smoke-induced emphysema in mice. *J. Clin. Invest.*, *114*, 9, 1248–1259, 2004.

138. Blake, D.J., Singh, A., Kombairaju, P., Malhotra, D., Mariani, T.J., Tuder, R.M., Gabrielson, E., Biswal, S., Deletion of Keap1 in the lung attenuates acute cigarette smoke-induced oxidative stress and inflammation. *Am. J. Respir. Cell Mol. Biol.*, *42*, 5, 524–536, 2010.
139. Sussan, T.E., Rangasamy, T., Blake, D.J., Malhotra, D., El-Haddad, H., Bedja, D., Yates, M.S., Kombairaju, P., Yamamoto, M., Liby, K.T., Sporn, M.B., Targeting Nrf2 with the triterpenoid CDDO-imidazolide attenuates cigarette smoke-induced emphysema and cardiac dysfunction in mice. *Proc. Natl. Acad. Sci.*, *106*, 1, 250–255, 2009.
140. Liu, Y., Lu, F., Kang, L., Wang, Z., Wang, Y., Pirfenidone attenuates bleomycin-induced pulmonary fibrosis in mice by regulating Nrf2/Bach1 equilibrium. *BMC Pulm. Med.*, *17*, 1–11, 2017.
141. Artaud-Macari, E., Goven, D., Brayer, S., Hamimi, A., Besnard, V., Marchal-Somme, J., Ali, Z.E., Crestani, B., Kerdine-Römer, S., Boutten, A., Bonay, M., Nuclear factor erythroid 2-related factor 2 nuclear translocation induces myofibroblasticdedifferentiation in idiopathic pulmonary fibrosis. *Antioxid. Redox Signaling*, *18*, 1, 66–79, 2013.
142. Hatwalne, M.S., Free radical scavengers in anaesthesiology and critical care. *Indian J. Anaesth.*, *56*, 3, 227, 2012.
143. Tsao, C.S., An overview of ascorbic acid chemistry and biochemistry, in: *Vitamin C in Health and Disease*, pp. 25–58, 1997.
144. Sporn, M.B. and Liby, K.T., NRF2 and cancer: the good, the bad and the importance of context. *Nat. Rev. Cancer*, *12*, 8, 564–571, 2012.
145. DeBlasi, J.M. and DeNicola, G.M., Dissecting the crosstalk between NRF2 signaling and metabolic processes in cancer. *Cancers*, *12*, 10, 3023, 2020.
146. Hammad, A., Namani, A., Elshaer, M., Wang, X.J., Tang, X., "NRF2 addiction" in lung cancer cells and its impact on cancer therapy. *Cancer Lett.*, *467*, 40–49, 2019.
147. Tao, S., Rojo de la Vega, M., Chapman, E., Ooi, A., Zhang, D.D., The effects of NRF2 modulation on the initiation and progression of chemically and genetically induced lung cancer. *Mol. Carcinog.*, *57*, 2, 182–192, 2018.
148. Kalantari, H. and Das, D.K., Physiological effects of resveratrol. *Biofactors*, *36*, 5, 401–406, 2010.
149. Shi, W., Zhai, C., Feng, W., Wang, J., Zhu, Y., Li, S., Wang, Q., Zhang, Q., Yan, X., Chai, L., Liu, P., Resveratrol inhibits monocrotaline-induced pulmonary arterial remodeling by suppression of SphK1-mediated NF-κB activation. *Life Sci.*, *210*, 140–149, 2018.
150. Csiszar, A., Labinskyy, N., Olson, S., Pinto, J.T., Gupte, S., Wu, J.M., Hu, F., Ballabh, P., Podlutsky, A., Losonczy, G., De Cabo, R., Resveratrol prevents monocrotaline-induced pulmonary hypertension in rats. *Hypertension*, *54*, 3, 668–675, 2009.
151. Javkhedkar, A.A., Quiroz, Y., Rodriguez-Iturbe, B., Vaziri, N.D., Lokhandwala, M.F., Banday, A.A., Resveratrol restored Nrf2 function, reduced renal inflammation, and mitigated hypertension in spontaneously hypertensive rats. *Am. J. Physiol. Regul. Integr. Comp. Physiol.*, *308*, 10, R840–R846, 2015.
152. Halliwell, B. and Whiteman, M., Antioxidant and prooxidant properties of vitamin C, in: *Antioxidants in Health and Disease Series*, pp. 59–74, 1997.
153. Winklhofer-Roob, B.M., Ellemunter, H., Frühwirth, M., Schlegel-Haueter, S.E., Khoschsorur, G., Van't Hof, M.A., Shmerling, D.H., Plasma vitamin C concentrations in patients with cystic fibrosis: evidence of associations with lung inflammation. *Am. J. Clin. Nutr.*, 65, 6, 1858–66, 1997 Jun 1.
154. Meister, A., On the antioxidant effects of ascorbic acid and glutathione. *Biochem. Pharmacol.*, *44*, 10, 1905–1915, 1992.
155. Washko, P.W., Wang, Y., Levine, M., Ascorbic acid recycling in human neutrophils. *J. Biol. Chem.*, *268*, 21, 15531–15535, 1993.

156. Fogarty, A., Lewis, S.A., Scrivener, S.L., Antoniak, M., Pacey, S., Pringle, M., Britton, J., Oral magnesium and vitamin C supplements in asthma: a parallel group randomized placebo-controlled trial. *Clin. Exp. Allergy*, *33*, 10, 1355–1359, 2003.
157. Fogarty, A., Lewis, S.A., Scrivener, S.L., Antoniak, M., Pacey, S., Pringle, M., Britton, J., Corticosteroid sparing effects of vitamin C and magnesium in asthma: a randomised trial. *Respir. Med.*, *100*, 1, 174–179, 2006.
158. Pirabbasi, E., Shahar, S., Manaf, Z.A., Rajab, N.F., Manap, R.A., Efficacy of ascorbic acid (vitamin C) and/N-acetylcysteine (NAC) supplementation on nutritional and antioxidant status of male chronic obstructive pulmonary disease (COPD) patients. *J. Nutr. Sci. Vitaminol.*, *62*, 1, 54–61, 2016.
159. Park, H.J., Byun, M.K., Kim, H.J., Kim, J.Y., Kim, Y.I., Yoo, K.H., Chun, E.M., Jung, J.Y., Lee, S.H., Ahn, C.M., Dietary vitamin C intake protects against COPD: the Korea National Health and Nutrition Examination Survey in 2012. *Int. J. Chron. Obstruct. Pulmon. Dis.*, 11, 2721–2728, 2016.
160. Riordan, N.H., Riordan, H.D., Casciari, J.P., Clinical and experimental experiences with intravenous vitamin C. *J. Orthomol. Med.*, *15*, 4, 201–213, 2000.
161. Traber, M.G. and Head, B., Vitamin E: How much is enough, too much and why! *Free Radical Biol. Med.*, *177*, 212–225, 2021.
162. Tanumihardjo, S.A., Vitamin A: biomarkers of nutrition for development. *Am. J. Clin. Nutr.*, *94*, 2, 658S–665, 2011.
163. Mastrangelo, D., Massai, L., Fioritoni, G., Coco, F.L., Vitamin C against cancer. *Vitam. C*, *1*, 648, 2017.
164. Nauman, G., Gray, J.C., Parkinson, R., Levine, M., Paller, C.J., Systematic review of intravenous ascorbate in cancer clinical trials. *Antioxidants*, *7*, 7, 89, 2018.
165. Kupari, M. and Rapola, J., Reversible pulmonary hypertension associated with vitamin C deficiency. *Chest*, *142*, 1, 225–227, 2012.
166. Litonjua, A.A., Fat-soluble vitamins and atopic disease: what is the evidence? *Proc. Nutr. Soc.*, *71*, 1, 67–74, 2012.
167. Pearson, P.J.K., Lewis, S.A., Britton, J., Fogarty, A., Vitamin E supplements in asthma: a parallel group randomized placebo controlled trial. *Thorax*, *59*, 8, 652–656, 2004.
168. Dunstan, J.A., Breckler, L., Hale, J., Lehmann, H., Franklin, P., Lyons, G., Ching, S.Y.L., Mori, T.A., Barden, A., Prescott, S.L., Supplementation with vitamins C, E, β-carotene and selenium has no effect on anti-oxidant status and immune responses in allergic adults: a randomized controlled trial. *Clin. Exp. Allergy*, *37*, 2, 180–187, 2007.
169. Romieu, I., Sienra-Monge, J.J., Ramírez-Aguilar, M., Téllez-Rojo, M.M., Moreno-Macías, H., Reyes-Ruiz, N.I., del Río-Navarro, B.E., Ruiz-Navarro, M.X., Hatch, G., Slade, R., Hernández-Avila, M., Antioxidant supplementation and lung functions among children with asthma exposed to high levels of air pollutants. *Am. J. Respir. Crit. Care Med.*, *166*, 5, 703–709, 2002.
170. Trenca, C.A., Koenig, J.Q., Williams, P.V., Dietary antioxidants and ozone-induced bronchial hyperresponsiveness in adults with asthma. *Arch. Environ. Health: Int. J.*, *56*, 3, 242–249, 2001.
171. Romieu, I., Meneses, F., Ramirez, M., Ruiz, S., Padilla, R.P., Sienra, J.J., Gerber, M., Grievink, L., Dekker, R., Walda, I., Brunekreef, B., Antioxidant supplementation and respiratory functions among workers exposed to high levels of ozone. *Am. J. Respir. Crit. Care Med.*, *158*, 1, 226–232, 1998.
172. Grievink, L., Zijlstra, A.G., Ke, X., Brunekreef, B., Double-blind intervention trial on modulation of ozone effects on pulmonary function by antioxidant supplements. *Am. J. Epidemiol.*, *149*, 4, 306–314, 1999.

173. Gref, A., Rautiainen, S., Gruzieva, O., Håkansson, N., Kull, I., Pershagen, G., Wickman, M., Wolk, A., Melén, E., Bergström, A., Dietary total antioxidant capacity in early school age and subsequent allergic disease. *Clin. Exp. Allergy*, *47*, 6, 751–759, 2017.
174. Arnér, E.S. and Holmgren, A., Physiological functions of thioredoxin and thioredoxin reductase. *Eur. J. Biochem.*, *267*, 20, 6102–6109, 2000.
175. Su, W.Y., Folz, R., Chen, J.S., Crapo, J.D., Chang, L.Y., Extracellular superoxide dismutase mRNA expressions in the human lung by *in situ* hybridization. *Am. J. Respir. Cell Mol. Biol.*, *16*, 2, 162–170, 1997.
176. Oury, T.D., Day, B.J., Crapo, J.D., Extracellular superoxide dismutase in vessels and airways of humans and baboons. *Free Radical Biol. Med.*, *20*, 7, 957–965, 1996.
177. Oury, T.D., Chang, L.Y., Marklund, S.L., Day, B.J., Crapo, J.D., Immunocytochemical localization of extracellular superoxide dismutase in human lung. *Lab. Invest.*, *70*, 6, 889–898, 1994.
178. van der Vliet, A., O'Neill, C.A., Cross, C.E., Koostra, J.M., Volz, W.G., Halliwell, B., Louie, S., Determination of low-molecular-mass antioxidant concentrations in human respiratory tract lining fluids. *Am. J. Physiol. Lung Cell. Mol. Physiol.*, *276*, 2, L289–L296, 1999.
179. Patil, J.S. and Sarasija, S., Pulmonary drug delivery strategies: A concise, systematic review. *Lung India: Off. Organ Indian Chest Soc.*, *29*, 1, 44, 2012.
180. Newman, S.P., Drug delivery to the lungs: Challenges and opportunities. *Ther. Deliv.*, *8*, 8, 647–661, 2017.
181. Patel, B., Gupta, N., Ahsan, F., *Barriers that inhaled particles encounter*, pp. 707–727, International Society for Aerosols in Medicine, Online Publication, Werne, North Rhine-Westphalia, Germany, 2015.
182. Patton, J.S., Brain, J.D., Davies, L.A., Fiegel, J., Gumbleton, M., Kim, K.J., Sakagami, M., Vanbever, R., Ehrhardt, C., The particle has landed—characterizing the fate of inhaled pharmaceuticals. *J. Aerosol Med. Pulm. Drug Deliv.*, *23*, S2, S-71, 2010.
183. Cipolla, D., Will pulmonary drug delivery for systemic application ever fulfill its rich promise? *Expert. Opin. Drug Deliv.*, *13*, 10, 1337–1340, 2016.
184. Sanchis, J., Gich, I., Pedersen, ST., Systematic Review of Errors in Inhaler Use. *CHEST*, *150*, 2, 394–406, 2016.
185. Sharma, P., Mehta, M., Dhanjal, D.S., Kaur, S., Gupta, G., Singh, H., Thangavelu, L., Rajeshkumar, S., Tambuwala, M., Bakshi, H.A., Chellappan, D.K., Emerging trends in the novel drug delivery approaches for the treatment of lung cancer. *Chem. Biol. Interact.*, *309*, 108–720, 2019.
186. Xu, Y., Liu, H., Song, L., Novel drug delivery systems targeting oxidative stress in chronic obstructive pulmonary disease: a review. *J. Nanobiotechnol.*, *18*, 1, 1–25, 2020.
187. Rathi, R., Nagpal, M., Kaur, M., Ballouria, P., Dutta, R., Aggarwal, G., Arora, S., Treatment of Tuberculosis in Nano Era: Recent Avenues. *Curr. Bioact. Compd.*, *18*, 7, 13–26, 2022.
188. Ourique, A.F., dos Santos Chaves, P., Souto, G.D., Pohlmann, A.R., Guterres, S.S., Beck, R.C.R., Redispersible liposomal-N-acetylcysteine powder for pulmonary administration: development, *in vitro* characterization and antioxidant activity. *Eur. J. Pharm. Sci.*, 65, 174–182, 2014.
189. Castellani, S., Trapani, A., Spagnoletta, A., Di Toma, L., Magrone, T., Di Gioia, S., Mandracchia, D., Trapani, G., Jirillo, E., Conese, M., Nanoparticle delivery of grape seed-derived proanthocyanidins to airway epithelial cells dampens oxidative stress and inflammation. *J. Transl. Med.*, *16*, 1–15, 2018.
190. Gil, D., Rodriguez, J., Ward, B., Vertegel, A., Ivanov, V., Reukov, V., Antioxidant activity of SOD and catalase conjugated with nanocrystalline ceria. *Bioengineering*, *4*, 1, 18, 2017.
191. Lee, W.H., Loo, C.Y., Young, P.M., Rohanizadeh, R., Traini, D., Curcumin nanoparticles attenuates inflammation caused by macrophages: effects of nanoparticles' surface charges. *J. Aerosol. Med. Pulm. Drug Deliv.*, *21*, A7–A7, 2014.

192. Lee, W.H., Loo, C.Y., Young, P.M., Rohanizadeh, R., Traini, D., Curcumin nanoparticles attenuate production of pro-inflammatory markers in lipopolysaccharide-induced macrophages. *Pharm. Res.*, *33*, 315–327, 2016.
193. Passi, M., Kumar, V., Packirisamy, G., Theranosticnanozyme: Silk fibroin based multifunctional nanocomposites to combat oxidative stress. *Mater. Sci. Eng., C*, *107*, 110255, 2020.
194. Jacobs, C. and Müller, R.H., Production and characterization of a budesonide nanosuspension for pulmonary administration. *Pharm. Res.*, *19*, 189–194, 2002.
195. Christofidou-Solomidou, M. and Muzykantov, V.R., Antioxidant strategies in respiratory medicine. *Treat. Respir. Med.*, *5*, 47–78, 2006.
196. Suwanpitak, K., Lim, L.Y., Singh, I., Sriamornsak, P., Thepsonthi, T., Huanbutta, K., Sangnim, T., Development of an Add-On Device Using 3D Printing for the Enhancement of Drug Administration Efficiency of Dry Powder Inhalers (Accuhaler). *Pharmaceutics*, *14*, 9, 1922, 2022.

8

Antioxidants in Aging

Harpreet Singh[1]*, Sourav Mohanto[2], Adrija Bhunia[2], Bhuvnesh Kumar Singh[3], Kanupriya Chauhan[3], Arvind Kumar[1], Amrita Mishra[4], Arun Kumar Mishra[5] and Alankar Shrivastav[6]

[1]*School of Pharmaceutical Sciences, IFTM University, Moradabad, India*
[2]*Department of Pharmaceutics, Yenepoya Pharmacy College & Research Centre, Yenepoya (Deemed to be University), Mangalore, Karnataka, India*
[3]*Faculty of Pharmacy, Moradabad Educational Trust, Moradabad, India*
[4]*School of Pharmaceutical Sciences, Delhi Pharmaceutical Sciences and Research University, New Delhi, India*
[5]*SOS School of Pharmacy, IFTM University, Moradabad, India*
[6]*Pharmacy Academy, IFTM University, Moradabad, India*

Abstract
The chapter provides a comprehensive overview of the relationship between oxidative stress and the aging process. Oxidative stress arises from an imbalance between the production of Reactive Oxygen Species (ROS) and the body's antioxidant defense system. The mechanisms of action of antioxidants are explored, including enzymatic and non-enzymatic antioxidants, metal-binding proteins, dietary sources, and nutritional supplements. The chapter also delves into the role of oxidative stress in cellular senescence and the aging process. Furthermore, the effectiveness of nutritional supplements in combating aging is examined. The chapter also highlights the impact of oxidative stress on skin health and aging, and discusses the use of topical antioxidants and anti-aging cosmetics. Lifestyle factors such as exercise, stress, sleep, smoking, and alcohol consumption are also explored as contributors to diminished antioxidant defense. The chapter concludes with a discussion on the potential for personalized antioxidant interventions and the development of novel compounds from plants, marine, microorganisms, and synthetic sources for future antioxidant research. Overall, this chapter provides valuable insights into the mechanisms of oxidative stress and the potential of antioxidant interventions in promoting healthy aging.

***Keywords*:** Aging, antioxidants, skin health, stress, reactive oxygen species, natural

8.1 Introduction

Aging is a normal biological process involving complex reactions that happens to all living things. It is characterized by a gradual decline in physiological functions as well as a loss of integrity at the cellular, tissue, and organ levels. Aging is characterized by a gradual

**Corresponding author*: harpreetproctor@rediffmail.com

Rakesh K. Sindhu, Inderbir Singh and M. Arockia Babu (eds.) Antioxidants: Nature's Defense Against Disease, (257–284)

decline in the body's ability to repair and regenerate itself, increasing vulnerability to various age-related diseases and, eventually, death [1]. The basic mechanism behind aging process istelomere shortening, which affects cell stability and replication capacity, as well as the accumulation of genetic mutations and DNA damage [2]. Furthermore, the efficiency of cellular processes, i.e., protein synthesis, mitochondrial function, and the removal of damaged molecules via autophagy get decreased when aging dominates. A combination of genetic, environmental, and lifestyle factors influence aging [3]. Although genetics can affect aging, an individual's environment also plays a role. Various factors, i.e., diet, physical activity, exposure to toxins, and chronic stress, can all impact aging [4].

Aging follows wide range of consequences that affect the physiological functions of the organ system of an organism. The consequences include loss of sensory perception, decreased muscle mass and strength, impaired cognitive functions, an increased susceptibility to chronic diseases such as cardiovascular disease, cancer, neurodegenerative disorders, and a reduced ability to recover from illness or injury [5]. At present, extensive scientific research is being carried out to better understand the mechanisms and processes involved in aging. Scientists have investigated aging theories such as oxidative stress theory, telomere theory, mitochondrial dysfunction theory, and the role of inflammation and cellular senescences. This suggests the scope of research to understand the underlying causes of aging and to develop interventions with an aim to promote healthy aging and prevent or delay age-related diseases [6]. The effects of aging on human health are substantial, ranging from changes in how the body functions to an increased vulnerability to age-related diseases. Loss of physical vitality and agility are prominent features of aging. Muscle mass declines with age, which results in weaker bodies and reduced mobility [7]. This can lead to problems with balance, coordination, and general mobility, increasing the risk of falls and injuries amongst the elderly. The immune system is another key aspect which is negatively impacted by time. The immune system's ability to detect and destroy foreign invaders declines with age, leaving the elderly more susceptible to illness and injury. The increase in the incidence of chronic diseases like cancer, cardiovascular disease, and autoimmune disorders has been linked to age-related declines in immune function [8].

The cardiovascular system is also severely affected by aging. Cardiac changes associated with aging, such as thickening of the cardiac muscle and reduced pumping efficiency, further contribute to cardiovascular issues in the elderly. In addition, the elder patients are at higher risk of developing neurodegenerative diseases like Alzheimer's and Parkinson's. Memory, focus and speed of thoughts are all areas that generally get decline with age [9]. The effects of cognitive decline on daily functioning and quality of life necessitate individualized approaches to treatment and care. Loss of bone density and an increased propensity for fractures and osteoporosis are additional skeletal system changes associated with aging. Bone loss gets accelerated by hormonal changes, especially in women during menopause. Joint pain, stiffness, and loss of mobility are all symptoms of age-related declines in joint health, caused by conditions such as osteoarthritis. The, mental and emotional wellbeing tend to shift as people age. Isolation, loneliness, and the loss of loved ones are all factors that can contribute to the development of depression and anxiety in the elderly. Retirement, financial worries, and a shifting sense of identity can all take a toll on a person's mental health and increases age [10]. The aging process is greatly influenced by oxidative stress, which is also linked to many age-related diseases. It happens when the body's antioxidant defense mechanisms are overwhelmed by the body's production of ROS, further depicted in Figure 8.1. ROS are byproducts of normal cellular metabolism that play key roles in various physiological processes. However, oxidative stress occurs when their production exceeds the body's ability to neutralize them. Oxidative stress rises with age because our cells become less efficient at dealing with ROS. Some of the causes of this phenomenon are the gradual weakening of antioxidant defenses, the disruption of mitochondrial function, and the accumulation of damaged cellular components. Damage to proteins, lipids, and DNA due to oxidative stress can interfere with normal cellular functions and has been linked to the aging process [11].

Inflammation is a common response to oxidative stress. ROS can set off inflammatory responses in the body, leading to chronic but, low-grade inflammation. Many age-related diseases, such as cardiovascular disease, neurodegenerative disorders, and some types of cancers, are characterized by inflammation. Oxidative stress is a major contributor to tissue damage and the aging process; therefore, it can be elucidated that chronic inflammation can keep the cycle going [12]. In discussions of molecular mechanisms, telomeres are crucial factors in the aging process. Telomeres, which serve as caps at the ends of chromosomes, are also negatively impacted by oxidative stress. With each cell division comes a shortening of the telomeres, but oxidative stress speeds up this process [13]. Telomere shortening is linked to cellular aging which causes an increased risk of developing age-related diseases. When telomeres are shortened due to oxidative stress, cells enter a senescent state in which they can no longer divide and function normally [13]. Furthermore, mitochondrial function, which is essential for cellular energy production, is negatively impacted by oxidative stress. The high metabolic activity and ROS production in mitochondria make the mitochondria especially vulnerable to oxidative damage. Damage to mitochondria over the time reduces the cell's ability to generate energy, which in turn increases oxidative stress and speeds up the aging process. The endogenous antioxidant defense systems of the body are

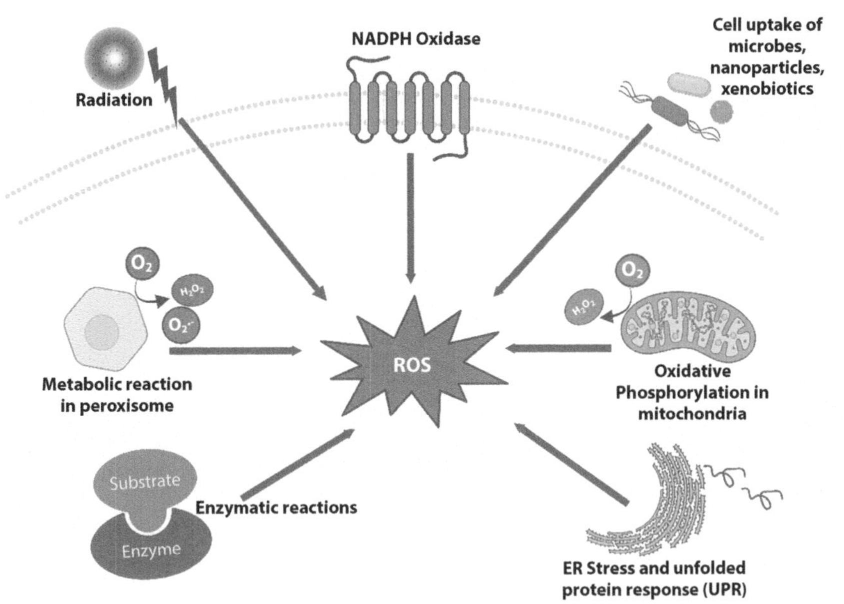

Figure 8.1 Sources of ROS generation. (*Created with Biorender*).

responsible for mitigating the harmful effects caused due to oxidative stress. Antioxidants prevent oxidative damage to cells by neutralizing ROS [14]. Cells become more susceptible to oxidative stress as age increases. Oxidative stress can be further exacerbated by factors such as poor diet, smoking, exposure to environmental pollutants, and chronic diseases. ROS may be one way to encourage healthy aging. Oxidative stress can be mitigated with the help of antioxidant-rich diets, regular physical exercise, and applying stress-reduction techniques. Exogenous antioxidants, such as those found in supplements or medical interventions, have also been investigated for their ability to mitigate the harmful effects caused by oxidative stress. However, extensive research is required to determine whether or not such interventions are effective and safe [15].

Antioxidants are crucial for preventing oxidative damage caused by ROS and maintaining cellular health. Antioxidants are bioactive compounds that neutralize ROS and other free radicals, preventing them from damaging cells and tissues. They function by donating an electron to unstable ROS, stabilizing them, and reducing their ability to initiate chain reactions that damage cellular components. There are numerous types of antioxidants, including naturally occurring enzymes example SOD, catalase and glutathione peroxidase. By converting the ROS into less harmful substances, these enzymes play a crucial role in the antioxidant defense mechanism. In addition to this, many dietary antioxidants are derived from fruits, vegetables, and other plant-based sources, such as vitamins (i.e., vitamin C, vitamin E, and beta-carotene), minerals (such as selenium and zinc), and phytochemicals (i.e., flavonoids and carotenoids) [16], have further been summarized in Table 8.1.

Antioxidants serve primarily to scavenge and neutralize ROS, thereby protecting cells from oxidative damage. They achieve this goal through various mechanisms. Some antioxidants react directly with ROS, stabilizing them and preventing their damaging effects. Vitamin C, for instance, donate electrons to free radicals, thereby effectively reduces their reactivity. Vitamin E functions as a lipid-soluble antioxidant, protecting cell membranes

Table 8.1 Summarized various endogenous and exogenous antioxidants derived from natural sources [17].

Endogenous antioxidants	Exogenous antioxidants
Enzymatic	**Mineral elements**
➤ Catalase	➤ Selenium
➤ Superoxide dismutase	➤ Zinc
➤ Glutathione	➤ Manganese
➤ Glutamyl trans peptidase	➤ Copper
Non-Enzymatic	**Nutritional**
➤ Uric acid	➤ Carotenoids
➤ Ubiquinone	➤ Vitamin C
➤ Tocopherol	➤ Vitamin E
➤ Retinol	➤ Phytochemicals
➤ Glutathione	**Supplementations**
➤ Melatonin	➤ Beta-Carotene, Coenzyme Q_{10}
➤ Nuclear Factor erythroid	➤ Lipoic acid, Glutathione
➤ 2-related factor 2 (NRF2)	➤ Phytoestrogens, Polyphenol

from lipid peroxidation caused by free radicals. It resides within cell membranes [18]. In addition to this, antioxidants have the capacity to regenerate other antioxidants, thereby enhancing their overall efficacy. Vitamin C, for instance, can regenerate vitamin E after it has neutralized a free radical, ensuring that vitamin E retains its antioxidant function. This interaction between antioxidants allows for a continuous cycle of oxidative damage protection [19]. Antioxidants not only combat oxidative damage directly but also modulate inflammation and gene-expression-related signaling pathways. They are able to regulate the activity of transcription factors such as nuclear factor-kappa B (NF-kB), which regulates the expression of genes involved in inflammation and immune response. By modulating these pathways, antioxidants can mitigate the deleterious effects of stress-induced chronic inflammation [20]. It is essential to note that the efficacy of antioxidants can vary depending on their location in the body and the ROS they target. Various antioxidants have preferences for specific cellular compartments and ROS types. For optimal protection against oxidative damage, a balanced consumption of various antioxidants through a varied diet is essential [21].

Antioxidants are necessary for preventing oxidative damage, but it is important to note that excessive antioxidant supplementation may not always be advantageous and may even have negative effects. The delicate balance between ROS and antioxidants is essential for normal cellular signaling and physiological processes. Disturbing this equilibrium by excessively elevating antioxidant levels may interfere with beneficial signaling pathways and have negative impacts [22].

8.2 Mechanisms of Action of Various Antioxidant

8.2.1 Enzymatic Antioxidants

Enzymatic antioxidants play a crucial role in combating oxidative stress and its associated effects on aging in living beings. Two important enzymatic antioxidants are superoxide dismutase (SOD) and catalase. These enzymes are involved in the neutralization of harmful ROS produced as byproducts of normal cellular metabolism. By comprehending their mechanisms, a better knowledge of their effects on aging can be gained [23]. The enzyme SOD converts the superoxide radical ($O_2\bullet^-$) into hydrogen peroxide (H_2O_2) and molecular oxygen (O_2). Extremely reactive, $O_2\bullet$- can damage cellular components such as deoxyribonucleic Acid (DNA), proteins and lipids. Through a process known as the fenton reaction, SOD prevents the formation of highly toxic hydroxyl radicals ($OH\bullet^-$) by converting $O_2\bullet^-$ into H_2O_2. It is well known that $OH\bullet^-$ is particularly damaging to cells and contribute to aging processes. Therefore, SOD is a crucial first line of defense against oxidative stress, reducing the likelihood of cellular damage and slowing the aging process [24, 25].

Catalase, on the other hand, is an enzyme that targets H_2O_2, a potentially harmful byproduct of numerous cellular reactions. Catalase converts $H_2O_2$to water (H_2O) and oxygen (O_2), deactivating this reactive compound. If left unchecked, H_2O_2 can cause oxidative damage to cells and tissues, leading to accelerate the aging process. Catalase helps to maintain cellular homeostasis and reduces the detrimental effects of oxidative stress on aging by facilitating the breakdown of H_2O_2 [26]. Thus, SOD and catalase in collaboration maintain the

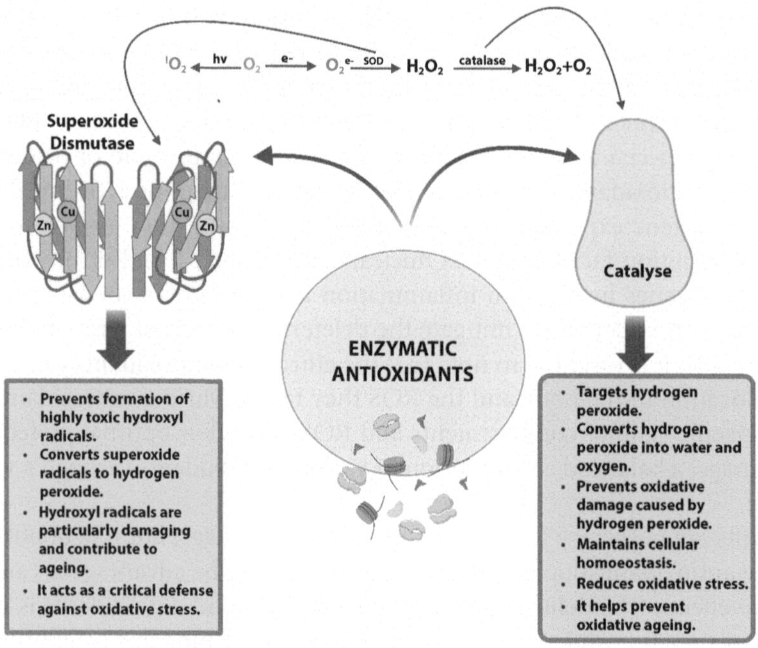

Figure 8.2 Mechanism of SOD and catalase as antioxidants. (*Created with Biorender*).

equilibrium of ROS within the cell. Their coordinated action reduces oxidative damage and its effects on aging. Nonetheless, as organisms age increases, the activity of these enzymatic antioxidants may diminish, resulting in an imbalance between ROS production and antioxidant defense mechanisms, further depicted the mechanisms in Figure 8.2. This imbalance can lead to an accumulation of oxidative damage and a speed-up of aging process [27]. In addition, it is important to note that environmental stressors, lifestyle choices, and genetic predispositions can influence the activity of enzymatic antioxidants and accelerate aging. For example, exposure to environmental toxins, ultraviolet radiation, and unhealthy behaviors such as smoking or a poor diet can increase oxidative stress, thereby overwhelming the enzymatic antioxidant defense system [28].

8.2.2 Non-Enzymatic Antioxidants

Non-enzymatic antioxidants, such as vitamin C, vitamin E and glutathione, are also important in reducing the effects of oxidative stress on aging in living beings. These antioxidants work through various mechanisms to scavenge and neutralize ROS and protect cells from damage. Comprehending the intricacies of their mechanisms can facilitate a comprehensive grasp of their influence on the aging process [29–31]. Vitamin C, also known as ascorbic acid, is a water-soluble antioxidant that acts as a free radical scavenger. It contributes electrons to the neutralization and stabilization of ROS such as $O_2\bullet^-$ and $OH\bullet^-$. Furthermore, vitamin C regenerates other antioxidants, like vitamin E, by converting their oxidized forms back to active states. This prevents the spread of oxidative damage and protects cellular components from oxidative stress, such as proteins, lipids and DNA. Furthermore, vitamin

C aids in the synthesis of collagen, a structural protein required for the integrity of tissues and organs. Aging is associated with collagen depletion and loss of connective tissue elasticity, and vitamin C's involvement in collagen synthesis contributes to its anti-aging effects [32].

Vitamin E is a group of fat-soluble compounds, the most biologically active of which is alpha-tocopherol. Vitamin E is mostly found in cellular membranes, where it functions as a lipid-soluble antioxidant. It prevents lipid peroxidation, a chain reaction caused by free radicals attacking polyunsaturated fatty acids in cell membranes, which causes cellular damage and aging. Vitamin E donates electrons to lipid radicals, halting the chain reaction and preventing oxidative damage to cell membranes. Vitamin E has also been shown to modulate signaling pathways involved in inflammation and cellular stress responses, which contributes to its anti-aging properties. Vitamin E helps to maintain cellular function and slows the aging process by preserving membrane integrity and reducing inflammation [33].

Glutathione is a tripeptide made up of three different amino acids: cysteine, glycine, and glutamic acid. It is an important intracellular antioxidant that participates in a variety of cellular processes, including ROS detoxification and antioxidant regeneration. Glutathione is a direct scavenger of free radicals, especially the $OH\bullet^{-}$, and it can regenerate vitamin C and vitamin E by reducing their oxidized forms. Furthermore, glutathione participates in the enzymatic detoxification of harmful compounds such as environmental toxins and heavy metals. These detoxification processes alleviate the burden of oxidative stress on cells and aid in the prevention of age-related damage [34].

Non-enzymatic antioxidants, such as vitamin C, vitamin E and glutathione, have been found to decline with age in the living beings. This decrease could be due to decreased biosynthesis, increased utilization, or faulty recycling mechanisms. As a result, cell antioxidant capacity declines and oxidative stress increases, contributing to accelerated aging processes. External factors such as poor nutrition, exposure to environmental toxins and lifestyle choices can also deplete antioxidant levels and worsen oxidative damage [35].

8.2.3 Metal-Binding Proteins and Their Role in Redox Regulation

Metal-binding proteins are important in redox regulation, which involves controlling oxidation-reduction (redox) reactions within living organisms. These proteins participate in the binding and coordination of metal ions such as iron, copper, zinc and manganese, and their interactions with these metal ions are required for their biological functions. Gaining insight into their mechanisms enables a more profound appreciation of their involvement in the processes of aging and redox homeostasis [36].

Metal-binding proteins act as redox catalysts in a variety of cellular processes, including energy production, DNA repair, and antioxidant defense. One example is hemoglobin, an iron-containing protein that transports oxygen throughout the body and is essential for cellular respiration. The iron ion in hemoglobin goes through reversible oxidation and reduction processes, allowing it to bind and release oxygen as needed. This iron redox regulation in hemoglobin is essential for efficient oxygen transport and energy production [37]. Metal-binding proteins contribute to redox regulation by maintaining metal ion homeostasis in addition to their catalytic roles. Through fenton chemistry, metal ions can participate in redox reactions and generate ROS. Excessive ROS can cause oxidative damage to cells and hasten aging. Metal-binding proteins aid in the sequestration and distribution of metal ions within cells, thereby preventing their accumulation and potential contribution to oxidative stress. Metal-binding proteins, i.e., metallothionein, bind to and regulate the levels of essential metals such as zinc and copper. Metallothioneins help to maintain redox balance and protect against oxidative damage by preventing the accumulation of free metal ions [38].

Metal-binding proteins also play a role in the formation and function of antioxidant enzymes. Metal ions are required as cofactors for the activity of many antioxidant enzymes, including SOD and catalase. These metal ions are required for these enzymes' catalytic mechanisms, which allow them to neutralize ROS and maintain redox homeostasis. Copper-zinc superoxide dismutase (CuZnSOD) and manganese superoxide dismutase (MnSOD), for example, catalyze the dismutation of $O_2\bullet^-$ into less harmful molecules by using copper and manganese ions, respectively. Metal-binding proteins help protect against oxidative stress and its effects on aging by promoting the activity of antioxidant enzymes [39]. Disruptions in metal-binding proteins and their interactions with metal ions, on the other hand, can cause redox imbalances and accelerate the aging process [40]. Metal ions can become dysregulated and accumulate in cells in some cases, resulting in the production of excessive ROS and oxidative damage. Iron accumulation in the brain, for example, has been linked to neurodegenerative diseases associated with aging, such as Alzheimer's and Parkinson's disease. Changes in the expression or activity of metal-binding proteins can also affect the function of antioxidant enzymes and other redox-related processes, contributing to oxidative stress and aging [41].

8.3 Antioxidants in the Diet

8.3.1 Analysis of Antioxidant-Rich Foods

Antioxidant-rich foods like fruits, vegetables, nuts, and seeds have gained attention for their potential health benefits. These foods contain a variety of antioxidant-containing compounds, including vitamins and minerals, summarized in Table 8.2. Understanding their potential protective effects against oxidative stress-related diseases, such as aging, requires an examination of the antioxidant content of these foods [42]. Several methods can be used to evaluate the antioxidant capacity of foods. The measurement of total antioxidant capacity (TAC) is a common method. TAC assays assess the cumulative antioxidant activity of a food sample by considering the contribution of multiple antioxidants present. These assays frequently employ colorimetric or fluorometric techniques to quantify the antioxidants' ability to scavenge free radicals or inhibit oxidative reactions [43].

Individual antioxidants in complex food matrices can be quantified using high-performance liquid chromatography (HPLC) coupled with various detection systems, such as UV-Visible spectroscopy or mass spectrometry. These methods provide comprehensive information regarding the concentration and composition of antioxidants in a particular food item [44]. In addition to quantitative analysis, qualitative evaluation of foods rich in antioxidants is also beneficial. This entails identifying the particular antioxidant compound present in a food sample. The identification and structural elucidation of antioxidants can be aided by techniques such as nuclear magnetic resonance (NMR) spectroscopy and gas

Table 8.2 Summarization of key effects of nutrients and diet in skin aging [46].

Nutrient/diet	Effects on skin
Water	Maintain skin internal balance and tissue function (e.g., aging and inflammation).
Protein	Constitution and repair of skin tissues (involved in protein synthesis and metabolism), mediation of skin physiological functions and supply of energy.
Copper	Involved in extracellular matrix, synthesis and stabilization of skin proteins, and angiogenesis.
Zinc	Participates in the proliferation and differentiation of epidermal keratinocytes.
Iron	Closely related to the activity of antioxidant enzymes in skin cells.
Selenium	Essential for the development and function of skin keratinocytes and related to skin antioxidant enzyme activity.
Vitamin A	Commonly used anti-aging ingredients prevent skin aging by regulating the expression of genes and matrix metalloproteinases.
Vitamin B	Associated with skin inflammation and pigmentation.
Vitamin C	Involved in skin collagen synthesis and elimination of intracellular reactive oxygen species.
Vitamin D	Reduces skin DNA damage, inflammation, and photo carcinogenesis.
Vitamin E	Prevent skin aging by inhibiting lipid peroxidation.
Fat	High fat is associated with skin inflammation, essential fatty acids are involved in skin lipid synthesis and metabolism.
Tobacco	Change skin cuticle thickness, accelerate skin pigmentation and skin necrosis.
Alcohol	Promote the proliferation of keratinocytes, change the skin permeability, destroy the barrier function of the skin, affect the skin lipid composition.
Sugar and Baked Goods	Associated with skin thickness, autophagy and inflammation.

chromatography-mass spectrometry (GC-MS). By understanding the types of antioxidants present, scientists are able to correlate them with potential health benefits and investigate their mechanisms of action [44, 45].

Analyzing foods rich in antioxidants involves more than measuring antioxidant content alone. Bioavailability and synergistic effects are important factors to consider. Some antioxidants may have limited bioavailability, meaning they may be poorly absorbed or utilized by the body. Assessing the bioavailability of antioxidants can shed light on their *in vitro* efficacy.

In addition, the synergistic interactions between various antioxidants present in a food matrix can boost antioxidant activity overall. These interactions can help reveal the potential additive or synergistic effects of consuming whole foods as opposed to isolated antioxidants [47].

8.3.2 Nutritional Supplements and Their Effectiveness in Aging

Nutritional supplements have gained popularity to promote healthy aging and reduce age-related decline. These supplements typically contain vitamins, minerals, antioxidants and other bioactive compounds believed to have positive effects on various aging-related factors. Although the use of nutritional supplements in aging is the subject of ongoing research, their efficacy and impact on the outcomes of aging are still being determined [48]. Antioxidants are a commonly used category of nutritional supplements for aging. Antioxidants, such as vitamin C, vitamin E and coenzyme Q10, are believed to mitigate oxidative stress and ROS-induced damage. Antioxidant supplements have demonstrated promising results in preclinical studies, but their efficacy in human trials has been inconsistent. Some studies have found potential benefits, while others have found no significant effects on the outcomes associated with aging. In addition, excessive consumption of antioxidant supplements may disrupt the delicate balance of ROS in the body and impair the body's natural defenses [49].

Some commonly-used nutritional supplements for anti-aging are resveratrol, curcumin, and polyphenols. Resveratrol, a compound found in red grapes and wine, has been the subject of extensive research due to its potential anti-aging properties. In animal models, it has been shown to activate longevity-related genes and pathways, but its effects on humans remain unknown [50]. Curcumin, a compound derived from turmeric, possesses anti-inflammatory and antioxidant properties and has shown promise in preclinical studies for treating age-related diseases. However, additional clinical research is required to establish its anti-aging efficacy. Polyphenols, which are found in foods such as green tea, berries and dark chocolate, have also been linked to potential anti-aging health benefits. They have antioxidant and anti-inflammatory properties and may improve cardiovascular health and cognitive performance. However, additional research is necessary to determine their direct effects on the aging process [27, 51–53].

It is important to note that the efficacy of nutritional supplements in aging can vary based on individual factors such as genetics, overall diet quality, lifestyle choices, and health status at the onset of aging. In addition, the quality and formulation of the supplements, as well as the correct dosage, have a significant impact on their potential health benefits. Before beginning any supplement regimen, it is recommended to consult a healthcare professional who can provide personalized recommendations based on individual needs and potential drug interactions [54].

The effectiveness of nutritional supplements in slowing the aging process and preventing age-related diseases remains under investigation. Although some dietary supplements have demonstrated promising results in preclinical studies, the evidence from human trials is frequently contradictory or inconclusive. Maintaining a well-balanced, nutrient-rich diet and adopting a healthy lifestyle continue to be essential for promoting healthy aging. To better comprehend the function of nutritional supplements and their optimal application in the context of aging, additional study is required [55].

8.4 The Role of Antioxidants in Cellular Senescence

8.4.1 Cellular Senescence and its Contribution to Aging

Cellular senescence, or irreversible growth arrest, is an important factor in the aging process. Senescence was first discovered as a mechanism to prevent the proliferation of damaged cells, but it has since been linked to many aspects of aging and age-related diseases. Senescent cells lose their ability to divide and perform normal cellular functions, resulting in a decline in tissue and organ function over time [56]. The impact of cellular senescence on tissue homeostasis is one of the most important contributions of cellular senescence to aging. Senescent cells naturally accumulate in various tissues throughout the body as a result of aging. The senescence-associated secretory phenotype (SASP) refers to the ability of these cells to secrete a variety of bioactive molecules, such as pro-inflammatory cytokines, growth factors, and matrix metalloproteinases [57]. The SASP has the potential to have both local and systemic effects, disrupting the surrounding tissue microenvironment and contributing to chronic low-grade inflammation, which is a hallmark of aging. Chronic inflammation, also known as inflammation, hastens the aging process and raises the risk of age-related diseases like cardiovascular disease, neurodegenerative disorders, and cancer [57, 58].

Cellular senescence affects tissue regenerative capacity in addition to promoting inflammation. The presence of senescent cells and SASP has a negative impact on stem cells, which are essential for tissue repair and regeneration. The SASP can impair the function of nearby stem cells, reducing their ability to regenerate. This impairment can lead to a loss of tissue homeostasis and a decrease in the ability of aged tissues to heal and regenerate in response to injury or stress [59]. Furthermore, the development of age-related pathologies has been linked to cellular senescence. Senescent cells can persist in tissues for extended periods of time, and their accumulation has been observed in a variety of age-related diseases such as atherosclerosis, osteoarthritis, and pulmonary fibrosis. By secreting harmful molecules, inducing chronic inflammation, and promoting tissue remodeling and fibrosis, these senescent cells contribute to tissue dysfunction [60].

8.4.2 Antioxidant Interventions to Counteract Cellular Senescence

Antioxidants are essential for neutralizing ROS and protecting cells from oxidative stress. They can scavenge free radicals and prevent cellular senescence caused by oxidative stress. Several studies have been conducted to investigate the effects of antioxidant interventions on cellular senescence, both *in vitro* and *in vivo*. Natural antioxidants, i.e., vitamin C, vitamin E, glutathione, and coenzyme Q10, have been shown to reduce ROS levels and delay cellular senescence in various cell types [61]. Furthermore, antioxidant-rich dietary interventions have shown promise in combating cellular senescence. A diet rich in fruits, vegetables, and whole grains contains a diverse range of antioxidants, including polyphenols, carotenoids, and flavonoids, all of which have potent antioxidant properties. Epidemiological studies have found an inverse relationship between antioxidant-rich diet consumption and cellular senescence markers, implying that dietary interventions can influence the aging process [62].

Aside from dietary interventions, antioxidant supplementation has been studied as a possible strategy to combat cellular senescence. However, the effectiveness of antioxidant supplements in preventing or reversing senescence-related phenotypes is debatable. While some studies found positive effects, others found no significant benefits or even negative effects. This inconsistency may be attributed to variables such as the timing and duration of antioxidant administration, the antioxidants used, and the context of cellular senescence being targeted [63]. Furthermore, the delicate balance between oxidative stress and cellular signaling pathways must be considered. Emerging evidence suggests that ROS play an important signaling role in cellular processes such as senescence. As a result, complete oxidative stress suppression may not be desirable because it may interfere with important physiological processes. The goal of optimal antioxidant interventions should be to restore redox balance while maintaining ROS's beneficial signaling functions [64].

8.4.3 Implications for Age-Related Diseases and Longevity

Antioxidant interventions to combat cellular senescence have important implications for age-related diseases and longevity. Age-related diseases, such as cardiovascular disease, neurodegenerative disorders, and cancer, are strongly linked to cellular senescence and oxidative stress. Antioxidant interventions have the potential to reduce age-related diseases and improve overall health by reducing oxidative damage and inhibiting cellular senescence [65].

Cardiovascular disease is the leading cause of death in the elderly. The accumulation of senescent cells and oxidative stress contribute to the development of atherosclerosis, arterial stiffness, and endothelial dysfunction. Antioxidant interventions can aid in the reduction of oxidative stress, the preservation of vascular function, and the prevention of cardiovascular disease. Several studies have found that antioxidants like vitamins C vitamin E, resveratrol and selenium can improve vascular health and slow the progression of atherosclerosis, promoting cardiovascular longevity in the long run [66].

Various neurodegenerative diseases, i.e., Alzheimer's and Parkinson's, are distinguished by the progressive loss of neurons and cognitive decline. The presence of senescent cells in the brain and oxidative stress both play important roles in the pathogenesis of these disorders. Antioxidant interventions have shown promise in reducing neuroinflammation, reducing oxidative damage and preserving neuronal function. Neuroprotective substances such as alpha-lipoic acid, curcumin and green tea catechins have been studied for their ability to delay the onset or slow the progression of neurodegenerative diseases [67].

Cancer incidence rises with age, and senescent cells, through their pro-inflammatory and tissue-remodeling effects, can contribute to tumor progression. Antioxidants that target cellular senescence have the potential to modulate the tumor microenvironment and inhibit cancer cell growth and spread. The use of antioxidant interventions in cancer prevention or treatment, however, is complicated because antioxidants can both protect cancer cells from oxidative damage and interfere with cancer therapies. To avoid promoting tumor growth while reducing senescence-associated inflammation and tissue dysfunction, a careful balance must be maintained [68–70]. The extension of a healthy lifespan, or longevity, is a primary goal in aging research. Antioxidant interventions have the potential to promote longevity and improve overall health by reducing oxidative stress and cellular senescence. Preclinical research using animal models has shown that senolytic therapy, or the targeted removal of senescent cells, can delay the onset of age-related pathologies and extend lifespan. Antioxidant interventions that delay or slow cellular senescence may provide similar benefits and promote healthy aging [71].

8.5 Antioxidants and Age-Related Neurodegenerative Diseases

8.5.1 Oxidative Stress and its Link to Neurodegeneration

The progressive loss of structure or function of neurons in the central nervous system is referred to as neurodegeneration. It is a defining characteristic of numerous neurodegenerative diseases, such as Alzheimer's disease, Parkinson's disease, and amyotrophic lateral sclerosis. Emerging evidence indicates a strong association between oxidative stress and the onset and progression of neurodegenerative disorders.

Multiple sources can generate oxidative stress in the context of neurodegeneration. The accumulation of abnormal proteins, such as beta-amyloid plaques in Alzheimer's disease and alpha-synuclein aggregates in Parkinson's disease, is a major contributor [72]. These abnormal proteins can induce the production of ROS and compromise the antioxidant defense mechanisms of neurons, resulting in oxidative damage. Moreover, mitochondrial dysfunction, which is frequently observed in neurodegenerative diseases, can exacerbate oxidative stress. Mitochondria are the primary source of ROS within cells, and when their function is compromised, they generate excessive ROS, which overwhelms the antioxidant capacity of the cell [73].

The detrimental effects of oxidative stress on neurons and other brain cells can directly damage lipids, proteins and DNA, among other cellular components. ROS-initiated lipid peroxidation can result in the disruption of cell membranes and compromise their integrity. Protein oxidation can result in misfolding and aggregation, contributing to the formation of neurodegenerative diseases' toxic protein aggregates. In addition, oxidative damage to DNA can result in mutations and genomic instability, which further compromise cellular function [73]. Oxidative stress has far-reaching effects on neurodegenerative diseases. Through the activation of immune cells and the release of pro-inflammatory molecules, oxidative stress can induce chronic inflammation, a defining feature of these conditions. In turn, inflammatory processes can exacerbate oxidative stress, perpetuating neurodegeneration in a vicious cycle. Additionally, oxidative stress-related neuronal damage can impair their functionality and ultimately cause cell death. This neuronal loss contributes to the progressive decline in cognitive and motor function seen in neurodegenerative disease patients [74].

Understanding the role of oxidative stress in neurodegeneration is essential for developing potential therapeutic strategies. Therapies based on antioxidants try to restore the balance between ROS production and antioxidant defense systems. This makes oxidative stress less harmful. Several antioxidants, such as vitamins C, vitamin E, coenzyme Q10 and polyphenols, have demonstrated promising results in preclinical studies and clinical trials. In addition, targeting the upstream factors that contribute to oxidative stress, such as abnormal protein accumulation or mitochondrial dysfunction, may provide additional therapeutic intervention opportunities [75], further summarized the neurodegeneration due to protein aggregations and mitochondrial dysfunction in Figure 8.3.

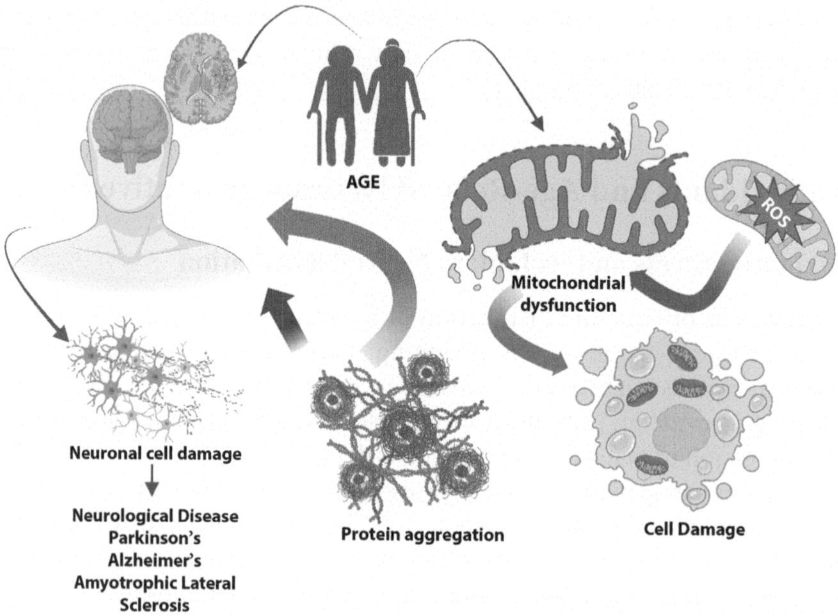

Figure 8.3 Age dependent neurodegeneration due to protein aggregation & mitochondrial dysfunction. (*Created with Biorender*).

8.5.2 Impact of Antioxidants to Reduce Oxidative Stress that Leads to Neurodegeneration

Antioxidants play a crucial role in reducing oxidative stress and protecting against neurodegeneration [76]. Various studies have shown that genetic and pharmacological manipulation of antioxidant defense mechanisms can modify neurodegeneration, indicating the causal role of oxidative stress in neurotoxicity [76]. The brain is particularly susceptible to oxidative insults due to its high oxygen consumption, abundant oxidation-sensitive lipids, and low antioxidant defense [77]. Antioxidants such as vitamin C, vitamin E, and coenzyme Q10 have been used in clinical trials to reduce oxidative stress in neurodegenerative diseases [77]. Melatonin has also been found to have neuroprotective effects by potentiation of brain antioxidants [78]. Additionally, phenolics and vitamin C in foods have been shown to reduce oxidative stress and protect neuronal cells against oxidative damage [79]. Overall, antioxidants have the potential to mitigate oxidative stress and prevent neurodegeneration [80].

8.6 Antioxidants and Skin Aging

8.6.1 The Impact of Oxidative Stress on Skin Health and Aging

Oxidative stress has a significant impact on skin health and is instrumental in the aging process. Constant exposure to environmental factors such as ultraviolet (UV) radiation, pollution, and toxins can generate ROS and induce oxidative stress in the skin.

Furthermore, endogenous factors such as metabolic processes and inflammation can contribute to oxidative stress in the skin [81]. Over the time, oxidative stress in the skin causes the accumulation of oxidative damage. ROS can directly damage vital components of the skin, including lipids, proteins, and DNA. ROS-caused lipid peroxidation can damage the integrity of the skin's lipid barrier, leading to more water loss through the epidermis and a weaker barrier function. This can lead to dryness and heightened sensitivity [81].

Proteins such as collagen and elastin are necessary for the structural integrity and elasticity of the skin. However, oxidative stress can induce protein oxidation, resulting in collagen and elastin fiber breakdown. This contributes to the loss of firmness and elasticity of the skin, resulting in the development of wrinkles and sagging skin. In addition, oxidative stress can activate matrix metalloproteinases, which further degrade collagen and elastin, thereby accelerating the aging process [82].

Oxidative stress-induced DNA damage can result in genetic mutations and cellular dysfunction in the skin. This accumulation of DNA damage impairs the skin's ability to repair itself over time, resulting in a slower rate of cell turnover and healing. Reduced skin regenerative capacity contributes to the appearance of age spots, uneven skin tone, and slowed wound healing [83, 84]. In addition, oxidative stress in the skin initiates an inflammatory response. During oxidative stress, the release of inflammatory mediators can perpetuate a chronic inflammatory state, leading to tissue damage and accelerating the aging process. Chronic skin inflammation is associated with the release of pro-inflammatory cytokines, an increase in matrix metalloproteinase levels, and the activation of immune cells. These processes contribute to the deterioration of skin components, the impairment of normal skin function, and the acceleration of aging [85].

8.6.2 Anti-Aging Cosmetic Products Containing Antioxidants

In the pursuit of youthful and radiant skin, anti-aging cosmetics containing antioxidants have gained popularity. These products are formulated to harness the power of antioxidants to combat oxidative stress and promote healthier, more resilient skin [86]. Antioxidant-rich cosmetic products typically contain a variety of substances with known antioxidant properties. These may consist of vitamins C and E, green tea extract, coenzyme Q10, resveratrol, and other botanical extracts. These antioxidants collaborate to neutralize harmful free radicals, protect against environmental damage, and strengthen the skin's natural defenses [87, 88].

Anti-aging products with antioxidants have the ability to combat oxidative stress as one of their primary benefits. Environmental factors like UV radiation, pollution, and lifestyle choices all contribute to premature aging. Antioxidants aid in preventing the oxidative stress-related harm that results in collagen degradation, DNA damage, and the development of wrinkles and fine lines by neutralizing these free radicals. Regular use of cosmetics containing antioxidants can reduce the appearance of aging and promote a more youthful complexion [89]. In addition to their protective properties, antioxidants in cosmetics support the skin's natural processes of repair and rejuvenation. Antioxidants, such as vitamin C, play a crucial role in collagen synthesis, which helps maintain the elasticity and firmness of the skin. Antioxidant-infused products can improve skin texture and reduce the appearance of wrinkles and sagging by stimulating collagen production [90].

Generally, antioxidants have been shown to have anti-inflammatory properties, which have multiple benefits for the skin. Chronic inflammation is linked to skin aging, and antioxidants can reduce inflammation and calm irritated skin. This is especially advantageous for individuals with sensitive or reactive skin. Antioxidant-containing products can help restore the skin's equilibrium and promote a healthier complexion by reducing inflammation [91]. It is important to consider formulation, stability, and concentration when selecting anti-aging cosmetic products containing antioxidants. Antioxidants can be susceptible to light, air, and heat, which can diminish their effectiveness. Therefore, the products that are properly packaged and formulated to preserve the stability of antioxidants should be used [92]. In addition, the product's antioxidant concentration must be sufficient to provide therapeutic benefits. Anti-aging cosmetic products containing antioxidants can improve skin health, but they are not a complete solution. A holistic approach to skincare, which includes sun protection, a healthy lifestyle, and a well-rounded skincare routine, is necessary for optimal results. Consistent and prolonged use of products containing antioxidants can help maintain the vitality of the skin and combat the effects of aging [92].

8.7 Lifestyle Factors and Antioxidant Defense

8.7.1 The Influence of Exercise on Antioxidant Capacity

Exercise has a substantial effect on the body's antioxidant capacity. Regular physical activity has been demonstrated to strengthen the body's antioxidant defenses and promote a more effective antioxidant response [93]. Increase in oxygen consumption and energy production during exercise, results in the production of ROS. The ROS act as signaling molecules that cause antioxidant enzymes like SOD, catalase and glutathione peroxidase to become more active. By inducing controlled oxidative stress, exercise stimulates the adaptive response of the antioxidant defense system, thereby enhancing its ability to neutralize and eliminate excess ROS. Regular exercise has also been shown to enhance mitochondrial function, which is essential for antioxidant defense [93]. Mitochondria are the body's primary source of ROS production, and mitochondrial dysfunction can increase oxidative stress. However, exercise promotes mitochondrial biogenesis and increases the activity of mitochondrial antioxidant enzymes. This improves mitochondrial function overall, reduces ROS production, and enhances the mitochondria's ability to scavenge and detoxify ROS [93].

Exercise has indirect effects on antioxidant capacity in addition to its direct effects on antioxidant defense systems. Physical activity improves blood circulation, which increases oxygen and nutrient delivery to tissues, including those involved in antioxidant defense. This enhanced blood flow facilitates the removal of metabolic waste and improves the delivery of antioxidants to various parts of the body. In contrast, physical activity has been shown to reduce chronic inflammation, which is associated with increased oxidative stress. By reducing inflammation, exercise alleviates oxidative stress and enhances the antioxidant capacity of the body [94]. The impact of physical activity on antioxidant capacity is not limited to its acute effects during exercise. It has been demonstrated that regular exercise induces long-lasting adaptations that result in sustained improvements in antioxidant defense systems. Individuals who engage in consistent exercise over time tend to have higher levels of endogenous antioxidants, such as glutathione, vitamin C and vitamin E, and higher antioxidant enzyme activity. These adaptations contribute to a stronger antioxidant defense system, which reduces the risk of oxidative damage and promotes overall health [95]. It is essential to recognize that the relationship between exercise and antioxidant capacity is complex and influenced by a number of factors, including the type, intensity, and duration of exercise [95]. Exercising at a moderate intensity has been shown to increase antioxidant capacity; however, excessive exercise or intense training without adequate recovery can tip the scales in favor of increased oxidative stress. To maximize the beneficial effects of exercise on antioxidant capacity, it is essential to find the optimal balance and allow for adequate recovery [96].

8.7.2 Effects of Stress and Sleep on Oxidative Stress Levels

Both stress and sleep have substantial effects on the body's oxidative stress levels. Psychological or physical stress can contribute to an increase in oxidative stress, whereas sufficient and quality sleep has a restorative effect on the body's antioxidants.

Chronic or prolonged stress may have a negative impact on oxidative stress levels. When the body experiences stress, it initiates the stress response, which involves the release of stress hormones such as cortisol. Elevated cortisol levels can increase the production of ROS and impair antioxidant defense mechanisms, resulting in an imbalance and elevated oxidative stress. In addition, stress can lead to behaviors that exacerbate oxidative stress, such as poor dietary choices, excessive alcohol consumption, and a lack of physical activity [97].

On the other hand, sleep is essential for maintaining the body's antioxidant balance. Various restorative processes occur during sleep, including tissue repair, hormone regulation, and metabolic waste elimination. Appropriate sleep duration and quality allow the body's antioxidant defenses to be restored and replenished. Sleep deprivation or poor sleep quality can disrupt these restorative processes and diminish the body's antioxidant capacity. According to research, sleep deprivation can increase oxidative stress markers and decrease antioxidant enzyme activity [98].

There is a bidirectional relationship between stress, sleep, and oxidative stress levels. Chronic stress can disrupt sleep patterns, leading to insomnia or sleep disturbances. In turn, these sleep disturbances can exacerbate oxidative stress levels. Sleep disruption impairs the body's ability to repair and restore itself, as well as its antioxidant defense mechanisms, resulting in oxidative damage accumulation [98]. Additionally, poor sleep can contribute to elevated stress levels, perpetuating the oxidative stress cycle. Maintaining optimal levels of oxidative stress requires managing stress and encouraging healthy sleep habits. Mindfulness, relaxation exercises, and engaging in activities that promote emotional well-being can aid in lowering stress levels and mitigating oxidative stress's negative effects. In addition, adopting good sleep hygiene practices, such as maintaining a regular sleep schedule, creating a conducive sleep environment, and practicing relaxation techniques prior to bedtime, can enhance sleep quality and strengthen the body's antioxidant defenses [99, 100].

8.7.3 Smoking, Alcohol, and Their Impact on the Body's Antioxidant System

Smoking and alcohol consumption have significant impacts on the body's antioxidant system. Chronic alcoholism has been shown to result in chronic oxidant stress in the alveolar space, leading to alterations in the systemic and pulmonary glutathione redox status [101].

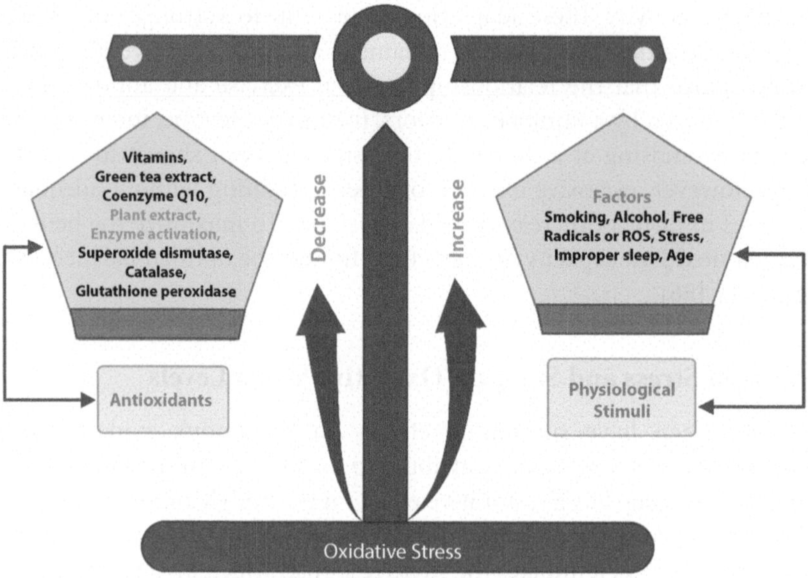

Figure 8.4 Balancing of oxidative stress. (*Created with Biorender*).

Smoking and alcohol are major factors that affect both the antioxidant system and the production of free radicals and reactive oxygen species. Studies have found that smoking decreases blood antioxidant capacity and increases lipid peroxidation, which can damage cell membrane lipids [102, 103]. Additionally, smoking has been associated with changes in salivary antioxidant levels, indicating systemic changes and effects on oral mucosa [104, 105]. Furthermore, smoking induces a potent antioxidant response and increases systemic production of glutathione [106]. Overall, both smoking and alcohol consumption have detrimental effects on the body's antioxidant system, leading to increased oxidative stress and potential damage to cells and tissues, depicted in Figure 8.4.

8.8 Recent Advancements in Antioxidants Research in Aging and Age-Related Diseases

8.8.1 The Potential of Personalized Antioxidant Interventions

Personalized antioxidant interventions hold great potential for addressing individual differences in antioxidant requirements and optimizing health outcomes. As the understanding of antioxidant systems and their interactions with cellular processes increases, researchers can develop more targeted and individualized antioxidant supplementation and intervention strategies [107]. The researchers are able to identify specific genetic variations, biomarkers, or metabolic pathways that influence an individual's antioxidant capacity or susceptibility to oxidative stress due to the development of technologies such as genomics, proteomics, and metabolomics. By incorporating this individualized data, the researchers are able to tailor antioxidant interventions to meet the unique needs of each individual, resulting in enhanced efficacy and outcomes. In addition, the concept of "nutrigenomics" permits the individualization of antioxidant interventions based on a person's unique genetic profile [107]. By identifying genetic variations associated with antioxidant metabolism or response, the researchers can determine the optimal antioxidant compounds, dosages, and routes of administration for each individual. Furthermore, wearable devices and biosensors that monitor multiple biomarkers in real-time provide new opportunities for personalized antioxidant interventions. These devices are able to monitor oxidative stress markers, such as ROS or antioxidant enzyme activity, and provide valuable information to guide individualized interventions [107].

8.8.2 Novel Antioxidant Compounds and Their Development

The development of novel antioxidant compounds remains an active field of study. To discover new antioxidant molecules with enhanced efficacy, bioavailability and specificity, scientists are examining various sources, including natural products, synthetic compounds, and even repurposing existing drugs. Various natural compounds, such as antioxidants derived from plants, continue to be an important source of novel compounds [108, 109]. Table 8.3 presents a compelling selection of plants renowned for their rich reservoir of antioxidants, offering a powerful resource for harnessing their health-boosting potential. Screening natural product libraries, botanical extracts, or traditional medicines can help identify antioxidant molecules with distinct chemical structures and bioactivities.

Furthermore, marine organisms and microorganisms are being investigated as possible sources of novel antioxidants. Synthetic compounds designed for their antioxidant properties are an additional area of study [112]. To target particular cellular processes or antioxidant pathways, medicinal chemists can develop molecules with optimized structures, stability, and bioavailability. These compounds can be tailored to treat specific diseases or conditions associated with oxidative stress. In addition, repurposing existing drugs with antioxidant properties is a cost-effective and time-efficient method for developing new antioxidant interventions [113]. Numerous FDA-approved drugs that were originally formulated for other purposes have been discovered to possess antioxidant activity. By repurposing these compounds, researchers can expedite their development for conditions associated with oxidative stress [113].

8.8.3 Outlook for Antioxidant Research in Aging and Age-Related Diseases

Understanding the underlying mechanisms of aging and age-associated diseases will continue to rely heavily on antioxidant research. Oxidative stress has been linked to a variety of age-related disorders, including cardiovascular diseases, neurodegenerative disorders, and cancer. Consequently, antioxidants hold great therapeutic potential for treating these conditions [114]. In the future, scientists will likely concentrate on elucidating the precise roles of particular antioxidants, antioxidant enzymes, and antioxidant pathways in aging and age-related diseases. This information will aid in the identification of novel intervention targets and guide the development of antioxidant-based therapies [115–117].

In addition, the interaction between antioxidants and other cellular processes, including inflammation, mitochondrial function, and cellular signaling, will be investigated in greater depth. Understanding the complex interaction between oxidative stress and these processes will enable the development of comprehensive therapeutic strategies that simultaneously

Table 8.3 Some of the plants commonly used as the potential source of antioxidants [110, 111].

Common name	Botanical name	Family	Parts to be used
Bengal quince	*Aegle marmelos*	Rutaceae	Fruit pulp
Onion	*Allium cepa*	Amaryllidaceae	Bulb
Indian aloe	*Aloe vera*	Xanthorrhoeaceae	Leaf
Satavar	*Asparagus racemosus*	Liliaceae	Shoot
Neem	*Azadirachta indica*	Meliaceae	Leaf
Brahmi	*Bacopa monniera*	Plantaginaceae	Leaf
Beet root	*Beta vulgaris*	Amaranthaceae	Root
Green tea	*Camellia sinensis*	Theaceae	Leaf
Tejpat	*Cinnamomum tamala*	Lauraceae	Leaf
Turmeric	*Curcuma longa*	Zingiberaceae	Rhizome
Akashabela	*Cuscuta reflexa*	Convolvulaceae	Stem
Carrot	*Daucus carota*	Apiaceae	Root
Amla/Emblic	*Emblica officinalis*	Euphorbiaceae	Fruit
River red gum	*Eucalyptus camaldulensis*	Myrtaceae	Leaf
Saunf	*Foeniculum vulgare*	Apiaceae	Fruit oil
Lavender	*Lavandula angustifolia*	Lamiaceae	Aerial parts
Aam	*Mangifera indica*	Anacardiaceae	Root
Curry tree	*Murraya koenigii*	Rutaceae	Leaf
Tulsi	*Ocimum sanctum*	Lamiaceae	Leaf
Common bean	*Phaseolus vulgaris*	Fabaceae	Fruit
Black pepper	*Piper nigrum*	Piperaceae	Fruit
Chinese plantain	*Plantago asiatica*	Plantaginaceae	Seed
Plums	*Prunus domestica*	Rosaceae	Fruit
Common sage	*Salvia officinalis*	Lamiaceae	Aerial parts
Sandalwood	*Santalum album*	Santalaceae	Bark
Black nightshade	*Solanum nigrum*	Solanaceae	Leaf
Potato	*Solanum tuberosum*	Solanaceae	Tuber
Behda	*Terminalia bellarica*	Combretaceae	Fruit
Ashwagandha	*Withania somnifera*	Solanaceae	Root, leaf and seed
Ginger	*Zingiber officinale*	Zingiberaceae	Rhizome

target multiple pathways [118]. Also, new technologies and methods, such as single-cell analysis, high-throughput screening, and computational modeling, will help the researchers learn more about how antioxidants work and find new ways to use them faster. These developments will enable more precise and efficient identification and evaluation of potential antioxidant compounds in preclinical and clinical settings [119].

8.9 Conclusion

Oxidative stress, caused by an imbalance between ROS and antioxidants, plays a significant role in the aging process. ROS can damage cellular components such as proteins, lipids, and DNA, leading to cellular dysfunction and increased risk of age-related diseases. Chronic inflammation, triggered by oxidative stress, further contributes to tissue damage and aging. Telomeres, which protect chromosomes, are also affected by oxidative stress, leading to cellular aging. Mitochondrial function, essential for energy production, is negatively impacted by oxidative stress, accelerating the aging process. Antioxidants, both endogenous and exogenous, play a crucial role in neutralizing ROS and preventing oxidative damage. Enzymatic antioxidants like SOD and catalase convert harmful ROS into less damaging substances. Non-enzymatic antioxidants like vitamin C, vitamin E, and glutathione scavenge and neutralize ROS, protecting cellular components. However, excessive antioxidant supplementation may disrupt the delicate balance between ROS and antioxidants, potentially interfering with beneficial cellular signaling pathways. Environmental factors, lifestyle choices, and genetic predispositions can influence the activity of antioxidants and accelerate aging. Overall, maintaining a balanced consumption of antioxidants through a varied diet is important for optimal protection against oxidative damage and healthy aging.

References

1. López-Otín, C., Blasco, M.A., Partridge, L., Serrano, M., Kroemer, G., The hallmarks of aging. *Cell*, 153, 6, 1194–217, 2013 Jun.
2. Guo, J., Huang, X., Dou, L., Yan, M., Shen, T., Tang, W. *et al.*, Aging and aging-related diseases: from molecular mechanisms to interventions and treatments. *Signal Transduct Target Ther.*, 7, 1, 391, 2022.
3. Terman, A., Kurz, T., Navratil, M., Arriaga, E.A., Brunk, U.T., Mitochondrial turnover and aging of long-lived postmitotic cells: the mitochondrial-lysosomal axis theory of aging. *Antioxid. Redox Signal*, 12, 4, 503–35, 2010 Apr.
4. Wojcik, M., Burzynska-Pedziwiatr, I., Wozniak, L.A., A review of natural and synthetic antioxidants important for health and longevity. *Curr. Med. Chem.*, 17, 28, 3262–88, 2010.
5. Frei, B., Stocker, R., Ames, B.N., Antioxidant defenses and lipid peroxidation in human blood plasma. *Proc. Natl. Acad. Sci. U.S.A.*, 85, 24, 9748–52, 1988 Dec.
6. Wayner, D.D.M., Burton, G.W., Ingold, K.U., Barclay, L.R.C., Locke, S.J., The relative contributions of vitamin E, urate, ascorbate and proteins to the total peroxyl radical-trapping antioxidant activity of human blood plasma. *Biochim. Biophys. Acta - Gen. Subj.*, 924, 3, 408–19, 1987.
7. Distefano, G. and Goodpaster, B.H., Effects of Exercise and Aging on Skeletal Muscle. *Cold Spring Harb. Perspect. Med.*, 8, 3, 1–15, 2018 Mar.

8. Babizhayev, M.A., Seguin, M.C., Gueyne, J., Evstigneeva, R.P., Ageyeva, E.A., Zheltukhina, G.A., L-carnosine (beta-alanyl-L-histidine) and carcinine (beta-alanylhistamine) act as natural antioxidants with hydroxyl-radical-scavenging and lipid-peroxidase activities. *Biochem J.*, 304, Pt 2, 509–16, 1994 Dec.
9. Elmadfa, I. and Meyer, A.L., Body composition, changing physiological functions and nutrient requirements of the elderly. *Ann. Nutr. Metab.*, 52 Suppl 1, 2–5, 2008.
10. Rizvi, S.I., Jha, R., Maurya, P.K., Erythrocyte plasma membrane redox system in human aging. *Rejuvenation Res.*, 9, 4, 470–4, 2006.
11. Oliver, C.N., Ahn, B.W., Moerman, E.J., Goldstein, S., Stadtman, E.R., Age-related changes in oxidized proteins. *J. Biol. Chem.*, 262, 12, 5488–91, 1987.
12. Stadtman, E.R., Protein oxidation and aging. *Free Radic. Res.*, 40, 12, 1250–8, 2006 Dec.
13. von Zglinicki, T., Oxidative stress shortens telomeres. *Trends Biochem. Sci*, 27, 7, 339–44, 2002 Jul.
14. Samaranayaka, A.G.P. and Li-Chan, E.C.Y., Food-derived peptidic antioxidants: A review of their production, assessment, and potential applications. *J. Funct. Foods*, 3, 4, 229–54, 2011.
15. Teo, R.D., Termini, J., Gray, H.B., Lanthanides: Applications in Cancer Diagnosis and Therapy. *J. Med. Chem.*, 59, 13, 6012–24, 2016 Jul.
16. T., J., Free radicals and antioxidants in normal physiological functions and human disease. *Int. J. Biochem. Cell Biol.*, 39, 44, 2007.
17. Brand, M.D., Mitochondrial generation of superoxide and hydrogen peroxide as the source of mitochondrial redox signaling. *Free Radic. Biol. Med.*, 100, 14–31, 2016 Nov.
18. Sies, H., Oxidative stress: a concept in redox biology and medicine. *Redox. Biol*, 4, 180–3, 2015.
19. Brawek, B., Löffler, M., Wagner, K., Huppertz, H.-J., Wendling, A.-S., Weyerbrock, A. *et al.*, Reactive oxygen species (ROS) in the human neocortex: role of aging and cognition. *Brain Res. Bull.*, 81, 4–5, 484–90, 2010 Mar.
20. Biswas, S.K., Does the Interdependence between Oxidative Stress and Inflammation Explain the Antioxidant Paradox? *Oxid. Med. Cell Longev.*, 2016, 5698931, 2016.
21. Buehler, B.A., The Free Radical Theory of Aging and Antioxidant Supplements: A Systematic Review. *J. Evidence-Based Complement Altern. Med.*, 17, 3, 218–20, 2012.
22. P-H., C., Free radicals, antioxidants in disease and health. *Int. J. Biomed. Sci. IJBS*, 4, 89, 2008.
23. Sampayo, J.N., Gill, M.S., Lithgow, G.J., Oxidative stress and aging – the use of superoxide dismutase/catalase mimetics to extend lifespan. *Biochem. Soc. Trans.*, 31, 6, 1305–7, 2003 Dec 1.
24. McCord, J.M. and Fridovich, I., Superoxide Dismutase. *J. Biol. Chem.*, 244, 22, 6049–55, 1969 Nov.
25. F., I., Superoxide radical and superoxide dismutases. *Ann. Rev. Biochem.*, 64, 97, 1995.
26. Carocho, M., Ferreira, I.C.F.R., Morales, P., Soković, M., Antioxidants and Prooxidants: Effects on Health and Aging 2018. *Oxid. Med. Cell Longev.*, 2019, 7971613, 2019.
27. Sohal, R.S. and Weindruch, R., Oxidative stress, caloric restriction, and aging. *Science*, 273, 5271, 59–63, 1996 Jul.
28. Chandrasekaran, A., Idelchik, M.D.P.S., Melendez, J.A., Redox control of senescence and age-related disease. *Redox. Biol.*, 11, 91–102, 2017 Apr.
29. Pehlivan, F.E., Vitamin C: An Antioxidant Agent, in: *Vitamin C*, A.H. Hamza (Ed.), IntechOpen, Rijeka, 2017.
30. Niki, E. and Noguchi, N., Dynamics of Antioxidant Action of Vitamin E. *Acc. Chem. Res.*, 37, 1, 45–51, 2004 Jan 1.
31. Hsu, J.L. and Sung, J.M., Antioxidant role of glutathione associated with accelerated aging and hydration of triploid watermelon seeds. *Physiol. Plant*, 100, 4, 967–74, 1997.
32. Chance, B., Sies, H., Boveris, A., Hydroperoxide metabolism in mammalian organs. *Physiol. Rev.*, 59, 3, 527–605, 1979 Jul.

33. Chang, S.K., Alasalvar, C., Shahidi, F., Review of dried fruits: Phytochemicals, antioxidant efficacies, and health benefits. *J. Funct. Foods*, 21, 113–32, 2016.
34. Warraich, U-E-A, Hussain, F., Kayani, H.U.R., Aging - Oxidative stress, antioxidants and computational modeling. *Heliyon*, 6, 5, e04107, 2020 May.
35. Hemilä, H. and Kaprio, J., Modification of the Effect of Vitamin E Supplementation on the Mortality of Male Smokers by Age and Dietary Vitamin C. *Am. J. Epidemiol.*, 169, 8, 946–53, 2009.
36. Chaves, A.C.S., Fraga, V.G., Guimarães, H.C., Teixeira, A.L., Barbosa, M.T., Carvalho M das, G. *et al.*, Estrogen receptor-alpha gene XbaI A > G polymorphism influences short-term cognitive decline in healthy oldest-old individuals. *Arq. Neuropsiquiatr.*, 75, 3, 172–5, 2017 Mar.
37. Bonaventura, C., Henkens, R., Alayash, A.I., Banerjee, S., Crumbliss, A.L., Molecular controls of the oxygenation and redox reactions of hemoglobin. *Antioxid. Redox Signal*, 18, 17, 2298–313, 2013 Jun.
38. Cheng, Z. and Ristow, M., Mitochondria and metabolic homeostasis. *Antioxid. Redox Signaling*, 19, 240–2, 2013, United States.
39. Honda, Y., Fujita, Y., Maruyama, H., Araki, Y., Ichihara, K., Sato, A. *et al.*, Lifespan-extending effects of royal jelly and its related substances on the nematode Caenorhabditis elegans. *PLoS One*, 6, 8, e23527, 2011.
40. Gaeta, A. and Hider, R.C., The crucial role of metal ions in neurodegeneration: the basis for a promising therapeutic strategy. *Br. J. Pharmacol.*, 146, 8, 1041–59, 2005 Dec.
41. Singh, A., Kukreti, R., Saso, L., Kukreti, S., Oxidative Stress: A Key Modulator in Neurodegenerative Diseases. *Molecules*, 24, 8, 1–20, 2019 Apr.
42. Le Bourg, E., Hormesis, aging and longevity. *Biochim. Biophys. Acta*, 1790, 10, 1030–9, 2009 Oct.
43. Carney, J.M., Starke-Reed, P.E., Oliver, C.N., Landum, R.W., Cheng, M.S., Wu, J.F. *et al.*, Reversal of age-related increase in brain protein oxidation, decrease in enzyme activity, and loss in temporal and spatial memory by chronic administration of the spin-trapping compound N-tert-butyl-alpha-phenylnitrone. *Proc. Natl. Acad. Sci. U.S.A.*, 88, 9, 3633–6, 1991 May.
44. Shetty, R.A., Forster, M.J., Sumien, N., Coenzyme Q(10) supplementation reverses age-related impairments in spatial learning and lowers protein oxidation. *Age (Dordr)*, 35, 5, 1821–34, 2013 Oct.
45. Partridge, L. and Gems, D., Mechanisms of ageing: public or private? *Nat. Rev. Genet.*, 3, 3, 165–75, 2002 Mar.
46. Inoue, S., Koya-Miyata, S., Ushio, S., Iwaki, K., Ikeda, M., Kurimoto, M., Royal Jelly prolongs the life span of C3H/HeJ mice: correlation with reduced DNA damage. *Exp. Gerontol.*, 38, 9, 965–9, 2003.
47. Biliński, T., Zadrąg-Tęcza, R., Bartosz, G., Hypertrophy hypothesis as an alternative explanation of the phenomenon of replicative aging of yeast. *FEMS Yeast Res.*, 12, 1, 97–101, 2012.
48. Zadrag, R., Bartosz, G., Bilinski, T., Is the yeast a relevant model for aging of multicellular organisms? An insight from the total lifespan of Saccharomyces cerevisiae. *Curr. Aging Sci.*, 1, 3, 159–65, 2008 Dec.
49. Michels, A.J. and Frei, B., Myths, artifacts, and fatal flaws: identifying limitations and opportunities in vitamin C research. *Nutrients*, 5, 12, 5161–92, 2013 Dec.
50. Zhang, L.-X., Li, C.-X., Kakar, M.U., Khan, M.S., Wu, P.-F., Amir, R.M. *et al.*, Resveratrol (RV): A pharmacological review and call for further research. *Biomed. Pharmacother.*, 143, 112164, 2021.
51. Pop, A., Fizeşan, I., Vlase, L., Rusu, M.E., Cherfan, J., Babota, M. *et al.*, Enhanced Recovery of Phenolic and Tocopherolic Compounds from Walnut (Juglans Regia L.) Male Flowers Based on

Process Optimization of Ultrasonic Assisted-Extraction: Phytochemical Profile and Biological Activities. *Antioxidants*, 10, 4, 1–25, 2021.

52. LB, E., Oxidative stress, aging and longevity in Drosophila melanogaster. *FEBS Lett.*, 498, 2–3, 183–6, 2001 Jun.
53. Xue, Y.-L., Ahiko, T., Miyakawa, T., Amino, H., Hu, F., Furihata, K. *et al.*, Isolation and Caenorhabditis elegans lifespan assay of flavonoids from onion. *J. Agric. Food Chem.*, 59, 11, 5927–34, 2011 Jun.
54. Vedeanu, N., Voica, C., Magdas, A., Kiss, B., Ştefan, M.-G., Simedrea, R. *et al.*, Subacute co-exposure to low doses of ruthenium(III) changes the distribution, excretion and biological effects of silver ions in rats. *Environ. Chem.*, 17, 163–72, 2019.
55. Rusu, M.E., Georgiu, C., Pop, A., Mocan, A., Kiss, B., Vostinaru, O. *et al.*, Antioxidant Effects of Walnut (Juglans regia L.) Kernel and Walnut Septum Extract in a D-Galactose-Induced Aging Model and in Naturally Aged Rats. *Antioxidants (Basel, Switzerland)*, 9, 5, 1–25, 2020 May.
56. Fizeşan, I., Rusu, M.E., Georgiu, C., Pop, A., Ştefan, M.-G., Muntean, D.-M. *et al.*, Antitussive, Antioxidant, and Anti-Inflammatory Effects of a Walnut (Juglans regia L.) Septum Extract Rich in Bioactive Compounds. *Antioxidants (Basel, Switzerland)*, 10, 1, 1–17, 2021 Jan.
57. Ghzaiel, I., Zarrouk, A., Nury, T., Libergoli, M., Florio, F., Hammouda, S. *et al.*, Antioxidant Properties and Cytoprotective Effect of Pistacia lentiscus L. Seed Oil against 7β-Hydroxycholesterol-Induced Toxicity in C2C12 Myoblasts: Reduction in Oxidative Stress, Mitochondrial and Peroxisomal Dysfunctions and Attenuation of Cell Death. *Antioxidants (Basel, Switzerland)*, 10, 11, 1–37, 2021 Nov.
58. Rusu, M.E., Mocan, A., Ferreira, I.C.F.R., Popa, D.-S., Health Benefits of Nut Consumption in Middle-Aged and Elderly Population. *Antioxidants (Basel, Switzerland)*, 8, 8, 1–37, 2019 Aug.
59. Ha, J.W. and Boo, Y.C., Siegesbeckiae Herba Extract and Chlorogenic Acid Ameliorate the Death of HaCaT Keratinocytes Exposed to Airborne Particulate Matter by Mitigating Oxidative Stress. *Antioxidants*, 10, 11, 1–20, 2021.
60. Childs, B.G., Durik, M., Baker, D.J., van Deursen, J.M., Cellular senescence in aging and age-related disease: from mechanisms to therapy. *Nat. Med.*, 21, 12, 1424–35, 2015 Dec.
61. Jo, S., Jung, Y.-S., Cho, Y.-R., Seo, J.-W., Lim, W.-C., Nam, T.-G. *et al.*, Oral Administration of Rosa gallica Prevents UVB-Induced Skin Aging through Targeting the c-Raf Signaling Axis. *Antioxidants (Basel, Switzerland)*, 10, 11, 1–14, 2021 Oct.
62. Correia, P., Araújo, P., Ribeiro, C., Oliveira, H., Pereira, A.R., Mateus, N. *et al.*, Anthocyanin-Related Pigments: Natural Allies for Skin Health Maintenance and Protection. *Antioxidants (Basel, Switzerland)*, 10, 7, 1–24, 2021 Jun.
63. Sawada, M. and Enesco, H.E., Vitamin E extends lifespan in the short-lived rotifer Asplanchna brightwelli. *Exp. Gerontol.*, 19, 3, 179–83, 1984.
64. Ishii, N., Senoo-Matsuda, N., Miyake, K., Yasuda, K., Ishii, T., Hartman, P.S. *et al.*, Coenzyme Q10 can prolong C. elegans lifespan by lowering oxidative stress. *Mech. Ageing Dev.*, 125, 1, 41–6, 2004 Jan.
65. Harrington, L.A. and Harley, C.B., Effect of vitamin E on lifespan and reproduction in Caenorhabditis elegans. *Mech. Ageing Dev.*, 43, 1, 71–8, 1988 Apr.
66. Zou, S., Sinclair, J., Wilson, M.A., Carey, J.R., Liedo, P., Oropeza, A. *et al.*, Comparative approaches to facilitate the discovery of prolongevity interventions: effects of tocopherols on lifespan of three invertebrate species. *Mech. Ageing Dev.*, 128, 2, 222–6, 2007 Feb.
67. Xie, A., Gao, J., Xu, L., Meng, D., Shared mechanisms of neurodegeneration in Alzheimer's disease and Parkinson's disease. *Biomed. Res. Int.*, 2014, 648740, 2014.
68. Park, S., Kim, B.-K., Park, S.-K., Supplementation with phosphatidylethanolamine confers anti-oxidant and anti-aging effects via hormesis and reduced insulin/IGF-1-like signaling in C. elegans. *Mech. Ageing Dev.*, 197, 111498, 2021.

69. Surco-Laos, F., Cabello, J., Gómez-Orte, E., González-Manzano, S., González-Paramás, A.M., Santos-Buelga, C. *et al.*, Effects of O-methylated metabolites of quercetin on oxidative stress, thermotolerance, lifespan and bioavailability on Caenorhabditis elegans. *Food Funct.*, 2, 8, 445–56, 2011 Aug.
70. Dueñas, M., Surco-Laos, F., González-Manzano, S., González-Paramás, A.M., Gómez-Orte, E., Cabello, J. *et al.*, Deglycosylation is a key step in biotransformation and lifespan effects of quercetin-3-O-glucoside in Caenorhabditis elegans. *Pharmacol. Res.*, 76, 41–8, 2013 Oct.
71. Al-Naggar, I.M.A., Kuchel, G.A., Xu, M., Senolytics: targeting senescent cells for age-associated diseases. *Curr. Mol. Biol. Rep.*, 6, 4, 161–72, 2020 Dec.
72. Pietsch, K., Saul, N., Chakrabarti, S., Stürzenbaum, S.R., Menzel, R., Steinberg, C.E.W., Hormetins, antioxidants and prooxidants: defining quercetin-, caffeic acid- and rosmarinic acid-mediated life extension in C. elegans. *Biogerontology*, 12, 4, 329–47, 2011 Aug.
73. Abbas, S. and Wink, M., Epigallocatechin gallate from green tea (Camellia sinensis) increases lifespan and stress resistance in Caenorhabditis elegans. *Planta Med.*, 75, 3, 216–21, 2009 Feb.
74. Zhang, L., Jie, G., Zhang, J., Zhao, B., Significant longevity-extending effects of EGCG on Caenorhabditis elegans under stress. *Free Radic. Biol. Med.*, 46, 3, 414–21, 2009.
75. Sayed, A.A.R., Ferulsinaic acid attenuation of advanced glycation end products extends the lifespan of Caenorhabditis elegans. *J. Pharm. Pharmacol.*, 63, 3, 423–8, 2011 Mar.
76. Dias-Santagata, D., Fulga, T.A., Duttaroy, A., Feany, M.B., Oxidative stress mediates tau-induced neurodegeneration in Drosophila. *J. Clin. Invest.*, 117, 1, 236–45, 2007.
77. Gao, H.-M., Zhou, H., Hong, J.-S., NADPH oxidases: novel therapeutic targets for neurodegenerative diseases. *Trends Pharmacol. Sci.*, 33, 6, 295–303, 2012 Jun 1.
78. Ikram, M., Park, H.Y., Ali, T., Kim, M.O., Melatonin as a Potential Regulator of Oxidative Stress, and Neuroinflammation: Mechanisms and Implications for the Management of Brain Injury-Induced Neurodegeneration. *J. Inflamm. Res.*, 14, 6251–64, 2021 Nov.
79. Jeong, H.-R., Kim, K.J., Lee, S.G., Cho, H.S., Cho, Y.-S., Kim, D.-O., Phenolic Profiles of Hardy Kiwifruits and Their Neuroprotective Effects on PC-12 and SH-SY5Y Cells against Oxidative Stress. *J. Microbiol. Biotechnol.*, 30, 6, 912–9, 2020 Jun.
80. Panahi, Y., Rajaee, S.M., Johnston, T.P., Sahebkar, A., Neuroprotective effects of antioxidants in the management of neurodegenerative disorders: A literature review. *J. Cell. Biochem.*, 120, 3, 2742–8, 2019.
81. Vardi, M., Levy, N.S., Levy, A.P., Vitamin E in the prevention of cardiovascular disease: the importance of proper patient selection. *J. Lipid Res.*, 54, 9, 2307–14, 2013 Sep.
82. Reiter, R.J., Mayo, J.C., Tan, D.-X., Sainz, R.M., Alatorre-Jimenez, M., Qin, L., Melatonin as an antioxidant: under promises but over delivers. *J. Pineal Res.*, 61, 3, 253–78, 2016 Oct.
83. Papaccio, F., Arino, A.D., Caputo, S., Bellei, B., Focus on the Contribution of Oxidative Stress in Skin Aging. *Antioxidants (Basel, Switzerland)*, 11, 6, 1–33, 2022 Jun.
84. Gowda, B.H.J., Mohanto, S., Singh, A., Bhunia, A., Abdelgawad, M.A., Ghosh, S. *et al.*, Nanoparticle-based therapeutic approaches for wound healing: a review of the state-of-the-art. *Mater. Today Chem.*, 27, 101319, 2023.
85. Pinzani, P., Petruzzi, E., Orlando, C., Stefanescu, A., Antonini, M.F., Serio, M. *et al.*, Reduced Serum Antioxidant Capacity In Healthy Centenarians. *Clin. Chem.*, 43, 5, 855a – 856, 1997.
86. Silva, S., Ferreira, M., Oliveira, A.S., Magalhães, C., Sousa, M.E., Pinto, M. *et al.*, Evolution of the use of antioxidants in anti-ageing cosmetics. *Int. J. Cosmet. Sci.*, 41, 4, 378–86, 2019.
87. Vincenzi, S., Tomasi, D., Gaiotti, F., Lovat, L., Giacosa, S., Torchio, F. *et al.*, Comparative Study of the Resveratrol Content of Twenty-one Italian Red Grape Varieties. *South African J. Enol. Vitic.*, 34, 30–5, 2013.

88. Lee, B.-J., Tseng, Y.-F., Yen, C.-H., Lin, P.-T., Effects of coenzyme Q10 supplementation (300 mg/day) on antioxidation and anti-inflammation in coronary artery disease patients during statins therapy: a randomized, placebo-controlled trial. *Nutr. J.*, 12, 1, 142, 2013 Nov.
89. Al-Horani, R.A., A Narrative Review of Exercise-Induced Oxidative Stress: Oxidative DNA Damage Underlined. *Open Sports Sci. J.*, 15, 1, 1–46, 2022 May 27.
90. Quindry, J.C., Stone, W.L., King, J., Broeder, C.E., The effects of acute exercise on neutrophils and plasma oxidative stress. *Med. Sci. Sports Exerc.*, 35, 7, 1139–45, 2003 Jul.
91. Michalak, M., Plant-Derived Antioxidants: Significance in Skin Health and the Ageing Process. *Int. J. Mol. Sci.*, 23, 2, 1–29, 2022 Jan.
92. Shanbhag, S., Nayak, A., Narayan, R., Nayak, U.Y., Anti-aging and Sunscreens: Paradigm Shift in Cosmetics. *Adv. Pharm. Bull.*, 9, 3, 348–59, 2019 Aug.
93. Kwon, K.-R., Alam, M.B., Park, J.-H., Kim, T.-H., Lee, S.-H., Attenuation of UVB-Induced Photo-Aging by Polyphenolic-Rich Spatholobus Suberectus Stem Extract Via Modulation of MAPK/AP-1/MMPs Signaling in Human Keratinocytes. *Nutrients*, 11, 6, 1–14, 2019 Jun.
94. Xiao, J., Liu, B., Zhuang, Y., Effects of rambutan (Nephelium lappaceum) peel phenolics and Leu-Ser-Gly-Tyr-Gly-Pro on hairless mice skin photoaging induced by ultraviolet irradiation. *Food Chem. Toxicol. Int. J. Publ. Br. Ind. Biol. Res. Assoc.*, 129, 30–7, 2019 Jul.
95. Higgins, M.R., Izadi, A., Kaviani, M., Antioxidants and Exercise Performance: With a Focus on Vitamin E and C Supplementation. *Int. J. Environ. Res. Public Health*, 17, 22, 1–26, 2020 Nov.
96. Kim, H.-I., Jeong, Y.-U., Kim, J.-H., Park, Y.-J., 3,5,6,7,8,3',4'-Heptamethoxyflavone, a Citrus Flavonoid, Inhibits Collagenase Activity and Induces Type I Procollagen Synthesis in HDFn Cells. *Int. J. Mol. Sci.*, 19, 2, 1–12, 2018 Feb.
97. Noh, D., Choi, J.G., Huh, E., Oh, M.S., Tectorigenin, a Flavonoid-Based Compound of Leopard Lily Rhizome, Attenuates UV-B-Induced Apoptosis and Collagen Degradation by Inhibiting Oxidative Stress in Human Keratinocytes. *Nutrients*, 10, 12, 1–10, 2018 Dec.
98. Vyazovskiy, V.V., Sleep, recovery, and metaregulation: explaining the benefits of sleep. *Nat. Sci. Sleep*, 7, 171–84, 2015.
99. Leonidis, A., Korozi, M., Sykianaki, E., Tsolakou, E., Kouroumalis, V., Ioannidi, D. *et al.*, Improving Stress Management and Sleep Hygiene in Intelligent Homes. *Sensors (Basel)*, 21, 7, 1–37, 2021 Mar.
100. Kanagasabai, T. and Ardern, C.I., Inflammation, Oxidative Stress, and Antioxidants Contribute to Selected Sleep Quality and Cardiometabolic Health Relationships: A Cross-Sectional Study. *Mediators Inflamm.*, 2015, 824589, 2015.
101. Yeh, M.Y., Burnham, E.L., Moss, M., Brown, L.A.S., Chronic alcoholism alters systemic and pulmonary glutathione redox status. *Am. J. Respir. Crit. Care Med.*, 176, 3, 270–6, 2007 Aug.
102. Bloomer, R.J., Decreased blood antioxidant capacity and increased lipid peroxidation in young cigarette smokers compared to nonsmokers: Impact of dietary intake. *Nutr. J.*, 6, 1, 39, 2007.
103. Garg, N., Singh, R., Dixit, J., Jain, A., Tewari, V., Levels of lipid peroxides and antioxidants in smokers and nonsmokers. *J. Periodontal Res.*, 41, 5, 405–10, 2006.
104. Ahmadi-Motamayel, F., Falsafi, P., Goodarzi, M.T., Poorolajal, J., Evaluation of salivary catalase, vitamin C, and alpha-amylase in smokers and non-smokers: a retrospective cohort study. *J. Oral Pathol. Med. Off. Publ. Int. Assoc. Oral Pathol. Am. Acad. Oral Pathol.*, 46, 5, 377–80, 2017 May.
105. Singh, S., Sharma, M., Rohilla, N., Salgotra, V., Kumar, V., Sharma, R.K., Assessment of Salivary Catalase, a-Amylase, and Cotinine Levels in Chronic Smokers: A Comparative Study. *J. Contemp. Dent. Pract.*, 19, 3, 253–256, 2018 Mar.
106. Ballesteros, A., Jiang, P., Summerfelt, A., Du, X., Chiappelli, J., O'Donnell, P. *et al.*, No evidence of exogenous origin for the abnormal glutathione redox state in schizophrenia. *Schizophr. Res.*, 146, 1–3, 184–9, 2013 May.

107. Polte, T. and Tyrrell, R.M., Involvement of lipid peroxidation and organic peroxides in UVA-induced matrix metalloproteinase-1 expression. *Free Radic. Biol. Med.*, 36, 12, 1566–74, 2004 Jun.
108. Parcheta, M., Świsłocka, R., Orzechowska, S., Akimowicz, M., Choińska, R., Lewandowski, W., Recent Developments in Effective Antioxidants: The Structure and Antioxidant Properties. *Mater (Basel, Switzerland)*, 14, 8, 1–24, 2021 Apr.
109. Marino, A., Battaglini, M., Moles, N., Ciofani, G., Natural Antioxidant Compounds as Potential Pharmaceutical Tools against Neurodegenerative Diseases. *ACS Omega*, 7, 30, 25974–90, 2022 Aug 2.
110. Reid, M.E., Duffield-Lillico, A.J., Garland, L., Turnbull, B.W., Clark, L.C., Marshall, J.R., Selenium supplementation and lung cancer incidence: an update of the nutritional prevention of cancer trial. *Cancer Epidemiol. Biomarkers Prev. Publ. Am. Assoc. Cancer Res. Cosponsored Am. Soc. Prev. Oncol.*, 11, 11, 1285–91, 2002 Nov.
111. Ramakrishnan, U., Prevalence of micronutrient malnutrition worldwide. *Nutr. Rev.*, 60, 5 Pt 2, S46–52, 2002 May.
112. Rutault, K., Alderman, C., Chain, B.M., Katz, D.R., Reactive oxygen species activate human peripheral blood dendritic cells. *Free Radic. Biol. Med.*, 26, 1–2, 232–8, 1999 Jan.
113. Song, H.Y., Ju, S.M., Goh, A.R., Kwon, D.-J., Choi, S.Y., Park, J., Suppression of TNF-alpha-induced MMP-9 expression by a cell-permeable superoxide dismutase in keratinocytes. *BMB Rep.*, 44, 7, 462–7, 2011 Jul.
114. Corsini, E., Galbiati, V., Nikitovic, D., Tsatsakis, A.M., Role of oxidative stress in chemical allergens induced skin cells activation. *Food Chem. Toxicol. Int. J. Publ. Br. Ind. Biol. Res. Assoc.*, 61, 74–81, 2013 Nov.
115. Kurutas, E.B., The importance of antioxidants which play the role in cellular response against oxidative/nitrosative stress: current state. *Nutr. J.*, 15, 1, 71, 2016 Jul.
116. Kushwah, N., Bora, K., Maurya, M., Pavlovich, M.C., Chen, J., Oxidative Stress and Antioxidants in Age-Related Macular Degeneration. *Antioxidants*, 12, 7, 1–28, 2023.
117. Tan, B.L., Norhaizan, M.E., Liew, W-P-P, Sulaiman Rahman, H., Antioxidant and Oxidative Stress: A Mutual Interplay in Age-Related Diseases. *Front Pharmacol.*, 9, 1162, 2018.
118. Naziroglu, M. and Kokcam, I., Antioxidants and lipid peroxidation status in the blood of patients with alopecia. *Cell Biochem. Funct.*, 18, 3, 169–73, 2000 Sep.
119. Yamamoto, Y. and Yasuoka, A., Welsh Onion Attenuates Hyperlipidemia in Rats Fed on High-Fat High-Sucrose Diet. *Biosci. Biotechnol. Biochem.*, 74, 2, 402–4, 2010.

9

Antioxidants Against Neurological Disorders

Jasjeet Kaur Narang[1]*, Anmol Dogra[2], Tajpreet Kaur[3], Ramandeep Singh Narang[4] and Amrit Pal Singh[5]

[1]*Department of Pharmaceutics, Khalsa College of Pharmacy, Amritsar, Punjab, India*
[2]*Department of Pharmaceutics, NIET Institute of Pharmacy, Greater Noida, Knowledge Park-II, Uttar Pradesh, India*
[3]*Department of Pharmacology, Khalsa College of Pharmacy, Amritsar, India*
[4]*Department of Oral and Maxillofacial Pathology, Sri Guru Ram Das Institute of Dental Sciences and Research, Amritsar, Punjab, India*
[5]*Department of Pharmaceutical Sciences, Guru Nanak Dev University, Amritsar, India*

Abstract
The advent and progression of neurological disorders has found to be related to oxidative stress. Reactive oxygen and nitrogen species play an instrumental role in triggering cascade of inflammatory and degenerative changes due to damage of proteins, cell membrane, mitochondria as well as DNA. Amyotrophic lateral sclerosis, Alzheimer's disease, stroke, Huntington's disease, trigeminal neuralgia and Parkinson's disease are few examples of diseases related to neuronal damage. The researchers globally have utilized both synthetic as well as natural antioxidants for controlling the progression of neurodegenerative changes in various pathologies of the brain as they have shown promising results in regaining the redox balance thereby alleviating oxidative stress. Exogeneous antioxidants used for the cure as well as neurodegenerative diseases management include hormones, vitamins, carotenes and polyphenols. In spite of showing encouraging results in preclinical investigations for neurodegenerative diseases, their clinical application is restricted to only adjunct therapy. The reasons attributed to poor clinical use include their poor bioavailability, administration of high and repeated doses for achieving a therapeutic effect as well as instability in some cases. In order to overcome these challenges the antioxidants are delivered by various nanotechnology based therapeutic systems. This chapter will provide explain the antioxidants role in management of neurodegenerative diseases besides providing information about the various dosage forms which are explored for the effective administration of these antioxidants for treating neurological diseases.

Keywords: Antioxidants, nanoformulations, neurological disorders, clinical trials

9.1 Introduction

The words "oxidative stress" were initially used by Helmut Sies [1] to describe a disbalance that might harm biological systems between the generation of oxidants and antioxidant

**Corresponding author*: jasjeet2975@yahoo.com

Rakesh K. Sindhu, Inderbir Singh and M. Arockia Babu (eds.) Antioxidants: Nature's Defense Against Disease, (285–368)

defenses. Ever since, the area of redox biology has advanced from concepts of stress related to oxidation in illness to redox signaling in physiology [2–4].

Neuropathologies, such as neurodegenerative illnesses and neurotrauma, are fundamentally influenced by oxidative stress. Secondary degeneration in neurotrauma instances is a chronic damage to the brain's neurons caused by a buildup of reactive oxygen compounds along with an imbalance in calcium ions [5]. Myelin, which plays an instrumental role in the signals generated by nerves, is also affected due to this excessive formation of oxygen species that are reactive (ROS). Neuronal damage resulting from myelin breakdown can be made worse by oligodendrocyte death (from excessive exposure to ROS). Compared to glial cells, oligodendrocytes have lower amounts of antioxidants (such glutathione and Manganese superoxide dismutase (MnSOD) and greater intracellular iron levels, which makes them more susceptible to ROS [5, 6]. The pathophysiology of neurodegenerative disorders has been linked to ROS.

Nonetheless, there is mounting evidence that ROS contribute to the acceleration of disease development in neurodegenerative disorders, even though the exact cause of many illnesses is still unknown. ROS generation is thought to be induced in frequently seen diseases of misfolded protein aggregation, e.g., Parkinson's disorder (PD). This inflammatory reaction is related to the development of disease [7, 8]. Additionally, the generation of ROS resulting from mitochondrial failure in neurodegenerative disorders intensifies inflammation and disrupts downstream redox signaling pathways, which have been linked to persistent apoptosis [9]. Given the inconsistent evidence about the impact of extrinsic therapeutic antioxidants on ROS generation in neurodegenerative illnesses, several research avenues are now being pursued to identify the specific mechanism by which ROS participate to the pathogenesis of neurodegenerative disorders [9, 10].

Free radicals contain a single electron in exterior orbit. Radicals which are free have positive impacts on living things and play a significant role in the genesis of life and biological evolution [11]. Oxygen radicals have an effect on numerous cellular metabolic activities, such as DNA transcription, signal transmission, and the regulation of aqueous guanylate cyclase activity. Nitric oxide (NO), a crucial signaling molecule, functions as a primary regulator of the unwinding and development of cells of the smooth muscle in the circulatory system [12].

Under normal cellular metabolism, unstable oxygen species related stress is inevitably produced by the respiration of mitochondria and a group of membrane-attached nicotinamide adenine dinucleotide phosphate hydrogen oxidases. Free radicals, or molecules with single electrons, and non-radical varieties make up the majority of ROS. Hydrogen peroxide, hydroxyl radical, and superoxide anion are the three primary types of reactive oxygen compounds in biology [13]. Because reactive oxygen compounds oxidize certain collocated proteins, they are required at low to moderate concentrations for physiologically and functional cellular functioning, which sets off a range of biochemical reactions related to cell survival, proliferation, differentiation, and metabolic flexibility [14]. However, large amounts of reactive oxygen compounds are dangerous because they oxidatively destroy vital cell components such proteins, lipids, and deoxyribonucleic acid, ultimately resulting in death and cell cycle halt [15–19].

When cells operate normally, endogenous antioxidants counteract reactive oxygen compounds given by mitochondrial activity. It should be noted that inbuilt antioxidant levels differ significantly throughout cell types. The measurement of the quantity of endogenous antioxidant is a challenging task even with tests like 2,2-diphenyl-1-picrylhydrazyl [20–22]. Nevertheless, techniques for measuring the quantity of ROS produced *in vivo* and *in vitro* exist [23–26].

Research on external antioxidants has been pushed to increase availability and lower the production of oxidative stress because of the variations in endogenous antioxidant levels in various cell types. Antioxidants can be categorized as endogenous (man-made or natural) or exogenous (natural). They are both in charge of eliminating free radicals, scavenging reactive oxygen compounds or their derivatives, preventing the creation of reactive oxygen compounds, and attaching the metal ions needed for ROS production catalysis [27]. Exogenous antioxidants can come from synthetic or natural sources. Superoxide dismutase [28], glutathione peroxidase [29], and catalase [30] are examples of enzymatic antioxidants. Direct-acting antioxidants, which are crucial for protecting against reactive oxygen compounds, are a type of non-enzymatic antioxidant. Vitamin C [30], alpha-lipoic acid [31], phenolic acids [32], α-tocopherol [33], and β-carotene [34] are among them. Turmeric [35], Poliglusam [36], buckytubes [37], polyurethane [38], ceria nanoaprticles [39], 5-Methoxy-N-acetyltryptamine [40], fullerene 60 [41], Querco-tannic acid [42], natural melanin [43], and 2,4-Disulfophenyl-N-tert-butylnitrone [44–47] are examples of synthetic antioxidants. A small percentage of these compounds are produced by the cell itself. Antioxidants with indirect actions mostly consist of chelating compounds that attach to redox metals to prevent the formation of harmful free radicals [48]. The role of various antioxidants against neurological disorders is shown in Figure 9.1 [49].

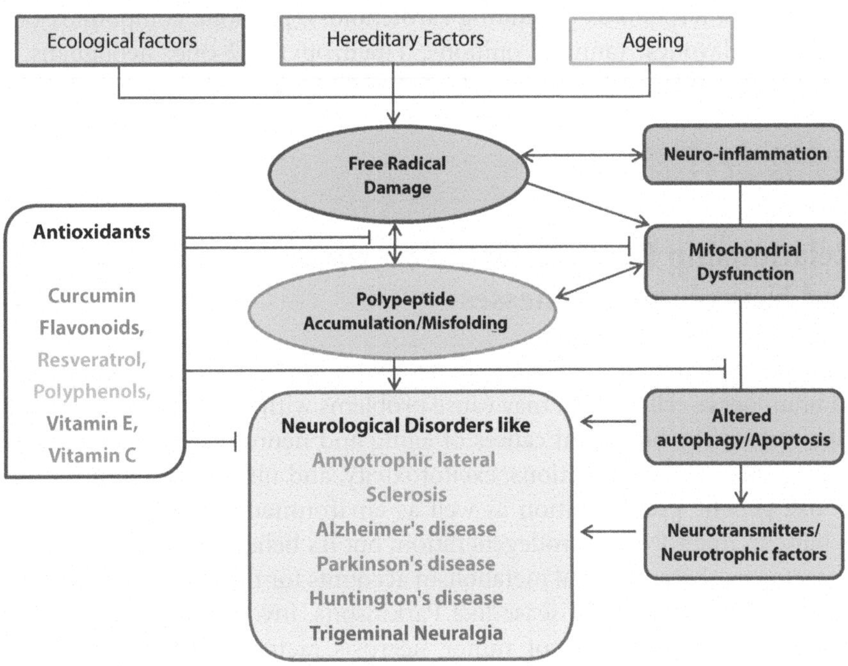

Figure 9.1 Role of various antioxidants used to treat neurological disorders [Modified version of Mittal *et al.* [49]].

9.2 Sources of Antioxidants

Most oxidants generated by cells seem to come from four endogenous sources [50]. During regular aerobic respiration, oxygen is consumed by the mitochondria and then reduced to generate -OH, O_2, as well as H_2O_2 byproducts [51]. Phagocytosis, which uses an oxidative surge of nitric oxide, oxygen, hydrogen peroxide as well as oxychloride is responsible for killing of cells infected by virus as well as bacteria [52]. Peroxisomes break down lipids such as fatty acids and other molecules of lipid that catalase then break down further, producing hydrogen peroxide as an aftereffect. Research indicates that some peroxide may be able to escape from degradation under some circumstances. This allows the peroxide to go inside other cell compartments and increase oxidative stress, which in turn damages DNA. Animal cytochrome P450 enzymes are a key component of the defensive mechanisms that shield the body from naturally occurring hazardous compounds derived from plants, which are the main source of dietary poisons. Although these enzymes can produce some oxidative intermediates that damage DNA, they are nevertheless protective against the acute harmful effects of foreign substances [53].

The human body receives a variety of antioxidants from its food, whether it is vegetarian or not. The most well-known antioxidants in food are coenzyme Q, ascorbic acid and α-vitamin E, as well as carotenoids. The primary sources of alpha-vitamin E are plant-based oils and flour germ. This fat-soluble nutrient is absorbed in the stomach and enters the plasma through lipoproteins. Beta-tocopherol is the primary prevalent as well as effective isomer of vitamin E among its eight organic state forms. The antioxidant vitamin E effectively prevents the lipid in the plasma membrane from peroxiding because it is lipid soluble [54, 55].

Many different types of molecules that scavenge free radicals can be found in plants, including vitamins, terpenoids (including carotenoids), phenolic compounds (acids containing phenols, flavones, tannins, quinons, 1-benzopyran-2-one, neolignans, stilbenes etc.), compounds containing nitrogen derivatives (Amine substances, alkaloidal compounds, betalamic acid, etc.), plus a few more naturally occurring metabolites with high antioxidant activity [56–59].

9.3 Relationship Between Oxidative Stress and Neurological Illnesses

One ailment that falls under the category of neurological illnesses is the degeneration of the spine and brain nerve cells, which may cause problems with sensation or can result in cognitive impairment. The biological causes of aging and neurological conditions have been linked to mitochondrial dysfunctions, excitotoxicity, and ultimately apoptosis. Numerous variables, like genetic predisposition as well as environmental exposure, have been suggested to have an impact on neurodegeneration; but it's believed that reactive metal abuse is a major factor, as aberrant metal metabolism accounts for the majority of symptoms [60].

The genesis of neurological disease like Parkinson's, involves the glial cells activation, besides having the involvement of tumor necrosis factor (α-TNF), cyclooxygenase-2, enzymes like astrocytic myeloperoxidase, inducible nitric oxide synthase, as well as reduced nicotinamide adenine dinucleotide phosphate oxidase [61]. Increased oxidative stress due to free radicals, which results in increased mitochondrial DNA mutation, disturbance of protein homeostasis, and nucleic acid instability, is principally responses to how Parkinson's disease develops [62]. Oxidative stress induced by free radicals in the substantia nigra is responsible for harming neurons involved in dopamine production.

As per reports, blood levels of 3-mononitrotyrosine, 8-Oxo-2'-deoxyguanosine, 8-oxoguanosine, 4-Hydroxy-2-nonenal, and Malonic aldehyde, increase in patients of Alzheimer's disease (AD). Variations in the production of antioxidant enzymes like superoxide dismutase as well as catalase have been connected to related AD [63]. Higher production of beta-amyloid in AD leads to mutations mostly in the precursor proteins of amyloid (APP), presenilin genes like (PS-1), and (PS-2). In humans, the amount of β-amyloid rises with age [64]. The scientists investigated this age-related beta-amyloid accumulation using mice with transgenic genes expressing the amyloid precursor protein. In mice that were 3, 6, 9, 12, and 14 months old, brain slices were examined. The 3-month-old mice's brains showed no beta-amyloid deposition, however, there was more accumulation in the neonatal cortex as well as hippocampus of the 12-month-old mice's brains. At 14 months, beta-amyloid was noticed in the brain and leptomeninges, tiny blood vessel walls, indicating that deposition of beta-amyloid in animals is age-related. Furthermore, compared to wild-type mice, it was discovered that the super oxygen dismutase enzyme's activity had significantly decreased, which led to oxidative stress, apoptosis, and mitochondrial malfunction. Therefore, it is possible to conclude that aging results in a rise in β-amyloid, which in turn causes the enzyme superoxide dismutase to become inactive. This raises ROS levels, which are associated to oxidative stress-related loss of neurons [64].

Muscle atrophy, weakness, and paralysis are the outcomes of amyotrophic lateral sclerosis, which are attributed to the loss of both the upper as well as lower motor neurons. Although the cause is unclear, mutations in the superoxide dismutase enzyme have been found [65]. The super oxygen dismutase-1 enzyme's activity is reduced by protein aggregates produced its mutations. The enzyme superoxide dismutase-1 is crucial for eliminating free oxygen radicals. The loss of enzyme function observed in ALS causes an enhanced build-up of oxygen free radicals, which in turn causes oxidation related stress.

9.4 Role of Antioxidants in Neurological Diseases

The role of antioxidants in the treatment of neurological diseases is significant. Table 9.1 provides an outline of the mechanism by which antioxidants protect against neurological diseases.

9.5 Therapeutic Strategies Using Antioxidants for Treatment of Neurological Disorder

9.5.1 Alzheimer's Disease

Alzheimer's disease (AD) impacts neurons over time, eventually leading to incurable dementia syndrome [126, 127]. AD is more frequent in elderly adults and increases in frequency with age [128]. Presently, 32.6 million individuals globally are afflicted, but in the absence

Table 9.1 Mechanisms of different antioxidants against various neurological diseases.

S. no.	Antioxidant	Mechanism	Reference
1.	class A scavenger receptor activator XD4 (W20/XD4-SPIONs)	It can stimulate the microglia's class A scavenger receptor (SR-A) and facilitate the phagocytosis of alpha-beta oligomers	[66]
2.	Pomegranate seed oil	Diminished levels of β-secretase and soluble amyloid protein procurer β (sAPPβ) maintain brain function by preventing neurons from dying	[67]
3.	Resveratrol	Activation of silent information regulator-1 (SIRT-1), which is mostly expressed in neurons in the adult mammalian brain. Resveratrol's activation of SIRT1 inhibits Aβ-induced microglial apoptosis and promotes better cognitive function	[68]
4.	Apolipoprotein E2	The formation of senile plaques and cerebral amyloid angiopathy (CAA), two prominent features of amyloid pathology in AD brains, are significantly influenced by APOE genotypes. The brains of AD patients have senile plaques that co-deposit ApoE, according to immunohistological findings	[69]
5.	Huperzine A	By boosting the activity of antioxidant enzymes such as glutathione peroxidase (GSH-Px), superoxide dismutase (SOD), and catalase (CAT), HupA can operate as an antioxidant in the Aβ-induced oxidative stress paradigm.	[70]
6.	Quercetin (QU) and rosmarinic acid (RA)	Quercetin may inhibit xanthine oxidase and lipid peroxidation *in vitro*, it may have a strong oxygen radical scavenging action. and Rosmarinic acid can block apoptotic indicators, reduce hyperphosphorylation of tau protein, and stop Aβ-induced memory loss (Iuvone *et al.*, 2006).	[71, 72]
7.	Rivastigmine	Acetylcholinesterase (AchE) inhibitors like rivastigmine have antioxidant qualities in addition to inhibiting AChE. as rivastigmine affected not only the glutathione levels in macrophages but also affected antioxidant enzymes activity.	[73]

(*Continued*)

Table 9.1 Mechanisms of different antioxidants against various neurological diseases. (*Continued*)

S. no.	Antioxidant	Mechanism	Reference
8.	Curcumin	Inhibition of generation of cytokines involved in inflammation, inducible nitric oxide synthase produced by NFκB as well as Cox-2.	[74]
9.	Piperine	Piperine, as anti-inflammatory agent, can prevent damage due to oxidation by squelching or suppressing oxygen species which are reactive, free radicals, and hydroxyl radicals.	[75]
10.	Thymoquinone rich fraction	By inhibiting mitochondrial dysfunction, it can aid in the prevention of glutamate-induced apoptosis.	[76]
11.	α-Bisabolol	It not only reduces the production of protein carbonyls , lipid peroxides as well as nitric oxides in Neuro-2a cells when treated with β-amyloid (Aβ)25–35 peptide, but it also inhibits acetylcholinesterase.	[77]
12.	Vitamin D-binding protein	It suppressed amyloidogenesis by raising serum levels of the amyloid peptide $A\beta_{1-40}$ and lowering Aβ levels in the brain [76]. Several studies proposed likely processes. In AD patients, 1α,25(OH)2D 3 may encourage macrophages to eat Aβ [77].	[78, 79]
13.	Memantine	It inhibits current flow through channels of receptor N-methyl-d-aspartate (NMDA), which have significant involvement in brain functioning.	[80]
14.	Zinc	Zinc is a redox-inert metal, but it has antioxidant properties because of the catalytic activity of copper/zinc-superoxide dismutase, membrane structure stabilization, protein sulfhydryl group protection, and increased expression of metallothionein, a metal-binding protein.	[81]

(*Continued*)

Table 9.1 Mechanisms of different antioxidants against various neurological diseases. (*Continued*)

S. no.	Antioxidant	Mechanism	Reference
15.	Tacrine	It's N-methyl-D-aspartate receptor (NMDAR) antagonist with limited affinity. Tacrine has been shown to have a number of cholinergic properties in addition to being an AChE inhibitor, such as increasing the production and release of acetylcholine (ACh) and modifying muscarinic and nicotinic receptors. Moreover, interactions between tacrine and monoaminergic systems have been demonstrated.	[82]
16.	Osthole	By altering oxidative stress, OST can cure CNS disorders and prevent glutamate-induced neurotoxicity and neuronal death.	[83]
17.	Morin hydrate	It acts by scavenging the free radicals	[84]
18.	Huperzine A (HA) and ligustrazine phosphate (LP)	By boosting the activity of glutathione peroxidase (GSH-Px), superoxide dismutase (SOD), and catalase (CAT), HupA plays a role as antioxidant. In a model of D-galactose-induced AD mice, ligustrazine dramatically increased cholinergic function of the hippocampus, attenuated damage due to oxidation, resulting in enhanced learning as well as enhanced memory.	[85, 86]
19.	Genistein	Genistein prevents the synthesis of Aβ by obstructing BACE1 (β-secretase enzyme).	[87]
20.	Melatonin	Melatonin's antioxidant and anti-amyloid qualities effectively shield neuronal cells from the harm caused by Aβ. It interacts with Aβ in a structure-dependent manner to prevent amyloid fibril formation in addition to inhibiting Aβ synthesis.	[88]
21.	Rosmarinic acid Cross-reacting material 197 (CRM197) and apolipoprotein E (ApoE)	A useful antioxidant that can help remove neurotoxic Peroxynitrite is rosmarinic acid (RA), which also has antiallergenic and anti-inflammatory properties.	[89]
22.	Berberine	Berberine can prevent Aβ-induced cortical neuronal apoptosis in rats by reducing reactive oxygen species and malondialdehyde formation	[90]

(*Continued*)

Table 9.1 Mechanisms of different antioxidants against various neurological diseases. (*Continued*)

S. no.	Antioxidant	Mechanism	Reference
23.	Galantamine hydrobromide	By inhibiting AChE selectively and reversibly, galantamine (Razadyne®) increases ACh levels in the cerebral cortex by delaying the breakdown of neurotransmitters and modifying nicotinic ACh receptors allosterically. ACh is further increased in presynaptic nerve terminals by this modulation. Moreover, glutamate and serotonin levels are raised by galantamine.	[91]
24.	Estradiol	The phenolic nature of the A ring reduces the amount of peroxide that mitochondria produce, thereby preventing oxidation.	[92]
25.	Basic fibroblast growth factor	Reduced inflammation and oxidative stress	[93]
26.	Silica	In cell lines, SiNPs reduced cell viability significantly, caused apoptosis of the cells, as well as affected internalized reactive oxygen species (ROS) levels in a dose-dependent manner.	[94]
27.	Sialic acid	A crucial nutrient required in microlevels for health and is involved in reducing brain oxidative stress.	[95]
28.	Ceria dioxide	The ability of cerium (CeO_2) nanoparticles to transition between the oxidation states of cerium(3+) and cerium(4+) allows them to act as potent and reusable ROS scavengers.	[96]
29.	Erythropoietin	Caspases Activation owing to the buildup of TNFα in AD patients, inhibition of apoptotic proteins, and a rise in the nuclear factor kappa-light-chain-enhancer function of NFκB.	[97]
30.	Naringenin	Antioxidant in nature, scavenging the free radical species	[98]
31.	Green tea polyphenol and (-)-epigallocatechin-3-	EGCG participates in PKC pathways by directly activating PKC through rapid phosphorylation, which enhances the positive, neuroprotective effects.	[99]

(*Continued*)

Table 9.1 Mechanisms of different antioxidants against various neurological diseases. (*Continued*)

S. no.	Antioxidant	Mechanism	Reference
32.	Donepezil	It is a powerful antioxidant that scavenges free radicals with about the same effectiveness as vitamin C	[100]
33.	Sesamol	Sesamol is a strong antioxidant and cytokine production inhibitor	[101]
34.	Ginsenoside Rg3 and Thioflavin T	A key component of ginseng, ginsenoside Rg3 is known to provide antioxidant effects by reducing freely present radicals which in turn exerts a crucial function in enhanced memory and cognition.	[102]
35.	N-acetylcysteine	It hydrolyzes quickly to produce cysteine, which can decrease highly oxidized GSH and replenish the whole GSH concentration. Additionally, preclinical research shows that NAC therapy reduces oxidative damage in AD mouse models	[103]
36.	Ropinirole hydrochloride	Increasing the frequency at which striatal neurons fire by activating D2-dopamine receptors selectively	[104]
37.	Selegiline	By increasing dopamine levels and bringing them back to normal, e helps to alleviate Parkinson's disease symptoms	[105]
38.	Catalase	It degrades hydrogen peroxide to produce harmless byproducts like oxygen and water	[106]
39.	Pramipexole dihydrochloride	A nonergot-based dopamine agonist called pramipexole is MAO-B inhibitors to successfully slow the course of Parkinson's disease.	[107]
40.	Ferulic acid	Ferulic acid and its additions have capability of binding with transition metals e.g. iron and acting as a secondary antioxidant. This stops harmful hydroxyl radicals from forming, which might cause peroxidation of cell membranes	[108]
41.	Dopamine	It is well known that dopamine functions as a potent antioxidant, scavenging free radicals to protect neurons from the damaging effects of stress induced by oxidation processes.	[109]

(*Continued*)

Table 9.1 Mechanisms of different antioxidants against various neurological diseases. (*Continued*)

S. no.	Antioxidant	Mechanism	Reference
42.	Gallic acid	It inhibits free radicals as well as peroxidation of lipids. It has properties to prevent inflammation and apoptosis as well as possesses ability to reduce glutathione.	[110]
43.	Hydroxytyrosol	The primary mechanisms investigated were antioxidant, ability to uptake free radicals, prevent inflammation and apoptosis.	[111]
44.	Graphene oxide	Reducing Aβ production besides increasing its destruction and promoting the transfer of Aβ to lysosomes by endosomal transport are the mechanisms by which it acts.	[112]
45.	β-Carotene	β-carotene was able to restore dopamine levels, oxidative stress markers (CAT, SOD, ROS, and TBARS), and increase motor function, memory, and survival	[113]
46.	Polydopamine (PDA) with selenocystine (SeCys)	To prevent the production of reactive oxygen species	[114]
47.	Fisetin	Lowering Amyloid Precursor Protein levels	[115]
48.	Poly(trehalose)	Trehalose partially reduces harmful oxidative stress and has an anti-inflammatory impact because it strengthens the body's natural antioxidant defense system, which is represented by the Nrf2 protein	[116]
49.	Boehmite	Antioxidant amino-nanozymes shown a remarkable ability to eliminate mutant huntingtin aggregates in addition to their strong antioxidant activity	[117]
50.	Ginger	According to the active bio-ingredient 6-shogaols' antioxidant activity, microglia cells may be inhibited when brief ischemia occurs	[118]
51.	Manganese	Involves hydrogen peroxide scavenging activity	[119]
52.	Verapamil and Riluzole	By scavenging hydrogen peroxide species	[120]
53.	FM19G11	It is a modulator of hypoxia-inducible factor which enhances self-renewal as well as generation of cells of the epithelium.	[121]

(*Continued*)

Table 9.1 Mechanisms of different antioxidants against various neurological diseases. (*Continued*)

S. no.	Antioxidant	Mechanism	Reference
54.	Leptin and Pioglitazone	Reported to prevent oxidation as well as inflammation of brain cells.	[122]
55.	Selenium and Crocin	Both act as antioxidant agents to scavenge the free radical species.	[123]
56.	(PEG-HCCs)	Acts by scavenging ROS superoxide ($O_2^{\bullet-}$) as well as hydroxyl ion.	[124]
57.	Capsaicin	It leads to reduced oxidative stress.	[125]

of novel therapies, this figure is expected to roughly quadruple every twenty years, with seventy eight million expected by 2030 which by 2050 is further expected to reach 139 million [129]. It causes psychological, social, and other cognitive abilities to deteriorate. Memory loss results from the death of brain cells caused by it [130]. The build-up of tau (τ) protein intraneuronal neurofibrillary tangles and Amyloid beta plaques extracellularly aggregate to cause AD, a neuropathological illness [131]. Aβ, produced by the division of amyloid precursor protein along with other type I membrane proteins, which is produced and found in considerable amounts in the plasma and cerebrospinal fluid of patients. The advancement of AD therapy has been made possible by the invention of nanotechnology. It has great potential in enhancing the way medications target the brain [132]. Nano-emulsions (NEs), dendrimers, nanostructured lipid carriers (NLCs), polymeric nanoparticles (PNPs), as well as solid lipid nanoparticles (SLNs), are nanoscale medication delivery systems explored over decades for the treatment of neurological disorders is the nano-capsule. To make certain that an appreciable amount of the medication reaches the brain, these nanoscale systems can be given intravenously, transdermally, or via intranasally.

The various antioxidants, which have been loaded in nanoformulations and found application in Alzheimer's disease treatment are given below:

9.5.1.1 Curcumin

Curcumin, an established antioxidant, has been utilized for treating Alzheimer's disease [74]. Curcumin's anti-amyloid as well as antioxidant properties make it an excellent line of treatment for Alzheimer's. The curcumin has been incorporated extensively in various nanoformulations for curing Alzheimer's disease, a few of which are given below:

Papadia *et al.*, 2017, developed multifunctional LUV liposomes. The development of multifunctional LUV liposomes, or mf-LIPs, involved the adherence of ligands to the surface for targeting the receptors (LDL as well as transferrin) of the blood-brain barrier (BBB) as well as a curcumin-lipid ligand (TREG) which had potential to target amyloid species. On hCMEC/D3 monolayers, mf-LIPs were assessed for targeting the brain as well as their capacity to prevent aggregation of Aβ-peptide. The transfer of mf-LIP via monolayers of hCMEC/D3 resembled BBB-LIPs, suggesting that their TREG bound to the surface did not lessen their capacity to target the brain. The inhibitory impact of mf-LIP on aggregation of Aβ was also comparable to that of TREG functionalized LIPs, indicating that the amyloid-specific ligand remains functional even when ligands meant for targeting the brain are present. Certain liposome types were shown to be more intact when curcumin-lipid was added, and the influence of serum proteins on the liposomes' ability to connect with brain endothelial cells was lessened. The *in vivo* data supported the *in vitro* findings. The authors concluded that the particular curcumin-lipid derivative had the potential to be a part of multifunctional LIPs with capability to target brain, which are meant to operate both as a therapeutic and as a diagnostic system for AD, according to the present data [133].

For curcumin to illicit its beneficial effects in treating AD, it must be capable to pass across the blood-brain barrier. The ultimate size of particles and stability have an impact on curcumin's bioavailability, which can be enhanced with the use of the most recent nanoparticle technology. In another study done by Cheng *et al.*, the authors formulated a stable formulation of curcumin nanoparticles for testing in Tg2576 mice, an AD model, and *in vitro*. By nanoprecipitation of polyethylene glycol-polylactic acid co-block polymer,. Curcumin as well as polyvinylpyrrolidone using a multi-inlet vortex mixer, and then using β-cyclodextrin to freeze dry, nanocurcumin in a dry form with less than 80 nm average size of particles was produced. Tg2576 mice were given either unformulated curcumin, powdered nanocurcumin, or a placebo orally for three months. Contextual fear conditioning tests and the radial arm maze were utilized for assessing memory both before and after therapy. When compared to placebo, nanocurcumin demonstrated a substantial (p equals to 0.04) improvement in the fear conditioning test's as compared to regular curcumin. In order to understand the processes thoroughly, both *in vivo* and *in vitro*, measurements were done of the pharmacokinetic parameters, density of amyloid plaque as well as penetration ability using Madin-Darby canine kidney cell monolayer. Compared to regular curcumin, nanocurcumin generated a plasma concentration of curcumin that was noticeably greater, with six times increased average residence period and area under the curve in the brain. For nanocurcumin, the P(app) values for tetrahydrocurcumin and curcumin were 1.6×10(-5) cm/s and 1.8×10(-6), respectively. The authors concluded that the developed nanosized curcumin dosage form was stable and possessed good efficacy in Tg2576 mice [134].

Reports suggest that, AD, exhibits decreased neurogenesis – the method of neuron production. By affecting the brain's ability to regenerate itself, for these situations, targeting natural neural stem cells (NSCs) to stimulate neurogenesis is reported to be a workable therapeutic approach. The neuroprotective agent curcumin has a low brain bioavailability. In a study reported in the literature, Tiwari *et al.*, stated that in comparison to uncoated bulk curcumin, subventricular zone and the hippocampus of adult rats, as well as NSC multiplication and neuronal differentiation *in vitro*, were all significantly stimulated by curcumin loaded PLGA nanoparticles (Cur-PLGA-NPs). Cur-PLGA-NPs were up taken by the hippocampus NSC, where they caused neurogenesis. Cur-PLGA-NPs upregulated genes linked to proliferation of cells (reelin, nestin, and Pax6) and neural growth (neuroligin, neuregulin, neuroD1, neurogenin and Stat3) significantly. Curcumin nanoparticles facilitated the development of neurons by Wnt and beta pathway activation, which is important in the regulating neurogenesis. These nanoparticles raised β-catenin's nuclear translocation, lowered levels of GSK-3β, and enhanced activity of TCF/LEF promoter as well as cyclin-D1. In a rat model of AD-like symptoms caused by amyloid beta, the formulation enhanced learning and improved memory by promoting neurogenesis. Research employing *in vitro* molecular docking suggested interaction of curcumin with Dkk, GSK-3β as well as Wif-1. These findings showed that curcumin loaded nanoparticles activated the canonical

Wnt/β-catenin pathway, thereby stimulating adult neurogenesis. The authors could thus potentially provide a strategy for treating neurological ailments like AD by stimulating self-healing process of the brain [135].

In another study done by Barbara *et al.*, the authors loaded curcumin in g7 ligand functionalized nanoparticles of polylactide-co-glycolic-acid to promote BBB bridging. Using primary hippocampal cell cultures, these NPs' potential toxicity, absorption, and—most importantly—their potential to impact Aβ pathophysiology *in vitro* were all investigated. The findings of authors indicated that the formed NPs did not appear to be hazardous, but they could cause a notable reduction in Aβ aggregates when they were loaded with curcumin [136].

The main symptoms of Alzheimer's disease, which has no known cure, are tau hyperphosphorylation and β-amyloid build up. Curcumin, often known as Cur, has demonstrated potential benefits due to its anti-inflammatory nature, anti-amyloid as well as antioxidant properties in Alzheimer's disease treatment. In another study done by Fan *et al.*, APP/PS1 Al transgenic mice as well as HT22 cells were given a novel brain-targeted poly(lactide-co-glycolide)-block-poly(ethylene glycol) (PLGA-PEG) nanoparticles, which were connected with the B6 peptide and loaded with Cur (PLGA-PEG-B6/Cur). The outcomes of the *in vitro* examinations, which comprised red blood cell lysis along with thromboelastography (TEG) analysis, flow cytometry (FCM), and dynamic light scattering (DLS), suggested that the formulation might improve Cur's blood compatibility, reduce its diameter, and enhance its cellular absorption. The results of the Morris Water Maze showed that PLGA-PEG-B6/Cur could considerably enhance spatial memory as well as learning ability of APP/PS1 mice in comparison to Cur native. *Ex vivo* assays like western blotting, immunostaining and Bielschowsky silver staining showed that PLGA-PEG-B6/Cur may decrease the generation of tau hyperphosphorylation and β-amyloid as well as its deposits in the hippocampal regions. Therefore, the authors suggested that the use nanoparticles of PLGA-PEG-B6/Cur in Alzheimer's disease treatment might be promising [137].

The build-up of plaques containing amyloid peptide (Aβ) is one defining pathology of AD. More research is being done on the creation of compounds that can target the amyloid pathology in tissue for both therapeutic and diagnostic applications. Due to its limited solubility, curcumin, a fluorescent chemical with a strong affinity for the Aβ peptide, cannot be used in therapeutic settings. In a study done Lazar *et al.*, formulated curcumin-conjugated nanoliposomes. The formulation inhibited the release of amyloid peptide, was non-toxic *in vitro*, and somewhat shielded against the toxicity caused by Aβ. APPxPS1 mice and AD patients' post-mortem examination of brain tissue showed markedly elevated Aβ accumulation. The authors concluded that curcumin-conjugated nanoliposomes might be helpful for diagnosing AD and administering medicine in a customized way [138].

In another study, curcumin loaded Se-PLGA nanospheres greatly enhanced the memory impairment in mice and lowered the quantity of amyloid-β in AD animal's samples of brain. Utilizing fluorescent microscopy, Alpha beta plaques and Curcumin loaded Se-PLGA nanospheres were shown to precisely link together. It was concluded that the cur-loaded Se-PLGA nanoformulation is a valuable delivery method for the targeted distribution and an efficient treatment for AD [139].

In another study, Mathew *et al.*, formulated PLGA coated curcumin nanoparticles which were water-soluble. Since the blood-brain barrier restricts targeting drugs for disorders with a cerebral origin, the authors combined the nanoparticle and Tet-1 peptide, having neuronal affinity as well as retrograde transport capabilities. The findings confirmed that PLGA nanoparticles encapsulated with curcumin non-cytotoxic, had anti-oxidative properties, and may break up amyloid clumps. Since curcumin's natural qualities are preserved when it is encapsulated in PLGA, PLGA-curcumin nanoparticles are medication with numerous applications in the AD treatment [140].

Curcumin has been extensively researched for its strong antioxidant properties, scavenger properties, besides its ability to protect nerve cells. However, two significant disadvantages of its use are its low bioavailability with restricted capacity to pass the blood-brain barrier. In an investigation conducted by Sadegh Malvajerd *et al.*, curcumin loaded nanostructured lipid carriers (NLCs) were developed to enhance its efficiency in curing Aβ-induced deficits in cognition in AD rats. The application of curcumin-loaded NLCs considerably raised both the serum levels and the accumulation rate of curcumin (505.76±38.4 ng/g^{-1} h) in the rat brain. Reduced oxidative stress markers in the hippocampus as well as an increase in spatial memory demonstrated the efficacious function of NLCs for curcumin transport to the brain. Additionally, histological investigations demonstrated Cur-NLCs' capacity to lessen Aβ in AD animals. Curcumin loading in NLCs was confirmed to be a strategy for enhanced brain delivery of curcumin with reduced neurological and memory deficits [141].

In another study conducted by Huang *et al.*, the authors formulated poly(lactide-co-glycolic acid) (PLGA) nanoparticles by conjugating the iron-mimic peptide brain targeting peptide CRT (cyclic CRTIGPSVC peptide), which targets the transferrin receptor (TfR), with curcumin and Aβ generation inhibitor S1 (PQVGHL peptide) to target the factors detrimental to the development of AD. Drug-loaded and CRT-conjugated PLGA nanoparticles were 128.6 nm and 139.8 nm in size, respectively. The Y-maze as well as new object recognition test findings showed that the PLGA nanoparticles greatly enhanced the transgenic AD mice's recognition and spatial memory. Additionally, in the AD mice brains, PLGA formulation significantly reduced the levels of Aβ, ROS, TNF-α, and IL-6 while increasing the number of synapses and the activity of super oxide dismutase (SOD). The CRT peptide modified-PLGA nanoparticles that co-delivered S1 and curcumin showed the most favorable impact on the AD mice treatment when compared to other PLGA nanoparticles [142].

In another study done by Malvajerd *et al.*, the authors formulated curcumin entrapped solid lipid nanoparticles (SLNs) along with nanostructured lipid carriers (NLCs) to increase brain accumulation. Multiobjective optimization techniques and experimental design were used to optimize the preparation procedures. Curcumin was shown to have an entrapment efficiency of 82% ± 0.49 in SLNs and 94% ± 0.74 in NLCs. Following intravenous (IV) administration of a 4 mg/kg curcumin, the pharmacokinetic studies demonstrated significantly higher curcumin levels in the brain in rats given curcumin-loaded NLCs (AUC_{0-t} = 505.76 ng/g h) in comparison to those administered with free curcumin (AUC_{0-t} = 0.00 ng/g h) as well as curcumin-loaded SLNs (AUC_{0-t} = 116.31 ng/g h) ($P < 0.005$). Differential scanning calorimetry as well as X-ray diffraction indicated that curcumin was distributed throughout the nanocarriers in an amorphous state. Images captured by scanning electron microscopy verified the nanometric spherical nanoparticles. The results confirmed the role of curcumin-loaded in the management of CNS disorders [143].

In study conducted by Desai *et al.*, pharmaceutical curcumin cocrystals were made and then were added to micellar nanocarriers for intranasal brain delivery. Using a molecular modeling technique, an antioxidant hydrophilic coformer was carefully chosen for cocrystals. A planetary ball mill was used to create the cocrystals, and 32 factorial design was used to improve the process. X-ray diffraction, differential scanning calorimetry, and Fourier-transform infrared spectroscopy studies were then used to characterize the product. The globule size of the cocrystal micelles was 28.79 ± 0.86 nm. Additionally, compared to curcumin, curcumin cocrystal as well as co-crystal micelles showed a considerably lower (p value <0.01) antioxidant activity, supporting higher antioxidant efficacy. Curcumin cocrystal micelles had an approximate 1.7-fold absolute bioavailability in *in vivo* experiments, with a Cmax of 1218.38 ± 58.11 ng/mL and a very large distribution in brain even after 6 hours of administration. Thus, curcumin cocrystal micellar nanocarriers were shown to have improved bioavailability, increased brain absorption, retention, and delayed elimination [144].

In another study done by Nasr *et al.*, the authors produced a mucoadhesive lipid nanoemulsion containing hyaluronic acid that co-encapsulated both curcumin and resveratrol, for the intranasal therapy of brain disorders. The spontaneous emulsification method was applied to develop the formulations, which were then examined for morphology, mucoadhesive strength, zeta potential, and particle size. The antioxidant capacity, safety on nasal mucosa, *in vitro* as well as *ex vivo* release of drugs, and quantification of the medications in brains of rats were all evaluated. After three months of storage, changes in the drug's concentration, zeta potential, particle size, and antioxidant potential were assessed in order to assess the stability of the material. The nanoemulsion formula using hyaluronic acid as the optimized ingredient showed a zeta potential of -23.9 ± 1.7 and a particle size of 115.2 ± 0.15. Comparing the formula to its non-mucoadhesive equivalent, it showed a spherical shape and much greater mucoadhesive strength. Additionally, the antioxidant capacity of the encapsulated polyphenols was maintained. The two medications were released under controlled diffusion for up to six hours, with *ex vivo* fluxes of 2.86 and 2.09 μg/cm(2) hr corresponding to resveratrol and curcumin, respectively, using sheep nasal mucosa. Additionally, the mucoadhesive nanoemulsion enhanced the brain levels of both polyphenolic drugs (resveratrol by approximately 7 times and curcumin by approximately 9 folds, respectively) while being safe for the nasal mucosa. The formulation demonstrated itself to be an effective carrier that not only improved the solubility and stability but also enhanced brain targeting efficiency of polyphenolic drugs [145].

Doggui *et al.*, encapsulated curcumin in a formulation of biodegradable nanoparticulates (Nps-Cur) based on poly (lactide-co-glycolide) (PLGA). A particle diameter of 80 to 120 nm was revealed by transmission electronic microscopy investigation and dynamic laser light scattering. Curcumin from Nps-Cur's *in vitro* release kinetics showed a two phase pattern, with an early exponential release followed by a gradual release. Fluorescence and confocal imaging were used to establish that Nps-Cur had internalized into the cells. The fluorescence was widely distributed throughout the cytoplasm and nucleus. The produced nanoformulation's biological activity and cellular toxicity were assessed. Human neuroblastoma SK-N-SH cells were not harmed by empty PLGA nanoparticles (Nps) or PLGA nanoparticles loaded with curcumin (Nps-Cur), according to cytotoxicity tests. Furthermore, Nps-Cur was successful in shielding SK-N-SH cells from H_2O_2, stop the rise in reactive oxygen compounds, and stop the H_2O_2-induced glutathione uptake. Remarkably, in the H_2O_2 presence, Nps-Cur could also stop the Nrf2 activation. When combined, these findings point to Nps-Cur as a possible medication delivery method to shield neurons from oxidative damage [146].

9.5.1.2 *Class A Scavenger Receptor Activator XD4 (W20/XD4-SPIONs)*

Alzheimer disease is a neurological illness that worsens with time. As of right now, there is no therapy to halt or decrease the progression of AD. Alpha-beta oligomers, as opposed to monomers or fibrils, are reported to be primary species causing neurotoxicity. Treatment strategies for AD that target Alpha-beta oligomers and encourage Aβ elimination may be very beneficial. Liu *et al.*, reported flexible nanoparticles formed by superparamagnetic iron oxide coupled with the Aβ oligomer-specific scFv antibody W20 and the category A-antioxidant activating receptor XD4 were developed. Apart from their diagnostic efficiency, the nanoparticles preserved the anti-Aβ properties of the scFv antibody specific to Aβ oligomer and the category A-antioxidant activating receptor by lowering Alpha-beta aggregation, decreasing Alpha-beta oligomers-induced cytotoxicity, and increasing microglia's cell engulfing of Alpha-beta. Nanoparticles dramatically reduced AD mice's neuropathology and reversed cognitive impairments when administered to amyloid precursor protein/PS1 mice. These results suggested that the anti-AD capabilities of developed nanoparticles exist and that they present a promising agent for the detection and treatment of AD in its early stages [147].

9.5.1.3 *Pomegranate Seed Oil/Extract*

It has been used as a vital antioxidant in various nanoformulations for treating Alzheimer's disease [67]. Cyclin-dependent kinase 5 (CDK5) is deregulated in diseases like Alzheimer's disease (AD) and is induced by coupling to the triggered calpain product p25. One metabolite of punicic acid (PA), the primary ingredient in pomegranate seed oil (PSO), is conjugated linoleic acid (CLA. Long-term therapy with *Nanopunica granatum* seed oil, a nanodroplet dosage form of pomegranate seed oil is reported to delay the sickness onset as well as mitochondrial damage in mice having inherited Creutzfeldt Jacob disease (CJD). The authors reported that CLA accumulates in the mice brains that are administered Nano-pomegranate seed oil, but not in brains treated with normal PSO. The long-lasting impact of Nano-pomegranate seed oil dosing on mental, biochemical processes and pathological outcomes were next investigated using 5XFAD mice as Alzheimer's disease models. The authors further demonstrated that in 5XFAD mice, Nano-p therapy reduced cognitive decline related to age as well as damage of mitochondria. Additionally, brains of mice treated with Nano-PSO showed enhanced expression of the crucial mitochondrial enzyme COX IV-1 and decreased Aβ and p25 accumulation. Based on the salient results, the authors revealed that administering Nano-PSO caused CLA to be targeted to the brain and proposed that the start of Alzheimer's disease and Creutzfeldt-Jacob disease may be delayed or prevented by this medication [148].

In another study done by Almuhayawi *et al.*, the authors using $AlCl_3$-induced AD rat model evaluated the preventive and therapeutic uses of pomegranate extract (PE) as well as nanoparticles encapsulated with pomegranate extract (PE nano). Following the synthesis of nanoparticles with 0.68% w/w PE load, seven groups of seven male Wistar rats were formed: Groups I, II, and III were the controls; Group IV got $AlCl_3$ (50 mg/kg) orally every day for four weeks; Group V was administered PE for two weeks; Group VI got PE nano for two weeks; and Group V and Group VI were given treatment following the cessation of $AlCl_3$ administration. Before $AlCl_3$ was given, Group VII got PE for two. Groups administered PE or nano PE showed a substantial increase in the levels of catalase, glutathione as well as total antioxidant activity in homogenates of brain. The thiobarbituric acid reactive substance, which is used to detect peroxidation of lipid, was shown to be considerably higher in the AD rats and to be lower in the groups that received PE nanotreatment or protection. The brains of AD rats showed neurofibrillary tangles and senile plaques, as well as hyaline degeneration and chromatolysis—the hallmarks of the disease—according to histopathological investigations conducted with hematoxylin and eosin, cresyl violet, and silver stains. Brains given treatment with PE or PE nano showed histological architecture restoration, Nissl granulesas well as limited emergence of markers of AD. To sum up, PE was more successful as a protectant than as a treatment to lessen the impacts of lipid peroxidation, antioxidants, and histological features in brains implicated with AD. However, the PE-encapsulated nanoparticles yielded outcomes comparable to the protective PE and enhanced the effectiveness of the active ingredients [149].

9.5.1.4 Resveratrol

Rajput *et al.*, developed an *in-situ* gel that was nanostructured and contained resveratrol for Alzheimer's disease therapy. The finished product when evaluated exhibited drug loading of 9.26 ± 3.79%, surface charge potential of -23 ± 3.79 mV, particle dimension of 132 ± 11.90 nm, polydispersity index of 0.209 ± 0.005, as well as entrapment efficiency of 74 ± 11.40%. The warm emulsification-probe sonication approach was utilized to create the formulation with resveratrol. After resveratrol nanostructured lipid carrier was added to gellan gum and xanthan gum, *in situ* gel was produced. In comparison to an *in-situ* gel based on resveratrol solution, the improved *in-situ* gel exhibited a five-fold enhancement in penetration across the nasal mucosa. Further rats treated with customized *in situ* gel demonstrated much improved memory performance than those given oral resveratrol solution. Given its improved memory function and increased permeability across nasal mucosa, for Alzheimer's disease, the developed formulation seemed to be a promising treatment strategy [150].

In another study done by Hao *et al.*, the authors formulated a composite *in situ* gel for difficult intranasal administration by combining nanoparticles with deacetylated gellan gum (DGG) matrix which was ionic-triggered. Nanosuspensions loaded with Resveratrol (Res-NSs) were dispersed in DGG matrix. Particle size, *in situ* gelation capabilities, textural characteristics and rheological features were two of the standards utilized for describing the *in situ* gel's characteristics. Studies were conducted on the brain-targeting efficacy and pharmacokinetics following intravenous and intranasal delivery, respectively. Additionally biological distribution and identification utilizing fluorescent microscopy and *in vivo* imaging tools were also done. The mixture with 0.6% w/v DGG had the required viscosity and solidified well. The rheology data were used to calculate the yield stress, thixotropy, and features of shear thinning in the DGG *in situ* gel. Comparing the *in situ* gel and DGG solution, the textured profile data showed that the *in situ* gel exhibited more adhesion and viscosity. The Higuchi mathematical model was employed in studies on *in vitro* penetration. The intranasal Res-NSs *in situ* gel preparation increased brain bioavailability 2.88 times, according to pharmacokinetic investigations. The nose-brain pathway was shown to be the direct route of delivery based on both the percentage of direct transportation of 78.18% and the drug targeted efficiency of 458.2%. The authors made a conclusion that the delivery system formulated, had the potential to be used intranasally to improve nasal mucosal permeability and extend the duration of the nasal cavity residency [151].

Lipid microparticles (LMs) carrying the neuroprotective polyphenol resveratrol were delivered to the brain via the nasal route. The lipid microscopic particles encapsulating resveratrol (LMs-Res) were developed by using melt emulsification with stearic acid serving as the lipid ingredient and phosphatidylcholine serving as the surfactant. A chitosan solution was added to the produced particles to create the Lipid microparticles-Res-Ch containing chitosan solution (1.75% w/v) and Lipid microparticles-Res-Ch-plus containing chitosan solution (8.75% w/v). The average size of the particles for LMs-Res-Ch, LMs-Res as well as LMs-Res-plus were 76.3 ± 5.2 μm, 68.5 ± 3.1 μm and 84.5 ± 8.1 μm,, indicating that they were suitable for nasal administration. The uncoated particles had a negative surface charge of (-12.7 ± 2.1 mV), for the LM-Res-Ch as well as LM-Res-Ch+ wrapped with chitosan, the surface charge was greater (24.0-± 4.7 and 44.6-± 3.1 mV, respectively). Studies on the penetration of human NCM460 cell monolayers showed that the existence of empty LMs, packed LMs-Res, packed LMs-Res-Ch, or free resveratrol did not affect the transepithelial electrical resistance values of these cells. Conversely, when LMs-Res-Ch-plus was present, the transepithelial electrical resistance values reduced to 41 ± 3 Ω cm^2 from 150 ± 7, which correlated with a notable rise in resveratrol's permeability as conducted by (Papp) (518 ± 8 × 10^{-4} cm/min to 750 ±98 × 10^{-4} cm/min). After the polyphenol was infused intravenously, there was no resveratrol in the rat CSF (cerebrospinal fluid), according to *in vivo* investigations. On the other hand, Resveratrol dissolved in water and delivered via nasal route, either as a chitosan solution or as uncoated Lipid microparticles-Res, led to resveratrol uptake in the cerebrospinal fluid, with maximum concentration values (1.30 ± 0.30 μg/ml and 0.79 ± 0.15 μg/ml), respectively, after 1 hour. On the other hand, the injection of an aqueous solution of Lipid microparticles-Res-Ch-plus resulted in a substantial enhancement in the amounts of drug reaching the CSF, with a maximum concentration (9.7 ± 1.9 μg/ml) after 1 hour. There was no generalized distribution throughout the considerable increase in CSF accessibility, suggesting a direct intranasal brain delivery [152].

In another study, the authors formulated Resveratrol, a potent antioxidant, loaded intranasal transferosomal mucoadhesive gel in order to improve its bioavailability. To make transferosomes containing RES, the technique of reverse evaporation-vortexing sonication was utilized. Using soy lecithin, permeation enhancers, and surfactants, 34 conclusive screening designs were used to create transferosomes. The formulation displayed sphere-like particles with a size of 83.79 ± 2.54 nm and an entrapment efficiency (EE%) 72.58 ± 4.51%. Following the preparation and assessment of mucoadhesive gels, optimal Resveratrol transferosomes were combined with a selected gel and FTIR spectroscopy, *in vitro and in vivo* studies were utilized to characterize them. An *in vivo* pharmacokinetic investigation and a histopathological analysis of the nasal mucosa were carried out. 65.87 ± 2.12% of the medication was released *in vitro* from the transferosomal gel, while 75.95 ± 3.19% was absorbed *in vivo*. A histopathological analysis verified the improved formula's safety. When opposed to the oral RES solution, the RES trans-gel showed a 2.15-fold boost in Cmax and a 22.5-fold rise in AUC(0-∞). The RES trans-gel led to an improvement in nasal safety and absorption through liver and intestinal metabolism [153].

Neuritic plaques are the histological markers of AD, which are connected to the amyloid-β peptide's (Alpha beta) aggregation. Many natural substances have been proposed as AD treatments. Resveratrol is one of these substances that has generated a lot of attention because of its neuroprotective qualities. However, after being injected intravenously, resveratrol is quickly broken down into glucuronic acid and sulfate in the hepatocytes and intestinal epithelial cells conjugations of the phenol-containing groups (within less than two hours), which are subsequently excreted. In another study done by Loureiro *et al.*, the authors confirmed that solid lipid nanoparticles (SLNs) treated with anti-transferrin receptor monoclonal antibody (OX26 mAb) might be a viable delivery system for the extract to enter the brain. Experiments on human brain-like cells called endothelial cells have demonstrated that the cellular consumption of OX26 SLNs is substantially higher in comparison to conventional solid lipid nanoparticles and solid lipid nanoparticles infused with an antibody which is unspecific. Consequently, it may concluded that these various SLNs have a greater capacity for transcytosis when functionalized with OX-26 [154].

9.5.1.5 Apolipoprotein E

It has been incorporated into nanoformulations as a rich antioxidant source for treating Alzheimer's disease [69]. In a study reported by Rodrigues *et al.*, the authors checked the possibility of treating AD using brain-targeted modified liposomes loaded with apolipoprotein E2 plasmids. This was accomplished by administering liposomes containing therapeutic genes that target the mice's brains systemically. The efficiency of targeting and transfecting bEnd.3, primary glial, and primary brain cells with produced liposomes were evaluated. The liposomal preparations ability to migrate across the blood-brain barrier (BBB) and afterward transfect basic neuronal cells was evaluated employing an *in vitro* triple co-culture BBB paradigm. After receiving a single intravenous liposomes injection with plasmid apolipoprotein E2 aimed at the brain, the authors measured the expression of apolipoprotein E2 in the brains of mice. The efficiency of the proposed liposomes' *in vitro* transfection was improved by dual surface modification. Research has demonstrated that transferrin and penetratin modified liposomes can be used for effectively transporting therapeutic genes across the blood-brain barrier. Mice's brains exhibited a significant ($p<0.05$) rise in levels of apolipoprotein E2 following intravenous injection of Tf-Pen-liposomes containing plasmid apolipoprotein E2. The findings demonstrated enhanced neurological targeting and delivery of genes by using liposome gene delivery methods based on dual ligands. The authors concluded that distribution of plasmid apolipoprotein E2 by transferrin-penetratin-modified liposomes has considerable promise for the therapy of AD [155].

Kuo *et al.*, grafted both apolipoprotein E (ApoE) and cross-reacting component 197 (CRM197) onto polyacrylamide-chitosan-poly(lactide-co-glycolide) nanoparticles loaded with rosmarinic acid with the goal to restore degenerated neurons and address the blood-brain barrier. CRM197-ApoE-RA-PAAM-CH-PLGA NPs were applied to RWA264.7 cells, Aβ-protected SK-N-MC cells, and human cerebral microvascular endothelial cells. The findings of the experiments showed that the grafting efficiency, zeta potential, and particle size of ApoE and CRM197 were all reduced with an increase in the weight % of PAAM. Furthermore, it was observed that surface DSPE-PEG(2000) may shield CRM197-ApoE-RA-PAAM-CH-PLGA NPs from RWA264.7 cell uptake. Propidium iodide and RA were observed to easily cross the blood-brain barrier when the quantities of ApoE and CRM197 elevated. This was due to a reduction in transendothelial electrical resistance. The SK-N-MC cells' viability was observed as follows: ApoE-RA-PAAM-CH-PLGA NPs > CRM197-RA-PAAM-CH-RA NPs > RA. Therefore, the authors concluded that in Alzheimer's disease cure, CRM197-ApoE-RA-PAAM-CH-PLGA NPs might be a useful delivery system for RA delivery to neurons that are Aβ-insulated [156].

9.5.1.6 *Huperzine A*

It has been incorporated as antioxidant in various nanoformulations for treating Alzheimer's disease [70]. A few examples reported in literature are given as follows: Yang *et al.*, synthesized and optimized Huperzine A encapsulated nanostructured lipid carriers (NLC), for successful AD cure using traditional medicine from China, which is reported to enhance cognitive ability in AD brain for reduction in oxidative stress. NLC was successfully created using an altered version of melted ultrasonication and high-pressure homogenization. Glycerol Triester of caprylic and capric acid a liquid lipid, Soybean phosphatidylcholine (Spc), a solid lipid, Cetyl Palmitate (CP), and Solutol HS15 were used for NLC formulation. To find the best formulation, a Box-Behnken approach with three components and three stages was employed. The formulation's ratio of lipid to drug, quantity of mixed lipid, and amount of emulsifier combination were all looked at as independent factors. The dependent variables were the size of the particles, drug concentration (DC), and efficiency of entrapment (EE). Properties of NLC including morphological variables like particle dimension, surface charge potential, efficiency of entrapment, loading of drug, and drug-releasing behavior were studied. The conclusion was that the proposed nanoparticles had approximately particles that were spherical with 120 nm mean size and potential difference of -22.93 +/- 0.91 mV. Hup A was released from NLC *in vitro* in stages, with a burst release observed at the beginning with a slow release spread over 96 hours. The results indicate that the developed Hup A encapsulated NLC system is a workable strategy for enhancing drug loading and for providing drug release in a controlled manner [157].

In another study done by Shi *et al.*, the author's formulated a new type of microemulsion patch that would enable the transdermal dispersion of ligustrazine phosphate (LP) and huperzine A (HA) at the same time. The pseudo-ternary phase diagrams were made employing the oil (oleic acid), the surfactant (Cremophor RH40), and the cosurfactant (ethanol). To improve penetration, the microemulsion was prepared with 1,8-cineole. The lamination approach was used to create the transdermal patches based on microemulsion. To evaluate the effectiveness of various transdermal patches and microemulsions in distributing ligustrazine phosphate and Huperzine A through the abdomen skin of rats, permeation investigations were carried out *in vitro*. The findings demonstrated that the transdermal patch had zero order penetration kinetics and that microemulsions increased HA and LP permeation rates in comparison to the control. According to the findings of the pharmacodynamic investigations, transdermal HA and LP combination treatment outperformed monotherapy in terms of its ability to prevent amnesia. After transdermal treatment at several dosages for nine consecutive days, Rats with scopolamine-induced forgetfulness were utilized to confirm the anti-amnesic effects, and the results indicated a dose-dependent pattern in the efficacy. In conclusion, transdermal patches based on microemulsions containing ligustrazine phosphate and huperzine A may provide a feasible means of preventing AD [158].

Meng *et al.*, produced intranasal mucoadhesive, brain targeted, lactoferrin (Lf)-conjugated N-trimethylated chitosan (TMC) polylactide-co-glycoside (PLGA) nanoparticles (NPs) transporting hupA (HupA-TMC-Lf NPs) for treating Alzheimer's disease. HupA-TMC-Lf NPs were produced by the emulsion-solvent evaporation method, and optimization was carried out using the Box-Behnken design. Investigations were done on the adhesion, surface charge potential, entrapment effectiveness of drug, size of particles, and *in vitro* release behavior. The uptake by cells was evaluated using fluorescence microscopy as well as flow cytometry. The MTT test was employed to assess the NPs'. The impact of nanoparticles administered intranasally on brain targeting was examined using an *in vivo* imaging system. Following intranasal administration, Hup-A NPs' biodistribution was ascertained using tandem mass spectrometry and liquid chromatography. The tailored HupA TMC-NPs-Lf demonstrated a 48-hour *in vitro* release, 73.8%±5.7% of drug entrapment efficiency, 0.229±0.078 polydispersity index, 153.2±13.7 nm particle size, and +35.6±5.2 mV zeta potential. The proportion of mucin adsorption on TMC-NPs-Lf was 86.9%±1.8%, which was more than the percentage adsorbed onto PLGA NPs of 32.1%±2.5%. In the 16 human bronchial cell line, HupA TMC-NPs-Lf showed less toxicity as compared to HupA solution. Cellular uptake tests, both qualitative and quantitative, revealed that in 16 human bronchial cell line and SH-SY5Y cells, TMC-NPs-Lf accumulated at a faster rate than non-targeted analogs. According to the results of *in vivo* imaging as compared to non-targeted NPs, Lf-TMC NPs had enhanced residence time and increased intensity of fluorescence in the brain. Following intranasal therapy, TMC-NPs-Lf increased the brain distribution of HupA; the drug aiming index values in the areas of the olfactory bulb of mouse, cerebrum (without the hippocampus), cerebellum, and hippocampal were observed to be 2.0, 1.6, 1.9, and 1.9, respectively. TMC-NPs-Lf demonstrated robust sustained-release activity, attachment and targeted capabilities, and a broad variety of potential uses as a carrier for nasal delivery [159].

9.5.1.7 Quercetin and Rosmarinic Acid

These two components are used widely in combination or alone as antioxidants [72], for curing AD. Quercetin is reported to protect against diseases such as Alzheimer's disease (AD), which are related to age. However, quercetin's poor oral bioavailability hinders its practical applicability.

In a study the authors employed apolipoprotein E-modified liposomes in conjunction with phosphatidic acid (PA) as a drug carrier system to enhance blood brain barrier transport of rosmarinic acid (RA) and quercetin (QU). Using an *in vitro* blood brain barrier model, the capacity of the dual-functioning liposomes to cross the blood-brain barrier and maintain high endothelial viability was shown to be superior to that of Tween 80 liposomes. Immunochemical staining pictures revealed that liposomes were able to transverse the blood-brain barrier. This was made possible by the strong attraction between apolipoprotein E and low-density lipoprotein receptors in brain microvascular endothelial cells, which also prevented $A\beta_{1\text{-}42}$-insulated SK-N-MC cells from undergoing apoptosis. In an *in vivo* AD model, liposomes inhibited the generation of Aβ plaque, acetylcholinesterase activity, and lipid neuroblastoma cells that display an epithelial cell type level. The authors confirmed that blood-brain barrier may be crossed, and QU and RA may be delivered to the brain via surface functionalization of liposomes with PA and apolipoprotein E, which would control $A\beta_{1\text{-}42}$-induced neurotoxicity [160].

In a study, the authors evaluated the potential of quercetin as an oral treatment for AD by encapsulating it in zein nanoparticles (NPQ), to increase the flavonoid's oral absorption and bioavailability. The authors administered SAMP8 mice with either given either NPQ (25 mg/kg every 48 hours) or a quercetin (Q; 25 mg/kg daily) solution for two months orally. There were no notable effects seen in animals for Q. Conversely, the SAMP8 mice's cognitive and memory deficits were ameliorated by the oral administration of NPQ. These findings seemed linked to a decrease in the expression of the GFAP marker, which is connected to astrocytes in the hippocampal region. Furthermore, quercetin in significant levels was detected in the cerebral cortex of nanoparticles treated mice. The results showed that zein nanoparticles can improve quercetin oral absorption and that the flavonoid may be used to treat AD etiology [161].

The reports in literature suggest that neuropathological alterations linked with AD that lead to malfunction followed by death of neurons seem to be primarily caused by chronic inflammatory processes. Glial cell activation and proliferation, in addition to a rise in output of free radicals and inflammatory mediators, serve as illustrations of the inflammation related responses seen in the Alzheimer's affected brain. In conjunction with glial cells, neurons can react to and play a role in changes related to neuroinflammation in the AD brain as precursors to inflammatory mediators. Since the brain is unable to break down excess cholesterol, it must be eliminated from the body as oxysterols, or products of cholesterol oxidation, in order to keep it from building up. Because of their neurotoxic effects, oxysterols seem to symbolize the lack of a link between abnormal metabolism of cholesterol and AD, one of the neurodegenerative disease's risk factors. The findings of another study demonstrated that three oxysterols that might have a role in the etiology of AD, 24-hydroxycholesterol, 27-hydroxycholesterol, and 7β-hydroxycholesterol, trigger the release of a few mediators that promote Toll-like receptor-4/cyclooxygenase-2/membrane-bound prostaglandin E synthase (TLR4/COX-2/mPGES-1), which mediates inflammation in human neuroblastoma SH-SY5Y cells; this implies that oxysterols may play a role in neuroinflammatory changes in AD. Quercetin, an anti-inflammatory flavonoid, was added to β-cyclodextrin-dodecylcarbonate nanoparticles, which significantly improved the flavonoid's inflammation reducing properties in SH-SY5Y cells in comparison to those of cells that were simply exposed to free quercetin. The objective of developing encapsulated quercetin nanoparticles was to increase its ability to cross the blood-brain barrier and enhance its target bioavailability. The results imply that this pharmaceutical delivery method could be a cutting-edge strategy aimed at slowing or stopping the progress of AD [162].

In another study done by Palle *et al.*, the authors assessed how well NQC protected against scopolamine-induced deficits in spatial memory when compared to free quercetin. NQC were made using the precipitation technique. For eight days in a row, the reference medications were quercetin, NQC (30 mg/kg p.o.), and rivastigmine at a dose of 2 mg/kg i.p. At the end of the treatment period, a single scopolamine injection (20 mg/kg; i.p.) caused memory deficits. After 30 minutes, the rats were killed, and brain homogenates utilized to assess the levels of several chemicals, including acetylcholinesterase (AchE), catalase, glutathione, malondialdehyde, and conditional avoidance and rectangular maze tests. Furthermore, histopathologic investigations were carried out. NQC was found to be less than 300 nm. When compared to the group given scopolamine, both the transfer delay and the conditioned avoidance reaction were significantly reduced by NQC ($p < 0.05$). A significant ($p < 0.05$) decrease in catalase and GSH levels in brain as well as an increase in MDA and acetylcholinesterase levels prior to taking NQC, which was similar to the results of the rivastigmine group. Rats treated with NQC had superior outcomes in all studies in comparison to the quercetin group, suggesting that NQC might be used as a prophylactic against AD progression. The authors concluded that quercetin nanoparticle approach may be utilized in the future to treat neurodegenerative illnesses in people [163].

It has been demonstrated that a connection between the beginning of Alzheimer's disease (AD) and the dysfunctional relationship of amyloid-β (Aβ) with excess metal ions exists. In another study done by Sun *et al.*, the authors studied PLGA-functionalized quercetin (PLGA@QT) NPs effects on $A\beta_{42}$ plaques/fibrils that have been blocked and deconstructed. *In vitro* experiments on SH-SY5Y cells demonstrated that the PLGA@QT NPs were not harmful. As expected, the human neuroblastoma cells SH-SY5Y exhibited concentration-dependent behavior with PLGA@QT NPs with concentration-dependent behavior. Furthermore, PLGA@QT NPs were observed to improve neuron cell survival and reduce the Zn^{2+}-$A\beta_{42}$ system's neurotoxicity. Behavior test findings showed that injecting APP/PS1 mice with PLGA@QT NPs, improved memory and cognitive deficits. The most encouraging discovery was that there were no indications of a negative impact on mice after administration of PLGA@QT NPs *in vivo* as determined by histological analysis in important organs. As a result, it was concluded that the produced nanoscale drug delivery carrier based on quercetin effectively lowered the negative effects and raised the therapeutic level. The results obtained were highly encouraging and provided solid confirmation of the efficacy of PLGA@QT NPs for biomedical application [164].

Rosmarinic acid (RA) loaded chitosan-coated mucoadhesive nanoemulsions were optimized by authors for nasal administration. The optimal proportion among the mixture's components that produced the smallest polydispersity index and droplet size, as well as the greatest ζ-potential and rosmarinic acid amount, was found using Box-Behnken design (BBD). The optimal values were 3.10 lecithin to oil phase ratio (weight by weight), 8.5% oil phase (weight/volume), and 0.1% chitosan final concentration (weight/volume). There were tests for *in vitro* release, physicochemical characterization, mucoadhesion evaluation, and permeation/retention. Rosmarinic acid nanoemulsions coated with chitosan showed enhanced rosmarinic acid penetration and retention across the swine mucosa of the nose, which also exhibited satisfactory physicochemical properties, extended drug release duration, robust mucoadhesive capability, and long-lasting penetration durations. Furthermore, fibroblast cell survival and necrosis were evaluated to examine the formulation's safety. After treatments for either 24 hours (3.125–50 μM) or 48 hours (3.125–25 μM), the formulations did not cause cytotoxicity. Overall, the data indicated that optimized formulation could be a successful delivery method for nasal therapy of RA intended to promote neuroprotective treatments [165].

9.5.1.8 Thymoquinone

A rich source of antioxidant, Thymoquinone, has been incorporated in nanoformulations used for curing AD [76], a few of which are given below:

Ismail *et al.*, determined how peroxidation of lipids, total antioxidant levels, memory and learning were affected by thymoquinone (TQ) and thymoquinone-rich fraction, expression of antioxidant genes, and levels of soluble beta-amyloid (Aβ) in rodents given a high-fat and cholesterol diet. A supercritical fluid extraction technology was used to extract the fraction rich in thymoquinone from Nigella sativa seeds. The resulting nanoemulsion was subsequently dubbed thymoquinone-rich fraction nanoemulsion. Meanwhile, thymoquinone was synthesized into a thymoquinone nanoemulsion via commercial acquisition. The thymoquinone-rich fraction and T conventional emulsions, respectively, were investigated as a point of comparison. The control medications included probucol, a medication that lowers cholesterol without statins and donepezil, a drug for AD, ranging from moderate to severe. The Sprague Dawley rats received oral gavage treatment every day for the final three months of their six-month treatment with the intervention groups after being fed HFCD.

Consequently, the HFCD-fed rats developed hypercholesterolemia, cognitive impairment, reduced levels of overall antioxidant status, elevated lipid peroxidation, soluble Alpha-beta plaques, and reduced antioxidant gene expression. Thymoquinone-rich fraction nanoemulsion showed similar findings to the other treatment groups, control drugs, and serum biomarker levels in memory and learning tests. In some way, thymoquinone-rich fraction nanoemulsion was more prevalent in brain indicators relative to the protein levels and expression of genes in the intervention groups as well as the control drugs. The authors concluded that adding thymoquinone-rich fraction nanoemulsion to an HFCD may help with the peroxidation of lipids, soluble Alpha-beta levels, memory impairment, overall antioxidant state, and the way antioxidant genes are expressed [166].

Strong phytoconstituent thymoquinone (TQ) from Nigella sativa is known for its neural activity through mechanisms based on antioxidant effects; however, due to its enhanced lipophilicity and ability to bind with plasma proteins, which reduces its bioavailability and eliminates any significant therapeutic effect, it has not been formulated into any conventional dosage form. In a unique endeavor, the authors synthesized Polysorbate-80 coated thymoquinone PLGA nanoparticles (P-80-TQN) for targeting the drug to brain and assessed their effectiveness in AD. The TQN was developed using a modified single-emulsion solvent-evaporation process, and Polysorbate-80 was then used to create brain-targeted P-80-TQN. The biochemical Super-Oxide Dismutase (SOD) enzymatic test and the behavioral Despair Test in Streptozotocin (STZ)-induced Alzheimer's mouse models were used to assess the efficacy of TQ delivered from P-80-TQN in AD at 5 mg/kg TQ equivalent. After giving mice 5 mg/kg equivalent P-80-TQN for 28 days, it raised functioning of SOD (8.33 ± 2.61 units/mg) and greatly decreased the IT (39.45 ± 3.32 s). While the *in vivo* TQ release from P-80-TQN was noted and justified as AUC (0-600mins) and Cmax, which were observed to be 4.5 and 4.3 times respectively, as well as Tmax which was thrice faster as compared to TQ normal, an *in vitro* biphasic release of TQ (59.10 ± 2.10) from P-80-TQN was seen over a 24-hour period. The histological investigation revealed decreased and solvated protein aggregates (<30 μm), demonstrating the efficacy of P-80-TQN against AD. The authors concluded that in AD-affected mice, TQ in P-80-TQN form reduced elevated levels of SOD [167].

9.5.1.9 *Memantine*

It has been incorporated as an antioxidant in various nanoformulations used for treating Alzheimer's disease [80]. Memantine (MEM), in-spite of being approved for AD, has not shown complete effectiveness for mild to severe Alzheimer's disease. This issue was effectively resolved by using polylactic-co-glycolic (PLGA) nanoparticles, which increased the drug's potency at the desired site while lowering adverse effects. The authors added memantine to polyethylene glycol (PEG)-surface-coated biodegradable poly(lactic acid) nanoparticles (PLGA) produced by the double emulsion approach. The MEM–PEG nanoparticles of PLGA had a mean particle size of 152.6 ± 0.5 nm, a zeta potential of − 22.4 mV, and a polydispersity index, PI < 0.1). The NPs' physicochemical characterization verified that the drug in crystalline form was distributed throughout the matrix of PLGA. Astrocytes cells and neuronal lines to MEM-PEG NPs of PLGA (bEnd.3) were observed to be cytotoxic in nature. Because memantine released from the NPs more slowly than it did from the free drug solution, fewer *in vivo* drug administrations were necessary. Nanoparticles demonstrated the potential to traverse the BBB. Transgenic APPswe/PS1dE9 mice were used for behavioral assessments, and the results showed that using MEM–PEG NPs of PLGA instead of the drug free solution, increased the effect of reducing memory impairment. Histopathological studies revealed that MEM-PEG NPs of PLGA decreased beta-amyloid plaques as well as inflammation with them. Memantine loaded NPs exhibited enhanced effectiveness in comparison to the drug in its free form for treating AD [168].

A memantine-loaded nanoemulsion intended for intranasal administration for bypassing the blood-brain barrier in AD therapy was formulated by Kaur *et al.* The homogenization and ultrasonication techniques were used to develop the nanoemulsion. The nanoemulsion was examined for antioxidant capability, *in vitro* release, and characterization. Technetium pertechnetate was used to radiolabel the memantine for the *in vivo* investigations. Approximately 11 nm of particle size and 99% transmittance were observed in the completed NE. 80% of the medication was released in simulated nasal fluid, according to *in vitro* release experiments. Antioxidative tests confirmed that the memantine encapsulation in a nanoemulsion maintained its antioxidant activity, with 98% cell survival. The intranasally treated rats' brains exhibited a greater absorption of the formulation, as demonstrated by gamma imaging and biodistribution data, with a percentage radioactivity of $3.6 \pm 0.18\%/g$ at 1.5 hours [169].

9.5.1.10 Tacrine

It has evolved as a good antioxidant [82]. Tacrine (TAC) was approved for use in therapy; nevertheless, because of hepatotoxicity from oral administration, it is imperative to minimize its adverse events.

PAMAM versions 4.0 and 4.5 dendrimers (DG4.0 and DG4.5) are used as single nanodrugs as well as drug delivery vehicles. In a study done by Igartúa *et al.*, the authors co-administered PAMAM dendrimer and TAC. FTIR, stability of drugs, and *in vitro* release of drugs were used to study the TAC and dendrimers interactions. Zebrafish larvae, Neuro-2a cell culture, and human red blood cells were used to evaluate the toxicological profile of dendrimer co-administered. Additionally, cell culture was used to investigate the anti-acetylcholinesterase action. It is possible to suspend DG4.0- and DG4.5-TAC without affecting the drug's ability to dissolve. The electrostatic nature of the way that tacrine and DG4.5 interacted was verified by Fourier transform Infrared spectroscopy along with *in vitro* release investigations. While the contemporaneous injection of DG4.5 decreased the cytotoxicity of tacrine on the Neuro-2a cell line, human red blood cells did not exhibit any adverse effects. Furthermore, to decrease the structural and hepatotoxic impacts of tacrine in zebrafish larvae, DG4.5-tacrine and DG4.0-tacrine were administered jointly *in vivo*. Since tacrine still has anti-acetylcholinesterase action when given with dendrimers, the decrease in toxicity did not result in a decrease in activity. In conclusion, co-administering tacrine in conjunction with DG4.0 and DG4.5 represents a novel treatment strategy, as it possesses comparable efficacy to the free medication while being less toxic and more biocompatible [170].

Jogani and coworkers prepared and characterized of tacrine's microemulsion/mucoadhesive microemulsion (TME/TMME) was done with the intention of evaluating the pharmacokinetic and pharmacological properties of the material for improving memory and focusing on the brain in mice with scopolamine-induced amnesia. The titration approach was used to produce and characterize the TME. Using 99m Tc as a marker, tacrine solution and formulations' biological distribution after parenteral and intranasal administration doses was assessed. Drug targeting effectiveness, direct medication delivery to the brain through nose, and pharmacokinetic characteristics were computed from the data. In order to verify the location of the medication in the rabbit brain, gamma scintigraphy was used. Selective nose-to-brain transfer was shown by shorter Tmax values of 60 min after intravenous dosing as opposed to intranasal injection (120 min). When compared to intranasal tacrine solution, the brain tacrine bioavailability following intranasal TMME was observed to be two times greater, suggesting a broader extent of drug distribution to the brain with intranasal TMME. Additionally, after intranasal medication treatment, rabbit brain scintigraphy demonstrated increased drug absorption into the brain. The findings showed that following intranasal TMME, tacrine was rapidly and extensively transported into the mouse brain, resulting in the quickest recovery of memory loss in scopolamine-induced amnesic mice. Therefore, the findings suggest that intranasal tacrine delivery has potential to treat AD patients [171].

9.5.1.11 Green Tea and Epigallocatechin-3-Gallate

These both antioxidants [99], have been incorporated in combination or alone in nanoformulations against Alzheimer's disease. Studies have confirmed the strong antioxidant, metal chelation, and anti-amyloidogenic effects of epigallocatechin-3-gallate (EGCG). However, because of its low bioavailability and stability, its therapeutic potential is constrained *in vivo*.

Green tea polyphenol (-)-epigallocatechin-3-gallate (EGCG) has recently shown encouraging preclinical results in mouse induced with Alzheimer's disease as well as in HIV-associated dementia (HAD); however, the translation into clinical use has proven difficult, mainly due to low bioavailability and ineffective central nervous system (CNS) delivery. Although EGCG is well known for its antioxidant qualities, the reports in literature have demonstrated that it can also upregulate alpha-secretase, which in turn enhances non-amyloidogenic processing of amyloid precursor protein (APP). This prevents the formation of brain beta amyloid plaque, which is a common finding in HIV infection and a hallmark of AD pathology. In a pilot study, the authors looked at how well one preformulation technique may increase EGCG's oral bioavailability. It has been observed that the formation of nanolipidic EGCG particles enhanced the alpha-secretase boosting ability of neural SweAPP N2a cells by 91% (P<001) *in vitro* and increased its oral *in vivo* bioavailability by more than double compared to free EGCG [172].

An investigation by Singh *et al.*, attempted to assess the neuroprotective function of EGCG nanoparticles (nanoEGCG) against $A\beta_{42}$ fibrillation produced by Al(III) *in vitro*. Following synthesis, the physiochemical characterization of nanoEGCG was carried out. The stability of nanoEGCG in artificial digestive fluids, as well as its *in vitro* release patterns and metal chelation capability, were assessed. Using Al(III) induction and nanoEGCG therapy, the anti-amyloidogenic ability of nanoEGCG on $A\beta_{42}$ secondary structure and its shape was assessed. Additionally, the impact of $A\beta_{42}$ on cellular toxicity was evaluated. NanoEGCG were formulated with a spherical to slightly elliptical particle shape with 300 nm diameter, exhibiting encapsulation effectiveness of 96%. When released in simulated gastric juices, the EGCG release from the nanoparticle happened steadily. Over time, nanoEGCG's metal chelation and antioxidant capacities were observed to be more than those of its free form. Treatment with nanoEGCG was seen to effectively suppress both Al(III)-induced $A\beta_{42}$

fibrillation and $A\beta_{42}$ itself. This was accomplished by producing fibrils and soluble $A\beta_{42}$ amorphous aggregates rather than insoluble $A\beta_{42}$ oligomers. When nanoEGCG was used as a treatment, a noticeable reduction in toxicity to the cells was observed. The authors concluded that EGCG nanoparticles can prevent $A\beta_{42}$ fibrillation caused by Al(III) and lessen its neurotoxicity *in vitro* [173].

The leading neurodegenerative illness, AD, is brought on by a build-up of amyloid-β (Aβ) plaque deposits in the brain. There is mounting evidence that by preventing Aβ aggregation, epigallocatechin-3-gallate (EGCG) can partially shield cells from Aβ-mediated neurotoxicity. In a study, the authors surface bonded EGCG to selenium nanoparticles (EGCG@Se) to lessen EGCG cytotoxicity at high doses and to understand the processes involved in Aβ aggregation and amyloid fibril disaggregation in a better way. The authors developed Tet-1-EGCG@Se. This was done because Tet-1 peptide is involved in not only in preventing oxidation but also causes neuroprotection, both of which play a crucial role in prevention and advent of AD. The findings demonstrated that Tet-1-EGCG@Se may efficiently impede Aβ fibrillation besides breaking down Aβ fibrils into safe deposits [174].

9.5.1.12 Rivastigmine

It has been used as antioxidant [73] therapy in AD. Owing to the enormous demand for one of the best drugs for treating Alzheimer's disease, rivastigmine, liposomal dispersion of the drug was created by Nageeb *et al.*, 2017. Lecithin, Didecyldimethyl ammonium bromide (DDAB), and polysorbate 80 were mixed in a 1:0.02:0.25 molar ratio with 2 mol% PEG-DSPE. Maintaining a stable formulation is a significant challenge when preparing liposomes, in order to overcome this challenge, the author developed electrostatic stealth (ESS) liposomes, which were aimed to boost liposomal stability by adding DDAB to offer electrical stability and PEG-DSPE to improve stability by steric hindrance. A medium-sized liposome of 478 ± 4.94 nm was produced, exhibiting an entrapment percentage of 48 ± 6.22 with a zeta potential of -8 ± 0.2 mV. After storing for three months, stability experiments revealed insignificant changes in terms of polydispersity index, particle dimension, surface charge potential, efficiency of entrapment, or *in vitro* release, which validated the effective development of steady liposomes. No histopathological alteration was seen in the nasal mucosa of sheep electrostatic stealth liposomes. In contrast to the drug solution, which only penetrated 28.7% of the sheep's nasal mucosa, 48.6% of the medication entered the animal when electrostatic stealth liposomes were administered. Following the completion of the *in vivo* pharmacokinetic studies on thirty-six rabbits, it was concluded that RHT intranasal solution had a relative bioavailability of 486% in brain, while the formulated electrostatic stealth had a relative bioavailability of 424.2% in blood [175].

9.5.1.13 Piperine

It is an alkaloid-based antioxidant which is utilized in AD [75]. Its application is however limited by its final delivery issues.

Etman *et al.*, prepared microemulsions (ME) as oral Piperine (PIP) nanocarrier for treating Alzheimer's disease and assessed their efficacy for the brain and other internal tissues. The authors enhanced PIP targeting to the brain by using bioactive surfactants at standard, safe dosages. Transcutol HP as a co-surfactant and Caproyl 90 as oil phase, Tween

80/Cremophor RH 40 as surfactant were the components of ME systems. The formulations that were generated had a particle diameter of less than 150 nm and a negative zeta potential. The *in vivo* findings demonstrated that ME had a stronger impact than free PIP. Toxicological data suggested that there could be nephrotoxicity from ME. When compared to free medication, oral microemulsion improved PIP's effectiveness and brain delivery, leading to a superior therapeutic result [176].

9.5.1.14 α-Bisabolol

It has been used as a potent antioxidant [77], to treat AD.

In a study done by Sathya *et al.*, the authors made SLNs loaded with α-bisabolol using the hot homogenization process and assessed their neuroprotective impact on Neuro-2a cells. Through DLS analysis, the physicochemical characterization of α-bisabolol-SLNs revealed a comparatively small size with a zeta-potential value which was stable. These findings suggest that α-bisabolol-SLNs will have improved drug delivery capabilities at the target location. α-bisabolol-SLNs had high stability and were amorphous in form, as demonstrated by the PXRD and FTIR study. α-Bisabolol-SLNs exhibited antioxidant potential and also markedly suppressed AChE. Moreover, the Neuro-2a cells were protected against the neurotoxicity and aggregation of Aβ, by the α-bisabolol-SLNs, as confirmed by the thioflavin T test, fluorescence microscopy, and FT-IR analysis. Therefore, the current study's findings demonstrated α-bisabolol's therapeutic effectiveness and bioavailability to treat AD had improved [177].

9.5.1.15 Vitamin D-Binding Protein

In a study, researcher incorporated vitamin D-binding protein [79] as an antioxidant in nanoformulation. Amyloid beta (Aβ) peptide aggregation and accumulation is thought considered be the primary cause behind Alzheimer's disease's (AD) pathogenesis. Vitamin D-binding (DBP) protein has the ability to decrease the accumulation and disintegration of Aβ. It is possible to control the release rate of therapeutic compounds from dosage form containing poly(PLGA) (D,L-lactic acid-co-glycolic acid), a biocompatible polymer. Using a PLGA-based drug delivery approach, the authors investigated the potential therapeutic benefits of nanoparticles of DBP-PLGA in mice overexpressing Alpha-beta (5XFAD). The characteristics of the resultant DBP-PLGA nanoparticle mixture were investigated when D-binding protein was incorporated into PLGA nanoparticles. The authors found that DBP-PLGA nanoparticles greatly suppressed Alpha-beta aggregation *in vitro* using a thioflavin-T test. Furthermore, the results discovered by authors suggested that the 5XFAD mice's Aβ buildup, neuroinflammation, neuronal death, and cognitive impairment were all markedly reduced by an intravenous infusion of DBP-PLGA nanoparticles. All things considered, the authors findings pointed to the possibility suggesting D-binding protein-PLGA nanoparticles could represent an effective treatment for AD management [178].

9.5.1.16 Zinc

It has been incorporated as antioxidant [81] and used in the treatment of AD. Alzheimer's disease (AD) has been linked to a regional dyshomeostasis of zinc ions surrounding amyloid aggregates which is attributed to the zinc accumulation in elderly plaques. Although, elevated brain zinc concentrations may help recover some pathological changes brought on by local zinc shortage, they may also encourage the beta-amyloid accumulation (Aβ plaques). Zinc may also enable in AD brains, the quantity of synapses regulates. In a study, zinc-loaded nanoparticles that were capable to reach the blood-brain barrier and transport zinc to the brain were formulated, the authors also investigated whether changing amounts of Zinc could preserve and cure AD. Using wild-type (WT) and APP23 mice, the authors assessed synapse loss, inflammation, and plaque load *in vivo*. In addition, the authors conducted behavioral analyses. The data demonstrated a considerable decrease in the size of the plaque and the impact on cytokines involved in inflammation, examples of which include Interleukin IL-6 and Interleukin IL-18 following a 14-day chronic administration of these nanoparticles. At the behavioral level, the authors were unable to identify any detrimental consequences of elevated zinc amounts in the neural system of APP23 mice. Further, it was observed that hyperlocomotion in animals was resumed. Thus, the authors concluded that intentional elevation of brain zinc levels might be beneficial in AD [179].

9.5.1.17 Osthole

It is a coumarin based antioxidant [83], explored for AD cure. An example of a coumarin chemical and possible Alzheimer's disease (AD) medication is osthole, or Ost. However, Ost's solubility, absorption, and poor blood-brain barrier permeability restrict its efficacy. The authors created osthole liposomes in this study using surface-modified CXCR4 abbreviated as CXCR4-osthole-Liposomes and explored the distribution of the liposomes in amyloid precursor protein-SH-SY5Y cells. Additionally, research (*in vitro* as well as *in vivo*) was done on the neuroprotective properties of CXCR4-osthole-Liposomes. The findings confirmed that CXCR4-osthole-Liposomes had a cytoprotective effect *in vitro* and enhanced intracellular uptake by APP-SH-SY5Y cells. The distribution of osthole in the brain revealed that CXCR4-osthole-Liposomes enhanced the amount of osthole in the brain and extended the mice's cycle duration. Furthermore, CXCR4-Osthole-Liposomes improved osthole's ability to relieve diseases associated with AD. These results imply that CXCR4-modified liposomes might function as an osthole carrier in Alzheimer's disease therapy [180].

9.5.1.18 Morin Hydrate

It has been reported to have antioxidant action, when given in nanoformulation against AD. In a study done by Sharma *et al.*, the authors formulated microemulsion containing morin hydrate for intranasal delivery to be used in Alzheimer's disease treatment. Following intranasal administration, optimized morin hydrate loaded microemulsion was shown to have greater blood and brain drug concentrations than plain morin hydrate. A microemulsion containing morin hydrate delivered intranasally caused a significant reduction ($p<0.05$) in the evaluated pharmacodynamic parameters as compared to a control group (Sham group). Wistar rats with dementia, which will caused by STZ showed a substantial improvement in memory following 21 days of daily chronic treatment with a microemulsion loaded with morin [181].

9.5.1.19 Genistein

There have been reports of neuroprotective and antioxidant properties for genistein [87]. Despite positive *in vitro* and *in vivo* results, genistein's therapeutic application is restricted by a variety of issues, including as limited solubility in water, rapid metabolic rate, and low bioavailability via the mouth.

Rassu *et al.*, developed sodium hexametaphosphate-based chitosan nanoparticles loaded with genistein for intranasal medication administration using ionic gelation technique. Following their *in vitro* description, the potential for cytotoxicity of the nanoparticles was investigated using PC12 cells. The authors found that sodium hexametaphosphate was a suitable substitute for established cross-linkers. The production process and composition of the formulation had an impact on the physical stability and properties of the nanoparticles. Compared to pure genistein, the created tiny and uniform nanoparticles of mean diameters of 200–300 nm, could enhance genistein diffusion through the mucosa of nose. The nasal fluid's pH was found to be comparable to the dispersions of nanoparticles, and the viability of PC12 cells was preserved [182].

9.5.1.20 Melatonin

Melatonin is used as an antioxidant used against Alzheimer's disease [88]. Exogenous melatonin (MT) has significant benefits in protecting neuronal cells in patients with Parkinson's and Alzheimer's disease.

In a study, the authors used central brain microdialysis to examine the nasal transport of MT as a polymeric gel solution to the brain in rats who have been given anesthesia. For nasal delivery, micronized MT solutions were produced utilizing polymeric compounds, carboxymethyl cellulose, and polyethylene glycol 400. The formulations' *in vitro* penetration was assessed using the three-dimensional EpiAirway(TM) tissue culture model. Brain microdialysis was used on male Wistar rats to examine the central brain transport of nasally given MT gel solutions into the olfactory bulb. Using electrochemical detection. The determination of the MT concentration of microdialysis specimens was conducted using the technique of high-performance liquid chromatography (HPLC). MT dosage forms delivered by the nose to the brain were compared to MT solution given intravenously. When compared to the control, PEG400, MT suspensions in vehicles with carboxy methyl cellulose and carbopol had substantially greater permeability over Epiairway (TM) ($P < 0.05$). Lastly, studies using microdialysis revealed that nasal MT therapy increased rat brain levels [183].

9.5.1.21 Berberine

It is an antioxidant used in AD [90]. Lohan *et al.*, produced multiwalled nanotubes of carbon (MWCNTs) coated with polysorbate and phospholipid that are loaded with berberine (BRB), to effectively treat Alzheimer's disease (AD). The optimized formulation demonstrated 186 nm of size of the particles and 96% of release of the drug. A 36 percentage of degree of carboxylation was found. Studies using FTIR and FESEM verified that polysorbate and phospholipid were coated on the side walls of MWCNTs. *In vivo* pharmacokinetic investigation conducted in rats confirmed a notable improvement over pure medication in terms of the absorption rate and extent of drugs in the cerebral cortex and blood. The Morris Maze test was used to confirm the increased performance efficiency. Furthermore, compared to other groups, the MWCNTs encapsulated with polysorbate and phospholipid demonstrated an improvement in memory performance between the 18th and 20th day which was significantly high. The authors concluded that BRB's polysorbate/phospholipid coated MWCNTs had potential for curing AD [184].

9.5.1.22 Galantamine Hydrobromide

It is an antioxidant used in AD [91]. However, its limited brain penetration causes a reduction in the target site's bioavailability. Misra *et al.*, formulated solid-lipid nanoparticules of galantamine hydrobromide using biodegradable and biocompatible components to overcome its drawbacks. The developed formulation had drug entrapment of 83.42 ± 0.63% at a size smaller than 100 nm. Over 90% of the drug exhibited *in vitro* from these spherically-shaped drug-loaded nanoparticles over the course of a controlled 24-hour period. When compared to a naïve medication, *in vivo* assessments showed that cognitive deficit mice were significantly capable of restoring their memory. The bioavailability of the created carriers was roughly double that of the unmodified medication. Consequently, solid-lipid nanoparticles of galantamine hydrobromide may be a useful drug administration system, especially for ailments like AD [185].

9.5.1.23 Estradiol

It has proved to be a potent antioxidant [92], due to a phenolic ring and explored for Alzheimer's disease treatment. Mittal *et al.*, created tween 80 coated polylactide-co-glycolide nanoparticles that would allow estradiol to be administered orally to the brain. Single emulsion was used to create estradiol-containing nanoparticles. They demonstrated stability in both artificial intestinal fluid (SIF) and artificial gastric fluid (SGF). When tween 80 encapsulated nanoparticles were given orally at a dosage of 0.2 mg/rat, the brain levels of estradiol of 1.969 ± 0.197 ng/gram tissue were achieved after 24 hours, which were considerably greater than that for uncoated formulation (1.105 ± 0.136 ng/gram tissue), indicating a major function for surface coating. The fact that these brain estradiol levels were very similar to those acquired after administration of an identical dose of drug suspension through a intramuscular route with 100% bioavailability (2.123 ± 0.370 ng/gram tissue) provided a proof that a greater amount of drug was reaching the brain after oral administration of formulation. Furthermore, the group that received nanoparticle treatment was able to stop the hippocampal area of the brain from expressing amyloid β-42 ($A\beta_{42}$) immune-reactivity [186].

9.5.1.24 Basic Fibroblast Growth Factor

It has reported to be a good antioxidant [93]. Nanoparticles of polyethylene glycol-polylactide-polyglycolide (PEG-PLGA) treated with lectins were formed by Zhang *et al.*, to enhance the transportation of drug to the brain after nose-to-brain administration. The formulation encapsulated basic fibroblast growth factor (bFGF) and transferred it to the brain using nanoparticles in conjunction with Solanum tuberosum lectin (STL). N-acetylglucosamine on the olfactory epithelial membrane is the only binding site for STL. The nanoparticles had a uniform size of particles and a negative surface charge potential. Using the radioisotope tracing technique, following intranasal administration, the formulations' distribution throughout the brain was assessed. The Mirror water maze test was used to evaluate the neuron's protective effects of intracerebroventricular doses of beta-amyloid$_{25\text{-}35}$ as well as ibotenic acid in rats. The memory and spatial cognition of AD rats in the STL-bFGF-NP group was improved as compared to the AD model group. The findings aligned with the rat hippocampal region's histology observations and the rat hippocampal choline acetyltransferase activity value. The histopathology examinations confirmed the safety of STL-bFGF-NP *in vivo*. The results clearly confirmed the efficacy of Solanum tuberosum lectin NP as a vehicle for the therapeutic introduction of proteins and peptides, such as bFGF, into the nervous system of the body [187].

9.5.1.25 Silica

The silica is used as an antioxidant against Alzheimer's disease [94]. The high absorption efficiency of silica nanoparticles (SiNPs) in the cells make them useful for BBB targeting. These nanoparticles, which aggregate in internal amyloid cells ($A\beta_{1\text{-}42}$), help to cure AD. The generated SiNPs have the dose-related capacity to lower cellular mortality and produce reactive oxygen species (ROS) inside the intracellular space. The two main features of AD are elevated phosphorylation and amyloid peptide build-up. SiNPs' effectiveness in reducing the development of $A\beta_{1\text{-}42}$ plaque and hyperphosphorylation was examined by Yang *et al.* [33]. Selenium (II), sodium selenite (VI), and sodium selenite (IV), trace minerals that serve as antioxidants, are essential for reducing the potential harm that reactive oxygen species (ROS) might do to cells. Among the various vital micronutrients with antioxidant qualities are selenite and selenium (Se). Elevated antioxidant effects have been demonstrated by Se-NPs and selenite NPs [188].

9.5.1.26 Sialic Acid

It is used against Alzheimer's disease, due to its antioxidant efficacy [95]. In a study reported, the authors synthesized selenium (Se) nanoparticles which were modified by sialic acid (SA) and coupled them with a different peptide, B6 peptide (B6-SA-SeNPs, an artificial version of selenoprotein), which has a high BBB permeability and might be employed as a potential nanomedicine to treat AD. Inductively linked plasma-atomic emission spectroscopy, flow cytometry analysis, and laser-scanning confocal microscopy ICP-AES revealed that cerebral endothelial cells absorbed B6-SA-SeNPs at an elevated cell level (bEnd.3). In a Transwell experiment employing an *in vitro* model of BBB, the migration effectiveness of B6-SA-SeNPs was evaluated. It confirmed that B6-SA-SeNPs penetrated the blood-brain barrier, where PC12 cells subsequently absorbed them. Furthermore, B6-SA-SeNPs were shown to have inhibitory effects on fragmentation of amyloid-beta peptide (Aβ) in PC12 as well as bEnd3 cells. B6-Sialic acid-Se nanoparticles were able to disintegrate produced Aβ strands into granular oligomers that are safe in addition to efficiently inhibiting Aβ aggregation. These findings showed that B6-SA-SeNPs would offer a potential strategy, for treating brain diseases using nanoparticles [189].

9.5.1.27 Ceria Dioxide

It has been incorporated into nanoformulation due to its antioxidant [96] potential and explored for treatment of AD. Mitochondrial oxidative stress is one of the primary pathogenic variables in Alzheimer's disease. Aberrant production of reactive oxygen compound may result in neural cell death brought on by mitochondrial malfunction. The ability of cerium (CeO_2) nanoparticles to transition between the oxidation states of cerium(3+) and cerium(4+) allows them to act as potent and reusable ROS scavengers. Thus, specifically targeting mitochondria with ceria nanoparticles may be a viable treatment option for neurological illnesses.

In a study done by Kwon *et al.*, the authors developed triphenylphosphonium-linked ceria nanoparticles, which were tested in a 5XFAD transgenic Alzheimer's disease mice model. The mice's morphological mitochondrial damage and reactive gliosis were lessened by the triphenylphosphonium-conjugated nanoparticles of ceria. The findings of the study suggested that triphenylphosphonium-conjugated nanoparticles of ceria might be a beneficial therapy for oxidative stress in mitochondria, which is associated with Alzheimer's disease [96].

9.5.1.28 Erythropoietin

It has been explored as an antioxidant [97] with AD treating potential. One promising neuroprotective options for treating neurological diseases like Alzheimer's dementia (AD) is erythropoietin (EPO), a hematopoietic factor. In the event of systemic treatment, EPO was incapable of crossing the blood-brain barrier entirely due to high molecular weight, hydrophilicity, as well as quick clearance from circulation.

In a study reported, the authors generated EPO-loaded Solid Lipid Nanoparticles (EPO-SLN) via a double emulsion solvent evaporation process (W1/O/W2) in order to get over this constraint. Dichloromethane (DCM), tween®80, span®80/span®60, glycerin monostearate (GMS), and solvent were selected as the lipid, internal phase surfactants, external aqueous phase surfactant, and solvent, respectively. Following physico-chemical analyses, the impact of EPO-SLN on the animal model of beta-amyloid-induced AD was studied. Using the Morris water maze test *in vivo* assessments, it was shown that the rats with cognitive deficits treated with EPO-SLN had much better memory than the rats treated with native medication. Furthermore, compared to free EPO, EPO-SLN was more efficient in reducing oxidative stress, the ADP/ATP ratio, and beta-amyloid plaque formation in the hippocampal regions. As a result, the proposed SLN was confirmed by authors to be a viable method for the efficient as well as safe EPO delivery for the treatment of AD [190].

9.5.1.29 Aβ Generation Inhibitor S1 (PQVGHL Peptide) and Curcumin

Vaz *et al.*, formulated nasal lipid nanocarriers loaded with CUR to efficiently reach brain in order to overcome such limitations. Using the hot solvent diffusion linked with the phase inversion temperature procedures, the lipid nanocarriers (NE) were created. The nanocarriers' morphological, physico-chemical, and *in vitro* drug release analyses were done. Using pig nasal mucosa, the CUR permeation and retention was examined in a Franz-type diffusion cell. Histopathological investigations and confocal laser scanning were also carried out. The findings demonstrated that the NE sizes had negative zeta potential and varied in size from 18 to 44 nm. The CUR concentration varied between 0.24 and 1.50 mg/mL, exhibiting a 99% encapsulation effectiveness. The CUR release profiles suggested a biphasic kinetics. The level of CUR-NE penetration into the nasal mucosa of the pig was greater than that of free CUR. The validity of these results was confirmed by confocal microscopy examination. Furthermore, a histological study revealed no harm on the nasal mucosa. The chance that CUR will permeate through the pig nasal mucosa is increased by such an encapsulation [191].

9.5.1.30 Ligustrazine Phosphate

It has been incorporated into nanoformulation due to its antioxidant [86] potential. Shi *et al.*, used ligustrazine phosphate (LP) to treat AD. The ethosomes for transdermal delivery for LP was formulated. For the *in vitro* permeation investigations, Franz-type diffusion cells along with confocal laser scanning microscopy were employed. Additionally, the Morris water maze test was used to check the behavioral performance of the scopolamine-induced amnesia rats with intention to evaluate the impact of the LP transdermal ethosomal system on AD. Rat brain levels of the lipid peroxidation product malondialdehyde (MDA) and the activity of antioxidant enzymes were also measured. The outcomes demonstrated that the LP ethosomal system's skin medication deposition and penetration capacity were both noticeably enhanced. Improvements in behavioral performance were also indirectly indicative of the antioxidant enzyme activities and MDA levels in the brains of the rats with amnesia being restored to normalcy by the LP transdermal ethosomal system. In conclusion, the authors stated that ethosomes may be the preferred vesicles for LP transdermal distribution, and LP may provide a viable therapeutic option in the battle against AD [192].

9.5.1.31 Naringenin

It is employed in AD [98], due to its antioxidant potential. A bioflavonoid with anti-inflammatory and antioxidant qualities, naringenin has neuroprotective benefits. Unfortunately, because it transports inefficiently across cellular membranes, its clinical use is restricted.

In a work reported, a human neuroblastoma cell line (SH-SY5Y) was used to assess the neuroprotective benefits of a produced naringenin nanoemulsion for treating β-amyloid-induced neurotoxicity. The optimized formulation of the naringenin-loaded nanoemulsion was investigated using photon correlation spectroscopy as well as transmission electron microscopy, yielding droplet sizes of 113.83 ± 3.35 nm and about 50 nm, respectively. The preparation exhibited a zeta potential, percentage transmittance and polydispersity index of 0.312 ± 0.003, 97.01%, 12.4 ± 1.05 respectively. The capacity of naringenin nanoemulsions to shield SH-SY5Y neuroblastoma cells from the neurotoxic effects of beta amyloid (Aβ) was used to measure their neuroprotective efficacy. It was also established that Aβ promoted the formation of total tau, phosphorylated tau (pT231), amyloid precursor protein (APP), β-secretase (BACE), and reactive oxygen compounds (ROS). The direct neurotoxic effects of Aβ on SH-SY5Y cells were considerably decreased by the naringenin-encapsulated nanoemulsion; this was correlated with a down-regulation of APP and BACE expression, suggesting a decrease in amyloidogenesis. Moreover, a reduction in phosphorylated tau in SH-SY5Y cells exposed to Aβ was observed. These observations confirmed that nanoemulsion laden with naringenin may be a potential therapy for Alzheimer's disease [98].

9.5.1.32 *Donepezil*

It has been employed for treating Alzheimer's disease [100], due to its antioxidant potential. Kaur *et al.*, formulated oil in water (o/w) nanoemulsion (NE) loaded with donepezil an antioxidant Labrasol (10%), cetyl pyridinium chloride (1% in 80% water), and glycerol (10%) were used to develop the NE, which had a drug concentration of 1 mg/ml. Particle size, zeta potential, and polydispersity index (PDI) were measured for the formulated NE. Studies on *in vitro* release were carried out to monitor medication release. Technetium pertechnetate (99mTc) tagged formulations were used in additional *in vivo* tests of developed NE on Sprague Dawley rats to study the intranasal route for brain delivery. The nanoemulsion had a zeta potential of -10.7 mV, a PDI of 0.084, and a particle size of 65.36 nm. A release of 99.22% in 4 hours in phosphate-buffered saline, 98% in 2 hours in artificial cerebrospinal fluid, and 96% in 2 hours in simulated nasal fluid were observed in *in vitro* release investigations. The NE's antioxidant and cytotoxic properties exhibited cytotoxicity, which was dependent on dose and percentage radical scavenging activity (%RSA). The proposed formulation did not affect the morphology of cells, as verified by giemsa staining photographs developed. Scintigrams confirmed the maximal NE intake by brain. It was concluded that created donepezil hydrochloride NE may offer a novel method for treatment of Alzheimer's disease by delivering medication [193].

9.5.1.33 *Sesamol*

A common characteristic of AD and other long-term neurological illnesses is neuroinflammation. Sesamol functions as both an antioxidant [101] and a strong inhibitor of cytokine synthesis. In an investigation, the authors determined if sesamol, when added to solid lipid nanoparticles (SLNs), could effectively address the cognitive abnormalities due to ICV-STZ in rats. Streptozotoçin (STZ)-induced intracerebroventricular (ICV) impairment of cognition has been extensively utilized as model of AD. Significant cognitive impairments were seen following ICV-STZ treatment, as measured by the Morris water maze and raised plus maze tasks. These deficits were accompanied by considerably higher blood TNF-α levels, changed acetylcholinesterase in the rat brain, and dramatically enhanced nitrodative stress. In ICV-STZ rats, a long-term therapy with sesamol and sesamol-loaded SLNs mitigated nitrodative stress and cytokine production while also dose-dependently restores cognitive impairments. Sesamol was effectively delivered to the brain via SLNs, as evidenced by a noticeably improved reduction in oxidative stress markers. The results showed that loading sesamol into SLNs is a useful strategy to reduce neuronal dysfunction and memory impairments caused by ICV-STZ [101].

9.5.1.34 *Ginsenoside Rg3 and Thioflavin T*

It has been employed in nanoformulations for treating AD [102], because to its antioxidant potential. Besides the increased production, the accumulation of β-amyloid (Aβ) peptide 1-42 ($A\beta_{1-42}$) is known to be the initiator of the pathogenic cascade that leads to cognitive impairment in AD. A key component of ginseng, ginsenoside Rg3 is believed to provide antioxidant effects by reducing radicals which are free and which are crucial in memory and cognition. Consequently, ginsenoside Rg3 seems like a good option to treat AD by acting as a neuroprotective agent. A unique approach to nanotherapy that increases the ginsenosides movement across the BBB to enhance their delivery to the brain would promote neuroprotection and prevent the build-up of plaques of Aβ and eventual degeneration of neurons. In a work, the authors synthesized and evaluated poly(lactic-co-glycolic acid) (PLGA) nanoparticles (NPs) encapsulating ginsenoside Rg3 along with Thioflavin T, and also investigated the NPs' neuroprotective potential as well as their capacity to traverse blood-brain barrier using an BBB *in vitro* model. Based on findings, the authors concluded that for the purpose of identifying and treating AD, our PLGA-Rg3 NPs present an innovative novel theranostic agent that can encapsulate natural nutraceuticals [102].

9.5.2 Parkinson's Disease

Parkinson's disease, the second commonly prevalent neuronal disease after AD, is characterized by increasing stiffness, bradykinesia, along with tremor. Pathologically, Lewy bodies and the loss of pigmented neurons in the substantia nigra in the midbrain are seen [194, 195]. As of 2006, about 4 million individuals worldwide, with an average age of 60, were living with Parkinson's disease (PD); the prevalence of PD is more in men as compared to women [196, 197]. Higher estrogen concentrations may be the cause of the smaller impact in females [198]. Numerous findings from post-mortem investigations showed that a number of processes, such as dysfunction of mitochondria, oxidative stress, inflammation of neurons, excitotoxicity, and the build-up of proteins which are misfolded as a result of proteasomal and autophagic disorders, are connected to apoptosis or necrosis [199].

The various antioxidants, which have been loaded in nanoformulations and used in Parkinson's disease are given below:

9.5.2.1 Resveratrol

Resveratrol (RSV) has demonstrated encouraging neuroprotective properties through its anti-inflammatory as well as antioxidant properties. It is employed in nanoformulations used to treat Parkinson's disease [68], due to its antioxidant potential. The BBB, which restricts its accessibilty to the CNS, is the main barrier which has to be circumvented for improving its efficacy.

Palle *et al.*, formulated resveratrol nanoparticles (NRSV) and assessed their efficacy in rats using rotenone-induced neurodegeneration model. The temperature-controlled antisolvent precipitation technique was used to develop the NRSV, which were then evaluated for size, shape, and characteristics during dissolution. Additionally, NRSV effects were compared with RSV, or free resveratrol. Four groups of animals were used: I was the control group; II was the rotenone (2 mg/kg s.c.); III was the RSV (40 mg/kg, p.o.) + rotenone group; and IV was the NRSV (40 mg/kg, p.o.) + rotenone group. For a duration of 35 days, the animals were given treatments 30 minutes before the administration of rotenone. The rotarod test and rearing behavior were used to quantify behavior 24 hours following the previous dosage. The midbrains of rats were removed in order to evaluate complex-I activity, oxidative markers like glutathione (GSH), catalase, and lipid peroxidation (LPO), and the tricarboxylic acid cycle enzymes. Furthermore, histological investigations were carried out. The findings demonstrated that long-term rotenone therapy resulted in mitochondrial malfunction, oxidative stress, motor impairments, and reduced rearing behavior. Additionally, in rats given rotenone, histological examination revealed neuronal damage. The current study's key discovery was that NRSV was much more effective than RSV therapy in reducing the behavioral abnormalities, histological and biochemical changes, oxidative stress, and dysfunction of mitochondria that are caused by rotenone in rats with PD [200].

Wang *et al.*, designed design and fabrication of Fe_3O_4-modified Res liposomes (Res-lips@Fe_3O_4), a magnetic targeting drug nanocarrier. Using magnetic resonance imaging, the fractional anisotropy (FA) values and T2 relaxation time of the rats treated with Res-lips@ Fe_3O_4 were determined. Res-lips@Fe_3O_4 demonstrated robust magnetic response qualities and good drug capability and stability. *In vitro*, Res-lips@Fe_3O_4 showed prolonged and gradual drug release. Under an external magnetic field, *in vivo* tests demonstrated that the Res-lips@Fe_3O_4 could efficiently penetrate the blood-brain barrier and raise the drug content at the target region. T2 relaxation duration and FA values also showed that the magnetic Res liposomes had the best therapeutic effect [201].

In Parkinson's disease (PD) inflicted mice, the neuronal protection effects of polysorbate 80 (PS80)-coated poly(lactide) nanoparticles loaded with resveratrol (RVT) were examined by da Rocha *et al.* The results were compared with bulk RVT. Mice (C57BL/6) were given single intraperitoneal injections of 1-methyl-4-phenyl-1,2,3,6-tetrahydropyridine (MPTP), a toxin that affects dopaminergic neurons and results in Parkinson's disease. Significant reduction in olfactory discrimination as well as social recognition recall were brought about by MPTP. Additionally, striatal oxidative stress and decreased expression of tyrosine hydroxylase were seen in the striatum. The behavioral and neurochemical alterations generated by MPTP were significantly protected against by drug entrapped nanoparticles. These findings suggested that poly(lactide)-nanoparticles coated with PS80 and loaded with might be a useful adjuvant treatment for Parkinson's disease [202].

In another study, in order to increase resveratrol diffusion into the brain, the authors developed RSV-entrapped nanoparticles of PLGA (RSV-PLGA-NPs) linked to lactoferrin (Lf) and evaluated the neuroprotective benefits of RSV in PD animal models. Compared to RSV-PLGA-NPs and free RSV, the Lf-conjugated RSV-PLGA-NPs (Lf-RSV-PLGA-NPs) showed improved localization into SH-SY5Y as well as human brain microvascular endothelial cells. Furthermore, attenuating the MPP+-induced formation of reactive oxygen compounds, lowering of membrane potential of mitochondria, and death of cells was more successful with Lf-RSV-PLGA-NPs as compared to RSV-PLGA-NPs and free RSV. Crucially, bioluminescent imaging investigations revealed that Lf conjugation selectively enhanced the RSV-PLGA-NPs accumulation in the brain. In the MPTP-induced PD model, the developed formulation significantly enhanced the neuroprotective effects of RSV. Therefore, Lf-RSV-PLGA-NPs are a potentially useful technique for enhancing RSV bioavailability and brain neuroprotection [203].

In another study Wang *et al.*, evaluated the protective impact of resveratrol generated from Polygonum cuspidatum in liposomal on PD rats' nigral cells. Through unilateral microinjection of 6-hydroxy dopamine in the striatum, a rat PD model was created. Next, for two weeks in a row, P. cuspidatum along with its liposomal formulation were given intraperitoneally daily once. The behaviors, ROS level, apoptotic cells, tyrosine hydroxylase positive cells, and total antioxidant capacity were observed. Following a 14-day oral treatment using resveratrol or resveratrol liposome (20 mg/kg per day), the findings demonstrated a significant improvement in the overall antioxidant potential of nigral tissues as well as a marked reduction in the abnormal rotational behavior, apoptosis, and nigral cell loss. Additionally, resveratrol liposomes had more significant benefits than resveratrol in its free form. The antioxidant and radical-scavenging qualities of resveratrol produced from P. cuspidatum as well as its liposomal formulation may contribute to the protection of the dopaminergic neurons in Parkinson's disease (PD) mice. Resveratrol liposome provides more effective protection than free resveratrol, possibly because of its higher bioavailability. As a result, it shows promise as a therapy for PD [204].

Pangeni *et al.*, formulated an oil-in-water (o/w) nanoemulsion by utilizing vitamin E:sefsol (1:1), Tween 80 and Transcutol P as the oil phase, surfactant, and co-surfactant respectively. A spontaneous emulsification technique was utilized to formulate the nanoemulsion, which was then homogenized under high pressure. Ternary phase diagrams were created to pinpoint the nanoemulsion's location. The produced nanoemulsions were examined for surface properties, *in vitro* studies, *ex vivo* release, globule size, zeta potential, refractive index, and viscosity. Spherical shaped particles having 102 ± 1.46 nm as diameter, 0.158 ± 0.02 polydispersity index, and -35 ± 0.02 mv zeta potential were observed in the formulation containing 150 mg ml(-1) of resveratrol after homogenization. After a 24-hour period, the pre-homogenized resveratrol suspension as well as nanoemulsion, and nanoemulsion after homogenization had cumulative percentage drug releases of 24.18 ± 2.30%, 54.32 ± 0.95%, and 88.57 ± 1.92%, respectively. Additionally, the release *ex vivo* revealed a cumulative percentage release of drug during 24 hours of 85.48 ± 1.34%. For the improved formulation, the DPPH assay revealed good scavenging efficiency in the antioxidant activity. Pharmacokinetic analyses revealed that the improved nanoemulsion was administered intranasally, resulting in a greater drug concentration in the brain (brain/blood ratio: 2.86 ± 0.70). The results of histopathology revealed that the groups receiving resveratrol nanoemulsion had less degenerative alterations. In the group treated with resveratrol nanoemulsion, the concentrations of GSH, SOD, and MDA were considerably lower and higher, respectively [205].

9.5.2.2 Ropinirole

In order to cure PD, ropinirole (RP), an antioxidant was administered using a novel drug delivery system made of poly (D,L-lactide-co-glycolide) (PLGA) nanoparticles (NPs). The optimal dosage form showed zero-order *in vitro* RP release for 5 days, mean encapsulation efficiency of 74.8 ± 8.2%, average size of particles less than 155 nm, and -14.25 ± 0.43 mV zeta potential. Male Wistar rats were administered daily intraperitoneal doses of the neurotoxic rotenone (2 mg/kg), which caused PD-like behavioral and neural alterations. Animals were given RP in saline (1 mg/kg/d for 35 days) or entrapped in PLGA NPs (amount of NPs equivalent to 1 mg/kg/d RP every 3 days for 35 days) once neurodegeneration had established (15 days). The results of behavioral tests and histology of brain and immunochemistry demonstrated that RP-loaded PLGA NPs could reverse neurodegeneration symptoms in the *in vivo* model examined [206].

In another study, the authors fabricated and assessed new surface-modified polymer-lipid hybrid nanoparticles (PLN) as durable delivery systems for ropinirole hydrochloride (ROPI HCl) administered intranasally. PLN were developed using the emulsification-solvent diffusion approach, and they were assessed for stability, mucosal toxicity, *ex vivo* permeation, *in vitro* diffusion, and physicochemical properties. The study utilized the Box-Behnken experimental design technique to study the impact of two independent variables on the produced PLN's particle size, zeta potential, and entrapment efficiency: the concentration of stearylamine and surfactant (Pluronic F-68). The optimum formulation sample was chosen using numerical optimization techniques, which was then validated using regression equations and three-dimensional response surface plots. The spherical shape and encapsulation of the drug's amorphous form within the PLN system were shown by DSC and SEM investigation. The nasal mucosa's integrity did not reveal any significant symptoms of degradation, and the PLN formulation showed excellent retention. The non-Fickian super case II diffusion mechanism with zero order kinetic model best described the pattern of release of the formulation. To evaluate the therapeutic effectiveness of the developed nasal PLN formulation, pharmacodynamic experiments were done. The findings suggested that the PLN had the potential to be a stable and safe carrier for ROPI HCl intranasal administration, particularly in the treatment of PD [207].

9.5.2.3 Naringenin

Oxidative stress is associated with Parkinson's disease. A strong antioxidant called naringenin (NAR) has been demonstrated to help with experimental Parkinson's disease. Due to its antioxidant potential [98], it has been incorporated in nanoformulation against PD. However, owing to its limited solubility in water, insufficient permeability, and less bioavailability brought on by gastrointestinal degradation, NAR's clinical development has been hindered.

Gaba *et al.*, formulated a naringenin (NRG) nanoemulsion (NE) encapsulating vitamin E- that can be directly used for nose to the brain to improve Parkinson's disease (PD) treatment. Multiple behavioral investigations (narrow beam test, grip strength test, muscle coordination test, forced swimming test, and akinesia test) in a rat model were used to evaluate the efficacy of tailored NE against PD. A droplet size of 38.70 ± 3.11 nm, a PDI of 0.14 ± 0.0024, a refractive index of 1.43 ± 0.01, a transmittance of 98.12 ± 0.07 %, a zeta potential of -27.4 ± 0.14 mV, and a viscosity of 19.67 ± 0.25 Pa s were measured in the optimized NE. Rats in which PD was induced by 6-OHDA, were successfully treated when given NRG NE intranasally in addition to levodopa, according to behavioral tests. The rats given treatment with NRG NE combined with levodopa via intranasal route had much lower malondialdehyde (MDA) levels, but enhanced quantities of GSH and SOD. Positive findings from the study suggest that the delivery method of NRG may be effective in Parkinson's disorder [208].

Md *et al.*, formulated chitosan nanoparticles (NAR NPs) containing naringenin for transport from the nose to the brain. In SH-SY5Y cell line the uptake, cellular toxicity, as well as neuroprotective effects of NAR NPs were investigated. Ionic gelation was used to develop NAR NPs, which were then examined using a zetasizer, transmission electron microscopy (TEM), and field emission microscopy (FESEM). The 24 hour *in vitro* release profile, zeta potential, entrapment efficiency, average particle size, polydispersity index (PDI), and 15.36 ± 2.05 mV were recorded as 87.6 ± 8.47 nm, 0.31 ± 0.04, 91.12 ± 2.99%, and 54.80 ± 4.22%, respectively. It was observed that NPs had a percentage of 67.90 ± 0.72% of NAR penetration through nasal mucosa. Fluorescence microscopy was used to confirm that the produced NPs were absorbed by cells. Reactive oxygen compounds (ROS) levels and viability tests were used to assess the neuroprotective efficacy of NAR NPs. In SH-SY5Y cells, NAR NPs exhibited improved neuroprotection and antioxidant activity against neurotoxicity caused by 6-OHDA [209].

In another study, after screening antioxidants using in silico and Drosophila experiments, Mani *et al.*, selected Naringenin to investigate its possible neuroprotective effects in mice after administration of rotenone (ROT). Naringenin was incorporated into solid-lipid nanoparticles (SLN) and assessed for improved brain bioavailability. Particle size average was 134.5 ± 20 nm, polydispersity index was 0.893, zeta potential was −11.9 mV, and entrapment efficiency was 89.87% ± 0.15%. With an R^2 of 0.9721 for the regression coefficient, the Higuchi model described the *in vitro* release profile when utilizing the dialysis procedure. Findings from biomarkers and behavioral observations led to theory that naringenin SLN, exhibits neuroprotection that may slow PD advent. The authors suggested that to fully grasp the potential of antioxidants in cutting-edge delivery methods and to encourage the large-scale, reasonably priced production of such formulations, more research is necessary [210].

9.5.2.4 Curcumin

It has anti-inflammatory, antioxidant, as well as a neuroprotective properties. In research on PD, rotenone that induces neurotoxicity has been investigated extensively. Bollimpelli *et al.*, evaluated the effect caused on dopaminergic cell line SK-N-SH by lactoferrin nanoparticles loaded with curcumin. These curcumin-loaded nanoparticles had a diameter of 43–60 nm and a hydrodynamic size of about 100 nm, respectively. 61.3% ± 2.4% was the encapsulation efficiency. Utilizing these nanoparticles, the cellular absorption of curcumin was verified via spectrofluorimetric analysis and confocal imaging. Compared to their soluble equivalent, the lactoferrin nanoparticles loaded with curcumin demonstrated higher intracellular drug absorption, prolonged retention, and improved neuroprotection. Viability tests and ROS level estimation were used to characterize neuroprotective activity. The authors came to conclusion that curcumin-loaded lactoferrin nanoparticles might offer a viable medication delivery method for preventing dopaminergic neuron damage [211].

9.5.2.5 Tannic Acid and Ferulic Acid

The brain's resident macrophages, or microglia, contribute to inflammation of neurons. Microglia in response to oxidative stress or some stimuli present outside cell get activated, which exacerbates inflammation of neurons. Furthermore, microglia, the first line of defense in the central nervous system, are important for the removal and destruction of αSYN. Despite the widespread recognition of microglia's involvement in neuronal degeneration diseases, limited treatment options have been explored to address activation of microglia. In this context, Zhao *et al.*, developed nanoparticles (NPs) that would administer antioxidants that block aggregation in order to improve aggregation of αSYN and lessen the pro-inflammatory microglial phenotype activation. Tannic acid (TA) and ferulic acid diacid with an adipic acid linker (FAA) were used as the shell and core to create NPs via flash nanoprecipitation. Using a Thioflavin T test, the NPs had a robust inhibitory impact on αSYN fibrillization, greatly reducing αSYN fibrillization *in vitro* in comparison to αSYN, which were untreated. Applying NPs to microglia resulted in a reduction of both intracellular αSYN oligomer formation and total αSYN internalization. Moreover, NP treatment reduced the *in vitro* release of pro-inflammatory cytokines TNF-α and IL-6 and inhibited the generation of reactive oxygen compounds and nitric oxide triggered by αSYN. Additionally, NP therapy dramatically lowered nuclear factor kappa B (NF-κB) nuclear translocation and reduced expression of Iba-1 in αSYN-challenged microglia. All things considered, the study establishes the preliminary study for a pathogenic protein aggregation and neuroinflammatory target for an antioxidant-based nanotherapeutic option in neurodegenerative illnesses [212].

9.5.2.6 *Cerium Oxide*

Due to its antioxidant potential [96], it has been used in Parkinson's disease. Cerium oxide nanoparticles, or CONPs, have been studied for treating Parkinson's disease (PD). Even in certain animal models of Parkinsonism and other neurodegenerative illnesses, cerium oxide nanoparticles (CeO_2NPs) demonstrated encouraging results.

The purpose of an investigation reported was to determine if treating rats with CeO_2NPs might improve their neurobehavioral and neurobiochemical deficiencies caused by 6-OHDA-induced neurotoxicity. 50 rats were injected left intrastriatally (IS) with either 6-OHDA (n = 40) or saline (control, n = 10). Neurobehavioral tests were used to confirm motor impairment three weeks after the injury. Then, for three weeks, 10 sick rats each were administered injection intraperitoneally of either a vehicle or CeO_2NPs (0.1, 0.5, or 1 mg/kg). Rats underwent behavioral evaluations before being killed for striatal biochemical tests. Assessments included stratothallic dopamine levels, oxidative stress indicators such as malondialdehyde (MDA) and total antioxidant capacity (TAC), and activity of caspase 3 as an apoptotic marker. The motor dysfunctions caused by 6-OHDA injection were varyingly improved by different dosages of CeO_2NPs in the open field, Rota Rod, and stepping tests. Furthermore, the 0.5 mg/kg dose of CeO_2NPs nearly corrected the neurobiochemical disturbances, but the 0.1 mg/kg dose was insufficient to change the striatal biochemical measures. One milligram per kilogram of CeO_2NPs was administered, and while apoptosis was reduced and striatal dopamine was somewhat improved, oxidative stress was not significantly affected. In treating 6-OHDA-induced Parkinsonian mice, the current work demonstrated a potential therapeutic role for CeO_2NPs and proposed their antioxidant and antiapoptotic properties as potential explanations for increased striatal dopamine levels and enhanced motor function [213].

In another investigation, rats with haloperidol-induced Parkinson's disease (PD) were given CONPs intranasally to help in reduction of the oxidative stress caused by free radicals. The ferric reducing antioxidant power (FRAP) test was used to evaluate the antioxidant capability of the CONPs *in vitro*. Using goat nasal mucosa, the penetration and local toxicity of the CONPs were assessed *ex vivo*. Rats were also used to study the acute local toxicity of intranasal CONPs. In order to evaluate CONPs' focused brain delivery, gamma scintigraphy was employed. Rats were used in acute toxicity tests to prove that intranasal CONPs are safe. Furthermore, the effectiveness of intranasal CONPs in a rats given haloperidol to induce PD was evaluated using the open field test, pole test, biochemical estimates, and brain histology. At 25 µg/mL, the produced CONPs exhibited the maximum antioxidant activity, according to the FRAP assay. The goat nasal mucus layers exhibited not only a deep but also uniform distribution of CONPs, as demonstrated by confocal imaging. After using improved CONPs, the goat nasal membrane showed no symptoms of irritation or damage. Acute toxicity research proved the safety of intranasal CONPs, while scintigraphy investigations in rats established their focused brain delivery. Rats administered intranasal CONPs exhibited a significantly substantial ($p < 0.001$) enhancement in locomotor activity in comparison to rats who were not treated, as demonstrated by the results of the open field and pole test. Additionally, the rat brain histopathology in the therapy group revealed more viable cells and less neurodegeneration. Following the intranasal administration of CONPs, there was a significant reduction in the amount of thiobarbituric acid reactive substances (TBARS), a significant enhancement in catalase (CAT), superoxide dismutase (SOD), and glutathione (GSH), and a significant reduction in the amounts of interleukin-6 (IL-6) and tumor necrosis factor-alpha (TNF-α). Furthermore, the dopamine content in the intranasal CONPs was substantially higher ($p < 0.001$) at 13.93 ± 0.85 ng/mg protein than in the rats treated with haloperidol (5.76 ± 0.70 ng/mg protein). Overall, the findings suggested that intranasal CONPs may be a safe treatment option for Parkinson's disease [214].

9.5.2.7 N-Acetylcysteine

One of the main factors influencing the onset and course of Parkinson's disease (PD) is oxidative stress. N-acetylcysteine (NAC), an antioxidant, [103] has drawn attention as a potential treatment for Parkinson's disease (PD); nevertheless, its short half-life, low bioavailability, and restricted brain penetration make it impractical.

The objective of a study reported was to utilize cellular rotenone PD model, to test nanocarriers which target mitochondria for the NAC delivery both alone and in conjunction with deferoxamine (DFO), a chelator of iron, and to evaluate the capacity of these nanocarriers to guard against oxidative stress. Dequalinium (DQA) and pluronic F68 (P68) nanocarriers were developed using an altered thin-film hydration technique. Ferrozine assays and ferritin immunoassays were used to quantify iron status and evaluate cell viability, respectively, in an MTT assay. Three assays were used to measure oxidative stress: the mitochondrial hydroxyl test, the thiobarbituric acid-reactive compounds assay, and a modified cellular antioxidant activity assay. The work showed that NAC and NAC + DFO can be successfully formulated into P68 + DQA nanocarriers for neuronal delivery for the first time. The authors concluded that these NAC nanocarriers show great promise for being moved to additional preclinical testing for Parkinson's disease [215].

9.5.2.8 Selegiline

An inhibitor of monoamine oxidase B (MAO-B), selegiline for Parkinson's disease. Due to its antioxidant potential [105], it has been explored for treating PD. However, because of its poor water solubility and substantial liver metabolism, it has 10% bioavailability after oral administration, which is the primary issue with oral administration. In a study, the authors developed a nanoemulsion containing selegiline for direct nasal administration to the brain, improving PD treatment. The production and optimization of nanoemulsion were done using a statistical multivariate method that applied a quality by design (QbD) methodology. The effectiveness of the improved formulation in treating Parkinson's disease was evaluated in Wistar rats using behavioral investigations. The polydispersity index (PDI), transmittance, zeta potential, viscosity, and droplet size of the selegiline nanoemulsion were determined to be, respectively, 61.43 ± 4.10 nm, 0.203 ± 0.005, 1.30 ± 0.01, 99.80 ± 0.04%, -34 mV, and 31.85 ± 0.24 mPas. Studies confirmed that the particles in nanoemulsion were spherical and that drug penetration was enhanced by 3.7 times when compared to drug solution. Behavior tests revealed that when selegiline nanoemulsion was given intranasally to treat rats with haloperidol-induced PD, there was a notable betterment of the rats' behavioral activities compared to when the medication was taken orally. It was stated that nanoemulsion appeared to be a potentially effective novel drug delivery vehicle for selegiline intranasal administration for Parkinson's disease therapy [216].

9.5.2.9 Catalase

In recent years, exosomes—naturally occurring nanosized vesicles—have garnered a lot of interest as potential drug delivery systems. Bilayers of lipids which occur naturally with a large number of sticky proteins which easily connect with membranes of the cells make up exosomes. Haney *et al.*, formulated a novel exosomal-based method of delivering catalase, an antioxidant, to treat Parkinson's disease (PD). The catalase-loaded exosomes (exoCAT) that were obtained had a size between 100 and 200 nm. Exosomes that were reconstituted by sonication, extrusion, or saponin permeabilization showed great efficiency of loading, prolonged release, and protection of catalase against degradation by protease. Neuronal cells absorbed exosomes easily *in vitro*. Exosomes existed in significant quantities in the brain of PD mice after intranasal delivery. ExoCAT significantly improved neuroprotection in both *in vitro* and *in vivo* Parkinson's disease models. All things considered, the authors concluded that exosome-based catalase compositions may prove to be an effective treatment for inflammatory and neurological conditions [217].

9.5.2.10 Pramipexole Dihydrochloride

In a study reported, ionic gelation was used to encapsulate pramipexole dihydrochloride, an antioxidant, into chitosan nanoparticles (P-CNs) for delivery via intranasal route in PD treatment. Particle size and entrapment efficiency of optimized P-CNs containing chitosan and sodium tripolyphosphate (6:1 v/v) were 292.5 nm ± 8.80 and 91.25% ± 0.95, respectively. After 24 hours, the diffusion of the P-CNs across the goat nasal mucosa and artificial membrane was found to be 93.32% ± 2.56 and 83.03% ± 3.48, respectively. The fact that P-CNs particles were spherical was shown by transmission electron microscopy, whereas scanning electron microphotographs revealed rough surface morphology. In pharmacodynamic investigations, the P-CN treatment group showed a lower motor activity and a better photoactometer score when compared to solution (nasal) or commercial tablets. P-CNs also markedly raised brain levels of dopamine and enhanced antioxidant effect as demonstrated by increased catalase and superoxide dismutase activity. Thus, it can be said that, intranasal distribution of Pramipexole-loaded chitosan nanoparticles demonstrated crucial *in vitro* features with better *in vivo* efficacy [218].

9.5.2.11 Ceria Dioxide

Due to its antioxidant potential [96], it has been used to treat PD. Reactive oxygen compounds (ROS)-induced oxidative stress is key elements in the development of many illnesses. However, a complete knowledge of the pathogenic consequences of ROS is limited by the absence of appropriate ways to scavenge ROS based on their cellular location.

Kwon *et al.*, checked three distinct kinds of ceria nanoparticles (NPs) may be used to selectively scavenge mitochondrial, within cell, and extracellular ROS, and how this might be useful in PD. According to authors findings, scavenging of within cell or ROS in mitochondria prevents lipid peroxidation and microglial activation while shielding the tyrosine hydroxylase (TH) present in the striata of mice used PD models. These findings highlight the critical roles that intracellular and mitochondrial ROS play in the development of PD. Based on findings of the study, the authors concluded that ceria NP systems should prove to be a valuable instrument in clarifying the roles of different ROS in illnesses [219].

9.5.2.12 Dopamine

To transfer dopamine to the brain, Pahuja *et al.*, formulated neurotransmitter containing PLGA nanoparticles (DA NPs). These nanoparticles demonstrated decreased quinone adduct production, decreased dopamine autoxidation, and a delayed and continuous release of dopamine in plasma. In the substantia nigra and striatum, areas impacted by Parkinson's disease, dopaminergic SH-SY5Y cells and dopaminergic neurons absorbed DA NPs. DA NP treatment did not result in decreased viability of cells or changes in morphology of SH-SY5Y cells, in contrast to cells treated with dopamine that displayed decreased viability. In rats given 6-hydroxydopamine (6-OHDA), the formulation was successful in traversing the blood-brain barrier and the endothelium of capillary in the striatum as well as substantia nigra. In the striatum of Parkinson induced rats, systemic intravenous injection of DA NPs resulted in markedly elevated quantities of dopamine as well as its metabolites and decreased super sensitivity of dopamine-D2 receptor. Furthermore, in 6-OHDA-induced parkinsonian mice, DA NPs dramatically corrected neurobehavioral deficits. In comparison to rats lesioned with 6-OHDA, dopamine administered by NPs did not result in increased ROS production, degeneration of dopaminergic neurons, or structural alterations in the striatum and substantia nigra. It's interesting to note that dopamine distribution by nanoformulation did not result in any rapid changes in the brain or any other peripheral organ, nor did it produce changes in blood pressure or heart rate. These findings imply that dopamine was supplied by NPs to the brain, which in turn prevented the toxicity caused by dopamine autoxidation and eventually corrected the neurochemical and behavioral abnormalities in parkinsonian rats [220].

9.5.2.13 Gallic Acid

Neurodegenerative illnesses, including PD, are largely caused by the 140-residue protein α-synuclein (αSN) aggregating. Gallic acid (GA), an antioxidant [110], has been shown in earlier studies to prevent αSN aggregation. Since endothelial cells are lipophilic, which prevents GA from being transported to the brain, the authors entrapped GA onto human serum albumin (PEI-HSA) nanoparticles which were coated with polyethylenimine and examined the characteristics of developed nanoparticles. The authors found that drug in PEI-HSA NPs (PEI-HSA-GA NPs) reduced the interaction of NPs' with αSN and stabilized αSN in the unfolded conformation. PEI-HSA-GA NPs suppressed αSN aggregation in a concentration-dependent way, to the same amount as free GA, while αSN aggregation rate increased when PEI-HSA NPs plus free GA (PEI-HSA NPs + GA) were present owing to the impact of PEI-HSA NPs. Furthermore, GA prevented PEI-HSA-GA NPs from interacting with calcein-filled vesicles, showing that adding GA can lessen PEI-HSA NPs' toxicity. Furthermore, PEI-HSA NPs reduced the quantity of hazardous oligomers while simultaneously preventing αSN oligomers from disrupting the membrane. Based on the findings, the authors concluded that PEI-HSA-GA NPs are viable options for effective GA brain delivery [221].

9.5.2.14 *Hydroxytyrosol*

In Parkinson's disease models, hydroxytyrosol (HT), a naturally occurring phenolic antioxidant [111], exhibits neuroprotective properties. However, HT does not enter the brain at the desired quantities needed to provide a clinical impact because of problems including fast metabolism. Mursaleen *et al.*, formulated micellar nanosized carriers using dequalinium (DQA) and Pluronic F68® (P68) that are ideal for delivering iron chelators and antioxidants to the brain. They further evaluated the physical properties and blood–brain barrier-crossing capacity of the P68 + DQA nanoformulations for HT taken individually or in conjunction with deferoxamine (DFO), a chelator of iron as well as their capacity to provide protection against rotenone in hCMEC/D3-SH-SY5Y cells. The HT and formulations containing showed significant encapsulation efficiencies (up to 97%) and were both smaller than 170 nm. The average blood–brain barrier (BBB) transfer of HT was improved by 50% by the P68 + DQA nanoformulation. In comparison to the equivalent free drug treatments, this led to significantly improved protection against cytotoxicity induced by rotenone and oxidative stress by 12% and 9%, respectively. The authors concluded that the formulation shielded against oxidative damage by delivering the formulated delivery system across a BBB model [222].

9.5.2.15 *Graphene Oxide*

In a study, the authors used the reserpine PD model to examine the effectiveness of graphene oxide nanoparticles (GONPs). GONPs have been used as a therapeutic treatment in numerous illnesses due to the degeneration of neurons at low doses. The authors evaluated the neurobehavioral changes in the PD reserpine model and looked into the antioxidant and neuroprotective properties of GONPs in this paradigm. Thirty male mice (n = 10) were divided into three groups: Res (reserpine 0.25 mg/kg); C (control); and Res + GONPs (reserpine 0.25 mg/kg with GONPs 25 mg/kg). According to the findings, the brain and brainstem exhibited a considerable rise in superoxide dismutase (SOD) level, catalase (CAT) level, and malondialdehyde (MDA) level in response to reserpine neurotoxicity, which also caused hypoactivity. The reserpine-induced hypoactivity that was accompanied by lower neuronal CAT and MDA levels was restored by GONPs. According to these results, GONPs may be used to prevent degeneration of neurons in the reserpine induced PD model by acting as an antioxidant agent [223].

9.5.2.16 *β-Carotene*

It is reported to possess antioxidant potential [113]. The Parkinson's disease model, Drosophila melanogaster, is useful for examining putative neuroprotective effects. Chaves *et al.*, divided four-day-old flies into four groups and gave different drugs for seven days: (1) control; (2) rotenone in food (500 μM); (3) β-carotene containing nanoparticles (20 μM); and (4) β-carotene containing nanoparticles with rotenone. The proportion of survival, food consumption, aversive phototaxis, open field, and geotaxis tests were then assessed. Following the behaviors, dopamine and acetylcholinesterase (AChE) in the fly heads were assessed, along with the reactive species (ROS) levels, thiobarbituric acid reactive substances (TBARS) levels, catalase (CAT) levels, and superoxide dismutase (SOD) levels. After being exposed to rotenone, nanoparticles loaded with β-carotene enhanced survival, memory, as well as motor function. They also restored levels of dopamine, activity of AChE, and markers for oxidative stress (CAT, SOD, ROS, and TBARS). It was concluded that nanoparticles loaded with β-carotene had a good impact in providing neuroprotection against damage brought on by the Parkinson-like illness model, suggesting that they may be used as a potential therapy [224].

9.5.2.17 Polydopamine (PDA) with Selenocystine

In the substantia nigra pars compacta (SNpc) of patients suffering from PD, neuromelanin-containing dopaminergic neurons degenerate because of the excess of reactive oxygen compounds (ROS) and cumulative oxidative stress. Melanin-like polydopamine (PDA) has been investigated for its capacity to eliminate reactive oxygen compounds (ROS) through a range of antioxidant enzyme mimic activities, such as catalase (CAT) and superoxide dismutase (SOD). This is because of its redox property. While the preservation of ROS metabolic equilibrium depends on glutathione peroxidase (GPx), few GPx-like nanozymes are being investigated *in vivo*. It is well known that selenocysteine is necessary for GPx's antioxidant function. Therefore, in order to create a nanocomposite (PDASeCys) which possesses GPx-like activity, Wang *et al.*, co-synthesized PDA and selenocystine (SeCys). According to the findings, the PDASeCys nanocomposite exhibited greater efficiency to scavenge free radicals and more GPx enzymatic activity as compared to PDA while having the same CAT and SOD enzymatic activities. In the 1-methyl-4-phenyl-pyridine ion (MPP+)-induced Parkinson's disease model, PDASeCys was successful in efficiently raising levels of intracellular GPx and shield SH-SY5Y neuronal cells from MPP+-induced oxidative stress. When the PDASeCys nanocomposite was injected into the mice substantia nigra (SN), it successfully prevented 1-methyl-4-phenyl-1,2,3,6-tetrahydropyridinium (MPTP)-induced Parkinson-related symptoms [225].

9.5.3 Huntington's Disease

The Huntington gene's aberrant CAG expansions are linked to Huntington's disease (HD), an autosomal neurological condition [226–228]. The gene located in the short arm of chromosome 4's exon 1 of the mutant HTT has polymorphic CAG repeats exceeding 39 [228–230]. In the end, the CAG mutations cause the polyglutamine (polyQ) tracts in the HTT protein to grow abnormally, which causes misfolding as well as the loss of function of protein [231, 232]. The main cause of the degeneration of medium spiny neurons (MSNs) in the striatum has been shown to be the polyQ expansion, which also gradually causes neurodegeneration in the hippocampus, cortex, hypothalamus, as well as brain stem [232–234]. The HD epidemiology indicates that while the illness is present everywhere, the prevalence of HD varies according to geographic location and genetic variety [234, 235]. Compared to the Asian population, the Caucasian population has a significantly greater incidence of HD. India accounts for a sizable portion of all HD cases in Asia, even though estimates indicate that HD is more common and is trending upward in North America,Western Europe, United Kingdom and Australia [236, 237].

Numerous theories have been put out to explain how clinical signs of neurodegeneration in HD occur. Among them, oxidative stress linked to polyQ expansion that triggers caspase-mediated death of neuronal cells is thought to be a possible source of pathological alterations in HD [234]. Because of their unstable reactive nature, free radicals are very reactive which contain unpaired electrons on their valence orbital that can cause a variety of molecular and cellular vulnerabilities [238, 239]. Reactive oxygen compounds and reactive nitrogen species (RNS) are the terms utilized to describe the most prevalent physiologically significant free radicals, which are superoxide (O−), hydroxyl (OH−), and nitric monoxide (NO) species [239, 240]. Enzymatic and nonenzymatic processes, , infections, electromagnetic radiation, immunological changes, intoxication, abnormal diet, malnutrition, and vitamin and trace element deficiencies, can all cause cells to become activated and produce free radicals [241].

The various antioxidants, which have been loaded in nanoformulations and used for HD:

9.5.3.1 Rosmarinic Acid

Bhatt *et al.*, investigated how solid lipid nanoparticles (SLNs) may be used as a medication delivery mechanism to improve ability of rosmarinic acid (RA) to target the brain after intranasal (i.n.) delivery. Using glycerol monostearate (GMS) as the lipid, tween 80 and soy lecithin as surfactants, and hydrogenated soy phosphatidyl choline (HSPC) as a stabilizer, the hot homogenization technique was utilized to prepare the SLN loaded with RA, which were then evaluated. The proposed formulation was administered nasally to Wistar rats, and behavioral and biochemical characteristics were studied thereafter. The tween 80-optimized RA-loaded SLNs (SLNPRT) had an average size of 149.2 ± 3.2 nm, a ZP of -38.27 mV, and an entrapment efficiency of 61.9 ± 2.2%. When compared to the control group, the rat treated with 3-NP exhibited a substantial increase in behavioral changes and oxidative damage. In rats treated with 3NP, SLNPRT therapy greatly reduced oxidative stress and improved behavioral impairments. However, SLNPRT administered by the nose had a far greater therapeutic effect than when administered intravenously. The brain drug concentration was 5.69 μg in the organ distribution study, and the Cmax, tmax, $t_{1/2}$, and AUC values were 1.5 h, 3.17 h, 1.505 μg/ml/h, and 0.284 μg/ml, respectively, in the pharmacokinetic research. The positive outcomes validated the creation of an optimal RA-loaded SLNs formulation, which is a potentially beneficial therapeutic formulation for the management of Huntington disease [242].

9.5.3.2 Curcumin

Finding compounds that support neuroprotective pathways might be a possible new curative strategy for diseases like Huntington's disease. Despite its low absorption, fast metabolism, systemic elimination, and restricted permeability across blood-brain barrier (BBB), curcumin is a well-established antioxidant and neuroprotective agent. Because of its antioxidant potential [74], it is used for the treating HD. Huntington's disease neurodegeneration's precise etiology is still unknown, however mitochondrial failure seems to be a key factor in the illness's pathophysiology. Consequently, methods to lessen mitochondrial deficiencies may offer a possible course of treatment.

Sandhir *et al.*, treated HD caused by 3-nitropropionic acid (3-NP) in rats by using curcumin loaded solid lipid nanoparticles (C-SLNs). Complex II activity was significantly reduced, according to the results of the succinate dehydrogenase (SDH) staining of the striatum and the MTT (3-(4,5-dimethylthiazolyl-2)-2,5-diphenyltetrazolium bromide) test. On the other hand, mice treated with C-SLN not only exhibited a substantial enhancement in levels of cytochrome but also mitochondrial complex activity. Moreover, superoxide dismutase activity and glutathione levels were recovered by C-SLNs. Furthermore, rats given C-SLNs showed a substantial decrease in reactive oxygen compounds, lipid peroxidation, protein carbonyls, and mitochondrial edema. The injection of C-SLNs to rats treated with 3-NP activated the nuclear factor-erythroid 2 antioxidant pathway, as demonstrated by the results of Western blot and quantitative PCR. Furthermore, rats treated with C-SLN exhibited a noteworthy enhancement in their neuromotor coordination in contrast to rats treated with 3-NP. Therefore, the outcomes of the study indicated that administration of C-SLNs might be a potential treatment strategy to improve HD patients' mitochondrial dysfunctions [243].

Pepe *et al.*, formulated curcumin-containing biodegradable nanoparticles that may improve the compound's transport and brain bioavailability. The natural fatty acids in the generated hyaluronan-based materials, which are capable of self-assembling into stable spherical nanoparticles, were chemically attached to the natural polysaccharide, since all of the starting materials for this study were derived from natural resources, the goal was to develop a potentially effective system to deliver curcumin. The aim was that, once the drug has been released at the designated site, the biopolymer will break down into nontoxic fragments prior to renal excretion. In HD *vitro* model, the results showed that curcumin loaded nanoparticles infiltrated the cells and decreased their apoptotic sensitivity [244].

9.5.3.3 Selenium

It has antioxidant potential [95], and therefore explored for treating Huntington's disease. Epigenetic, oxidative stress, metabolic, as well as dietary variables possess a role in the gradual loss of brain function associated with Huntington's disease (HD), an incurable condition. One approach to HD treatment is targeting the suppression of protein aggregation, although this approach has not shown good results. Studies report that people suffering from HD have lower brain amounts of selenium (Se). Improvements in Se homeostasis may decrease loss of neurons as well as dysfunction.

Based on transgenic HD models of Caenorhabditis worms (C. elegans), Cong *et al.*, used selenium loaded nanoparticles (NPs) (Nano-Se) for the treatment of HD illness. Nano-Se NPs at low doses alleviated behavioral dysfunction, dramatically decreased neuronal death, and protected C. elegans from damage due to stress. The mechanism at molecular level also demonstrated that Nano-Se decreased the expression of histone deacetylase family members at the level of mRNA, decreased oxidative stress, and saved huntingtin proteins from aggregation. The findings indicated that Nano-Se had a lot of promise for treating Huntington's illness. The authors concluded that method by which Nano-Se NPs restore brain functions and shield against damage under stressful circumstances will be advantageous for the treatment of HD illness [245].

9.5.3.4 *Fisetin*

There have been several reports of the pharmacological actions of fisetin (FS). Due to its antioxidant potential [115], it has been incorporated in nanoformulation and used for treating Huntington's disease. However, it is not widely used because to its poor oral bioavailability and water solubility.

Kumar *et al.*, attempted to increase fisetin's oral bioavailability since the pharmacological efficacy is significantly influenced by its solubility and bioavailability. The authors formulated a fisetin self-nanoemulsifying drug delivery system (SNEDDS). Rats in which PD was induced by rotenone were used as the subject of pharmacokinetic and pharmacodynamic investigations involving developed SNEDDS. The analysis of pharmacokinetics confirmed that SNEDDS enhanced the mean residence time of the drug in plasma and enhanced the oral bioavailability of FS. The results of ELISA and biochemical estimation supported the significantly better neuroprotection in rats given treatment with FS entrapped SNEDDS in comparison to rats treated with FS alone. According to this study, FS's oral bioavailability was enhanced by SNEDDS, which in turn increased the drug's neuroprotective effects in a PD-affected rat model. It also raises the possibility of using FS-SNEDDS to effectively control Parkinson's disease [115].

9.5.3.5 *Poly(trehalose)*

The main goals of prevention and treatment approaches for different neurodegenerative disorders are to reduce the toxicity caused by protein aggregates, remove aggregates of protein plaques from the brain, and limit protein fibrillation. Singh *et al.*, formulated poly(trehalose) nanoparticles that could stop polyglutamine aggregation in a HD mice, lessen the cytotoxicity resulting from such aggregation, and suppress aggregation of amyloid/polyglutamine in extracellular and intracellular environments. The nanoparticles consisted of a 6 nm core of iron oxide and a polymer shell with approximately 5–12 weight percent covalently attached to trehalose. Their hydrodynamic diameter was 20–30 nm. The engineered poly(trehalose) nanoparticles exhibited 1000–10000 times greater efficacy as compared to molecular trehalose in terms of suppressing mutant huntington aggregates present in the HD mice brain, preventing the polyglutamine-containing mutant huntingtin protein aggregation in model neuronal cells, and inhibiting protein fibrillation in extracellular space. The authors reported that effective brain targeting, entrance inside neuronal cells, and inhibition of mutant huntingtin aggregation depend on zwitterionic charge on the surface of trehalose nanoparticles with a multivalency (number of trehalose molecules per nanoparticle) of around 80–200. The study demonstrated that, in contrast to molecular trehalose, which requires millimollar to molar concentrations, nanoscale trehalose can provide extremely effective anti-amyloidogenic activity at micromolar concentration. The authors concluded that this strategy can be used *in vivo* to treat neurodegenerative illnesses caused by protein aggregation [246].

9.5.3.6 *Thymoquinone*

Its antioxidant potential [76], makes it a potential candidate for treating Huntington's disease. In HD, oxidative stress, excitotoxicity, neuroinflammation, and defective gene transcription are the main disease processes that result in striatal cell loss and aberrant motor behavior. In an investigation done, homogeneous pathology akin to HD was inferred by the utilization of 3-Nitropropionic acid (3-NP), a strong mitochondrial toxin. In the investigation, the 14-day administration of 3-NP was found to foreshadow neuroinflammation, motor impairments, glial cell activation, and stimulation of N-methyl-D-aspartate (NMDA) receptors. By encapsulating the physiologically active substance thymoquinone in solid lipid nanoparticles, a colloidal carrier, the therapeutic approach in the study was modified. When 3-NP-induced animals were given 80 mg/kg b.w. of suspension of thymoquinone (TQ-S) or 10 and 20 mg/kg b.w. of thymoquinone encapsulated solid lipid nanoparticles (TQ-SLNs), their ATPases functioned significantly ($P < 0.01$) better than the TQ-S (40 mg/kg b.w.) treated group. Treatment with TQ-SLNs (10 and 20 mg/kg) also reduced the upregulation of pro-inflammatory cytokines, p-p65 NFκB nuclear translocation, and glial fibrillary acidic protein (GFAP) in mice exposed to 3-NP. Additionally, compared to the TQ-S treated group, the TQ-SLNs therapy enhanced tyrosine hydroxylase (TH) immune reactive neurons, desensitized NR2B-subtype NMDA receptor, and lessened motor abnormalities in rats intoxicated with 3-NP. The findings therefore indicated that therapy with lesser dosages of thymoquinone nanoformulation as compared to thymoquinone solution can effectively terminate the course of 3-NP caused HD in the male Wistar rat striatum [247].

9.5.3.7 *Boehmite*

It has antioxidant potential [117]. To address some of the main causes of Huntington disease, a new amino-nanozyme based on boehmite nanoparticles (BNPs) functionalized with tetra-azapyridinophane (L1) has been developed. L1 has remarkable SOD activity when it forms Cu^{2+} complexes. Additionally, when grafted onto BNPs, it demonstrates mitoROS scavenging capabilities and the capacity to disintegrate mutant huntingtin deposits within cells [248].

9.5.3.8 *Ginger*

Due to its antioxidant potential [118], it has been explored for treating Huntington's disease. By the formulation of chitosan nanoparticles loaded with ginger and assessment of their potential for neuroprotection after treatment with 3-Nitropropionic acid (3-NP) induced HD rats, the aim of a study reported was to boost the bioavailability and brain transport of ginger. Using the ionic gelation process, five distinct formulations (F1-F5) of chitosan nanoparticles loaded with ginger were formulated. Formulations F1 and F3 were selected for physicochemical evaluation based on their release. Rats with Huntington's disease (HD) model were utilized for assessing the neuroprotection offered by formulations F1 and F3 in comparison to ginger extract using behavioral, biochemical, and neurochemical tests. Formulations F1 and F3 exhibited notable and almost identical controlled release. The spherical nanoparticles in Formulation F1 had a negative zeta potential and an ideal size range. Rats given chitosan nanoformulations loaded with ginger showed improved gait, mobility, grip strength, and memory compared to rats given ginger extract, according to the behavioral evaluation. Additionally, biochemical and neurochemical investigations demonstrated that, as compared to ginger extract, ginger-loaded chitosan nanoformulations significantly reduced oxidative stress markers ($p<0.05$). Compared to ginger extract, the ginger nanoformulations significantly reduced the production of free radicals, which boosted the antioxidant enzymes activity. When ginger nanoformulations were given to Wistar rats, their memory and cognition significantly enhanced compared to when ginger extract was given to the same rats ($p<0.05$ because of the suppression of the acetylcholine esterase). The study found that because of their nano size, which allows them to travel across blood-brain barrier and ultimately increases the bioavailability of ginger, chitosan nanoparticles loaded with ginger had a stronger neuroprotective impact than its extract [249].

9.5.3.9 *Manganese*

It is reported to have antioxidant potential [119]. Adhikari *et al.*, demonstrated that a biocompatible nanometric substance based on citrate functionalized manganese (C-Mn_3O_4 NP) may be readily incorporated into the cellular multienzyme cascade for H_2O_2 scavenging, while also successfully mimicking the physiological milieu's glutathione peroxidase enzyme. The mechanism of the activity of the nanozyme is revealed by a thorough computational analysis. The authors also demonstrated the C-Mn_3O_4 nanozyme's *in vivo* therapeutic effectiveness in a preclinical model of HD, a common progressive neurological illness for which no cure exists. The authors demonstrated that, in a preclinical animal model of neurodegenerative Huntington's disease, citrate-functionalized spherical Mn_3O_4 nanoparticles may successfully imitate the glutathione peroxidase (GPX) enzyme without the drawbacks of modern nanozymes. Given that Mn is a necessary micronutrient for mammals and that citrate, a stabilizing ligand, aids in the nanoparticles' ability to travel across blood-brain barrier. It has been demonstrated by the authors that the nanozyme is readily integrated into the cellular antioxidant enzyme cascade. When compared to previous published nanozymes, the cascade's nanozyme's specificity and effectiveness were much greater [250].

9.5.4 Amyotrophic Lateral Sclerosis (ALS)

Lou Gehrig's disease, or ALS, is an age-related degenerative disease of the motor neurons that affects the brainstem, spinal cord, and motor cortex. It is deadly. ALS can be either familial (FALS) or sporadic (SALS). Sporadic ALS has an unknown molecular origin, but it is an incurable illness with a median life expectancy of 3–5 years following detection, while some people may survive considerably longer [251, 252]. Although the term "ALS" is frequently used imprecisely, it officially only refers to a condition that affects the anterior horn cells in addition to the pyramidal tract [253]. Because of this, there are around 5 cases of ALS for every 100,000 people in the US. The term "motor neuron diseases" (MND) can refer to a group of clinically related disorders that show a technically distinct pattern of motor neuron degeneration. Ten to fifteen percent of instances of ALS are familial, or inherited. Twenty to thirty percent of this proportion can be attributed to changes in the antioxidant Cu, Zn-superoxide dismutase (SOD1). In separate FALS-affected families, more than 90 distinct SOD1 mutations have been discovered [253, 254]. Hereditary ALS, with very few exceptions, spreads dominantly, suggesting a gain of toxicity as opposed to a reduction of enzyme activity. Although some mutations, such as the G93A and A4V alterations, indicate a faster rate of illness development, the clinical characteristics of person with FALS are nearly identical to those of an individual with SALS.

A neurodegenerative condition called amyotrophic lateral sclerosis is characterized by gradual muscular weakness brought on by the loss of upper and lower motor neurons. It is believed to be brought on by environmental, age-related, and genetic factors (up to 20 genes connected) [255]. Early pathophysiological process in amyotrophic lateral sclerosis is malfunction of mitochondria; these changes may be found in the mitochondria at several levels, such as ATP generation and respiration, calcium handling, dynamics, and apoptotic signaling [256]. The disease's associated mutant proteins build up in mitochondria and harm them. Therapeutic approaches for significant amyotrophic lateral sclerosis have centered on apoptotic inhibition pathways, ROS reduction, and mitochondrial biogenesis [257, 258]. Transgenic SOD1 mice are the traditional *in vivo* genetic model of amyotrophic lateral sclerosis [258, 259]. N-acetyl-L-tryptophan treatment postponed the beginning of the illness, increased survival, and improved loss of motor neuron, atrophy, as well as performance in transgenic mice in addition to reducing inflammation [260]. The results of the investigation showed that caspase-3 activation was inhibited, Bcl-xL levels were elevated, and cytochrome c/smac/AIF release was decreased [260]. Delivery of medicines related with antioxidant capabilities, [261–264], exhibited minimal impact on span of life whereas creatine protected motor neurons and increased the lives of SOD1 mice by 20% [256, 257]. Riluzole [256] is the only medication given license by the FDA to cure amyotrophic lateral sclerosis. It works by reducing the formation of ROS through glutathione [256].

The various antioxidants, which have been loaded in nanoformulations and used for Amyotrophic lateral sclerosis are:

9.5.4.1 Curcumin

A study reported the development of a carrier-in-carrier dosage form that was injected intravenously. The device was mediated by micelle-loaded mesenchymal stromal cells (MSCs). The secret to first order targeting and second order delivery is MSCs' natural capacity to penetrate wounded tissues, including the CNS and other tissues which are damaged. Because of their size (about 7 nm in diameter), inulin-D-alfa-tocopherol succinate micelles (INVITE M) may dissolve extremely hydrophobic medications and enter cells rapidly. The work showed that the concentration-dependent loading of the micelles loaded with curcumin (INVITE MC), which were sterilized by filtering, achieved its highest loading in MSCs in a matter of minutes. MSCs showed clear signs of cytotoxicity when "naked" curcumin was employed, but INVITE micelles shielded them from this impact. INVITE MC-entrapped MSCs could also release the medication that has been entrapped. The study provided compelling evidence in favor of the carrier-in-carrier approach's viability for treating ALS [263].

The investigation done by Ahmadi *et al.*, aimed to evaluate not only the safety but also the effectiveness of nanosized curcumin as an antioxidant and anti-inflammatory medication in people suffering from amyotrophic lateral sclerosis (ALS). The authors carried out a 12-month, randomized, double-blind, placebo-controlled study at an Iranian neurological referral clinic. A 1:1 randomization process was used to assign eligible individuals with a confirmed or probable diagnosis of ALS to be administered either nanocurcumin (80 mg daily) or a placebo. The randomization list was created with an automated random number generator. The distribution of treatments was hidden from both the study team and the patients. The main result was survival, and an event was one of mechanical ventilation reliance or death. Every patient who was administered at least one dosage of the study medication was selected for the intention-to-treat analysis. 54 participants in all were randomly assigned to be administered either a placebo (n = 27) or nanocurcumin (n = 27). Events happened in six patients (22.2%) belonging to the placebo group and 1 patient (3.7%) belonging to the nanocurcumin group after a 12-month period. A significant difference in the survival curves of the study groups' was found using Kaplan-Meier analysis ($p = 0.036$). Regarding all other outcome variables, there were no noteworthy variations between the groups. There were no recorded significant side effects or treatment-related fatalities. There were no patient withdrawals due to adverse medication events. The findings imply that curcumin in nano size is safe and may increase the likelihood of survival when used in addition to other ALS treatments, particularly for individuals who already have bulbar symptoms [264].

9.5.4.2 Riluzole

The therapy of amyotrophic lateral sclerosis (ALS) presently involves two authorized drugs, one of which is riluzole. Due to its antioxidant potential [120], it has been explored for treating Amyotrophic lateral Sclerosis. Nevertheless, the blood-brain barrier's (BBB) efflux transporters impose barrier on riluzole's brain disposition owing to it being a substrate of P-glycoprotein (P-gp).

Yang *et al.*, suggested using verapamil, a P-gp inhibitor, to decrease efflux pumps in order to create a liposomal co-delivery system that might effectively distribute riluzole to mind cells. Lipid film hydrolysis was used to create cocktail liposomes containing riluzole and verapamil. The polydispersity index (PDI) of cocktail liposomes was 0.272 ± 0.017, and their average size of particles was 194.3 ± 6.0 nm. The cocktail liposomes exhibited encapsulation efficiencies of 86.0 ± 1.4% for verapamil and 85.6 ± 1.1% for riluzole. After 8 hours at 37°C in PBS, 78.4 ± 6.2% riluzole and 76.7 ± 3.8% verapamil were released from cocktail liposomes. After three months, at 4 °C, the average size of liposomes did not exhibit any significant change. In brain endothelial bEND.3 cells, verapamil containing liposomes suppressed the levels of P-gp as determined by western blotting in a dose as well as time dependent manner. In astrocytes C8D1A and bEND.3, more drug efflux transporters were found, which were stimulated by either hydrogen peroxide (H_2O_2) or tumor necrosis factor (TNF-α). The formulated liposome treatments resulted in restoration of accumulations of riluzole and the fluorescent dye rhodamine 123 in bEND.3 cells. It suggested that the co-delivery liposomes inhibitory potential against P-gp may facilitate the transfer of both P-gp substrates. The authors concluded that to improve riluzole pharmacosistance and enhance ALS treatment, liposomes co-loaded with verapamil and riluzole may be utilized [120].

9.5.4.3 FM19G11

The proliferative restorative attempts of ependymal stem progenitor cells (epSPCs), which are ordinarily quiescent in the spinal cord, are linked to neurodegeneration in ALS animals. Therefore, controlling epSPC proliferation might be a useful tactic to halt neurodegeneration. Recent research has shown that the hypoxia-inducible factor modulator FM19G11 promotes the proliferation and growth as well as self-renewal of epithelial cells. Marcuzzo *et al.*, explored if gold nanoparticles (NPs) loaded with FM19G11 could influence the processes of self-renewal and proliferation in epSPCs isolated from G93A-SOD1 mice at the beginning of the illness. After treating with NPs loaded with FM19G11 the authors found increased transcriptional and protein levels of SOX2, OCT4, AKT1, and AKT3, important genes linked to pluripotency, self-renewal, and proliferation, in G93A-SOD1 epSPCs. In treated cells, the researchers also saw an upregulation of the mitochondrial uncoupling protein (UCP) gene. The expression of the cell cycle-related microRNA (miR)-19a and its target gene PTEN in G93A-SOD1 epSPCs was likewise impacted by FM19G11-loaded NPs treatment. Overall, the results demonstrated the profound influence of NPs loaded with FM19G11 on the cellular pathways responsible for self-renewal and proliferation in G93A-SOD1 epSPCs, which would stimulate the development of fresh, specialized methods to halt the advancement of ALS illness [265].

9.5.4.4 Cerium Oxide

Due to its antioxidant potential [96], it has been explored for treating Amyotrophic lateral Sclerosis. Reactive oxygen and nitrogen species are neutralized using cerium oxide nanoparticles, or CeNPs. Given the oxidative stress involvement in both the human and SOD1G93A animal models of ALS, the researchers in a study investigated the potential benefits of administering CeNPs to enhance survival and mitigate severity of disease in SOD1G93A transgenic mice. When CeNPs were administered two times a week intravenously, a week to SOD1G93A mice, beginning at the outset of muscular weakening, muscle function was retained, and both male and female lifespan was enhanced. Following the start of CeNP therapy, the mice's median survival was 33.0 ± 3.7 days (N = 20), compared to just 22.0 ± 2.5 days (N = 27; P = 0.022) for those treated with vehicle, control injections. Given that these citrate–EDTA stabilized CeNPs demonstrated catalase and oxidase activity in both *in vitro* models of ischemia oxidative stress and cell-free systems, the authors speculated that antioxidant activity serves as the survival strategy that prolongs life in SOD1G93A animals [266].

9.5.4.5 Mesoporous Silica

In a study, the authors looked at how nanomedicine might be used to create a single platform by using mesoporous silica nanoparticles (MSNs) for treating an animal model of ALS using a combination of drugs like leptin, which provide neuroprotection, and pioglitazone, which acts against inflammation. These drugs have already shown promise in treating other neurodegenerative diseases. The objective of the researchers was to investigate the ability of functionalized mesoporous materials as anti-ALS drugs by employing MSNs as nanometric carriers for the leptin/pioglitazone (MSN-LEP-PIO) medication cocktail. The suggested medication, MSN-LEP-PIO, was tested *in vivo* using the TDP-43 proteinopathy mouse model (TDP-43A315T animals). Body weight loss was examined, and motor function was evaluated using the rotarod test. TDP-43A315T mice and wild-type (WT) mice showed a consistent decrease in weight of body as well as coordination of motor functions. However, the illness progressed more slowly and the motor function significantly improved, suggesting that TDP-43A315T animals given treatment with MSN-LEP-PIO appeared to require reduced energy in the latter stages of ALS symptoms. All of these findings seem to support the effectiveness of the dosage forms *in vivo* and the value of applying them to ALS [267].

9.5.5 Multiple Sclerosis

A complex neurodegenerative disease, multiple sclerosis (MS) affects people from early adulthood to age forty [268]. It is characterized by chronic inflammation of the peripheral and central nervous systems, oxidative stress, disruption of the blood–brain barrier, and subsequent demyelination and damage to neurons and axons [269]. The etiology of multiple sclerosis (MS) is not entirely understood, but it is widely recognized to be associated with inflammation marked by the loss of oligodendrocytes progressively, which causes demyelination, decreased remyelination, and decreased survival of axons. Additionally, activation of microglia and antigen-presenting cells (APCs) can affect immune responses, which in turn causes the autoimmune MS disease to spread [270]. Four distinct clinical courses—the relapsing-remitting (RR), primary progressive (PP), secondary progressive (SP), and progressive-relapsing (PR)—define the clinical stages of MS, which alternate between remission and relapse and gradually worsen as the illness progresses. RRMS, clinically isolated syndrome (CIS), radiologically isolated syndrome (RIS), PPMS, and SPMS are among the current multiple sclerosis phenotypic categories [271]. The primary form of the disease's manifestation, RRMS, is marked by oxidative stress and inflammatory cytokines during relapses. It is not certain that it will improve during the remission phase, and it can progress to progressive MS as neurological deficits, particularly damage to axons, worsen [272].

The exact cause of multiple sclerosis (MS) is unknown, but it most probably is an outcome of a complex interplay between genetic, autoimmune, and environment related factors that disrupt the immune system. These factors include disruptions to the BBB brought on by immune system changes that allow T cells to enter the CNS and release pro-inflammatory cytokines, which in turn trigger macrophage activation and white matter inflammation that destroys the myelin sheath [273]. Th1 and Th17 in particular, which are CD4+ T-helper (Th) cells, indicate the production of autoantibodies by inducing B lymphocytes and lowering the T/B lymphocyte ratio. Natural killer cells, CD4+, and CD8+ cells upregulate the immune responses, while regulatory T cells (T-regs) suppress the autoimmune responses derived from T and B cells [274]. Different kinds of leucocytes may improve permeability of BBB by activating different peripheral cells, such as platelets [275]. After migrating into the CNS, myelin reactive CD4+ T cells exhibit a crucial function in demyelination. Naive T lymphocytes, on the other hand, mature into Th1 and Th17 and secrete pro-inflammatory cytokines. The main mechanism for activating T cells is antigen presentation on dendritic cells via major histocompatibility complex II [276]. B-lymphocytes may potentially play a role in the etiology of MS by generating antibodies against the myelin basic protein.

The various antioxidants, which have been loaded in nanoformulations and used for Multiple sclerosis treatment are given below:

9.5.5.1 Pomegranate Seed Oil

The CNS is affected by MS, a chronic illness characterized by inflammation that is linked to demyelination, degeneration of neurons, and oxidative stress sensitivity. In a study, mice with experimentally generated experimental autoimmune encephalomyelitis (EAE), a well-established MS model, were given a nanodroplet formulation of pomegranate seed oil (PSO), called Nano-PSO. Punicic acid, a special kind of polyunsaturated fatty acid that is regarded as one of the most potent naturally occurring antioxidant, is present in increased amounts in PSO. The authors demonstrated that although mice induced with EAE and given natural PSO showed some decrease in disease, this positive impact was greatly amplified after EAE mice were given Nano-PSO of a particular nanosize at substantially reduced oil amounts. The severe neurological disease's hallmarks, demyelination as well as lipid oxidation, were significantly decreased in the sick animal brains after receiving Nano-PSO treatment, according to pathological investigations. The authors suggested that individuals with demyelinating illnesses might benefit from the use of unique dosage forms containing antioxidants which are present naturally, such Nano-PSO. The results point out that lipid oxidation might play a key factor in the mechanistic aspects of demyelination and neurodegeneration [277].

9.5.5.2 Selenium-Crocin

It is believed that between 40 and 60 percent of persons with multiple sclerosis (MS) have cognitive impairment. The amount of data supporting the antioxidant effects of crocin and selenium on cognition is growing. Rezaeimanesh *et al.*, examined the impact of crocin-selenium nanoparticles (Cor@SeNs) on oxidative stress indicators and cognition in patients suffering with RS in their study for the first time. A randomized clinical study with triple blinding was carried out with sixty MS patients. Block randomization was used to randomly assign the individuals in a 1:1 ratio to the Cor@SeNs or placebo group. The subjects were given Cor@SeNs capsules, which contained 5.74 mg crocin and 55 mcg selenium, or placebo capsules for a duration of 12 weeks. The Persian version of the Brief International Cognitive Assessment for MS (BICAMS) battery was used to evaluate cognitive abilities. Colorimetric kits were used to measure the serum levels of malondialdehyde (MDA), glutathione reductase (GR) activity, and total antioxidant capacity (TAC). The data was analyzed by SPSS version 26. The range that was deemed significant was $P<0.05$. In the intervention and placebo groups, the mean ± SD of TAC change was 0.03 ± 0.07 mM and -0.03 ± 0.09 mM, respectively (Time × group effect P: 0.01; effect size: 0.10). The results indicated that there was a substantial and increasing temporal impact of intervention on the California Verbal Learning Test, Second Edition (CVLT-II) ($P < 0.01$; effect size: 0.29), CVLT-II-delay ($P < 0.01$; effect size: 0.29), and the Symbol Digit Modalities Test (SDMT) ($P < 0.01$; effect size: 0.18). Furthermore, both groups saw a substantial decrease in GR activity over time as a result of the intervention ($P < 0.01$; effect size: 0.20). The findings indicated that the Cor@SeNs treatment significantly improved TAC. During their investigation, the researchers also saw a considerable increase in both groups' cognitive performance. The Cor@SeNs group showed a greater degree of change in cognition and antioxidant indicators of serum than the placebo group [123].

9.5.5.3 PEG-HCC

Due to its radical scavenging property, it has been incorporated into nanoformulations used to treat against multiple sclerosis. T lymphocyte-mediated autoimmune disorders like multiple sclerosis are now treated mostly with broad-spectrum immunosuppressants, which have the potential to cause negative side effects. Antioxidants offer an alternate treatment strategy for autoimmune illnesses, although dietary antioxidants are not enough to fulfill this function. Carbon nanoparticle antioxidants are more effective than dietary and endogenous antioxidants at scavenging reactive oxygen compounds (ROS). Carbon nanoparticles' affinity for particular cell types also indicates a new strategy for targeting cells. An investigation revealed that T lymphocytes preferentially internalize poly(ethylene glycol)-functionalized hydrophilic carbon clusters (PEG-HCCs), which are recognized ROS superoxide ($O2^-$) and hydroxyl radical scavengers. Using this selectivity, T cell activation was prevented without impairing macrophages' ability to perform their primary role as antigen-presenting cells, which is essential for T cell activation. The researchers showed that PEG-HCCs are beneficial *in vivo* in decreasing T lymphocyte-mediated inflammation [278].

9.5.6 Trigeminal Neuralgia

The most prevalent kind of neuropathic pain that affects the different trigeminal nerve branches is called trigeminal neuralgia (TN). Unilateral acute, firing, lancinating, electric shock-like, scorching, and agonizing pain assaults are its defining feature. As a result, TN is regarded as among the most excruciating human ailments. However, the fundamental causes of TN remain unclear, and pharmacological or surgical therapies frequently have unsatisfactory results [278]. Therefore, it is imperative to create novel strategies and find fresh substances to treat TN.

Prior research on treating neuropathic pain has concentrated on the main sensory neurons and how they affect spinal dorsal horn neurons' activity [279, 280]. Trigeminal neuralgia is thought to be mostly caused by both astrocytes and microglia that are persistently active and reside in the spine trigeminal nucleus, according to new research [281, 282]. Pro-inflammatory cytokines, including tumor necrosis factor-α, interleukin-one beta (IL-1β), and interleukin six (IL-6), are released by activated glial cells in the CNS and increase pain hypersensitivity [283–286] beside being involved in the pathophysiology of neuropathic pain [287].

Between four to thirteen persons per 100,000 year are thought to be affected [288–291]. Trigeminal neuralgia, as per the definition by the International Association for the Study of Pain, includes abrupt, typically unilateral, intense, short stabbing spells that reoccur in one or more trigeminal nerve branches [283]. Pain is localized to the region innervated by one or more trigeminal nerve branches and is typically characterized as stabbing, paroxysmal, scorching, or evoking an electric shock. About three-quarters of instances involve both the maxillary and mandibular branches, whereas roughly 60% of cases involve only one of the branches. Conversely, less than 4% of patients have an impact on the ophthalmic branch [284]. Trigeminal discomfort is more likely to develop in older people, usually those over 50 [285]. Older People are risk indicator for this condition. The female-to-male ratio is around 2-3:1, indicating a greater occurrence in women [286, 291]. Trigeminal nerve innervation typically contains trigger sites, which are the source of pain episodes. A few instances of triggers for pain episodes are mild facial contact, cleaning one's teeth, and using the masticatory and muscles of the face for speech or feeding.

The key to effective therapy and a positive result is a proper diagnosis. Since therapy for trigeminal autonomic cephalalgias (such as paroxysmal hemicrania, cluster migraines, SUNCT, and) differs significantly from those associated with other cranial and facial pain syndromes or chronic idiopathic face pain, it is crucial to distinguish between the two conditions. Patients who only have first-division trigeminal pain may also have symptomatic TN (STN) [292], which is more challenging to treat [293, 294]. STN might be caused by a tumor or multiple sclerosis. When eliminating individuals with microvascular conflict, routine magnetic resonance imaging (MRI) of the head can identify structural reasons in a maximum of 15% (95% confidence interval 11–20%) of cases. For the purpose of diagnosing STN patients, the reflex of blinking and other trigeminal response tests are highly accurate. A 94% combined sensitivity (95% CI 91–97%) and a 87% combined specificity (95% CI 77–93%) were demonstrated by five separate investigations. Insufficient data was found to differentiate between classic TN and STN using evoked potentials (combined specificity 64%, 95% CI 56–71%; combined sensitivity 84%, 95% CI 73–92%) [295, 296].

The antioxidants used in TN are given below:

9.5.6.1 *Capsaicin*

In a study, the researchers manufactured nanoparticles of capsaicin and also changed them to gel form, so that their therapeutic effects on rats with trigeminal neuralgia may be seen. Using PLGA as the carrier and bovine serum albumin as the stabilizer, capsaicin -loaded polylactic-co-glycolic acid/Capsaicin nanoparticles were developed by the emulsion solvent evaporation technique. A highly sensitive surface charge potential analyzer, Fourier transform infrared spectroscopy, scanning electron microscopy, and nanoparticle size were used for the evaluation of nanoparticles. Utilizing differential scanning calorimetry, the characteristics of the nanoparticles were examined. After that, Carbopol 934 was used to convert the Cap nanoparticles into a gel. Using high-performance liquid chromatography, the loading capacity of drug, *in vitro* release study and encapsulation efficiency were ascertained. The polymer dispersity index of PLGA/Capsaicin Nanoparticles at 17 mg was 0.270±0.018; the surface charge potential was 33.56±2.61 mV; the drug loading capacity was 57.56±0.05%); the encapsulation efficiency was 90.26±0.90)%; and the maximum cumulative release rate was 77.19%. The NPs had a regular spherical shape with particle size of 389.58±12.02 nm. Trigeminal pain threshold was considerably raised ($P<0.05$) in rats when 3 %, 0.03% and 0.3%, PLGA/Capsaicin Nanoparticles gels were applied to the surgical site. When comparing the treatment group to the model group, there was a significant drop in the blood levels of prostaglandin E2 (PGE2) and substance P (SP). The authors confirmed the therapeutic benefits of cap nanoparticles gel on trigeminal neuralgia of rat [125].

Table 9.2 compiles the various nanoformulations loaded with antioxidants and explored for different neurodegenerative ailments.

Table 9.2 Various nanoformulations of antioxidants against neurodegenerative ailments.

S. no.	Disease	Antioxidant	Nanoformulation	Reference
1.	**Alzheimer's disease**	Curcumin	Multifunctional LUV liposomes	[133]
2.		Curcumin	Nanoparticles	[134]
3.		Curcumin	PLGA Nanoparticles	[135]
4.		Curcumin	PLGA Nanoparticles	[136]
5.		Curcumin	PLGA-PEG-B6	[137]

(*Continued*)

Table 9.2 Various nanoformulations of antioxidants against neurodegenerative ailments. (*Continued*)

S. no.	Disease	Antioxidant	Nanoformulation	Reference
6.		Curcumin	Nanoliposomes	[138]
7.		Curcumin	PLGA based selenium Nanoparticles	[139]
8.		Curcumin	PLGA Nanoparticles	[140]
9.		Curcumin	Nanolipid carriers	[141]
10.		Curcumin	PLGA Nanoparticles	[142]
11.		Curcumin	SLN Nanoparticles	[143]
12.		Curcumin	Cocrystal micellar nanocarriers	[144]
		Resveratrol and Curcumin	Mucoadhesive lipidic nanoemulsion	[145]
13.		Curcumin	PLGA Nanoparticles	[146]
14.		class A scavenger receptor activator XD4 (W20/ XD4-SPIONs)	Superparamagnetic iron oxide nanoparticle	[147]
15.		Pomegranate seed oil	Nanodroplet formulation	[148]
16.		Pomegranate extract	Nanoparticles	[149]
17.		Resveratrol	Nanostructured lipid carrier	[150]
18.		Resveratrol	Nanosuspension	[151]
19.		Resveratrol	Lipid microparticles	[152]
20.		Resveratrol	transferosomal mucoadhesive gel	[153]
21.		Resveratrol	Solid lipid nanoparticles	[154]
22.		Apolipoprotein E2	Liposomes	[155]
23.		Cross-reacting material 197 (CRM197) and apolipoprotein E (ApoE)	RA-PAAM-CH-PLGA	[156]
24.		Huperzine A	Nanostructured lipid carrier	[157]

(*Continued*)

Table 9.2 Various nanoformulations of antioxidants against neurodegenerative ailments. (*Continued*)

S. no.	Disease	Antioxidant	Nanoformulation	Reference
		Huperzine A (HA) and ligustrazine phosphate (LP)	Microemulsion	[158]
25.		Huperzine A	PLGA Nanoparticles	[159]
26.		Quercetin (QU) and rosmarinic acid (RA)	Liposomes	[160]
27.		Quercetin	Zein nanoparticles	[161]
28.		Quercetin	Nanoparticles	[162]
29.		Quercetin	Nanoparticles	[163]
30.		Quercetin	PLGA Nanoparticles	[164]
31.		Rosmarinic acid (RA)	Mucoadhesive chitosan-coated nanoemulsions	[165]
32.		thymoquinone rich fraction	Nanoemulsion	[166]
33.		Thymoquinone	PLGA Nanoparticles	[167]
34.		Memantine	PLGA Nanoparticles	[168]
35.		Memantine	Nanoemulsion	[169]
36.		Tacrine	PAMAM dendrimer	[170]
37.		Tacrine	mucoadhesive microemulsion	[171]
38.		Green tea polyphenol and (-)-epigallocatechin-3-	Nanolipid	[172]
39.		Epigallocatechin-3- gallate	Nanoparticles	[173]
40.		Epigallocatechin-3-gallate	Selenium Nanoparticles	[174]
41.		Rivastigmine	Liposomes	[175]
42.		Piperine	Microemulsion	[176]
43.		α-bisabolol	Solid lipid nanoparticles	[177]
44.		Vitamin D-binding protein	PLGA Nanoparticles	[178]
45.		Zinc	Nanoparticles	[179]
46.		Osthole	Liposomes	[180]

(*Continued*)

Table 9.2 Various nanoformulations of antioxidants against neurodegenerative ailments. (*Continued*)

S. no.	Disease	Antioxidant	Nanoformulation	Reference
47.		Morin hydrate	Microemulsion	[181]
48.		Genistein	Chitosan Nanoparticles	[182]
49.		Melatonin	Polymeric gel suspension	[183]
50.		Berberine	Multiwalled carbon nanotubes	[184]
51.		Galantamine hydrobromide	Solid lipid nanoparticles	[185]
52.		Estradiol	PLGA Nanoparticles	[186]
53.		Basic fibroblast growth factor	PLGA-PEG Nanoparticles	[187]
54.		Silica	Nanoparticles	[188]
55.		Sialic acid	Selenium Nanoparticles	[189]
56.		Ceria dioxide	Ceria Nanoparticles	[96]
57.		Erythropoietin	Solid lipid nanoparticles	[190]
58.		Aβ generation inhibitor S1 (PQVGHL peptide) and curcumin	PLGA Nanoparticles	[191]
59.		Ligustrazine phosphate	Ethosomes	[192]
60.		Naringenin	Nanoemulsion	[98]
61.		Donepezil	Nanoemulsion	[193]
62.		Sesamol	Solid lipid nanoparticles	[101]
63.		Ginsenoside Rg3 and Thioflavin T	PLGA Nanoparticles	[102]
67.	**Parkinson's disease**	Resveratrol	Nanoparticles	[200]
		Fe_3O_4-modified Resveratrol	Liposomes	[201]
		Resveratrol (RVT)-loaded polysorbate 80	Polylactide Nanoparticles	[202]
68.		Resveratrol	PLGA Nanoparticles	[203]
69.		Resveratrol	Liposomes	[204]

(*Continued*)

Table 9.2 Various nanoformulations of antioxidants against neurodegenerative ailments. (*Continued*)

S. no.	Disease	Antioxidant	Nanoformulation	Reference
70.		Resveratrol	Nanoemulsion	[205]
71.		Ropinirole hydrochloride	Polymer-lipid hybrid nanoparticles	[206]
72.		Ropinirole	PLGA Nanoparticles	[207]
73.		Naringenin	Nanoemulsion	[208]
74.		Naringenin	Chitosan Nanoparticles	[209]
75.		Naringenin	Solid Lipid Nanoparticles	[210]
76.		Curcumin	Lactoferrin nanoparticles	[211]
77.		Ferulic acid diacid with an adipic acid linker (FAA) and tannic acid (TA)	Nanoparticles	[212]
78.		Cerium oxide	Nanoparticles	[213]
79.		Cerium oxide	Nanoparticles	[214]
80.		N-acetylcysteine	Nanocarrier	[215]
81.		Selegiline	Nanoparticles	[216]
82.		Catalase	Exosomes	[217]
83.		Pramipexole dihydrochloride	Chitosan Nanoparticles	[218]
84.		Ceria dioxide	Nanoparticles	[219]
85.		Dopamine	Nanoparticles	[220]
86.		Gallic acid	Polyethylenimine-coated human serum albumin (PEI-HSA) Nanoparticles	[221]
87.		Hydroxytyrosol	Micellar Nanocarrier	[221]
88.		Graphene oxide	Nanoparticles	[223]
89.		β-carotene	Nanoparticles	[224]
90.		Polydopamine (PDA) with selenocystine (SeCys)	Nanocomposite	[225]

(*Continued*)

Table 9.2 Various nanoformulations of antioxidants against neurodegenerative ailments. (*Continued*)

S. no.	Disease	Antioxidant	Nanoformulation	Reference
95.	**Huntington's disease**	Rosmarinic acid	Solid Lipid Nanoparticles	[242]
96.		Curcumin	Solid Lipid Nanoparticles	[243]
97.		Curcumin	Hyaluronan-based Nanoparticles	[244]
98.		Selenium	Nanoparticles	[245]
99.		Fisetin	Self-nanoemulsifying drug delivery system	[115]
101.		Poly(trehalose)	Nanoparticles	[246]
102.		Thymoquinone	Solid Lipid Nanoparticles	[247]
103.		Boehmite	Nanoparticles	[248]
104.		Ginger	Chitosan Nanoparticles	[249]
105.		Manganese	Nanoparticles	[250]
107.	**Amyotrophic lateral sclerosis**	Curcumin	Micelles	[263]
		Curcumin	Nanocarrier	[264]
108.		Verapamil and Riluzole	Cocktail liposomes	[120]
109		FM19G11	Gold Nanoparticles	[265]
111.		Cerium oxide	Nanoparticles	[266]
112.		Mesoporous	Silica Nanoparticles	[267]
113.	**Multiple Sclerosis**	Pomegranate seed oil	Nanodroplets	[277]
114.		Selenium and Crocin	Nanoparticles	[123]
115.		(PEG-HCCs)	Nanoparticles	[124]
116.	**Trigeminal Neuralgia**	Capsaicin	PLGA-Nanoparticles	[125]

9.6 Clinical Trials of Nanoformulations Containing Antioxidants Against Neurological Disorders

Antioxidant treatment has new opportunities due to the crucial function that ROS has in certain neurodegenerative ailments and neurological illnesses. Only a small number of antioxidant medications, including idebenone, edaravone, and dimethyl fumarate, have received approval for treating neurological disorders in humans, with the majority of medications still undergoing preclinical and clinical testing [297].

Sina Curcumin, or nanocurcumin (curcumin encapsulated in nanomicelles), is an efficient antioxidant nanoformulation whose immunomodulatory action has been used to treat MS. A clinical trial conducted recently [298] gave patients suffering from MS curcumin loaded in polymeric based nanoparticles (NanoCurc) or a placebo every day for six months. Blood samples were subjected to real-time PCR to analyze the miRNAs level, transcription factors, proinflammatory cytokines, and miRNA-dependent targets' gene expression. ELISA tests were then performed to examine the release of cytokines. Interestingly, inflammatory miRNAs (miR-145, miR-132, and miR-16), STAT1, NF-κB, and AP-1 were considerably less expressed in MS patients treated with nanocurcumin. These findings suggested that nanocurcumin could be able to reduce MS patients' neuroinflammation [298].

9.7 Conclusion

Numerous studies in the literature demonstrate the importance of oxidative stress as well as ROS production in the pathophysiology of neurodegenerative diseases. Antioxidants have emerged as novel therapeutics for treatment of neurological diseases. However, there are still several bioavailability issues associated with their use that need to be resolved in the case of administration systemically, including the BBB's pathway, which poses a significant barrier to brain targeting. Besides these, other problems associated with antioxidants include their low aqueous solubility, their instability in the gastrointestinal tract, and potential metabolism [299]. However, in spite of all problems, incorporating antioxidants in nano-formulations has shown encouraging results in treating neurological disorders.

References

1. Sies, H., Oxidative stress: introductory remarks In: *Oxidative Stress.* Sies. H., editor. pp. 1–8, Academic Press, London: 1985.
2. Flohé, L., Looking back at the early stages of redox biology. *Antioxidants*, 9, 1254, 2020.
3. Sies, H., Hydrogen peroxide as a central redox signaling molecule in physiological oxidative stress: oxidative eustress. *Redox Biol.*, 11, 613–619, 2017.
4. Sies, H., Berndt, C., Jones, D.P., Oxidative stress. *Annu. Rev. Biochem.*, 86, 715–748, 2017.
5. Milbourn, H.R., Toomey, L.M., Gavriel, N., Gray, C.G.G., Gough, A.H., Fehily, B., Giacci, M.K., Fitzgerald, M., Limiting Oxidative Stress Following Neurotrauma with a Combination of Ion Channel Inhibitors. *Discov. Med.*, 23, 361–369, 2017.
6. Giacci, M.K., Bartlett, C.A., Smith, N.M., Iyer, K.S., Toomey, L.M., Jiang, H., Guagliardo, P., Kilburn, M.R., Fitzgerald, M., Oligodendroglia Are Particularly Vulnerable to Oxidative Damage after Neurotrauma. *In Vivo. J. Neurosci.*, 38, 6491–6504, 2018.
7. Yeung, A.W.K., Tzvetkov, N.T., Georgieva, M.G., Ognyanov, I.V., Kordos, K., Jóźwik, A., Kühl, T., Perry, G., Petralia, M.C., Mazzon, E., Atanasov, A.G., Reactive Oxygen Species and Their Impact in Neurodegenerative Diseases: Literature Landscape Analysis. *Antioxid. Redox Signal.*, 34, 5, 402–420, 2021.

8. Liu, Z., Zhou, T., Ziegler, A.C., Dimitrion, P., Zuo, L., Oxidative Stress in Neurodegenerative Diseases: From Molecular Mechanisms to Clinical Applications. *Oxid. Med. Cell. Longev.*, 2017, e2525967, 2017.
9. Lee, Y.M., He, W., Liou, Y.C., The Redox Language in Neurodegenerative Diseases: Oxidative Post-Translational Modifications by Hydrogen Peroxide. *Cell Death Dis.*, 12, 1–13, 2021.
10. Halliwell, B., Role of Free Radicals in the Neurodegenerative Diseases: Therapeutic Implications for Antioxidant Treatment. *Drugs Aging*, 18, 685–716, 2001.
11. McCord, J.M., The evolution of free radicals and oxidative stress. *Am. J. Med.*, 108, 652, 2000.
12. Zheng, M. and Storz, G., Redox sensing by prokaryotic transcription factors. *Biochem. Pharmacol.*, 59, 1–6, 2000.
13. Birben, E., Sahiner, U.M., Sackesen, C., Erzurum, S., Kalayci, O., Oxidative Stress and Antioxidant Defense. *World Allergy Organ. J.*, 5, 9–19, 2012.
14. Chandel, N.S. and Tuveson, D.A., The Promise and Perils of Antioxidants for Cancer Patients. *N. Engl. J. Med.*, 371, 177–178, 2014.
15. Krumova, K. and Cosa, G., Chapter 1: Overview of Reactive Oxygen Species, in: *Singlet Oxygen: Applications in Biosciences and Nanosciences*, pp. 1–21, RSC Publishing, London, UK, 2016.
16. Agarwal, V. and Chatterjee, K., Recent Advances in the Field of Transition Metal Dichalcogenides for Biomedical Applications. *Nanoscale*, 10, 16365–16397, 2018.
17. Zuo, L., Zhou, T., Pannell, B.K., Ziegler, A.C., Best, T.M., Biological and Physiological Role of Reactive Oxygen Species—The Good, the Bad and the Ugly. *Acta Physiol.*, 214, 329–348, 2015.
18. Samantaray, P.K., Indrakumar, S., Chatterjee, K., Agarwal, V., Bose, S., Template-Free Hierarchical MoS2 Foam as a Sustainable 'Green' Scavenger of Heavy Metals and Bacteria in Point of Use Water Purification. *Nanoscale Adv.*, 2, 2824–2834, 2020.
19. Saikolappan, S., Kumar, B., Shishodia, G., Koul, S., Koul, H.K., Reactive Oxygen Species and Cancer: A Complex Interaction. *Cancer Lett.*, 452, 132–143, 2019.
20. Shah, R., Farmer, L.A., Zilka, O., Kessel, A.T.M.V., Pratt, D.A., Beyond, D.P.P.H., Use of Fluorescence-Enabled Inhibited Autoxidation to Predict Oxidative Cell Death Rescue. *Cell Chem. Biol.*, 26, 1594–1607.e7, 2019.
21. Schaich, K.M., Tian, X., Xie, J., Hurdles and Pitfalls in Measuring Antioxidant Efficacy: A Critical Evaluation of ABTS, DPPH, and. *J. Funct. Foods*, 14, 111–125, 2015.
22. Kraus, R.L., Pasieczny, R., Lariosa-Willingham, K., Turner, M.S., Jiang, A., Trauger, J.W., Antioxidant Properties of Minocycline: Neuroprotection in an Oxidative Stress Assay and Direct Radical-Scavenging Activity. *J. Neurochem.*, 94, 819–827, 2005.
23. Agarwal, V., Tjandra, E.S., Iyer, K.S., Humfrey, B., Fear, M., Wood, F.M., Dunlop, S., Raston, C.L., Evaluating the Effects of Nacre on Human Skin and Scar Cells in Culture. *Toxicol. Res.*, 3, 223–227, 2014.
24. Yamato, M., Egashira, T., Utsumi, H., Application of *in Vivo* ESR Spectroscopy to Measurement of Cerebrovascular ROS Generation in Stroke. *Free Radic. Biol. Med.*, 35, 1619–1631, 2003.
25. Woolley, J.F., Stanicka, J., Cotter, T.G., Recent Advances in Reactive Oxygen Species Measurement in Biological Systems. *Trends Biochem. Sci*, 38, 556–565, 2013.
26. Fernández-Castro, P., Vallejo, M., San Román, M.F., Ortiz, I., Insight on the Fundamentals of Advanced Oxidation Processes. Role and Review of the Determination Methods of Reactive Oxygen Species. *J. Chem. Technol. Biotechnol.*, 90, 796–820, 2015.
27. Arrigoni, O. and De Tullio, M.C., Ascorbic Acid: Much More than Just an Antioxidant. *Biochim. Biophys. Acta*, 1569, 1–9, 2002.
28. Eleutherio, E.C.A., Silva Magalhães, R.S., de Araújo Brasil, A., Monteiro Neto, J.R., de Holanda Paranhos, L., SOD1, More than Just an Antioxidant. *Arch. Biochem. Biophys.*, 697, 108701, 2021.

29. Cardoso, B.R., Hare, D.J., Bush, A.I., Roberts, B.R., Glutathione Peroxidase 4: A New Player in Neurodegeneration? *Mol. Psychiatry*, 22, 328–335, 2017.
30. Glorieux, C. and Calderon, P.B., Catalase, a Remarkable Enzyme: Targeting the Oldest Antioxidant Enzyme to Find a New Cancer Treatment Approach. *Biol. Chem.*, 398, 1095–1108, 2017.
31. Packer, L., Witt, E.H., Tritschler, H.J., Alpha-Lipoic Acid as a Biological Antioxidant. *Free Radic. Biol. Med.*, 19, 227–250, 1995.
32. Dragsted, L.O., Antioxidant Actions of Polyphenols in Humans. *Int. J. Vitam. Nutr. Res.*, 73, 112–119, 2003.
33. Jiang, Q., Natural Forms of Vitamin E: Metabolism, Antioxidant, and Anti-Inflammatory Activities and Their Role in Disease Prevention and Therapy. *Free Radic. Biol. Med.*, 72, 76–90, 2014.
34. Uttara, B., Singh, A.V., Zamboni, P., Mahajan, R.T., Oxidative Stress and Neurodegenerative Diseases: A Review of Upstream and Downstream Antioxidant Therapeutic Options. *Curr. Neuropharmacol.*, 7, 65–74, 2009.
35. Menon, V.P. and Sudheer, A.R., Antioxidant and Anti-Inflammatory Properties of Curcumin. *Adv. Exp. Med. Biol.*, 595, 105–125, 2007.
36. AbdEl-Hack, M.E., El-Saadony, M.T., Shafi, M.E., Zabermawi, N.M., Arif, M., Batiha, G.E., Khafaga, A.F., Abd El-Hakim, Y.M., Al-Sagheer, A.A., Antimicrobial and Antioxidant Properties of Chitosan and Its Derivatives and Their Applications: A Review. *Int. J. Biol. Macromol.*, 164, 2726–2744, 2020.
37. Jahromi, H.K., Farzin, A., Hasanzadeh, E., Barough, S.E., Mahmoodi, N., Najafabadi, M.R.H., Farahani, M.S., Mansoori, K., Shirian, S., Ai, J., Enhanced Sciatic Nerve Regeneration by Poly-L-Lactic Acid/Multi-Wall Carbon Nanotube Neural Guidance Conduit Containing Schwann Cells and Curcumin Encapsulated Chitosan Nanoparticles in Rat. *Mater. Sci. Eng., C*, 109, 110564, 2020.
38. Singh, A., Shiekh, P.A., Das, M., Seppälä, J., Kumar, A., Aligned Chitosan-Gelatin Cryogel-Filled Polyurethane Nerve Guidance Channel for Neural Tissue Engineering: Fabrication, Characterization, and *In Vitro* Evaluation. *Biomacromolecules*, 20, 662–673, 2019.
39. Rubio, L., Marcos, R., Hernández, A., Nanoceria Acts as Antioxidant in Tumoral and Transformed Cells. *Chem. Biol. Interact.*, 291, 7–15, 2018.
40. Reiter, R.J., Rosales-Corral, S., Tan, D.X., Jou, M.J., Galano, A., Xu, B., Melatonin as a Mitochondria-Targeted Antioxidant: One of Evolution's Best Ideas. *Cell. Mol. Life Sci.*, 74, 3863–3881, 2017.
41. Sergeeva, V., Kraevaya, O., Ershova, E., Kameneva, L., Malinovskaya, E., Dolgikh, O., Konkova, M., Voronov, I., Zhilenkov, A., Veiko, N., Troshin, P., Kutsev, S., Kostyuk, S., Antioxidant Properties of Fullerene Derivatives Depend on Their Chemical Structure: A Study of Two Fullerene Derivatives on HELFs. *Oxid. Med. Cell. Longev.*, 2019, 4398695, 2019.
42. Türkan, F., Taslimi, P., Saltan, F.Z., Tannic Acid as a Natural Antioxidant Compound: Discovery of a Potent Metabolic Enzyme Inhibitor for a New Therapeutic Approach in Diabetes and Alzheimer's Disease. *J. Biochem. Mol. Toxicol.*, 33, e22340, 2019.
43. Moreno-García, A., Kun, A., Calero, M., Calero, O., The Neuromelanin Paradox and Its Dual Role in Oxidative Stress and Neurodegeneration. *Antioxidants*, 10, 124, 2021.
44. Floyd, R.A., Kopke, R.D., Choi, C.H., Foster, S.B., Doblas, S., Towner, R.A., Nitrones as Therapeutics. *Free Radic. Biol. Med.*, 45, 1361–1374, 2008.
45. Piotrowska, D.G., Mediavilla, L., Cuarental, L., Głowacka, I.E., Marco-Contelles, J., Hadjipavlou-Litina, D., López-Muñoz, F., Oset-Gasque, M.J., Synthesis and Neuroprotective Properties of N-Substituted C-Dialkoxyphosphorylated Nitrones. *ACS Omega*, 4, 8581–8587, 2019.

46. Chioua, M., Salgado-Ramos, M., Diez-Iriepa, D., Escobar-Peso, A., Iriepa, I., Hadjipavlou-Litina, D., Martínez-Alonso, E., Alcázar, A., Marco-Contelles, J., Novel Quinolylnitrones Combining Neuroprotective and Antioxidant Properties. *ACS Chem. Neurosci.*, 10, 2703–2706, 2019.
47. Ayuso, M.I., Martínez-Alonso, E., Chioua, M., Escobar-Peso, A., Gonzalo-Gobernado, R., Montaner, J., Marco-Contelles, J., Alcázar, A., Quinolinyl Nitrone RP19 Induces Neuroprotection after Transient Brain Ischemia. *ACS Chem. Neurosci.*, 8, 2202–2213, 2017.
48. Gilgun-Sherki, Y., Melamed, E., Offen, D., Oxidative stress induced-neurodegenerative diseases: the need for antioxidants that penetrate the blood brain barrier. *Neuropharmacology*, 40, 959–975, 2001.
49. Mittal, P., Dhankhar, S., Chauhan, S., Garg, N., Bhattacharya, T., Ali, M., Chaudhary, A.A., Rudayni, H.A., Al-Zharani, M., Ahmad, W., Khan, S., Singh, T.G., Mujwar, S., A Review on Natural Antioxidants for Their Role in the Treatment of Parkinson's Disease. *Pharmaceuticals*, 16, 7, 908, 2023.
50. Angelo, D., Erene, M., Rakez, K., Saskia, C., Milton, I.P., Charles, G.G., Calcium dysregulation and membrane disruption as a ubiquitous neurotoxic mechanism of soluble amyloid Oligomers. *J. Biol. Chem.*, 280, 17, 172–94, 2005.
51. Ashraf, V., Franco, G., Syed, I., Zbigniew, B., Syed, A., Possible mechanism for the neuroprotective effects of l-Carnitine on methamphetamine-evoked neurotoxicity. *Ann. N. Y. Acad. Sci.*, 993, 197–207, 2006.
52. Barry, H., Antioxidants in human health and disease. *Annu. Rev. Nut.*, 16, 33–50, 1996.
53. Ames, B.N., Shigenaga, M.K., Hagen, T.M., Oxidants, antioxidants, and the degenerative diseases of aging. *Proc. Natl. Acad. Sci.*, 90, 17, 7915–22, 1993.
54. Burton, G.W. and Ingold, K.U., Mechanisms of antioxidant action: preventive and chain-breaking antioxidants in, in: *CRC Handbook of Free Radicals and Antioxidants in Biomedicine*, A. Miguel, A.T. Quintanilha, H. Weber (Eds.), pp. 29–43, CRC Press Inc, Boca Raton, FL, 1990.
55. Burton, G.W. and Ingold, K.U., Vitamin E as an *in vitro* and *in vivo* antioxidant. *Ann. N. Y. Acad. Sci.*, 570, 7–22, 1989.
56. Cai, Y.Z., Sun, M., Corke, H., Antioxidant activity of betalains from plants of the amaranthaceae. *J. Agric. Food. Chem.*, 51, 8, 2288–2294, 2003.
57. Cotelle, N., Bernier, J.L., Catteau, J.P., Pommery, J., Wallet, J.C., Gaydou, E.M., Antioxidant properties of hydroxyflavones. *Free Rad. Bio. Med.*, 20, 1, 35–43, 1996.
58. Velioglu, Y.S., Mazza, G., Gao, L., Oomah, B.D., Antioxidant activity and total phenolics in selected fruits, vegetables, and grain products. *J. Agric. Food. Chem.*, 46, 10, 4113–4117, 1998.
59. Zheng, W. and Wang, S.Y., Antioxidant activity and phenolic compounds in selected herbs. *J. Agric. Food. Chem.*, 49, 11, 5165–5170, 2001.
60. Mark, P.M., Metal-catalyzed disruption of membrane protein and lipid signaling in the pathogenesis of neurodegenerative Disorders. *Ann. N. Y. Acad. Sci.*, 1012, 37–50, 2004.
61. Hirsch, E.C. and Hunot, S., Neuroinflammation in Parkinson's Disease: A Target for Neuroprotection? *Lancet Neurol.*, 8, 382–397, 2009.
62. Cacabelos, R., Parkinson's Disease: From Pathogenesis to Pharmacogenomics. *Int. J. Mol. Sci.*, 18, 551, 2017.
63. Chen, Z. and Zhong, C., Oxidative Stress in Alzheimer's Disease. *Neurosci. Bull.*, 30, 271–281, 2014.
64. Anantharaman, M., Tangpong, J., Keller, J.N., Murphy, M.P., Markesbery, W.R., Kiningham, K.K., St Clair, D.K., Beta-Amyloid Mediated Nitration of Manganese Superoxide Dismutase: Implication for Oxidative Stress in a APPNLH/NLH X PS-1P264L/P264LDouble Knock-in Mouse Model of Alzheimer's Disease. *Am. J. Pathol.*, 168, 1608–1618, 2006.

65. Niedzielska, E., Smaga, I., Gawlik, M., Moniczewski, A., Stankowicz, P., Pera, J., Filip, M., Oxidative Stress in Neurodegenerative Diseases. *Mol. Neurobiol.*, 53, 4094–4125, 2016.
66. Zhang, H., Su, Y.J., Zhou, W.W., Wang, S.W., Xu, P.X., Yu, X.L., Liu, R.T., Activated scavenger receptor A promotes glial internalization of abeta. *PLoS One*, 9, e94197, 2014.
67. Ullah, A., Khan, A., Ahmed, S., Irfan, H.M., Hafiz, A.A., Jabeen, K., Alruwaili, M., Alotaibi, M., Al Naeem, W., Khan, A., Al-Harrasi, A., A review of pomegranate supplementation: A promising remedial avenue for Alzheimer's disease. *Heliyon*, 15, 9(11), e22483, 2023.
68. Chen, J., Zhou, Y., Mueller-Steiner, S., Chen, L.F., Kwon, H., Yi, S., Mucke, L., Gan, L., SIRT1 protects against microglia-dependent amyloid-β toxicity through inhibiting NF-κB signaling. *J. Biol. Chem.*, 280, 48, 40364–40374, 2005.
69. Ellis, R.J., Olichney, J.M., Thal, L.J., Mirra, S.S., Morris, J.C., Beekly, D., Heyman, A., Cerebral amyloid angiopathy in the brains of patients with Alzheimer's disease: the CERAD experience, Part XV. *Neurology*, 46, 6, 1592–6, 1996.
70. Xiao, X.Q., Wang, R., Han, Y.F., Tang, X.C., Protective effects of huperzine A on beta-amyloid (25–35) induced oxidative injury in rat pheochromocytoma cells. *Neurosci. Lett.*, 286, 155–158, 2000a.
71. Fiorani, M., Guidarelli, A., Blasa, M., Azzolini, C., Candiracci, M., Piatti, E., Cantoni, O., Mitochondria accumulate large amounts of quercetin: prevention of mitochondrial damage and release upon oxidation of the extramitochondrial fraction of the flavonoid. *J. Nutr. Biochem.*, 21, 5, 397–404, 2010.
72. Iuvone, T., Filippis, D.D., Esposito, G., D'Amico, A., Izzo, A.A., The spice sage and its active ingredient rosmarinic acid protect PC12 cells from amyloid-beta peptide-induced neurotoxicity. *J. Pharmacol. Exp. Ther.*, 317, 1143e1149, 2006.
73. Goschorska, M., Gutowska, I., Baranowska-Bosiacka, I., Piotrowska, K., Metryka, E., Safranow, K., Chlubek, D., Influence of Acetylcholinesterase Inhibitors Used in Alzheimer's Disease Treatment on the Activity of Antioxidant Enzymes and the Concentration of Glutathione in THP-1 Macrophages under Fluoride-Induced Oxidative Stress. *Int. J. Environ. Res. Public Health*, 16, 1, 10, 2018.
74. Lim, G.P., Chu, T., Yang, F., Beech, W., Frautschy, S.A., Cole, G.M., The curry spice curcumin reduces oxidative damage and amyloid pathology in an Alzheimer transgenic mouse. *J. Neurosci.*, 21, 8370–8377, 2001.
75. Mittal, R. and Gupta, R.L., *In vitro* antioxidant activity of piperine. *Methods Find Exp. Clin. Pharmacol.*, 22, 5, 271–4, 2000.
76. Pottoo, F.H., Ibrahim, A.M., Alammar, A., Alsinan, R., Aleid, M., Alshehhi, A., Alshehri, M., Mishra, S., Alhajri, N., Thymoquinone: Review of Its Potential in the Treatment of Neurological Diseases. *Pharmaceuticals (Basel)*, 15, 4, 408, 2022.
77. Jeyakumar, M., Sathya, S., Gandhi, S., Tharra, P., Aarthy, M., Jaya Balan, D., Kiruthiga, C., Baire, B., Singh, S.K., Devi, K.P., α-bisabolol β-D-fucopyranoside inhibits β-amyloid (Aβ)25–35 induced oxidative stress in Neuro-2a cells via antioxidant approaches. *Process Biochem.*, 121, 493–503, 2022.
78. Miller, B.J., Whisner, C.M., Johnston, C.S., Vitamin D supplementation appears to increase plasma Aβ40 in vitamin D insufficient older adults: a pilot randomized controlled trial. *J. Alzheimers Dis.*, 52, 843–847, 2016.
79. Masoumi, A., Goldenson, B., Ghirmai, S., Avagyan, H., Zaghi, J., Abel, K., Zheng, X., Espinosa-Jeffrey, A., Mahanian, M., Liu, P.T., Hewison, M., Mizwickie, M., Cashman, J., Fiala, M., 1alpha,25-dihydroxyvitamin D3 interacts with curcuminoids to stimulate amyloid-beta clearance by macrophages of Alzheimer's disease patients. *J. Alzheimers Dis.*, 17, 3, 703–17, 2009.
80. Johnson, J.W. and Kotermanski, S.E., Mechanism of action of memantine. *Curr. Opin. Pharmacol.*, 6, 1, 61–7, 2006.

81. Lee, S.R., Critical Role of Zinc as Either an Antioxidant or a Prooxidant in Cellular Systems. *Oxid. Med. Cell. Longev.*, 9156285, 1–11, 2018.
82. Fridjonsdottir, E., Vallianatou, T., Mantas, I., Shariatgorji, R., Nilsson, A., Schembri, L.S., Odell, L.R., Svenningsson, P., Andrén, P.E., Region-Specific and Age-Dependent Multitarget Effects of Acetylcholinesterase Inhibitor Tacrine on Comprehensive Neurotransmitter Systems. *ACS Chem. Biol.*, 17, 1, 147–158, 2022.
83. Chu, Q., Zhu, Y., Cao, T., Zhang, Y., Chang, Z., Liu, Y., Lu, J., Zhang, Y., Studies on the Neuroprotection of Osthole on Glutamate-Induced Apoptotic Cells and an Alzheimer's Disease Mouse Model via Modulation Oxidative Stress. *Appl. Biochem. Biotechnol.*, 190, 2, 634–644, 2020.
84. Rajput, S.A., Wang, X., Yan, H.C., Morin hydrate: A comprehensive review on novel natural dietary bioactive compound with versatile biological and pharmacological potential. *Biomed. Pharmacother.*, 138, 111511, 2021.
85. Xiao, X.Q., Wang, R., Han, Y.F., Tang, X.C., Protective effects of huperzine A on beta-amyloid (25–35) induced oxidative injury in rat pheochromocytoma cells. *Neurosci. Lett.*, 286, 155–158, 2000a.
86. Zhao, L., Wei, M.J., He, M., Jin, W.B., Zhao, H.S., Yao, W.F., The effects of tetramethylpyrazine on learning and memory abilities of mice with Alzheimer disease and its possible mechanism. *Chin. Pharmacol. Bull.*, 24, 1088–1092, 2008.
87. Li, R., He, P., Cui, J., Staufenbiel, M., Harada, N., Shen, Y., Brain endogenous estrogen levels determine responses to estrogen replacement therapy via regulation of BACE1 and NEP in female Alzheimer's transgenic mice. *Mol. Neurobiol.*, 47, 3, 857–67, 2013.
88. Lin, L., Huang, Q.X., Yang, S.S., Chu, J., Wang, J.Z., Tian, Q., Melatonin in Alzheimer's disease. *Int. J. Mol. Sci.*, 14575-93, 14(7), 2013.
89. Alkam, T., Nitta, A., Mizoguchi, H., Itoh, A., Nabeshima, T., A natural scavenger of peroxynitrites, rosmarinic acid, protects against impairment of memory induced by Abeta (25-35). *Behav. Brain Res.*, 180, 2, 139–45, 2007.
90. Luo, T., Jiang, W., Kong, Y., Li, S., He, F., Xu, J., Wang, H.Q., The protective effects of jatrorrhizine on β-amyloid (25-35)-induced neurotoxicity in rat cortical neurons. *CNS Neurol. Disord. Drug Targets*, 11, 8, 1030–7, 2012.
91. Seltzer, B., Galantamine-ER for the treatment of mild-to-moderate Alzheimer's disease. *Clin. Interv. Aging*, 5, 1–6, 2010.
92. Rink, S.M., Yarema, K.J., Solomon, M.S., Paige, L.A., Tadayoni-Rebek, B.M., Essigmann, J.M., Croy, R.G., Synthesis and biological activity of DNA damaging agents that form decoy binding sites for the estrogen receptor. *Proc. Natl. Acad. Sci. U.S.A.*, 93, 26, 15063–8, 1996.
93. Takami, K., Matsuo, A., Terai, K., Walker, D.G., McGeer, E.G., McGeer, P.L., Fibroblast growth factor receptor-1 expression in the cortex and hippocampus in Alzheimer's disease. *Brain Res.*, 802, 89–97, 1998.
94. Yang, X., He, C., Li, J., Chen, H., Ma, Q., Sui, X., Tian, S., Ying, M., Zhang, Q., Luo, Y., Zhuang, Z., Liu, J., Uptake of silica nanoparticles: neurotoxicity and Alzheimer-like pathology in human SK-N-SH and mouse neuro2a neuroblastoma cells. *Toxicol. Lett.*, 229, 1, 240–9, 2014. 17.
95. Peng, D.G., Zhang, J.S., Liu, Q.L., Taylor, E.W., Size effect of elemental selenium nanoparticles (Nano-Se) at supranutritional levels on selenium accumulation and glutathione S-transferase activity. *J. Inorg. Biochem.*, 101, 1457–63, 2007.
96. Kwon, H.J., Cha, M.Y., Kim, D., Kim, D.K., Soh, M., Shin, K., Hyeon, T., Mook-Jung, I., Mitochondria-Targeting Ceria Nanoparticles as Antioxidants for Alzheimer's Disease. *ACS Nano*, 10, 2860–2870, 2016.

97. Hernández, C.C., Burgos, C.F., Gajardo, A.H., Silva-Grecchi, T., Gavilan, J., Toledo, J.R., Fuentealba, J., Neuroprotective effects of erythropoietin on neurodegenerative and ischemic brain diseases: the role of erythropoietin receptor. *Neural Regen. Res.*, 12, 9, 1381–1389, 2017.
98. Md, S., Gan, S.Y., Haw, Y.H., Ho, C.L., Wong, S., Choudhury, H., *In Vitro* neuroprotective effects of naringenin nanoemulsion against-amyloid toxicity through the regulation of amyloidogenesis and tau phosphorylation. *Int. J. Biol. Macromol.*, 118, 1211–1219, 2018.
99. Mandel, S.A., Amit, T., Kalfon, L., Reznichenko, L., Youdim, M.B., Targeting multiple neurodegenerative diseases etiologies with multimodal-acting green tea catechins. *J. Nutr.*, 138, 1578S–1583S, 2008.
100. Munishamappa, V., Vijayakumar, A.E., Rajathilagam, T., Evaluation of the antioxidant activity of donepezil-*in vitro* study. *Natl. J. Physiol. Pharm. Pharmacol.*, 9, 2, 108–110, 2019.
101. Sachdeva, A.K., Misra, S., Pal Kaur, I., Chopra, K., Neuroprotective potential of sesamol and its loaded solid lipid nanoparticles in ICV-STZ-induced cognitive deficits: behavioral and biochemical evidence. *Eur. J. Pharmacol.*, 747, 132–40, 2015.
102. Aalinkeel, R., Kutscher, H.L., Singh, A., Cwiklinski, K., Khechen, N., Schwartz, S.A., Prasad, P.N., Mahajan, S.D., Neuroprotective effects of a biodegradable poly(lactic-co-glycolic acid)-ginsenoside Rg3 nanoformulation: a potential nanotherapy for Alzheimer's disease? *J. Drug Target.*, 26, 2, 182–193, 2018.
103. Tchantchou, F., Graves, M., Rogers, E., Ortiz, D., Shea, T.B., N-acteyl cysteine alleviates oxidative damage to central nervous system of ApoE-deficient mice following folate and vitamin E-deficiency. *J. Alzheimer's Dis.*, 7, 135–138, 2005.
104. Fukuzaki, K., Kamenosono, T., Nagata, R., Effects of ropinirole onvarious parkinsonian models in mice, rats, and cynomolgus monkeys. *Pharmacol. Biochem. Behav.*, 65, 503–8, 2000.
105. Lecht, S., Haroutiunian, S., Hoffman, A., Lazarovici, P., Rasagiline—a novel MAO B inhibitor in Parkinson's disease therapy. *Ther. Clin. Risk Manag.*, 3, 467–74, 2017.
106. Nandi, A., Yan, L.J., Jana, C.K., Das, N., Role of Catalase in Oxidative Stress- and Age-Associated Degenerative Diseases. *Oxid. Med. Cell. Longev.*, 2019, 9613090, 2019.
107. Jing, X.Z., Yang, H.J., Taximaimaiti, R., Wang, X.P., Advances in the Therapeutic Use of Non-Ergot Dopamine Agonists in the Treatment of Motor and Non-Motor Symptoms of Parkinson's Disease. *Curr. Neuropharmacol.*, 21, 5, 1224–1240, 2023.
108. Pawlowska, E., Szczepanska, J., Koskela, A., Kaarniranta, K., Blasiak, J., Dietary Polyphenols in Age-Related Macular Degeneration: Protection against Oxidative Stress and Beyond. *Oxid. Med. Cell Longev.*, 2019, 9682318, 2019.
109. Iuga, C., Alvarez-Idaboy, J.R., Vivier-Bunge, A., ROS initiated oxidation of dopamine under oxidative stress conditions in aqueous and lipidic environments. *J. Phys. Chem. B*, 115, 42, 12234–46, 2011.
110. Kawabata, K., Kawai, Y., Terao, J., Suppressive Effect of Quercetin on Acute Stress-induced Hypothalamic-pituitary-adrenal Axis Response in Wistar Rats. *J. Nutr. Biochem.*, 21, 5, 374–380, 2010.
111. Khalatbary, A.R., Olive oil phenols and neuroprotection. *Nutr. Neurosci.*, 16, 6, 243–9, 2013.
112. Zhang, J., Zhu, S., Jin, P., Huang, Y., Dai, Q., Zhu, Q., Wei, P., Yang, Z., Zhang, L., Liu, H., Xu, G., Chen, L., Gu, E., Zhang, Y., Wen, L., Liu, X., Graphene oxide improves postoperative cognitive dysfunction by maximally alleviating amyloid beta burden in mice. *Theranostics*, 10, 26, 11908–11920, 2020.
113. Chaves, N.S.G., Janner, D.E., Poetini, M.R., Fernandes, E.J., de Almeida, F.P., Musachio, E.A.S., Reginaldo, J.C., Dahleh, M.M.M., de Carvalho, A.S., Leimann, F.V., Gonçalves, O.H., Ramborger, B.P., Roehrs, R., Prigol, M., Guerra, G.P., β-carotene-loaded nanoparticles protect against neuromotor damage, oxidative stress, and dopamine deficits in a model of Parkinson's

disease in Drosophila melanogaster. *Comp. Biochem. Physiol. C. Toxicol. Pharmacol.*, 268, 109615, 2023.

114. Wang, W., Zheng, J., Zhou, H., Liu, Q., Jia, L., Zhang, X., Ge, D., Shi, W., Sun, Y., Polydopamine-Based Nanocomposite as a Biomimetic Antioxidant with a Variety of Enzymatic Activities for Parkinson's Disease. *ACS Appl. Mater. Interfaces*, 14, 29, 32901–32913, 2022.
115. Kumar, R., Kumar, R., Khurana, N., Singh, S.K., Khurana, S., Verma, S., Sharma, N., Kapoor, B., Vyas, M., Khursheed, R., Awasthi, A., Kaur, J., Corrie, L., Enhanced oral bioavailability and neuroprotective effect of fisetin through its SNEDDS against rotenone-induced Parkinson's disease rat model. *Food Chem. Toxicol.*, 144, 111590, 2020.
116. Pupyshev, A.B., Klyushnik, T.P., Akopyan, A.A., Singh, S.K., Tikhonova, M.A., Disaccharide trehalose in experimental therapies for neurodegenerative disorders: Molecular targets and translational potential. *Pharmacol. Res.*, 183, 106373, 2022.
117. Camarena, A.M., Merino, M., Sánchez-Sánchez, A.V., Blasco, S., Llinares, J.M., Mullor, J.L., An antioxidant boehmite amino-nanozyme able to disaggregate Huntington's inclusion bodies. *Chem. Commun.*, 32, 5021–5024, 2022.
118. Bakr, A.F., Abdelgayed, S.S., EL-Tawil, O.S., Bakeer, A.M., Ginger extract and ginger nanoparticles; characterization and applications. *Int. J. Vet. Sci.*, 9, 203–9, 2020.
119. Adhikari, A., Mondal, S., Das, M., Biswas, P., Pal, U., Darbar, S., Bhattacharya, S.S., Pal, D., Saha-Dasgupta, T., Das, A.K., Mallick, A.K., Pal, S.K., Incorporation of a Biocompatible Nanozyme in Cellular Antioxidant Enzyme Cascade Reverses Huntington's Like Disorder in Preclinical Model. *Adv. Healthc. Mater.*, 10, 7, e2001736, 2021.
120. Yang, T., Ferrill, L., Gallant, L., McGillicuddy, S., Fernandes, T., Schields, N., Bai, S., Verapamil and riluzole cocktail liposomes overcome pharmacoresistance by inhibiting P-glycoprotein in brain endothelial and astrocyte cells: A potent approach to treat amyotrophic lateral sclerosis. *Eur. J. Pharm. Sci.*, 120, 30–39, 2018.
121. Marcuzzo, S., Isaia, D., Bonanno, S., Malacarne, C., Cavalcante, P., Zacheo, A., Laquintana, V., Denora, N., Sanavio, B., Salvati, E., Andreozzi, P., Stellacci, F., Krol, S., Mellado-López, M., Mantegazza, R., Moreno-Manzano, V., Bernasconi, P., FM19G11-Loaded Gold Nanoparticles Enhance the Proliferation and Self-Renewal of Ependymal Stem Progenitor Cells Derived from ALS Mice. *Cells.*, 8, 3, 279, 2019.
122. Alhowail, A., Alsikhan, R., Alsaud, M., Aldubayan, M., Rabbani, S.I., Protective Effects of Pioglitazone on Cognitive Impairment and the Underlying Mechanisms: A Review of Literature. *Drug Des. Devel. Ther.*, 16, 2919–2931, 2022.
123. Rezaeimanesh, N., Rafiee, P., Saeedi, R., Khosravian, P., Sahraian, M.A., Eskandarieh, S., Moghadasi, A.N., Jahromi, S.R., The effect of crocin-selenium nanoparticles on the cognition and oxidative stress markers of multiple sclerosis patients: a randomized triple-blinded placebo-controlled clinical trial. *Biometals*, 37, 2, 305–319, 2023.
124. Redwan, H., Samuel, E.L.G., Sikkema, W.K.A., Nilewski, L.G., Lee, T., Tanner, M.R., Khan, F.S., Porter, P.C., Tajhya, R.B., Patel, R.S., Inoue, T., Pautler, R.G., Corry, D.B., Tour, J.M., Beeton, C., Preferential uptake of antioxidant carbon nanoparticles by T lymphocytes for immunomodulation. *Sci. Rep.*, 6, 33808, 2016.
125. Xiong, Y., Zheng, X., Chen, J., Chen, L., Jin, Y., Wang, P., Preparation of Capsaicin Nanoparticles and Their Effect on Treatment of Trigeminal Neuralgia in Rats. *J. Oral Sci. Res.*, 39, 10, 911–916, 2023.
126. Ferri, C.P., Prince, M., Brayne, C., Brodaty, H., Fratiglioni, L., Ganguli, M., Hall, K., Hasegawa, K., Hendrie, H., Huang, Y., Jorm, A., Mathers, C., Menezes, P.R., Rimmer, E., Scazufca, M., Alzheimer's Disease International. Global prevalence of dementia: a Delphi consensus study. *Lancet*, 366, 9503, 2112–7, 2005.

127. Goedert, M. and Spillantini, M.G., A Century of Alzheimer's Disease. *Science*, 314, 777–781, 2006.
128. Sloane, P.D., Zimmerman, S., Suchindran, C., Reed, P., Wang, L., Boustani, M., Sudha, S., The Public Health Impact of Alzheimer's Disease, 2000–2050: Potential Implication of Treatment Advances. *Annu. Rev. Public Health*, 23, 213–231, 2002.
129. Gustavsson, A., Norton, N., Fast, T., Frölich, L., Georges, J., Holzapfel, D., Kirabali, T., Krolak-Salmon, P., Rossini, P.M., Ferretti, M.T., Lanman, L., Chadha, A.S., van der Flier, W.M., Global estimates on the number of persons across the Alzheimer's disease continuum. *Alzheimer's Dement.*, 19, 658–670, 2022.
130. Anand, A., Arya, M., Kaithwas, G., Singh, G., Saraf, S.A., Sucrose stearate as a biosurfactant for development of rivastigmine containing nanostructured lipid carriers and assessment of its activity against dementia in C. elegans model. *J. Drug Deliv. Sci. Technol.*, 49, 219–226, 2018.
131. Montine, T.J., Phelps, C.H., Beach, T.G., Bigio, E.H., Cairns, N.J., Dickson, D.W., Duyckaerts, C., Frosch, M.P., Masliah, E., Mirra, S.S., Nelson, P.T., Schneider, J.A., Thal, D.R., Trojanowski, J.Q., Vinters, H.V., Hyman, B.T., National Institute on Aging–Alzheimer's Association guidelines for the neuropathologic assessment of Alzheimer's disease: A practical approach. *Acta Neuropathol.*, 123, 1–11, 2011.
132. Mudshinge, S.R., Deore, A.B., Patil, S., Bhalgat, C.M., Nanoparticles: Emerging carriers for drug delivery. *Saudi Pharm. J.*, 19, 129–141, 2011.
133. Papadia, K., Giannou, A.D., Markoutsa, E., Bigot, C., Vanhoute, G., Mourtas, S., Van der Linded, A., Stathopoulos, G.T., Antimisiaris, S.G., Multifunctional LUV liposomes decorated for BBB and amyloid targeting-B. *In vivo* brain targeting potential in wild-type and APP/PS1 mice. *Eur. J. Pharmaceut. Sci.*, 102, 180–187, 2017.
134. Cheng, K.K., Yeung, C.F., Ho, S.W., Chow, S.F., Chow, A.H., Baum, L., Highly stabilized curcumin nanoparticles tested in an *in vitro* blood-brain barrier model and in Alzheimer's disease Tg2576 mice. *AAPS J.*, 15, 2, 324–36, 2013.
135. Tiwari, S.K., Agarwal, S., Seth, B., Yadav, A., Nair, S., Bhatnagar, P., Karmakar, M., Kumari, M., Chauhan, L.K., Patel, D.K., Srivastava, V., Singh, D., Gupta, S.K., Tripathi, A., Chaturvedi, R.K., Gupta, K.C., Curcumin-loaded nanoparticles potently induce adult neurogenesis and reverse cognitive deficits in Alzheimer's disease model via canonical Wnt/β-catenin pathway. *ACS Nano*, 8, 1, 76–103, 2014.
136. Barbara, R., Belletti, D., Pederzoli, F., Masoni, M., Keller, J., Ballestrazzi, A., Vandelli, M.A., Tosi, G., Grabrucker, A.M., Novel Curcumin loaded nanoparticles engineered for Blood-Brain Barrier crossing and able to disrupt Abeta aggregates. *Int. J. Pharm.*, 526, 1-2, 413–424, 2017.
137. Fan, S., Zheng, Y., Liu, X., Fang, W., Chen, X., Liao, W., Jing, X., Lei, M., Tao, E., Ma, Q., Zhang, X., Guo, R., Liu, J., Curcumin-loaded PLGA-PEG nanoparticles conjugated with B6 peptide for potential use in Alzheimer's disease. *Drug Deliv.*, 25, 1, 1091–1102, 2018.
138. Lazar, A.N., Mourtas, S., Youssef, I., Parizot, C., Dauphin, A., Delatour, B., Antimisiaris, S.G., Duyckaerts, C., Curcumin-conjugated nanoliposomes with high affinity for Aβ deposits: possible applications to Alzheimer disease. *Nanomedicine*, 9, 5, 712–21, 2013.
139. Huo, X., Zhang, Y., Jin, X., Li, Y., Zhang, L., A novel synthesis of selenium nanoparticles encapsulated PLGA nanospheres with curcumin molecules for the inhibition of amyloid aggregation in Alzheimer's disease. *J. Photochem. Photobiol. B Biol.*, 190, 98–102, 2019.
140. Mathew, A., Fukuda, T., Nagaoka, Y., Hasumura, T., Morimoto, H., Yoshida, Y., Maekawa, T., Venugopal, K., Kumar, D.S., Curcumin loaded-PLGA nanoparticles conjugated with Tet-1 peptide for potential use in Alzheimer's disease. *PLoS One*, 7, 3, e32616, 2012.
141. Sadegh Malvajerd, S., Izadi, Z., Azadi, A., Kurd, M., Derakhshankhah, H., Sharifzadeh, M., Akbari Javar, H., Hamidi, M., Neuroprotective Potential of Curcumin-Loaded Nanostructured

Lipid Carrier in an Animal Model of Alzheimer's Disease: Behavioral and Biochemical Evidence. *J. Alzheimers Dis.*, 69, 3, 671–686, 2019.

142. Huang, N., Lu, S., Liu, X.G., Zhu, J., Wang, Y.J., Liu, R.T., PLGA nanoparticles modified with a BBB-penetrating peptide co-delivering Aβ generation inhibitor and curcumin attenuate memory deficits and neuropathology in Alzheimer's disease mice. *Oncotarget.*, 8, 46, 81001–81013, 2017.
143. Malvajerd, S.S., Azadi, A., Izadi, Z., Kurd, M., Dara, T., Dibaei, M., Zadeh, M.S., Javar, H.A., Hamidi, M., Brain Delivery of Curcumin Using Solid Lipid Nanoparticles and Nanostructured Lipid Carriers: Preparation, Optimization, and Pharmacokinetic Evaluation. *ACS Chem. Neurosci.*, 10, 728–739, 2019.
144. Desai, P.P. and Patravale, V.B., Curcumin Cocrystal Micelles-Multifunctional Nanocomposites for Management of Neurodegenerative Ailments. *J. Pharm. Sci.*, 107, 4, 1143–1156, 2018.
145. Nasr, M., Development of an optimized hyaluronic acid-based lipidic nanoemulsion co-encapsulating two polyphenols for nose to brain delivery. *Drug Deliv.*, 23, 4, 1444–52, 2016.
146. Doggui, S., Sahni, J.K., Arseneault, M., Dao, L., Ramassamy, C., Neuronal uptake and neuroprotective effect of curcumin-loaded PLGA nanoparticles on the human SK-N-SH cell line. *J. Alzheimers Dis.*, 30, 2, 377–92, 2012.
147. Liu, X.G., Zhang, L., Lu, S., Liu, D.Q., Huang, Y.R., Zhu, J., Zhou, W.W., Yu, X.L., Liu, R.T., Superparamagnetic iron oxide nanoparticles conjugated with Aβ oligomer-specific scFv antibody and class Aβ scavenger receptor activator show therapeutic potentials for Alzheimer's Disease. *J. Nanobiotechnol.*, 18, 160, 2020.
148. Binyamin, O., Nitzan, K., Frid, K., Ungar, Y., Rosenmann, H., Gabizon, R., Brain Targeting of 9c,11t-Conjugated Linoleic Acid, a Natural Calpain Inhibitor, Preserves Memory and Reduces Aβ and P25 Accumulation in 5XFAD Mice. *Sci. Rep.*, 9, 1, 18437, 2019.
149. Almuhayawi, M.S., Ramadan, W.S., Harakeh, S., Al Jaouni, S.K., Bharali, D.J., Mousa, S.A., Almuhayawi, S.M., The potential role of pomegranate and its nano-formulations on cerebral neurons in aluminum chloride induced Alzheimer rat model. *Saudi J. Biol. Sci.*, 27, 1710–1716, 2020.
150. Rajput, A., Bariya, A., Allam, A., Othman, S., Butani, S.B., *In situ* nanostructured hydrogel of resveratrol for brain targeting: *in vitro-in vivo* characterization. *Drug Deliv. Transl. Res.*, 8, 5, 1460–1470, 2018.
151. Hao, J., Zhao, J., Zhang, S., Tong, T., Zhuang, Q., Jin, K., Chen, W., Tang, H., Fabrication of an ionic-sensitive in situ gel loaded with resveratrol nanosuspensions intended for direct nose-to-brain delivery. *Colloids Surf., B*, 147, 376–386, 2016.
152. Trotta, V., Pavan, B., Ferraro, L., Beggiato, S., Traini, D., Reis, L.G.D., Traini, D., Dalpiaz, A., Brain targeting of resveratrol by nasal administration of chitosan-coated lipid microparticles. *Eur. J. Pharm. Biopharm.*, 127, 250–259, 2018.
153. Salem, H.F., Kharshoum, R.M., Abou-Taleb, H.A., Naguib, D.M., Nanosized Transferosome-Based Intranasal In Situ Gel for Brain Targeting of Resveratrol: Formulation, Optimization *In Vitro* Evaluation, and *In Vivo* Pharmacokinetic Study. *AAPS Pharm. Sci. Tech.*, 20, 181, 2019.
154. Loureiro, J.A., Andrade, S., Duarte, A., Neves, A.R., Queiroz, J.F., Nunes, C., Sevin, E., Fenart, L., Gosselet, F., Coelho, M.A., Pereira, M.C., Resveratrol and Grape Extract-loaded Solid Lipid Nanoparticles for the Treatment of Alzheimer's Disease. *Molecules*, 22, 2, 277, 2017.
155. dos Santos Rodrigues, B., Kanekiyo, T., Singh, J., ApoE-2 Brain-Targeted Gene Therapy Through Transferrin and Penetratin Tagged Liposomal Nanoparticles. *Pharm. Res.*, 36, 161, 2019.
156. Kuo, Y.C. and Rajesh, R., Targeted delivery of rosmarinic acid across the blood-brain barrier for neuronal rescue using polyacrylamide-chitosan-poly(lactide-co-glycolide) nanoparticles with surface cross-reacting material 197 and apolipoprotein E. *Int. J. Pharm.*, 528, 1–2, 228–241, 2017.

157. Yang, C.R., Zhao, X.L., Hu, H.Y., Li, K.X., Sun, X., Li, L., Chen, D.W., Preparation, optimization and characteristic of huperzine a loaded nanostructured lipid carriers. *Chem. Pharm. Bull. (Tokyo)*, 58, 5, 656–61, 2010.
158. Shi, J., Cong, W., Wang, Y., Liu, Q., Luo, G., Microemulsion-based patch for transdermal delivery of huperzine A and ligustrazine phosphate in treatment of Alzheimer's disease. *Drug Dev. Ind. Pharm.*, 38, 752–761, 2021.
159. Meng, Q., Wang, A., Hua, H., Jiang, Y., Wang, Y., Mu, H., Wu, Z., Sun, K., Intranasal delivery of Huperzine A to the brain using lactoferrin-conjugated N-trimethylated chitosan surface-modified PLGA nanoparticles for treatment of Alzheimer's disease. *Int. J. Nanomed.*, 13, 705–718, 2018.
160. Kuo, Y.C., Chen, I.Y., Rajesh, R., Use of functionalized liposomes loaded with antioxidants to permeate the blood–brain barrier and inhibit β-amyloid-induced neurodegeneration in the brain. *J. Taiwan Inst. Chem. Eng.*, 87, 1–14, 2018.
161. Moreno, L.C.G.E.I., Puerta, E., Suárez-Santiago, J.E., Santos-Magalhães, N.S., Ramirez, M.J., Irache, J.M., Effect of the oral administration of nanoencapsulated quercetin on a mouse model of Alzheimer's disease. *Int. J. Pharm.*, 517, 1–2, 50–57, 2017.
162. Testa, G., Gamba, P., Badilli, U., Gargiulo, S., Maina, M., Guina, T., Calfapietra, S., Biasi, F., Cavalli, R., Poli, G., Leonarduzzi, G., Loading into nanoparticles improves quercetin's efficacy in preventing neuroinflammation induced by oxysterols. *PLoS One*, 9, 5, e96795, 2014.
163. Palle, S. and Neerati, P., Quercetin nanoparticles attenuates scopolamine induced spatial memory deficits and pathological damages in rats. *Bull. Fac. Pharm. Cairo Univ.*, 55, 1, 101–106, 2017.
164. Sun, D., Li, N., Zhang, W., Zhao, Z., Mou, Z., Huang, D., Liu, J., Wang, W., Design of PLGA-functionalized quercetin nanoparticles for potential use in Alzheimer's disease. *Colloids Surf., B*, 148, 116–129, 2016.
165. Fachel, F.N.S., Medeiros-Neves, B., Prá, M.D., Schuh, R.S., Veras, K.S., Bassani, V.L., Koester, L.S., Henriques, A.T., Braganhol, E., Teixeira, H.F., Box-Behnken design optimization of mucoadhesive chitosan-coated nanoemulsions for rosmarinic acid nasal delivery—*In vitro* studies. *Carbohydr. Polym.*, 199, 572–582, 2018.
166. Ismail, N., Ismail, M., Azmi, N.H., Bakar, M.F.A., Yida, Z., Stanslas, J., Sani, D., Basri, H., Abdullah, M.A., Beneficial effects of TQRF and TQ nano- and conventional emulsions on memory deficit, lipid peroxidation, total antioxidant status, antioxidants genes expression and soluble Aβ levels in high fat-cholesterol diet induced rats. *Chem. Biol. Interact.*, 275, 61–73, 2017.
167. Yusuf, M., Khan, M., Alrobaian, M.M., Alghamdi, S.A., Warsi, M.H., Sultana, S., Khan, R.A., Brain targeted Polysorbate-80 coated PLGA thymoquinone nanoparticles for the treatment of Alzheimer's disease, with biomechanistic insights. *J. Drug Deliv. Sci. Technol.*, 61, 102214, 2021.
168. Sánchez-López, E., Ettcheto, M., Egea, M.A., Espina, M., Cano, A., Calpena, A.C., Camins, A., Carmona-Ule, N., Silva, A.M., Souto, E.B., Garcia, M.L., Memantine loaded PLGA PEGylated nanoparticles for Alzheimer's disease: *In Vitro* and *In Vivo* characterization. *J. Nanobiotechnol.*, 16, 32, 2018.
169. Kaur, A., Nigam, K., Srivastava, S., Tyagi, A., Dang, S., Memantine nanoemulsion: A new approach to treat Alzheimer's disease. *J. Microencapsul.*, 37, 355–365, 2020.
170. Igartúa, D.E., Martinez, C.S., Del, V., Alonso, S., Prieto, M.J., Combined Therapy for Alzheimer's Disease: Tacrine and PAMAM Dendrimers Co-Administration Reduces the Side Effects of the Drug without Modifying its Activity. *AAPS Pharm. Sci. Tech.*, 21, 3, 110, 2020.
171. Jogani, V.V., Shah, P.J., Mishra, P., Mishra, A.K., Misra, A., Intranasal Mucoadhesive Microemulsion of Tacrine to Improve Brain Targeting. *Alzheimer Dis. Assoc. Disord.*, 22, 116–124, 2008.

172. Smith, A., Giunta, B., Bickford, P.C., Fountain, M., Tan, J., Shytle, R.D., Nanolipidic particles improve the bioavailability and alpha-secretase inducing ability of epigallocatechin-3-gallate (EGCG) for the treatment of Alzheimer's disease. *Int. J. Pharm.*, 389, 1–2, 207–12, 2010.
173. Singh, N.A., Mandal, A.K.A., Khan, Z.A., Inhibition of Al(III)-Induced $A\beta_{42}$ Fibrillation and Reduction of Neurotoxicity by Epigallocatechin-3-Gallate Nanoparticles. *J. Biomed. Nanotechnol.*, 14, 6, 1147–1158, 2018.
174. Zhang, J., Zhou, X., Yu, Q., Yang, L., Sun, D., Zhou, Y., Liu, J., Epigallocatechin-3-gallate (EGCG)-stabilized selenium nanoparticles coated with Tet-1 peptide to reduce amyloid-β aggregation and cytotoxicity. *ACS Appl. Mater. Interfaces*, 6, 11, 8475–87, 2014.
175. Nageeb El-Helaly, S., Abd Elbary, A., Kassem, M.A., El-Nabarawi, M.A., Electrosteric stealth Rivastigmine loaded liposomes for brain targeting: preparation, characterization, *ex vivo*, bio-distribution and *in vivo* pharmacokinetic studies. *Drug Deliv.*, 24, 1, 692–700, 2017.
176. Etman, S.M., Elnaggar, Y.S.R., Abdelmonsif, D.A., Abdallah, O.Y., Oral brain-targeted microemulsion for enhanced piperine delivery in alzheimer's disease therapy: *in vitro* appraisal, *in vivo* activity, and nanotoxicity. *AAPS Pharm. Sci. Tech.*, 19, 3698–3711, 2018.
177. Sathya, S., Shanmuganathan, B., Manirathinam, G., Ruckmani, K., Devi, K.P., α-Bisabolol loaded solid lipid nanoparticles attenuates Aβ aggregation and protects Neuro-2a cells from Aβ induced neurotoxicity. *J. Mol. Liq.*, 264, 431–441, 2018.
178. Jeon, S.G., Cha, M.Y., Kim, J.I., Hwang, T.W., Kim, K.A., Kim, T.H., Song, K.C., Kim, J.J., Moon, M., Vitamin D-binding protein-loaded PLGA nanoparticles suppress Alzheimer's disease-related pathology in 5XFAD mice. *Nanomed. Nanotechnol. Biol. Med.*, 17, 297–307, 2019.
179. Vilella, A., Belletti, D., Sauer, A.K., Hagmeyer, S., Sarowar, T., Masoni, M., Stasiak, N., Mulvihill, J.J., Ruozi, B., Forni, F., Vandelli, M.A., Tosi, G., Zoli, M., Grabrucker, A.M., Reduced plaque size and inflammation in the APP23 mouse model for Alzheimer's disease after chronic application of polymeric nanoparticles for CNS targeted zinc delivery. *J. Trace Ele. Med. Biol.*, 49, 210–221, 2018.
180. Ni, Y.N., Kong, L., Xiao, H.H., Wu, Y.T., Liang, X.C., Lin, Y., Li, W.Y., Deng, Y., Li, Y., Shi, Y., Cheng, L., Li, H.Y., Ju, R.J., Yang, J.X., Multifunctional osthole liposomes and brain targeting functionality with potential applications in a mouse model of Alzheimer's disease. *J. Liposome Res.*, 31, 3, 267–278, 2021.
181. Sharma, D., Singh, M., Kumar, P., Vikram, V., Mishra, N., Development and characterization of morin hydrate loaded microemulsion for the management of Alzheimer's disease. *Artif. Cells Nanomed. Biotechnol.*, 45, 8, 1620–1630, 2017.
182. Rassu, G., Porcu, E.P., Fancello, S., Obinu, A., Senes, N., Galleri, G., Migheli, R., Gavini, E., Giunchedi, P., Intranasal Delivery of Genistein-Loaded Nanoparticles as a Potential Preventive System against Neurodegenerative Disorders. *Pharmaceutics*, 11, 8, 2018.
183. Jayachandra Babu, R., Dayal, P.P., Pawar, K., Singh, M., Nose-to-brain transport of melatonin from polymer gel suspensions: a microdialysis study in rats. *J. Drug Target.*, 19, 9, 731–40, 2011.
184. Lohan, S., Raza, K., Mehta, S.K., Bhatti, G.K., Saini, S., Singh, B., Anti-Alzheimer's potential of berberine using surface decorated multi-walled carbon nanotubes: A preclinical evidence. *Int. J. Pharm.*, 530, 1–2, 263–278, 2017.
185. Misra, S., Chopra, K., Sinha, V.R., Medhi, B., Galantamine-loaded solid-lipid nanoparticles for enhanced brain delivery: preparation, characterization, *in vitro* and *in vivo* evaluations. *Drug Deliv.*, 23, 4, 1434–43, 2016.
186. Mittal, G., Carswell, H., Brett, R., Currie, S., Kumar, M.N., Development and evaluation of polymer nanoparticles for oral delivery of estradiol to rat brain in a model of Alzheimer's pathology. *J. Control Release*, 150, 2, 220–8, 2011.

187. Zhang, C., Chen, J., Feng, C., Shao, X., Liu, Q., Zhang, Q., Pang, Z., Jiang, X., Intranasal nanoparticles of basic fibroblast growth factor for brain delivery to treat Alzheimer's disease. *Int. J. Pharm.*, 461, 1–2, 192–202, 2014.
188. Yang, X., He, C., Li, J., Chen, H., Ma, Q., Sui, X., Tian, S., Ying, M., Zhang, Q., Luo, Y., Zhuang, Z., Liu, J., Uptake of silica nanoparticles: Neurotoxicity and Alzheimer-like pathology in human SK-N-SH and mouse neuro2a neuroblastoma cells. *Toxicol. Lett.*, 229, 240–249, 2014.
189. Yin, T., Yang, L., Liu, Y., Zhou, X., Sun, J., Liu, J., Sialic acid (SA)-modified selenium nanoparticles coated with a high blood–brain barrier permeability peptide-B6 peptide for potential use in Alzheimer's disease. *Acta Biomater.*, 25, 172–183, 2015.
190. Dara, T., Vatanara, A., Sharifzadeh, M., Khani, S., Vakilinezhad, M.A., Vakhshiteh, F., Meybodi, M.N., Malvajerd, S.S., Hassani, S., Mosaddegh, M.H., Improvement of memory deficits in the rat model of Alzheimer's disease by erythropoietin-loaded solid lipid nanoparticles. *Neurobiol. Learn. Mem.*, 166, 107082, 2019.
191. Vaz, G.R., Hädrich, G., Bidone, J., Rodrigues, J.L., Falkembach, M.C., Putaux, J.L., Hort, M.A., Monserrat, J.M., Varela Junior, A.S., Teixeira, H.F., Muccillo-Baisch, A.L., Horn, A.P., Dora, C.L., Development of Nasal Lipid Nanocarriers Containing Curcumin for Brain Targeting. *J. Alzheimers Dis.*, 59, 3, 961–974, 2017.
192. Shi, J., Wang, Y., Luo, G., Ligustrazine Phosphate Ethosomes for Treatment of Alzheimer's Disease, *In Vitro* and in Animal Model Studies. *AAPS Pharm. Sci. Tech.*, 13, 485–492, 2012.
193. Kaur, A., Nigam, K., Bhatnagar, I., Sukhpal, H., Awasthy, S., Shankar, S., Tyagi, A., Dang, S., Treatment of Alzheimer's diseases using donepezil nanoemulsion: An intranasal approach. *Drug Deliv. Transl. Res.*, 10, 1862–1875, 2020.
194. Lin, M.T. and Beal, M.F., Mitochondrial dysfunction and oxidative stress in neurodegenerative diseases. *Nature.*, 443, 787, 2006.
195. Lotharius, J. and Brundin, P., Pathogenesis of Parkinson's disease: dopamine, vesicles and α-synuclein. *Nat. Rev. Neurosci.*, 3, 932, 2002.
196. De Lau, L.M. and Breteler, M.M., Epidemiology of Parkinson's disease. *Lancet Neurol.*, 5, 525–35, 2006.
197. Bose, A. and Beal, M.F., Mitochondrial dysfunction in Parkinson's disease. *J. Neurochem.*, 139, 216–31, 2016.
198. Hirsch, L., Jette, N., Frolkis, A., Steeves, T., Pringsheim, T., The incidence of Parkinson's disease: a systematic review and meta-analysis. *Neuroepidemiology.*, 46, 292–300, 2016.
199. Esteves, A.R., Arduíno, D.M., Swerdlow, R.H., Oliveira, C.R., Cardoso, S.M., Oxidative stress involvement in α-synuclein oligomerization in Parkinson's disease cybrids. *Antioxid. Redox Signal.*, 11, 439–48, 2009.
200. Palle, S. and Neerati, P., Improved neuroprotective effect of resveratrol nanoparticles as evinced by abrogation of rotenone-induced behavioral deficits and oxidative and mitochondrial dysfunctions in rat model of Parkinson's disease. *Naunyn. Schmiedebergs. Arch. Pharmacol.*, 391, 4, 445–453, 2018.
201. Wang, M., Li, L., Zhang, X., Liu, Y., Zhu, R., Liu, L., Fang, Y., Gao, Z., Gao, D., Magnetic Resveratrol Liposomes as a New Theranostic Platform for Magnetic Resonance Imaging Guided Parkinson's Disease Targeting Therapy. *ACS Sustain. Chem. Eng.*, 6, 12, 17124–17133, 2018.
202. da Rocha Lindner, G., Bonfanti Santos, D., Colle, D., Gasnhar Moreira, E.L., Daniel Prediger, R., Farina, M., Khalil, N.M., Mara Mainardes, R., Improved neuroprotective effects of resveratrol-loaded polysorbate 80-coated poly(lactide) nanoparticles in MPTP-induced Parkinsonism. *Nanomed. (Lond.)*, 10, 7, 1127–38, 2015.

203. Katila, N., Duwa, R., Bhurtel, S., Khanal, S., Maharjan, S., Jeong, J.H., Lee, S., Choi, D.Y., Yook, S., Enhancement of blood-brain barrier penetration and the neuroprotective effect of resveratrol. *J. Control Release*, 346, 1–19, 2022.
204. Wang, Y., Xu, H., Fu, Q., Ma, R., Xiang, J., Protective effect of resveratrol derived from Polygonum cuspidatum and its liposomal form on nigral cells in Parkinsonian rats. *J. Neurol. Sci.*, 304, 1-2, 29–34, 2011.
205. Pangeni, R., Sharma, S., Mustafa, G., Ali, J., Baboota, S., Vitamin E loaded resveratrol nanoemulsion for brain targeting for the treatment of Parkinson's disease by reducing oxidative stress. *Nanotechnology*, 25, 485102, 2014.
206. Barcia, E., Boeva, L., García-García, L., Slowing, K., Fernández-Carballido, A., Casanova, Y., Negro, S., Nanotechnology-based drug delivery of ropinirole for Parkinson's disease. *Drug Deliv.*, 24, 1, 1112–1123, 2017.
207. Pardeshi, C.V., Belgamwar, V.S., Tekade, A.R., Surana, S.J., Novel surface modified polymer-lipid hybrid nanoparticles as intranasal carriers for ropinirole hydrochloride: *in vitro*, *ex vivo* and *in vivo* pharmacodynamic evaluation. *J. Mater. Sci. - Mater. Med.*, 24, 9, 2101–15, 2013.
208. Gaba, B., Khan, T., Haider, F., Alam, T., Baboota, S., Parvez, S., Ali, J., Vitamin E Loaded Naringenin Nanoemulsion via Intranasal Delivery for the Management of Oxidative Stress in a 6-OHDA Parkinson's Disease Model. *Biomed Res. Int.*, 2019, 2382563, 2019.
209. Md, S., Alhakamy, N.A., Aldawsari, H.M., Asfour, H.Z., Neuroprotective and Antioxidant Effect of Naringenin-Loaded Nanoparticles for Nose-to-Brain Delivery. *Brain Sci.*, 9, 10, 275, 2019.
210. Mani, M., Balasubramanian, S., Manikandan, K.R., Kulandaivel, B., Neuroprotective potential of Naringenin loaded solid-lipid nanoparticles against Rotenone-induced Parkinson's disease model. *J. Appl. Pharm. Sci.*, 11, 02, 019–028, 2021.
211. Bollimpelli, V.S., Kumar, P., Kumari, S., Kondapi, A.K., Neuroprotective effect of curcumin-loaded lactoferrin nano particles against rotenone induced neurotoxicity. *Neurochem. Int.*, 95, 37–45, 2016.
212. Zhao, N., Yang, X., Calvelli, H.R., Cao, Y., Francis, N.L., Chmielowski, R.A., Joseph, L.B., Pang, Z.P., Uhrich, K.E., Baum, J., Moghe, P.V., Antioxidant Nanoparticles for Concerted Inhibition of α-Synuclein Fibrillization, and Attenuation of Microglial Intracellular Aggregation and Activation. *Front. Bioeng. Biotechnol.*, 8, 112, 2020.
213. Hegazy, M.A., Maklad, H.M., Samy, D.M., Abdelmonsif, D.A., El Sabaa, B.M., Elnozahy, F.Y., Cerium oxide nanoparticles could ameliorate behavioral and neurochemical impairments in 6-hydroxydopamine induced Parkinson's disease in rats. *Neurochem. Int.*, 108, 361–371, 2017.
214. Mohammad, Khan, U.A., Warsi, M.H., Alkreathy, H.M., Karim, S., Jain, G.K., Ali, A., Intranasal cerium oxide nanoparticles improves locomotor activity and reduces oxidative stress and neuroinflammation in haloperidol-induced parkinsonism in rats. *Front. Pharmacol.*, 14, 1188470, 2023.
215. Mursaleen, L., Noble, B., Chan, S.H.Y., Somavarapu, S., Zariwala, M.G., N-Acetylcysteine Nanocarriers Protect against Oxidative Stress in a Cellular Model of Parkinson's Disease. *Antioxid. (Basel)*, 9, 7, 600, 2020.
216. Kumar, S., Ali, J., Baboota, S., Design Expert(®) supported optimization and predictive analysis of selegiline nanoemulsion via the olfactory region with enhanced behavioural performance in Parkinson's disease. *Nanotechnology*, 27, 43, 435101, 2016.
217. Haney, M.J., Klyachko, N.L., Zhao, Y., Gupta, R., Plotnikova, E.G., He, Z., Patel, T., Piroyan, A., Sokolsky, M., Kabanov, A.V., Batrakova, E.V., Exosomes as drug delivery vehicles for Parkinson's disease therapy. *J. Control. Release*, 207, 18–30, 2015.

218. Raj, R., Wairkar, S., Sridhar, V., Gaud, R., Pramipexole dihydrochloride loaded chitosan nanoparticles for nose to brain delivery: Development, characterization and *in vivo* anti-Parkinson activity. *Int. J. Biol. Macromol.*, 109, 27–35, 2018.
219. Kwon, H.J., Kim, D., Seo, K., Kim, Y.G., Han, S.I., Kang, T., Soh, M., Hyeon, T., Ceria Nanoparticle Systems for Selective Scavenging of Mitochondrial, Intracellular, and Extracellular Reactive Oxygen Species in Parkinson's Disease. *Angew. Chem. Int. Ed. Engl.*, 57, 30, 9408–9412, 2018.
220. Pahuja, R., Seth, K., Shukla, A., Shukla, R.K., Bhatnagar, P., Chauhan, L.K.S., Saxena, P.N., Arun, J., Chaudhari, B.P., Patel, D.K., Singh, S.P., Shukla, R., Khanna, V.K., Kumar, P., Chaturvedi, R.K., Gupta, K.C., Correction to Trans-Blood Brain Barrier Delivery of Dopamine-Loaded Nanoparticles Reverses Functional Deficits in Parkinsonian Rats. *ACS Nano*, 13, 7, 8490, 2019.
221. Beigi, H.M., Morshedi, D., Shojaosadati, S.A., Pedersen, J.N., Marvian, A.T., Aliakbari, F., Christiansen, G., Pedersena, J.S., Otzen, D.E., Gallic acid loaded onto polyethylenimine-coated human serum albumin nanoparticles (PEI-HSA-GA NPs) stabilizes α-synuclein in the unfolded conformation and inhibits aggregation. *RSC Adv.*, 6, 85312–85323, 2016.
222. Mursaleen, L., Noble, B., Somavarapu, S., Zariwala, M.G., Micellar Nanocarriers of Hydroxytyrosol Are Protective against Parkinson's Related Oxidative Stress in an *In Vitro* hCMEC/D3-SH-SY5Y Co-Culture System. *Antioxidants (Basel)*, 10, 6, 887, 2021.
223. Oukhrib, M., Tamegart, L., Assafi, A., Hejji, L., Azzouz, A., Villarejo, L.P., Haida, M., Mohamed, C., Gamrani, H., Effects of graphene oxide nanoparticles administration against reserpine-induced neurobehavioral damage and oxidative stress in an animal model of Parkinson's disease. *Food Chem. Toxicol.*, 178, 113904, 2023.
224. Chaves, N.S.G., Janner, D.E., Poetini, M.R., Fernandes, E.J., de Almeida, F.P., Musachio, E.A.S., Reginaldo, J.C., Dahleh, M.M.M., de Carvalho, A.S., Leimann, F.V., Gonçalves, O.H., Ramborger, B.P., Roehrs, R., Prigol, M., Guerra, G.P., β-carotene-loaded nanoparticles protect against neuromotor damage, oxidative stress, and dopamine deficits in a model of Parkinson's disease in Drosophila melanogaster. *Comp. Biochem. Physiol. C. Toxicol. Pharmacol.*, 268, 109615, 2023.
225. Wang, W., Zheng, J., Zhou, H., Liu, Q., Jia, L., Zhang, X., Ge, D., Shi, W., Sun, Y., Polydopamine-Based Nanocomposite as a Biomimetic Antioxidant with a Variety of Enzymatic Activities for Parkinson's Disease. *ACS Appl. Mater. Interfaces*, 14, 29, 32901–32913, 2022.
226. Vonsattel, J.P.G. and DiFiglia, M., Huntington disease. *J. Neuropathol. Exp. Neurol.*, 57, 5, 369–384, 1998.
227. Bonilla, E., Huntington disease. A review. *Investigacion. Clinica.*, 41, 2, 117–141, 2000.
228. A novel gene containing a trinucleotide repeat that is expanded and unstable on Huntington's disease chromosomes. *The Huntington's Disease Collaborative Research Group Cell*, 72, 6, 971–83, 1993.
229. ACMG/ASHG statement, Laboratory guidelines for Huntington disease genetic testing. The American College of Medical Genetics/American Society of Human Genetics Huntington Disease Genetic Testing Working Group. *Am. J. Hum. Genet.*, 62, 5, 1243–7, 1998.
230. Gusella, J.F., Gibbons, K., Hobbs, W., The G8 locus linked to Huntington's disease. *Am. J. Hum. Genet.*, 36, 139S, 1984.
231. Gidalevitz, T., Ben-Zvi, A., Ho, K.H., Brignull, H.R., Morimoto, R.I., Progressive disruption of cellular protein folding in models of polyglutamine diseases. *Science*, 311, 5766, 1471–1474, 2006.
232. DiFiglia, M., Sapp, E., Chase, K.O., Davies, S.W., Bates, G.P., Vonsattel, J.P., Aronin, N., Aggregation of huntingtin in neuronal intranuclear inclusions and dystrophic neurites in brain. *Science.*, 277, 5334, 1990–3, 1997.

233. Paulsen, J.S., Magnotta, V.A., Mikos, A.E., Paulson, H.L., Penziner, E., Andreasen, N.C., Nopoulos, P.C., Brain structure in preclinical Huntington's disease. *Biol. Psychiatry*, 59, 1, 57–63, 2006.
234. Walker, F.O., Huntington's disease. *Lancet*, 369, 9557, 218–228, 2007.
235. Warby, S.C., Visscher, H., Collins, J.A., Doty, C.N., Carter, C., Butland, S.L., Hayden, A.R., Kanazawa, I., Ross, C.J., Hayden, M.R., HTT haplotypes contribute to differences in Huntington disease prevalence between Europe and East Asia. *Eur. J. Hum. Genet.*, 19, 5, 561–6, 2011.
236. Rawlins, M.D., Wexler, N.S., Wexler, A.R., Tabrizi, S.J., Douglas, I., Stephen, J.W.E., Smeeth, L., The prevalence of Huntington's disease. *Neuroepidemiology*, 46, 2, 144–153, 2016.
237. Moily, N.S., Kota, L.N., Anjanappa, R.M., Venugopal, S., Vaidyanathan, R., Pal, P., Purushottam, M., Jain, S., Kandasamy, M., Trinucleotide repeats and haplotypes at the huntingtin locus in an Indian sample overlaps with European haplogroup a. *PLoS Curr.*, 6, 2014. https://www.ncbi.nlm.nih.gov/pmc/articles/PMC4205232/
238. Trushina, E. and McMurray, C.T., Oxidative stress and mitochondrial dysfunction in neurodegenerative diseases. *Neuroscience*, 145, 4, 1233–1248, 2007.
239. Freeman, B.A. and Crapo, J.D., Biology of disease. Free radicals and tissue injury. *Lab. Invest.*, 47, 5, 412–426, 1982.
240. Emerit, J., Edeas, M., Bricaire, F., Neurodegenerative diseases and oxidative stress. *Biomed. Pharmacother.*, 58, 1, 39–46, 2004.
241. Finkel, T. and Holbrook, N.J., Oxidants, oxidative stress and the biology of ageing. *Nature*, 408, 6809, 239–247, 2000.
242. Bhatt, R., Singh, D., Prakash, A., Mishra, N., Development, characterization and nasal delivery of rosmarinic acid-loaded solid lipid nanoparticles for the effective management of Huntington's disease. *Drug Deliv.*, 22, 931–939, 2014.
243. Sandhir, R., Yadav, A., Mehrotra, A., Sunkaria, A., Singh, A., Sharma, S., Curcumin nanoparticles attenuate neurochemical and neurobehavioral deficits in experimental model of Huntington's disease. *Neuromol. Med.*, 16, 1, 106–18, 2014.
244. Pepe, G., Calce, E., Verdoliva, V., Saviano, M., Maglione, V., Di Pardo, A., De Luca, S., Curcumin-Loaded Nanoparticles Based on Amphiphilic Hyaluronan-Conjugate Explored as Targeting Delivery System for Neurodegenerative Disorders. *Int. J. Mol. Sci.*, 21, 22, 8846, 2020.
245. Cong, W., Bai, R., Li, Y.F., Wang, L., Chen, C., Selenium Nanoparticles as an Efficient Nanomedicine for the Therapy of Huntington's Disease. *ACS Appl. Mater. Interfaces*, 11, 38, 34725–34735, 2019.
246. Debnath, K., Pradhan, N., Singh, B.K., Jana, N.R., Jana, N.R., Poly (trehalose) nanoparticles prevent amyloid aggregation and suppress polyglutamine aggregation in a Huntington's Disease model mouse. *ACS Appl. Mater. Interfaces*, 9, 28, 24126–24139, 2017.
247. Ramachandran, S. and Thangarajan, S., Thymoquinone loaded solid lipid nanoparticles counteracts 3-Nitropropionic acid induced motor impairments and neuroinflammation in rat model of Huntington's disease. *Metab. Brain Dis.*, 33, 5, 1459–1470, 2018.
248. Álvaro, M.-C., Merino, M., Ana Virginia, S.-S., Blasco, S., Llinares, J.M., Mullorb, J.L., Enrique, G.-E., An antioxidant boehmite amino-nanozyme able to disaggregate Huntington's inclusion bodies. *Chem. Commun.*, 58, 5021–5024, 2022.
249. Akila, R.M. and Shaji, D.M., Ginger loaded chitosan nanoparticles for the management of 3–Nitropropionic acid-induced Huntington's Disease-like symptoms in male wistar rats. *Int. J. Pharm. Pharm. Sci.*, 14, 1, 28–36, 2022.
250. Adhikari, A., Mondal, S., Das, M., Biswas, P., Pal, U., Darbar, S., Bhattacharya, S.S., Pal, D., Saha-Dasgupta, T., Das, A.K., Mallick, A.K., Pal, S.K., Incorporation of a Biocompatible Nanozyme in Cellular Antioxidant Enzyme Cascade Reverses Huntington's Like Disorder in Preclinical Model. *Adv. Healthc. Mater.*, 10, 7, e2001736, 2021.

251. Cudkowicz, M.E., Kenna-Yasek, D., Sapp, P.E., Chin, W., Geller, B., Hayden, D.L., Schoenfeld, D.A., Hosler, B.A., Horvitz, H.R., Brown, R.H., Epidemiology of mutations in superoxide dismutase in amyotrophic lateral sclerosis. *Ann. Neurol.*, 41, 210–221, 1997.
252. Orrell, R.W., Amyotrophic lateral sclerosis: copper/zinc superoxide dismutase (SOD1) gene mutations. *Neuromuscul. Disord.*, 10, 63–68, 2000.
253. Masrori, P. and van Damme, P., Amyotrophic lateral sclerosis: A clinical review. *Eur. J. Neurol.*, 27, 1918–1929, 2020.
254. Smith, E.F., Shaw, P.J., de Vos, K.J., The role of mitochondria in amyotrophic lateral sclerosis. *Neurosci. Lett.*, 710, 132933, 2019.
255. Wu, Y., Chen, M., Jiang, J., Mitochondrial dysfunction in neurodegenerative diseases and drug targets via apoptotic signaling. *Mitochondrion.*, 49, 35–45, 2019.
256. Kostic, V., Jackson-Lewis, V., de Bilbao, F., Dubois-Dauphin, M., Przedborski, S., Bcl-2: Prolonging life in a transgenic mouse model of familial amyotrophic lateral sclerosis. *Science*, 277, 559–562, 1997.
257. Zhu, S., Stavrovskaya, I.G., Drozda, M., Kim, B.Y.S., Ona, V., Li, M., Sarang, S., Liu, A.S., Hartley, D.M., Wu, D.C., Gullans, S., Ferrante, R.J., Przedborski, S., Kristal, B.S., Friedlander, R.M., Minocycline inhibits cytochrome c release and delays progression of amyotrophic lateral sclerosis in mice. *Nature*, 417, 74–78, 2002.
258. Li, W., Fotinos, A., Wu, Q., Chen, Y., Zhu, Y., Baranov, S., Tu, Y., Zhou, E.W., Sinha, B., Kristal, B.S., Wang, X., N-acetyl-L-tryptophan delays disease onset and extends survival in an amyotrophic lateral sclerosis transgenic mouse model. *Neurobiol. Dis.*, 80, 93–103, 2015.
259. Cheah, B.C. and Kiernan, M.C., Dexpramipexole, the R(+) enantiomer of pramipexole, for the potential treatment of amyotrophic lateral sclerosis. *IDrugs*, 13, 911–920, 2010.
260. Li, X., Zhan, J., Hou, Y., Hou, Y., Chen, S., Luo, D., Luan, J., Wang, L., Lin, D., Coenzyme Q10 Regulation of Apoptosis and Oxidative Stress in H_2O_2 Induced BMSC Death by Modulating the Nrf-2/NQO-1 Signaling Pathway and Its Application in a Model of Spinal Cord Injury. *Oxid. Med. Cell. Longev.*, 2019, 6493081, 2019.
261. Kaviani, M., Keshtkar, S., Azarpira, N., Aghdaei, M.H., Geramizadeh, B., Karimi, M.H., Shamsaeefar, A., Motazedian, N., Nikeghbalian, S., Al-Abdullah, I.H., Ghahremani, M.S., Cytoprotective effects of olesoxime on isolated human pancreatic islets in order to attenuate apoptotic pathway. *Biomed. Pharm.*, 112, 108674, 2019.
262. Klopstock, T., Elstner, M., Bender, A., Creatine in mouse models of neurodegeneration and aging. *Amino Acids*, 40, 1297–1303, 2011.
263. Tripodo, G., Chlapanidas, T., Perteghella, S., Vigani, B., Mandracchia, D., Trapani, A., Galuzzi, M., Tosca, M.C., Antonioli, B., Gaetani, P., Marazzi, M., Torre, M.L., Mesenchymal stromal cells loading curcumin-INVITE-micelles: a drug delivery system for neurodegenerative diseases. *Colloids Surf., B.*, 125, 300–8, 2015.
264. Ahmadi, M., Agah, E., Nafissi, S., Jaafari, M.R., Harirchian, M.H., Sarraf, P., Faghihi-Kashani, S., Hosseini, S.J., Ghoreishi, A., Aghamollaii, V., Hosseini, M., Tafakhori, A., Safety and Efficacy of Nanocurcumin as Add-On Therapy to Riluzole in Patients With Amyotrophic Lateral Sclerosis: A Pilot Randomized Clinical Trial. *Neurotherapeutics*, 15, 2, 430–438, 2018.
265. Marcuzzo, S., Isaia, D., Bonanno, S., Malacarne, C., Cavalcante, P., Zacheo, A., Laquintana, V., Denora, N., Sanavio, B., Salvati, E., Andreozzi, P., Stellacci, F., Krol, S., Mellado-López, M., Mantegazza, R., Moreno-Manzano, V., Bernasconi, P., FM19G11-Loaded Gold Nanoparticles Enhance the Proliferation and Self-Renewal of Ependymal Stem Progenitor Cells Derived from ALS Mice. *Cells*, 8, 3, 279, 2019.
266. DeCoteau, W., Heckman, K.L., Estevez, A.Y., Reed, K.J., Costanzo, W., Sandford, D., Studlack, P., Clauss, J., Nichols, E., Lipps, J., Parker, M., Hays-Erlichman, B., Leiter, J.C., Erlichman, J.S., Cerium oxide nanoparticles with antioxidant properties ameliorate strength and prolong life in

mouse model of amyotrophic lateral sclerosis. *Nanomed. Nanotechnol. Biol. Med.*, 12, 8, 2311–2320, 2016.
267. Díaz-García, D., Ferrer-Donato, A., Méndez-Arriaga, J.M., Cabrera-Pinto, M., Díaz-Sánchez, M., Prashar, S., Carmen, M., Fernandez-Martos, and Santiago Gómez-Ruiz, Design of Mesoporous Silica Nanoparticles for the Treatment of Amyotrophic Lateral Sclerosis (ALS) with a Therapeutic Cocktail Based on Leptin and Pioglitazone. *ACS Biomater. Sci. Eng.*, 8, 11, 4838–4849, 2022.
268. Goldenberg, M.M., Multiple sclerosis review. *Pharm. Ther.*, 37, 175–184, 2012.
269. Ohl, K., Tenbrock, K., Kipp, M., Oxidative stress in multiple sclerosis: Central and peripheral mode of action. *Exp. Neurol.*, 277, 58–67, 2016.
270. Kempuraj, D., Thangavel, R., Natteru, P.A., Selvakumar, G.P., Saeed, D., Zahoor, H., Zaheer, S., Iyer, S.S., Zaheer, A., Neuroinflammation induces neurodegeneration. *J. Neurol. Neurosurg. Spine*, 1, 1003, 2016.
271. Lublin, F.D., Reingold, S.C., Cohen, J.A., Cutter, G.R., Sørensen, S.P., Thompson, A.J., Wolinsky, J.S., Balcer, L.J., Banwell, B., Barkhof, F., Bebo, B., Jr, Calabresi, P.A., Clanet, M., Comi, G., Fox, R.J., Freedman, M.S., Goodman, A.D., Inglese, M., Kappos, L., Kieseier, B.C., Lincoln, J.A., Lubetzki, C., Miller, A.E., Montalban, X., O'Connor, P.W., Petkau, J., Pozzilli, C., Rudick, R.A., Sormani, M.P., Stüve, O., Waubant, E., Polman, C.H., Defining the clinical course of multiple sclerosis. The 2013 revisions. *Neurology*, 83, 278–286, 2014.
272. Dutta, R. and Trapp, B.D., Relapsing and progressive forms of multiple sclerosis—Insights from pathology. *Curr. Opin. Neurol.*, 27, 271–278, 2014.
273. Miller, E.D., Dziedzic, A., Saluk-Bijak, J., Bijak, M.A., Review of Various Antioxidant Compounds and their Potential Utility as Complementary Therapy in Multiple Sclerosis. *Nutrients.*, 11, 1528, 2019.
274. Yadav, S.K., Mindur, J.E., Ito, K., Dhib-Jalbut, S., Advances in the immunopathogenesis of multiple sclerosis. *Curr. Opin. Neurol.*, 28, 206–219, 2015.
275. Sun, Y. and Langer, H.F., Platelets, Thromboinflammation and Neurovascular Disease. *Front. Immunol.*, 13, 843404, 2022.
276. Raphael, I., Nalawade, S., Eagar, T.N., Forsthuber, T.G., T cell subsets and their signature cytokines in autoimmune and inflammatory diseases. *Cytokine*, 74, 5–17, 2015.
277. Binyamin, O., Larush, L., Frid, K., Keller, G., Friedman-Levi, Y., Ovadia, H., Abramsky, O., Magdassi, S., Gabizon, R., Treatment of a multiple sclerosis animal model by a novel nanodrop formulation of a natural antioxidant. *Int. J. Nanomed.*, 10, 7165–74, 2015.
278. Cheshire, W.P., Trigeminal neuralgia: for one nerve a multitude of treatments. *Expert Rev. Neurother.*, 7, 1565–79, 2007.
279. Hokfelt, T., Zhang, X., Wiesenfeld-Hallin, Z., Messenger plasticity in primary sensory neurons following axotomy and its functional implications. *Trends Neurosci.*, 17, 22–30, 1994.
280. Alvares, D. and Fitzgerald, M., Building blocks of pain: the regulation of key molecules in spinal sensory neurones during development and following peripheral axotomy. *Pain*, 6, S71–85, 1999.
281. Dieb, W. and Hafidi, A., Astrocytes are involved in trigeminal dynamic mechanical allodynia: potential role of D-serine. *J. Dent. Res.*, 92, 808–13, 2013.
282. Shibuta, K., Suzuki, I., Shinoda, M., Tsuboi, Y., Honda, K., Shimizu, N., Sessle, B.J., Iwata, K., Organization of hyperactive microglial cells in trigeminal spinal subnucleus caudalis and upper cervical spinal cord associated with orofacial neuropathic pain. *Brain Res.*, 1451, 74–86, 2012.
283. DeLeo, J.A. and Yezierski, R.P., The role of neuroinflammation and neuroimmune activation in persistent pain. *Pain*, 90, 1–6, 2001.
284. Watkins, L.R., Milligan, E.D., Maier, S.F., Glial activation: a driving force for pathological pain. *Trends Neurosci.*, 24, 450–5, 2001.

285. Watkins, L.R., Milligan, E.D., Maier, S.F., Spinal cord glia: new players in pain. *Pain*, 93, 201–5, 2001.
286. Hanisch, U.K., Microglia as a source and target of cytokines. *Glia*, 40, 140–55, 2002.
287. Moalem, G. and Tracey, D.J., Immune and inflammatory mechanisms in neuropathic pain. *Brain Res. Rev.*, 51, 240–64, 2006.
288. Yoshimasu, F., Kurland, L.T., Elveback, L.R., Tic douloureux in Rochester, Minnesota, 1945-1969. *Neurology*, 22, 952–956, 1972.
289. Katusic, S., Beard, C.M., Bergstralh, E., Kurland, L.T., Incidence and clinical features of trigeminal neuralgia, Rochester, Minnesota, 1945-1984. *Ann. Neurol.*, 27, 89–95, 1990.
290. Obermann, M., Treatment options in trigeminal neuralgia. *Ther. Adv. Neurol. Disord.*, 3, 107–115, 2010.
291. Jones, M.R., Urits, I., Ehrhardt, K.P., Cefalu, J.N., Kendrick, J.B., Park, D.J., Cornett, E.M., Kaye, A.D., Viswanath, O., A comprehensive review of trigeminal neuralgia. *Curr. Pain Headache Rep.*, 23, 74, 2019.
292. Cohen, A.S., Matharu, M.S., Goadsby, P.J., Short-lasting unilateral neuralgiform headache attacks with conjunctival injection and tearing (SUNCT) or cranial autonomic features (SUNA)–a prospective clinical study of SUNCT and SUNA. *Brain*, 129, 10, 2746–60, 2006.
293. Zebenholzer, K., Wöber, C., Vigl, M., Wessely, P., Wöber-Bingöl, C., Facial pain and the second edition of the International Classification of Headache Disorders. *Headache*, 46, 2, 259–63, 2006.
294. Zakrzewska, J.M., Diagnosis and Differential Diagnosis of Trigeminal Neuralgia. *Clin. J. Pain*, 18, 1, 14–21, 2002.
295. Cruccu, G., Gronseth, G., Alksne, J., Argoff, C., Brainin, M., Burchiel, K., Nurmikko, T., Zakrzewska, J.M., American Academy of Neurology Society; European Federation of Neurological Society. AAN-EFNS guidelines on trigeminal neuralgia management. *Eur. J. Neurol.*, 15, 10, 1013–28, 2008.
296. Gronseth, G., Cruccu, G., Alksne, J., Argoff, C., Brainin, M., Burchiel, K., Nurmikko, T., Zakrzewska, J.M., Practice parameter: the diagnostic evaluation and treatment of trigeminal neuralgia (an evidence-based review): report of the Quality Standards Subcommittee of the American Academy of Neurology and the European Federation of Neurological Societies. *Neurology*, 71, 15, 1183–90, 2008.
297. Sandhir, R., Yadav, A., Sunkaria, A., Singhal, N., Nano-antioxidants: An emerging strategy for intervention against neurodegenerative conditions. *Neurochem. Int.*, 89, 209–26, 2015.
298. Dolati, S., Ahmadi, M., Aghebti-Maleki, L., Nikmaram, A., Marofi, F., Rikhtegar, R., Ayromlou, H., Yousefi, M., Nanocurcumin is a potential novel therapy for multiple sclerosis by influencing inflammatory mediators. *Pharmacol. Rep.*, 70, 6, 1158–1167, 2018.
299. Bonferoni, M.C., Rassu, G., Gavini, E., Sorrenti, M., Catenacci, L., Giunchedi, P., Nose-to-Brain Delivery of Antioxidants as a Potential Tool for the Therapy of Neurological Diseases. *Pharmaceutics*, 12, 12, 1246, 2020.

10

Role of Antioxidants for the Treatment of Metabolic Disorders

Sanshita[1], Mayank Sharma[2], Payal Dande[2], Piyush Ghode[2], Atul A. Shirkhedkar[3], Amit Chaudhary[4] and Inderbir Singh[1*]

[1]Chitkara College of Pharmacy, Chitkara University, Punjab, India
[2]School of Pharmacy & Technology Management, SVKM's NMIMS, Shirpur, Dhule, Maharashtra, India
[3]R.C. Patel Institute of Pharmaceutical Education and Research, Shirpur Dist: Dhule, Maharashtra, India
[4]Chitkara University School of Pharmacy, Chitkara University, Himachal Pradesh, India

Abstract
Metabolic disorders are a group of various medical conditions affecting metabolic processes in the body and result from abnormal biochemical processes interfering with the proper utilization of nutrients. They can be divided into two major categories including inherited or acquired metabolic disorders. The inherited metabolic disorders (e.g., phenylketonuria, Wilson's disease, etc.) are less prevalent as compared to the acquired metabolic disorders (e.g., obesity, diabetes mellitus, etc.). In fact, acquired metabolic disorders have become a major health concern worldwide. These disorders are a result of the oxidative stress that results from an imbalance between the generation of reactive oxygen species and the body's antioxidant defense against them. There are a variety of herbal drugs and natural products known for their antioxidant activity. These drugs comprise different chemical constituents that can reduce oxidative stress by different mechanisms, such as free radical scavenging, inhibition of lipid peroxidation, or metal ion reduction. Most of these constituents are polyphenols, flavonoids, tannins, or carotenoids in nature. In the current chapter, an examination is conducted on acquired metabolic disorders, including obesity, diabetes, and cardiovascular diseases, with respect to their prevalence, risk factors, diagnosis, complications, management strategies, and the function of oxidative stress. Subsequently, it elucidates the function of antioxidants in metabolic disorders. Eventually, it also discusses the primary natural sources of drugs, including their antioxidant components and, when possible, the specific mechanisms by which they exhibit antioxidant effects. Thus, this chapter serves as a valuable resource, offering information on various natural products as antioxidants and insights into their molecular mechanisms for relieving oxidative stress, paving the way for the integration of antioxidant drugs in the evolving landscape of metabolic disorder treatment.

***Keywords*:** Antioxidants, metabolic disorders, obesity, diabetes, cardiovascular diseases, polyphenols

**Corresponding author*: inderbir.singh@chitkara.edu.in

Rakesh K. Sindhu, Inderbir Singh and M. Arockia Babu (eds.) Antioxidants: Nature's Defense Against Disease, (369–410)

10.1 Introduction

10.1.1 Metabolic Disorders

Metabolic disorders comprise of diverse medical conditions that impact the body's metabolic processes. These disorders result from abnormal chemical processes that interfere with the body's ability to properly process and utilize nutrients, resulting in an imbalance of essential substances crucial for maintaining good health.

10.1.2 Classification of Metabolic Disorders

In general, metabolic disorders are grouped into inherited metabolic disorders (IMDs) and acquired metabolic disorders (AMDs) and have been represented in Figure 10.1.

10.1.2.1 Inherited Metabolic Disorders (IMDs)

IMDs also known as inborn errors in metabolism are caused due to genetic defects in the proteins or enzymes involved in metabolism that lead to disruptions in the normal biochemical processes. The International Classification of Inherited Metabolic Disorders, or ICIMD – includes 1,450 disorders [1], with ongoing discoveries. Examples of these disorders include metabolic myopathies, characterized by a deficiency or dysfunction of essential metabolites involved in muscle contraction [2]. These myopathies can be further classified into glycogen storage disease (GSD; glycogenosis), disorders of lipid metabolism, and mitochondrial myopathies [3]. Other examples of IMDs include organic acidopathies,

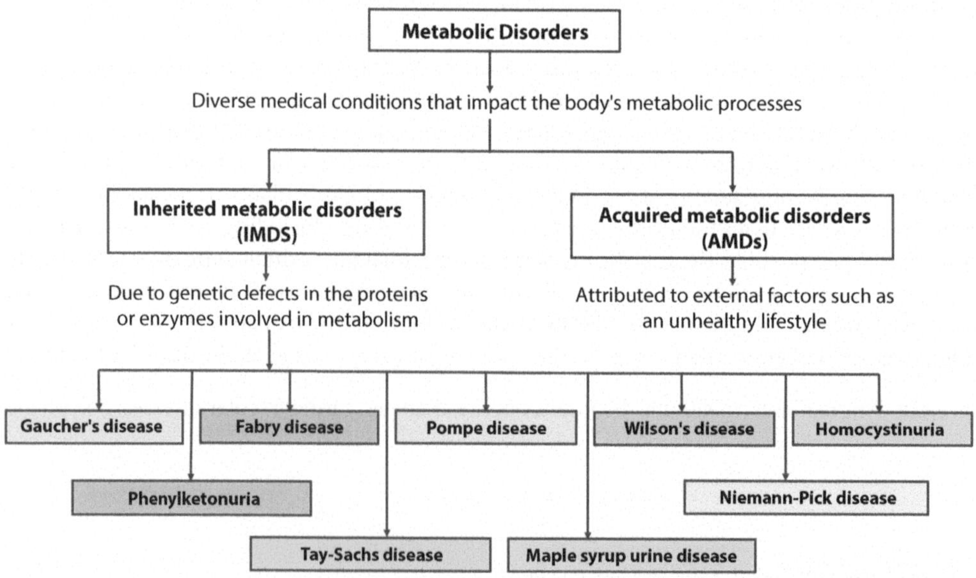

Figure 10.1 Types of metabolic disorders.

amino acidopathies, fatty acid disorders, and lysosomal storage diseases [4, 5] and Wilson's disease [6].

10.1.2.1.1 Prevalence of IMDs

According to an estimate from Centre for Global Health Research, The Usher Institute for Population Health Sciences and Informatics, University of Edinburgh, Scotland, where they conducted a systematic literature review of birth prevalence and case fatality of IMDs globally from 1980 to 2017, the global birth prevalence of all-cause IEM to be 50.9 per 100 000 live births [7]. Although the specific prevalence of individual IMDs can vary greatly and can have a significant impact due to their cumulative incidence.

10.1.2.1.2 Risk Factors for IMDs

These disorders can result from enzyme deficiencies or defects in the biosynthesis or degradation of glycoconjugates [8], autonomic dysfunction and impaired glucose homeostasis [9]. Symptoms of inherited metabolic disorders can vary widely, but may include muscle weakness, myalgia, rhabdomyolysis, lethargy, weight loss, jaundice, seizures, and other multisystemic manifestations. Since, generally, all the IMDs are considered as orphan diseases, their appropriate treatment depends majorly on early detection and accurate diagnosis. Proteomics, particularly tandem mass spectrometry, has emerged as a promising tool for studying IMDs, aiding in early diagnosis, identifying biomarkers, and understanding the underlying pathophysiology.

10.1.2.1.3 Types of Inherited Metabolic Disorders (IMDs)

Phenylketonuria (PKU): is a disorder of amino acid metabolism. It caused by phenylalanine hydroxylase deficiency that leads to the accumulation of phenylalanine and its metabolites in the blood and brain, causing intellectual disability, seizures, behavioral problems, and skin disorders.

Galactosemia: is a disorder of carbohydrate metabolism that is caused by a deficiency of galactose-1-phosphate uridyltransferase, leading to the accumulation of galactose and its metabolites in the blood and tissues, causing liver damage, cataracts, intellectual disability, and other complications.

Maple syrup urine disease (MSUD): is a disorder of amino acid metabolism which is caused by a deficiency of the branched-chain alpha-keto acid dehydrogenase complex, resulting in the accumulation of branched-chain amino acids and their metabolites in the blood and tissues, causing neurological problems, developmental delays, and other complications.

Gaucher's disease: is a disorder of lipid metabolism that results due to glucocerebrosidase deficiency, causing the accumulation of glucocerebroside in the spleen, liver, and bone marrow, causing anemia, thrombocytopenia, hepatosplenomegaly, and bone pain.

Tay-Sachs disease: is a disorder of lipid metabolism, which is caused by a deficiency of the hexosaminidase A, causing progressive neurological deterioration, blindness, and death due to the accumulation of gangliosides in the brain and nervous system, causing progressive neurological deterioration, blindness and death.

Niemann-Pick disease: It is a group of lipid metabolism disorders caused by a deficiency of the sphingomyelinase, which leads to the accumulation of sphingomyelin in the spleen, liver, and brain, causing hepatosplenomegaly, neurological problems, and other complications.
Fabry disease: It is an X-linked recessive disorder of lipid metabolism resulting from a deficiency of alpha-galactosidase A, causing the accumulation of globotriaosylceramide in the blood vessels and organs, causing kidney failure, heart disease, and other complications.
Homocystinuria: It is a disorder of amino acid metabolism caused by a deficiency of the enzymes involved in the metabolism of methionine, which leads to the accumulation of homocysteine and its metabolites in the blood and tissues, causing intellectual disability, skeletal abnormalities, and other complications.
Wilson's disease: It is a disorder related to metal accumulation caused by a mutation in the ATP7B gene, which leads to the increased concentration of copper in the liver, brain, and other organs, causing liver disease, neurological problems, and other complications.
Pompe disease: It is a disorder of glycogen metabolism caused by a deficiency of acid alpha-glucosidase, which leads to the glycogen accumulation in the muscles and other tissues, causing muscle weakness, respiratory problems, and other complications.

10.1.2.2 Acquired Metabolic Disorders (AMDs)

AMDs can be attributed to external factors such as an unhealthy lifestyle and excessive consumption of calories and are becoming more common worldwide. The most prevalent AMDs are diabetes, obesity, inflammatory bowel disease and metabolic syndrome [10]. Among these, metabolic syndrome is a collection of metabolic abnormalities that elevate the chances of developing cardiovascular disease, stroke, and type 2 diabetes. The five primary components of metabolic syndrome comprise of obesity of the abdomen, elevated blood pressure, heightened fasting blood sugar, abnormal cholesterol or triglyceride levels, and inflammatory markers. Other AMDs constitute the deficiency of ornithine transcarbamylase, porphyria, hemochromatosis, insufficiency of alpha 1 antitrypsin, uremia, and nutritional problems caused by vitamin deficiencies [6, 11, 12], Non-Alcoholic Fatty Liver Disease, gout [13], Osteoporosis [14], Acquired Hypo or hyperthyroidism, and Cushing's syndrome.

10.1.2.2.1 Prevalence of AMDs

As the AMDs contain different diseases, the prevalence of different diseases may vary across the globe. Although as, the metabolic syndrome encompasses the majority of these diseases, a data for prevalence of metabolic syndrome can show the gravity of the situation worldwide. According to some studies [15–17] the worldwide prevalence of metabolic syndrome ranges from <10% to a roaring 84% and can dependent upon the region, environment (urban or rural) age, sex, race or ethnicity.

10.1.2.2.2 Risk Factors for AMDs

The acquired metabolic disorders can result from abnormalities in carbohydrate metabolism, lipid metabolism, and amino acid metabolism [18], as well as factors such as bariatric surgery, exposure to organic chemicals, consumption of alcohol, repetitive need of blood transfusions, excessive iron supplements through oral route, hepatitis C, and infectious Tropical Pulmonary Eosinophilia [19].

10.1.3 Clinical Features of Metabolic Diseases

There may be a range of clinical features exhibited by metabolic disorders. Among these are increased blood pressure, abdominal adiposity, irregular lipid profiles, insulin resistance, and impaired glucose tolerance. Further, there may be some specific physical signs and symptoms that can be recognized by trained professionals. The impact of these disease conditions on the body can result in atherosclerotic coronary disease, diabetes mellitus, as well as vascular and neurological complications such as stroke [20]. Modifications in cerebral metabolism, including changes in glucose, ketone bodies, and amino acid metabolism, have been observed in individuals with both Alzheimer's disease and Parkinson's disease [21]. Moreover, the cutaneous signs and symptoms that may be observed in IMDs include vascular lesions, ichthyosis, anomalous pigmentation, and nail irregularities [21].

10.2 Correlation of Metabolic Disorders and Oxidative Stress and ROS

Reactive Oxygen Species and Oxidative Stress

Reactive oxygen species (ROS) are extremely reactive molecules due to presence of at least one unpaired electron on oxygen. They are produced particularly during cellular respiration, as a result of use of oxygen by cells to produce adenosine triphosphate (ATP). Although produced ROS play a pivotal role in normal cellular signaling and defense mechanisms, their uncontrolled release can lead to complication known as oxidative stress. Some of the contributing factors for generation of ROS in the body are inclusive of mitochondrial respiration, inflammation, environmental factors, lifestyle factors, ultraviolet (UV) radiation, drug and toxin metabolism, ischemia-reperfusion injury and the presence of metal ions such as iron and copper (Figure 10.2). The most abundant ROS include superoxide radicals, hydrogen peroxide and hydroxyl radicals. A number of factors may lead to an increase in ROS in the body including age, unhealthy diet, smoking and alcohol consumption, environmental exposures, and some chronic conditions such as diabetes, cardiovascular diseases, and cancer etc. On the other hand, excessive production of ROS may lead to DNA damage, lipid peroxidation resulting in cell death, damage to tissues, protein oxidation leading to alteration in their structure and function, and mitochondrial dysfunction responsible for disrupting cellular energy production.

The pernicious effects of ROS are countered by the antioxidant mechanism in the body by their neutralization and regulation. Antioxidants are the compounds that can hand out electrons or hydrogen atoms, leading to stabilization of the free radicals. The body has a variety of defense mechanisms for maintenance of balance amidst ROS production and their elimination. Some key antioxidant systems in the body encompass enzymatic antioxidants (superoxide dismutase (SOD), glutathione peroxidase etc.), non-enzymatic antioxidants (vitamins, ubiquinone (coenzyme Q10) etc.), metal binding proteins (Ferritin, metallothionein), Thioredoxin and Thioredoxin Reductase System, bilirubin and Nrf2-Keap1 Pathway.

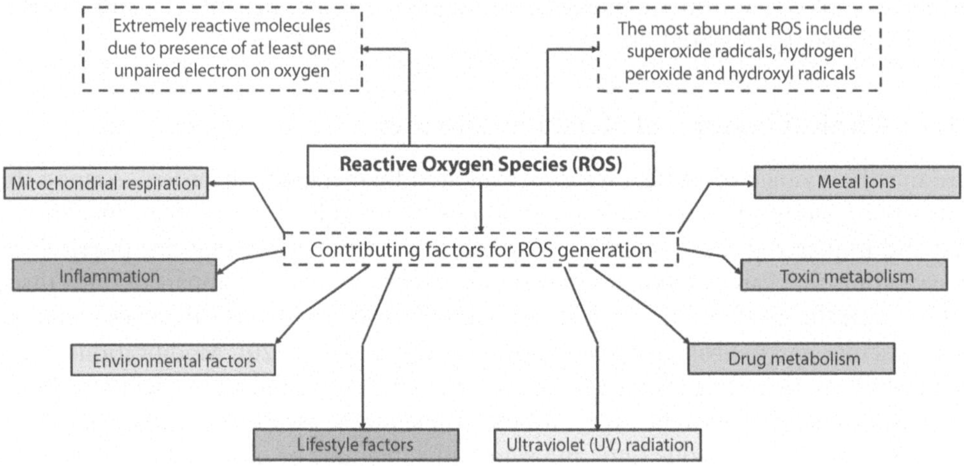

Figure 10.2 Contributing factors for ROS generation.

The equity between ROS production and the antioxidant defense mechanisms is critical to maintain the cellular health. A well-balanced diet rich in fruits, vegetables, and other sources of antioxidants, along with a healthy lifestyle are the key factors in management of this balance. Physiologically, this balance is maintained through the combined action of intrinsic and extrinsic antioxidants, biomarkers for measuring oxidative damage, and the antioxidant defense systems in various tissues and organelles. When this delicate harmony between ROS and antioxidants is shifted towards ROS, the condition is known as oxidative stress that is responsible for the emergence of oxidative damage of the biomacromolecules. Oxidative stress may lead to the pathogenesis of various diseases, including atherosclerosis, neurodegenerative diseases, cancer, inflammatory diseases, and AMDs. However, oxidative stress also has a physiological role in cell signaling and adaptation.

10.3 Role of Oxidative Stress in Metabolic Disorders

Oxidative stress is meticulously correlated with metabolic disorders, such as obesity, insulin resistance, type 2 diabetes, and cardiovascular complications. Rani *et al.* have depicted that oxidative stress plays a significant portrayal in the pathogenesis of metabolic syndrome [22]. According to Furukawa *et al.*, production of ROS increases selectively in adipose tissue as observed in obese mice, abetted by extended expression of NADPH oxidase and weakened expression of antioxidative enzymes. Increased oxidative stress in accumulated fat is an early instigator of metabolic syndrome, that causes un-regulated manufacture of adipocytokines, including adiponectin, plasminogen activator inhibitor-1, IL-6, and monocyte chemotactic protein-1. Thus making the redox state in adipose tissue a likely pragmatic therapeutic target for metabolic syndrome associated with obesity [23]. Nutritional stress that may be precipitated due to high fat and high carbohydrate diet also boosts up oxidative stress as conspicuous by heightened lipid peroxidation products, protein carbonylation, and decreased antioxidant system [22]. In turn, this nutritional stress contributes to the pathogenesis of diabetes by disrupting cell cycle regulation, impairing DNA repair, and promoting inflammation. Increased oxidative stress is considered a potential common factor in various cardiovascular diseases. Oxidative stress is particularly implicated in conditions like atherosclerosis, hypertension, myocardial infarction, ischemia/reperfusion injury, and heart failure [24]. Thus, the oxidative stress conditions can lead to a variety of AMDs. Some AMDs with their prevalence and risk factors are described below:

10.3.1 Obesity

Obesity is a complex metabolic disorder that has been characterized by anomalistic and excessive buildup of adipose tissue, resulting in a body mass index (BMI) of 30 or higher. It stems from a multifaceted interplay of genetic, environmental, and behavioral factors, impacting various physiological pathways [25]. It is associated with heightened risks of comorbidities like cardiovascular diseases, insulin resistance, and systemic inflammation, comprehensive interventions encompassing dietary, lifestyle, and medical approaches are necessary for effective management [26].

10.3.1.1 Prevalence of Obesity

Obesity has reached epidemic proportions globally, with a substantial rise in prevalence over the past few decades. The World Health Organization (WHO) estimates that over 650 million adults worldwide are obese, and this number is expected to escalate further in the coming years. The prevalence is not limited to developed nations; it has become a significant health concern in low- and middle-income countries as well [27].

10.3.1.2 Risk Factors for Obesity

Numerous interrelated factors contribute to the development of obesity, encompassing genetic, environmental, and behavioral influences. Genetic predisposition, characterized by a familial aggregation of obesity, underscores the heritability of this condition. Environmental factors such as sedentary lifestyles, increased calorie intake, and the ubiquity of high-calorie, low-nutrient foods contribute significantly [28].

Socioeconomic status plays a pivotal role, with individuals from lower-income groups facing heightened susceptibility. Moreover, psychological factors, including stress and depression, can contribute to overeating and weight gain. The intricate interplay of these factors underscores the multifactorial nature of obesity etiology [29].

10.3.1.3 Diagnosis

The diagnosis of obesity is primarily based on the measurement of body mass index (BMI), calculated by dividing an individual's weight in kilograms by the square of their height in meters. The WHO classifies individuals with a BMI of 30 or above as obese. However, BMI may not be a perfect indicator, as it does not account for variations in body composition [30].

In addition to BMI, other assessments such as waist circumference and body fat percentage may provide a more comprehensive evaluation of adiposity. Clinical evaluation, including a detailed medical history and physical examination, is crucial to identify underlying conditions contributing to obesity and to assess associated risk factors [31, 32].

10.3.1.4 Complications

Obesity is intricately linked to a myriad of health complications that extend beyond its physical manifestation. Cardiovascular diseases, type 2 diabetes, hypertension, and dyslipidemia are among the prominent comorbidities associated with obesity. Inflammatory processes induced by adipose tissue contribute to insulin resistance and chronic low-grade inflammation, exacerbating metabolic dysfunction. Joint diseases, sleep apnea, and reproductive disorders further underscore the diverse complications of obesity [33].

Psychosocial consequences, such as depression and diminished quality of life, add to the holistic impact of this condition [34, 35].

10.3.1.5 Treatment and Prevention

Addressing obesity necessitates a comprehensive, multidisciplinary approach. Lifestyle modifications involving dietary interventions and increased physical activity form the cornerstone of obesity management. Behavioral therapy, pharmacotherapy, and, in severe cases, bariatric surgery are viable options for individuals with obesity and associated comorbidities [36, 37].

Preventive strategies must focus on fostering a health-promoting environment, edifying healthy eating habits, and encouraging regular physical activity [38]. Public health initiatives targeting awareness and education are crucial in stemming the rising tide of obesity on a population level. Obesity presents a complex and multifaceted challenge with far-reaching consequences [39, 40].

10.3.1.6 Role of Oxidative Stress in Obesity

Oxidative stress is a pivotal factor in obesity-related disorders, particularly cardiovascular health [41]. Obesity induces an expansion of adipose tissue, significantly contributing to oxidative stress by manufacturing of ROS through metabolic processes [42]. The chronic low-grade inflammation characteristic of obesity further stimulates ROS production, establishing a synergistic relationship [43]. Insulin resistance, a hallmark of obesity, plays a crucial role in the oxidative stress mechanism by disrupting insulin signaling and dysregulating glucose metabolism. Additionally, obesity is linked to impaired mitochondrial function, intensifying ROS production and oxidative stress [44].

Impact of Oxidative Stress in Obesity include damaged insulin receptors and disrupted signaling pathways, resulting in insulin resistance and type 2 diabetes. It also damages the endothelial lining of blood vessels, impairing vascular function and increasing the risk of cardiovascular complications [43].

ROS further contribute to the oxidation of low-density lipoprotein (LDL) cholesterol, fostering the formation of atherosclerotic plaques and elevating the risk of cardiovascular diseases [45, 46].

Lifestyle modifications, incorporating dietary changes, physical activity, and antioxidant-rich interventions, are crucial for managing oxidative stress in individuals with obesity. Recognizing the pivotal role of oxidative stress in metabolic disorders is imperative for formulating effective prevention and treatment strategies [47].

Lifestyle interventions such as weight loss, physical activity, and adherence to antioxidant-rich diets emerge as crucial strategies for mitigating oxidative stress in obesity [48–51].

10.3.2 Diabetes

Diabetes mellitus, a chronic disorder characterized by high blood sugar, arises from defects in insulin secretion, insulin action, or both [52].

Insulin, the key regulator of blood sugar levels, facilitates its uptake into cells for energy. Impaired insulin function disrupts this balance, causing hyperglycemia and its associated complications. While various forms of diabetes exist, each with its unique characteristics, a common thread emerges: the intricate interplay between diabetes and oxidative stress [53].

10.3.2.1 Prevalence

Diabetes has reached epidemic proportions, affecting millions worldwide. According to the International Diabetes Federation (IDF), an estimated 537 million individuals were living with diabetes in 2021, and this number is premeditated to rise to 643 million by 2030. The prevalence varies across regions, with an increasing trend observed in both developed and developing nations [54–56].

10.3.2.2 Risk Factors

Understanding the multifaceted risk factors for diabetes is pivotal for effective prevention and management. Genetics play a significant role, with a familial predisposition increasing susceptibility. Lifestyle factors such as sedentary behavior, unhealthy dietary patterns, and obesity amplify the risk. Additionally, age, ethnicity, and gestational diabetes contribute to the intricate tapestry of risk factors [57–59].

10.3.2.3 Diagnosis

Accurate diagnosis is fairly important for timely intervention. Fasting blood glucose, oral glucose tolerance tests, and HbA1c levels serve as crucial diagnostic tools. In recent years, advancements in glycemic monitoring technology, including continuous glucose monitoring systems, have enhanced precision in diabetes diagnosis and management. A multidisciplinary approach, involving clinical, laboratory, and imaging assessments, aids in a comprehensive evaluation of the disease [60, 61].

10.3.2.4 Complications

Diabetes exerts its deleterious effects on multiple organ systems, culminating in a spectrum of complications. Microvascular complications include retinopathy, nephropathy, and neuropathy, while macrovascular complications manifest as cardiovascular diseases, stroke, and peripheral vascular disease. Chronic hyperglycemia can cause the production of advanced glycation end-products (AGEs), fostering inflammation and oxidative stress, thereby contributing to the pathogenesis of complications [62–64].

10.3.2.5 Role of Oxidative Stress in Diabetes

Oxidative stress, a situation where there occurs an imbalance between ROS and the body's capacity to neutralize and counterbalance them, significantly contributes to diabetes development and progression. Chronic exposure to elevated ROS levels impairs the function of pancreatic beta cells, responsible for insulin production, leading to reduced insulin secretion and ultimately contributing to diabetes development [53].

Furthermore, ROS can disrupt insulin signaling pathways, hindering the normal cellular response to insulin, and decreasing glucose uptake by peripheral tissues, resulting in elevated blood sugar levels. This, coupled with oxidative stress-induced inflammation, exacerbates insulin resistance, creating a vicious cycle that fuels diabetes progression. Oxidative stress also damages endothelial cells lining blood vessels, contributing to endothelial dysfunction, characterized by reduced nitric oxide production. This dysfunction paves the way for the precipitation of diabetic vascular complications, encompassing both microvascular and macrovascular diseases [65–67].

The intricate link between oxidative stress and inflammation is evident in diabetes. Elevated ROS levels trigger inflammatory pathways, leading to the release of pro-inflammatory cytokines. Chronic inflammation, a key player in the pathogenesis of diabetes, further promotes insulin resistance and accelerates the destruction of pancreatic beta cells [68]. Additionally, oxidative stress impairs mitochondrial function, the energy-producing organelles within cells, leading to reduced energy production and increased ROS generation within the cell. This mitochondrial dysfunction contributes to cellular damage and apoptosis, processes implicated in diabetes progression. While the body possesses an antioxidant defense system to neutralize excess ROS and maintain redox balance, in diabetes, this system may be compromised, hindering its ability to adequately counteract oxidative stress. Genetic factors, lifestyle choices, and environmental factors can all influence the efficiency of these antioxidant defense mechanisms [69–71].

Understanding the linkage between oxidative stress and diabetes highlights the importance of targeted therapeutic strategies. Antioxidant interventions, lifestyle modifications, and pharmacological agents aimed at mitigating oxidative stress are actively being explored as potential avenues for preventing or managing diabetes and its complications. Ongoing research holds the promise of unraveling the precise mechanisms underlying this complex relationship, paving the way for the development of more effective strategies to combat oxidative stress and improve the lives of individuals with diabetes [72].

10.3.3 Cardiovascular Diseases

CVDs are a heterogeneous group of disorders affecting the structure and/or function of the heart and blood vessels, leading to impaired blood circulation and subsequent tissue damage. This encompasses a diverse spectrum of pathologies, including atherosclerosis, coronary artery disease, cerebrovascular disease, peripheral artery disease, aortic disease, rheumatic heart disease, cardiomyopathies, and arrhythmias. These conditions can manifest with various symptoms, ranging from angina pectoris (chest pain) and claudication (leg pain) to myocardial infarction (heart attack), stroke, and heart failure. Understanding the underlying pathophysiology of CVDs, including the role of risk factors like hyperlipidemia, hypertension, diabetes, and smoking, is crucial for developing effective prevention and treatment strategies [73].

10.3.3.1 Prevalence

Cardiovascular diseases (CVDs) represent a pervasive global health issue, with their prevalence exhibiting an alarming upward trajectory. CVDs are the well-known causes of death worldwide, claiming an estimated 17.9 million lives each year. That is about 32% of all global deaths, making CVDs a major public health concern in every corner of the world. Nearly 523 million people were living with CVDs in 2019, a significant increase from 271 million in 1990; 85% of CVD deaths are due to heart attacks and strokes. Out of the 17 million premature deaths (under the age of 70) due to noncommunicable diseases in 2019, 38% were caused by CVDs.

High systolic blood pressure is the preeminent modifiable risk factor for CVDs, responsible for over 10 million deaths annually. Epidemiological studies consistently underscore the substantial burden of CVDs on public health, emphasizing the urgent need for effective prevention and management strategies [74, 75].

10.3.3.2 Risk Factors

The multifactorial nature of CVDs is underscored by a diverse range of risk factors, both modifiable and non-modifiable. Several risk factors contribute to the development and progression of cardiovascular diseases, posing significant threats to public health. One primary risk factor is age, as the likelihood of CVD increases with advancing years. Additionally, gender plays a role, with men generally facing a higher risk than premenopausal women; however, this gender difference tends to narrow after menopause [76].

Genetics also contribute, as individuals with a family history of cardiovascular problems are more susceptible. Lifestyle choices significantly impact cardiovascular health, with poor diet, shortfall of physical activity, smoking, and redundant consumption of alcohol all identified as modifiable risk factors. High blood pressure and elevated cholesterol levels are key contributors, placing strain on the cardiovascular system. Diabetes, both type 1 and type 2, heightens the risk, as the condition can lead to damaged blood vessels.

Furthermore, obesity, particularly abdominal obesity, is related with an increased likelihood of developing CVD. Chronic stress and inadequate sleep also play roles in raising cardiovascular risk. Addressing these multifaceted risk factors through lifestyle modifications, medication, and preventive measures is crucial for reducing the prevalence and impact of cardiovascular diseases on global health [76, 77].

10.3.3.3 Diagnosis of Cardiovascular Diseases

Accurate and timely diagnosis is pivotal for effective management of CVDs. The diagnosis of cardiovascular diseases (CVD) involves a combination of medical history assessment, physical examinations, and various diagnostic tests. Patient history is a crucial initial step, where healthcare professionals inquire about risk factors, symptoms, and family medical history. A thorough physical examination helps detect visible signs of cardiovascular issues, such as abnormal heart sounds or irregular pulses. To supplement these initial steps, several diagnostic tests play pivotal roles. Electrocardiography (ECG or EKG) records the heart's electrical activity, aiding in the identification of irregular rhythms or signs of myocardial infarction. Blood tests measure biomarkers like troponin and creatine kinase-MB, indicating heart muscle damage. Imaging techniques, such as echocardiography, utilize sound waves to produce detailed images of the heart's structure and function, providing valuable information about valve function, chamber size, and wall motion [78].

Stress tests, often performed on a treadmill or using medication, evaluate the heart's response to physical exertion. Angiography, including coronary angiography, involves injecting contrast dye into blood vessels to visualize blockages or abnormalities. Advanced technologies like cardiac CT scans and MRI offer detailed images of the heart and surrounding structures. Ultimately, the integration of these diagnostic modalities enables healthcare professionals to comprehensively assess the cardiovascular system, facilitating accurate diagnosis and tailored treatment plans for individuals at risk or affected by cardiovascular diseases.

10.3.3.4 Complications

Complications stemming from CVDs are both diverse and severe, significantly impacting patient outcomes. Myocardial infarction, heart failure, arrhythmias, and ischemic strokes are common consequences of uncontrolled cardiovascular pathology. Among these, heart failure emerges as a prevalent complication, marked by compromised cardiac contractility and inadequate systemic perfusion. Arrhythmias, stemming from aberrations in the heart's electrical conduction system, manifest as irregular heart rhythms, precipitating hemodynamic disturbances and elevating the risk of sudden cardiac arrest [79, 80].

Atherosclerosis, a primary etiological factor in CVD, engenders thrombotic events wherein blood clots impede coronary or cerebral circulation, culminating in myocardial infarctions or strokes. Chronic kidney disease ensues from prolonged vascular impairment, as diminished renal perfusion compromises nephron function. Peripheral artery disease, characterized by occlusive arterial lesions, precipitates claudication and, in severe cases, necessitates limb amputation [81–83].

Aneurysm formation, a consequence of weakened arterial walls, introduces the peril of rupture and life-threatening hemorrhage. Furthermore, the intricate interplay between CVD and cognitive decline underscores the vulnerability of the brain to compromised perfusion. This intricate web of complications necessitates a holistic approach, emphasizing not only the management of primary cardiovascular pathology but also the meticulous mitigation of associated risk factors to curtail the expansive ramifications of CVD on diverse organ systems. These complications not only compromise an individual's quality of life but also contribute to increased morbidity and mortality rates [84–86].

10.3.3.5 Management Strategies

The management of CVDs necessitates a multifaceted approach, encompassing lifestyle modifications, pharmacological interventions, and, when warranted, invasive procedures.

Lifestyle changes, including adherence to heart-healthy diets, regular physical activity, smoking cessation, and moderation of alcohol intake, form the foundation of prevention and management efforts. Pharmacological interventions commonly include medications targeting hypertension, dyslipidemia, and antiplatelet therapy. Invasive procedures, such as coronary angioplasty or bypass surgery, may be considered in advanced cases, highlighting the importance of personalized treatment plans [87].

10.3.3.6 Role of Oxidative Stress in Cardiovascular Diseases

The role of oxidative stress in cardiovascular diseases (CVD) is substantial. It arises from an imbalance between the production of reactive oxygen species (ROS) and the body's antioxidant defense system. In physiological conditions, cells meticulously maintain a delicate redox balance to regulate essential signaling pathways [24]. However, disruptions in this equilibrium, characterized by either excessive ROS production or diminished antioxidant capacity, can trigger oxidative stress [88]. Within the cardiovascular system, oxidative stress detrimentally affects endothelial function, fosters inflammation, induces smooth muscle cell proliferation, and contributes to the formation of atherosclerotic plaques [89]. Additionally, oxidative stress is implicated in myocardial ischemia-reperfusion injury and the progression of heart failure [90].

At the molecular level, ROS play a modulatory role in key signaling pathways involved in vascular homeostasis, mediated by nuclear factor-κB (NF-κB), mitogen-activated protein kinases (MAPKs), and endothelial nitric oxide synthase (eNOS). These signaling cascades influence inflammatory responses, cell proliferation, and vascular tone, collectively shaping the cardiovascular phenotype. Moreover, oxidative stress can initiate lipid peroxidation, further contributing to the precipitation of atherosclerosis. The recognition of the pivotal role of oxidative stress in CVD pathogenesis has prompted the exploration of therapeutic strategies targeting redox imbalance [91, 92].

ROS have the capacity to disrupt cell signaling, DNA, lipids, and proteins, leading to inflammation and apoptosis. Mitochondria, being the primary source of ROS production, play a crucial role in the pathogenesis of CVD [93]. Increased ROS production diminishes nitric oxide (NO) bioavailability, resulting in vasoconstriction and hypertension. ROS also contributes to the development of atherosclerotic plaque [94]. Additionally, oxidative stress-induced endothelial dysfunction is a major player in the etiology of CVD [24], reducing NO bioavailability, inducing inflammation, and impairing mitochondrial function [95]. Mitochondrial dysfunction stemming from oxidative stress exposure further contributes to cardiovascular diseases. In summary, oxidative stress disrupts cellular processes and significantly contributes to the development and progression of cardiovascular diseases [88, 96, 97].

10.4 Role of Antioxidants in Metabolic Disorders

Antioxidants are reducing agents that inhibit the oxidation of other molecules. These compounds decrease the oxidative stress through different mechanism. Some antioxidant compounds can scavenge the free radicals while other can inhibit their activity or expression. The antioxidant can also inhibit the activity or expression of enzymes responsible for the generation of free radicals while few can help in enhancing the activity or expression of intracellular antioxidant enzymes. As oxidative stress and ROS play a modulatory role in several metabolic disorders, use of external antioxidant compounds proves useful approach in the prevention and treatment of such complications. They can be included in the nutritional regimen or added as an adjuvant in management of the oxidative stress related to metabolic dysfunctions.

Apart from vitamins and minerals of edible food, there are important phytochemicals in plant drugs and inedible parts of vegetative food that are known to have significant antioxidant activity. These phytochemicals include polyphenols like flavonoids, tannins, coumarins, chalcones, phenolic acids, and anthocyanins. Besides polyphenols, terpenoids, polysaccharides and few important glycosides have also shown free radical scavenging activity through different mechanisms. Some of them have also been evaluated for their role in various metabolic disorders. Some antioxidants useful in metabolic disorders have been represented in Table 10.1.

10.5 Conclusion

Metabolism is the powerhouse of numerous biochemical processes pivotal in the production and utilization of energy for the normal functioning of the human body. Certain inherited deficiencies and lifestyle induced metabolic abnormalities lead to different types of metabolic disorders. Metabolic disorders are increasing worldwide leading to increased rates of mortality especially in case of young adults due to various inherited and acquired metabolic dysfunction. These metabolic disorders if not managed can lead to precipitation of other significant pathological conditions [201].

The lifestyle abnormalities increase the production of reactive oxygen species causing oxidative stress that leads to various metabolic disorders. The risk of such complicated metabolic disorders like diabetes mellitus (Type 2) [202], cardio metabolic dysfunction, Obesity, dyslipidemia and non-alcoholic fatty liver disease are raising the global concern.

Current research is therefore investigating the role of antioxidants in the prevention and management of such disorders. Along with important vitamins, minerals and nutrients, the antioxidant phytochemicals from various plant drugs, edible and inedible components have shown to play a significant role in the prevention and management of these metabolic disorders. This chapter focuses on the role of oxidative stress in various metabolic disorders and the significance of antioxidant phytochemicals such as polyphenols, flavonoids, tannins, carotenoids in the management and treatment of several growing metabolic disorders. Further investigations to their molecular mechanism and role will help in including them as active drug ingredients in preventive care and as an adjuvant for their synergistic effect in therapeutic management schemes.

Table 10.1 Antioxidants in metabolic disorders.

S. no.	Natural drug source	Parts used	Antioxidant constituents	Pharmacological activity/mechanism of action	Ref.
1.	*Allium sativum* *(Garlic)*	bulb	Allicin, diallyl disulphide and diallyl trisulphide	Serum total cholesterol, elevated LDL and an elevation in LDL oxidation, reduce H2O2 induced DNA damage	[98–100]
2.	*Achyroclinesatureioides* *(Macela)*	Inflorescence/Flowers	Flower extract (quercitrin, isoquercitrin, quercetin and caffeic acid, rutin, quercetin, 3-O-methylquercetin, and luteolin) phenolic compounds (gallic acid, ellagic acid, chlorogenic acid, caffeic acid, catechin, epicatechin, Kaempferol, Luteolin), terpenoids in the essential oils, coumarins and phloroglucinol derivatives	Flower infusion showed significant cumulative antioxidant adequacy in DPPH• and ABTS+• scavenging, FRAP, and thiobarbituric acid reactive substances (TBA-RS) assay. Both extract, as well as paramount phytochemicals standards illustrated protection against lipid peroxidation in rat brains. *In silico* toxic predictions depicted that the important compounds found in the extract had relatively decreased probability for toxic risk. The isolated major compounds protected homogenates of rat brains from lipid damage caused due to $OH^{\bullet}$	[101, 102]

(*Continued*)

Table 10.1 Antioxidants in metabolic disorders. (*Continued*)

S. no.	Natural drug source	Parts used	Antioxidant constituents	Pharmacological activity/mechanism of action	Ref.
3.	*Angelica sylvestris* (*Angelica*)	Essential oil	Essential oil (limonene (75.3%) and α-pinene (9.6%), Fruit essential oil contain as (α-pinene (23.2%) and β-phellandrene (34.5%) as major components, Root essential oil contain majorly sesquiterpenes, major compounds were spathulenol (12.4%), germacrene D (10.6%), α-humulene (7.6%) in the leave essential oil; elemol (5.4%), 10-epi-γ-eudesmol (5.4%), spathulenol (4.8%) in the root oil; α-pinene (42.0) and β -phellandrene (25.5) in the flower and α-pinene (23.2%) and β-phellandrene (34.5%) in the fruit essential oil, respectively. LC-MS/MS evaluation showed chlorgenic acid in roots, benzoid acid in seeds, ferulic acid in stem and hesperidin in leaf extract	EO exhibited significant antioxidant potential ranging from 1.48 to 2.07 mg VitC/g for 30 μL/mL of tested concentration in ABTS assay. IC50 values in DPPH assay lied between 59.895 and 149.498 mg/mL of concentrations that were tested. Fruit essential oil exhibited highest antioxidant activity at 10mg/ml concentration. In another study the leaf methanolic extract and essential oil showed significant antioxidant potential in (DPPH·) (ABTS·+), and (FRAP) assay	[103–105]

(*Continued*)

Table 10.1 Antioxidants in metabolic disorders. (*Continued*)

S. no.	Natural drug source	Parts used	Antioxidant constituents	Pharmacological activity/mechanism of action	Ref.
4.	*Apium Graveolens (Celery)*	Roots & leaves	Appin, apigenin, luteolin, kaempferol, limonene, selinene, coumarin glycosides, flavonoids, and vitamins A and C, caffeic acid, p-coumaric acid, ferulic acid,	Celery root, specially its leaves possesses the property to eliminate OH and DPPH (2,2-diphenyl-1-picrylhydrazyl) radicals and the plant also decreases the magnitude of liposomal peroxidation, decreased glutathione content, antioxidative capacity (FRAP) in liver homogenate while leaf juice lead to an elevation in GSH content. The active constituent called Apiin was isolated from leaves and found to exhibit a noteworthy inhibiting activity on MDA and LPF. While it promoted TAOC and significantly elevated the activities of SOD, GSH-Px, and CAT.	[106]
5.	*Arachis hypogaea (Peanut)*	Peanut skin & roots	Peanut skin contain phenolic acid (ferulic acid, coumaric acid, and chlorogenic acid), flavonoid (epicatechin, catechin, and Resveratrol) and anthocyanins majorly as cyanidin-3-O-sophoroside(C3So) and cyanidin-3-O-sambubioside (C3Sa) Roots (quinones, stilbenoids, and flavones and flavonoid glucosides)	Red peanut skin extract showed higher antioxidant activity in ABTS, DPPH radical scavenging activities, and reducing power (FRAP) assay. The antioxidant activity is attributed the large number of aromatic hydroxyl groups. Aqueous ethanol and methanol extract showed high ABTS+ radical scavenging activity and DPPH radical scavenging activity respectively	[107, 108]

(*Continued*)

Table 10.1 Antioxidants in metabolic disorders. (*Continued*)

S. no.	Natural drug source	Parts used	Antioxidant constituents	Pharmacological activity/mechanism of action	Ref.
6.	*Arctium lappa (Burdock)*	Roots & leaves	Roots contain polysaccharides that are mainly composed of glucose, rhamnose, arabinose, and galactose. Roots also contain arctigenin, quercetin, chlorogenic acid and caffeic acid compounds, flavonoids and phenolics (quercetin, rutin, chlorogenic acid, caffeoylquinic acid derivatives, and caffeic acid 7-O-apiosyl-glucoside, apigenin 7-O-glucoside, apigenin 7-O-rutinoside, morin, kaempferol, kaempferol 3,7-O-diglucoside, and quercetin 3-O-rhamnoside), polysaccharide(fructan) in roots, Sesquiterpene lactone (Onopordopicrin) in leaves	The roots' derived polysaccharides exhibited potent antioxidant activity *in vitro*, with IC50 values of 1.732 mg/ml for ·OH radical scavenging, 0.29 mg/ml for DPPH radical scavenging, and 0.15 mg/ml for superoxide anion scavenging. In another study, aqueous ethanol extract of roots containing flavonoid exhibited strongest free radical scavenging activity. Aqueous root extract depicted scavenging influence on superoxide and H_2O_2 by promoting termination of free-radical reactions and ROS quenching. Also showed antioxidant, anti-inflammatory, anti-diabetic, gastroprotective, and hepatoprotective activity	[109–114]
7.	*Arctostaphylos alpina (Bearberry)*	Berries	Cyanidin-3- O-galactoside (cy-gal), flavonols, phenolic compounds, sugars, fruit acids and dietary fiber.		[115]
8.	*Arctostaphylos uva-ursi L. Spreng. (Bearberry)*	Leaves	Phenolic compounds, flavonoids, arbutin, methylarbutin, hyperoside, corilagin, and methylartutin, ursolic, tannic and gallic acid.	The leaf extract exhibited robust antioxidant capability in scavenging radicals, reducing power assay, liposome model, scavenging hydroxyl radicals (HO). Additionally, it demonstrated potential as protectors against lipid oxidation and biological ageing of tissues.	[116]
9.	*Armoracia rusticana (Horseradish)*	Roots	Phenolic acid (syringic acid, ferulic acid, caffeic acid) and flavonoids (epicatechin, kaempferol-3-O-rutinoside and kaempferol-3-O-glucoside)	Fresh extract of roots showed highest value of TEAC determined by ABTS and DPPH assays while sous-vide treated root extract showed anti-AGEs activity by increasing polyphenolic compounds which captured α-dicarbonyl compounds	[117]

(*Continued*)

Table 10.1 Antioxidants in metabolic disorders. (*Continued*)

S. no.	Natural drug source	Parts used	Antioxidant constituents	Pharmacological activity/mechanism of action	Ref.
10.	*Artemisia dracunculus (Tarragon)*	Aerial parts & Essential oil	Essential oil contains p-allylanisole (40–85%), sabinene (35%), methyl eugenol (25%), and elemicin (up to 57%) Other compounds present in the oil in concentrations greater than 10% are: terpinen-4-ol, Ocimene (E)-beta (7.46%), Ocimene (Z)-beta (6.24%) and Limonene (1.42%), and trans-anethole, α-phellandrene, β-phellandrene (Z)-artemidin, capillene.	hydro-alcoholic extract of aerial parts of Artemisia dracunculus showed strong antioxidant activity in FRAP, DPPH and ABTS assays and also exhibited hepatoprotective activity	[118, 119]
11.	*Artemisia vulgaris (Mugwort)*	Roots and leaves	Leaf contains saponins, glycosides, flavanoids, protein, triterpenoids and roots contain Thujone, Flavonoids, Triterpenes, and Coumarin	Methanolic leaves extract has a potent antioxidants in DPPH and Fecl3 reducing power assay while root extracts showed strong scavenging activity in DPPH*- radical and Fe3+ - reducing power assay	[120–122]
12.	*Astragalus membranaceus (Astragalus)*	Roots	Astragalosides, flavonoids and polysaccharides,	Phytoconstituents Astragalus roots showed significant protection against heart, brain, kidney, intestine, liver and lung injury in various models of oxidative stress-related disease. In another study the root extract showed SOD-like activity, tyrosinase activity, nitrite scavenging activity and EDA (electron donating ability) increasing superoxide dismutase and decreasing lipid peroxide activity	[123–125]
13.	*Avena sativa (Oats)*	Oat bran	Vitamin E (tocols), phytic acid, phenolic compounds, beta-carotene, polyphenols, flavonoids and avenanthramides also contain flavonoids, sterols Beta-glucan, Beta-D-glucans, Polysaccharides	Alleviating oxidative stress-induced cellular damage in the human body by eliminating reactive oxygen species also contributes to mitigating cardiovascular risks.	[126, 127]

(*Continued*)

Table 10.1 Antioxidants in metabolic disorders. (*Continued*)

S. no.	Natural drug source	Parts used	Antioxidant constituents	Pharmacological activity/mechanism of action	Ref.
14.	*Camellia sinensis (Tea and Green Tea)*		Tea catechins (epigallocatechin gallate (EGCG), epigallocatechin (EGC), epicatechin gallate (ECG), gallocatechin (GC), epicatechin (EC), and catechin)	Fresh tea leaves extract showed potent antioxidant by scavenging free radicals in DPPH assay, exhibited potent inhibition of lipid peroxidation in Ferric Thiocyanate (FTC) assay and exhibited the ability to provide protection against CCl4 intoxication induced oxidative stress.	[128]
15.	*Capsicum annuum (Bell peppers)*		Capsicum antioxidant lycopene, Carotene, Like lycopene, Para-coumaric acid, Vitamin C	Mediated immune responses, Antibacterial	
16.	*Carica papaya (Papaya)*	Fruit, leaves, fruit peel and seeds	Leaves (papain, cystatin, chymopapain, tocopherol, phenolic acids, cyanogenic glucosides, glucosinolates, and vitamin C) Fruit peel (gallic acid (18.06 μg/g), caffeic acid (29.28 μg/g), p-coumaric acid (38.16 μg/g), ferulic acid (95.46 μg/g), and quercetin (3.17 μg/g)) Fruit pulp (Lycopene)	Water and Methanol extract of leaves depicted radical scavenging ability in DPPH. Papaya peel extract showed demonstrated radical scavenging activity in DPPH, FRAP and ABTS assays also showed decrease in SOD, CAT, GPx, GR activity, and GSH levels Fully ripened fruit pulp showed the maximal antioxidant capability to scavenge free radicals.	[129–131]
17.	*Cinnamomum verum (Cinnamon)*	Bark, leaf Essential Oil	Cinnamic acid, Cinnamaldehyde, eugenol, carvacrol, cinnamic acetate, thymol, coumarin	Showed protective effect against H2O2 damage, by increasing GPx activity, improve hyperlipidemia; possibly by decreasing the production of cholesterol, and inhibiting lipid peroxidation. Robust inhibitory action was also visible against β-carotene discoloration. Also depicted potential to suppress lipid oxidation reaction, and as a acting as a preservative for food.	[132–134]

(*Continued*)

Table 10.1 Antioxidants in metabolic disorders. (*Continued*)

S. no.	Natural drug source	Parts used	Antioxidant constituents	Pharmacological activity/mechanism of action	Ref.
18.	*Citrus sinensis (Citrus)*	Fruit pulp, Fruit Peel, seed oil	Fruit juice conatin flavonoids, Limonoids citric acid, Vitamin C, Flavonoids Fruit peel contains gallic acid, protocatechuic acid, 4-hydroxybenzoic acid, caffeic acid and ferulic acid and flavonoids like catechin, epigallocatechin, vitexin, rutin, luteolin and apigenin. Seed oil contain carotenoids, phenolic compounds, α-tocopherol and phytosterols	Fruit peel showed potential antioxidant activity in DPPH, FRAP and ORAC assay Seed oil showed free radical scavenging in DPPH assay	[135, 136]
19.	*Coriandrum sativum (Coriander)*	Fruit, leaves and roots	Fruit contains, phenolic acids, flavonoids, Terpenes linalool, Pinene, chlorogenic and gallic acids	Methanolic extracts of fruit presented exceptional DPPH radical scavenging and reducing power activity	[137]
20.	*Crataegus monogyna (Hawthorn)*	Vegetative and reproductive parts	Phenolic compounds, proanthocyanidins, catechins, flavonoids, and phenolic acids	The ethanol extract of leaves depicted antioxidant potential IC50 =22.50 μg/ml. Ethanolic flower extract of Crataegus monogyna Jacq. Showed considerable radical scavenging in ABTS and CUPRAC method. In another study, unripe fruits and flowers revealed paramount antioxidant activity. antioxidant and potential cardiovascular activity	[138–140]

(*Continued*)

Table 10.1 Antioxidants in metabolic disorders. (*Continued*)

S. no.	Natural drug source	Parts used	Antioxidant constituents	Pharmacological activity/mechanism of action	Ref.
21.	*Crocus Sativus (Saffron)*	Dried stigma of the flowers	carotenoids (crocins, Picrocrocins, crocetin, safranal), anthocyanins, flavonoids, vitamins (riboflavin and thiamine),	Crocin acts as an antioxidant by neutralizing free radicals, particularly superoxide anions, thereby protecting cells from oxidative stress. Crocetin reduced lipid peroxidation caused by reactive oxygen species (ROS) in primary hepatocytes of rats and by BαP in mice. Crocetin reduced the production of malondialdehyde (MDA), which serves as a marker for lipid peroxidation caused by reactive oxygen species (ROS). Crocin decreased the amount of cholesterol in hyperlipidemic rats with 2 months feeding of excessive cholesterol [44]. A separate investigation discovered that crocin exhibits potent triglyceridemic and cholesterolemic reduction effects in quails suffering from coronary artery disease [22]. Additional investigation confirmed that crocetin has the ability to decrease the levels of serum, total cholesterol, and malondialdehyde. It also prevents the decrease of nitric oxide in the serum of hyperlipidemic-diet quails [39]. In the elucidation of the hypolipidemic mechanism of crocin, Sheng *et al.* [44] indicated that crocin inhibited the absorption of dietary fat and cholesterol.	[14, 18, 20–22, 34–47, 61, 62, 141]

(*Continued*)

Table 10.1 Antioxidants in metabolic disorders. (*Continued*)

S. no.	Natural drug source	Parts used	Antioxidant constituents	Pharmacological activity/mechanism of action	Ref.
22.	*Curcuma longa (Turmeric)*	Rhizome	Curcumin, demethoxycurcumin, bisdemethoxycurcumin, bisabolone-9-one (1), 4-methyllene-5-hydroxybisabola-2,10-diene-9-one, turmeronol B, 5-hydroxy-1,7-bis(4-hydroxy-3-methoxyphenyl)-1-hepten-3-one, 3-hydroxy-1,7-bis(4-hydroxyphenyl)-6-hepten-1,5-dione, cyclobisdemethoxycurcumin	The methanolic rhizome extract showed significant radical scavenging capability in DPPH (IC_{50}: 26.4 μg/mL), ORAC (14,090 μmol Trolox equivalent/g extract), reducing power absorbance (0.33) and hydroxyl radical scavenging activity (IC_{50}: 7.4 μg/mL).	[142, 143]
23.	*Cymbopogon nardus (Lemongrass)*		Citronellol, Geraniol, Geranyl acetate, nerol, elemol, delta-elemene Cariophyllene Germacrene D and Naphthalene.	The essential oil showed stronger radical-scavenging capacity IC50 = 1.38 ± 0.04 μ g/mL than quercetin(IC50 = 6.56 ± 0.06 μ g/mL).	[144]
24.	*Daucus carota subsp. Sativus (Carrot)*	Roots orange, red, yellow and black	Orange carrots contain β-carotene and α-carotene; red carrots has additional lycopene; yellow carrots contain xanthophylls and lutein,; black/purple carrots have anthocyanins	All carrot extract showed antioxidant activity in the following order (black carrot > rainbow carrot > red carrot ≈ orange carrot ≈ yellow carrot) in FRAP, CUPRAC and ABTS assay.	[145]
25.	*Eucalyptus globulus (Eucalyptus)*	Leaves, Essential oil	Leaves contain salicylic acid β-D-glucuronide, chlorogenic acid, epicatechin, 2″-O-galloylhyperin, isoquercitrin, isorhapontin, quercitrin, and quercetin-3-O-glucuronide. Essential oil of leaves contain eucalyptol (1,8-cineole)	30–50% ethanolic extract of leaves gave best antioxidant results in DPPH and ABTS assay. EO of Eucalyptus species demonstrated strong antioxidant applicability and decreased the oxidative cellular stress by modulation of antioxidant enzymes, glutamate-cysteine ligase (GCL) and heme oxygenase-1 (Hmox-1) in *in vitro* assays.	[146, 147]

(*Continued*)

Table 10.1 Antioxidants in metabolic disorders. (*Continued*)

S. no.	Natural drug source	Parts used	Antioxidant constituents	Pharmacological activity/mechanism of action	Ref.
26.	*Ginkgo Biloba (Maidenhair tree)*	Leaves	Flavonoids and polyphenolic acids such as quercetin-3-O-rutinoside, Chlorogenic acid, syringic acid, protocatechuic acid, caffeic acid and coumaric acid, epicatechin, catechin, coumarin and apigenin.	Aqueous acetone extract of leaves depicted highest antioxidant potential in DPPH; TEAC – Trolox equivalent antioxidant capacity; MRP – molybdenum reducing power assay. In another study, aqueous infusions from Ginkgo biloba leaves showed antioxidant activity in FRAP assay	[148, 149]
27.	*Glycyrrhiza glabra (Licorice)*	Roots	Flavonoids and isoflavone (Glabradin, glabridin, hispaglabridin A, and 30-hydroxy-4-O-methylglabridin,glabrene, glabrol, licoflavonol, licoricone, formononetin, phaseollinisoflavan, hispaglabridin A and B, 3-hydroxy glabrol, 3-methoxy glabridin) licochalcone B & D	The methanolic extracts of licorice roots showed the protection from oxidation in DPPH assay	[150]
28.	*Hedera rhombea (Ivy)*		Triterpenoid Saponins (Alpha-hederin, hederasaponin-C, hederacolchisides-E and –F)	The triterpenoid saponin extract showed saponins strong antioxidant potential in 1,1-di-phenyl-2-picryl-hydrazyl (DPPH.) free radical scavenging, total antioxidant activity, reducing power, superoxide anion radical scavenging, hydrogen peroxide scavenging, and metal chelating activities at a concentration of 75 pg/mL,	[151]
29.	*Inula helenium (Elecampane)*	Roots	Phenolic acids (caffeic, chlorogenic, dicaffeoylquinic, hydroxibenzoic), terpenes (alantolactone) and flavonoids (epicatechin, catechin gallate, ferulic acid-4-O-glucoside, dihydroquercetinpentosylrutinoside, kaempherol-7-O-dipentoside, quercetin-3-O-β-glucopyranoside)	Chloroform and ethyl acetate fraction of dried roots of *I. helenium* showed strong antioxidant activity correlating to its phenolic and flavonoid content.	[152]

(*Continued*)

Table 10.1 Antioxidants in metabolic disorders. (*Continued*)

S. no.	Natural drug source	Parts used	Antioxidant constituents	Pharmacological activity/mechanism of action	Ref.
30.	*Laurus nobilis (Bay laurel)*	Essential oil of leaves	1,8-cineole, 1-p-menthen-8-ethyl acetate, linalool and sabinene, while the seed oil was characterized by β-ocimene, 1,8-cineole, α-pinene and β-pinene	The leaf essential oil exhibited the most potent antioxidant activity in the β-carotene/linoleic acid system, with an IC_{50} value of 35.6 μg mL^{-1}	[153]
31.	*Leonurus cardiac (Hoarhound)*	Aerial parts including leaves and flowers	Alkaloid leonurine, ursolic acid	Increase in mitochondrial oxidation Attenuation of ROS generation in mitochondria	[154–156]
32.	*Mentha pulegium (Pennyroya)*	Essential oils from leaves and stems	chrysoeriol, kaempferol, and luteolin	Free radical scavenging property, thereby reducing oxidative stress	[157, 158]
33.	*Mentha spicata (Mint)*	Aerial parts including leaves and stems	oxygenated monoterpenes and monoterpene hydrocarbons present in the essential oil	DPPH, ABTS radical scavenging, and ferric-reducing power assay	[159, 160]
34.	*Myristica fragrans (Nutmeg)*	Seeds	dehydrodiisoeugenol, malabaricone B, and malabaricone C present in the methanol, ethanol, and acetone extracts of Myristica fragrans seeds.	DPPH, ABTS, superoxide radical and hydroxyl radical scavenging	[161]
35.	*Ocimum sanctum (Holy Basil)*	Leaves	O. sanctum leaf extracts in butanol and ethylacetate with high polyphenolics and flavonoids, had strong antioxidant potential Phenolic compounds such as such as cirsilineol, cirsimaritin, isothymusin, apigenin, rosmarinic acid, and eugenol and flavonoids are responsible for antioxidant activity.	1. Inhibition of lipid peroxidation by Aqueous extracts of Ocimum sanctum in rats which is responsible for generation of ROS 2. DPPH, ABTS and Phosphomolybdate assay	[162, 163]
36.	*Olea europaea (Olive)*	Leaves (Major source of antioxidants)	Oleuropein, a bitter glycoside; Phenolic compounds	Antioxidant activity was determined through DDPH assay, reducing power and nitric acid scavenging activity	[164]
37.	*Origanum majorana (Marjoram)*	Leaves	Borneol, Camphor, Pinene	Antiviral, Antibacterial, Antifungal effects	

(*Continued*)

Table 10.1 Antioxidants in metabolic disorders. (*Continued*)

S. no.	Natural drug source	Parts used	Antioxidant constituents	Pharmacological activity/mechanism of action	Ref.
38.	*Origanum vulgare (Oregano)*	Shrub	Carvacrol and thymol	Ethanolic extract of the plant shows antioxidant activity through free radical scavenging property.	[165]
39.	*Oryza sativa (Rice)*	Grain	g-oryzanol (e.g. 24-methylenecycloartanyl trans-ferulate), Anthocyanins (e.g. peonidin-3-O-glucoside, malvidin, cyanidin-3-O-glucoside, and petunidin-3-O-glucoside)	Free radical scavenging	[166]
40.	*Papaver somniferum (Opium poppy)*	Seeds, leaves	Alkaloids (berberine and macranthine), Phenolic compounds	1. Free radical scavenging, 2. Inhibition of lipid peroxidation, 3. Enhancing the activity of antioxidant enzymes, such as superoxide dismutase (SOD) and catalase	[167]
41.	*Phaseolus Vulgaris (Beans)*	Seeds	Flavonoids (kaempferol and quercetin), Anthocyanins (cyanidin 3-glucoside)	Free radical scavenging	[168]
42.	*Phyllanthus emblica L. (Amla)*	Fruit	Ascorbic acid, gallic acid, ellagic acid, and phyllembin, phenoic acids (4-hydroxybenzoic acid, coumaric acid, protocatechuic acid), flavones (apigenin, luteolin, and myricetin), ellagitannins (chebulinic acid, chebulagic acid, emblicanin A and B, pedunculagin, phyllanemblinins A–F, and punigluconin)	Improved cell viability, GSH, CAT, and SOD levels and reduced lipid peroxidation, ROS level	[169]
43.	*Pimenta dioica (Allspice)*	Berry	Essential oil	High free radical scavenging of radicals such as DPPH, ABTS and superoxide anion	[170]
44.	*Pimpinella anisum (Anise)*	Seed	Essential oil	Neutralization of free radicals	[171]

(*Continued*)

Table 10.1 Antioxidants in metabolic disorders. (*Continued*)

S. no.	Natural drug source	Parts used	Antioxidant constituents	Pharmacological activity/mechanism of action	Ref.
45.	*Piper betle (Betel leaf)*	Leave	Phenolic compound hydroxychavicol is particularly responsible for antioxidant activity	1. Free radical scavenging 2. inhibiting metal-catalyzed oxidation reactions 3. Reducing the formation of lipid peroxides	[172]
46.	*Piper nigrum (Black pepper)*	Seeds	Piperine, phenolic compounds, essential oil	1. Free radical scavenging 2. Inhibition of lipid peroxidation	[173]
47.	*Punica granatum (pomegranate)*	Fruit Peel, Seed	Phenolic compounds, flavonoids	Water and ethanol extracts of punica granatum exhibit antioxidant activity through Free radical scavenging properties. This activity is determined through scavenging of DPPH and ABTS, TPTZ reducing, Fe^{3+} reducing and Cu^{2+} reducing methods	[174]
48.	*Quercus robur (Oak)*	Leaves, twigs, and acorns	Phenolic acids, flavonoids, tannins	1. Water extract: lipid peroxidation 2. 80% ethanol: ARP (Antioxidant Radical Power) determinations and FRAP (Ferric Reducing Antioxidant Power) 3. Acidic ethanol (0.1 moldm3 HCl in EtOH): determining total phenol (TPC) and tannin (TAC) content	[175, 176]
49.	*Rosmarinus officinalis (Rosemary)*	Leaves, essential oil	Myrcene, 1,8-cineole, α-pinene and β-pinene	Free radical scavenging of superoxide anions and hydroxyl radicals. Rosemary essential oil also exhibits iron-chelating ability.	[177, 178]
50.	*Rubus occidentalis (Raspberry)*	Berries, fresh leaves	Anthocyanin pigments, Polyphenols, especially ellagic acid	Methanol extract was used for determination of antioxidant activity through scavenging of DPPH, Fe^{3+} reducing and Cu^{2+} reducing methods	[179–181]
51.	*Rubus ursinus (California Blackberry, marionberry)*	Berries	Phenolic compounds, such as anthocyanins (e.g. delphinidin and cyaniding), flavonols, and ellagitannins	Free radical scavenging. The evaluation of antioxidant activity has been performed through DPPH radical scavenging.	[179]

(*Continued*)

Table 10.1 Antioxidants in metabolic disorders. (*Continued*)

S. no.	Natural drug source	Parts used	Antioxidant constituents	Pharmacological activity/mechanism of action	Ref.
52.	*Serenoa repens (Saw palmetto)*	Seeds	The fermented saw palmetto seed extract of Serenoa repens contains phenolic compounds that exhibit antioxidant activity	Free radical scavenging	[182]
53.	*Schisandra chinensis (Schizandra)*	Fruits	The dibenzocyclooctadiene lignin, schizandrin is considered as the major antioxidant. Other antioxidants include phenolic acids (chlorogenic, gentisic, p-hydroxybenzoic, p-coumaric, protocatechuic, syringic, and salicylic acids) and flavonoids (quercetin, isoquercitrin, rutin, and hyperoside)	Free radical scavenging and activation of the cellular antioxidant defense system components	[183]
54.	*Sesamum indicum (Sesame)*	Seed	Sesame seed as well as seed coats extract exhibited high phenolic (sesamol, Sesamin and sesaminol) and lignin content	free radical scavenging, inhibition of lipid peroxidation, enhanced antioxidant enzyme activity, and chelation of metal ions	[184–186]
55.	*Syzygiumcumini (Indian Blackberry, Java plum, Malabar Plum)*	Fruit pulp, seed and leaves	Fruit extract contain anthocyanins (malvidin (44.4%), petunidin (24.2%), delphinidin (20.3%), cyanidin (6.6%), and peonidin (2.2%), Seed extract contain phenolics, flavonoids, and tannin (oxalic acid, tannic acid, gallic acid, epicatechin, and catechin hydrate) leaf extracts (ferulic acid and catechin,)	*S. cumini* seed extract illustrated significantly superior antioxidant capability as compared from ascorbic acid in ABTS free radical scavenging assay. Catechin from leaf extract showed high a peroxyl radical scavenging activity. Fruits and seed extract showed strong radical scavenging activity and also inhibited the enzymes, alpha-amylase and intestinal alpha-glucosidase and also show hypoglycemic activity	[187, 188]

(*Continued*)

Table 10.1 Antioxidants in metabolic disorders. (*Continued*)

S. no.	Natural drug source	Parts used	Antioxidant constituents	Pharmacological activity/mechanism of action	Ref.
56.	*Spinacia oleracea (Spinach)*	Leaves	flavonoids, p-coumaric acid derivatives, The water extract also contains phenolic alkaloids that exhibit antioxidant activity.	Spinach extracts have been shown to have *in vitro* antioxidant activity, as measured by total polyphenol content (TPC), oxygen radical absorbance capacity (ORAC), cellular antioxidant activity (CAA), and antigenotoxic activity	[189, 190]
57.	*Terminalia chebula (Harde)*	Fruit	The aqueous extract exhibits the presence of phenolic compounds, tannins, flavonoids, and gallic acid	The antioxidant activity of the aqueous extract of terminalia chebula fruit was observed through Trolox equivalent antioxidant capacity (TEAC) assay, Superoxide anion generation inhibition assay, Nitric oxide radical scavenging assay, Singlet oxygen scavenging assay and Hypochlorous acid scavenging assay	[191]
58.	*Theobroma cacao (Cocoa)*	Seeds	Polyphenos, specifically monomeric flavan-3-ols, are responsible for antioxidant activity. Further, these compounds can undergo structural changes during processing. These changes, may lead to new compounds, which can also contribute to the antioxidant properties of cocoa.	Free radical scavenging, metal ion chelation	[192]
59.	*Thymus Vulgaris*		thymol, apigenin, rosmarinic acid, and carvacrol rosmarenic acid, methyl rosmarenate, caffeic acid, cinnamic acid, chlorogenic acid and quinic acid as phenolic acids, besides some flavonoids such as ferulic acid, apigenin, luteolin and quercetin.	Methanol extract of thyme exhibited strong antioxidative activity, than α-tocopherol and BHA in DPPH free radical scavenging method. In another study, thyme–sage extract showed lowest EC50 (DPPH•, 55.51 μg·ml−1) and the highest FRAP value (95.51 mM Fe (II) mg−1 extract). The essential oil of thymus (Vulgaris) and thymus (Vulgaris) silver nanoparticles showed strong radical scavenging activity in DPPH assay	[193]

(*Continued*)

Table 10.1 Antioxidants in metabolic disorders. (*Continued*)

S. no.	Natural drug source	Parts used	Antioxidant constituents	Pharmacological activity/mechanism of action	Ref.
60.	*Urtica thunbergiana (Nettle, Urtica)*	Whole plant	The ethanolic extract contains active phenolic compounds, such as caffeic acid (CA) and chlorogenic acid (CGA)	Free radical scavenging	[194]
61.	*Vaccinium corymbosum (Blueberry)*	Fruit	Anthocyanin and phenolic compounds, malvidin, malvidin-3-glucoside, and malvidin-3-galactoside	Scavenging the ROS and preventing the formation of reactive free radicals by upregulating the SOD and HO-1 protein expression levels.	[195]
62.	*Vaccinium oxycoccos (Cranberry)*	Berries	Polyphenolic compounds including anthocyanins, proanthocyanins, and flavonols, especially quercetin	Polyphenols act as potent antioxidants by scavenging free radicals	[196]
63.	*Vaccinium myrtillus (Bilberry)*	Berries	Anthocyanins (delphinidin, petunidin, cyanidin, peonidin, and malvidin)	Causes ROS quenching on HepG2 cells.	[197]
64.	*Withaniasomnifera (Ashwagandha)*	Roots	Withaferin A, Withanolides	Suppress oxidative and inflammatory responses by suppressing LPS-induced production of NO and promoting the working of Nrf2/HO-1 pathway,	[198, 199]
65.	*Zingiber officinale (Ginger)*	Rhizome	Ginger oleo resin, 6-Gingerol, 6-Shogaol, phenylpropoinoids	Reducing the generation and content of ROS, Decreasing the level of MDA, reducing the lipid peroxidation, increasing the activities of antioxidant enzymes and the level of GSH, Increasing Nrf2 activity and the level of GSTP1, Increasing the intracellular GSH/GSSG ratio	[200]

References

1. Ferreira, C.R., Rahman, S., Keller, M., Zschocke, J., An international classification of inherited metabolic disorders (ICIMD). *J. Inherit. Metab. Dis.*, *44*, 164–177, 2021.
2. Angelini, C., Burlina, A., Blau, N., Ferreira, C.R., Clinical and biochemical footprints of inherited metabolic disorders: X. Metabolic myopathies. *Mol. Genet. Metab.*, *137*, 213–222, 2022.
3. Vorgerd, M. and Deschauer, M., *Em Neuromuscular Disorders*, Elsevier, Amsterdam, Netherlands, 2022.
4. Zakir, F., Mohapatra, S., Farooq, U., Mirza, M.A., Iqbal, Z., *Em Drug Delivery Systems for Metabolic Disorders*, Elsevier, 2022.
5. Abdelsattar, S., Obada, M., El-Hawy, M.A., Abd El Naby, S.A., Zaki, O.K., Elsaid, H., Inherited metabolic disorders in a cohort of Egyptian children. *Egypt. Liver J.*, *12*, 20, 2022.
6. Pardo Lameda, I.L. and Koch, T.R., *Em Liver Diseases*, Springer International Publishing: Cham, Switzerland, 2020.
7. Waters, D., Adeloye, D., Woolham, D., Wastnedge, E., Patel, S., Rudan, I., Global birth prevalence and mortality from inborn errors of metabolism: a systematic analysis of the evidence. *J. Glob. Health*, *8*, 021102, 2018.
8. Pakanova, Z., Nemcovic, M., Ziburova, J., Mucha, J., Salingova, A., Sebova, C., Jurickova, K., Barath, P., Inherited metabolic disorders of glycoconjugate metabolism. *Bratislava Med. J.*, *121*, 760–766, 2020.
9. Ramachandran, R., *Em Postural Tachycardia Syndrome*, Springer International Publishing: Cham, Gewerbestrasse, Cham, Switzerland, 2021.
10. Tabatabaei-Malazy, O., Larijani, B., Abdollahi, M., Targeting metabolic disorders by natural products. *J. Diabetes Metab. Disord.*, *14*, 57, 2015.
11. Wu, G., *Em Animal Agriculture*, Elsevier, Amsterdam, Netherlands, 2020.
12. Bassam, B.A. and Bertorini, T.E., *Em Neuromuscular Disorders*, Elsevier, Amsterdam, Netherlands, 2022.
13. Yan, M. and Zhai, Q., Sperm tsRNAs and acquired metabolic disorders. *J. Endocrinol.*, *230*, F13–F18, 2016.
14. Resch, H., *Osteoporosis and Related Bone Metabolic Disease*, MDPI, Basel, Switzerland, 2023.
15. Desroches, S. and Lamarche, B., The evolving definitions and increasing prevalence of the metabolic syndrome. *Appl. Physiol. Nutr. Metab.*, *32*, 23–32, 2007.
16. Kolovou, G.D., Anagnostopoulou, K.K., Salpea, K.D., Mikhailidis, D.P., The Prevalence of Metabolic Syndrome in Various Populations. *Am. J. Med. Sci.*, *333*, 362–371, 2007.
17. Manaf, M.R.A., Nawi, A.M., Tauhid, N.M., Othman, H., Rahman, M.R.A., Yusoff, H.M., Safian, N., Ng, P.Y., Manaf, Z.A., Kadir, N.B.A., Yanasegaran, K., Basir, S.M.A., Ramakrishnappa, S., Ganasegeran, K., Prevalence of metabolic syndrome and its associated risk factors among staffs in a Malaysian public university. *Sci. Rep.*, *11*, 8132, 2021.
18. Akkol, E.K. and Aschner, M., *Em The Role of Phytonutrients in Metabolic Disorders*, Elsevier, Amsterdam, Netherlands, 2022.
19. Sherry, E.B., Lee, P., Choi, I.-Y., *In Vivo* NMR Studies of the Brain with Hereditary or Acquired Metabolic Disorders. *Neurochem. Res.*, *40*, 2647–2685, 2015.
20. Nathan, V., Therapeutics in metabolic diseases, in: *Genome Editing in Cardiovascular and Metabolic Diseases. Advances in Experimental Medicine and Biology*, vol. 1396, J. Xiao (Eds.), Springer, Singapore, 2023.
21. Ferreira, C.R., Martinelli, D., Blau, N., Clinical and biochemical footprints of inherited metabolic diseases. VI. Metabolic dermatoses. *Mol. Genet. Metab.*, *134*, 87–95, 2021.
22. Rani, V., Deep, G., Singh, R.K., Palle, K., Yadav, U.C.S., Oxidative stress and metabolic disorders: Pathogenesis and therapeutic strategies. *Life Sci.*, *148*, 183–193, 2016.

23. Furukawa, S., Fujita, T., Shimabukuro, M., Iwaki, M., Yamada, Y., Nakajima, Y., Nakayama, O., Makishima, M., Matsuda, M., Shimomura, I., Increased oxidative stress in obesity and its impact on metabolic syndrome. *J. Clin. Invest.*, *114*, 1752–1761, 2004.
24. Dubois-Deruy, E., Peugnet, V., Turkieh, A., Pinet, F., Oxidative Stress in Cardiovascular Diseases. *Antioxidants*, *9*, 864, 2020.
25. Singla, P., Metabolic effects of obesity: A review. *World J. Diabetes*, *1*, 76, 2010.
26. Ärnlöv, J., Sundström, J., Ingelsson, E., Lind, L., Impact of BMI and the Metabolic Syndrome on the Risk of Diabetes in Middle-Aged Men. *Diabetes Care*, *34*, 61–65, 2011.
27. Racette, S.B., Deusinger, S.S., Deusinger, R.H., Obesity: Overview of Prevalence, Etiology, and Treatment. *Phys. Ther.*, *83*, 276–288, 2003.
28. Pillon, N.J., Loos, R.J.F., Marshall, S.M., Zierath, J.R., Metabolic consequences of obesity and type 2 diabetes: Balancing genes and environment for personalized care. *Cell*, *184*, 1530–1544, 2021.
29. Sheikh, A.B., Nasrullah, A., Haq, S., Akhtar, A., Ghazanfar, H., Nasir, A., Afzal, R.M., Bukhari, M.M., Chaudhary, A.Y., Naqvi, S.W., The Interplay of Genetics and Environmental Factors in the Development of Obesity. *Cureus*, 9, 7, e1435, 2017.
30. Ciemins, E.L., Joshi, V., Cuddeback, J.K., Kushner, R.F., Horn, D.B., Garvey, W.T., Diagnosing Obesity as a First Step to Weight Loss: An Observational Study. *Obesity*, *28*, 2305–2309, 2020.
31. Purnell, J.Q., Definitions, Classification, and Epidemiology of Obesity. *Endotext*, 2023. http://www.ncbi.nlm.nih.gov/pubmed/25905390.
32. Obesity - Diagnosis and treatment - Mayo Clinic, [s.d.], https://www.mayoclinic.org/diseases-conditions/obesity/diagnosis-treatment/drc-20375749.
33. Fruh, S.M., Obesity. *J. Am. Assoc. Nurse Pract.*, *29*, S3–S14, 2017.
34. Clinical Problems Caused by Obesity - Endotext - NCBI Bookshelf, , [s.d.], https://www.ncbi.nlm.nih.gov/books/NBK278973/.
35. Malnick, S.D.H., The medical complications of obesity. *QJM*, *99*, 565–579, 2006.
36. Foster, D., Sanchez-Collins, S., Cheskin, L.J., Multidisciplinary Team-Based Obesity Treatment in Patients With Diabetes: Current Practices and the State of the Science. *Diabetes Spectr.*, *30*, 244–249, 2017.
37. Yu, B., Chen, Y., Qin, H., Chen, Q., Wang, J., Chen, P., Using multi-disciplinary teams to treat obese patients helps improve clinical efficacy: the general practitioner's perspective. *Am. J. Transl. Res.*, *13*, 2571–2580, 2021. http://www.ncbi.nlm.nih.gov/pubmed/34017416.
38. Bright, D., O'Hare, K., Beesley, R., Tapp, H., Tipping the scales: Provider perspectives on a multi-disciplinary approach to obesity. *Exp. Biol. Med.*, *244*, 183–192, 2019.
39. Cochrane, A.J., Dick, B., King, N.A., Hills, A.P., Kavanagh, D.J., Developing dimensions for a multicomponent multidisciplinary approach to obesity management: a qualitative study. *BMC Public Health*, *17*, 814, 2017.
40. Frank, A., A Multidisciplinary Approach to Obesity Management. *J. Am. Diet. Assoc.*, *98*, S44–S48, 1998.
41. Manna, P. and Jain, S.K., Obesity, Oxidative Stress, Adipose Tissue Dysfunction, and the Associated Health Risks: Causes and Therapeutic Strategies. *Metab. Syndr. Relat. Disord.*, *13*, 423–444, 2015.
42. Rotariu, D., Babes, E.E., Tit, D.M., Moisi, M., Bustea, C., Stoicescu, M., Radu, A.-F., Vesa, C.M., Behl, T., Bungau, A.F., Bungau, S.G., Oxidative stress - Complex pathological issues concerning the hallmark of cardiovascular and metabolic disorders. *Biomed. Pharmacother.*, *152*, 113238, 2022.
43. De Marchi, E., Baldassari, F., Bononi, A., Wieckowski, M.R., Pinton, P., Oxidative Stress in Cardiovascular Diseases and Obesity: Role of p66Shc and Protein Kinase C. *Oxid. Med. Cell. Longev.*, *2013*, 1–11, 2013.

44. Niemann, B., Rohrbach, S., Miller, M.R., Newby, D.E., Fuster, V., Kovacic, J.C., Oxidative Stress and Cardiovascular Risk: Obesity, Diabetes, Smoking, and Pollution. *J. Am. Coll. Cardiol.*, *70*, 230–251, 2017.
45. Marseglia, L., Manti, S., D'Angelo, G., Nicotera, A., Parisi, E., Di Rosa, G., Gitto, E., Arrigo, T., Oxidative Stress in Obesity: A Critical Component in Human Diseases. *Int. J. Mol. Sci.*, *16*, 378–400, 2014.
46. Huang, C.-J., McAllister, M.J., Slusher, A.L., Webb, H.E., Mock, J.T., Acevedo, E.O., Obesity-Related Oxidative Stress: the Impact of Physical Activity and Diet Manipulation. *Sport. Med. - Open*, *1*, 32, 2015.
47. Świątkiewicz, I., Wróblewski, M., Nuszkiewicz, J., Sutkowy, P., Wróblewska, J., Woźniak, A., The Role of Oxidative Stress Enhanced by Adiposity in Cardiometabolic Diseases. *Int. J. Mol. Sci.*, *24*, 6382, 2023.
48. Jakubiak, G.K., Osadnik, K., Lejawa, M., Kasperczyk, S., Osadnik, T., Pawlas, N., Oxidative Stress in Association with Metabolic Health and Obesity in Young Adults. *Oxid. Med. Cell. Longev.*, *2021*, 1–19, 2021.
49. Sharifi-Rad, M., Anil Kumar, N.V., Zucca, P., Varoni, E.M., Dini, L., Panzarini, E., Rajkovic, J., Tsouh Fokou, P.V., Azzini, E., Peluso, I., Prakash Mishra, A., Nigam, M., El Rayess, Y., Beyrouthy, M.E., Polito, L., Iriti, M., Martins, N., Martorell, M., Docea, A.O., Setzer, W.N., Calina, D., Cho, W.C., Sharifi-Rad, J., Lifestyle, Oxidative Stress, and Antioxidants: Back and Forth in the Pathophysiology of Chronic Diseases. *Front. Physiol.*, *11*, 694, 2020.
50. Sharebiani, H., Keramat, S., Chavoshan, A., Fazeli, B., Stanek, A., The Influence of Antioxidants on Oxidative Stress-Induced Vascular Aging in Obesity. *Antioxidants*, *12*, 1295, 2023.
51. Barteková, M., Adameová, A., Görbe, A., Ferenczyová, K., Pecháňová, O., Lazou, A., Dhalla, N.S., Ferdinandy, P., Giricz, Z., Natural and synthetic antioxidants targeting cardiac oxidative stress and redox signaling in cardiometabolic diseases. *Free Radic. Biol. Med.*, *169*, 446–477, 2021.
52. Asmat, U., Abad, K., Ismail, K., Diabetes mellitus and oxidative stress-A concise review. *Saudi Pharm. J.*, *24*, 547–553, 2016.
53. Giacco, F. and Brownlee, M., Oxidative stress and diabetic complications. *Circ. Res.*, *107*, 1058–70, 2010, http://www.ncbi.nlm.nih.gov/pubmed/21030723.
54. Darenskaya, M.A., Kolesnikova, L., II, Kolesnikov, S., II, Oxidative Stress: Pathogenetic Role in Diabetes Mellitus and Its Complications and Therapeutic Approaches to Correction. *Bull. Exp. Biol. Med.*, *171*, 179–189, 2021.
55. Pradeepa, R. and Mohan, V., Epidemiology of type 2 diabetes in India. *Indian J. Ophthalmol.*, *69*, 2932, 2021.
56. Diabetes, [s.d.], https://www.who.int/news-room/fact-sheets/detail/diabetes.
57. Smith, S.C., Multiple Risk Factors for Cardiovascular Disease and Diabetes Mellitus. *Am. J. Med.*, *120*, S3–S11, 2007.
58. Alam, S., Hasan, M.K., Neaz, S., Hussain, N., Hossain, M.F., Rahman, T., Diabetes Mellitus: Insights from Epidemiology, Biochemistry, Risk Factors, Diagnosis, Complications and Comprehensive Management. *Diabetology*, *2*, 36–50, 2021.
59. Barrett-Connor, E. and Khaw, K.-T., Diabetes mellitus: An independent risk factor for stroke? *Am. J. Epidemiol.*, *128*, 116–123, 1988.
60. Reddy, N., Verma, N., Dungan, K., Monitoring Technologies- Continuous Glucose Monitoring, Mobile Technology, Biomarkers of Glycemic Control. *Endotext*, 2023, http://www.ncbi.nlm.nih.gov/pubmed/25905275.
61. Bain, S., Cummings, M.H., McKay, G.A., Multidisciplinary Approach to Management and Care of Patients with Type 2 Diabetes Mellitus. *EMJ Diabetes*, 7, 73–81, 2019.

62. Deshpande, A.D., Harris-Hayes, M., Schootman, M., Epidemiology of Diabetes and Diabetes-Related Complications. *Phys. Ther.*, *88*, 1254–1264, 2008.
63. Forbes, J.M. and Cooper, M.E., Mechanisms of Diabetic Complications. *Physiol. Rev.*, *93*, 137–188, 2013.
64. Banday, M.Z., Sameer, A.S., Nissar, S., Pathophysiology of diabetes: An overview. *Avicenna J. Med.*, *10*, 174–188, 2020.
65. Tiwari, B.K., Pandey, K.B., Abidi, A.B., Rizvi, S., II, Markers of Oxidative Stress during Diabetes Mellitus. *J. Biomarkers*, *2013*, 1–8, 2013.
66. Bhatti, J.S., Sehrawat, A., Mishra, J., Sidhu, I.S., Navik, U., Khullar, N., Kumar, S., Bhatti, G.K., Reddy, P.H., Oxidative stress in the pathophysiology of type 2 diabetes and related complications: Current therapeutics strategies and future perspectives. *Free Radic. Biol. Med.*, *184*, 114–134, 2022.
67. Yousef, H., Khandoker, A.H., Feng, S.F., Helf, C., Jelinek, H.F., Inflammation, oxidative stress and mitochondrial dysfunction in the progression of type II diabetes mellitus with coexisting hypertension. *Front. Endocrinol. (Lausanne)*, *14*, 1173402, 2023.
68. Iacobini, C., Vitale, M., Pesce, C., Pugliese, G., Menini, S., Diabetic Complications and Oxidative Stress: A 20-Year Voyage Back in Time and Back to the Future. *Antioxidants*, *10*, 727, 2021.
69. Matough, F.A., Budin, S.B., Hamid, Z.A., Alwahaibi, N., Mohamed, J., The Role of Oxidative Stress and Antioxidants in Diabetic Complications = دور الإجهاد الـ تأكسدي و المواد المضادة للأكسدة في مضاعفات مرض السكري, *Sultan Qaboos Univ. Med. J.*, *12*, 5–18, 2012.
70. Bajaj, S. and Khan, A., Antioxidants and diabetes. *Indian J. Endocrinol. Metab.*, *16*, S267–71, 2012, http://www.ncbi.nlm.nih.gov/pubmed/23565396.
71. Rahimi-Madiseh, M., Malekpour-Tehrani, A., Bahmani, M., Rafieian-Kopaei, M., The research and development on the antioxidants in prevention of diabetic complications. *Asian Pac. J. Trop. Med.*, *9*, 825–831, 2016.
72. An, Y., Xu, B., Wan, S., Ma, X., Long, Y., Xu, Y., Jiang, Z., The role of oxidative stress in diabetes mellitus-induced vascular endothelial dysfunction. *Cardiovasc. Diabetol.*, *22*, 237, 2023.
73. Frąk, W., Wojtasińska, A.; Lisińska, W.; Młynarska, E., Franczyk, B., Rysz, J., Pathophysiology of Cardiovascular Diseases: New Insights into Molecular Mechanisms of Atherosclerosis, Arterial Hypertension, and Coronary Artery Disease. *Biomedicines*, *10*, 1938, 2022.
74. Cardiovascular diseases (CVDs), [s.d.], https://www.who.int/news-room/fact-sheets/detail/cardiovascular-diseases-(cvds).
75. Deshpande, A.D., Harris-Hayes, M., Schootman, M., Epidemiology of diabetes and diabetes-related complications. *Phys. Ther.*, *88*, 1254–64, 2008.
76. Teo, K.K. and Rafiq, T., Cardiovascular Risk Factors and Prevention: A Perspective From Developing Countries. *Can. J. Cardiol*, *37*, 733–743, 2021.
77. Rippe, J.M., Lifestyle Strategies for Risk Factor Reduction, Prevention, and Treatment of Cardiovascular Disease. *Am. J. Lifestyle Med.*, *13*, 204–212, 2019, http://www.ncbi.nlm.nih.gov/pubmed/30800027.
78. Heart disease - Diagnosis and treatment - Mayo Clinic, [s.d.], https://www.mayoclinic.org/diseases-conditions/heart-disease/diagnosis-treatment/drc-20353124.
79. Farahmand, F., Lou, H., Singal, P.K., Oxidative stress in cardiovascular complications of diabetes, in: *Atherosclerosis, Hypertension and Diabetes. Progressing in Experimental Cardiology*, G.N. Pierce, M. Nagano, P. Zahradka, N.S. Dhalla (Eds.), vol. 8, Springer, Boston, MA, 2003.
80. Mezzetti, A., Oxidative stress and cardiovascular complications in diabetes: isoprostanes as new markers on an old paradigm. *Cardiovasc. Res.*, *47*, 475–488, 2000.
81. Lusis, A.J., Atherosclerosis. *Nature*, *407*, 233–241, 2000.
82. Badimon, L., Padró, T., Vilahur, G., Atherosclerosis, platelets and thrombosis in acute ischaemic heart disease. *Eur. Hear. J. Acute Cardiovasc. Care*, *1*, 60–74, 2012.

83. Song, R., Xu, H., Dintica, C.S., Pan, K.-Y., Qi, X., Buchman, A.S., Bennett, D.A., Xu, W., Associations Between Cardiovascular Risk, Structural Brain Changes, and Cognitive Decline. *J. Am. Coll. Cardiol.*, *75*, 2525–2534, 2020.
84. Jia, R., Wang, Q., Huang, H., Yang, Y., Chung, Y.F., Liang, T., Cardiovascular disease risk models and dementia or cognitive decline: a systematic review. *Front. Aging Neurosci.*, *15*, 1257367, 2023.
85. van Nieuwkerk, A.C., Delewi, R., Wolters, F.J., Muller, M., Daemen, M., Biessels, G.J., Cognitive Impairment in Patients With Cardiac Disease: Implications for Clinical Practice. *Stroke*, *54*, 2181–2191, 2023.
86. Wang, Y., Zhang, H., Liu, L., Li, Z., Zhou, Y., Wei, J., Xu, Y., Zhou, Y., Tang, Y., Cognitive function and cardiovascular health in the elderly: network analysis based on hypertension, diabetes, cerebrovascular disease, and coronary heart disease. *Front. Aging Neurosci.*, *15*, 1229559, 2023.
87. Jain, K.K., Personalized Management of Cardiovascular Disorders. *Med. Princ. Pract.*, *26*, 399-414, 2017.
88. Senoner, T. and Dichtl, W., Oxidative Stress in Cardiovascular Diseases: Still a Therapeutic Target? *Nutrients*, *11*, 2090, 2019, http://www.ncbi.nlm.nih.gov/pubmed/31487802.
89. Madamanchi, N.R., Vendrov, A., Runge, M.S., Oxidative Stress and Vascular Disease. *Arterioscler. Thromb. Vasc. Biol.*, *25*, 29–38, 2005.
90. Kayama, Y., Raaz, U., Jagger, A., Adam, M., Schellinger, I.N., Sakamoto, M., Suzuki, H., Toyama, K., Spin, J.M., Tsao, P.S., Diabetic Cardiovascular Disease Induced by Oxidative Stress. *Int. J. Mol. Sci.*, *16*, 25234–63, 2015, http://www.ncbi.nlm.nih.gov/pubmed/26512646.
91. Dhalla, N.S., Temsah, R.M., Netticadan, T., Role of oxidative stress in cardiovascular diseases. *J. Hypertens.*, *18*, 655–673, 2000.
92. Cervantes Gracia, K., Llanas-Cornejo, D., Husi, H., CVD and Oxidative Stress. *J. Clin. Med.*, *6*, 22, 2017.
93. Farías, J., Molina, V., Carrasco, R., Zepeda, A., Figueroa, E., Letelier, P., Castillo, R., Antioxidant Therapeutic Strategies for Cardiovascular Conditions Associated with Oxidative Stress. *Nutrients*, *9*, 966, 2017.
94. Chen, J., Ye, Z., Wang, X., Chang, J., Yang, M., Zhong, H., Hong, F., Yang, S., Nitric oxide bioavailability dysfunction involves in atherosclerosis. *Biomed. Pharmacother.*, *97*, 423–428, 2018.
95. Lubos, E., Role of oxidative stress and nitric oxide in atherothrombosis. *Front. Biosci.*, 13, 5323, 2008.
96. Moris, D., Spartalis, M., Spartalis, E., Karachaliou, G.-S., Karaolanis, G., II, Tsourouflis, G., Tsilimigras, D., II, Tzatzaki, E., Theocharis, S., The role of reactive oxygen species in the pathophysiology of cardiovascular diseases and the clinical significance of myocardial redox. *Ann. Transl. Med.*, *5*, 326–326, 2017.
97. Salekeen, R., Haider, A.N., Akhter, F., Billah, M.M., Islam, M.E., Didarul Islam, K.M., Lipid oxidation in pathophysiology of atherosclerosis: Current understanding and therapeutic strategies. *Int. J. Cardiol. Cardiovasc. Risk Prev.*, *14*, 200143, 2022.
98. Mashhadi, N.S., Ghiasvand, R., Askari, G., Hariri, M., Darvishi, L., Mofid, M.R., Anti-oxidative and anti-inflammatory effects of ginger in health and physical activity: review of current evidence. *Int. J. Prev. Med.*, *4*, S36–42, 2013, http://www.ncbi.nlm.nih.gov/pubmed/23717767.
99. Bozin, B., Mimica-Dukic, N., Samojlik, I., Goran, A., Igic, R., Phenolics as antioxidants in garlic (Allium sativum L., Alliaceae). *Food Chem.*, *111*, 925–929, 2008.
100. Capasso, A., Antioxidant action and therapeutic efficacy of Allium sativum L. *Molecules*, *18*, 690-700, 2013, http://www.ncbi.nlm.nih.gov/pubmed/23292331.
101. Salgueiro, A.C.F., Folmer, V., da Rosa, H.S., Costa, M.T., Boligon, A.A., Paula, F.R., Roos, D.H., Puntel, G.O., *In vitro* and *in silico* antioxidant and toxicological activities of Achyrocline satureioides. *J. Ethnopharmacol.*, *194*, 6–14, 2016.

102. Arredondo, M., Blasina, F., Echeverry, C., Morquio, A., Ferreira, M., Abin-Carriquiry, J., Lafon, L., Dajas, F., Cytoprotection by Achyrocline satureioides (Lam) D.C. and some of its main flavonoids against oxidative stress. *J. Ethnopharmacol.*, *91*, 13–20, 2004.
103. Stanković, N.; Mihajilov-Krstev, T.; Zlatković, B.; Stankov-Jovanović, V.; Kocić, B.; Čomić, L., Antibacterial and antioxidant activity of wild-growing *Angelica* species (Apiaceae) from Balkan Peninsula against human pathogenic bacteria. *J. Essent. Oil Res.*, *32*, 464–473, 2020.
104. Ağalar, H.G., Göger, F., Demirci, B., Malyer, P., Kirimer, Dr. H., Angelica Sylvestris Var, N., Sylvestris L.: Essential oils and antioxidant activity evaluation. *Eskişehir Tech. Univ. J. Sci. Technol. A - Appl. Sci. Eng.*, *21*, 39–48, 2020.
105. Karaçelik, A. A.; Türkuçar, S. A.; Karaköse, M., Phytochemical Composition and Biological Activities of *Angelica sylvestris* L. var. *stenoptera* Avé-Lall ex Boiss.: An Endangered Medicinal Plant of Northeast Turkey. *Chem. Biodivers.*, *19*, e202200552, 2022.
106. Kooti, W. and Daraei, N.A., Review of the Antioxidant Activity of Celery (*Apium graveolens* L). *J. Evid. Based. Complementary Altern. Med.*, *22*, 1029–1034, 2017.
107. Kim, M.Y., Kim, H., Lee, Y., Kim, M.H., Lee, J.Y., Kang, M.S., Koo, B.C., Lee, B.W., Antioxidant and anti-inflammatory effects of Peanut (*Arachishypogaea L.*) skin extracts of various cultivars in oxidative-damaged HepG2 cells and LPS-induced raw 264.7 macrophages. *Food Sci. Nutr.*, *9*, 973–984, 2021.
108. Ullah, S., Hussain, S.A., Shaukat, F., Hameed, A., Yang, W., Song, Y., Antioxidant Potential and the Characterization of *Arachis hypogaea* Roots. *Biomed Res. Int.*, *2019*, 1–9, 2019.
109. Gao, D., Chen, H., Li, H., Yang, X., Guo, X., Zhang, Y., Ma, J., Yang, J., Ma, S., Extraction, structural characterization, and antioxidant activity of polysaccharides derived from Arctium lappa L. *Front. Nutr.*, *10*, 1149137, 2023.
110. Predes, F.S., Ruiz, A.L., Carvalho, J.E., Foglio, M.A., Dolder, H., Antioxidative and *in vitro* antiproliferative activity of Arctium lappa root extracts. *BMC Complement. Altern. Med.*, *11*, 25, 2011.
111. Yosri, N., Alsharif, S.M., Xiao, J., Musharraf, S.G., Zhao, C., Saeed, A., Gao, R., Said, N.S., Di Minno, A., Daglia, M., Guo, Z., Khalifa, S.A.M., El-Seedi, H.R., Arctium lappa (Burdock): Insights from ethnopharmacology potential, chemical constituents, clinical studies, pharmacological utility and nanomedicine. *Biomed. Pharmacother.*, *158*, 114104, 2023.
112. Duh, P., Antioxidant activity of burdock (*Arctium lappa* Linné): Its scavenging effect on free-radical and active oxygen. *J. Am. Oil Chem. Soc.*, *75*, 455–461, 1998.
113. Liu, W., Wang, J., Zhang, Z., Xu, J., Xie, Z., Slavin, M., Gao, X., *In vitro* and *in vivo* antioxidant activity of a fructan from the roots of. *Int. J. Biol. Macromol.*, *65*, 446–453, 2014.
114. El Khatib, N., Morel, S., Hugon, G., Rapior, S., Carnac, G., Saint, N., Identification of a Sesquiterpene Lactone from Arctium lappa Leaves with Antioxidant Activity in Primary Human Muscle Cells. *Molecules*, *26*, 1328, 2021.
115. Yang, W., Kortesniemi, M., Yang, B., Zheng, J., Enzymatic Acylation of Anthocyanins Isolated from Alpine Bearberry (*Arctostaphylos alpina*) and Lipophilic Properties, Thermostability, and Antioxidant Capacity of the Derivatives. *J. Agric. Food. Chem.*, *66*, 2909–2916, 2018.
116. Mohd Azman, N., Gallego, M., Segovia, F., Abdullah, S., Shaarani, S., Almajano Pablos, M., Study of the Properties of Bearberry Leaf Extract as a Natural Antioxidant in Model Foods. *Antioxidants*, *5*, 11, 2016.
117. Karafyllaki, D., Narwojsz, A., Kurp, L., Sawicki, T., Effects of different processing methods on the polyphenolic compounds profile and the antioxidant and anti-glycaemic properties of horseradish roots (Armoracia rusticana). *Eur. Food Res. Technol.*, *249*, 1739–1747, 2023.
118. Behbahani, B.A., Shahidi, F., Yazdi, F.T., Mortazavi, S.A., Mohebbi, M., Antioxidant activity and antimicrobial effect of tarragon (Artemisia dracunculus) extract and chemical composition of its essential oil. *J. Food Meas. Charact.*, *11*, 847–863, 2017.

119. Ekiert, H., Świątkowska, J., Knut, E., Klin, P., Rzepiela, A., Tomczyk, M., Szopa, A., Artemisia dracunculus (Tarragon): A Review of Its Traditional Uses, Phytochemistry and Pharmacology. *Front. Pharmacol.*, *12*, 653993, 2021.
120. Sharma, K.R. and Adhikari, S., Phytochemical analysis and biological activities of *Artemisia vulgaris* grown in different altitudes of Nepal. *Int. J. Food Prop.*, *26*, 414–427, 2023.
121. Thangjam, N.M., Taijong, J., Kumar, A., Phytochemical and pharmacological activities of methanol extract of Artemisia vulgaris L. leaves. *Clin. Phytosci.*, *6*, 72, 2020.
122. Matvieieva, N., Drobot, K., Duplij, V., Ratushniak, Y., Shakhovsky, A., Kyrpa-Nesmiian, T., Mickevičius, S., Brindza, J., Flavonoid content and antioxidant activity of *Artemisia vulgaris* L. "hairy" roots. *Prep. Biochem. Biotechnol.*, *49*, 82–87, 2019.
123. Shahzad, M., Shabbir, A., Wojcikowski, K., Wohlmuth, H., C. Gobe, G., The Antioxidant Effects of Radix Astragali (Astragalus membranaceus and Related Species) in Protecting Tissues from Injury and Disease. *Curr. Drug Targets*, *17*, 1331–1340, 2016.
124. Hu, H.C., Zhang, W., Xiong, P.Y., Song, L., Jia, B., Liu, X.L., Anti-inflammatory and antioxidant activity of astragalus polysaccharide in ulcerative colitis: A systematic review and meta-analysis of animal studies. *Front. Pharmacol.*, *13*, 1043236, 2022.
125. Kim, J.Y., Antioxidant Effects of Astragalus membranaceus Extracts with Extraction Conditions. *J. Korean Soc. Appl. Biol. Chem.*, *52*, 702–707, 2009.
126. Peterson, D.M., Oat Antioxidants. *J. Cereal Sci.*, *33*, 115–129, 2001.
127. Kim, I.-S., Hwang, C.-W., Yang, W.-S., Kim, C.-H., Multiple Antioxidative and Bioactive Molecules of Oats (Avena sativa L.) in Human Health. *Antioxidants*, *10*, 1454, 2021.
128. Thitimuta, S., Pithayanukul, P., Nithitanakool, S., Bavovada, R., Leanpolchareanchai, J., Saparpakorn, P., Camellia sinensis L. Extract and Its Potential Beneficial Effects in Antioxidant, Anti-Inflammatory, Anti-Hepatotoxic, and Anti-Tyrosinase Activities. *Molecules*, *22*, 401, 2017.
129. Sharma, A., Sharma, R., Sharma, M., Kumar, M., Barbhai, M.D., Lorenzo, J.M., Sharma, S., Samota, M.K., Atanassova, M., Caruso, G., Naushad, M., Radha, Chandran, D., Prakash, P., Hasan, M., Rais, N., Dey, A., Mahato, D.K., Dhumal, S., Singh, S., Senapathy, M., Rajalingam, S., Visvanathan, M., Saleena, L.A.K., Mekhemar, M., Carica papaya L. Leaves: Deciphering Its Antioxidant Bioactives, Biological Activities, Innovative Products, and Safety Aspects. *Oxid. Med. Cell. Longev.*, *2022*, 1–20, 2022.
130. Addai, Z.R., Abdullah, A., Abd.Mutalib, S., Musa, K.H., Douqan, E.M.A., Antioxidant Activity and Physicochemical Properties of Mature Papaya Fruit (Carica papaya L. cv. Eksotika). *Adv. J. Food Sci. Technol.*, *5*, 859–865, 2013.
131. Insanu, M., Nayaka, N. M. D. M. W., Solihin, L., Wirasutisna, K.R., Pramastya, H., Fidrianny, I., Antioxidant activities and phytochemicals of polar, semi-polar, and nonpolar extracts of used and unused parts of Carica papaya fruit. *Biocatal. Agric. Biotechnol.*, *39*, 102270, 2022.
132. Chandula Weerasekera, A., Samarasinghe, K., Krishantha Sameera de Zoysa, H., Chathuranga Bamunuarachchige, T., Yashasvi Waisundara, V., *Antioxidants - Benefits, Sources, Mechanisms of Action*, IntechOpen, London, UK, 2021, Em.
133. Alizadeh Behbahani, B., Falah, F., Lavi Arab, F., Vasiee, M., Tabatabaee Yazdi, F., Chemical Composition and Antioxidant, Antimicrobial, and Antiproliferative Activities of *Cinnamomum zeylanicum* Bark Essential Oil. *Evidence-Based Complement. Altern. Med.*, *2020*, 1–8, 2020.
134. Pagliari, S., Forcella, M., Lonati, E., Sacco, G., Romaniello, F., Rovellini, P., Fusi, P., Palestini, P., Campone, L., Labra, M., Bulbarelli, A., Bruni, I., Antioxidant and Anti-Inflammatory Effect of Cinnamon (Cinnamomum verum J. Presl) Bark Extract after *In Vitro* Digestion Simulation. *Foods*, *12*, 452, 2023.
135. Liew, S.S., Ho, W.Y., Yeap, S.K., Sharifudin, S.A., Bin Phytochemical composition and *in vitro* antioxidant activities of *Citrus sinensis* peel extracts. *PeerJ*, *6*, e5331, 2018.

136. Jorge, N., da Silva, A.C., Aranha, C.P.M., Antioxidant activity of oils extracted from orange (Citrus sinensis) seeds. *An. Acad. Bras. Cienc.*, *88*, 951–958, 2016.
137. Msaada, K., Ben Jemia, M., Salem, N., Bachrouch, O., Sriti, J., Tammar, S., Bettaieb, I., Jabri, I., Kefi, S., Limam, F., Marzouk, B., Antioxidant activity of methanolic extracts from three coriander (Coriandrum sativum L.) fruit varieties. *Arab. J. Chem.*, *10*, S3176–S3183, 2017.
138. Bahorun, T., Trotin, F., Pommery, J., Vasseur, J., Pinkas, M., Antioxidant Activities of *Crataegus monogyna* Extracts. *Planta Med.*, *60*, 323–328, 1994.
139. Belabdelli, F., Bekhti, N., Piras, A., Benhafsa, F.M., Ilham, M., Adil, S., Anes, L., Chemical composition, antioxidant and antibacterial activity of *Crataegus monogyna* leaves' extracts. *Nat. Prod. Res.*, *36*, 3234–3239, 2022.
140. Konyalioglu, S., Cebe, G.E., Aktar, S., Antioxidant activity of Crataegus Monogyna L flowers. *Free Radic. Biol. Med.*, *108*, S56, 2017.
141. Mashmoul, M., Azlan, A., Khaza'ai, H., Yusof, B., Noor, S., Saffron: A Natural Potent Antioxidant as a Promising Anti-Obesity Drug. *Antioxidants*, *2*, 293–308, 2013.
142. Akter, J., Hossain, M.A., Takara, K., Islam, M.Z., Hou, D.-X., Antioxidant activity of different species and varieties of turmeric (Curcuma spp): Isolation of active compounds. *Comp. Biochem. Physiol. Part C Toxicol. Pharmacol.*, *215*, 9-17, 2019.
143. Quirós-Fallas, M., II, Vargas-Huertas, F., Quesada-Mora, S., Azofeifa-Cordero, G., Wilhelm-Romero, K., Vásquez-Castro, F., Alvarado-Corella, D., Sánchez-Kopper, A., Navarro-Hoyos, M., Polyphenolic HRMS Characterization, Contents and Antioxidant Activity of Curcuma longa Rhizomes from Costa Rica. *Antioxidants*, *11*, 620, 2022.
144. Tanya, C.Y., Kolo, S.M.D., Karnelasatri, K., Antimicrobial and antioxidant properties of Cymbopogon Nardus L (citronella grass) oil from Kefamemamu, Timor Tengah Utara Regency, Indonesia, in: *AIP Conference Proceedings*, vol. 2391, no. 1, AIP Publishing, Surakarta, Indonesia, 2022.
145. Singh, B.K., Koley, T.K., Maurya, A., Singh, P.M., Singh, B., Phytochemical and antioxidative potential of orange, red, yellow, rainbow and black coloured tropical carrots (Daucus carota subsp. sativus Schubl. & Martens). *Physiol. Mol. Biol. Plants*, *24*, 899–907, 2018.
146. Park, J.Y., Kim, J.Y., Son, Y.G., Kang, S.D., Lee, S.W., Kim, K.D., Kim, J.Y., Characterization of Chemical Composition and Antioxidant Activity of Eucalyptus globulus Leaves under Different Extraction Conditions. *Appl. Sci.*, *13*, 9984, 2023.
147. Polito, F., Fratianni, F., Nazzaro, F., Amri, I., Kouki, H., Khammassi, M., Hamrouni, L., Malaspina, P., Cornara, L., Khedhri, S., Romano, B., Maresca, D.C., Ianaro, A., Ercolano, G., De Feo, V., Essential Oil Composition, Antioxidant Activity and Leaf Micromorphology of Five Tunisian Eucalyptus Species. *Antioxidants*, *12*, 867, 2023.
148. Ražná, K.; Sawinska, Z.; Ivanišová, E.; Vukovic, N.; Terentjeva, M.; Stričík, M.; Kowalczewski, P. Ł.; Hlavačková, L.; Rovná, K.; Žiarovská, J.; Kačániová, M., Properties of Ginkgo biloba L.: Antioxidant Characterization, Antimicrobial Activities, and Genomic MicroRNA Based Marker Fingerprints. *Int. J. Mol. Sci.*, *21*, 3087, 2020.
149. Biernacka, P., Felisiak, K., Adamska, I., Śnieg, M.; Podsiadło, C., Effect of Brewing Conditions on Antioxidant Properties of Ginkgo biloba Leaves Infusion. *Antioxidants*, *12*, 1455, 2023.
150. Karahan, F., Avsar, C., Ozyigit, I., II, Berber, I., Antimicrobial and antioxidant activities of medicinal plant *Glycyrrhiza glabra* var. *glandulifera* from different habitats. *Biotechnol. Biotechnol. Equip.*, *30*, 797–804, 2016.
151. Gülçin, İ., Mshvildadze, V., Gepdiremen, A., Elias, R., Antioxidant Activity of Saponins Isolated from Ivy: α-Hederin, Hederasaponin-C, Hederacolchiside-E and Hederacolchiside-F. *Planta Med.*, *70*, 561–563, 2004.
152. Spiridon, I., Nechita, C., Niculaua, M., Silion, M., Armatu, A., Teacă, C.-A.; Bodîrlău, R., Antioxidant and chemical properties of Inula helenium root extracts. *Open Chem.*, *11*, 1699–1709, 2013.

153. Saab, A.M., Tundis, R., Loizzo, M.R., Lampronti, I., Borgatti, M., Gambari, R., Menichini, F., Esseily, F., Menichini, F., Antioxidant and antiproliferative activity of *Laurus nobilis* L. (Lauraceae) leaves and seeds essential oils against K562 human chronic myelogenous leukaemia cells. *Nat. Prod. Res.*, *26*, 1741–1745, 2012.
154. Fierascu, R.C., Fierascu, I., Ortan, A., Fierascu, I.C., Anuta, V., Velescu, B.S., Pituru, S.M., Dinu-Pirvu, C.E., *Leonurus cardiaca* L. as a Source of Bioactive Compounds: An Update of the European Medicines Agency Assessment Report (2010). *Biomed Res. Int.*, *2019*, 1–13, 2019.
155. Bernatoniene, J., Kopustinskiene, D.M., Jakstas, V., Majiene, D., Baniene, R., Kuršvietiene, L., Masteikova, R., Savickas, A., Toleikis, A., Trumbeckaite, S., The effect of Leonurus cardiaca herb extract and some of its flavonoids on mitochondrial oxidative phosphorylation in the heart. *Planta Med.*, *80*, 525–32, 2014, http://www.ncbi.nlm.nih.gov/pubmed/24841965.
156. Liu, Y., Duan, C., Chen, H., Wang, C., Liu, X., Qiu, M., Tang, H., Zhang, F., Zhou, X., Yang, J., Inhibition of COX-2/mPGES-1 and 5-LOX in macrophages by leonurine ameliorates monosodium urate crystal-induced inflammation. *Toxicol. Appl. Pharmacol.*, *351*, 1–11, 2018, http://www.ncbi.nlm.nih.gov/pubmed/29763636.
157. Al-Rajhi, A.M.H., Qanash, H., Almuhayawi, M.S., Al Jaouni, S.K., Bakri, M.M., Ganash, M., Salama, H.M., Selim, S., Abdelghany, T.M., Molecular Interaction Studies and Phytochemical Characterization of Mentha pulegium L. Constituents with Multiple Biological Utilities as Antioxidant, Antimicrobial, Anticancer and Anti-Hemolytic Agents. *Molecules*, *27*, 4824, 2022.
158. Aimad, A., Sanae, R., Anas, F., Abdelfattah, E.M., Bourhia, M., Mohammad Salamatullah, A., Alzahrani, A., Alyahya, H.K., Albadr, N.A., Abdelkrim, A., El Barnossi, A., Noureddine, E., Chemical Characterization and Antioxidant, Antimicrobial, and Insecticidal Properties of Essential Oil from Mentha pulegium L. *Evidence-Based Complement. Altern. Med.*, *2021*, 1–12, 2021.
159. El Menyiy, N., Mrabti, H.N., El Omari, N., Bakili, A.E., Bakrim, S., Mekkaoui, M., Balahbib, A., Amiri-Ardekani, E., Ullah, R., Alqahtani, A.S., Shahat, A.A., Bouyahya, A., Medicinal Uses, Phytochemistry, Pharmacology, and Toxicology of Mentha spicata. *Evidence-Based Complement. Altern. Med.*, *2022*, 1–32, 2022.
160. Bardaweel, S.K., Bakchiche, B., ALSalamat, H.A., Rezzoug, M., Gherib, A., Flamini, G., Chemical composition, antioxidant, antimicrobial and Antiproliferative activities of essential oil of Mentha spicata L. (Lamiaceae) from Algerian Saharan atlas. *BMC Complement. Altern. Med.*, *18*, 201, 2018.
161. Li, C.-W., Chu, Y.-C., Huang, C.-Y., Fu, S.-L., Chen, J.-J., Evaluation of Antioxidant and Anti-α-glucosidase Activities of Various Solvent Extracts and Major Bioactive Components from the Seeds of Myristica fragrans. *Molecules*, *25*, 5198, 2020.
162. Suanarunsawat, T., Devakul Na Ayutthaya, W., Songsak, T., Thirawarapan, S., Poungshompoo, S., Lipid-Lowering and Antioxidative Activities of Aqueous Extracts of *Ocimum sanctum* L. Leaves in Rats Fed with a High-Cholesterol Diet. *Oxid. Med. Cell. Longev.*, *2011*, 1–9, 2011.
163. Chaudhary, A., Sharma, S., Mittal, A., Gupta, S., Dua, A., Phytochemical and antioxidant profiling of Ocimum sanctum. *J. Food Sci. Technol.*, *57*, 3852–3863, 2020.
164. Mansour, H.M.M., Zeitoun, A.A., Abd-Rabou, H.S., El Enshasy, H.A., Dailin, D.J., Zeitoun, M.A.A., El-Sohaimy, S.A., Antioxidant and Anti-Diabetic Properties of Olive (Olea europaea) Leaf Extracts: *In Vitro* and *In Vivo* Evaluation. *Antioxidants*, *12*, 1275, 2023.
165. Coccimiglio, J., Alipour, M., Jiang, Z.-H., Gottardo, C., Suntres, Z., Antioxidant, Antibacterial, and Cytotoxic Activities of the Ethanolic *Origanum vulgare* Extract and Its Major Constituents. *Oxid. Med. Cell. Longev.*, *2016*, 1–8, 2016.
166. Goufo, P. and Trindade, H., Rice antioxidants: phenolic acids, flavonoids, anthocyanins, proanthocyanidins, tocopherols, tocotrienols, Î³ -oryzanol, and phytic acid. *Food Sci. Nutr.*, *2*, 75–104, 2014.

167. Butnariu, M., Quispe, C., Herrera-Bravo, J., Pentea, M., Sarac, I., Küşümler, A.S., Özçelik, B., Painuli, S., Semwal, P., Imran, M., Gondal, T.A., Emamzadeh-Yazdi, S., Lapava, N., Yousaf, Z., Kumar, M., Eid, A.H., Al-Dhaheri, Y., Suleria, H.A.R., del Mar Contreras, M., Sharifi-Rad, J., Cho, W.C., Papaver Plants: Current Insights on Phytochemical and Nutritional Composition Along with Biotechnological Applications. *Oxid. Med. Cell. Longev., 2022*, 1–23, 2022.
168. Alcázar-Valle, M.; Lugo-Cervantes, E.; Mojica, L.; Morales-Hernández, N.; Reyes-Ramírez, H.; Enríquez-Vara, J. N.; García-Morales, S., Bioactive Compounds, Antioxidant Activity, and Antinutritional Content of Legumes: A Comparison between Four Phaseolus Species. *Molecules, 25*, 3528, 2020.
169. Gul, M., Liu, Z.W., Iahtisham-Ul-haq, Rabail, R., Faheem, F., Walayat, N., Nawaz, A., Shabbir, M.A., Munekata, P.E.S., Lorenzo, J.M., Aadil, R.M., Functional and Nutraceutical Significance of Amla (Phyllanthus emblica L.): A Review. *Antioxidants, 11*, 816, 2022, http://www.ncbi.nlm.nih.gov/pubmed/35624683.
170. Padmakumari, K.P., Sasidharan, I., Sreekumar, M.M., Composition and antioxidant activity of essential oil of pimento (*Pimenta dioica* (L) Merr.) from Jamaica. *Nat. Prod. Res., 25*, 152–160, 2011.
171. Shojaii, A. and Abdollahi Fard, M., Review of Pharmacological Properties and Chemical Constituents of *Pimpinella anisum. ISRN Pharm., 2012*, 1–8, 2012.
172. Carsono, N., Tumilaar, S.G., Kurnia, D., Latipudin, D., Satari, M.H.A., Review of Bioactive Compounds and Antioxidant Activity Properties of Piper Species. *Molecules, 27*, 6774, 2022.
173. Lee, J.-G., Chae, Y., Shin, Y., Kim, Y.-J., Chemical composition and antioxidant capacity of black pepper pericarp. *Appl. Biol. Chem., 63*, 35, 2020.
174. Karagecili, H., İzol, E., Kirecci, E., Gulcin, İ., Determination of Antioxidant, Anti-Alzheimer, Antidiabetic, Antiglaucoma and Antimicrobial Effects of Zivzik Pomegranate (Punica granatum)â€"A Chemical Profiling by LC-MS/MS. *Life, 13*, 735, 2023.
175. Popović, B. M.; Štajner, D.; Ždero, R.; Orlović, S.; Galić, Z., Antioxidant Characterization of Oak Extracts Combining Spectrophotometric Assays and Chemometrics. *Sci. World J., 2013*, 1–8, 2013.
176. Othón-Díaz, E. D.; Fimbres-García, J. O.; Flores-Sauceda, M.; Silva-Espinoza, B. A.; López-Martínez, L. X.; Bernal-Mercado, A. T.; Ayala-Zavala, J. F., Antioxidants in Oak (Quercus sp.): Potential Application to Reduce Oxidative Rancidity in Foods. *Antioxidants, 12*, 861, 2023.
177. Li Pomi, F., Papa, V., Borgia, F., Vaccaro, M., Allegra, A., Cicero, N., Gangemi, S., Rosmarinus officinalis and Skin: Antioxidant Activity and Possible Therapeutical Role in Cutaneous Diseases. *Antioxidants, 12*, 680, 2023.
178. Rašković, A.; Milanović, I.; Pavlović, N.; Ćebović, T.; Vukmirović, S.; Mikov, M., Antioxidant activity of rosemary (Rosmarinus officinalis L.) essential oil and its hepatoprotective potential. *BMC Complement. Altern. Med., 14*, 225, 2014.
179. Wada, L. and Ou, B., Antioxidant Activity and Phenolic Content of Oregon Caneberries. *J. Agric. Food. Chem., 50*, 3495–3500, 2002.
180. Wang, S.Y. and Lin, H.-S., Antioxidant Activity in Fruits and Leaves of Blackberry, Raspberry, and Strawberry Varies with Cultivar and Developmental Stage. *J. Agric. Food. Chem., 48*, 140–146, 2000.
181. Kostecka-Gugała, A.; Ledwożyw-Smoleń, I.; Augustynowicz, J.; Wyżgolik, G.; Kruczek, M.; Kaszycki, P. Antioxidant properties of fruits of raspberry and blackberry grown in central Europe. *Open Chem.*, 13, 000010151520150143, 2015.
182. Abdel-Aty, A.M., Barakat, A.Z., Bassuiny, R., II, Mohamed, S.A., Antioxidant-polyphenols of saw palmetto seeds: statistical optimized production and improved functional properties under solid-state fermentation by Trichoderma reesei. *J. Food Meas. Charact., 17*, 1132–1143, 2023.

183. Kopustinskiene, D.M. and Bernatoniene, J., Antioxidant Effects of Schisandra chinensis Fruits and Their Active Constituents. *Antioxidants, 10*, 620, 2021.
184. Andargie, M., Vinas, M., Rathgeb, A., Möller, E., Karlovsky, P., Lignans of Sesame (Sesamum indicum L.): A Comprehensive Review. *Molecules, 26*, 883, 2021.
185. Morsy, M.K., Sami, R., Algarni, E., Al-Mushhin, A.A.M., Benajiba, N., A., A., Almasoudi, A.G., Mekawi, E., Phytochemical Profile and Antioxidant Activity of Sesame Seed (Sesamum indicum) By-Products for Stability and Shelf Life Improvement of Refined Olive Oil. *Antioxidants, 11*, 338, 2022.
186. Wu, M.-S., Aquino, L.B.B., Barbaza, M.Y.U., Hsieh, C.-L., De Castro-Cruz, K.A., Yang, L.-L., Tsai, P.-W., Anti-Inflammatory and Anticancer Properties of Bioactive Compounds from Sesamum indicum L.-A Review. *Molecules, 24*, 4426, 2019.
187. Ahmed, S., Jahan, I.A., Hossain, M.H., Ahmed, K.S., Rahman, M., Zzaman, W., Hoque, M.M., Bioactive compounds, antioxidant properties and phenolic profile of pulp and seed of Syzygium cumini. *J. Food Meas. Charact., 15*, 1991–1999, 2021.
188. Ruan, Z.P., Zhang, L.L., Lin, Y.M., Evaluation of the antioxidant activity of Syzygium cumini leaves. *Molecules, 13*, 2545–56, 2008, http://www.ncbi.nlm.nih.gov/pubmed/18927517.
189. Bergman, M., Varshavsky, L., Gottlieb, H.E., Grossman, S., The antioxidant activity of aqueous spinach extract: chemical identification of active fractions. *Phytochemistry, 58*, 143–152, 2001.
190. Ko, S.-H., Park, J.-H., Kim, S.-Y., Lee, S.W., Chun, S.-S., Park, E., Antioxidant Effects of Spinach (Spinacia oleracea L.) Supplementation in Hyperlipidemic Rats. *Prev. Nutr. Food Sci., 19*, 19–26, 2014.
191. Hazra, B., Sarkar, R., Biswas, S., Mandal, N., Comparative study of the antioxidant and reactive oxygen species scavenging properties in the extracts of the fruits of Terminalia chebula, Terminalia belerica and Emblica officinalis. *BMC Complement. Altern. Med., 10*, 20, 2010.
192. Oracz, J. and Żyżelewicz, D., Antioxidants in Cocoa. *Antioxidants, 9*, 1230, 2020.
193. Roby, M.H.H., Sarhan, M.A., Selim, K.A.-H., Khalel, K., II, Evaluation of antioxidant activity, total phenols and phenolic compounds in thyme (Thymus vulgaris L.), sage (Salvia officinalis L.), and marjoram (Origanum majorana L.) extracts. *Ind. Crops Prod., 43*, 827–831, 2013.
194. Ngo, H.T.T., Fang, M., Hwang, E., Kim, Y., Park, B., Seo, S.A., Do, N.Q., Nguyen, Q.T.N., Yi, T.-H., Inhibitory Effects of Urtica thunbergiana Ethanol Extract on Atopic Dermatitis-Induced NC/Nga Mice. *Antioxidants, 9*, 197, 2020.
195. Huang, W., Hutabarat, R.P., Chai, Z., Zheng, T., Zhang, W., Li, D., Antioxidant Blueberry Anthocyanins Induce Vasodilation via PI3K/Akt Signaling Pathway in High-Glucose-Induced Human Umbilical Vein Endothelial Cells. *Int. J. Mol. Sci., 21*, 575, 2020, http://www.ncbi.nlm.nih.gov/pubmed/32106617.
196. Jurikova, T., Skrovankova, S., Mlcek, J., Balla, S., Snopek, L., Bioactive Compounds, Antioxidant Activity, and Biological Effects of European Cranberry (Vaccinium oxycoccos). *Molecules, 24*, 24, 2018.
197. Vigliante, I., Mannino, G., Maffei, M.E., OxiCyan®, a phytocomplex of bilberry (Vaccinium myrtillus) and spirulina (Spirulina platensis), exerts both direct antioxidant activity and modulation of ARE/Nrf2 pathway in HepG2 cells. *J. Funct. Foods, 61*, 103508, 2019.
198. Shahzad, M., Shabbir, A., Wojcikowski, K., Wohlmuth, H., Gobe, G.C., The Antioxidant Effects of Radix Astragali (Astragalus membranaceus and Related Species) in Protecting Tissues from Injury and Disease. *Curr. Drug Targets, 17*, 1331–40, 2016, http://www.ncbi.nlm.nih.gov/pubmed/26343107.
199. Sun, G.Y., Li, R., Cui, J., Hannink, M., Gu, Z., Fritsche, K.L., Lubahn, D.B., Simonyi, A., Withania somnifera and Its Withanolides Attenuate Oxidative and Inflammatory Responses and Up-Regulate Antioxidant Responses in BV-2 Microglial Cells. *Neuromolecular Med., 18*, 241-52, 2016, http://www.ncbi.nlm.nih.gov/pubmed/27209361.

200. Mao, Q.-Q., Xu, X.-Y., Cao, S.-Y., Gan, R.-Y., Corke, H., Beta, T., Li, H.-B., Bioactive Compounds and Bioactivities of Ginger (Zingiber officinale Roscoe). *Foods (Basel, Switzerland)*, *8*, 185, 2019, http://www.ncbi.nlm.nih.gov/pubmed/31151279.
201. Kumar, S., Behl, T., Sachdeva, M., Sehgal, A., Kumari, S., Kumar, A., Kaur, G., Yadav, H.N., Bungau, S., Implicating the effect of ketogenic diet as a preventive measure to obesity and diabetes mellitus. *Life Sci.*, *264*, 118661, 2021.
202. Arora, A., Behl, T., Sehgal, A., Singh, S., Sharma, N., Bhatia, S., Sobarzo-Sanchez, E., Bungau, S., Unravelling the involvement of gut microbiota in type 2 diabetes mellitus. *Life Sci.*, *273*, 119311, 2021.

11

Hepatoprotective Potential of Antioxidants in Medicinal Plants

Sayed Mohammed Firdous*, Sahabaj Ali Khan and Amritangshu Maity

Department of Pharmacology, Calcutta Institute of Pharmaceutical Technology & AHS, Uluberia, Howrah, West Bengal, India

Abstract

The liver plays a vital role in maintaining overall health by regulating various metabolic processes and detoxifying harmful substances. However, it is highly susceptible to damage caused by oxidative stress, which can lead to the development of liver diseases. Antioxidants have emerged as potential therapeutic agents for protecting the liver against oxidative damage and preserving its normal functions. The chapter begins by discussing the mechanisms underlying liver damage and the role of oxidative stress in liver diseases. It then explores the concept of antioxidants and their ability to scavenge free radicals and inhibit oxidative stress. The focus then shifts to medicinal plants that are rich in antioxidants and have been traditionally used for liver ailments.

Several well-known medicinal plants are highlighted, including *Silybum marianum* (milk thistle), Curcuma longa (turmeric), *Phyllanthus niruri* (stonebreaker), *Andrographis paniculata* (King of Bitters), *Picrorhiza kurroa* (Katuki), and Glycyrrhiza glabra (Licorice). Furthermore, it discusses various experimental models and clinical studies that have investigated the efficacy of these antioxidants in liver disease management. Additionally, the chapter addresses the challenges associated with the use of antioxidants in medicinal plants, such as standardization, quality control, and potential herb-drug interactions.

In conclusion, this book chapter provides a comprehensive overview of the hepatoprotective potential of antioxidants derived from medicinal plants. It underscores the importance of these natural compounds in preventing and managing liver diseases. The information presented here will be valuable to researchers, healthcare professionals, and individuals seeking alternative and complementary approaches to liver health and disease prevention.

***Keywords*:** Antioxidants, hepatoprotective, herb-drug interaction, oxidative stress, medicinal plants

11.1 Introduction

The liver assumes an important role in the regulation of several physiological activities. It plays a crucial role in almost all metabolic processes associated with growth, immunity,

**Corresponding author*: firdous.oncology@gmail.com

Rakesh K. Sindhu, Inderbir Singh and M. Arockia Babu (eds.) Antioxidants: Nature's Defense Against Disease, (411–440)

nutrition delivery, energy generation, and reproduction. One of the primary roles of the liver is the metabolic processing of carbohydrates, proteins, and lipids [1]. Additionally, the liver assumes a vital role in the storage of glycogen, hormones, vitamins, iron, minerals, and several other chemicals [2]. Moreover, it serves as the principal site for the accumulation of toxins assimilated from the gastrointestinal tract and plays a vital part in the degradation and excretion of toxins, including pharmaceuticals and many exogenous chemical compounds. The liver plays an essential part in the detoxification and metabolic transformation of several harmful compounds, facilitating their elimination via the urinary system or the colon through bile secretion. The liver also serves a pivotal function in the metabolic processes of several pharmaceutical substances. Upon entering the liver, medicines undergo enzymatic conversion, resulting in the formation of either active metabolites or inactive forms. Pharmaceutical substances undergo metabolic processes predominantly inside the hepatic system, facilitated by a specific set of enzymes referred to as cytochrome P-450 [3, 4]. Hence, the liver is susceptible to the detrimental effects of toxic substances due to its intricate physiological processes.

Hepatic diseases contribute to the yearly death of thousands of individuals worldwide. Approximately two million individuals die to liver disease annually, with one million deaths attributed to complications arising from cirrhosis and the remaining one million resulting from viral hepatitis and hepatocellular carcinoma [5]. Despite notable advancements in contemporary medicine, there is still a lack of pharmaceutical interventions that are entirely efficacious in enhancing hepatic function, ensuring comprehensive organ protection, or aiding in the regeneration of hepatocytes. Steroids, antiviral medicines, and vaccines are often used in the management and prophylaxis of liver diseases. However, it is important to note that these interventions are not only associated with high costs, but they also carry the risk of adverse consequences. Multiple studies have shown the potential of medicinal plants and their phytochemical components in providing protection against a range of hepatic diseases [6, 7]. Therefore, significant focus has been dedicated to the discovery of medicinal plants with antioxidant, preventative, and therapeutic properties against liver disorders.

11.1.1 Overview of Liver Function and Importance

The liver is an essential organ inside the human body, playing a pivotal role in several physiological processes that contribute to metabolism, immune response, digestion, detoxification, and vitamin storage, among other vital tasks. As a result of these physiological processes, liver disorders persist as significant challenges to public health, exerting a global impact. Despite significant advancements in contemporary medical science, there is a lack of pharmaceutical interventions that exhibit comprehensive efficacy in stimulating hepatic activity, providing total organ protection, or facilitating hepatic cell regeneration. Therefore, it is essential to determine pharmaceutical alternatives for the management of hepatic disorders, with the objective of these alternatives exhibiting enhanced efficacy and reduced toxicity. The utilization of certain botanical species and the ingestion of various types of fruits have been integral to the field of human healthcare. Extensive scientific research has demonstrated that the identified plants and fruits possess advantageous properties that can be attributed to the existence of chemical compounds known as phytochemicals [2].

11.1.2 Oxidative Stress and Liver Damage

Oxidative stress is a physiological condition that arises from the disparity between the production of free radicals and the availability of antioxidants inside the human body. It has been associated with over 200 disorders [8]. Reactive oxygen species (ROS) are characterized by their inherent instability, since they possess a lone electron in their outer shell. These molecules are known to have a significant influence on the process of tissue destruction. The most prevalent examples are nitric oxide (NO•), superoxide anion (•O2$^-$), hydroxyl radical (•OH), and hypochlorous acid (HOCl) [9]. Cellular damage occurs as a result of the interaction between these substances and macromolecules, including proteins, lipids, and DNA. The impact of ROS on proteins is mostly facilitated by the first alteration of certain amino acid residues, including cysteine, methionine, histidine, tyrosine, and tryptophan. This modification subsequently leads to protein breakdown and structural alterations.

Lipids, particularly phospholipids, are a crucial constituent of cell membranes and organelles such as the nucleus and mitochondria. ROS target polyunsaturated fatty acids present in cell membranes, leading to cell death via a process known as lipid peroxidation. This process generates reactive aldehydes, specifically malondialdehyde (MDA) and 4-hydroxynonenal (HNE), which in turn facilitate the formation of protein cross-linkages. Additionally, this ROS-induced damage results in the depletion of intracellular reduced glutathione (GSH), the promotion of peroxide generation, activation of the epidermal growth factor receptor, and stimulation of fibronectin formation [10, 11]. Hydrogen peroxide and malondialdehyde are recognized as harmful second messengers involved in the signaling of oxidative stress. As such, they have significant importance as biomarkers in the study of oxidative damage.

ROS have the potential to induce persistent harm to DNA molecules, which includes the rupture of DNA strands, modification of bases such as the conversion of deoxyguanosine to 8-hydroxy-20-deoxyguanosine, deletion of nucleotides, diverse modifications in the organic bases of nucleotides, and the formation of cross-links with proteins [12]. Furthermore, ROS have the ability to oxidize both nuclear and mitochondrial DNA molecules. However, it has been shown that mitochondrial DNA is more susceptible to oxidative damage compared to nuclear DNA, with a vulnerability that is estimated to be 10-20 times higher. The bulk of these DNA alterations are associated with various health conditions such as cancer, aging, neurodegenerative diseases, immunological disorders, and cardiovascular problems. In addition, they have been shown to induce acute respiratory distress syndrome, disseminated intravascular coagulation, sepsis, chronic gastrointestinal inflammation, and hepatic problems [13]. ROS are also involved in the regulation of several kinases and transcription factors, including nuclear factor-kappa B (NF-κB). This regulation is associated with the modulation of proinflammatory indicators, including as tumor necrosis factor-alpha (TNF-α), interleukin 6 (IL6), and prostaglandin E2 (PGE2), which are involved in the inflammatory response [14].

11.1.3 Role of Antioxidants in Liver Protection

Both enzymatic and nonenzymatic processes have a role in mitigating or reducing the harmful consequences of ROS [41]. Antioxidant enzymes are a significant component of the antioxidant defense system, since they are essential in the removal of ROS.

Glutathione peroxidases (GPs), catalase (CAT), superoxide dismutases (SODs), glutathione S-transferase (GST), and quinone reductase (QR) are enzymes that play a role in the direct detoxification of reactive oxygen species (ROS). GPs, CAT, and SODs are recognized as major enzymes, whereas glutathione S-transferases and quinone reductase are classified as secondary enzymes. SODs play a crucial role in the detoxification of harmful superoxide radicals, while catalases and glutathione peroxidases contribute to the elimination of hydrogen peroxides [15]. Nicotinamide adenine dinucleotide phosphate (NADPH) and glutathione are examples of nonenzymatic antioxidants that play a crucial role in the antioxidant defense system. They play a significant role in the elimination of ROS. To provide more clarity, NADPH is involved in the glycolytic pathway (GP), whereas ROS generated during mitochondrial respiratory chain activity are eliminated by GSH. Furthermore, other phytochemical substances, which are nonenzymatic antioxidants, including α-tocopherol (often referred to as vitamin E) and ascorbic acid (also known as vitamin C), are essential in the process of detoxifying ROS [16].

Figure 11.1: This study investigates the impact of carbon tetrachloride (CCl4)-induced reactive oxygen species (ROS) on the defense mechanisms of antioxidant enzymes and phytochemical substances possessing antioxidant capabilities. The presence of antioxidant enzymes such as xanthine oxidase (XO), superoxide dismutase (SOD), catalase (CAT), glutathione peroxidase (GP), glutathione reductase (GR), glucose 6 phosphate dehydrogenase (G6PD), glutathione S-transferase (GST), and quinone reductase (QR), in addition

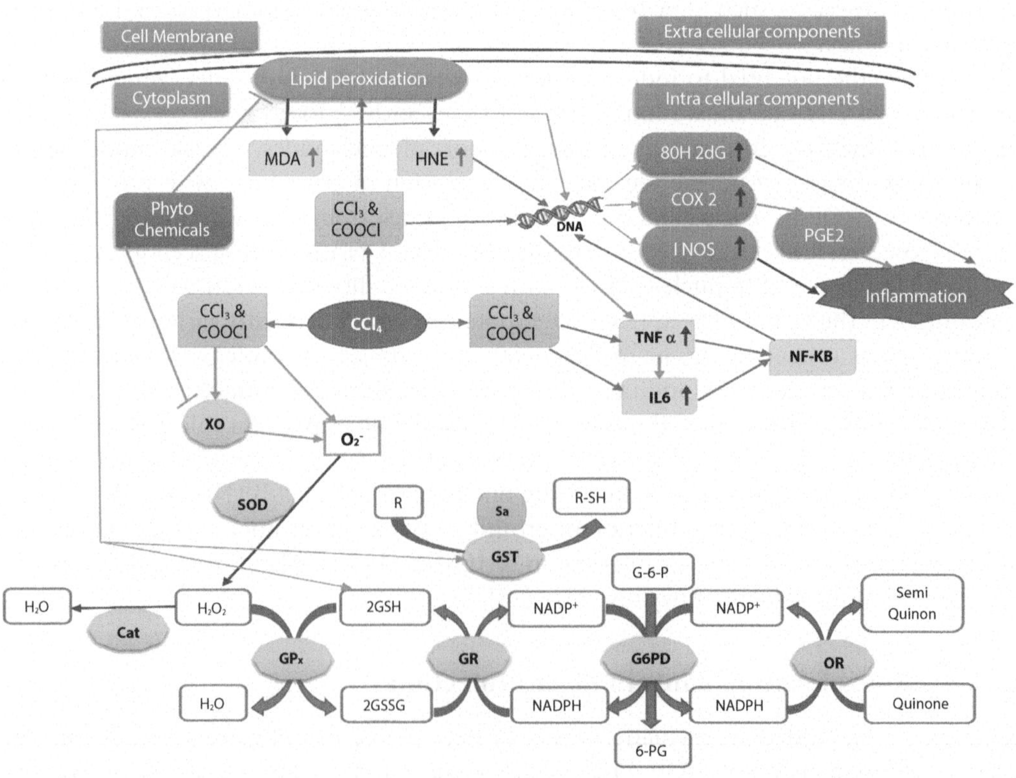

Figure 11.1 The impact of ROS from CCl4 on antioxidant enzymes and phytochemical substances.

to phytochemical compounds, has been shown to reduce the detrimental effects of reactive oxygen species (ROS). This is achieved by reducing the production of malondialdehyde (MDA), increasing the levels of reduced glutathione (GSH), and minimizing the excessive generation of 4-hydroxynonenal (HNE) protein adducts and 8-hydroxy-2′-deoxyguanosine (8OHdG). Furthermore, these enzymes and compounds have been found to upregulate the expression of tumor necrosis factor-alpha (TNF-α), interleukin 6 (IL6), and prostaglandin E2 (PGE2).

Oxidative stress arises from the inhibition or deficiency of antioxidant enzymes, resulting in the exacerbation of cellular damage and lysis. Antioxidant defense systems play a crucial role in mitigating the generation of free radicals via the scavenging of oxidants, the conversion of detrimental free radicals into less toxic compounds, and the suppression of inflammatory mediators and subsequent deleterious metabolites. The aforementioned defensive mechanisms collaborate in safeguarding the human organism from the detrimental consequences of oxidative stress [17].

11.2 Understanding Antioxidants

11.2.1 Definition of Antioxidants

The term "antioxidant" may be defined as a substance that acts against the process of oxidation. An antioxidant is defined as any chemical that, when present at low quantities relative to an oxidizable substrate, effectively retards or inhibits the oxidation process of such substrate. Antioxidants have a crucial function in the preservation of food quality and the maintenance of human health [18].

11.2.2 Mechanisms of Antioxidants

In general, antioxidants serve to protect against oxidative damage caused by free radicals via a variety of processes, as outlined in the following sections.

11.2.2.1 Use of Preventive Antioxidants

Reactive oxygen species, including H_2O_2, O_2, and $OH^\bullet$, are generated in an irreversible manner as a byproduct of metabolic processes. Consequently, several methodologies have been thoroughly examined in order to reduce the detrimental effects exacerbated by oxidative stress. The production of intracellular antioxidant enzymes inside the cell serves as a vital protective mechanism against the creation of free radicals. SOD, CAT, GPx, GR, GST, thioredoxin reductase and hemeoxygenase are well recognized as crucial enzymes involved in antioxidant defense mechanisms. Superoxide dismutases facilitate the conversion of molecular oxygen (O_2) into hydrogen peroxide (H_2O_2), which then undergoes transformation into water (H_2O) via the actions of CAT, GPx, and the Fenton reaction. Consequently, the transformation of two noxious species results in the production of a benign substance [19].

During the process of metabolism, peroxides are generated and then removed by the combined action of GST and GPX. The GRd enzyme plays a regulatory role in maintaining the balance between reduced GSH and oxidized glutathione (GSSG), with the ratio of GSH to GSSG acting as a recognized indicator of oxidative stress. The activity of GR is crucial in the elevation of GSH levels, hence preserving the oxidoreductive state inside the organism [20]. Therefore, the involvement of oxidative stress has been shown in the development and manifestation of autism complex disorder. A recent research conducted a comparison between persons with autism and a control group, revealing a drop in the ratio of GSH/GSSG and an increase in the formation of free reaction facilitates the conversion of H_2O_2 to the highly reactive $OH^•$, which then undergoes oxidation of Fe^{2+} to Fe^{3+} resulting in the formation of water.

Figure 11.2: The antioxidant enzyme system, known as O_2, is converted into H_2O_2 by the action of the SOD enzyme. The resulting hydrogen peroxide (H_2O_2) is transformed into water by the actions of catalase (CAT) and glutathione peroxidase (GPx). Through this process, two noxious species, $O_2^•$ and H_2O_2, undergo a transformation into the innocuous substance known as water. The enzyme GPx facilitates the neutralization of H_2O_2 by extracting hydrogen atoms from two molecules of glutathione (GSH), resulting in the formation of two water molecules and oxidized glutathione (GSSG). Glutathione reductase (GR) is responsible for the regeneration of reduced GSH from its oxidized form, GSSG. CAT, a crucial component of enzymatic defense mechanisms, facilitates the conversion of H_2O_2) into water.

The Fenton Radicals in the cells of those with autism, as compared to the control group. Furthermore, it should be noted that the enzyme GPx is distributed constantly across the cell, while the enzyme CAT is mostly localized within peroxisomes. The brain has a higher sensitivity to oxidative stress caused by free radicals. Notably, the activity of GPx in the

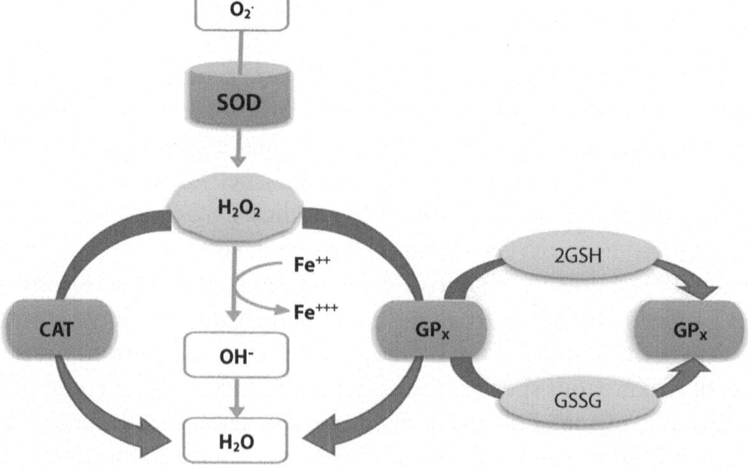

Figure 11.2 Antioxidant enzyme system.

brain is sevenfold greater than that of CAT. Furthermore, the liver, kidney, and erythrocytes are the primary sites where CAT exhibits its greatest levels, facilitating the decomposition of a significant portion of H_2O_2 [21].

11.2.2.2 Neutralizers of Free Radicals

11.2.2.2.1 The Process of Scavenging Superoxide and Other ROS

Superoxide ($O2\bullet^-$), a prevalent free radical inside cells, is implicated in several detrimental changes that are often accompanied with a deficiency in antioxidants and an increase in peroxidative activities. Although molecular oxygen (O_2) is not inherently reactive towards biomolecules, it plays a facilitating role in the generation of more reactive species such as hydroxyl radicals ($OH^\bullet$) and peroxynitrite ($ONOO^-$).Oxygen molecules, denoted as O_2, are produced in significant amounts inside phagocytes by the action of the Nicotinamide adenin dinucleotide phosphate oxidase (NADPH) oxidase enzyme during the process of pathogen elimination. Furthermore, it should be noted that it is a direct result of the process of mitochondrial respiration.

11.2.2.2.2 The Process of Scavenging Hydroxyl Radical and Other ROS

The hydroxyl radical ($OH\bullet$) is a more reactive and more harmful radical when compared to other radical species, particularly in its impact on biological components such as DNA, lipids, and proteins. In a broad context, the formation of $OH^\bullet$ is often attributed to the Fe^{2+} or $Cu+/H_2O_2$ Fenton reaction system. This process involves incubating $FeSO_4$ and H_2O_2 in an aqueous solution. Hence, the antioxidative function of antioxidants involves the scavenging of hydroxyl radicals ($OH^\bullet$) by either direct scavenging or the inhibition of $OH^\bullet$ formation through chelation of free metal ions or conversion of hydrogen peroxide (H_2O_2) into non-toxic molecules [22].

11.2.2.3 Inhibitors of Enzymes that Generate Free Radicals

According to available reports, many enzymes have been identified as the primary origins of free radicals in diverse physiological and pathological contexts. NADPH oxidases are a class of enzymes that are associated with the plasma membrane. These enzymes possess the capability to facilitate the transfer of a single electron from the cytosolic donor NADPH to an extracellular oxygen molecule, resulting in the formation of $O_2^{\bullet-}$ Uric acid is synthesized by the enzymatic activity of xanthine oxidase, which facilitates the oxidation of hypoxanthine and xanthine to produce uric acid, along with the generation of $O_2^{\bullet-}$ and H_2O_2. This process leads to an elevation in the oxidation state inside an organism [23]. Furthermore, molecular oxygen (O_2) is produced as a secondary product during mitochondrial respiration, as well as by several enzymes such as NADH oxidase, monooxygenases, and cyclooxygenases. O_2 is physiologically considered to be very harmful and is synthesized in substantial quantities by the enzyme NADPH oxidase. This production serves the purpose of facilitating oxygen-dependent systems that are responsible for eliminating invading pathogens. The respiratory burst is a crucial mechanism for regulating the synthesis of reactive oxygen derivatives in order to protect the organism against invading pathogens, while minimizing any significant impairment of tissue functioning. However, an overabundance of ROS might increase oxidative stress, leading to the oxidation of low density lipoprotein (LDL). A correlation has been discovered between heightened phagocytic NADPH oxidase activities and enhanced levels of circulating oxidized LDL in individuals with metabolic syndrome. Consequently, the regulation of NADPH oxidase to inhibit excessive formation of ROS and the administration of antioxidants have been identified as effective approaches in mitigating the detrimental impact of oxidative stress in individuals undergoing hemodialysis [24]. In recent years, there has been a growing body of research indicating that many natural antioxidants possess the ability to block enzymes that facilitate the formation of ROS. Additionally, these antioxidants have shown promise in the development of novel therapeutic treatments for disorders associated with oxidative stress.

11.2.2.4 Lipid Peroxidation Protection

Peroxidation of lipids (unsaturated fatty acids, glycolipids, cholesterol, cholesterol ester, phospholipids, etc.) is known as lipid peroxidation. Unsaturated fatty acids are easily damaged by ROS because they contain many double bonds and methylene-CH2-groups, which contain highly reactive hydrogen atoms [25]. Antioxidants may immediately react with peroxide radicals and quench them, thereby halting the chain reaction. Different types of chronic illnesses, including cancer and atherosclerosis, have been linked to lipid peroxidation and DNA damage. Antioxidants may eliminate ROS and peroxide radicals, preventing or curing many diseases. Lipid peroxidation is where researchers have focused their efforts in order to identify natural antioxidants and better understand how they work. Antioxidant research against free radical increased lipid peroxidation has been considered in several systems, including lipids, red blood cells, and LDL. The antioxidant activity of these polyphenols is very sensitive to the molecular structure of the compound, the reaction's initial circumstances, and the medium's microenvironment.

11.2.2.5 Prevention of Damage to DNA

In biological systems, the hydroxyl radical ($OH^{\bullet}$) and peroxynitrite anion ($ONOO^{-}$) radicals, which are generated from the reaction between nitric oxide and superoxide anion ($O2^{\bullet -}$), have the capability to directly interact with plasmid DNA macromolecules. This interaction results in the cleavage of a single DNA strand, therefore causing oxidative damage to the DNA. Neurodegenerative and cardiovascular illnesses, cancer, and the aging process are closely linked to cellular apoptosis and genetic mutation resulting from DNA damage. As a result, there has been a significant focus on the investigation and use of DNA or plasmid damage as experimental models for the examination and characterization of antioxidants [26]. The present work investigates the DNA damage resulting from the interaction between Cu^{+} ions and hydroxyl radicals (OH•) in a metal-free plasmid DNA system. This interaction is examined in the presence of Cu^{2+} ions, ascorbic acid, and H_2O_2 at a pH level of 7. The process involves the reduction of Cu^{2+} to Cu^{+} *in situ* using ascorbic acid. The $OH^{\bullet}$ generated from the reaction between Cu^{2+} and H_2O_2 induces the cleavage of a single DNA strand, resulting in the unwinding of the typically supercoiled plasmid DNA.

11.2.3 Types of Antioxidants and Their Sources

Antioxidants delay or stop the oxidation of other compounds, therefore facilitating their actions. Antioxidants were initially studied for their biological function in inhibiting the oxidation of unsaturated lipids.

The discovery of vitamins A, C, and E and the discovery of vitamin E's function in preventing lipid peroxidation [27] were, nevertheless, pivotal in the development of our knowledge of antioxidants and their importance to all forms of life. The figure (Figure 11.3) displays the categorization of antioxidants along with typical instances. Enzymatic and nonenzymatic antioxidants are the two most common categories. There is a wide range of compounds with a variety of mechanisms of action, sites of action, and outcomes. Because of its unique characteristics, each plays a specific function in the body. The most efficient antioxidant defense is provided by a network of interacting antioxidant enzymes including catalase (CAT), glutathione peroxidase (GPx), and glutathione reductase (GR). Vitamin C, E, coenzyme Q, carotenoids, glutathione, and trace elements are all examples of low-molecular-weight antioxidants that play a role in neutralizing reactive radicals. Glutathione, ubiquinone, albumin and metallothioneins, and uric acid are examples of endogenous antioxidants [28], but the vast majority come from dietary sources, either plant-based (flavonoids, phenolic acids, carotenoids, stilbenes, coumarins, lignans, organosulfur compounds,

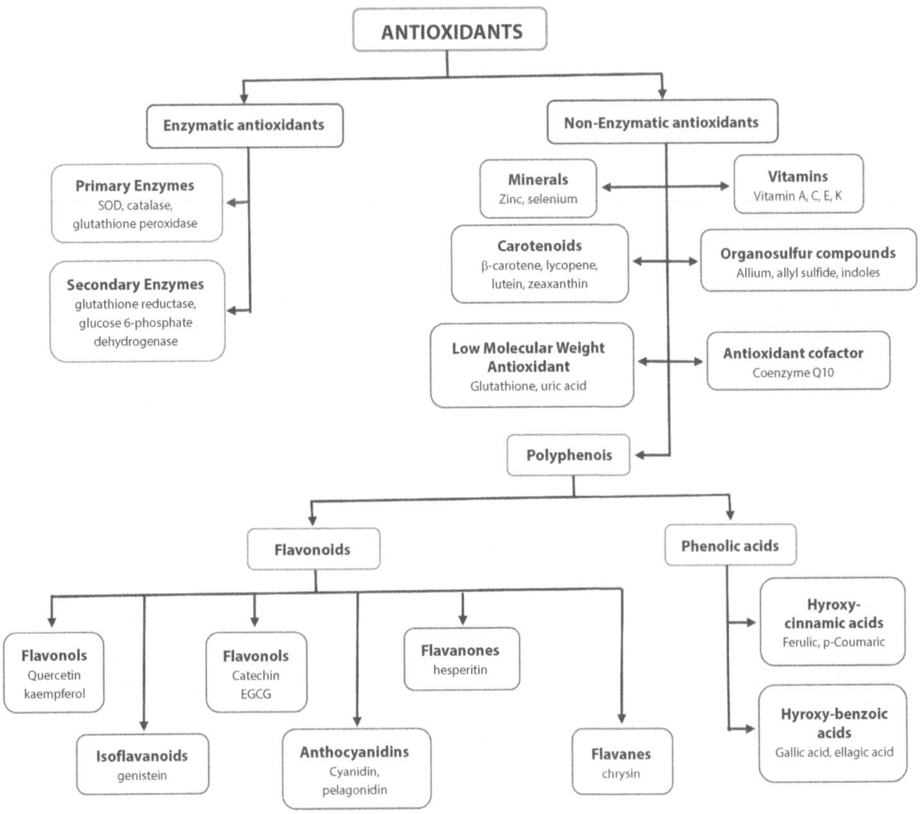

Figure 11.3 Classification of antioxidants.

vitamins) or mineral-based (selenium, zinc, manganese). Exogenous antioxidants are required when the body's naturally occurring antioxidants, which play a role in free radical defense, are insufficient to fend off reactive oxygen species (ROS). Both prokaryotes and eukaryotes have the capacity to produce beneficial compounds.

It is now possible to prevent or treat diseases caused by ROS using one of the many naturally occurring antioxidants that have been identified and described. A diet that incorporates a variety of fruits, tea, wine, vegetables, and grains might be regarded as a substantial reservoir of antioxidants. Certain medications, in addition to their intended therapeutic benefits, have been shown to possess antioxidant properties. For instance, captopril, which falls under the category of angiotensin-converting enzyme (ACE) inhibitors, as well as N-acetylcysteine and dihydropyridine calcium antagonists, have been identified as having antioxidant effects [29]. Nevertheless, the quantities used in the treatment fail to exhibit antioxidant action *in vivo.* Microorganisms, such as actinomycetes, bacteria, cyanobacteria, fungus, and lichens [30], are recognized as significant contributors of antioxidants and other bioactive chemicals. In contrast to plants, these organisms exhibit rapid growth rates when subjected to precise environmental controls, making them a very desirable reservoir of naturally occurring bioactive compounds for use in many industrial sectors such as food production, medicines, nutraceuticals, and agriculture.

11.2.4 Interplay Between Free Radicals and Antioxidants

The role of oxidative stress in the development of several diseases in humans has garnered significant scientific interest. Oxidative stress arises from a state of disequilibrium between the production of free radicals and the availability of antioxidants. Free radicals, which are well recognized as significant contributors to the aging process and the development of diseases, have the potential to cause harm to cells. Antioxidants play a crucial role in mitigating the adverse consequences of free radical-induced harm, serving as the primary protective barrier. Their multifaceted methods of action are vital for preserving overall well-being. The many categories of antioxidants include those that are produced internally by the cells of the body, as well as external agents such as dietary supplements. The development of antioxidant insufficiency may occur due to a reduction in antioxidant consumption, diminished production of endogenous enzymes, or an elevation in antioxidant usage. The practice of supplementing with antioxidants has gained popularity as a means to enhance protection against free radicals and preserve good bodily function [31].

11.3 Role of Nano-Antioxidants in Liver Function

11.3.1 Antioxidants Encapsulated in Nanoparticles (NP-Antioxidant)

Low amounts of ROS preserve the body's regular metabolism and shield it from bacterial viruses under healthy settings. A change in signal pathways and the development of cancer may result from a disparity between ROS and antioxidants [32]. In order to cure cancers, antioxidants can reduce ROS in tumors or tumor microenvironments. Nevertheless, issues like rapid blood circulation, low membrane permeability, low water solubility, and lysosome/endosome degradation significantly reduce their effectiveness. As previously indicated, nanocarrier technologies may be able to address these drawbacks. Nano-antioxidants are antioxidants encapsulated in nanocarriers [33].

11.3.2 Using Antioxidants to Treat Liver Disorders

By scavenging reactive oxygen species (ROS), antioxidants lower levels of ROS within cells and guard against oxidative damage. Several substances, including vitamin E, resveratrol, curcumin, and ebselen (an analog of glutathione peroxidase), have been seen to exhibit antioxidant properties. These substances can be employed as external antioxidants in order to penetrate tumor cells and lower ROS. It has been demonstrated that consuming antioxidants in moderation lowers the possibility of cancer and slows its spread [34].

11.3.3 Benefits of Antioxidant Treatment Mediated by Nanocarriers

The half-life of Nanoparticles (NP) is extended by hydrophilic surface changes, giving it enough time for reaching the tumor location. The tumor's surrounding defective microvascular endothelium allows the nanoparticles to reach the tumor matrix, where they become trapped since lymphatic vessel reflux is blocked. The passive localization of nanoparticles to solid tumors is accomplished through the fundamental concept of increased permeability and retention (PRE) effects. Nanoparticles bigger than 2 nm are blocked from passing through the typical endothelial cell gap due to tight junctions [35].

Small molecules are frequently employed as antioxidant agents in the treatment of cancer, but their inability to dissolve in water and generalized expulsion by the vascular endothelial mechanism and the mononuclear phagocytosis mechanism restrict their use in clinical settings. Owing to their tiny size, these molecules were primarily eliminated by metabolism in the liver and kidney, which also results in harmful side effects that exacerbate liver and kidney damage. This is also the primary cause of small molecule medicines' short biological half-lives and ineffectiveness as a therapeutic agent. These tiny molecules can be contained by nanomaterials, which can shield medications from rapid elimination, an acid-base environment, and enzyme deterioration. In addition to PEGylation, surface modification, and stimuli-responsiveness, nanocarriers offer a solution to the aforementioned issues [36].

11.4 Medicinal Plants with Hepatoprotective Properties

11.4.1 *Silybum marianum* (Milk Thistle)

11.4.1.1 Phytochemical Composition

Silybum marianum seeds are composed of 26.05% oil, 4.48% moisture, 1.93% ash, 5.48% crude fiber, 87.2% carbs, and 23% total proteins. Apigenin, silybonol, proteins, betaine, fixed oil, and free fatty acids are the active substances found in milk thistle seeds.

Silymarin is the primary bioactive substance in the medicinal plant milk thistle. Various flavonolignans, including silybinin A and B (SBN A&B), isosilybin A and B (ISBN A&B), silychristin (SCN), and silydianin (SDN), are combined to form silymarin. It is a popular Chinese herb. The milk thistle has silymarin in its seeds, fruit, and leaves, although the silymarin concentration is highest in the seeds. The milk thistle's fruit contains varying amounts of silymarin depending on the variety, region, and environment where it grows. Owing to the inclusion of phytoconstituents such antioxidants and total phenolics, milk thistle is utilized as medicine to treat a variety of disorders [37].

Due to the high oil content of milk thistle, silymarin extraction cannot be completed in a single step. Since oil is a byproduct of silymarin industrial production, it should be removed from the seeds prior to silymarin extraction. Extraction is the first and most crucial stage in recovering and purifying silymarin from *Silybum marianum* seeds [38].

11.4.1.2 Hepatoprotective Mechanisms

The main component obtained from *S. marianum* was silymarin, which showed hepatoprotective effects by lowering ALT, AST, TC, TG, HDL-C, and LDL-C levels, decreasing IL-6, mitogen-activated protein kinase 1 (MAPK1), Caspase 3, p53, and vascular endothelial growth factor A (VEGF-A) protein expressions, and raising AKT1 protein expression. The current investigation revealed novel details and offered a potential rationale for the molecular processes that protect *S. marianum* against NAFLD [39].

11.4.1.3 Experimental and Clinical Evidence

For about 2000 years, milk thistle derivatives have been utilized as herbal treatments. The extracts remain frequently utilized to manage chronic liver disorders and to safeguard the liver against poisons. Milk thistle extracts may also have anti-cancer, anti-diabetic, and cardioprotective properties, according to recent preclinical and clinical investigations. In this article, milk thistle clinical trials from the last five years are reviewed, along with pharmacokinetic and toxicity investigations, herb-drug interactions, and additional risk factors. The effects of milk thistle for people with liver disorders, cancer, hepatitis C, HIV, diabetes, and hypercholesterolemia have been investigated in numerous trials. The curative properties of milk thistle against several cancer types has shown promising findings, and ongoing studies will add to the body of knowledge concerning this impact. The long-awaited proof of milk thistle's effectiveness in treating liver illnesses will also be provided by newly determined doses and improvements in the quality and uniformity of this plant. Milk thistle preparations are widely known for being harmless and well accepted, and any hazardous or unfavorable effects seen in the clinical trials under consideration appear to be insignificant. Research on milk thistle has a bright future, and to further support the herb's safety and effectiveness, randomized clinical investigations comparing milk thistle to placebo may be required [40].

11.4.2 Curcuma Longa (Turmeric)

11.4.2.1 Active Constituents

Asian nations including India value the spice curcuma longa, also known as turmeric. Three curcuminoids, curcumin (diferuloylmethane, the main component responsible for turmeric's yellow color), demethoxycurcumin, and bisdemethoxycurcumin, are the main active components of turmeric. Volatile oils (such as zingiberene, atlantone, and tomerol) also have pharmacological effects. In addition, turmeric contains carbohydrates, proteins, and resins. Turmeric is a strong antioxidant and has significant anti-inflammatory qualities. It's possible that turmeric's anti-inflammatory properties are directly tied to its anticancer properties [41].

11.4.2.2 Antioxidant Effects on the Liver

Alcohol, drugs, viral infections, environmental pollutants, and dietary components can all cause liver damage, which can then progress into non-alcoholic steatohepatitis, non-alcoholic liver disease, liver fibrosis, and cirrhosis. Oxidative stress is thought to be a major contributing factor in this process. Even despite the significant advancements in the treatment of liver illness over the last 30 years, thousands of individuals globally continue to have chronic or acute liver conditions. Among the most widely used natural compounds, curcumin has several liver-protecting shielding properties. The current study aims to conduct a thorough assessment of the pharmacological effects, molecular mechanisms, and clinical data of curcumin as a potential drug for both the avoidance and cure of oxidative-associated liver disorders [42].

These methods include inhibiting the production of pro-inflammatory cytokines, the products of lipid peroxidation, the activation of phosphoinositide 3-kinase (PI3K)/Akt and hepatic stellate cells, as well as improving cellular reactions to oxidative stress, such as nuclear factor erythroid 2-related factor 2 (Nrf2), SOD, CAT, GSH, GPx, and GR. When taken as a whole, curcumin utilizes the phenolic, -diketone, and methoxy groups to act as an antagonist of free radicals over the action of distinct types of ROS [43].

11.4.2.3 Preclinical and Clinical Studies

Curcumin has been used as a spice, nutritional supplement, and component of numerous traditional Asian remedies for many years. Curcumin demonstrates a wide spectrum of pharmacological effects against many chronic diseases, especially type II diabetes, rheumatoid arthritis, multiple sclerosis, Alzheimer's disease, and atherosclerosis, according to recent preclinical investigations. Additionally, it prevents thrombosis, lowers platelet aggregation, and obstructs HIV propagation. Curcumin also promotes wound healing and guards against fibrosis, liver damage, cataract development, and liver toxicity. Last but not least, curcumin's potential as a preventative and therapeutic agent for a wide range of cancers, which includes gastrointestinal, melanoma, genitourinary, breast, lung, hematological, the head and neck, neurological, and sarcoma, has been thoroughly investigated [44].

11.4.2.4 Curcumin Nanoformulations

The main goal of the nanoformulations covered in this part is to improve curcumin's solubilization while also safeguarding it against hydrolysis-induced inactivation. The formulation must effectively prepare and load curcumin while holding onto it for the necessary amount of time. While some formulations offer additional methods for cellular transport or intracellular release, some formulations aim for a sustained release of curcumin.

11.4.2.5 *In Vitro Studies of Curcumin Nanoformulations*

Numerous types of cancer cell lines have been used to examine the cytotoxic property of curcumin nanoformulations. Due to the lengthy stimulation cell populations to elevated fixed dosages of curcumin (whether it's in its free form or as a nanoformulation), which are not always correlated with the concentrations attained *in vivo*, interpreting the significance of the findings is frequently challenging.

11.4.2.6 *In Vivo Studies of Curcumin Nanoformulations: Kinetics and Efficacy*

Numerous animal studies have examined the pharmacokinetics, biodistribution, and therapeutic efficiency of various curcumin nanoformulations in order to gain knowledge about the potential benefits of these frameworks for the treatment of various diseases. Numerous studies have demonstrated that when curcumin nanoformulations were administered orally or intravenously, the area under the concentration-time curve (AUC) was greater than when curcumin in its free form was given.

11.4.2.7 *Clinical Trials*

Curcumin has been studied in human clinical trials for many years, both in its free form and as nanoformulations, and it has demonstrated beneficial effects to individuals with colorectal cancer, pancreatic cancer, breast cancer, and multiple myeloma. Due to the limited bioavailability of curcumin, it is often taken through the oral route as capsules filled with the substance's powder [45].

11.4.3 ***Phyllanthus niruri* (Stonebreaker)**

11.4.3.1 *Phytochemicals in Phyllanthus niruri*

Flavonoids

Rutin: Rutin, a flavonol glycoside made up of the flavonol quercetin and the disaccharide rutinose, is one of more than 2000 different known compounds that are included in the broad group of phenolic secondary metabolites of plants. Rutin is crucial because it helps persons with coronary artery disease or high blood pressure by strengthening capillaries. Recent research has revealed that rutins are potent anti-oxidants that battle free radicals, which are thought to be the cause of up to 90% of all human diseases, including cancer, arteriosclerosis, stroke, and senility.

Quercetin: Due to its capacity to reinforce and regulate the permeability of blood vessel walls, especially capillaries, quercetin is significant in nutrition. This bioactive substance exhibits anti-aggregant, anticancer, antifungal (particularly anti-dermatophytic), anti-feedant, anti-glaucomic, anti-inflammatory, anti-oxidant, antiseptic, and anti-spasmodic activity.

Quercitrin: The flavone glycoside quercitrin, also known as quercetin 3-O-alpha-L-rhamnopyranoside, possesses anti-leishmanial, anti-nociceptive, and anti-inflammatory properties.

Astragalin: A flavanone, astragalin has diuretic properties. By boosting the manufacture of antibodies, accelerating phagocytosis, and raising the number of macrophages, it can also

control the immunological capabilities of both humans and animals. Astragalin reduces histamine release from cultivated KU812 cells, a human basophil cell line, and latent cutaneous anaphylaxis passive cutaneous anaphylaxis (PCA) responses in mice. The substance may lessen the incidence and extent of dermatitis as well as the levels of IgE in the serum in NC/Nga atopic mice. On histological examination of the skin of mice treated with D-N-galactosamine lipopolysaccharide, there was a reduced inflammatory cell infiltrate.
Catechin: *P. niruri* tissue culture tissue cultures yielded gallocatechin. According to studies, catechins promoted apoptosis and inhibited the development of human colon and liver epithelial cancer cells. In the presence of minor quantities of metals, especially Cu^{2+}, catechins exhibit pro-oxidative action and bactericidal properties [46].

Terpenes
Limonene: Monoterpenoid limonene has numerous uses in medicine and pharmaceuticals. The Diels-Alder reaction can be used to create the limonene molecule from two isoprene molecules. It exists in two mirror reflections of one another. Orange oil contains a significant amount of limonene. It has been documented that limonene has anti-carcinogenic effects in models of liver tumors. It has also been reported that limonene causes a spike in diclofenac absorption in hairless rats, indicating that limonene is a potent topical treatment for both epidermal and sub-dermal wounds.
p-Cymene: In the monoterpene family, p-cymene also has a variety of biological effects, including antioxidant and anti-microbial activity. Thyme, an essential oil that contains p-cymene, was found in a recent study to have strong anti-oxidant properties by blocking the oxidation of α-tocopherol. The paper disc diffusion method was used to assess p-Cymene's antimicrobial abilities, and the results showed that it had good antimicrobial activity. Chloro (p-cymene) ruthenium complexes, which are synthetic p-cymene derivatives, showed *in vitro* anti-cancer action against the human breast cancer cell line (MDAMB-435).
Lupeol: The pentacyclic triterpene lupeol can be produced through protosterol carbocation, which is a backbone rearrangement reaction. By blocking tyrosyl phosphorylation of a 45 kDa protein found in human neutrophils, it reduces the production of superoxide. It has been demonstrated that the substance possesses anti-inflammatory and anti-tumor properties. More intriguingly, it has been discovered that lupeol and its derivatives imitate the action of the whole-plant extract by preventing calcium oxalate (CaOx) crystal accumulation in experimental urolithiasis. Lupeol, a potent epidermis chemopreventive agent, can reduce the toxicity of benzoyl peroxide on the skin. Additionally, it has the power to strengthen the liver's antioxidant defenses against cadmium-induced damage in rats [47].

11.4.3.2 Hepatoprotective Mechanisms

A common tropical plant called *Phyllanthus niruri* L. is utilized in Ayurvedic medicine to treat liver and renal problems. The goal of the current investigation is to discover the *P. niruri* extract that is most effective at hepatoprotection and to use a bio-guided technique to pinpoint the key ingredients that provide this effect. *P. niruri* aerial components were extracted independently using ethanol concentrations of 80%, 50%, and 70%. In clone-9 and Hepg2 cells, the extracts' cytoprotective efficacy was assessed against CCl4-induced hepatotoxicity. To isolate the active components, the aqueous extract (AE) was fractionated under the direction of a bioassay [42]. The DPPH (1, 1-diphenyl-2-picrylhydrazyl) radical elimination technique and ferric reducing antioxidant power (FRAP) were used to measure antioxidant activity. After determining its LD50, the *in vivo* liver-protective efficacy of AE was assessed in CCl4-induced liver damage in rats at various doses. The amounts of alanine aminotransferase (ALT) and aspartate aminotransferase (AST) contrary to CCl4 injuries were significantly reduced after pretreatment of clone-9 and Hepg2 with various concentrations of AE (1, 0.1, 0.01 mg/ml), and the activity of the natural antioxidants glutathione (GSH) and superoxide dismutase (SOD) was restored toward normalization. The fractionation of AE produced fractions I through IV. *In vitro* hepatoprotective action was significantly present in fractions I, II, and IV. Seven compounds were obtained after I, II, and IV were purified: corilagins C1 and C2, isocorilagins C2 and C3, brevifolins C3 and C4, quercetins C4 and C5, and gallic acids C6 and C7. The hepatoprotective potency of molecules C1, C2, C5, and C7 was highest (p 0.001), whilst C3, C4, and C6 showed a moderate (p 0.001) activity. The AE demonstrated significant antioxidant FRAP (79.352 2.88 mM Ferrous equivalent) and DPPH (IC50 11.6 2 g/ml) activity. AST, ALT, alkaline phosphatase (ALP), lactate dehydrogenase (LDH), total cholesterol (TC), triglycerides (TG), total bilirubin (TB), glucose, total proteins (TP), urea, and creatinine levels that were elevated by CCl4 were normalized in rats after *in vivo* administration of AE (25, 50, 100, and 200 mg/kg). Additionally, AE reduced the expression of TNF-α, NF-κB, IL-6, IL-8, IL-10, and COX-2, and it strongly counteracted the effects of CCl4 on antioxidant enzymes such as SOD, CAT, GR, and GSP. The hepatoprotective action of AE was also validated by the histological analysis. *P. niruri* isolates shown strong hepatoprotective properties towards CCl4-induced liver damage in clone-9 and Hepg2 cell lines by reducing lipid peroxidation and retaining glutathione in its reduced state. Due to their phenolic makeup and consequent antioxidative potential, this is the case [48].

11.4.3.3 Evidence from Studies

A common tropical plant called *Phyllanthus niruri L.* is utilized in Ayurvedic medicine to treat liver and renal problems. The goal of the current investigation is to discover the *P. niruri* extract that is most effective at hepatoprotection and to use a bio-guided technique to pinpoint the key ingredients that provide this effect. *P. niruri* aerial components were extracted independently using ethanol concentrations of 80%, 50%, and 70%. In clone-9 and Hepg2 cells, the extracts' cytoprotective efficacy was assessed against CCl4-induced hepatotoxicity. Primary screening should consider a plant's *in vitro* cytotoxic and hepatoprotective potential using Hepg2, as this model is excellent for examining *in vitro* xenobiotic absorption and liver toxicity. Reactive oxygen species (ROS) were produced by CCl4 in a time-dependent manner. Lipid peroxidation was also induced, reaching its peak after 24 hours of incubation. In order to investigate the hepatoprotective impact of AE towards CCl4-induced toxicity, clone-9 and Hepg2 cells were treated with CCl4 for 24 hours [49].

11.4.4 Other Medicinal Plants with Hepatoprotective Potential

11.4.4.1 Andrographis paniculata (King of Bitters)

Andrographis paniculata is referred to as the "King of the Bitters". Substantial antioxidant and hepatoprotective activity have been reported for the plant's aerial parts' ethanol extract. The hepatic enzymes Alanine aminotransferase (ALT) and Aspartate aminotransferase (AST) were restored to 75 and 14.5% of their original levels when the plant extract was pretreated against CCl4 challenged rats at a dose of 300 mg/kg b.w. Animals exposed to *A. paniculata* extract had their MDA levels reduced by 40% and their GSH levels raised by 46%. There was a 22 to 94% significant restoration of the CAT, GPX, QR, GST, GR, and G6PD activities in the hepatic tissues. Furthermore, the plant's solvent extract exposure prevented the degenerative changes brought on by CCl4 treatment, including lipid alterations, cellular hypertrophy, necrotic cells, inflammatory cell infiltration, and sinusoidal dilatation [50].

11.4.4.2 Picrorhiza kurroa (Katuki)

The well-known herb *Picrorhiza kurroa* (*P. kurroa*; Family Scrophulariaceae), utilized in the Indian traditional Ayurvedic medical system, has been used to treat liver issues and is a key component of numerous herbal remedies for the treatment of liver conditions [51]. Picroliv is a standardized combination of iridoid glucosides that was extracted from *P. kurroa*'s roots and rhizomes. It has been employed as a hepatoprotective treatment for conditions like jaundice because it contains at least 60% of a 1:1.5 mixture of picroside I and kutkoside [52]. Additionally, *P. kurroa* extract has been shown to have nitric oxide scavenging, cardioprotective, anti-cancer, anti-diabetic, and anti-viral effects. Picroliv, a substance present in *P. kurroa*, has recently been shown to protect rat livers from damage brought on by ethanol [53].

11.4.4.3 Glycyrrhiza glabra (Licorice)

Liquorice, also known as *Glycyrrhiza glabra L.* (Fabaceae), is one of the more well-known medicinal plants. *G. glabra* is a stiff herb or small shrub that can reach a height of 1 m and has pinnate leaves. A loose inflorescence has axillary spikes of purple to pale grayish blue flowers. An oblong pod with several seeds makes up the fruit. The CCl4-induced liver damage model is frequently used for the investigation of the hepatoprotective activity of medications and plant materials because studies have shown that the changes associated with CCl4-induced liver damage are similar to those associated with acute viral hepatitis, many chemical substances (like Isoniazid, Rifampicin, Pyrazinamide, Non-steroidal anti-inflammatory drugs (NSAIDs), etc.) caused hepatopathy and oxidative stress [54]. On the carp hepatocyte injury brought on by carbon tetrachloride CCl4, *G. glabra* extract shown hepatoprotective and antioxidant activities. Animals with CCl4-induced hepatic damage are protected against oxidative stress by the combination of *Silybum marianum* and *G. glabra*. Licorice had a protective effect in mice against alcohol-induced liver damage, which was mediated through oxidative stress and inflammation. This investigation's goal was to assess *G. glabra* extracts' potential hepatoprotective properties [55].

11.5 Clinical Studies on Medicinal Plants and Liver Diseases

11.5.1 Non-Alcoholic Fatty Liver Disease (NAFLD)

The prevalence of non-alcoholic fatty liver disease (NAFLD), a common chronic liver condition, is high in developed nations. One of the main contributors to chronic liver disease and cryptogenic cirrhosis has been identified as NAFLD. A greater risk of developing NAFLD exists in those with obesity, insulin resistance, diabetes mellitus, hyperlipidemia, and hypertension. For the treatment of NAFLD, the herbal medicine has had significant positive effects on the reduction of inflammation and steatosis. It has been discovered that these impacts involved the various lipid metabolism and inflammatory regulation pathways [56].

Bile Acid Synthesis Involves in Herbal Medicine Modulation of Lipid Metabolism

The liver produces bile acids, which function as a biological detergent to metabolize lipids and cholesterol into bile [57].

Figure 11.4: provides crucial lipid mechanisms of action into the anti-NAFLD benefits of herbal therapy. The often prescribed herbal remedy increased lipolysis with a focus on fatty acid-oxidation, which involves oxidative stress and mitochondrial activity, while inhibiting fatty acid or diet-induced lipogenesis via the sterol regulatory element binding protein 1c

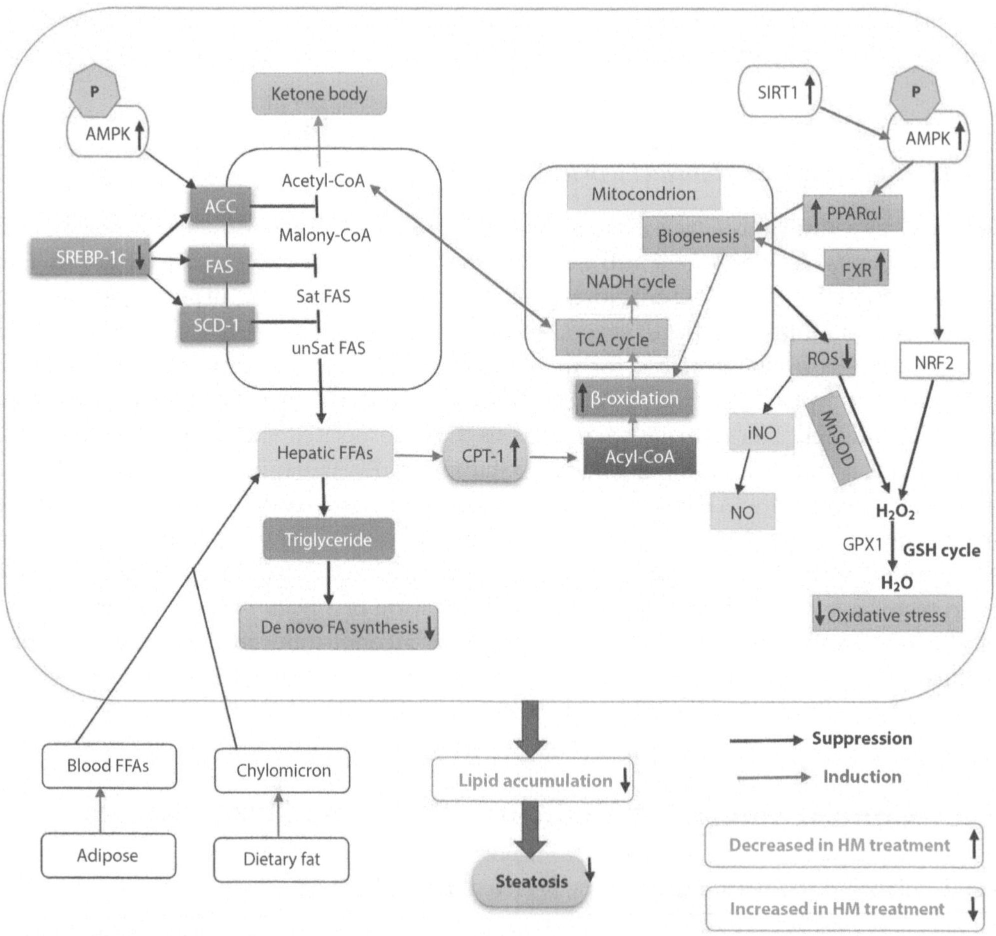

Figure 11.4 The lipid mechanistic understanding of herbal medicine's anti-NAFLD actions.

(SREBP-1c) pathway. By modifying the AMP-activated protein kinase (AMPK) signaling pathway, the fundamental mechanism of action could potentially be accomplished [58].

Herbal Medicine Ameliorates the Hepatic Inflammation

Hepatic inflammation encourages lipid accumulation and gets redistributed from adipose to hepatic tissue and fuels inflammatory damage to the liver. NAFLD begins as hepatic steatosis and progresses to steatohepatitis and fibrosis as a result of hepatic inflammation. Treatment for NAFLD must first reduce heaptic inflammation. The protective effects of herbal medicine against the development of steatohepatitis from hepatic steatosis have been demonstrated, and the underlying mechanism has been shown to be associated with preventing the inflammatory signaling pathway, in Figure 11.5. This regulates dyslipidemia and enhances liver function in NAFLD [59].

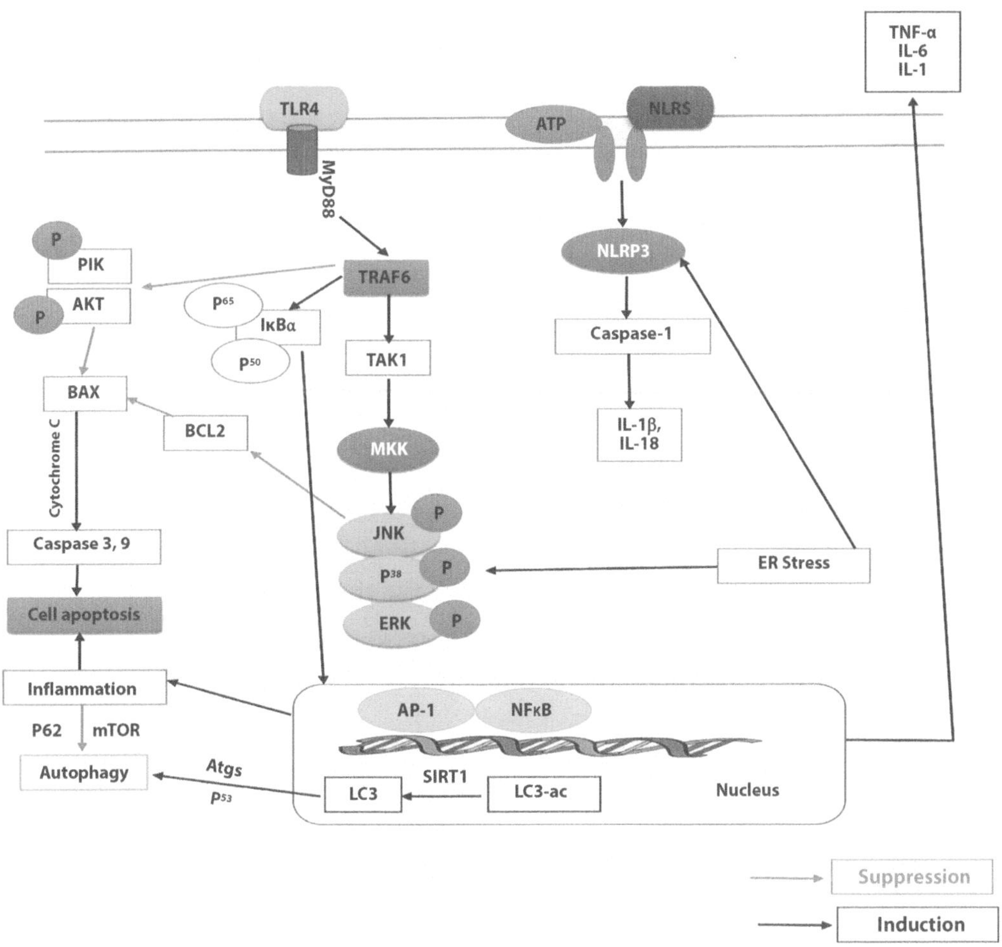

Figure 11.5 The inflammatory mechanistic understanding of herbal medicine's anti-NAFLD actions.

Herbal Medicine Improve Hepatic Lipid Metabolism

Hepatic steatosis is mostly brought on by a lipid overload. Excessive supply of free fatty acids (FFAs) to the liver causes the intrahepatic pool of FFAs to grow in a form of triglycerides. Fat accumulation causes hepatic lipotoxicity, which causes the production of pro-inflammatory cytokines by liver cells, the induction of oxidative stress, the activation of hepatic stellate cells, and finally the development of inflammatory injury to the liver. Enhancing the metabolism of fatty acids is a useful strategy for treating NAFLD, and both clinical and preclinical investigations of the disease have looked at the effectiveness of herbal remedies that target fatty acid metabolism [60].

Oxidative Stress Action Involves in Herbal Medicine Modulation of Lipid Metabolism

Oxidative stress is a reflection of a disparity between the generation of reactive species and antioxidant protection, which can result in liver damage as NAFLD progresses. Reactive oxygen species (ROS) are influenced by the lipid metabolic disease; specifically, fatty acid-oxidation appears to produce increased ROS in NAFLD. Herbal medicine's ability to decrease cholesterol and act against oxidative stress are related [61].

11.5.2 Alcoholic Liver Disease (ALD)

A global health issue without a viable therapeutic solution is alcoholic liver disease (ALD), which is caused by prolonged excessive alcohol consumption. As oxidative stress, inflammation, and increased fatty acid production all play a role in ALD, it is now more well

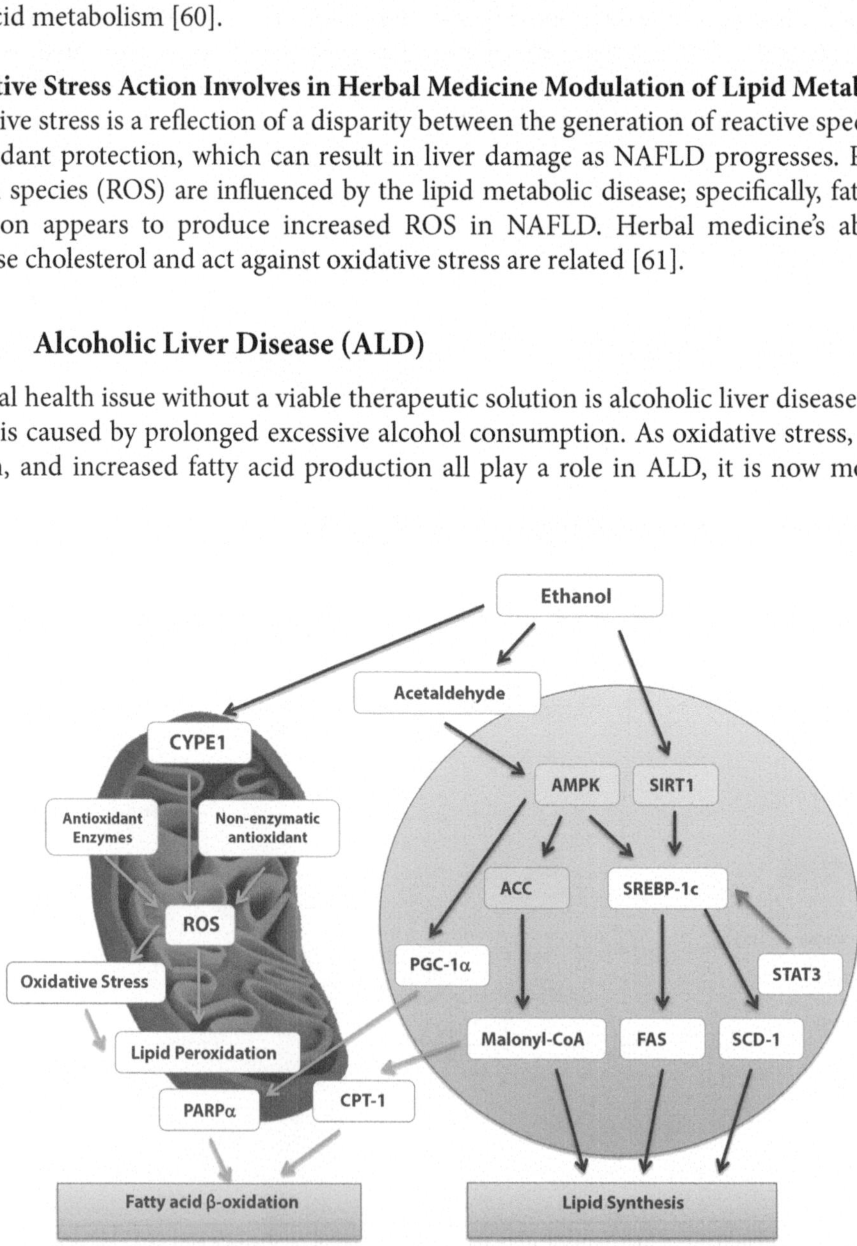

Figure 11.6 Biological targets leading to ALD and targets for herbal remedies.

recognized as a complex and diverse disease process. Due to their numerous therapeutic targets and little harmful side effects, herbal medications have received a lot of interest in the past ten years as prospective therapeutics in both the management and prevention of ALD. A number of plants, including *Cnidium monnieri* (L.) *Cusson* (Apiaceae), *Curcuma longa* L. (Zingiberaceae), *Pueraria lobata* (Willd.), *Ohwi* (Leguminosae), etc., have been demonstrated to be extremely effective and are currently used frequently in China to treat ALD either used alone or in combination [62].

Figure 11.6 presents a summary of the relevant biological targets that contribute to the onset of ALD and treated by herbal remedies. The antioxidation, anti-inflammatory, reduction of de novo lipid production, and promotion of mitochondrial fatty acid-oxidation in the liver may all play a role in the therapeutic effects of herbal medications in ALD [63].

11.5.3 Viral Hepatitis

Hepatitis C is a contagious liver infection brought on by the hepatitis C virus. This virus has a persistent infection in over 170 million persons worldwide. Serious consequences including liver cirrhosis, liver cancer, or liver failure are linked to the infection. Hepatitis C has been treated using herbal medications [64].

Individuals with chronic hepatitis were subjected to ten different herbal treatments in ten randomized trials. The data indicates some herbal medications when administered with interferon might have an impact on the elimination of hepatitis C virus (HCV) RNA and on the normalization of liver enzymes, however the current comprehensive analysis identified no substantial antiviral benefit of herbal remedies when compared with placebo. Due to the fact that the majority of the beneficial results came from research studies with poor methodological quality, there is, however, no clear evidence for the usefulness of these herbal remedies for treating chronic hepatitis C. It is not recommended to utilize herbal remedies for infection by the hepatitis C virus outside carefully planned, randomized clinical trials [65].

11.5.4 Drug-Induced Liver Injury (DILI)

Synthetic medicines and other xenobiotics can cause drug-induced liver injury (DILI), a serious consequence that can cause aberrant liver enzyme patterns in clinical tests and ultimately hepatic dysfunction. There are two categories for DILI: intrinsic and idiosyncratic. In both preclinical and clinical models, intrinsic DILI is defined as a medication that causes liver damage that is predictable and dose-dependent. For instance, the medication acetaminophen, which is biotransformed by the hepatocytes and catalyzed by cytochrome P450, produces an extra reactive/toxic intermediate. Intrinsic DILI is caused by three different mechanisms: aberrant mitochondrial activity, oxidative stress, and modified bile acid balance. All of the functions of the cells depend on the production of ATP by mitochondria [66]. Through the integration of signaling networks, mitochondria also play a variety of roles in metabolic processes. Therefore, the mitochondrial will activate the death receptor if there is injury to the signals. As a result, it has been a key target for a medicine to produce mitochondrial malfunction by reducing the level of ATP generation, which eventually results in cell necrosis and functional loss of cells [67]. Individuals' idiosyncratic DILI, which is less frequent, has a more varied appearance and a less distinct dose-toxicity connection [68]. Idiosyncratic DILI is typically unpredictable and uncontrolled, may be dose-related,

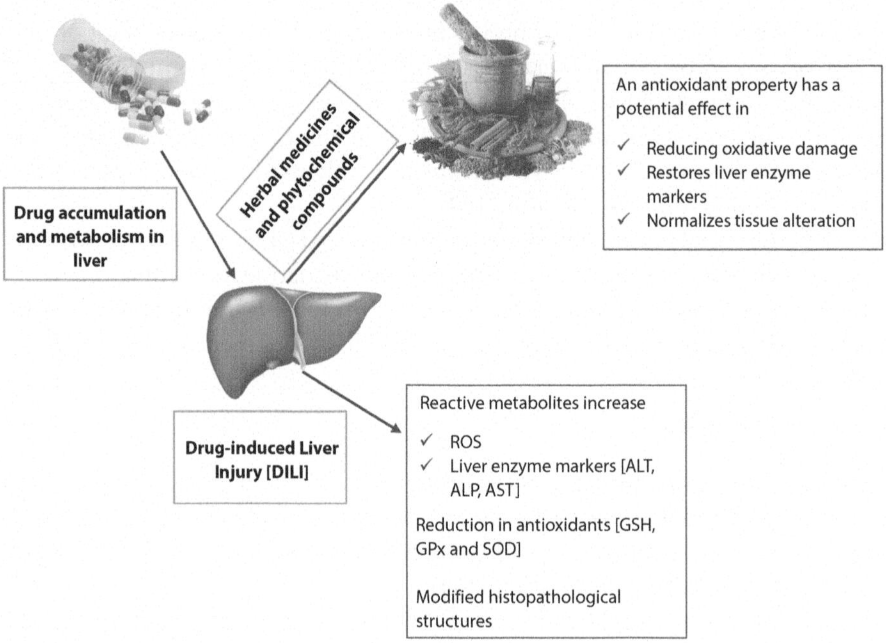

Figure 11.7 Liver damage and herbal plants' preventive effects.

Table 11.1 Anti-hepatotoxic drugs based on herbal plants with their mechanism of action.

Sl. no.	Botanical name (Family)	Solvent (plant parts)	MOA
1.	*Cuscuta campestris* (Cuscutaceae)	Ethanol (Whole plant)	Decreased the level of AST, ALT, cholesterol, triglyceride and MDA increased SOD, GPx, and GSH activities.
2.	*Marrubium vulgare* (Lamiaceae)	Aqueous (Leaves)	Decreases ALT, ALP, AST, LDH and TBARS increases SOD, GPx and CAT.
3.	*Euclea natalensis* (Ebenaceae)	Ethanol (Shoots)	Decreases ALT, and hepatic damage and scavenge free radicals. No alterations in liver cells.
4.	*Mammea africana* (Guttiferae)	Ethanol (Stem bark)	Decreased ALT, ALP, AST, total cholesterol, total and direct bilirubin, increased albumin, total protein, GPx, SOD, GSH and CAT.

and develops after a latent period [69]. Additionally, signs of an allergic idiosyncratic DILI include an adaptive immune response, which includes skin allergies, fever, the production of autoantibodies, and eosinophilia.

Herbs that preserve the liver are frequently used by people as therapeutic plants in case of DILI since they are readily accessible and more affordable than traditional medicinal medications on the market. Their mechanism is given in Figure 11.7.

Recently, studies on herbal plants that prevent drug-induced hepatotoxicity have been done. They are enlisted in Table 11.1.

11.6 Challenges and Limitations in Utilizing Medicinal Plants as Hepatoprotective Agents

11.6.1 Standardization and Quality Control

Because the majority of studies solely consider their antihepatotoxic effects, the current body of literature is insufficient to evaluate the safety of the majority of hepatoprotective and liver-regenerating herbs and herbal products. However, previous studies on rats show that administering 0.5 g/kg of A. membranaceus extracts intraperitoneally for 30 days did not result in any negative side effects, whereas excessive amounts (1 g/kg) of membranaceus extracts of roots caused mutagenicity in mice when administered straight into the stomach lining [70]. In a different investigation, Sato *et al.* found no detectable toxicity of glycyrrhizin at different doses on PLC/PRF/5 cells *in vitro* [71]. When administered to rat liver cells *in vitro* at a concentration of 1 mg/ml, p-Methoxybenzoic acid from *C. spinosa* was reported by Gadgoli and Mishra to be harmless [72]. This substantiates the traditional medical system's assertions. In a similar vein, silymarin has been demonstrated to be nontoxic and free of adverse reactions even at large doses [73]. Although some herbal remedies have the potential to cure both acute and chronic liver problems, premarketing drug testing and pharmacovigilance are still necessary, just like with any other medication. As of now, there is not enough data to support the use of herbal remedies for curing chronic liver illnesses outside of scientific studies [74].

11.6.2 Herb-Drug Interactions

The use of herbs as medicine is widespread throughout the world. Combining herbal and conventional treatment can be risky, especially in comorbid, hepato-insufficient, and frail individuals who are more likely to have hazardous herb-drug interactions (HDI) and neurotoxicity. This study uses data extraction and computational techniques to highlight the possibility of HDI of 15 internationally renowned plant species. It was discovered that several plant species imitate warfarin. Hypoglycemia was brought on by *M. charantia* phytochemicals. *G. biloba* and *M. chamomila* both exhibit anticoagulant properties. *S. hispanica* lowers blood sugar levels after meals. *A. sativum* can increase the clearing of anticancer drugs, whilst *R. officinalis* has been shown to hinder the expulsion of anticancer substrates. *P. ginseng* can affect how blood clots [75]. Numerous herbal metabolites, including ursolic and rosmarinic acid, among others, are likely or probably inhibitors of various CYP450 enzymes, according to a cross-linking of biological molecules and in silico data. So, it is advised to avoid mixing medications with plant species/metabolites that have the same pharmacological properties/metabolizing enzymes as a particular drug. Herbal remedies should only be taken in small amounts for a brief length of time under the guidance of a healthcare professional, even if they are mixed with other medications. This is to prevent any potential negative and dangerous side effects [76].

11.6.3 Dosage and Formulation Issues

Issues in herbal formulation

1. WHO defines herbal dosage forms as products generated from herbs in physical form, such as liquid, solid, or semi-solid goods, containing or not containing excipients, in a specific formulation (for example, decoctions, pills, ointment, etc.).
2. The current toxicological, epidemiological, and other information about herbal formulations is perplexing.
3. The authenticity of botanical ingredients is challenging. Documenting pharmacological, toxicological, and clinical information is laborious.
4. When dealing with herbal formulations, it might be challenging to adhere to pharmacovigilance rules.
5. Research into herb-drug interactions is necessary.
6. It is difficult to ensure consistency, safety, and effectiveness.
7. Clinical trials for herbal products face a number of obstacles [77].

11.7 Future Perspectives and Research Directions

11.7.1 Unexplored Medicinal Plants with Hepatoprotective Properties

1. Crypto-tanshinone

a. **Botanical source:** *Salvia miltiorrhiza* Bunge.
b. **Potential target/mechanism of action:** inhibits the production of sterol regulated component binding protein 1, its nuclear translocation, and the subsequent transactivation of targeted genes associated with fatty acid biosynthesis in dose-dependent manner in primary cultured rat hepatocytes, protecting hepatocytes from lipopolysaccharide- and ethanol-induced cell death.

2. Glycyrrhizin

a. **Botanical source:** Licorice (*Glycyrrhiza sp.)*
b. **Potential target/mechanism of action:** The administration of glycyrrhizin to PLC/PRF/5 cells reduced the release of hepatitis B surface antigen (HBsAg) into the cell culture medium, leading researchers to believe that glycyrrhizin alters hepatocyte surface characteristics and intracellular transport.

c. Glycyrrhizin, when given intraperitoneally to mice, decreases the liver damage caused by lipopolysaccharide and D-galactosamine by reducing inflammatory reactions and the generation of IL-18 [78].

11.7.2 Combination Therapies with Antioxidants and Conventional Medications

Combination therapy is a therapeutic approach that combines a variety of oral antioxidants with one another and, occasionally, with additional pharmacological agents and/or treatment modalities (such as intralesional and physical therapy) in order to provide superior outcomes to single-drug or single-mode treatment. Naturally, this is not a novel concept in medicine as a combination therapy has been employed in the past, for example, in the chemotherapy used to treat cancer and tuberculosis [79]. The present research attempts to make it clear the manner in which the connection among different antioxidants used in therapies for PD may substantially boost the possibility of therapeutic success, despite the fact that we have prior published multiple research papers explaining the use of combined therapy for the treatment of patients with Peyronie's disease (PD) [80].

11.7.3 Nanotechnology and Delivery Systems for Enhanced Efficacy

Since ancient times, herbal treatments have been utilized extensively around the world. Clarification of the chemical makeup and biological properties of a number of therapeutic plant products has been made possible by advances in phytochemical and phytopharmacological studies. The availability of active chemicals determines the efficacy of numerous varieties of medicinal plants. The majority of an extract's biologically active components, that include flavonoids, tannins, and terpenoids, easily dissolve in water but poorly absorbed due to their inability to pass through cell membrane lipids, their excessively large molecular size, or their poor absorption, which reduces their bioavailability and effectiveness. These barriers prevent some extracts from being used in clinical settings [81]. It has been suggested that herbal medicine and nanotechnology be combined since nano structured systems may be able to enhance the benefits of plant extracts, lowering dosage requirements, minimizing adverse effects, and enhancing activity. Nanosystems are able to route the active ingredient to the intended site of action during the whole treatment time while delivering it at a sufficient concentration. These prerequisites are not met by conventional therapy. This study's objective is to review herbal remedies and drug delivery methods based on nanotechnology [82].

11.8 Conclusion

From the above studies, we have come to know about the hepatoprotective potential of antioxidants derived from medicinal plants and the importance of these natural compounds in preventing and curing liver diseases. We have also learnt about the mechanisms causing liver damage and the role of oxidative stress in it supported with various experimental models and clinical study designs. This have also helped us to gain insight into various problems faced while using the medicinal plants as hepatoprotective agents. Researchers, medical professionals, and anyone looking for alternative and complementary methods to liver health and disease prevention will find the material offered here to be useful. Further research that elaborate and elucidate the actions of the antioxidants mentioned in this chapter are welcomed.

References

1. Oyinloye, B.E., Osunsanmi, F.O., Ajiboye, B.O., Ojo, O.A., Kappo, A.P., Modulatory effect of methanol extract of Piper guineense in CCl4-induced hepatotoxicity in male rats. *Int. J. Environ. Res. Public Health*, 14, 9, 955, 2017.
2. Madrigal-Santillán, E., Madrigal-Bujaidar, E., Álvarez-González, I., Sumaya-Martínez, M.T., Gutiérrez-Salinas, J., Bautista, M., Morales-González, A., y González-Rubio, M.G., JL, Aguilar-Faisal, Morales-González, J.A., Review of natural products with hepatoprotective effects. *World J. Gastroenterol. WJG*, 20, 40, 14787, 2014.
3. Okaiyeto, K., Nwodo, U., Mabinya, L., Okoh, A., A review on some medicinal plants with hepatoprotective effects. *Pharmacogn. Rev.*, 12, 24, 186–99, 2018.
4. Corsini, A. and Bortolini, M., Drug-induced liver injury: The role of drug metabolism and transport. *J. Clin. Pharmacol.*, 53, 5, 463–74, 2013.
5. Asrani, S.K., Larson, J.J., Yawn, B., Therneau, T.M., Kim, W.R., Underestimation of liver-related mortality in the United States. *Gastroenterology*, 145, 2, 375–82, 2013.
6. Ong, H.C., Faezah, A.W., Milow, P., Medicinal plants used by the jah hut orang asli at kampung pos penderas, Pahang, Malaysia. *Stud. Ethnomed.*, 6, 1, 11–5, 2012.
7. Shah, M.D., D'souza, U.J., Iqbal, M., The potential protective effect of Commelina nudiflora L. against carbon tetrachloride (CCl4)-induced hepatotoxicity in rats, mediated by suppression of oxidative stress and inflammation. *Environ. Health Prev. Med.*, 22, 1, 1–9, 2017.
8. Miller, J.K., Brzezinska-Slebodzinska, E., Madsen, F.C., Oxidative stress, antioxidants, and animal function. *J. Dairy Sci.*, 76, 9, 2812–23, 1993.
9. Abd Ellah, M.R., The role of liver biopsy in detection of hepatic oxidative stress. *Vet. Med. Int.*, 12, 4, 347–9, 2011.
10. Birben, E., Sahiner, U.M., Sackesen, C., Erzurum, S., Kalayci, O., Oxidative stress and antioxidant defense. *World Allergy Organ. J.*, 5, 9–19, 2012.
11. Tounsi, N., Djerdjouri, B., Bouzid, C., Bentabak, K., Correlation of adenosine deaminase operating under nitro-oxidative stress with tumor and vascularization in patients with advanced gallbladder carcinoma. *J. Appl. Biomed.*, 17, 3, 175–83, 2019.
12. Srinivas, U.S., Tan, B.W., Vellayappan, B.A., Jeyasekharan, A.D., ROS and the DNA damage response in cancer. *Redox Biol.*, 25, 101084, 2019.
13. Halliwell, B. and Gutteridge, J., Oxygen toxicity, oxygen radicals, transition metals and disease. *Biochem. J.*, 219, 1, 1, 1984.
14. Ranneh, Y., Ali, F., Akim, A.M., Hamid, H.A., Khazaai, H., Fadel, A., Crosstalk between reactive oxygen species and pro-inflammatory markers in developing various chronic diseases: a review. *Appl. Biol. Chem.*, 60, 3, 327–38, 2017.
15. Wu, D. and Cederbaum, A.I., Alcohol, oxidative stress, and free radical damage. *Alcohol Res. Health*, 27, 4, 277, 2003.
16. Nimse, S.B. and Pal, D., Free radicals, natural antioxidants, and their reaction mechanisms. *RSC Adv.*, 5, 35, 27986–8006, 2015.
17. Halliwell, B., Biochemistry of oxidative stress. *Biochem. Soc. Trans.*, 35, 5, 1147–50, 2007.
18. Sehwag, S. and Das, M., Antioxidant activity: an overview. *J. Food Sci. Technol.*, 2, 3, 1–1, 2013.

19. Pandey, K.B. and Rizvi, S.I., Markers of oxidative stress in erythrocytes and plasma during aging in humans. *Oxid. Med. Cell. Longev.*, 3, 1, 2–12, 2010.
20. Shih, P.H., Yeh, C.T., Yen, G.C., Anthocyanins induce the activation of phase II enzymes through the antioxidant response element pathway against oxidative stress-induced apoptosis. *J. Agric. Food. Chem.*, 55, 23, 9427–35, 2007.
21. Marklund, S.L., Westman, N.G., Lundgren, E., Roos, G., Copper-and zinc-containing superoxide dismutase, manganese-containing superoxide dismutase, catalase, and glutathione peroxidase in normal and neoplastic human cell lines and normal human tissues. *Cancer Res.*, 42, 5, 1955–61, 1982.
22. Krishnamurthy, P. and Wadhwani, A., Antioxidant enzymes and human health. *Antioxid. Enzyme*, 1, 3–18, 2012.
23. Schröder, K., Vecchione, C., Jung, O., Schreiber, J.G., Shiri-Sverdlov, R., van Gorp, P.J., Busse, R., Brandes, R.P., Xanthine oxidase inhibitor tungsten prevents the development of atherosclerosis in ApoE knockout mice fed a Western-type diet. *Free Radical Biol. Med.*, 41, 9, 1353–60, 2006.
24. Castilla, P., Dávalos, A., Teruel, J.L., Cerrato, F., Fernández-Lucas, M., Merino, J.L., Sánchez-Martín, C.C., Ortuño, J., Lasunción, M.A., Comparative effects of dietary supplementation with red grape juice and vitamin E on production of superoxide by circulating neutrophil NADPH oxidase in hemodialysis patients. *Am. J. Clin. Nutr.*, 87, 4, 1053–61, 2008.
25. Porter, N.A., Caldwell, S.E., Mills, K.A., Mechanisms of free radical oxidation of unsaturated lipids. *Lipids*, 30, 4, 277–90, 1995.
26. Zhao, F., Tang, Y.Z., Liu, Z.Q., Protective effect of icariin on DNA against radical-induced oxidative damage. *J. Pharm. Pharmacol.*, 59, 12, 1729–32, 2007.
27. German, J.B., Food processing and lipid oxidation. *Impact Process. Food Saf.*, 459, 23–50, 1999.
28. Shi, H., Noguchi, N., Niki, E., Comparative study on dynamics of antioxidative action of α-tocopheryl hydroquinone, ubiquinol, and α-tocopherol against lipid peroxidation. *Free Radical Biol. Med.*, 27, 3-4, 334–46, 1999.
29. Decramer, M., Rutten-van Mölken, M., Dekhuijzen, P.R., Troosters, T., van Herwaarden, C., Pellegrino, R., van Schayck, C.O., Olivieri, D., Del Donno, M., De Backer, W., Lankhorst, I., Effects of N-acetylcysteine on outcomes in chronic obstructive pulmonary disease (Bronchitis Randomized on NAC Cost-Utility Study, BRONCUS): a randomised placebo-controlled trial. *Lancet*, 365, 9470, 1552–60, 2005.
30. Chandra, P., Sharma, R.K., Arora, D.S., Antioxidant compounds from microbial sources: A review. *Food Res. Int.*, 129, 108849, 2020.
31. Aziz, M.A., Diab, A.S., Mohammed, A.A., Antioxidant categories and mode of action. *Antioxidants*, 5, 3–22, 2019.
32. Yang, Y., Wang, Y., Wei, S., Wang, X., Zhang, J., Effects and Mechanisms of Non-Thermal Plasma-Mediated ROS and Its Applications in Animal Husbandry and Biomedicine. *Int. J. Mol. Sci.*, 2;24, 21, 15889, 2023.
33. Gupta, A. and Sharma, B., Oxidative stress biomarkers in a living cell. *MOJ Toxicol.*, 7, 1, 38–43, 2023.
34. Tarin, M., Babaie, M., Eshghi, H., Matin, M.M., Saljooghi, A.S., Elesclomol, a copper-transporting therapeutic agent targeting mitochondria: from discovery to its novel applications. *J. Transl. Med.*, 20;21, 1, 745, 2023.
35. Parveen, S., Gupta, P., Kumar, S., Banerjee, M., Lipid polymer hybrid nanoparticles as potent vehicles for drug delivery in cancer therapeutics. *Med. Drug Discov.*, 30, 100165, 2023.
36. Guo, X., Wu, M., Deng, Y., Liu, Y., Liu, Y., Xu, J., Redox-Responsive Lipidic Prodrug Nano-Delivery System Improves Antitumor Effect of Curcumin Derivative C210. *Pharmaceutics*, 19;15, 5, 1546, 2023.

37. Aziz, M., Saeed, F., Ahmad, N., Ahmad, A., Afzaal, M., Hussain, S., Mohamed, A.A., Alamri, M.S., Anjum, F.M., Biochemical profile of milk thistle (Silybum Marianum L.) with special reference to silymarin content. *Food Sci. Nutr.*, 9, 1, 244–50, 2021.
38. Anthony, K.P. and Saleh, M.A., Free radical scavenging and antioxidant activities of silymarin components. *Antioxidants*, 2, 4, 398–407, 2013.
39. Jiang, G., Sun, C., Wang, X., Mei, J., Li, C., Zhan, H., Liao, Y., Zhu, Y., Mao, J., Hepatoprotective mechanism of Silybum marianum on nonalcoholic fatty liver disease based on network pharmacology and experimental verification. *Bioengineered*, 13, 3, 5216–35, 2022.
40. Tamayo, C. and Diamond, S., Review of clinical trials evaluating safety and efficacy of milk thistle (Silybum marianum [L.] Gaertn.). *Intgr. Cancer Ther.*, 6, 2, 146–57, 2007.
41. Hewlings, S.J. and Kalman, D.S., Curcumin: A review of its effects on human health. *Foods*, 6, 10, 92, 2017.
42. Cichoż-Lach, H. and Michalak, A., Oxidative stress as a crucial factor in liver diseases. *World J. Gastroenterol.: WJG*, 20, 25, 8082, 2014.
43. Farzaei, M.H., Zobeiri, M., Parvizi, F., El-Senduny, F.F., Marmouzi, I., Coy-Barrera, E., Naseri, R., Nabavi, S.M., Rahimi, R., Abdollahi, M., Curcumin in liver diseases: a systematic review of the cellular mechanisms of oxidative stress and clinical perspective. *Nutrients*, 10, 7, 855, 2018.
44. Kaushik, S., Masand, N., Iyer, M.R., Patil, V.M., Preclinical to Clinical Profile of Curcuma longa as Antidiabetic Therapeutics. *Curr. Top. Med. Chem.*, 23, 1, 2267–76, 2023.
45. Aggarwal, B.B., Kumar, A., Bharti, A.C., Anticancer potential of curcumin: preclinical and clinical studies. *Anticancer Res.*, 23, 1/A, 363–98, 2003.
46. Bagalkotkar, G., Sagineedu, S.R., Saad, M.S., Stanslas, J., Phytochemicals from Phyllanthus niruri Linn. and their pharmacological properties: a review. *J. Pharm. Pharmacol.*, 58, 12, 1559–70, 2006.
47. Orhan, D.D., Orhan, N., Ergun, E., Ergun, F., Hepatoprotective effect of Vitis vinifera L. leaves on carbon tetrachloride-induced acute liver damage in rats. *J. Ethnopharmacol.*, 112, 1, 145–51, 2007.
48. Ezzat, M.I., Okba, M.M., Ahmed, S.H., El-Banna, H.A., Prince, A., Mohamed, S.O., Ezzat, S.M., In-depth hepatoprotective mechanistic study of Phyllanthus niruri: *In vitro* and *in vivo* studies and its chemical characterization. *PLoS One*, 15, 1, e0226185, 2020.
49. Krithika, R., Mohankumar, R., Verma, R.J., Shrivastav, P.S., Mohamad, I.L., Gunasekaran, P., Narasimhan, S., Isolation, characterization and antioxidative effect of phyllanthin against CCl4-induced toxicity in HepG2 cell line. *Chem. Biol. Interact.*, 181, 3, 351–8, 2009.
50. Koh, P.H., Mokhtar, R.A., Iqbal, M., Andrographis paniculata ameliorates carbon tetrachloride (CCl4)-dependent hepatic damage and toxicity: Diminution of oxidative stress. *Redox Rep.*, 16, 3, 134–43, 2011.
51. Rastogi, R., Saksena, S., Garg, N.K., Kapoor, N.K., Agarwal, D.P., Dhawan, B.N., Picroliv Protects Against Alcohol-Induced Chronic Hepatotoxicity in Rats1. *Planta Med.*, 62, 03, 283–5, 1996.
52. Shukla, B., Visen, P.K., Patnaik, G.K., Dhawan, B.N., Choleretic effect of picroliv, the hepatoprotective principle of Picrorhiza kurroa1. *Planta Med.*, 57, 01, 29–3, 1991.
53. Visen, P.K., Shukla, B., Patnaik, G.K., Kaul, S., Kapoor, N.K., Dhawan, B.N., Hepatoprotective activity of picroliv, the active principle of Picrorhiza kurrooa, on rat hepatocytes against paracetamol toxicity. *Drug Dev. Res.*, 22, 3, 209–19, 1991.
54. Asl, M.N. and Hosseinzadeh, H., Review of pharmacological effects of Glycyrrhiza sp. and its bioactive compounds. *Phytother. Res.: Int. J. Devoted Pharmacol. Toxicol. Eval. Nat. Prod. Deriv.*, 22, 6, 709–24, 2008.
55. Kumar, A. and Dora, J., Review on Glycyrrhiza glabra (Liquorice). *J. Pharm. Sci. Innov. (JPSI)*, 1, 2, 1–4, 2012.

56. Xu, Y., Guo, W., Zhang, C., Chen, F., Tan, H.Y., Li, S., Wang, N., Feng, Y., Herbal medicine in the treatment of non-alcoholic fatty liver diseases-efficacy, action mechanism, and clinical application. *Front. Pharmacol.*, 11, 601, 2020.
57. Li, H., Ying, H., Hu, A., Hu, Y., Li, D., Therapeutic effect of gypenosides on nonalcoholic steatohepatitis via regulating hepatic lipogenesis and fatty acid oxidation. *Biol. Pharm. Bull.*, 40, 5, 650–7, 2017.
58. Ding, J., Zhang, B., Wang, P.J., He, G.N., Wei, D.M., Ding, J.L., Wang, B.C., Analysis on mechanisms and medication rules of herbal prescriptions for nonalcoholic fatty liver disease based on methods of data mining and biological information. *Zhongguo Zhong yao za zhi= Zhongguo zhongyao zazhi= China J. Chin. Materia Med.*, 44, 8, 1689–95, 2019.
59. Marra, F. and Lotersztajn, S., Pathophysiology of NASH: perspectives for a targeted treatment. *Curr. Pharm. Des.*, 19, 29, 5250–69, 2013.
60. Choi, J.Y., Kwon, E.Y., Choi, M.S., Elucidation of the metabolic and transcriptional responses of an oriental herbal medicine, bangpungtongseong-san, to nonalcoholic fatty liver disease in diet-induced obese mice. *J. Med. Food*, 22, 9, 928–36, 2019.
61. Albano, E., Oxidative mechanisms in the pathogenesis of alcoholic liver disease. *Mol. Aspects Med.*, 29, 1-2, 9–16, 2008.
62. Gao, B. and Bataller, R., Alcoholic liver disease: pathogenesis and new therapeutic targets. *Gastroenterology*, 141, 5, 1572–85, 2011.
63. Choo, Q.L., Kuo, G., Weiner, A.J., Overby, L.R., Bradley, D.W., Houghton, M., Isolation of a cDNA clone derived from a blood-borne non-A, non-B viral hepatitis genome. *Science*, 244, 4902, 359–62, 1989.
64. Farci, P., Choo, Q.L., Kuo, G., AJ, Weiner, LR, Overby, DW, B., Houghton, M., Isolation of a cDNA clone derived from a blood-borne non-A, non-B viral hepatitis genome [Science 1989; 244: 359–362]. *J. Hepatol.*, 36, 5, 582–5, 2002.
65. Aleo, M.D., Luo, Y., Swiss, R., Bonin, P.D., Potter, D.M., Will, Y., Human drug-induced liver injury severity is highly associated with dual inhibition of liver mitochondrial function and bile salt export pump. *Hepatology*, 60, 3, 1015–22, 2014.
66. Will, Y. and Dykens, J., Mitochondrial toxicity assessment in industry–a decade of technology development and insight. *Expert Opin. Drug Metab. Toxicol.*, 10, 8, 1061–7, 2014.
67. Chalasani, N.P., Hayashi, P.H., Bonkovsky, H.L., Navarro, V.J., Lee, W.M., Fontana, R.J., Practice Parameters Committee of the American College of Gastroenterology ACG Clinical Guideline: the diagnosis and management of idiosyncratic drug-induced liver injury. *Off. J. Am. Coll. Gastroenterol. ACG*, 109, 7, 950–66, 2014.
68. Hussaini, S.H. and Farrington, E.A., Idiosyncratic drug-induced liver injury: an overview. *Expert Opin. Drug Saf.*, 6, 6, 673–84, 2007.
69. Zhang, J., Xie, X., Li, C., Fu, P., Systematic review of the renal protective effect of Astragalus membranaceus (root) on diabetic nephropathy in animal models. *J. Ethnopharmacol.*, 126, 2, 189–96, 2009.
70. Sato, H., Goto, W., Yamamura, J.I., Kurokawa, M., Kageyama, S., Takahara, T., Watanabe, A., Shiraki, K., Therapeutic basis of glycyrrhizin on chronic hepatitis B. *Antiviral Res.*, 30, 2-3, 171–7, 1996.
71. Gadgoli, C. and Mishra, S.H., Antihepatotoxic activity of p-methoxy benzoic acid from Capparis spinosa. *J. Ethnopharmacol.*, 66, 2, 187–92, 1999.
72. Upadhyay, G., Kumar, A., Singh, M.P., Effect of silymarin on pyrogallol-and rifampicin-induced hepatotoxicity in mouse. *Eur. J. Pharmacol.*, 565, 1-3, 190–201, 2007.
73. Stickel, F., Patsenker, E., Schuppan, D., Herbal hepatotoxicity. *J. Hepatol.*, 43, 5, 901–10, 2005.
74. Suroowan, S., Abdallah, H.H., Mahomoodally, M.F., Herb-drug interactions and toxicity: Underscoring potential mechanisms and forecasting clinically relevant interactions induced

by common phytoconstituents via data mining and computational approaches. *Food Chem. Toxicol.*, 156, 112432, 2021.

75. Colalto, C., Herbal interactions on absorption of drugs: Mechanisms of action and clinical risk assessment. *Pharmacol. Res.*, 62, 3, 207–27, 2010.
76. Preussmann, R., *Hepatocarcinogens as potential risk for human liver cancer*, pp. 11–29, MTP Press, Lancaster, UK, 1978.
77. Ali, M., Khan, T., Fatima, K., Ali, Q.U., Ovais, M., Khalil, A.T., Ullah, I., Raza, A., Shinwari, Z.K., Idrees, M., Selected hepatoprotective herbal medicines: Evidence from ethnomedicinal applications, animal models, and possible mechanism of actions. *Phytother. Res.*, 32, 2, 199–215, 2018.
78. Burnett, A.L., Nitric oxide induces oxidative stress and mediates cytotoxicity to human cavernosal cells in culture. *Int. J. Impotence Res.*, 13, 6, 360–65, 2001.
79. Paulis, G., Barletta, D., Turchi, P., Vitarelli, A., Dachille, G., Fabiani, A., Gennaro, R., Efficacy and safety evaluation of pentoxifylline associated with other antioxidants in medical treatment of Peyronie's disease: a case-control study. *Res. Rep. Urol.*, 1–10, 2015.
80. Roy, M., Roy, A., Rustagi, S., Pandey, N., An Overview of Nanomaterial Applications in Pharmacology. *Biomed Res. Int.*, 1–23, 2023.
81. Bonifacio, B.V., da Silva, P.B., Ramos, M.A., Negri, K.M., Bauab, T.M., Chorilli, M., Nanotechnology-based drug delivery systems and herbal medicines: a review. *Int. J. Nanomed.*, 1-5, 2014.
82. Kumar, M., Keshwania, P., Chopra, S., Mahmood, S., Bhatia, A., Therapeutic potential of nanocarrier-mediated delivery of phytoconstituents for wound healing: their current status and future perspective. *AAPS PharmSciTech.*, 20;24, 6, 155, 2023.

12

Antioxidant Effects of Medicinal Plants for the Treatment of Epilepsy

Sayed Mohammed Firdous*, Shouvik Mallik and Bijoy Paria

Department of Pharmacology, Calcutta Institute of Pharmaceutical Technology & AHS, Uluberia, Howrah, West Bengal, India

Abstract

Epilepsy is a chronic neurological disorder characterized by recurrent seizures, affecting millions of individuals worldwide. Despite the availability of various antiepileptic drugs (AEDs), a substantial number of patients continue to experience inadequate seizure control and adverse effects. Consequently, there is a growing interest in exploring alternative treatment options, including medicinal plants, to complement conventional therapies and enhance patient outcomes. This book chapter aims to provide an overview of the antioxidant effects of medicinal plants in epilepsy. Oxidative stress, resulting from an imbalance between reactive oxygen species (ROS) production and the body's antioxidant defense system, has been implicated in the pathogenesis and progression of epilepsy. Medicinal plants rich in natural antioxidants offer a promising avenue for managing epilepsy by modulating oxidative stress. The chapter begins by discussing the fundamental concepts of epilepsy and oxidative stress, highlighting the intricate interplay between these two phenomena. It then delves into the diverse classes of medicinal plants known for their antioxidant properties and potential application in epilepsy management. Various plant extracts and phytochemicals, including flavonoids, phenolics, terpenoids, and alkaloids, are examined for their ability to scavenge free radicals, inhibit lipid peroxidation, and enhance endogenous antioxidant defenses.

Furthermore, the chapter explores the mechanistic insights behind the antioxidant effects of medicinal plants, such as the activation of the nuclear factor erythroid 2-related factor 2 (Nrf2) pathway, modulation of mitochondrial function, and inhibition of pro-inflammatory cytokines. The potential synergy between antioxidant compounds and conventional AEDs is also discussed, highlighting the prospect of combination therapies. Finally, the chapter addresses the challenges and future directions in the field of medicinal plants as antioxidants in epilepsy. Standardization of herbal preparations, identification of active compounds, and rigorous clinical trials are crucial for establishing their safety, efficacy, and optimal therapeutic strategies.

In summary, this book chapter provides a comprehensive exploration of the antioxidant effects of medicinal plants in epilepsy. It emphasizes the potential of these natural compounds as adjunctive therapies, opening new avenues for personalized treatment approaches that can improve seizure control and enhance the overall well-being of individuals living with epilepsy.

**Corresponding author*: firdous.oncology@gmail.com

Rakesh K. Sindhu, Inderbir Singh and M. Arockia Babu (eds.) Antioxidants: Nature's Defense Against Disease, (441–490)

***Keywords*:** Epilepsy, reactive oxygen species, antioxidant, phytochemicals, nuclear factor erythroid 2-related factor 2

12.1 Introduction

Epilepsy is a neurological condition that has severe neurobiological, psychological, cognitive, and social repercussions. Repeated seizures result from overexcitation of a particular set of neurons and produce a variety of momentary clinical symptoms and laboratory results. The creation of the epileptogenic foci, which has uncontrolled hyperexcitability due to partially sustained depolarization of cellular membranes, is fundamental to the development of the condition [1]. Prolonged depolarization occurs when there is an imbalance in the polarization of the neural system, favoring hyperpolarizing forces. This imbalance can be attributed to increased excitation, reduced inhibition, and other alterations within the neurological system. This elucidates the fundamental process underlying the onset and spread of a seizure in an individual without any pre-existing conditions [2]. There exists empirical data suggesting that the generation of reactive species or the diminished functionality of antioxidant systems may contribute to various manifestations of epilepsy, as well as an elevated propensity for recurrent epileptic episodes.

12.1.1 Background

There is data indicating that oxidative stress is a significant factor in acute neurological injuries, including persistent seizures, stroke, and traumatic brain damage. Additionally, it has been implicated in the development of neurodegenerative diseases such as Parkinson's and Alzheimer's disease. Undoubtedly, the occurrence of oxidative stress during and following seizures plays a crucial role in both immediate and prolonged excitotoxic neuronal demise [3]. Oxidative stress arises as a consequence of an equilibrium in the oxidation state, which is caused by an excessive generation of free radicals or their byproducts, such as reactive oxygen species (ROS), as well as a reduction in the quantities or functionality of cellular antioxidant mechanisms [4]. Oxidative stress can therefore be reduced by inhibiting the production of ROS, by the use of exogenous antioxidants or by promoting endogenous antioxidant systems, in particular through the activation of the transcription factor nuclear factor erythroid 2-related factor 2 (Nrf2). Our new findings demonstrate that the suppression of epilepsy development in rat models of temporal lobe epilepsy (TLE) may be achieved by activating Nrf2 through the regulation of Kelch ECH associating protein 1 (KEAP1), either alone or in conjunction with an exogenous antioxidant [5]. It was postulated that a potentially more efficacious approach would involve the simultaneous activation of endogenous antioxidant mechanisms and inhibition of seizure-induced reactive oxygen species (ROS) generation. The major intracellular sources of reactive oxygen species (ROS) are mitochondria and enzymes that produce ROS, such as NADPH oxidase (NOX). In earlier studies, it was postulated that mitochondria play a significant role in the production of reactive oxygen species (ROS) during epileptiform activity [6]. Nevertheless, new findings from our research team and other scientific groups have demonstrated that NOX serves as the primary catalyst for the formation of reactive oxygen species (ROS) during seizure-like activity. This process has been found to play a crucial role in the progression of neurodegeneration and the occurrence of cell death observed in several animal models of epilepsy [7]. The presence of persistent oxidative stress in acquired and hereditary epilepsy implies that the implementation of an efficient antioxidant approach may provide disease-modifying benefits in established epilepsy. In this study, we demonstrate that the combination of NOX inhibition and the enhancement of cellular antioxidant defenses by boosting Nrf2 availability is an effective approach for avoiding epileptogenesis and altering the course of chronic epilepsy [7].

12.1.2 Significance of Antioxidant in Epilepsy Treatment

Epilepsy, a chronic neurological disease that is common in adults and children alike, can be a serious problem. Antiepileptic drugs (AED) are available in a variety of forms, but 30% of the population still has seizures. The monotherapy treatment is ineffective for this group [8]. These individuals require a more assertive strategy since the use of a single therapeutic intervention is ineffective in managing epileptic seizures. The efficacy of polytherapy might be limited. The significance of adverse responses, namely neurological disturbances, mental and behavioral modifications, and metabolic abnormalities, is becoming more prominent. Hence, there exists a pressing demand for the development of antiepileptic drugs (AEDs) that are more tolerated and have enhanced efficacy, particularly within this specific cohort of individuals [9]. In the context of metabolic processes, a considerable number of antiepileptic drugs (AEDs), particularly those belonging to the older generation of AEDs such as PB, phenytoin (PHT), carbamazepine (CBZ), and valproic acid (VPA), generate reactive metabolites. These metabolites have the ability to form covalent bonds with various endogenous macromolecules or enhance the production of reactive oxygen species (ROS), leading to oxidative harm and subsequent toxicity. In order to substantiate this notion, considerable research has been conducted to assess the impact of epilepsy and antiepileptic drugs (AEDs) on the production of free radicals. These studies consistently indicate a potential association between either epilepsy or AEDs and the development of oxidative stress [10, 11]. Hence, this review provides a comprehensive overview of the underlying justification for including antioxidants as supplementary treatment in order to regulate the impact of free radical-induced oxidative stress in epilepsy. Additionally, it assesses the current body of research that examines the effectiveness of various antioxidant treatments.

12.2 Epilepsy: An Overview

12.2.1 Definition and Classification

A "clinical epileptic seizure" is described as abnormally synchronized neuronal activity in the brain resulting in temporary clinical indications or symptoms [12]. The elements of an epileptic seizure are

(1) Mode of onset and termination,
(2) Abnormally enhanced synchrony, and
(3) Clinical manifestations [12].

Given that it is a temporary event, a seizure should have a distinct beginning and end. Because of the postictal interval, the clinical cessation of the seizure is usually less evident. Clinical indicators might vary greatly, but a clinical seizure must have at least one clinical sign. The most recent update to this definition of a clinical epileptic seizure occurred in 2005, although the International League Against Epilepsy (ILAE) modified seizure classifications in 2014, this definition was not changed.

"Epilepsy" is defined as a disorder in which a person has

(1) Two or more unprovoked or reflex seizures that are more than 24 hours apart,
(2) A single unprovoked or reflex seizure in a person who has a 60% chance of having another seizure during the next ten years, or
(3) An epilepsy syndrome [13].

This definition of "epilepsy" was codified in 2014. The previous definition of "epilepsy" (from 2005) included only the first criterion noted here. Regarding the second criterion, the two things that classify a person as having a 60% risk of having a second seizure in the next 10 years are

(1) Findings on brain imaging that have epileptogenic potential or
(2) Epileptiform activity on an electroencephalogram (EEG).

A finding on brain imaging that is abnormal but does not have epileptiform potential (e.g., an arachnoid cyst, diffuse atrophy) coupled with a single unprovoked seizure would not meet the criteria for a diagnosis of epilepsy. The cut-off of 60% was chosen because this is the lower limit of the confidence interval for someone with two unprovoked seizures to have a third seizure [14]. The definition of "epilepsy" was modified to reflect clinical practice. Epileptologists commonly prescribed ASMs to patients who had a single seizure and an EEG showing epileptiform potentials, but they also had to inform the patient that they did not formally have epilepsy (yet). Given this discrepancy, it was decided that the definition should be changed. This type of patient can now be advised that they have epilepsy and need therapy. This change in the definition of "epilepsy" also reduced confusion among those unfamiliar with epilepsy about when to treat seizures with medications, increasing the likelihood that a patient would receive seizure prevention treatment after a single seizure when they were known to be at high risk for further seizures.

The Long-Term Prognosis of Epilepsy (PRO-LONG) study compared patients with epilepsy diagnosed using the 2014 diagnosis to those diagnosed using the old definition. This study found that those identified under the new definition had an 83.6% chance of having seizures again after 10 years [15]. Based on the findings of the Beretta *et al.* study, it is obvious that failing to diagnose persons with epilepsy after their first seizure who had an abnormal study (e.g., EEG, MRI, computed tomography [CT]) was a mistake. The delay in epilepsy diagnosis led in

(1) Patient noncompliance with seizure drugs due to a lack of comprehension of the possibility of having another seizure and
(2) Physician failure to treat.

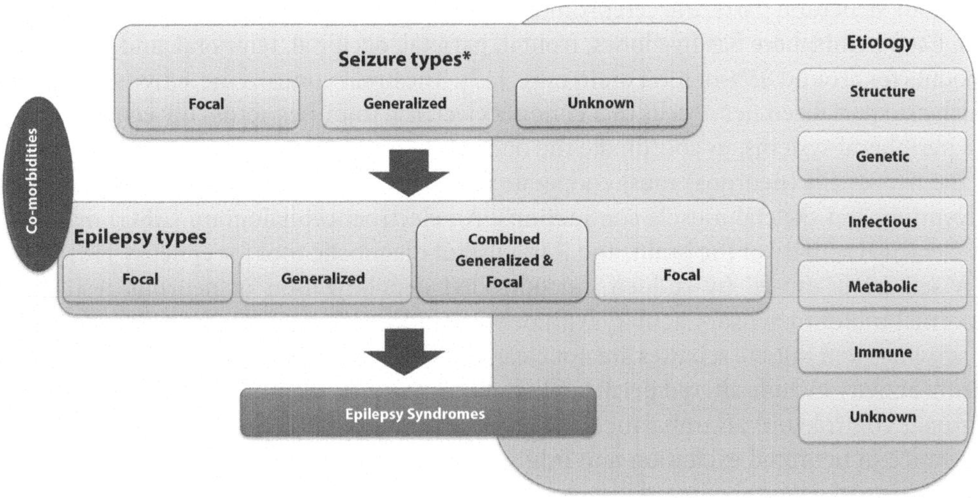

Figure 12.1 ILAE classification of epilepsies.

An additional change to the definition of "epilepsy" in 2014 was to define "epilepsy" as a disease instead of a disorder, as it was previously defined. This change was made as it was felt that patients could better understand a "disease" than a "disorder," and that they would take the diagnosis of a "disease" more seriously. Entities such as heart disease, diabetes, hypothyroidism, etc. are diseases, not disorders. Because epilepsy should be considered with similar gravity, it was renamed as a disease. Notably, there was some concern that defining epilepsy as a disease would increase stigma for patients, but it was felt that the benefits resulting from the change outweighed this potential concern.

Finally, the definition of "epilepsy" was changed to allow for its resolution. Previously, if you had epilepsy, you always had epilepsy. According to the current definition, epilepsy is considered resolved if

(1) No seizures have occurred for at least 10 years and
(2) No medications to treat epilepsy have been taken for at least the previous 5 years.

Epilepsy is also deemed resolved if the person has an age-dependent epilepsy syndrome and is over the age at which it resolves.

According to the ILAE Epilepsy is classified into 4 types (Focal, Generalized, Combined epilepsy), and seizure types are divided into 2 types (Focal and Generalized) (Figure 12.1).

12.2.2 Pathophysiology of Epilepsy

Seizures occur when there is abnormal synchronous neuronal activity in a portion of the brain or throughout the brain, when networks are irregularly established, or when networks are disrupted by a structural, viral, or metabolic disturbance. The most prevalent causes of seizures in children are hereditary, perinatal traumas, and cortical development abnormalities [16]. The brain is a delicate organ that is shielded by the stiff bone of the skull. It is made up of four structural parts, the cerebrum, the diencephalon, the brainstem, and the cerebellum. Each hemisphere has five lobes, frontal, parietal, occipital, temporal, and insular, and accounts for around 83% of total brain mass [17]. The brain controls the activity of neurons, which are specialized nerve cells that conduct electrical impulses across the central nervous and peripheral systems, to coordinate the operations of the body.

The nerve cells (neurons) must cooperate in order to control cognitive activity, sensory perception, and skeletal muscle contractions. An electroencephalogram (EEG) may assess the electrical activity of the brain, and it is a useful diagnostic tool for epilepsy [18].

A seizure is caused by sudden, uncontrolled depolarization of neurons resulting in abnormal motor or sensory activity with or without loss of consciousness. The exact cellular activities that initiate seizures are not clear, however, theories of specific mechanisms for seizure activity include altered permeability of the neuronal membrane, reduced inhibitory neuronal control, and an imbalance in neurotransmitters [19]. A seizure is caused by an imbalance in neuronal excitation and inhibition caused by hyperexcitation and hypersynchronization of a neural network.

Changes in ions and neurotransmitters cause hyperexcitation, and the loss of inhibitory neurons causes hyper synchronicity, which can cause a partial or generalized seizure depending on the spread of activity [21].

Because of hypoxia, alkalosis, hypoglycemia, and aberrant neurotransmitter characteristics, neuronal membrane properties might change, causing huge amounts of neurotransmitters to be released at the synapse and therefore promoting seizure. Clinical symptoms of aberrant neuronal discharge include changes in motor function, sensory, and autonomic function, awareness, and behavior [20].

The neurons in the epileptogenic center have a lower stimulation threshold and are hyperexcitable. Irritable neurons are quickly activated by physiological changes, and situations like as exhaustion or lack of sleep, stress, fever, and constipation may lower the seizure threshold. The focused cells excite surrounding normal cells, causing the activity to spread [19]. Physical stressors such as loud noise and bright light can cause epilepsy in some persons, as can metabolic stimuli such as excessive fluid retention, hypoglycemia, medication changes, alkalosis, and abrupt withdrawal from sedatives and alcohol [19].

The interval immediately following an epileptic seizure is known as the postictal state, and it can involve headache, disorientation, dysphasia, memory loss, short-term paralysis, or profound sleep [20].

12.3 Oxidative Stress in Epilepsy

Oxidative stress arises as a result of an elevated load of oxidants that surpasses the capability of the body's inherent antioxidants and repair mechanisms. Alternatively, it can occur due to a reduction in the body's inherent antioxidants and repair mechanisms, which are unable to adequately counterbalance the typical load of oxidants [22]. Glutamate, functioning as an excitatory neurotransmitter in the central nervous system (CNS) and potentially exhibiting excitotoxic effects, particularly at elevated levels, is considered to be a significant contributor to the formation of oxidative stress [23]. The heightened production of reactive oxygen species (ROS) leads to an augmented level of intracellular calcium ions (Ca^{2+}), which is observed in alterations of neuroplasticity and the occurrence of neuronal death produced by seizures, either via necrosis or apoptosis [24]. Elevated levels of intracellular calcium, which persist even throughout the chronic stage of epilepsy and are thus essential for the perpetuation of recurring seizures, might impact the recycling of GABA A receptors and hence modify neuronal excitability [25]. Furthermore, the elevation of intracellular calcium ions (Ca^{2+}) has the potential to modulate several cellular processes, including but not limited to gene transcription, protein production and degradation, neurogenesis, neuronal sprouting, and other essential physiological activities of cells [26].

12.3.1 Mitochondrial Oxidative Stress and Epilepsy

12.3.1.1 Mitochondrial ROS Homeostasis

Mitochondria serve as the primary locations for the generation of reactive oxygen species (ROS) within the cellular environment. This is primarily due to the substantial production of superoxide (O2) as a byproduct of various biochemical processes occurring within the mitochondria, such as electron leakage from the electron transport chain (ETC) during mitochondrial oxidative phosphorylation (OXPHOS) and the tricarboxylic acid (TCA) cycle. The prevailing consensus in the scientific community is that the respiratory chain is the primary source of reactive oxygen species (ROS) production within mitochondria. It has long been established that two of the respiratory chain complexes, specifically complexes I (CI) and CIII, are implicated in the generation of O^{2-} [27]. Mitochondrial reactive oxygen species (mtROS) are also generated by several proteins and complexes inside the mitochondrial matrix, such as enzymes involved in the tricarboxylic acid (TCA) cycle, namely aconitase, pyruvate dehydrogenase, and α-ketoglutarate dehydrogenase [28, 29]. Furthermore, a multitude of biochemical events occurring inside mitochondria have the capability to generate reactive oxygen species (ROS). These reactions involve enzymes such as glycerol-3-phosphate dehydrogenase, cytochrome P450, monoamine oxidase, and cytochrome b5 reductase. The close proximity of mitochondrial proteins, lipids, and DNA to reactive oxygen species (ROS) suggests that they are likely to be the main targets of oxidative damage in times of stress. This leads to the formation of a detrimental cycle of mitochondrial free radical injury. Furthermore, the regulation of mitochondrial redox state is not only dependent on the formation of mitochondrial reactive oxygen species (mtROS), but is also influenced by the presence of mitochondria-associated antioxidant defenses. In mitochondria, SOD2, peroxiredoxin 3 (Prdx3), and thioredoxin 2 (Trx2) often provide antioxidative functions [30, 31]. Physiologically, the maintenance of mitochondrial reactive oxygen species (mtROS) equilibrium is achieved by the regulation of both free radical production and scavenging mechanisms. However, in the presence of pathological diseases such as epilepsy, there is an increase in the generation of reactive oxygen species (ROS), particularly in the mitochondria of the brain. This excessive ROS production leads to oxidative damage.

12.3.2 Role of Mitochondrial Oxidative Stress in Epilepsy

12.3.2.1 mtDNA Damage

Mitochondria possess an independent genome known as mtDNA, with a total of 37 genes. This genetic material is exclusively transmitted from the maternal lineage. After exposure to reactive oxygen species (ROS), DNA has the potential to undergo damage [32]. In contrast to nucleus DNA (nDNA), mitochondrial DNA (mtDNA) is more susceptible to damage because it lacks the protective histones and is located in close proximity to the electron transport chain, which is the primary source of O^{2-}generation [33]. The direct evidence is derived from the observation that the levels of oxidized bases in mitochondrial DNA (mtDNA) are two to three times higher compared to nuclear DNA (nDNA). The encoding of essential components involved in oxidative phosphorylation (OXPHOS) by mitochondrial DNA (mtDNA) is widely recognized. This process enables mitochondria to deliver energy and maintain the equilibrium of mitochondrial reactive oxygen species (mtROS). The presence of oxidative mitochondrial DNA (mtDNA) damage is a contributing factor to impairments in oxidative phosphorylation (OXPHOS), resulting in inadequate cellular energy supply. This, in turn, leads to an excessive generation of reactive oxygen species (ROS), so establishing a self-perpetuating loop [34]. Oxidative damage to mitochondrial DNA (mtDNA), specifically targeting components of the electron transport chain (ETC), is believed to have a role in the development and advancement of several mitochondrial diseases in humans [35]. The prevailing forms of mitochondrial DNA (mtDNA) damage that contribute to the generation of reactive oxygen species (ROS) mostly consist of point mutations resulting from oxidative alterations of purines and pyrimidines, as well as gene deletions. Two significant mutagenesis outcomes occur as a result of reactive oxygen species (ROS) attack, the generation of 8-hydroxyguanine (8-OH-G) and structural damage, including single and double strand breaks [36]. This suggests that mitochondrial DNA (mtDNA) is especially susceptible to oxidative damage caused by reactive oxygen species (ROS).

Assessment of mitochondrial DNA (mtDNA) damage can also be conducted by the examination of thymine glycol (TG) and 8-hydroxydeoxyguanosine (8-OHdG), which are both indicative of oxidative lesions in DNA [37]. The accurate assessment of oxidative DNA damage is generally based on the ratio between steady-state levels of 8-OHdG and TG. Previous studies have shown that there is a time-dependent elevation in 8-OHdG/2dG in mitochondria after the occurrence of epileptic seizures produced by kainic acid (KA-). Likewise, these modifications exhibit a correlation with the generation of mitochondrial oxygen and are accompanied by a temporary reduction in the repair enzyme 8-oxoguanine glycosylase (Ogg1) responsible for repairing damaged mitochondrial DNA [38.]

12.3.2.2 Mitophagy

Yoshinori Ohsumi got the Nobel Prize in Physiology or Medicine in 2016 in recognition of his contributions to the understanding of the molecular mechanisms underlying autophagy [39]. Mitophagy, a kind of autophagy, refers to the targeted elimination of malfunctioning or surplus mitochondria, a process crucial for the preservation of mitochondrial equilibrium and cellular viability. The most widely recognized regulators associated with mitophagy are autophagy-related 32 protein (Atg32), NIP3-like protein X (NIX, sometimes referred to as BNIP3L), parkin, and PTEN-induced potential kinase protein 1 (PINK1) [40]. Research has demonstrated that it may be triggered by several stresses, including as hypoxia, food deprivation, and heightened oxidative phosphorylation activity. It is specifically induced by reactive oxygen species (ROS) in the context of nerve growth factor deficiency, rapamycin treatment, or starvation [41]. As noted above, mitochondria are considered ROS-generating organelles. On the one hand, induction of mitophagy by mtROS under physical conditions is beneficial for eliminating damaged mitochondria. On the other hand, excessive ROS-triggered oxidative stress due to mitochondrial dysfunction promotes alteration and leads to onset of multiple diseases. The brain is particularly susceptible to oxidative stress due to the high amount of oxygen consumption and lipid-rich content [42]. Hence, it can be characterized by either uncontrolled or insufficient mitophagy, as a consequence of mitochondrial abnormalities induced by oxidative stress, and possesses significant implications for the development of neurological disorders, including epilepsy. After newborn seizures, it has been seen that the expression of mitophagy markers PINK, dynamin-related protein 1 (Drp1), and polyhydroxybutyrate is considerably reduced. However, the administration of leptin has been found to effectively restore aberrant mitophagy and alleviate cognitive deficits associated with seizures [43].

12.3.2.3 Apoptosis

As previously discussed, oxidative stress plays a crucial role in the activation of many cellular death mechanisms, including apoptosis [44]. The mitochondria are known to have significant involvement in the process of apoptosis, mostly due to their heightened susceptibility to oxidative stress. Research has demonstrated that the initiation of the mitochondrial transition pore results in the liberation of proapoptotic substances, such as cytochrome c, into the cytoplasm [45]. The aforementioned procedure subsequently triggers the caspase-dependent mitochondrial pathway, ultimately enhancing the expression of apoptotic protease-activating factor-1 (Apaf-1) and procaspase-9, so promoting the assembly of the apoptosome [46]. Multiple variables have been documented to have a role in regulating mitochondrial oxidative stress and cellular apoptosis in epilepsy. During epileptiform activity generated by low levels of Mg^{2+}, there is an observed increase in mitochondrial Ca^{2+} buildup, which in turn stimulates the synthesis of NADH and reactive oxygen species (ROS) [47]. The process of mitochondrial fission has been found to be associated with oxidative stress, apoptosis, and several neurological disorders. In the rat model of epilepsy induced by pilocarpine, there is a notable increase in mitochondrial fission. However, when the mitochondrial fission protein Drp1 is inhibited by the compound mitochondrial division inhibitor 1 (mdivi-1), there is an enhancement in superoxide dismutase (SOD) activity. Additionally, the expressions of cytochrome c and caspase-3 are reduced, and there is an increase in neuron survival [48]. According to these findings, there is a direct correlation between the start of epileptic convulsions and mitochondrial fission, and inhibiting mitochondrial fission prevents neuronal death by blocking the mtROS/cytochrome c pathway. Results from the Stokman research group showed that loss of NLRX1 increased oxidative stress and apoptosis in epithelial cells during ischemia-reperfusion injury. The mitochondria-localized receptor, NLRX1, was also previously discovered to protect against mitochondrial damage and epithelial cell apoptosis in an oxidative stress-dependent manner [49]. It suggests that NLRX1 plays a crucial role in the regulation of mitochondrial activity and the avoidance of oxidative stress and apoptosis in tissue damage, while it is yet unknown whether NLRX1 promotes oxidative stress-associated apoptosis in the brain mitochondria after epilepsy.

12.3.2.4 Inflammation

The relationship between the reactive oxygen species (ROS) and the inflammatory response appears to be interconnected. Research has demonstrated that oxidative stress-induced mitochondrial dysfunction, which is related to reactive oxygen species (ROS), contributes to the exacerbation of inflammation in airway smooth muscle and cardiac tissue [50]. The mitigation of inflammation and the facilitation of organ recovery can be achieved by the focused scavenging of mitochondrial reactive oxygen species (ROS) using MitoVitE or MitoQ. These mitochondria-targeted antioxidants inhibit oxidative stress, restore proteasomal function, decrease mitochondrial uptake of Tfr iron, and ultimately help preserve mitochondrial cell permeability. These mitochondria-targeted anti-oxidants inhibit oxidative stress, restore proteasomal function, decrease the uptake of Tfr iron into mitochondria and ultimately help preserve the mitochondrial cell permeability [51]. The findings suggest that mitochondrial oxidative stress plays a significant role in contributing to tissue inflammation. After the occurrence of a reactive oxygen species (ROS) burst inside the mitochondria, the NACHT, LRR, and PYD domains-containing protein 3 (NLRP3) inflammasome may become activated. This activation then initiates the synthesis of the proinflammatory cytokine IL-1β, which can result in the impairment of organ function. The restoration of mitochondrial activity has the ability to inhibit the formation of reactive oxygen species (ROS) and reverse the inflammatory response [52]. The activation of NLRP3 by mtROS may be attributed to the oxidative impact on mtDNA, which results in the capacity to induce inflammation. Furthermore, the imbalance in the NAD+/NADH redox system caused by the decrease in Complex I (CI) activity also promotes the activation of the NLRP3 inflammasome [53]. This implies that the occurrence of mitochondrial oxidative damage contributes to the development of a pro-inflammatory condition and the initiation of illness etiopathogenesis. While there is currently no conclusive data to directly support the role of mitochondrial oxidative stress in contributing to inflammation in epilepsy, it is worth noting that therapy with CoQ10 has demonstrated clear neuroprotective effects against pentylenetetrazol- (PTZ-) induced kindling. These effects are believed to be attributed to the suppression of oxidative damage and neuroinflammation [54].

12.3.3 Implication of Oxidative Stress in Epileptogenesis

ROS/RNS, neurodegeneration, neurogenesis, and the reduction of the seizure threshold are widely acknowledged as the primary indicators of epileptogenesis (Figure 12.2) [55].

The primary source of data on this process is derived mostly from experimental models of partial complex seizures and postmortem examinations of individuals diagnosed with temporal lobe epilepsy. Currently, the most accurate models of epileptogenesis in rats are those generated by kainate or pilocarpine. Epileptogenesis is defined as the seizure-free interval that occurs between the initial damage to the brain and the occurrence of the first spontaneous seizure. During this period, a series of brain alterations occur, ultimately resulting in the occurrence of spontaneous recurring seizures. The latent phase, which can last for several years, provides an opportunity for therapeutic intervention to avoid or mitigate the onset of seizures. The primary precipitating events commonly observed in individuals diagnosed with temporal lobe epilepsy are traumatic brain injury, stroke, intracerebral infections, and intoxications [56]. Epileptogenesis is characterized by significant neuronal changes, such as increased production of reactive oxygen species (ROS) and reactive nitrogen species (RNS), as well as the release of pro-inflammatory mediators. These alterations contribute to neurodegeneration, abnormal neurogenesis, reorganization of surviving neuronal circuits, maladaptive changes in dendritic plasticity, and the sprouting of axons within excitatory pathways.

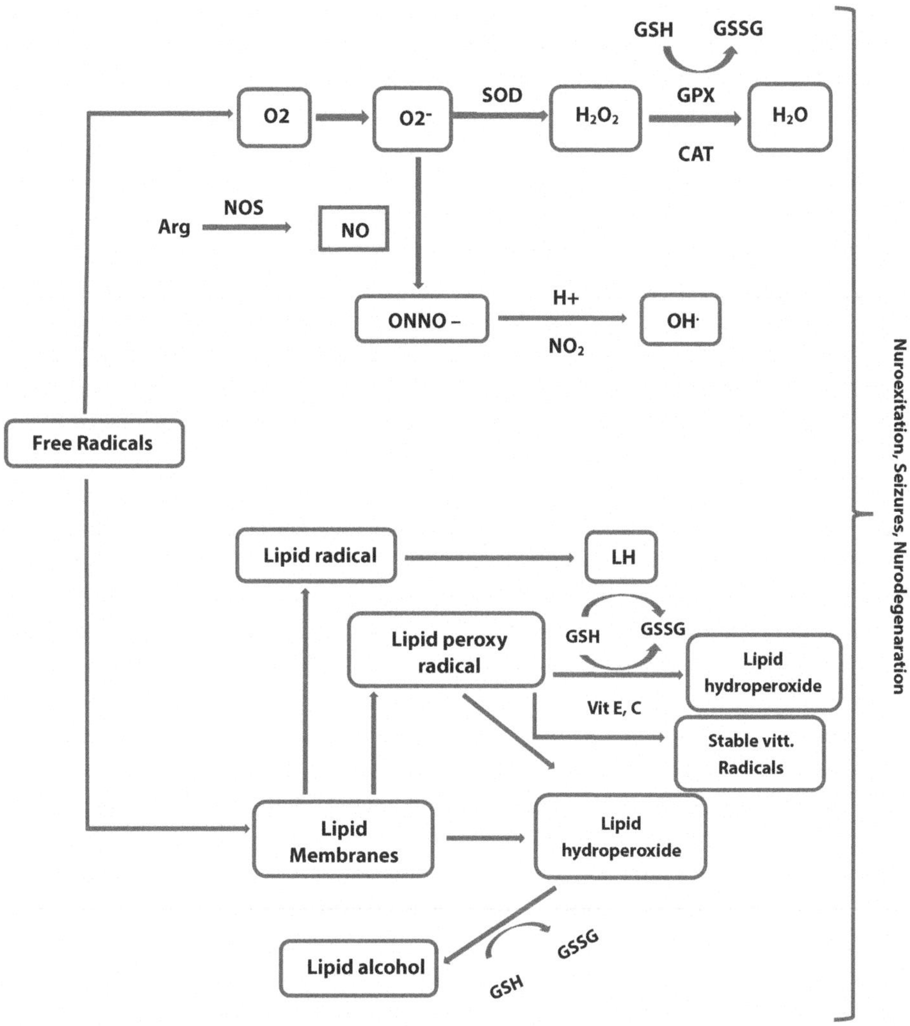

Figure 12.2 Summary of the mechanisms involved in oxidative stress.

Additionally, there is a loss of inhibitory interneurons, reactive gliosis, damage to the blood-brain barrier, and reconfiguration of the extracellular matrix and neuronal cytoarchitecture. Consequently, heightened brain excitability ultimately gives rise to neuronal hypersynchronization, escalated epileptiform spiking, and the occurrence of spontaneous recurring seizures [57]. The mouse model of temporal lobe epilepsy caused by Theiler's murine encephalomyelitis virus provided evidence of a strong association between two key factors in the development of epilepsy, neuroinflammation and an imbalance in redox status. Neuroinflammation is characterized by severe astrogliosis and microgliosis. Astrocytes, when stimulated by nuclear factor erythroid regulated factor 2 (Nrf2), a transcription factor located in the cytoplasm, generate antioxidants including superoxide dismutase, catalase, glutathione peroxidase, and GSH (Glutathione) [58]. The occurrence of seizures is accompanied by impaired functioning of glutamate transporters (GLT-1, GLAST) in astrocytes, which is well recognized as a significant contributing factor to the development of epilepsy in humans. The excess glutamate is then released extracellularly via a process that is reliant on calcium. Glutamine synthase is responsible for the conversion of glutamate into glutamine. Following the occurrence of a first seizure, there is a temporary elevation in the activity of glutamine synthase in nonreactive astrocytes. However, this activity subsequently declines in reactive astrocytes throughout the latent phase of epileptogenesis. The presence of elevated levels of unmetabolized glutamate in the extracellular space leads to heightened neuronal excitability and the occurrence of spontaneous recurring seizures . Furthermore, reactive astrocytes play a significant role in the production of proinflammatory mediators, including IL-1β, IL-6, IL-10, TNFα, and interferons. The significant contribution of pleiotropic effects of reactive oxygen species/reactive nitrogen species (ROS/RNS) and proinflammatory cytokines to the pathogenesis of epilepsy is well-established [59]. Interleukin-1β (IL-1β) and tumor necrosis factor α (TNFα) have the capacity to modify excitability by influencing the expression and activity of receptors. Interleukin-1 beta (IL-1β) has been shown to induce the activation of N-methyl-D-aspartate (NMDA) receptors, whilst tumor necrosis factor-alpha (TNFα) has been seen to modulate the functioning of α-amino-3-hydroxy-5-methyl-4-isoxazolepropionic The augmentation of calcium ion (Ca2+) influx via glutamate receptors is a contributing factor to both neuronal hyperexcitability and degeneration [60]. Interleukin-6 (IL-6), fibrinogen, and high-sensitivity C-reactive protein (hsCRP) were detected in the serum, cerebrospinal fluid, and brain tissues of individuals diagnosed with epilepsy. The pilocarpine model of status epilepticus in rats exhibited elevated levels of IL-1β, IL-6, COX-2, TNFα, and the inflammatory transcription factor nuclear factor kappa-light chain-enhancer of activated B cells (NFκB) in the hippocampus and piriform cortex regions. The activation of NF-κB, a crucial factor in neural plasticity and memory functions, occurs in response to pro-inflammatory cytokines such as IL-1β, IFN-γ, and TNF-α, as well as heightened excitatory synaptic transmission [61]. Notably, there was a strong correlation observed between alterations in cytokine expression and seizure activity. The concentrations of pro-inflammatory mediators exhibited a peak immediately following the first shock, subsequently reverted to baseline levels during the latent period, and subsequently increased once again during the phase characterized by chronic seizures. This observed trend aligns with alterations in reactive oxygen species (ROS) levels, the extent of mitochondrial impairment, and the occurrence of apoptotic cellular demise. Conversely, apoptosis exacerbates inflammatory reactions [62]. Fyn, a member of the Src family of tyrosine kinases (SFKs), is a neuroinflammatory mediator that plays a role in regulating cell proliferation. The experimental findings demonstrated that the inhibition of the Fyn pathway resulted in a reduction in inflammation and the formation of reactive nitrogen species (RNS) in the hippocampus. The efficacy of Saracatinib, a selective Src family kinase (SFK) inhibitor, in the prevention of epileptogenesis, following kainic acid-induced status epilepticus was seen in a study involving rats, with a success rate of 50%. In the remaining animal subjects, the intervention resulted in a decrease in the occurrence of spontaneous recurring seizures and epileptiform spiking. Thyroid hormones also have a crucial function in the processes of epileptogenesis and neuroregeneration. Molecular investigations have provided data indicating that a deficiency in thyroid hormones leads to mitochondrial damage, induces oxidative stress, and hampers the functionality of GABAergic neurons [63]. Astrocytes play a crucial role in the formation and maintenance of the blood-brain barrier. Astrocytes respond to brain damage by increasing the synthesis of matrix metalloproteinase (MMP-9), an endopeptidase. This heightened production of MMP-9 has the effect of further compromising the integrity of the blood-brain barrier, resulting in increased permeability to fluid and blood proteins. Consequently, albumins stimulate astrocytes by means of the p38 MAPK signaling pathway, leading to an increased production of inflammatory mediators. In contrast, astrocytes, due to their proximity to synaptic terminals and blood arteries, fulfill a supportive role for neurons by actively removing excess synaptic glutamate and potassium, while also supplying them with glucose [64]. The primary responsibility of nonreactive microglia, which is the resident macrophages in the brain, is the engulfment of cell debris. Following the first seizure event, microglia play a role in many processes including neuroinflammation, neuroprotection, neurogenesis, and neurodegeneration. These substances possess neurotrophic, anti-inflammatory, and cytotoxic properties, while also stimulating the Nox2 complex and enhancing the generation of reactive oxygen species (ROS) (Figure 12.2). Activated microglial cells migrate toward the site of damage and secrete tissue plasminogen activator (tPA), which is a recognized indicator of chronic epilepsy and plays a role in the phenomenon of mossy fiber sprouting. During the process of epileptogenesis, it has been observed that microglia undergo activation around 30 minutes after the occurrence of an initial insult, thereafter migrating towards the synaptic terminals. In this context, the individuals are actively involved in the process of dendritic pruning, which refers to the elimination of unnecessary connections between neurons. Additionally, they contribute to the preservation of neuronal viability by generating trophic agents, which are substances that promote cell growth and survival [65]. The initial form of epilepsy shown to have a connection with oxidative stress is known as myoclonic epilepsy with ragged red fibers. The causal cause of this phenomenon is a mutation occurring in mitochondrial DNA, which subsequently results in an elevation of reactive oxygen species (ROS) levels. This increase in ROS leads to damage in the respiratory chain and a subsequent reduction in adenosine triphosphate (ATP) production [66]. In preclinical investigations, it was revealed that the pilocarpine and kainate models of temporal lobe epilepsy had an elevated concentration of reactive oxygen species/reactive nitrogen species (ROS/RNS) and a diminished action of glutathione (GSH) [66]. In the brains of rats, the occurrence of status epilepticus produced by kainate was seen to be accompanied by the manifestation of tricyclic acid cycle problems. These disorders were characterized by a decrease in the availability of reducing components such as NADH, NAD cytochrome c reductase, and FADH2, as well as decreased synthesis of ATP. Furthermore, the administration of kainic acid leads to an elevation in the influx of intracellular calcium ions, hence triggering the activation of several enzymes, such as nitric oxide synthase (NOS). The initiation of induced nitrative stress resulted in a significant lipid peroxidation process, leading to eventual harm to both cellular membranes and organelles. In addition to hippocampus apoptosis, there were observed changes in gene expression and protein synthesis. In the present study, an increase in the expression of nNOS/iNOS and a biphasic alteration in the levels of nitric oxide and 3-nitrotyrosine (3NT) in the hippocampus were detected within the same model. The concentrations of NO and 3NT exhibited a quick rise following the induction of status epilepticus (SE). Subsequently, these concentrations returned to their baseline levels two days after the SE event. However, they increased once again and remained elevated for a duration of six weeks, which coincided with the epileptogenic phase. The presence of surviving neurons was found [67]. Posttraumatic epilepsy is a condition characterized by the occurrence of seizures following a traumatic brain injury. One of the factors that might trigger seizures in this condition is the extravasation of blood, which leads to the release of Fe^{3+} from red blood cells. This release of Fe^{3+} stimulates the creation of reactive oxygen species, specifically superoxide, hydroxyl, and hydrogen peroxide. The process of lipid peroxidation generated by ROS (reactive oxygen

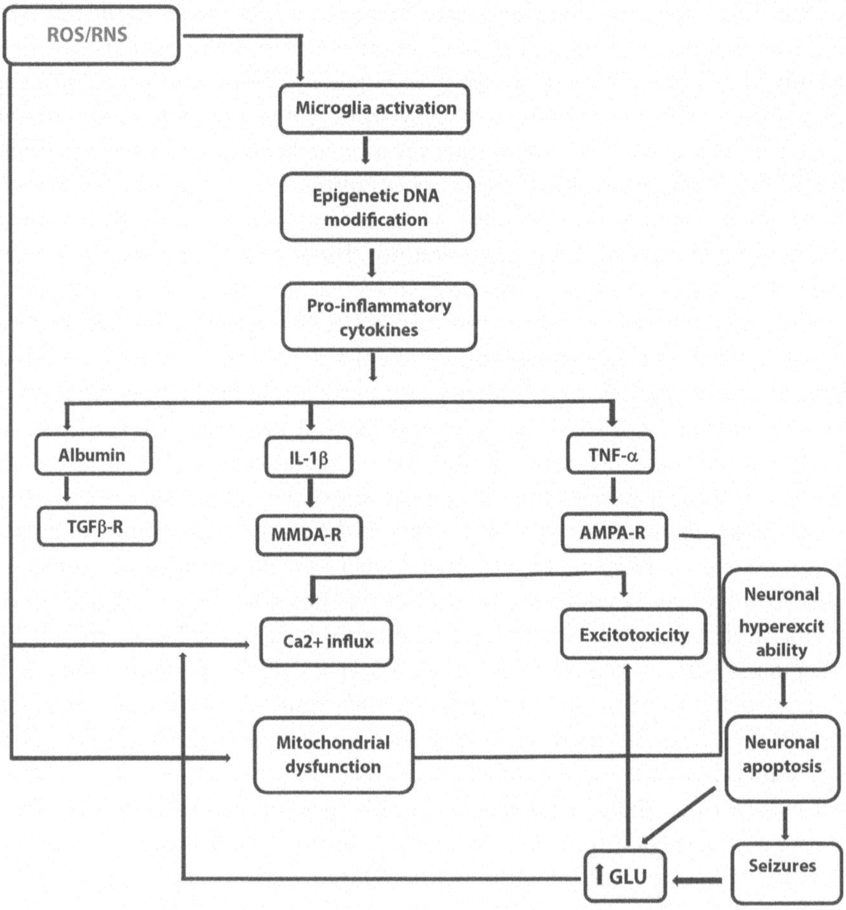

Figure 12.3 The role of oxidative stress and neuroinflammation in the etiology of seizures.

species) leads to the breakdown of cellular membranes, which subsequently triggers an elevated release of excitatory amino acids. In addition, hydroxyl radicals have been seen to stimulate the synthesis of methylguanidine, a naturally occurring convulsant that is produced from creatinine. Posttranslational modifications caused by ribonucleic acid signaling (RNS) may also play a role in the pathogenesis of posttraumatic epilepsy (Figure 12.3) [68].

12.4 Mechanisms of Antioxidant Effects in Medicinal Plants

12.4.1 Scavenging of Reactive Oxygen Species (ROS)

Since ROS are extremely hazardous, they can interact with lipids, proteins, and nucleic acids in cells to induce lipid peroxidation, membrane degradation, and enzyme inactivation, which can have an impact on a variety of physiological processes as well as cell viability. However, plants have developed a wide range of sophisticated defense mechanisms to maintain low ROS levels and prevent the negative effects of overly high ROS concentrations [69]. In plants, different cellular enzymatic and nonenzymatic systems quickly detoxify ROS. When there is a severe imbalance between the generation of ROS and the antioxidative defense, oxidative stress results [70]. Different soluble (ASH, GSH) and membrane (a-tocopherol) chemicals, as well as enzymes like SOD(Superoxide dismutases), APX(Ascorbate Peroxidase), GPX(Glutathione peroxidase), CAT(Catalase), MDHAR(Monodehydroascorbate reductase), DHAR(Dehydroascorbate reductase), and glutathione reductase (GR)) make up the antioxidant defense system [69].

12.4.1.1 Enzymatic Antioxidants

The most powerful intracellular enzymatic antioxidants are SODs, which are present in all aerobic species and subcellular compartments susceptible to oxidative stress caused by ROS. By dismutating superoxide to H_2O_2, superoxide dismutases serve as the first line of defense against the harmful effects of ROS [71, 72]. The upregulation of SODs is associated with the prevention of oxidative stress brought on by biotic and abiotic stressors and is essential for plant survival in adverse conditions [71]. Significant increase in SOD activity under salt stress has been observed in various plants, namely, mulberry [73], *Cicer arietinum* [74], and *Lycopersicon esculentum* [75], The effects of salt and drought stress on Glycyrrhiza uralensis Fisch were studied, and it was shown that SOD activity dramatically increased. However, under salt stress alone, an extra Mn-SOD isoenzyme was also found. 54j 3 Generation and Scavenging of Reactive Oxygen Species in Plants under Stress *Hordeum vulgare* [76], *A. thaliana* [77], *Oryza sativa* [78], *Triticum aestivum* [79], *Brassica juncea* [80], *Vigna mungo* and *C. arietinum* [81] have all shown increased SOD activity after Cd treatment.

Similarly, increased SOD activity following drought stress was also observed in three cultivars of *Phaseolus vulgaris* [82], *Alternanthera philoxeroides* [83] and *O. sativa* [84]. Following H2O2 scavenging, APX, GPX, and CAT are all involved, however, unlike CAT, APX needs an ascorbate and GSH regeneration pathway called the ascorbate-glutathione cycle. By oxidizing ascorbate to MDA, which can then be renewed by MDA reductase (MDAR) using NAD(P)H as reducing equivalents, APX detoxifies H_2O_2 to H_2O. Dehydroascorbate can spontaneously dismutate from MDA. Dehydroascorbate reductase (DHAR), which is activated by the oxidation of GSH to GSSG, mediates ascorbate regeneration. Finally, by employing NAD(P)H as a reducing agent, GR may regenerate GSH from GSSG [85]. Although the activity of the chloroplastic APX increased in plants under mild drought stress compared to control-grown plants, it decreased under more severe drought stress. Similar to APX, GPX detoxifies H_2O_2 to H_2O as well, but it does so by using GSH as a reducing agent directly. The GPX cycle is completed when GR regenerates GSH from GSSG(Oxidized glutathione) [85]. Abiotic stress increases GPX activity in cultivars of *Capsicum annuum* plants [86], but decreases in roots and causes no significant change in the leaves of Cdexposed *Pisum sativum* plants [87]. When under stress, tetrameric heme-containing enzymes called CATs are essential for ROS detoxification because they have the ability to dismutate H_2O_2 directly into H_2O and O_2 [88]. CAT is important in the removal of H_2O_2 generated in peroxisomes by oxidases involved in b-oxidation of fatty acids, photorespiration, and purine catabolism [85]. Unpredictable response of CAT activity under metal stress has been reported by various researchers. Under Cd stress, CAT activity declined in *Glycine max* [87], *Phragmites australis* [88], *C. annuum* [86], and *A. thaliana* [89], whereas its activity increased in *O. sativa* [78], *B. juncea* [80], *T. aestivum* [79], *C. arietinum* [81] and

V. mungo roots Under non-heat shock circumstances, pretreatment of rice seedlings with H2O2 increased CAT activity and shielded rice seedlings from ensuing Cd stress. According to reports, under salt stress, a sizable rise in CAT activity was seen in C. arietinum leaves [91] and roots [74]. It was reported that a decrease in CAT activity in *A. doliolum* under NaCl and Cu^{2+} stress [90, 92]. The levels of superoxide, H_2O_2, and hydroxyl radicals in a cell determine the severity of oxidative stress. Therefore, controlling the quantities of hazardous ROS in a cell will depend on the balance of SOD, APX, and CAT activities. The balance of scavenging enzymes can be altered to bring about compensatory mechanisms. Scavenging enzymes, such as APX and GPX, were increased in plants, for instance, when CAT activity was decreased. There may also be unanticipated outcomes [72]. Plants missing both APX and CAT were less vulnerable to oxidative stress than plants with reduced CAT. Because these plants' photosynthetic activity was reduced, a drop in APX and CAT could prevent ROS generation in the chloroplasts [72]. Additionally found in mitochondria and peroxisomes, MDHAR coupled with APX scavenges H_2O_2 there [93]. MDHAR activity was shown to be increased in Cd-exposed Pinus sylvestris and decreased in Cd-exposed poplar hybrids (*Populusx Canescens*) [94]. DHAR restores ASH from its oxidized state and controls the cellular ASH redox state, which is essential for allowing cells to withstand a variety of abiotic stressors without producing ROS [85]. Additionally, it has been discovered that DHAR overexpression increases plants' resistance to a variety of abiotic stressors. Plants overexpressing DHAR demonstrated Al stress resistance by maintaining high levels of ASH and drought and ozone stress tolerance in tobacco. Guaiacol peroxidase (GPOX) produces indole-3-acetic acid (IAA) and it plays an important role in the production of lignin and by scavenging H_2O_2defense against biotic stresses [85]. GPOX mainly prefer electron donors that are aromatic such as guaiacol, and pyrogallol, these are usually oxidizes ascorbate at the rate of around 1% that of guaiacol [95]. Depending on plant species and stress conditions GPOX's activity varies considerably. It was discovered that it increased in Cd-exposed plants of *T. aestivum* [96], *A. thaliana* [97], and *C. demersum* [98].

Flavoprotein oxidoreductase (GR), a possible enzyme of the ASH-GSH cycle, maintains the decreased status of GSH, which is crucial for the defense mechanism against ROS. GSH and GR play a crucial role in determining the tolerance of a plant under various stresses. In *C. annuum* [80], *A. thaliana* [99], *V. mungo* [80], *T. aestivum* [79], and *B. juncea* [80], it was discovered that GR activity increased in the presence of Cd. It is known that plant GSTs play roles in the detoxification of herbicides, hormone homeostasis, anthocyanin vacuolar sequestration, tyrosine metabolism, hydroperoxide detoxification, regulation of apoptosis, and responses to biotic and abiotic stressors. Under salinity stress, it was observed, an increased GST activity in roots of *L. esculentum* plants. When the salinity tolerance of drought-tolerant sorghum cv [75].

12.4.1.2 Nonenzymatic Antioxidants

The most prevalent, potent, and water-soluble antioxidant that reduces or eliminates ROS damage to plants is ascorbic acid [100]. Because of its capacity to contribute electrons in a variety of enzymatic and nonenzymatic activities, ASH is regarded as a highly potent ROS scavenger. By immediately scavenging the O_2 and OH as well as generating α-tocopherol from the tocopherol radical, it can preserve membranes [85]. Mutants with altered glutathione content [101] or decreased ascorbic acid levels [102] are hypersensitive to stress.

Ascorbate is oxidized to MDA and DHA whereas GSH is oxidized by ROS to generate GSSG. GSSG, MDA, and DHA can be decreased by the ascorbate-glutathione cycle, regenerating GSH and ascorbate [72]. Plants respond to chilling, heat shock, pathogen attack, and drought stress by increasing the activity of GSH biosynthetic enzymes and GSH levels [103]. It was reported that during Cd exposure, oxidized ascorbate content was increased in *H. vulgare* plants [104]. Under conditions of strong light and dryness, it was shown that *P. asperata* seedlings had a significantly higher ASH content [105]. The UV-B stress in *C. auriculata* seedlings was said to dramatically enhance the amount of ASH and DHA as well as GSH/GSSG. On the other hand, during Cd stress, a decrease in ASH content was also noticed in the roots and nodules of *G. max* [87]. Additionally, Cd reduced the ASH level in *Cucumis sativus* chloroplast, *A. thaliana*, and *P. sativum* leaves, but not in *Populusx Canescens* roots [106]. GSH is required to keep cells in their normal reduced condition in order to combat the debilitating effects of oxidative stress brought on by ROS [107]. It is a potential scavenger of O_2, H_2O_2 [108], and the most dangerous ROS such as OH [109]. The use of GSH as a stress marker is strongly supported by its function in the antioxidant defense system. However, the amount of cellular GSH varies significantly under abiotic stressors, which has a significant impact on how well it functions as an antioxidant. Furthermore, there is compelling evidence linking a high GSH concentration to plant resistance to oxidative stress brought on by metals. In *P. sativum* [110], *Sedum alfredii* [111], and *V. mungo* [112]. Rising Cd concentrations have been accompanied by rising GSH concentrations. For ROS scavenging in cells, a high ratio of reduced to oxidized ascorbate and GSH is necessary. By employing NADPH as a reducing power, GR, MDAR, and DHAR keep the antioxidants' reduced states stable [113, 114]. Additionally, the overall equilibrium between the various antioxidants needs to be closely regulated. Cells with increased glutathione biosynthesis in chloroplasts exhibit oxidative stress damage, presumably as a result of alterations in the total redox state of chloroplasts, highlighting the significance of this equilibrium [101]. In biomembranes, where they serve both antioxidant and nonantioxidant roles and may scavenge ROS and lipid radicals, tocopherols are regarded as key antioxidant [115]. Tocopherols are regarded as all-purpose antioxidants that protect membrane integrity by quenching or scavenging ROS like O_2. It has been demonstrated that tocopherols inhibit the chain propagation step in lipid autooxidation, making them efficient free radical traps. Additionally, it has been calculated that one α-tocopherol molecule can, by resonance energy transfer, scavenge up to 120 O_2 molecules [116]. Recently, it has been discovered that for the synthesis of tocopherols , oxidative stress activates the expression of genes that are responsible in higher plants [117]. Following trizole treatment, higher concentrations of a-tocopherol and ASH were observed in tomato plants, which may help protect membranes from oxidative damage and increase chilling tolerance [118]. Many professionals have also found that plants' tocopherol levels rise when they are under water stress [118].

12.4.2 Inhibition of Lipid Peroxidation

Atherosclerosis, cataracts, rheumatoid arthritis, and neurological disorders are only a few of the illnesses that have been linked to lipid peroxidation. As a result, antioxidants' role has drawn a lot of attention. Adaptive responses, repair of damage, excretion of harmful oxidation products, and avoidance of the synthesis of active oxidants are the four categories into which antioxidant defenses can be separated. By blocking either the activation or reaction of an enzyme, enzymatic lipid oxidation can be prevented [119]. It is possible to slow down chain initiation, chain growth, and/or chain termination in order to prevent free radical-mediated lipid peroxidation. By, for instance, dampening UV radiation and quenching the singlet oxygen itself by carotenoids, lipid peroxidation caused by singlet oxygen may be prevented. Radical-scavenging vitamins and medications, both natural and synthetic, have been investigated. Reactivity to radicals, the fate of radicals produced by the antioxidant, absorption, distribution, localization, mobility, and interactions with other antioxidants are some of the elements that affect the antioxidant activity *in vivo* [120]. The overall defense network against oxidative stress may also depend on activities like phase induction that are independent of antioxidant action. One of the interesting characteristics of the biological system is its heterogeneity. The antioxidants are localized in the aqueous or lipid phase. The lipophilic compartment of membranes and lipoproteins is distributed with the lipophilic antioxidants. Typical lipophilic and hydrophilic antioxidants are vitamin E and vitamin C, respectively. The aqueous phase is where vitamin C dwells and scavenges radicals [119]. The most thoroughly researched lipophilic antioxidant *in vivo* and one that is appropriate for examining antioxidant function *in vivo* is vitamin E. To stop chain propagation, it scavenges lipid peroxyl radicals, species that carry chains in lipid peroxidation [121]. There are eight isoforms of vitamin E, however, the α-tocopherol transfer protein in the liver binds only α-tocopherol (α-TOH) preferentially and it enhances its bioavailability [122]. It was shown that the distribution of lipids is directly inversely correlated with the subcellular distribution of both α-tocopherol and α-tocotrienol in nuclear, mitochondrial, microsomal, and cytosolic fractions [123]. With a methyl side chain in place of the phenyl group, 2,2,5,7,8-Pentamethyl-6-chromanol (PMC), an analogue of α-TOH, exerts roughly the same antioxidant activity as α-TOH in an organic solution, but has higher activity against lipid peroxidation in the membranes and LDL due to higher mobility within and between membranes and particles [121]. The water-lipid interface is home to the phenolic hydrogen at position 6 of α-TOH, the active site. The effectiveness of radical scavenging was shown to decline as the radical penetrated further into the membrane's core and LDL [124]. Recently, it was reported that α-TOH inhibited the cholesteryl ester oxidation is inhibited by in the LDL core more effectively than it did phospholipid oxidation at the surface monolayer [125]. Vitamin E was found to be a prooxidant against LDL oxidation in 1992 [126] and Stocker and his colleagues have extensively debated this effect, which is known as tocopherol-mediated peroxidation (TMP). According to this plan, α-TOH transfers radicals from the aqueous phase into the LDL particle (phase transfer effect), and the ensuing α-tocopheroxyl (α-TO) radical attacks the bisallylic hydrogen of PUFA in LDL (chain transfer effect), causing the chain reaction to continue. In reality, α-TOH works as a prooxidant, and numerous investigations on the *in vitro* oxidation of LDL have demonstrated the role of TMP [127]. It should be noted that while this TMP may be significant for isolated LDL oxidation, it is not relevant for plasma, because vitamin C quickly lowers the α-TO radical. Accordingly, it has been demonstrated that vitamin E and vitamin C together effectively reduce the lipid peroxidation of LDL and plasma, however plasma oxidation does still occur even in the absence of vitamin E. It is safe to say that α-TOH does not function *in vivo* as a prooxidant. The results of numerous large-scale, long-term intervention studies do not always support the beneficial effect of vitamin E and have raised some doubts on the oxidative modification hypothesis, despite the fact that it is now widely accepted that the oxidative modification of LDL is an important initial event in the progression of atherosclerosis and that antioxidants should be effective for prevention of atherosclerosis [128]. It should be noted that numerous oxidants can modify LDL through a variety of methods, and that vitamin E efficiently suppresses some of these pathways but not others.

12.4.3 Enhancement of Endogenous Antioxidant Defense Systems

Plants are subjected to a variety of challenges in the field due to the current state of the global climate. Plant development and sustainable crop production are significantly impacted by water stress (drought and waterlogging), salinity, metal/metalloid toxicity, severe temperatures, xenobiotics, and other abiotic stresses [129]. The cellular metabolism of plants naturally produces a variety of reactive oxygen species (ROS), including free radicals (superoxide anion, O_2•, hydroperoxyl radical, HO_2, alkoxy radical, RO•, and hydroxyl radical,•OH) and non-radical molecules (hydrogen peroxide, H_2O_2, and singlet oxygen, $^1O_2$). Though under stress, ROS are overproduced by plant cells. Because ROS are so reactive, they interfere with plant metabolism and seriously harm vital cellular elements such proteins, lipids, DNA, and others [130]. Consequently, this throws off the equilibrium between regular ROS production and antioxidant activity, which causes oxidative stress in plants [129]. The primary adaptive response to oxidative stress in plants is to increase the antioxidant defense system's capacity [131]. Low-molecular-weight nonenzymatic antioxidants such as ascorbic acid (AsA), glutathione (GSH), α-tocopherol, phenolic compounds, flavonoids, alkaloids, and nonprotein amino acids, together with other antioxidant enzymes, are primarily responsible for maintaining the system [129]. Ascorbate peroxidase (APX), catalase (CAT), monodehydroascorbate reductase (MDHAR), dehydroascorbate reductase (DHAR), glutathione reductase (GR), glutathione peroxidase (GPX), and glutathione S-transferase (GST) are just a few of the antioxidant enzymes linked to this defense system [132]. For example, SOD removes O_2•–, CAT converts H_2O_2 into H_2O and O_2, POD scavenges H_2O_2 in the vacuoles, GST combines GSH with the electrophilic or hydrophobic compounds, and MDHAR and DHAR control the ascorbate pool [133]. The GST and GPX enzymes transform additional xenobiotics and electrophilic chemicals into less dangerous molecules, which are then sequestered in extracellular areas [134]. Because it postpones planned cell death, plants in stressed conditions benefit greatly from having an antioxidant defense system. Lack of adequate antioxidant enzymes in plants causes cellular organelles to malfunction, which leads to lipid peroxidation, protein oxidation-related damage, the disintegration of DNA molecules and nucleic acids, and various enzyme inhibitions [135]. The antioxidant defense system delays oxidation and regulates DNA and nucleic acid damages under stress to ensure effective ROS detoxification, decreased lipid peroxidation in membranes, and the prevention of protein damage. The ability of the plants to withstand stress is thus enabled by the overall cellular protection provided [136]. ROS generation in plant cells has both negative and positive impacts. However, the precise mechanisms of ROS-mediated stress reduction remain to be fully understood [137]. The status of the signaling mechanisms/pathways, consequences, and management of ROS within stressed plant cells were all summarized. Additionally, they talked about how modern approaches like molecular priming, systems biology, phenomics, and crop modeling have advanced, and how these techniques can minimize oxidative damage and divert ROS into signaling pathways [137]. One of the plant's adaptive responses to abiotic stress linked to ROS signaling is cross-tolerance. found that in resistant chickpea genotypes, cold acclimation noticeably increased enzymatic (SOD, CAT, APX, GR) and non-enzymatic (AsA and GSH) activities, which reduced low-temperature-induced leaf damage during cold stress. Utilizing this knowledge will help focus efforts on developing cold tolerance [138]. It was discovered that the phytohormone-producing, plant-growth-promoting rhizobacteria *Acinetobacter bouvetii* P1 recovered the sunflower growth when exposed to Cr^{+6} by boosting the host's antioxidant system and inducing a higher production of enzyme antioxidants, including CAT, APX, SOD, and POD. The formation of nonenzymatic antioxidants such as flavonoids, phenolics, proline (Pro), and GSH was also increased by P1 [139]. The antioxidant system is stimulated by the presence of signaling molecules including nitric oxide (NO), hydrogen sulfide (H_2S), and H_2O_2. Some halophytes and plant species that can withstand stress have demonstrated improved capacity for the synthesis of stress-related signaling molecules like NO [136]. It was observed that the antioxidant defense, which was controlled by NO, was strongly connected with the increased salt tolerance in the halophyte *Kandelia obovata*. Sodium nitroprusside (SNP) treatment of salt-stressed plants boosted endogenous NO levels, decreased ion toxicity, improved nutritional homeostasis, and further enhanced Pro levels. SNP treatment enhanced the CAT, APX, MDHAR, and DHAR antioxidant enzymes' activity. However, NO scavengers or inhibitors therapy negated these advantageous SNP effects and made them worse [140]. It was found that alfalfa plants suffering from iron (Fe) deficiency may be maintained by the NO donor. Fe-homeostasis and oxidative state were restored in iron-deficient alfalfa by an exogenous NO donor. With regard to this, the rise in antioxidant genes and their associated enzymes (Fe-SOD, APX) in response to SNP treatment implies that Fe-SOD and APX are important factors in alfalfa's decreased oxidative stress and H_2O_2 buildup. Additionally, the increased expression of genes associated with the AsA-GSH pathway (GR and MDAR) in Fe deficit with SNP suggests that NO is linked to an improved antioxidant defense against Fe deficiency stress [141].

12.5 Medicinal Plant with Antioxidant Effect on Epilepsy Treatment

At today, it is well acknowledged that the formation of reactive oxygen species (ROS) is considered to be a pivotal process that underlying the mechanism by which epileptic activity exerts detrimental effects on the brain. The observed modifications in the neural network of the brain are indicative of underlying transformations that may ultimately give rise to a state characterized by repeated and unprovoked seizures. Consequently, the implementation of neuroprotective measures may potentially impede or halt the development of epilepsy, thereby reducing the intensity and frequency of seizures. Additionally, this approach has the potential to enhance the effectiveness of existing antiepileptic drug (AED) treatments and mitigate the well-documented issue of AED resistance [142, 143]. The investigation of compounds with antioxidant properties has been intensively conducted to explore their potential as anti-epileptogenic therapies, since the neutralization of heightened reactive oxygen species (ROS) production during seizures relies on the efficacy of antioxidant defense mechanisms [144]. The compounds mentioned encompass both endogenous antioxidants, such as α-lipoic acid, GSH, melatonin, ubiquinone (coenzyme Q10), and vitamin A, as well as exogenous substances with anticonvulsive and neuroprotective properties, such as α-tocopherol, ascorbic acid, curcumin, N-acetylcysteine, omega-3 fatty acids, and resveratrol. Additionally, there are novel synthetic radical scavengers, including aspalatone, EPC-K1, EUK-134, MnTBAP, and tempol [145]. Neuroprotective effects are exerted by these substances, which are often classified as molecules with the capacity to quench or diminish highly reactive oxygen species (ROS). These compounds can therefore protect the brain from oxidative stress, as was seen in some animal models of seizure activity [146, 147]. It is well known that standard AEDs do not provide enough control over seizures. In addition, the use of these AEDs for an extended period of time considerably increases oxidative stress and puts the user at risk for cognitive impairment. It's possible that the present AEDs are unable to restore the body's natural equilibrium between oxidants and antioxidant defense systems, which might be the root cause of both seizures and cognitive impairment [148]. In people who have epilepsy, there is a possibility that increased oxidative stress brought on by antiepileptic drugs (AEDs), in addition to the fact that epilepsy itself can cause oxidative stress, contributes to the onset of seizures and cognitive impairment Hence, the capacity of antioxidants to impede the onset of seizures and reduce signs of oxidative stress provides more evidence for their substantial involvement in purported antiepileptic properties. There is a limited amount of research available addressing the clinical effects of antiepileptic drugs (AEDs) on cognitive impairment. Several antiepileptic drugs (AEDs), such as phenobarbital (PB), valproic acid (VPA), and carbamazepine (CBZ), have been demonstrated to markedly disturb the crucial equilibrium between oxidants and antioxidants. Previous studies have demonstrated that CBZ and lamotrigine can induce a significant decrease in performance on learning and memory assessments [149]. In order to validate these findings, it has been reported that there is a notable decrease in cognitive functioning following treatment with CBZ in individuals who have just been diagnosed with epilepsy. Additionally, it has been observed that the administration of PB and CBZ resulted in notable oxidative stress, leading to the decline of cognitive functions such as learning and memory in rats over a period of 21 days. The administration of PB resulted in a greater degree of cognitive impairment when compared to CBZ [150]. Hence, the capacity of antioxidants to impede the onset of seizures and reduce signs of oxidative stress provides more evidence to indicate their substantial role in potentially possessing antiepileptic properties. There is a limited amount of available research about the clinical effects of antiepileptic drugs (AEDs) on cognitive impairment. Several antiepileptic drugs (AEDs), such as phenobarbital (PB), valproic acid (VPA), and carbamazepine (CBZ), have been found to considerably disturb the crucial equilibrium between oxidants and antioxidants [151].

12.5.1 Flavonoid-Rich Medicinal Plant

Several ongoing scientific investigations are now focused on the isolation and characterization of the bioactive constituents found in medicinal plants. These constituents have the potential to serve as "lead" molecules in the creation of new drugs. The vegetal kingdom has been identified as a significant reservoir of bioactive compounds that belong to a distinct class of phytochemicals. These compounds have the ability to interact with various targets [152]. Table 12.1 shows the list of flavonoids with its sources and targeted area. The metabolites exhibit many biological actions, such as anti-inflammatory, cytostatic, antiviral characteristics, and antinociceptive activity. In addition, it is widely recognized that some flavonoids and their glycosides include anxiolytic, sedative, and anticonvulsant properties that affect the

Table 12.1 Flavonoids with their target area.

Flavonoids	Source	Targeted area
Chrysin(5,7-dihydroxyflavone)	Honey, Propolis, *Diaphragma juglandis fruits*	The compound exhibited a modest level of affinity for the central benzodiazepine receptors and demonstrated anticonvulsant effects in animal models [165].
Apigenin (5,7,4′ trihydroxyflavone)	*Tanacetum parthenium*	The flavone in question has been identified as a benzodiazepine ligand with central activity, eliciting effects on the central nervous system upon systemic treatment. In the experimental paradigm involving picrotoxin-induced convulsions in rats [166].
Vitexin (5,7,4-trihydroxyflavone-8-glucoside)	*Passiflora* sp., bamboo leaves	The compound possesses several pharmacological characteristics, manifesting its anticonvulsant capabilities by prolonging the onset of seizures in PTZ-induced minimum clonic seizures (MCS) and generalized tonic-clonic seizures (GTCS). This action is likely achieved by binding to the benzodiazepine site of the GABAA receptor complex [167].
Luteolin (3′,4′,5,7-tetrahydroxyflavone)	Celery, parsley, broccoli, oni on leaves	The direct antioxidant impact of luteolin remains undisclosed, nevertheless, the observed elevation of endothelial nitric oxide synthase (eNOS) in the brain and other tissues, along with the reduction of inducible nitric oxide synthase (iNOS), suggests that luteolin has an indirect antioxidant effect [168].
Quercetin (3,3′,4′,5,7-pentahydroxyflavone)	Fruits, vegetables, tea, olive oil	There is a potential for quercetin to exert modulation on GABAA receptors and have antagonistic effects on NMDA receptors [169].
Rutin (3′,4′,5,7-tetrahydrohyflavone-3-rhamnoglucosidc)	Buckwheat, apples, and black tea	The observed results may perhaps be attributed to the antioxidative properties of rutin. There exists a potential for rutin to act as a ligand for benzodiazepine receptors by means of positive allosteric regulation of the GABAA receptor [170].
Epigallocatechin-3-gallate (EGCG)	green tea	It has been discovered to be a powerful antioxidant and may lessen oxidative brain injury during PTZ-induced kindling. The putative antioxidant impact of EGCG may contribute to its ability to protect against PTZ-induced kindling [171].
6-Hydroxyflavone (6HF)	*Barleria prionitis*	Benzodiazepine compounds exhibit a modest binding affinity to the benzodiazepine site on GABAA receptors. 6HF is classified as a partial agonist of GABAA receptors [172].

central nervous system (CNS) [153]. The neuroprotective effects of phenylpropanoids generated from caffeic acid have been observed. Polyphenolics, a class of antioxidants characterized by the presence of a polyphenolic substructure, encompass both flavonoid and nonflavonoid polyphenolics [154]. In recent times, there has been a notable surge in scientific and clinical interest surrounding polyphenolics due to their ability to provide protection against damage caused by free radicals [155]. Flavonoids are found in many plant-based sources such as fruits, vegetables, nuts, plant-derived drinks, traditional eastern treatments, and nutritional supplements including herbs. Antioxidants are recognized for their ability to safeguard cells from oxidative damage. The biological activities of flavonoids are predominantly influenced by their structural characteristics, resulting in a diverse array of bioactivities . Over 5000 distinct flavonoids have been identified and a significant number of them have been characterized in terms of their pharmacological attributes. The sedative-hypnotic effects of flavonoids derived from medicinal plants, such as valerian (*Valeriana officinalis*), chamomile (*Matricaria recutita*), and kava-kava (*Piper methysticum*), are attributed to their positive allosteric regulation of GABAA receptors [156]. Flavonoids from several chemical groups exhibit varied degrees of inhibition on kinases and phosphatases, which are enzymes responsible for phosphorylating and dephosphorylating crucial proteins involved in signaling transduction pathways. These pathways play a pivotal role in regulating oxidative stress, inflammation, and cell viability [157]. The anticonvulsant effect of these metabolites has been documented in previous studies. Previous research has demonstrated the anticonvulsant effects of rutin, quercetin, and isoquercitrin on experimental models of epilepsy. Experimental data has unequivocally proven that flavonoids possess antiepileptic properties through their modulation of the GABAA-Cl-channel complex. This effect can be attributed to the structural similarity between flavonoids and benzodiazepines [158].

12.5.2 Phenolic Compounds in Medicinal Plants

Polyphenols are a class of bioactive chemicals derived from plants that often exhibit strong reactive oxygen species (ROS) scavenging abilities. Numerous studies have demonstrated that polyphenols, such as curcumin, resveratrol, and lycopene, have protective effects against brain damage caused by epilepsy through the mitigation of oxidative stress originating from mitochondria [159]. Curcumin has been seen to provide protection to mitochondria against lipid peroxidation and cell death generated by reactive oxygen species (ROS). Additionally, curcumin has demonstrated neuroprotective benefits in the context of epileptic convulsions and cognitive impairments [160]. Furthermore, the compound known as resveratrol, derived from grapefruit, has been found to effectively mitigate brain damage caused by seizures by lowering levels of oxygen (O2) and lipid peroxidation. There is further evidence to suggest that lycopene obtained from tomatoes possesses the ability to mitigate the severity of seizures in a mouse model of epilepsy caused by PTZ. This effect is achieved by the reduction of oxidative damage, as seen by a notable drop in malondialdehyde (MDA) levels, as well as the enhancement of the activities of CI, CII, and CIV [161]. Taken together, the findings suggest that polyphenols, which are naturally occurring chemicals, hold the potential to provide neuroprotection against epileptic disease.

12.5.3 Alkaloids as Antioxidant in Epilepsy

12.5.3.1 Therapeutic Efficacy of Berberine (BBR) in Epilepsy

Epilepsy is a neurological condition defined by the occurrence of seizures. While there are several antiepileptic medicines (AEDs) now available, their administration is associated with a wide range of adverse effects, and many AEDs fail to effectively control seizures in certain people. This phenomenon serves to heighten the motivation of researchers in their pursuit of exploring phytotherapy as a potential means to mitigate occurrences of epilepsy [162]. Research findings have indicated that the utilization of *Berberis vulgaris* extract has potential advantages in the management of epilepsy and convulsions [163]. Research has demonstrated a correlation between the pathological manifestation of epilepsy and the excitotoxicity of NMDAR. Berberine (BBR) has been suggested to have therapeutic potential by preventing the overactivation of extra synaptic N-methyl-D-aspartate receptors (NMDARs). Previous studies have indicated that BBR can affect the neurotransmitter system and function as an antagonist of NMDARs. Nevertheless, further research is required to investigate the potential of BBR as an antagonist of NMDAR [164].

12.5.3.2 Therapeutic Potential of Montanine in Epilepsy

Research findings have indicated that GABAergic neurotransmission plays a significant role in the modulation of stress, anxiety, and epilepsy. There have been reports of disruption in GABAergic neurotransmission in individuals with anxiety that is comorbid with epilepsy [173]. According to experimental research findings, it has been observed that montanine has a protective effect against convulsions via altering several neurotransmitter systems, including the GABAA receptors. There has been a proposition positing that PTZ is associated with cognitive impairments that are correlated with seizures [174]. Research findings have indicated that GABAergic neurotransmission plays a significant role in the manifestation of stress, anxiety, and epilepsy. There have been reports of disruption in GABAergic neurotransmission in people with anxiety that is comorbid with epilepsy [175]. According to experimental investigations, it has been shown that montanine has a protective effect against convulsions through altering neurotransmitter systems, namely the GABAA receptors. There has been a suggestion that the administration of PTZ is associated with cognitive impairments that are correlated with the occurrence of seizures [176].

12.5.4 Therapeutic Efficacy of Piperine in Epilepsy

Piperine (PIP) is a primary alkaloid that is present in both long pepper (Piper longum) and black pepper (Piper nigrum). The plant in question is classified under the Piperaceae family. According to reports, piperine exhibits a wide range of pharmacological effects, including anti-inflammatory, antifungal, insecticidal, antihypertensive, antipyretic, antitumor, and analgesic properties. Additionally, piperine has been found to possess various neuroprotective properties, such as antioxidant, antidepressant, and anticonvulsant activities [177]. The PIP intervention exerts its effects on the neurotransmitter systems, specifically serotonin, GABA, and norepinephrine. By increasing the levels of serotonin and GABA in the cortical and hippocampus regions, PIP is able to effectively postpone the occurrence of tonic-clonic seizures. The suggestion has been made that malfunctioning of GABA receptors may serve as the fundamental etiology of epilepsy [178]. Electrophysiological investigations have shown the fundamental mechanism by which PIP functions as an anticonvulsant drug, owing to its capacity to oppose sodium/calcium channels [179].

12.6 The Synergy Between Antioxidants and Conventional Antiepileptic Drugs (AEDs)

12.6.1 Mechanistic Interactions Between Antioxidants and AEDs

AEDs are linked to a variety of pharmacological interactions, such as protein-binding displacement and the activation and inhibition of hepatic enzymes. A small subset of medications with poor therapeutic indices have their metabolism altered by hepatic induction brought on by AEDs. Potential issues could be reduced with appropriate clinical monitoring and anticipation of induction interactions. Epoxide hydrolase, CYP2C enzymes, and UGT enzymes are all broadly inhibited by valproate. Felbamate inhibits CYP2C19 substrates while inducing CYP3A4. Topiramate exclusively inhibits the substrates of CYP2C19. Tiagabine, etosuximide, gabapentin, and vigabatrin don't either increase or inhibit drug metabolism. Competition at the enzyme site is typically what inhibits hepatic enzymes. Clinicians are able to anticipate potential interactions thanks to their understanding of the precise metabolic enzymes involved in the metabolism of AEDs [180]. Drug interactions in patients using AEDs are a frequent side effect of treatment. If conventional dosages are left unchanged, AED stimulation of the metabolism of medications with limited therapeutic ranges will result in patients obtaining inadequate amounts of concurrent therapy. They are the main contributor to clinically important medication interactions because of the quick onset and unpredictability of inhibitory drug interactions. Drug interactions in patients using AEDs are a frequent side effect of treatment. If conventional dosages are left unchanged, AED stimulation of the metabolism of medications with limited therapeutic ranges will result in patients obtaining inadequate amounts of concurrent therapy. Inhibitory medication interactions are the main reason for clinically important drug interactions because of their quick onset and unpredictability. Toxic interactions that involve AED inhibition are frequent. A greater understanding of the precise CYP isoenzymes involved in the metabolism of AEDs has made it possible to qualitatively predict interactions between new AEDs and new non-AEDs. Additionally, the most recent association between the results of *in vitro* studies and clinical interaction studies evaluating topiramate and felbamate has shown the significance of in vitro studies in foretelling and guiding clinical interaction investigations. The pharmacist can play a critical role in foreseeing and subsequently preventing medication interactions by having a thorough understanding of the function of certain metabolic enzymes involved in the hepatic metabolism of the AEDs [180].

12.6.2 Clinical Evidence of Combined Therapy

The situation regarding clinical evidence in support of "rational polytherapy" is similarly sparse [181]. There are assertions in the literature that certain combinations, such as a sodium channel blocker and a GABA-ergic drug [182] or one with several modes of action, are particularly effective [183]. The only evidence for synergism, however, is again with sodium valproate and lamotrigine. In a pragmatic experiment, Brodie and colleagues attempted to replace lamotrigine as monotherapy in patients receiving less-than-optimal care from carbamazepine, phenytoin, or sodium valproate [184]. The pharmacokinetic interactions between these medications were taken into account while adjusting the lamotrigine dose regimens, resulting in comparable circulating lamotrigine concentrations for all three combinations.

Other helpful combinations that have been proposed in the literature are primarily based on anecdotal evidence from patient groups of a small size or studies with a small sample size. Valproate along with ethosuximide is one of these for absence seizures [185]. These include carbamazepine with valproate or vigabatrin for partial seizures [186], phenobarbital with phenytoin for generalized tonic-clonic seizures [187], vigabatrin with tiagabine for partial seizures [188], and lamotrigine with topiramate for a variety of seizure types[189]. It's interesting to note that in a multicenter double-blind, placebo-controlled, randomized, parallel group trial of rufinamide 1600 mg twice daily in adults and adolescents with refractory partial seizures, there was a 12% decrease in monthly seizure frequency in the 96 patients who were established on carbamazepine compared with a 29% seizure reduction ($p = 0.05$) in the 60 patients taking a regimen that did not contain carbamazepine [190]. This finding may indicate that rufinamide and carbamazepine affect voltage-dependent sodium channels similarly. Recent placebo-controlled dose ranging adjunctive trials with lacosamide have shown that although both groups showed a dose-dependent statistically significant improvement over placebo, response to this medication was less impressive in patients taking traditional sodium blockers than in those taking AEDs with alternative mechanisms of action [191]. Combining medications that block voltage-dependent sodium channels increases the risk of neurotoxic side effects such as ataxia, diplopia, and dizziness [190, 192, 193]. In a pooled review of clinical trial data with supplemental lacosamide, a similar scenario, specifically related to a higher chance of dizziness, has been identified in patients already receiving traditional sodium channel blockers compared to those taking other AEDs [191]. There is a significantly higher incidence of allergic reaction when lamotrigine is introduced to sodium valproate-dependent individuals than when the drug is first used as a monotherapy [194]. This is thought to be a result of the valproate's inhibitory action on lamotrigine metabolism, which leads to greater amounts.

This is thought to be a result of the valproate's inhibitory action on lamotrigine metabolism, which leads to greater amounts.

12.7 Challenges and Future Direction

The review presents information that suggests a link between mitochondrial oxidative stress and malfunction, and the initiation and advancement of epilepsy. The neuroprotective benefits of antioxidants that specifically target mitochondrial oxidative stress have been found to be significant in mitigating epileptic conditions. Numerous experimental research have been undertaken on antioxidants that demonstrate consistent seizure-suppressing activities, despite the comprehensive understanding of the intricate chemical pathways via which these molecules safeguard the brain against epileptic seizures. Indeed, many antioxidants related with mitochondria have been evaluated for their potential therapeutic effects in individuals diagnosed with epilepsy . The collective findings presented in this study provide compelling evidence that medicines aimed at mitigating mitochondrial oxidative stress have potential as neuroprotective agents against the onset of seizures and/or the development of epilepsy.

There are further points that require clarification. Firstly, extensive investigations have demonstrated the correlation between mitochondrial oxidative damage and the development of epilepsy. However, the regulatory mechanism underlying the mitochondrial dysfunction caused by oxidation in the context of pathological conditions is intricate, necessitating further exploration in ongoing research. The clarification of this subject is advantageous for the advancement of new antioxidants and other nonpharmacological therapies, including as dietary supplements and gene editing. Furthermore, it is imperative to conduct extensive clinical trials to thoroughly examine the bioavailability and blood-brain barrier (BBB) penetration, as previously discussed. Additionally, it is crucial to implement multi-center clinical investigations with a substantial sample size in order to establish the advantageous effects of these antioxidants on the long-term outcomes of epilepsy [195, 196].

12.7.1 Standardization of Herbal Preparation

12.7.1.1 Need for Standardization

In recent years, there has been a notable increase in the level of interest around the preservation and viability of Ayurvedic medicinal practices. From a worldwide standpoint, there is a noticeable trend towards the utilization of herbal therapy, since the limitations and risks associated with modern medicine have become increasingly evident. It is worth noting that a significant proportion of Ayurvedic formulations are derived from botanical sources.

The primary obligation of regulatory agencies is to ensure that customers receive medication that provides a guarantee of efficacy and safety. The key factors to consider in this context are purity, safety, potency, and efficacy. The regulatory authorities fulfill this responsibility by strictly adhering to prescribed quality standards for both raw materials and finished products as outlined in pharmacopoeias. They exercise control over manufacturing processes through the use of formularies and ensure compliance with statutory-imposed "Good Manufacturing Practices." These procedures may be used to all types of medications, regardless of whether they are part of the contemporary system of medicine or one of the older systems such as Ayurvedic medicine. Regrettably, the Ayurvedic pharmacopoeias and formulary now do not meet contemporary standards. Contemporary medical practices are subject to ongoing metabolic alterations and advancements in the domains of purity, safety, and effectiveness. Therefore, the preservation of the quality of Ayurvedic treatment rests solely on the shoulders of the manufacturer [197].

12.7.1.2 The Standardization and Quality Control of Herbal Crude Drugs is a Crucial Aspect in Ensuring Their Reliability and Efficacy

According to the World Health Organization (WHO), accounting refers to the comprehensive examination of crude drugs, encompassing various aspects such as the selection and handling of raw materials, evaluation of safety, efficacy, and stability of the final product, documentation of safety and risk based on past experiences, and the provision of product information to consumers as well as product promotion [198].

Macro and Microscopic Examination: To facilitate the identification of the appropriate variety and the detection of potential adulterants.

Foreign Organic Matter: The extraction process involves the elimination of extraneous substances, leaving behind just the source plant material to obtain the medicine in its pure form.

Ash Values: The assessment of the identification and purity of crude drugs involves the use of many criteria, such as total ash, sulphated ash, water-soluble ash, and acid insoluble ash, among others.

Moisture Content: The assessment of moisture content serves as a preventive measure against product deterioration.

Extractive Values: The following data provides an estimation of the chemical composition of the crude medication.

Crude Fiber: In order to ascertain an abundance of lignocellulosic material The parameters used to assess the level of purity.

Qualitative Chemical Evaluation: This study focuses on the identification and characterization of crude drugs in relation to their phytochemical constituents.

Chromatographic Examination: The identification of a crude medication can be accomplished by utilizing the principal chemical ingredient as a marker.

Qualitative Chemical Evaluation: Criteria for estimating the quantity of the primary class of elements.

Toxicological Studies: The reduction of pesticide residue, potentially harmful materials, and microbial count is a crucial technique to mitigate their impact on the final product.

Physical Evaluation

Every monograph has comprehensive botanical, macroscopic, and microscopic accounts of the physical attributes of each plant, which serve to ensure both its identification and purity. Accurate identification of materials is visually documented by comprehensive drawings and photographic photographs accompanying each description.

Microscopic Evaluation

The comprehensive and precise characterization of plant material necessitates the utilization of both physical and chemical testing. The utilization of microscopic studies in the study of plants is of great significance in verifying the authenticity of the material and serves as the first screening method for detecting contaminants. Many makers of herbal goods do not possess the necessary employees with expertise in quality control to effectively evaluate the microscopic aspects of plant identification and purity. The Ayurvedic Herbal Pharmacopoeia (AHP) comprehensively evaluates herbal items in accordance with existing literature and the AHP guidelines. The utilization of verified trade mark authenticated materials serves as a means to ensure the authenticity and integrity of test materials, therefore establishing a reliable method for verifying their identification. Ideally, it is recommended that submitted samples be presented in their complete or partially intact (trimmed) state to facilitate microscopic examination. Although, a significant amount of information may be derived from the microscopic analysis of powders.

Chemical Evaluation

The process of chemical evaluation encompasses the procedures of isolation, identification, and purification. It involves the analysis of drugs to determine the effectiveness of active components derived from plant and animal sources. Chemical tests, such as color reaction tests, are employed to ascertain the identity of the drug substance and detect any potential adulteration.

Analytical Method

The requirement for suitable analytical procedures to ascertain identification, purity, and relative potency is of utmost importance in adhering to any monograph standard. There exists a wide array of analytical procedures at one's disposal. Although, determining the most suitable option to utilize may sometimes pose a challenge. The main objective of the Analytic Hierarchy Process (AHP) is to offer a range of techniques for the identification and evaluation of various characteristics of the botanical specimen in a suitable manner.

Purity Determination

Every monograph has criteria for purity and other qualitative evaluations, such as, when applicable, foreign matter, ash, acid-insoluble ash, moisture content, loss of moisture on drying, and extractives.

High performance thin layer chromatography (HPTLC) is a significant method for assessing the quality of botanical materials. This technique enables the efficient and cost-effective examination of a wide range of chemicals. Furthermore, the utilization of several samples within a single analysis might lead to a significant reduction in analytical time. High-performance thin-layer chromatography (HPTLC) enables the examination of a plant's whole profile by utilizing several wavelengths of light, beyond the level of detail generally obtained with narrower analytical methods.

Quantitative Analysis

When deemed suitable, the most suitable quantitative analytical technique, together with its corresponding chromatograms, will be supplied. The main objective of the method(s) is to offer validated techniques for quantifying the compound(s) that exhibit the highest correlation with pharmacological activity or qualitative markers. These correlations are determined by analyzing primary pharmacological literature, examining constituent declarations in product labeling, and conducting a survey of experts in the field. The approaches Selection Committee will choose the method(s) from the main analytical literature, giving precedence to compendial approaches if they are accessible. In the present context, validation entails a minimum requirement of conducting a two-laboratory validation employing identical processes, samples, and reference standards.

The key criteria for evaluating the suitability of a technique encompass the precision of the results, efficiency, robustness, relevance to a broad range of industrial practices, and the absence of hazardous chemicals and solvents. In order to facilitate harmonization, priority is accorded to methodologies that have previously been endorsed by official pharmacopoeias or AOAC International. When deemed relevant, it is imperative to run comparison tests in order to ascertain the best suitable method(s) among those that are accessible. The validation procedure encompasses several key aspects, including standard precision, linearity, sample accuracy through the use of repeat samples, sample linearity, selectivity (including co-elution and sensitivity to analyte degradation), retention durations, and limits of detection. Additional techniques that might potentially contribute to the industry may be incorporated or referenced in the monograph, while they are not obligatory for adhering to the book's guidelines.

12.7.2 Identification of Active Compound

12.7.2.1 Techniques of Isolation and Purification of Bioactive Molecules from Plants

The technology of purifying and isolating bioactive chemicals from plants has experienced significant advancements in recent years [199]. This contemporary methodology provides the capacity to concurrently progress and make accessible several sophisticated bioassays, while also offering accurate methods for isolating, separating, and purifying substances. The primary objective in the pursuit of bioactive chemicals is identifying a suitable approach for screening the source material to determine its bioactivity, including attributes such as antioxidant, antibacterial, or cytotoxic properties. This endeavor is further guided by the need for methods that exhibit simplicity, specificity, and expeditiousness [200]. *In vitro* methodologies are often preferred over *in vivo* tests due to the costliness and time-consuming nature of animal research, as well as the potential for ethical problems. Several variables contribute to the challenge of establishing definitive methodologies or protocols for the isolation and characterization of certain bioactive compounds. The variation in phytochemical composition within plants can be attributed to the presence of distinct tissues, each capable of producing unique substances. Furthermore, the bioactive phytochemicals exhibit a wide range of chemical structures and physicochemical capabilities [201]. The acquisition and compilation of plant materials are regarded as fundamental stages in the process of isolating and characterizing a bioactive phytochemical. The subsequent phase is the recovery of ethno-botanical data to identify potential bioactive compounds. Subsequently, the active components accountable for the bioactivity might be isolated and purified by the utilization of different solvents. The utilization of column chromatographic methods is a viable approach for the extraction and refinement of bioactive substances. The utilization of advanced analytical techniques, such as High Pressure Liquid Chromatography (HPLC), expedites the purifying process of bioactive molecules. Various spectroscopic methods, including as UV-visible, Infrared (IR), Nuclear Magnetic Resonance (NMR), and mass spectroscopy, are capable of discerning the purified components [202].

12.7.2.2 Purification of Bioactive Molecule

Paper thin-layer and column chromatography technologies have been employed to separate and purify several bioactive compounds. Column chromatography and thin-layer chromatography (TLC) continue to be widely utilized in scientific research and industry primarily because of their practicality, cost-effectiveness, and the wide range of stationary phases they provide. Silica, alumina, cellulose, and polyamide have been identified as the most advantageous materials for the separation of phytochemicals. The presence of intricate phytochemicals in plant materials poses a significant challenge in achieving effective separation. Hence, the utilization of numerous mobile phases to enhance polarity becomes advantageous in achieving separations of significant value. Thin-layer chromatography (TLC) has traditionally been employed for the analysis of chemical fractions obtained from column chromatography. Silica gel column chromatography and thin-layer chromatography (TLC) have been employed as techniques for the separation of bioactive compounds, utilizing various analytical instruments [203].

12.7.2.3 UV-Visible Spectroscopy

UV-visible spectroscopy is a widely utilized technique in the field of analytical chemistry for the purpose of qualitative analysis and identification of certain component classes in both pure and biological mixtures. UV-visible spectroscopy is commonly employed for quantitative investigation due to the strong chromophoric properties shown by aromatic compounds within the UV region. The identification of natural chemicals can be accomplished by the utilization of UV-visible spectroscopy .Phenolic chemicals, such as anthocyanins, tannins, polymer dyes, and phenols, have been shown to form complexes with iron, which have been identified using ultraviolet/visible (UV-Vis) spectroscopy. Furthermore, it has been observed that spectroscopic UV-Vis methods exhibit lower selectivity and provide insights into the overall composition of polyphenolic compounds. UV-Vis spectroscopy was employed to ascertain the concentrations of several compounds in the sample, including the total phenolic extract at a wavelength of 280 nm, flavones at 320 nm, phenolic acids at 360 nm, and the total anthocyanins at 520 nm. This approach is characterized by its efficiency in terms of time consumption and cost-effectiveness when compared to alternative procedures [204].

12.7.2.4 Infrared Spectroscopy

When infrared radiation traverses an organic substance, certain frequencies are prone to absorption, while others are capable of passing through the sample without undergoing any absorption. The phenomenon of infrared absorption is associated with the alterations in molecular vibrations that occur upon exposure to infrared light. Hence, infrared spectroscopy may be fundamentally characterized as a form of vibrational spectroscopy. Various types of chemical bonds, including C–C, C=C, C≡C, C–O, C=O, O–H, and N–H, exhibit a wide range of vibrational frequencies. The identification of certain types of bonds inside an organic molecule can be accomplished by observing the distinctive absorption band in the infrared spectrum associated with these bonds [205]. Fourier Transform Infrared Spectroscopy (FTIR) is a technique widely employed in analytical chemistry for the purpose of identifying chemical ingredients and elucidating the structural compounds with a high level of precision. Fourier Transform Infrared Spectroscopy (FTIR) provides a quick and non-invasive method for analyzing and identifying the unique characteristics of herbal extracts or powders.

12.7.2.5 Nuclear Magnetic Resonance Spectroscopy

Nuclear Magnetic Resonance (NMR) is largely concerned with the magnetic characteristics shown by certain atomic nuclei, particularly those of the hydrogen atom (proton), carbon, and a carbon isotope. The utilization of NMR spectroscopy has facilitated extensive investigations by researchers in the field of molecular analysis. This technique allows for the detection and characterization of distinct magnetic nuclei, providing valuable insights on the spatial arrangement of these nuclei inside the molecule under study. Additionally, it will serve to illustrate the presence of atoms in adjacent groups. In conclusion, the number of atoms present in each of these habitats may be determined. Previous studies have employed preparative or semi-preparative thin-layer chromatography, liquid chromatography, and column chromatography techniques to separate and isolate distinct phenols. Subsequently, the structures of these phenols are identified by offline nuclear magnetic resonance (NMR) analysis [205].

12.7.2.6 Mass Spectroscopy for Chemical Compound Identification

In the field of mass spectrometry, organic molecules undergo bombardment by either electrons or lasers, resulting in their conversion into charged ions that possess significant levels of energy. A mass spectrum is a graphical representation depicting the proportional abundance of fragmented ions as a function of their mass-to-charge ratio. Mass spectrometry is a very accurate analytical technique that may be employed to measure the relative molecular mass (also known as molecular weight) of a compound. Additionally, by identifying the locations where the molecule has undergone fragmentation, it is possible to ascertain the precise molecular formula of the compound [206]. In prior research, the extraction and purification of bioactive compounds derived from pith were accomplished by a series of techniques including bioactivity-guided solvent extraction, column chromatography, and high-performance liquid chromatography (HPLC) [207]. The structure of the bioactive molecule was characterized using several spectroscopic approaches, including UV-visible, IR, NMR, and mass spectroscopy. In addition, it is possible to subject compounds to hydrolysis and afterwards analyze their derivatives for characterization purposes. The application of tandem mass spectrometry (MS) in mass spectrometry yields substantial data that facilitates the determination of complex structures. Hence, the integration of high-performance liquid chromatography (HPLC) with mass spectrometry (MS) enables expeditious and precise determination of chemical constituents inside medicinal plants, particularly in cases when a pure reference substance is not accessible [208, 209]. In recent times, the utilization of LC/MS has witnessed significant prevalence in the examination of phenolic compounds. The utilization of electrospray ionization (ESI) is favored in scientific research due to its notable capacity for efficient ionization of phenolic compounds.

12.7.3 Personalized Treatment Approaches

Epilepsy is a pathological disorder characterized by the occurrence of repeated, or potentially recurring, seizures resulting from an abnormal surge of electrical activity inside a specific cluster of neurons in the brain [210]. Currently, the therapeutic options available for epilepsy management are predominantly comprised of a diverse array of antiepileptic medicines (AEDs) that operate through various mechanisms of action. However, it is important to note that these medications primarily aim to mitigate symptoms, including seizures, rather than offer a curative solution. The treatment exhibits limited efficacy in a significant proportion of individuals and may occasionally exacerbate seizure activity or elicit unfavorable responses [211]. The diverse origins of epilepsy, the extensive range of distinct syndromes and seizure classifications, together with the varying individual responses to antiepileptic drugs (AEDs), continue to present significant challenges in the management of this disorder [212]. Furthermore, it is important to note that the potential negative consequences of medication administration might be significant and pose a risk to an individual's life. Additionally, many antiepileptic drugs (AEDs) have the potential to exacerbate the management of seizures and potentially trigger the occurrence of previously unseen seizure manifestations. The current consensus supports the notion that genetic variables play a significant role in accounting for the variation observed among individuals in their reaction to antiepileptic drugs (AEDs). This variability may be attributed to mutations occurring in various genes, which can impact drug pharmacokinetics, drug pharmacodynamics, or even contribute to the development of epilepsy. Furthermore, empirical research has demonstrated that the occurrence of epilepsy is associated with alterations in brain structure and function, which can be attributed to the involvement of epigenetic processes [213]. The concept of precision medicine endeavors to elucidate the customization of therapeutic interventions that ideally necessitate a focus on the specific biological mechanisms behind the development and progression of a certain illness. The field of precision medicine has made significant advancements in the realm of oncology, particularly in the area of cancer precision medicine. This specific branch of precision medicine, which was initiated by Barack Obama and received substantial financing, has emerged as a notable achievement in the pursuit of enhanced medical precision. Epilepsy presents a compelling and promising avenue for tailoring treatments due to several factors. Firstly, it has a significant global prevalence, affecting approximately 1% of individuals at the age of 20 and 3% at the age of 75. Furthermore, a considerable number of patients continue to experience seizures or encounter adverse reactions to existing medications. Additionally, the genetic underpinnings of numerous epileptic syndromes have been extensively investigated, with ongoing discoveries of novel genes [214].

12.7.3.1 Precision Therapy

The management of epilepsy continues to rely predominantly on empirical scientific approaches, and the selection of medications for individuals with epilepsy cannot be exclusively guided by an understanding of their underlying mechanisms of action. The efficacy of individualized therapy is constrained by the wide range of clinical phenotypes and the diverse etiology that underlies the condition. Recent advancements in the field of genetics mechanisms, neuroimaging studies, and epilepsy neurobiology have yielded valuable insights regarding the selection of pharmaceutical treatments. These findings have paved the way for a new era in epilepsy treatment, wherein patients can receive therapies tailored to the underlying causes of their conditions [215]. The latest anti-epileptic drugs (AEDs) have several therapeutic benefits in contrast to conventional and previous iterations of therapy. Notably, they exhibit a reduced risk profile of adverse effects and much fewer instances of medication interactions. Nevertheless, the current approach to therapy remains predominantly empirical, and the ability to prescribe medication based on the mechanism of action specific to each patient is typically lacking. One inherent issue is the limited understanding or lack of comprehensive clarity on the primary mechanisms of action and biochemical impacts exhibited by pharmaceutical substances. This might be attributed, in part, to the limited knowledge of the specific molecular targets of existing antiepileptic drugs (AEDs). Hence, the process of medication discovery lacks specificity and instead depends on advancements made through the utilization of *in vivo* seizure models. Moreover, as a result of these models, conventional antiepileptic treatments are restricted to managing the symptoms of epilepsy rather than avoiding the occurrence of epileptogenic episodes. It is quite probable that a comprehensive comprehension of epileptogenesis will be important in the development of innovative treatment strategies. Research is mostly focused on investigating the mechanisms of action and the molecular processes that illnesses undergo [216]. The utilization of particular molecular diagnostic tools in epileptic syndrome enables the verification of genetic factors, which may significantly enhance the accuracy of prognosis determination and the assessment of epileptic seizure recurrence [217]. Besides, it provides a valuable chance to get more comprehensive data on focused therapeutic interventions and the advancement of tailored pharmaceuticals. The significance of enhanced knowledge has great value, particularly in those diagnosed with epileptic encephalopathies. This category encompasses a set of neurodevelopmental illnesses defined by pronounced epileptic activity, which is accompanied by a decline in neurological development [218]. Contemporary pharmaceutical interventions primarily attenuate neural excitability through the modulation of ion channels and neurotransmitter receptors. The latest procurements have unveiled other avenues that demonstrate distinct processes, including synaptic vesicle traffic, mammalian target of rapamycin (mTOR) signaling, chromatin remodeling, and transcription. As a result, novel therapeutic approaches have emerged [219].

The genetic basis of epilepsy remains highly intricate, since various mutations in distinct genes can give rise to identical syndromes. Conversely, mutations in a single gene, such as SCN1A, can exhibit a broad spectrum of symptoms, spanning from febrile seizures to severe epileptic encephalopathies [220].

12.7.3.2 Drug Use

The explanation for the varying responses, including both positive and negative (paradoxical), to various antiepileptic drugs (AEDs) can be attributed, at least partially, to a certain type of epilepsy. The clinical manifestation of Dravet syndrome may be exacerbated by the administration of carbamazepine and phenytoin since this disorder is attributed to genetic abnormalities in the sodium channel gene (SCN1A), and both medications impede the functioning of the aforementioned channel by means of blockade. On the other hand, the therapeutic approach of choice for the epileptic condition linked to mutations in SCN8A (a sodium channel gene) and KCNQ2 genes is the inhibition of sodium channels [221]. Lacosamide (LTG) has been identified as a well-established antagonist of both sodium channels and N-type calcium channels, leading to its utilization in several medical contexts. However, the application of LTG has generated significant debate and controversy. Certain studies have posited its role in the worsening of seizures, thereby prompting research efforts to discourage its utilization in individuals diagnosed with Dravet syndrome. Conversely, alternative research posits a favorable impact on certain individuals diagnosed with Dravet syndrome. The observed positive impact may be elucidated through the mechanism that entails the activation of cyclic nucleotide channels by hyperpolarization processes [222].

Currently, the accepted treatment for Dravet syndrome involves the administration of a combination of three medications, namely valproate, clobazam, and frequently stiripentol. In relation to stiripentol, it is noteworthy that this particular drug has been subjected to a significant randomized controlled trial in the treatment of Dravet syndrome, specifically when administered in conjunction with valproate and clobazam. However, it is widely acknowledged that the utilization of stiripentol, valproate, and clobazam may result in the occurrence of severe adverse effects. Furthermore, it is commonly recognized that while stiripentol has the capacity to decrease the frequency of critical episodes, it does not entirely eradicate them [223].

12.8 Conclusion

12.8.1 Summary of Key Findings

The current review made it abundantly evident that antioxidants are essential for combating both oxidative stress and some degenerative illnesses. Medicinal plants have been considered a potential factor in this situation. Because 70–80% of the world's population is thought to be unable to afford modern medication, using medicinal herbs can be a significant source of free radicals. The review demonstrates that the majority of plant-based extracts and chemicals are helpful in reducing the symptoms of numerous diseases, additional research is required to realize the full therapeutic potential. Additionally, it is necessary to isolate bioactive components from medicinal plants and investigate how they affect a variety of health issues, including the prevention or treatment of diseases caused by free radicals. Therefore, such information could be exploited as cost-effective measure for the health of well-being.

12.8.2 Implications for Epilepsy Management

One of three main processes may underlie the action of antiepileptic medications, (i) lowering the electrical excitability of cell membranes, especially (by blocking) the voltage-dependent sodium channels that are in charge of the inward current that generates an action potential, (ii) enhancing GABA-mediated synaptic inhibition, by inhibiting GABA transaminase or by medications with direct GABA agonist properties, the result is increased membrane permeability to chloride ion, which lowers cell excitability, and (iii) inhibiting T-type glutamate. Here are some typical contraindications.

12.8.2.1 Epilepsy and Oral Contraceptives

Antiepileptic medications such carbamazepine, phenytoin, barbiturates, topiramate, and oxcarbazepine might cause steroid metabolizing enzymes to be activated, which can interfere with hormonal contraception. As a result, people taking antiepileptic medications need a larger dose of an oral contraceptive that contains estrogen (50 mg/day).

12.8.2.2 Pregnancy and Epilepsy

Problems with managing epilepsy during pregnancy could affect both the mother and the fetus. Women with epilepsy have higher rates of spontaneous abortion and stillbirth. Due to the risk of anoxia and metabolic condition, patients should have their seizure disorder thoroughly investigated and treated before becoming pregnant with the lowest possible dose. Patients should be urged to take oral vitamin K and folic acid supplements since they are likely innocuous and should not be eliminated. because folic acid deficiency is a risk factor for neural tube abnormalities and several antiepileptic medications influence folic acid metabolism. Antiepileptic medications that cause hepatic enzyme induction decrease the mother's vitamin K levels, which can exacerbate any postpartum hemorrhage.

12.8.2.3 Breast Feeding

Except barbiturates and ethosuximide, antiepileptics typically pass into breast milk in low concentrations, and breastfeeding is regarded as safe when administered in standard doses. The benzodiazepine and barbiturate drugs may cause problems for newborn sedation, and since ethosuximide is transported in considerable levels in breast milk, breastfeeding should be avoided.

12.8.2.4 Epilepsy in Children

Children are treated for fits in the same way that adults are, however they may react differently and get agitated, such as when given sodium valproate or phenobarbital. If febrile convulsions have occurred, a significant epilepsy medication (such as phenobarbitone, 3–4 mg/kg/day) may be administered consistently until the kid is 5 years old. However, prolonged medication therapy, such as that with phenobarbital or phenytoin, may hinder the development of cognitive function.

Thus, the effectiveness of an anticonvulsant medication for particular seizure types and epilepsy is the main factor considered when choosing one. Although most patients have satisfactory seizure control, a considerable portion of epilepsy patients still have uncontrollable or drug-resistant epilepsy despite receiving prompt treatment and the recommended daily dosage of an effective anticonvulsant. Thus, compared to the current antiepileptic treatments, there is a need for novel medications with better adverse effects and tolerance profiles, even at the sacrifice of efficacy.

Although there are many therapies available, a lot of work is put into developing new methods. Many of these methods center on understanding the genetic, cellular, and molecular causes of hyperexcitability, these revelations promise to reveal precise targets for cutting-edge treatments.

12.8.3 Future Perspectives and Recommendations

Both contemporary western medicine and folk or traditional medicine around the world use herbal medicine for epilepsy therapy. The existing literature contains studies on hundreds of herbal remedies. Some popular herbal medications have been evaluated in our study because it is difficult to describe all herbal treatments for epilepsy in one document.

These statistics are largely based on studies using animals. As stated in the data above, none of them, to our knowledge, have strong proof for therapeutic medicine, despite the antiepileptic effects of some of them demonstrated in animal research being widely established. Additionally, the effectiveness and toxicity of the herbal treatment should be examined using evidence. But there is a difficult path to follow.

Invariably produced as a by-product by plants during various metabolic processes in different cellular compartments such as chloroplast, mitochondria, peroxisomes, cytosol, plasma membrane, and apoplastic space, ROS are partially reduced or activated derivatives of oxygen ($^1O_2$, O_2, H_2O_2, OH, and HO_2) that are highly reactive and toxic and can cause oxidative stress and cell death. Cell viability can be impacted by ROS because they can interact with biological components (lipids, proteins, and nucleic acids) and result in lipid peroxidation, protein oxidation, DNA damage, membrane damage, and enzyme deactivation. The equilibrium between ROS generation and scavenging is disturbed by a variety of biotic and abiotic stresses, such as plant diseases, drought, salinity, extreme temperatures, excess light, pollutants, xenobiotics, ultraviolet radiation, ozone, hypoxia, nutrient deficiency, and so on. This causes an overproduction and higher accumulation of ROS, which in turn causes oxidative stress. High levels of ROS in plant cells negatively impact a number of biological processes by oxidizing proteins, lipids, and nucleic acids, which ultimately leads to cell death. One of the main factors contributing to low crop yield across the globe is regarded to be ROS toxicity, which results from numerous biotic and abiotic stress factors. Plants need ROS as second messengers in signal transduction cascade in a variety of physiological activities, therefore they cannot be entirely eradicated. It is still not clearly understood how this dual role is controlled in plants, however, it is quite clear that the steady-state level of ROS in cells needs to be tightly regulated. Additional genomic, proteomic, and metabolomics research along with recently developed technologies may provide light on the networks underlying several ROS-related plant activities. The identification of genes that can be used to manipulate/transform ROS-related metabolisms in plants for the development of superior transgenic crop plants resistant to biotic and abiotic stress may result from a deep and comprehensive understanding of the ROS gene network.

References

1. Liu, W., Xu, J., Zhang, L., Li, F., Zhang, L., Tai, Z., Yang, J., Zhang, H., Tuo, J., Yu, C., Xu, Z., Research progress on correlations between trace element levels and epilepsy. *Front. Cell Dev. Biol.*, 11, 1167626, 2023.
2. Olowe, R., Sandouka, S., Saadi, A., Shekh-Ahmad, T., Approaches for reactive oxygen species and oxidative stress quantification in epilepsy. *Antioxidants*, 14,9, 10, 990, 2020.
3. Koh, S., Dupuis, N., Auvin, S., Ketogenic diet and neuroinflammation. *Epilepsy Res.*, 167, 106454, 2020.
4. Kaltsas, A., Zikopoulos, A., Moustakli, E., Zachariou, A., Tsirka, G., Tsiampali, C., Palapela, N., Sofikitis, N., Dimitriadis, F., The Silent Threat to Women's Fertility, Uncovering the Devastating Effects of Oxidative Stress. *Antioxidants*, 12, 8, 1490, 2023.
5. Meng, L. and Wu, G., Recent advances in small molecules for improving mitochondrial disorders. *RSC Adv.*, 13, 30, 20476–85, 2023.

6. Kim, S.K., Lee, G.Y., Kim, S.K., Kwon, Y.J., Seo, E.B., Lee, H., Lee, S.H., Kim, S.J., Lee, S., Ye, S.K., Protective Effects of Repetitive Transcranial Magnetic Stimulation Against Streptozotocin-Induced Alzheimer's Disease. *Mol. Neurobiol.*, 61, 1–7, 2023 Sep 27.
7. Zilberter, Y., Tabuena, D.R., Zilberter, M., NOX-induced oxidative stress is a primary trigger of major neurodegenerative disorders. *Prog. Neurobiol.*, 231, 102539, 2023 Oct 12.
8. Sun, Y., Chen, J., Shi, X., Li, Z., Wan, L., Yan, H., Chen, Y., Wang, J., Wang, J., Zou, L., Reiter, R., Safety and efficacy of melatonin supplementation as an add-on treatment for infantile epileptic spasms syndrome, A randomized, placebo-controlled, double-blind trial. *J. Pineal Res.*, 76, e12922, 2023 Nov 1.
9. Ogunsuyi, O.B., Aro, P.O., Umar, H.I., Oboh, G., Administration of Curcumin plus Treadmill Exercise Modulate Some Neuronal Enzyme Activities, Redox Markers and BDNF mRNA Expression in Pilocarpine-Induced Epileptic Seizure in Rats. *Neurochem. J.*, 17, 3, 467–76, 2023 Sep.
10. Verrotti, A., Basciani, F., Trotta, D., Pomilio, M.P., Morgese, G., Chiarelli, F., Serum copper, zinc, selenium, glutathione peroxidase and superoxide dismutase levels in epileptic children before and after 1 year of sodium valproate and carbamazepine therapy. *Epilepsy Res.*, 1,48, 1-2, 71–5, 2002.
11. Hu, X., Gong, S., He, Q., Wu, J.L., Li, N., Less is more, A new perspective for toxicity of emerging contaminants by structures, protein adducts and proteomics. *TrAC, Trends Anal. Chem.*, 117289, 2023.
12. Fisher, R.S., Boas, W.V., Blume, W., Elger, C., Genton, P., Lee, P., Engel Jr, J., Epileptic seizures and epilepsy, definitions proposed by the International League Against Epilepsy (ILAE) and the International Bureau for Epilepsy (IBE). *Epilepsia*, 46, 4, 470–2, 2003.
13. Asadi-Pooya, A.A., Brigo, F., Lattanzi, S., Blumcke, I., Adult epilepsy. *Lancet*, 402, 10399, 412–24, 2023.
14. Jomaa, N., Nasreddine, W., Hmeimess, G., Beaini, S., Beydoun, A., Hotait, M., Wazne, J., Beydoun, A., Risk of Recurrence in Patients with an Unprovoked Tonic-Clonic Seizure and Generalized Epileptiform Discharges on EEG. *Epilepsia*, 64, 2153–2163, 2023.
15. Zhu, H., Wang, W., Li, Y., Molecular mechanism and regulation of autophagy and its potential role in epilepsy. *Cells*, 11, 17, 2621, 2022.
16. Aaberg, K.M., Surén, P., Søraas, C.L., Bakken, I.J., Lossius, M.I., Stoltenberg, C., Chin, R., Seizures, syndromes, and etiologies in childhood epilepsy, The International League Against Epilepsy 1981, 1989, and 2017 classifications used in a population-based cohort. *Epilepsia*, 58, 11, 1880–91, 2017.
17. McCabe, C., Feeney, A., Basa, M., Eustace-Cook, J., McCann, M., Nurses knowledge, attitudes and education needs towards acute pain management in hospital settings, A meta-analysis. *J. Clin. Nurs.*, 32, 4325–4336, 2023 Jan 12.
18. Girma, B., Nigussie, J., Tamir, T., Bekele, E., Public knowledge toward Epilepsy and its determinants in Ethiopia, A systematic review and meta-analysis. *Epilepsy & Behav.*, 133, 108764, 2022.
19. VanMeter, K.C. and Hubert, R.J., Pathophysiology for the Health Professions E-Book. *Elsevier Health Sci.*, 2022.
20. Fonferko-Shadrach, B., Lacey, A.S., White, C.P., Powell, H.R., Sawhney, I.M., Lyons, R.A., Smith, P.E., Kerr, M.P., Rees, M.I., Pickrell, W.O., Validating epilepsy diagnoses in routinely collected data. *Seizure*, 1, 52, 195–8, 2017.
21. Lawal, M., Omobayo, H., Lawal, K., Epilepsy, pathophysiology, clinical manifestations and treatment options. *Br. J. Neurosci. Nurs.*, 2,14, 2, 58–72, 2018.
22. Halliwell, B. and Gutteridge, J.M., *Free radicals in biology and medicine*, Oxford University Press, USA, 2015.

23. Wilson, D.M., Cookson, M.R., Van Den Bosch, L., Zetterberg, H., Holtzman, D.M., Dewachter, I., Hallmarks of neurodegenerative diseases. *Cell*, 186, 4, 693–714, 2023 Feb 16.
24. Torres-Martinez, N., Chabardes, S., Mitrofanis, J., Lights for epilepsy, can photobiomodulation reduce seizures and offer neuroprotection? *Neural Regener. Res.*, 18, 7, 1423, 2023 Jul.
25. Upaganlawar, A.B., Wankhede, N.L., Kale, M.B., Umare, M.D., Sehgal, A., Singh, S., Bhatia, S., Al-Harrasi, A., Najda, A., Nurzyńska-Wierdak, R., Bungau, S., Interweaving epilepsy and neurodegeneration, Vitamin E as a treatment approach. *Biomed. Pharmacother.*, 143, 112146, 2021.
26. Nguyen, T., Al-Juboori, M.H., Walerstein, J., Xiong, W., Jin, X., Impaired glutamate receptor function underlies early activity loss of ipsilesional motor cortex after closed-head mild traumatic brain injury. *J. Neurotrauma*, 38, 14, 2018–29, 2021.
27. Chen, Q., Nan, Y., Yang, Y., Xiao, Z., Liu, M., Huang, J., Xiang, Y., Long, X., Zhao, T., Wang, X., Huang, Q., Nanodrugs alleviate acute kidney injury, Manipulate RONS at kidney. *Bioact. Mater.*, 22, 141–67, 2023.
28. Hajam, Y.A., Rani, R., Ganie, S.Y., Sheikh, T.A., Javaid, D., Qadri, S.S., Pramodh, S., Alsulimani, A., Alkhanani, M.F., Harakeh, S., Hussain, A., Oxidative stress in human pathology and aging, Molecular mechanisms and perspectives. *Cells*, 11, 3, 552, 2022.
29. Morris, G., Gevezova, M., Sarafian, V., Maes, M., Redox regulation of the immune response. *Cell. Mol. Immunol.*, 19, 10, 1079–101, 2022.
30. Pearson-Smith, J.N. and Patel, M., Metabolic dysfunction and oxidative stress in epilepsy. *Int. J. Mol. Sci.*, 18, 11, 2365, 2017.
31. Yang, N., Guan, Q.W., Chen, F.H., Xia, Q.X., Yin, X.X., Zhou, H.H., Mao, X.Y., Antioxidants targeting mitochondrial oxidative stress, Promising neuroprotectants for epilepsy. *Oxid. Med. Cell. Longevity*, 2020, 6687185, 2020 Oct.
32. Covarrubias, A.J., Perrone, R., Grozio, A., Verdin, E., NAD+ metabolism and its roles in cellular processes during ageing. *Nat. Rev. Mol. Cell Biol.*, 22, 2, 119–41, 2021.
33. Hahn, A. and Zuryn, S., The cellular mitochondrial genome landscape in disease. *Trends Cell Biol.*, 29, 3, 227–40, 2019.
34. Chen, K., Lu, P., Beeraka, N.M., Sukocheva, O.A., Madhunapantula, S.V., Liu, J., Sinelnikov, M.Y., Nikolenko, V.N., Bulygin, K.V., Mikhaleva, L.M., Reshetov, I.V., Mitochondrial mutations and mitoepigenetics, Focus on regulation of oxidative stress-induced responses in breast cancers, in: *In Seminars in Cancer Biology*, vol. 83, pp. 556–569, Academic Press, United States of America, 2022 Aug.
35. Shemiakova, T., Ivanova, E., Grechko, A.V., Gerasimova, E.V., Sobenin, I.A., Orekhov, A.N., Mitochondrial dysfunction and DNA damage in the context of pathogenesis of atherosclerosis. *Biomedicines*, 8, 6, 166, 2020.
36. Kaarniranta, K., Pawlowska, E., Szczepanska, J., Jablkowska, A., Blasiak, J., Role of mitochondrial DNA damage in ROS-mediated pathogenesis of age-related macular degeneration (AMD). *Int. J. Mol. Sci.*, 14, 20, 10, 2374, 2019.
37. Hou, J., Yin, W., Li, P., Huang, Y., Wan, Y., Hu, C., Xu, T., Cheng, J., Wang, L., Yu, Z., Yuan, J., Effect of exposure to phthalates on association of polycyclic aromatic hydrocarbons with 8-hydroxy-2′-deoxyguanosine. *Sci. Total Environ.*, 15, 691, 378–92, 2019.
38. Armada-Moreira, A., Gomes, J.I., Pina, C.C., Savchak, O.K., Gonçalves-Ribeiro, J., Rei, N., Pinto, S., Morais, T.P., Martins, R.S., Ribeiro, F.F., Sebastião, A.M., Going the extra (synaptic) mile, excitotoxicity as the road toward neurodegenerative diseases. *Front. Cell. Neurosci.*, 14, 90, 2020.
39. Vesela, B., Svandova, E., Ramesova, A., Kratochvilova, A., Tucker, A.S., Matalova, E., Caspase inhibition affects the expression of autophagy-related molecules in chondrocytes. *Cartilage*, 13, 2_suppl, 956S–68S, 2021.
40. Rho, J.M. and Boison, D., The metabolic basis of epilepsy. *Nat. Rev. Neurol.*, 18, 6, 333–47, 2022.

41. Li, W., He, P., Huang, Y., Li, Y.F., Lu, J., Li, M., Kurihara, H., Luo, Z., Meng, T., Onishi, M., Ma, C., Selective autophagy of intracellular organelles, recent research advances. *Theranostics*, 11, 1, 222, 2021.
42. Sies, H. and Jones, D.P., Reactive oxygen species (ROS) as pleiotropic physiological signalling agents. *Nat. Rev. Mol. Cell Biol.*, 21, 7, 363–83, 2020.
43. Li, Y., Jin, M., Yin, X., Zhou, B., Ni, H., Effects of leptin treatment immediately after neonatal seizures on serum clusterin and VEGF levels and brain oxidative stress-related proteins and neurobehavioral phenotypes. *Epilepsy Behav.*, 138, 109016, 2023.
44. Mao, X.Y., Zhou, H.H., Jin, W.L., Redox-related neuronal death and crosstalk as drug targets, focus on epilepsy. *Front. Neurosci.*, 13, 512, 2019.
45. D'arcy, M.S., Cell death, a review of the major forms of apoptosis, necrosis and autophagy. *Cell Biol. Int.*, 43, 6, 582–92, 2019.
46. ToksoySARITAŞ, C.K.Z., Börü, U., Demiral, G.Z., Görücü, F., Bülbül, A., Demirel, H.H., Koç, Y., Investigation of the protective effect of anzer propolis in cerebral ischemia-reperfusion injury. *Eur. Rev. Med. Pharmacol. Sci.*, 27, 17, 8004–12, 2023.
47. Korczowska-Łącka, I., Słowikowski, B., Piekut, T., Hurła, M., Banaszek, N., Szymanowicz, O., Jagodziński, P.P., Kozubski, W., Permoda-Pachuta, A., Dorszewska, J., Disorders of Endogenous and Exogenous Antioxidants in Neurological Diseases. *Antioxidants*, 12, 10, 1811, 2023.
48. Qiu, X., Cao, L., Yang, X., Zhao, X., Liu, X., Han, Y., Xue, Y., Jiang, H., Chi, Z., Role of mitochondrial fission in neuronal injury in pilocarpine-induced epileptic rats. *Neuroscience*, 245, 157–65, 2013.
49. Komada, T. and Muruve, D.A., *Nat. Rev. Nephrol.*, 15, 8, 501–20, The role of inflammasomes in kidney disease2019.
50. Muri, J. and Kopf, M., Redox regulation of immunometabolism. *Nat. Rev. Immunol.*, 21, 6, 363–81, 2019.
51. Marchi, S., Guilbaud, E., Tait, S.W., Yamazaki, T., Galluzzi, L., Mitochondrial control of inflammation. *Nat. Rev. Immunol.*, 23, 3, 159–73, 2023.
52. Kelley, N., Jeltema, D., Duan, Y., He, Y., The NLRP3 inflammasome, an overview of mechanisms of activation and regulation. *Int. J. Mol. Sci.*, 20, 13, 3328, 2019.
53. Aggarwal, V., Tuli, H.S., Varol, A., Thakral, F., Yerer, M.B., Sak, K., Varol, M., Jain, A., Khan, M.A., Sethi, G., Role of reactive oxygen species in cancer progression, molecular mechanisms and recent advancements. *Biomolecules*, 9, 11, 735, 2019.
54. Xing, J., Han, D., Xu, D., Li, X., Sun, L., CREB protects against temporal lobe epilepsy associated with cognitive impairment by controlling oxidative neuronal damage. *Neurodegenerative Dis.*, 19, 5-6, 225–371, 2020.
55. Waldbaum, S. and Patel, M., Mitochondrial dysfunction and oxidative stress, a contributing link to acquired epilepsy? *J. Bioenerg. Biomembr.*, 42, 449–55, 2010.
56. Puttachary, S., Sharma, S., Stark, S., Thippeswamy, T., Seizure-induced oxidative stress in temporal lobe epilepsy. *Biomed Res. Int.*, 2015, 745612, 2015 Oct.
57. Kalozoumi, G., Kel-Margoulis, O., Vafiadaki, E., Greenberg, D., Bernard, H., Soreq, H., Depaulis, A., Sanoudou, D., Glial responses during epileptogenesis in Mus musculus point to potential therapeutic targets. *PLoS One*, 13, 8, e0201742, 2018.
58. McElroy, P.B., Liang, L.P., Day, B.J., Patel, M., Scavenging reactive oxygen species inhibits status epilepticus-induced neuroinflammation. *Exp. Neurol.*, 298, 13–22, 2017.
59. Borowicz-Reutt, K.K. and Czuczwar, S.J., Role of oxidative stress in epileptogenesis and potential implications for therapy. *Pharmacol. Rep.*, 72, 5, 1218–26, 2020.
60. Amini, E., Rezaei, M., Mohamed Ibrahim, N., Golpich, M., Ghasemi, R., Mohamed, Z., Raymond, A.A., Dargahi, L., Ahmadiani, A., A molecular approach to epilepsy management,

from current therapeutic methods to preconditioning efforts. *Mol. Neurobiol.*, 52, 492–513, 2015.

61. Sies, H., Belousov, V.V., Chandel, N.S., Davies, M.J., Jones, D.P., Mann, G.E., Murphy, M.P., Yamamoto, M., Winterbourn, C., Defining roles of specific reactive oxygen species (ROS) in cell biology and physiology. *Nat. Rev. Mol. Cell Biol.*, 23, 7, 499–515, 2022.
62. Ali, N.H., Al-Kuraishy, H.M., Al-Gareeb, A.I., Alnaaim, S.A., Alexiou, A., Papadakis, M., Saad, H.M., Batiha, G.E., Autophagy and autophagy signaling in Epilepsy, possible role of autophagy activator. *Mol. Med.*, 29, 1, 142, 2023.
63. Tamijani, S.M., Karimi, B., Amini, E., Golpich, M., Dargahi, L., Ali, R.A., Ibrahim, N.M., Mohamed, Z., Ghasemi, R., Ahmadiani, A., Thyroid hormones, possible roles in epilepsy pathology. *Seizure*, 31, 155–64, 2015.
64. Vezzani, A., Ravizza, T., Bedner, P., Aronica, E., Steinhäuser, C., Boison, D., Astrocytes in the initiation and progression of epilepsy. *Nat. Rev. Neurol.*, 12, 707–22, 2022.
65. Vezzani, A., Balosso, S., Ravizza, T., Neuroinflammatory pathways as treatment targets and biomarkers in epilepsy. *Nat. Rev. Neurol.*, 15, 8, 459–72, 2019.
66. Walters, G.C. and Usachev, Y.M., Mitochondrial calcium cycling in neuronal function and neurodegeneration. *Front. Cell Dev. Biol.*, 11, 1094356, 2023.
67. Atanasova, M., Petkova, Z., Pechlivanova, D., Dragomirova, P., Blazhev, A., Tchekalarova, J., Strain-dependent effects of long-term treatment with melatonin on kainic acid-induced status epilepticus, oxidative stress and the expression of heat shock proteins. *Pharmacol. Biochem. Behav.*, 111, 44–50, 2013.
68. Kobylarek, D., Iwanowski, P., Lewandowska, Z., Limphaibool, N., Szafranek, S., Labrzycka, A., Kozubski, W., Advances in the potential biomarkers of epilepsy. *Front. Neurol.*, 10, 685, 2019.
69. Woźniak, A., Bednarski, W., Dancewicz, K., Gabryś, B., Borowiak-Sobkowiak, B., Bocianowski, J., Samardakiewicz, S., Rucińska-Sobkowiak, R., Morkunas, I., Oxidative stress links response to lead and Acyrthosiphon pisum in Pisum sativum L. *J. Plant Physiol.*, 240, 152996, 2019.
70. Szymańska, R. and Strzałka, K., Reactive oxygen species in plants—production, deactivation and role in signal transduction. *POSTĘPY BIOCHEMII.*, 190, 182–190, 2008.
71. Hasanuzzaman, M., Bhuyan, M.B., Zulfiqar, F., Raza, A., Mohsin, S.M., Mahmud, J.A., Fujita, M., Fotopoulos, V., Reactive oxygen species and antioxidant defense in plants under abiotic stress, Revisiting the crucial role of a universal defense regulator. *Antioxidants*, 9, 8, 681, 2020.
72. Sachdev, S., Ansari, S.A., Ansari, M.I., Fujita, M., Hasanuzzaman, M., Abiotic stress and reactive oxygen species, Generation, signaling, and defense mechanisms. *Antioxidants*, 10, 2, 277, 2021.
73. Abbasi, Q., Pourakbar, L., Moghaddam, S.S., Potential role of apple wood biochar in mitigating mercury toxicity in corn (Zea mays L.). *Ecotoxicol. Environ. Saf.*, 267, 115619, 2023.
74. Rajput, V.D., Harish, Singh, RK, Verma, KK, Sharma, L., Quiroz-Figueroa, FR, Meena, M., Gour, VS, Minkina, T., Sushkova, S., Mandzhieva, S., Recent developments in enzymatic antioxidant defence mechanism in plants with special reference to abiotic stress. *Biology*, 10, 4, 267, 2021.
75. Chandran, H., Meena, M., Swapnil, P., Plant growth-promoting rhizobacteria as a green alternative for sustainable agriculture. *Sustainability*, 13, 19, 10986, 2021.
76. Guo, T., Zhang, G., Zhou, M., Wu, F., Chen, J., Effects of aluminum and cadmium toxicity on growth and antioxidant enzyme activities of two barley genotypes with different Al resistance. *Plant Soil*, 258, 241–8, 2004.
77. Skórzyńska-Polit, E., Dra̧żkiewicz, M., Krupa, Z., The activity of the antioxidative system in cadmium-treated Arabidopsis thaliana. *Biol. Plant.*, 47, 1, 71–8, 2003.
78. Hsu, Y.T. and Kao, C.H., Cadmium toxicity is reduced by nitric oxide in rice leaves. *Plant Growth Regul.*, 42, 227–38, 2004.

79. El Rasafi, T., Oukarroum, A., Haddioui, A., Song, H., Kwon, E.E., Bolan, N., Tack, F.M., Sebastian, A., Prasad, M.N., Rinklebe, J., Cadmium stress in plants,A critical review of the effects, mechanisms, and tolerance strategies. *Crit. Rev. Environ. Sci. Technol.*, 52, 5, 675–726, 2022.
80. Shahid, M., Dumat, C., Khalid, S., Niazi, N.K., Antunes, P.M., Cadmium bioavailability, uptake, toxicity and detoxification in soil-plant system. *Rev. Environ. Contam. Toxicol.*, 241, 73–137, 2017.
81. Haider, F.U., Liqun, C., Coulter, J.A., Cheema, S.A., Wu, J., Zhang, R., Wenjun, M., Farooq, M., Cadmium toxicity in plants, Impacts and remediation strategies. *Ecotoxicol. Environ. Saf.*, 211, 111887, 2021.
82. Zhao, L., Bai, T., Wei, H., Gardea-Torresdey, J.L., Keller, A., White, J.C., Nanobiotechnology-based strategies for enhanced crop stress resilience. *Nat. Food*, 3, 10, 829–36, 2022.
83. Liu, Z., Wu, X., Hou, L., Ji, S., Zhang, Y., Fan, W., Li, T., Zhang, L., Liu, P., Yang, L., Effects of cadmium on transcription, physiology, and ultrastructure of two tobacco cultivars. *Sci. Total Environ.*, 869, 161751, 2023.
84. Isah, T., Stress and defense responses in plant secondary metabolites production. *Biol. Res.*, 52, 2019.
85. Ghani, M.I., Saleem, S., Rather, S.A., Rehmani, M.S., Alamri, S., Rajput, V.D., Kalaji, H.M., Saleem, N., Sial, T.A., Liu, M., Foliar application of zinc oxide nanoparticles, An effective strategy to mitigate drought stress in cucumber seedling by modulating antioxidant defense system and osmolytes accumulation. *Chemosphere*, 289, 133202, 2022.
86. León, A.M., Palma, J.M., Corpas, F.J., Gómez, M., Romero-Puertas, M.C., Chatterjee, D., Mateos, R.M., del Río, L.A., Sandalio, L.M., Antioxidative enzymes in cultivars of pepper plants with different sensitivity to cadmium. *Plant Physiol. Biochem.*, 40, 10, 813–20, 2002.
87. Balestrasse, K.B., Gardey, L., Gallego, S.M., Tomaro, M.L., Response of antioxidant defence system in soybean nodules and roots subjected to cadmium stress. *Funct. Plant Biol.*, 28, 6, 497–504, 2001.
88. Iannelli, M.A., Pietrini, F., Fiore, L., Petrilli, L., Massacci, A., Antioxidant response to cadmium in Phragmites australis plants. *Plant Physiol. Biochem.*, 40, 11, 977–82, 2002.
89. Cho, U.H. and Seo, N.H., Oxidative stress in Arabidopsis thaliana exposed to cadmium is due to hydrogen peroxide accumulation. *Plant Sci.*, 168, 1, 113–20, 2005.
90. Garg, N. and Manchanda, G.J., ROS generation in plants, boon or bane? *Plant Biosyst.*, 143, 1, 81–96, 2009.
91. E yidogan, F. and Öz, M.T., Effect of salinity on antioxidant responses of chickpea seedlings. *Acta Physiol. Plant.*, 29, 485–93, 2007.
92. Sonmez, M.C., Ozgur, R., Uzilday, B., Reactive oxygen species, connecting eustress, hormesis, and allostasis in plants. *Plant Stress*, 8, 100164, 2023.
93. Gonzales, M.M., Garbarino, V.R., Pollet, E., Palavicini, J.P., Kellogg, D.L., Kraig, E., Orr, M.E., Biological aging processes underlying cognitive decline and neurodegenerative disease. *J. Clin. Invest.*, 132, 10, 2022 May 16.
94. Patel, M. and Parida, A.K., Salinity alleviates arsenic stress-induced oxidative damage via antioxidative defense and metabolic adjustment in the root of the halophyte Salvadora persica. *Planta*, 258, 6, 109, 2023 Dec.
95. Asada, K., The water-water cycle in chloroplasts, scavenging of active oxygens and dissipation of excess photons. *Annu. Rev. Plant Biol.*, 50, 1, 601–39, 1999.
96. Wang, B., Yuan, X., Zhu, Y., Wang, Y., Li, D., He, J., Xiao, Y., Low-level cadmium exposure induced hormesis in peppermint young plant by constantly activating antioxidant activity based on physiological and transcriptomic analyses. *Front. Plant Sci.*, 14, 1088285, 2023.

97. Milone, M.T., Sgherri, C., Clijsters, H., Navari-Izzo, F., Antioxidative responses of wheat treated with realistic concentration of cadmium. *Environ. Exp. Bot.*, 50, 3, 265–76, 2003.
98. Aravind, P. and Prasad, M.N., Zinc alleviates cadmium-induced oxidative stress in Ceratophyllum demersum L., a free floating freshwater macrophyte. *Plant Physiol. Biochem.*, 41, 4, 391–7, 2003.
99. Pan, Y., Wu, L.J., Yu, Z.L., Effect of salt and drought stress on antioxidant enzymes activities and SOD isoenzymes of liquorice (Glycyrrhiza uralensis Fisch). *Plant Growth Regul.*, 49, 157–65, 2006.
100. Demidchik, V., Mechanisms of oxidative stress in plants, from classical chemistry to cell biology. *Environ. Exp. Bot.*, 109, 212–28, 2015 Jan 1.
101. Creissen, G., Firmin, J., Fryer, M., Kular, B., Leyland, N., Reynolds, H., Pastori, G., Wellburn, F., Baker, N., Wellburn, A., Mullineaux, P., Elevated glutathione biosynthetic capacity in the chloroplasts of transgenic tobacco plants paradoxically causes increased oxidative stress. *Plant Cell*, 11, 7, 1277–91, 1999.
102. Mittler, R., Oxidative stress, antioxidants and stress tolerance. *Trends Plant Sci.*, 7, 9, 405–10, 2002.
103. Bilgin, D.D., ROS, oxidative stress and engineering resistance in higher plants, in: *Reactive oxygen species and antioxidants in higher plants*, pp. 205–27, CRC, Boca Raton, FL, 2010.
104. Demirevska-Kepova, K., Simova-Stoilova, L., Stoyanova, Z.P., Feller, U., Cadmium stress in barley, growth, leaf pigment, and protein composition and detoxification of reactive oxygen species. *J. Plant Nutr.*, 29, 3, 451–68, 2006.
105. Yang, Y., Han, C., Liu, Q., Lin, B., Wang, J., Effect of drought and low light on growth and enzymatic antioxidant system of Picea asperata seedlings. *Acta Physiol. Plant.*, 30, 433–40, 2008.
106. Romero-Puertas, M.C., Corpas, F.J., Rodríguez-Serrano, M., Gómez, M., Del Río, L.A., Sandalio, L.M., Differential expression and regulation of antioxidative enzymes by cadmium in pea plants. *J. Plant Physiol.*, 164, 10, 1346–57, 2007.
107. Meyer, A.J., The integration of glutathione homeostasis and redox signaling. *J. Plant Physiol.*, 165, 13, 1390–403, 2008.
108. Briviba, K., Klotz, L.O., Sies, H., Toxic and signaling effects of photochemically or chemically generated singlet oxygen in biological systems. *Biol. Chem.*, 378, 11, 1259–65, 1997.
109. Larson, R.A., The antioxidants of higher plants. *Phytochemistry*, 27, 4, 969–78, 1988.
110. Metwally, A., Safronova, V.I., Belimov, A.A., Dietz, K.J., Genotypic variation of the response to cadmium toxicity in Pisum sativum L. *J. Exp. Bot.*, 56, 409, 167–78, 2005.
111. Sun, Q., Ye, Z.H., Wang, X.R., Wong, M.H., Cadmium hyperaccumulation leads to an increase of glutathione rather than phytochelatins in the cadmium hyperaccumulator Sedum alfredii. *J. Plant Physiol.*, 164, 11, 1489–98, 2007.
112. Asada, K. and Takahashi, M., *Production and Scavenging of Active Oxygenin Photosynthesis*, D.J. Kyle, C.B. Osmond, C.J. Arntzen (Eds.), pp. 227–287, Elsevier, 1987.
113. Tsugane, K., Kobayashi, K., Niwa, Y., Ohba, Y., Wada, K., Kobayashi, H., A recessive Arabidopsis mutant that grows photoautotrophically under salt stress shows enhanced active oxygen detoxification. *Plant Cell*, 11, 7, 1195–206, 1999.
114. Holländer-Czytko, H., Grabowski, J., Sandorf, I., Weckermann, K., Weiler, E.W., Tocopherol content and activities of tyrosine aminotransferase and cystine lyase in Arabidopsis under stress conditions. *J. Plant Physiol.*, 162, 7, 767–70, 2005 Jul 1.
115. Munné-Bosch, S., The role of α-tocopherol in plant stress tolerance. *J. Plant Physiol.*, 162, 7, 743–8, 2005.
116. Gang, W., Zhen-Kuan, W., Yong-Xiang, W., Li-Ye, C., Hong-Bo, S., The mutual responses of higher plants to environment, physiological and microbiological aspects. *Colloids Surf., B*, 59, 2, 113–9, 2007.

117. Shao, H.B., Chu, L.Y., Wu, G., Zhang, J.H., Lu, Z.H., Hu, Y.C., Changes of some anti-oxidative physiological indices under soil water deficits among 10 wheat (Triticum aestivum L.) genotypes at tillering stage. *Colloids Surf., B, Biointerfaces*, 54, 2, 143–9, 2007.
118. Niki, E., Yoshida, Y., Saito, Y., Noguchi, N., Lipid peroxidation, mechanisms, inhibition, and biological effects. *Biochem. Biophys. Res. Commun.*, 338, 1, 668–76, 2005.
119. Niki, E. and Noguchi, N., Dynamics of antioxidant action of vitamin E. *Acc. Chem. Res.*, 37, 1, 45–51, 2004.
120. Galli F, Azzi, A., Birringer, M., Cook-Mills, J.M., Eggersdorfer, M., Frank, J., Cruciani, G., Lorkowski, S., Özer, N.K., Vitamin, E., Emerging aspects and new directions. *Free Radical Biol. Med.*, 102, 16–36, 2017.
121. Saito, Y., Yoshida, Y., Nishio, K., Hayakawa, M., Niki, E., Characterization of cellular uptake and distribution of vitamin E. *Ann. N. Y. Acad. Sci.*, 1031, 1, 368–75, 2004.
122. Niki, E., Assessment of antioxidant capacity *in vitro* and *in vivo*. *Free Radical Biol. Med.*, 49, 4, 503–15, 2010.
123. Yu, Y., Yan, Y., Niu, F., Wang, Y., Chen, X., Su, G., Liu, Y., Zhao, X., Qian, L., Liu, P., Xiong, Y., Ferroptosis, A cell death connecting oxidative stress, inflammation and cardiovascular diseases. *Cell Death Discov.*, 7, 1, 193, 2021.
124. Upston, J.M., Kritharides, L., Stocker, R., The role of vitamin E in atherosclerosis. *Prog. Lipid Res.*, 42, 5, 405–22, 2003.
125. Stocker, R. and Keaney Jr., J.F., Role of oxidative modifications in atherosclerosis. *Physiol. Rev.*, 84, 4, 1381–478, 2004.
126. Hasanuzzaman, M., Bhuyan, M.B., Zulfiqar, F., Raza, A., Mohsin, S.M., Mahmud, J.A., Fujita, M., Fotopoulos, V., Reactive oxygen species and antioxidant defense in plants under abiotic stress, Revisiting the crucial role of a universal defense regulator. *Antioxidants*, 9, 8, 681, 2020.
127. Raja, V., Majeed, U., Kang, H., Andrabi, K.I., John, R., Abiotic stress, Interplay between ROS, hormones and MAPKs. *Environ. Exp. Bot.*, 137, 142–57, 2017.
128. Fatma, M., Iqbal, N., Sehar, Z., Alyemeni, M.N., Kaushik, P., Khan, N.A., Ahmad, P., Methyl jasmonate protects the PS II system by maintaining the stability of chloroplast D1 protein and accelerating enzymatic antioxidants in heat-stressed wheat plants. *Antioxidants*, 10, 8, 1216, 2021.
129. Hasanuzzaman, M., Mahmud, J.A., Anee, T.I., Nahar, K., Islam, M.T., Drought stress tolerance in wheat, Omics approaches in understanding and enhancing antioxidant defense, in: *Abiotic stress-mediated sensing and signaling in plants, an omics perspective*, pp. 267–307, 2018.
130. Nahar, K., Rhaman, M.S., Parvin, K., Bardhan, K., Marques, D.N., García-Caparrós, P., Hasanuzzaman, M., Arsenic-induced oxidative stress and antioxidant defense in plants. *Stresses*, 2, 2, 179–209, 2022.
131. Hasanuzzaman, M., Raihan, M.R., Khojah, E., Samra, B.N., Fujita, M., Nahar, K., Biochar and chitosan regulate antioxidant defense and methylglyoxal detoxification systems and enhance salt tolerance in jute (Corchorus olitorius L.). *Antioxidants*, 10, 12, 2017, 2021.
132. Dumanović, J., Nepovimova, E., Natić, M., Kuča, K., Jaćević, V., The significance of reactive oxygen species and antioxidant defense system in plants,A concise overview. *Front. Plant Sci.*, 11, 552969, 2021.
133. Fujita, M. and Hasanuzzaman, M., Approaches to enhancing antioxidant defense in plants. *Antioxidants*, 11, 5, 925, 2022.
134. Sachdev, S., Ansari, S.A., Ansari, M.I., Fujita, M., Hasanuzzaman, M., Abiotic stress and reactive oxygen species, Generation, signaling, and defense mechanisms. *Antioxidants*, 10, 2, 277, 2021.
135. Rani, A., Kiran, A., Sharma, K.D., Prasad, P.V., Jha, U.C., Siddique, K.H., Nayyar, H., Cold tolerance during the reproductive phase in chickpea (Cicer arietinum L.) is associated with

superior cold acclimation ability involving antioxidants and cryoprotective solutes in anthers and ovules. *Antioxidants*, 10, 11, 1693, 2021.

136. Qadir, M., Hussain, A., Hamayun, M., Shah, M., Iqbal, A., Irshad, M., Ahmad, A., Lodhi, M.A., Lee, I.J., Phytohormones producing Acinetobacter bouvetii P1 mitigates chromate stress in sunflower by provoking host antioxidant response. *Antioxidants*, 10, 12, 1868, 2021.
137. Hasanuzzaman, M., Inafuku, M., Nahar, K., Fujita, M., Oku, H., Nitric oxide regulates plant growth, physiology, antioxidant defense, and ion homeostasis to confer salt tolerance in the mangrove species, Kandelia obovata. *Antioxidants*, 10, 4, 611, 2021 Apr 16.
138. Rahman, M.A., Kabir, A.H., Song, Y., Lee, S.H., Hasanuzzaman, M., Lee, K.W., Nitric oxide prevents Fe deficiency-induced photosynthetic disturbance, and oxidative stress in alfalfa by regulating Fe acquisition and antioxidant defense. *Antioxidants*, 10, 10, 1556, 2021.
139. Santos, I.M., Tomé, A.D., Saldanha, G.B., Ferreira, P.M., Militão, G.C., De Freitas, R.M., Oxidative stress in the hippocampus during experimental seizures can be ameliorated with the antioxidant ascorbic acid. *Oxid. Med. Cell. Longevity*, 2, 4, 214–21, 2009.
140. Rowles, J. and Olsen, M., Perspectives on the development of antioxidant antiepileptogenic agents. *Mini-Rev. Med. Chem.*, 12, 10, 1015–27, 2012.
141. Devi, P.U., Manocha, A., Vohora, D., Seizures, antiepileptics, antioxidants and oxidative stress, an insight for researchers. *Expert Opin. Pharmacother.*, 9, 18, 3169–77, 2008.
142. Sarker, U. and Oba, S., The response of salinity stress-induced A. tricolor to growth, anatomy, physiology, non-enzymatic and enzymatic antioxidants. *Front. Plant Sci.*, 11, 559876, 2021.
143. Hossain, M.N., Sarker, U., Raihan, M.S., Al-Huqail, A.A., Siddiqui, M.H., Oba, S., Influence of salinity stress on color parameters, leaf pigmentation, polyphenol and flavonoid contents, and antioxidant activity of Amaranthus lividus leafy vegetables. *Molecules*, 27, 6, 1821, 2022.
144. Sarker, U. and Oba, S., Phenolic profiles and antioxidant activities in selected drought-tolerant leafy vegetable amaranth. *Sci. Rep.*, 10, 1, 18287, 2020.
145. Martinc, B., Grabnar, I., Vovk, T., Antioxidants as a preventive treatment for epileptic process, a review of the current status. *Curr. Neuropharmacol.*, 12, 6, 527–50, 2014.
146. Tomé, A.D., Ferreira, P.M., Freitas, R.M., Inhibitory action of antioxidants (ascorbic acid or α-tocopherol) on seizures and brain damage induced by pilocarpine in rats. *Arquivos de Neuro-Psiquiatria*, 68, 355–61, 2010.
147. Mehla, J., Reeta, K.H., Gupta, P., Gupta, Y.K., Protective effect of curcumin against seizures and cognitive impairment in a pentylenetetrazole-kindled epileptic rat model. *Life Sci.*, 87, 19-22, 596–603, 2010.
148. Hamed, S.A., Abdellah, M.M., El-Melegy, N., Blood levels of trace elements, electrolytes, and oxidative stress/antioxidant systems in epileptic patients. *J. Pharmacol. Sci.*, 96, 4, 465–73, 2004.
149. Arora, T., Mehta, A.K., Sharma, K.K., Mediratta, P.K., Banerjee, B.D., Garg, G.R., Sharma, A.K., Effect of carbamazepine and lamotrigine on cognitive function and oxidative stress in brain during chemical epileptogenesis in rats. *Basic Clin. Physiol. Pharmacol.*, 106, 5, 372–7, 2010.
150. Sudha, K., Rao, A.V., Rao, A., Oxidative stress and antioxidants in epilepsy. *Clin. Chim. Acta*, 303, 1-2, 19–24, 2001.
151. Aycicek, A. and Iscan, A., The effects of carbamazepine, valproic acid and phenobarbital on the oxidative and antioxidative balance in epileptic children. *Eur. Neurol.*, 57, 2, 65–9, 2007.
152. Diniz, T.C., Silva, J.C., Lima-Saraiva, S.R., Ribeiro, F.P., Pacheco, A.G., de Freitas, R.M., Quintans-Júnior, L.J., Quintans, J.D., Mendes, R.L., Almeida, J.R., The role of flavonoids on oxidative stress in epilepsy. *Oxid. Med. Cell. Longevity*, 2015, 171756, 2015 Oct.
153. Estrada-Reyes, R., Martínez-Vázquez, M., Gallegos-Solís, A., Heinze, G., Moreno, J., Depressant effects of Clinopodium mexicanum Benth Govaerts (Lamiaceae) on the central nervous system. *J. Ethnopharmacol.*, 130, 1, 1–8, 2010.

154. Shin, E.J., Jeong, J.H., Chung, Y.H., Kim, W.K., Ko, K.H., Bach, J.H., Hong, J.S., Yoneda, Y., Kim, H.C., Role of oxidative stress in epileptic seizures. *Neurochem. Int.*, 59, 2, 122–37, 2011.
155. Ramalingam, R., Nath, A.R., Madhavi, B.B., Nagulu, M., Balasubramaniam, A., Free radical scavenging and antiepileptic activity of Leucas lanata. *J. Pharm. Res.*, 6, 3, 368–72, 2013.
156. Shrestha, S., Park, J.H., Lee, D.Y., Cho, J.G., Cho, S., Yang, H.J., Yong, H.I., Yoon, M.S., Han, D.S., Baek, N.I., Rhus parviflora and its biflavonoid constituent, rhusflavone, induce sleep through the positive allosteric modulation of GABAA-benzodiazepine receptors. *J. Ethnopharmacol.*, 142, 1, 213–20, 2012.
157. Keddy, P.G., Dunlop, K., Warford, J., Samson, M.L., Jones, Q.R., Rupasinghe, H.V., Robertson, G.S., Neuroprotective and anti-inflammatory effects of the flavonoid-enriched fraction AF4 in a mouse model of hypoxic-ischemic brain injury. *PLoS One*, 7, 12, e51324, 2012.
158. Rasouli, H., Farzaei, M.H., Khodarahmi, R., Polyphenols and their benefits,A review. *Int. J. Food Prop.*, 20, sup2, 1700–41, 2017.
159. Russo, G.L., Tedesco, I., Spagnuolo, C., Russo, M., Antioxidant polyphenols in cancer treatment, friend, foe or foil? *Semin. Cancer Biol.*, 46, 1–13, 2017.
160. Kaur, H., Bal, A., Sandhir, R., Curcumin supplementation improves mitochondrial and behavioral deficits in experimental model of chronic epilepsy. *Pharmacol. Biochem. Behav.*, 125, 55–64, 2014.
161. Folbergrová, J., Ješina, P., Kubová, H., Otáhal, J., Effect of resveratrol on oxidative stress and mitochondrial dysfunction in immature brain during epileptogenesis. *Mol. Neurobiol.*, 55, 7512–22, 2018.
162. Mojarad, T.B. and Roghani, M., The anticonvulsant and antioxidant effects of berberine in kainate-induced temporal lobe epilepsy in rats. *Basic Clin. Neurosci.*, 5, 2, 124, 2014.
163. Jiang, W., Wei, W., Gaertig, M.A., Li, S., Li, X.J., Therapeutic effect of berberine on Huntington's disease transgenic mouse model. *PLoS One*, 10, 7, e0134142, 2015.
164. Cortes, N., Posada-Duque, R.A., Alvarez, R., Alzate, F., Berkov, S., Cardona-Gómez, G.P., Osorio, E., Neuroprotective activity and acetylcholinesterase inhibition of five Amaryllidaceae species, a comparative study. 122, 42–50, 2015.AU: Please provide journal title.
165. Liu, X.G., Lu, X., Gao, W., Li, P., Yang, H., Structure, synthesis, biosynthesis, and activity of the characteristic compounds from Ginkgo biloba L. *Nat. Prod. Rep.*, 39, 3, 474–511, 2022.
166. Jäger, A.K., Krydsfeldt, K., Rasmussen, H.B., Bioassay-guided isolation of apigenin with GABA-benzodiazepine activity from Tanacetum parthenium. *Phytother. Res.*, 23, 11, 1642–4, 2009.
167. Abbasi, E., Nassiri-Asl, M., Shafeei, M., Sheikhi, M., Neuroprotective effects of vitexin, a flavonoid, on pentylenetetrazole-induced seizure in rats. *Chem. Biol. Drug Des.*, 80, 2, 274–8, 2012.
168. Taheri, Y., Sharifi-Rad, J., Antika, G., Yılmaz, Y.B., Tumer, T.B., Abuhamdah, S., Chandra, S., Saklani, S., Kılıç, C.S., Sestito, S., Daştan, S.D., Paving luteolin therapeutic potentialities and agro-food-pharma applications, emphasis on in vivo pharmacological effects and bioavailability traits. *Oxid. Med. Cell. Longevity*, 2021, 1987588, 2021.
169. Choudhary, N., Bijjem, K.R., Kalia, A.N., Antiepileptic potential of flavonoids fraction from the leaves of Anisomeles malabarica. *J. Ethnopharmacol.*, 135, 2, 238–42, 2011.
170. Nassiri-Asl, M., Shariati-Rad, S., Zamansoltani, F., Anticonvulsive effects of intracerebroventricular administration of rutin in rats. *Prog. Neuro-Psychopharmacol. Biol. Psychiatry*, 32, 4, 989–93, 2008.
171. Xie, T., Wang, W.P., Mao, Z.F., Qu, Z.Z., Luan, S.Q., Jia, L.J., Kan, M.C., Effects of epigallocatechin-3-gallate on pentylenetetrazole-induced kindling, cognitive impairment and oxidative stress in rats. *Neurosci. Lett.*, 516, 2, 237–41, 2012.
172. Ren, L., Wang, F., Xu, Z., Chan, W.M., Zhao, C., Xue, H., GABAA receptor subtype selectivity underlying anxiolytic effect of 6-hydroxyflavone. *Biochem. Pharmacol.*, 79, 9, 1337–44, 2010.

173. de Andrade, J.P., Giordani, R.B., Torras-Claveria, L., Pigni, N.B., Berkov, S., Font-Bardia, M., Calvet, T., Konrath, E., Bueno, K., Sachett, L.G., Dutilh, J.H., Brazilian Amaryllidaceae as a source of acetylcholinesterase inhibitory alkaloids. *Phytochem. Rev.*, 15, 147–60, 2016.
174. Rocha, L., Alonso-Vanegas, M., Martínez-Juárez, I.E., Orozco-Suárez, S., Escalante-Santiago, D., Feria-Romero, I.A., Zavala-Tecuapetla, C., Cisneros-Franco, J.M., Buentello-García, R.M., Cienfuegos, J., GABAergic alterations in neocortex of patients with pharmacoresistant temporal lobe epilepsy can explain the comorbidity of anxiety and depression, the potential impact of clinical factors. *Front. Cell. Neurosci.*, *8*, 442, 2015.
175. Hussain, G., Rasul, A., Anwar, H., Aziz, N., Razzaq, A., Wei, W., Ali, M., Li, J., Li, X., Role of plant derived alkaloids and their mechanism in neurodegenerative disorders. *Int. J. Biol. Sci.*, 14, 3, 341, 2018.
176. Rocha, L., Alonso-Vanegas, M., Martínez-Juárez, I.E., Orozco-Suárez, S., Escalante-Santiago, D., Feria-Romero, I.A., Zavala-Tecuapetla, C., Cisneros-Franco, J.M., Buentello-García, R.M., Cienfuegos, J., GABAergic alterations in neocortex of patients with pharmacoresistant temporal lobe epilepsy can explain the comorbidity of anxiety and depression, the potential impact of clinical factors. *Front. Cell. Neurosci.*, 8, 442, 2015.
177. Yeggoni, D.P., Rachamallu, A., Kallubai, M., Subramanyam, R., Cytotoxicity and comparative binding mechanism of piperine with human serum albumin and α-1-acid glycoprotein. *J. Biomol. Struct. Dyn.*, 33, 6, 1336–51, 2015.
178. Goswami, P.K., Samant, M.A., Srivastava, R.S., Snake venom, anti-snake venom & potential of snake venom. *Int. J. Pharm. Pharm. Sci.*, 6, 5, 4–7, 2014.
179. Mikkilineni, S., Cantuti-Castelvetri, I., Cahill, C.M., Balliedier, A., Greig, N.H., Rogers, J.T., The anticholinesterase phenserine and its enantiomer posiphen as 5′ untranslated-region-directed translation blockers of the Parkinson's alpha synuclein expression, in: *Parkinson's Disease*, 2012.
180. Ren, T., *Biopharmaceutics and Pharmacokinetics Characterization of Piperine and Its Pharmacokinetics and Pharmacodynamics Interactions with Carbamazepine in Anti-epileptic Treatment*, The Chinese University of Hong Kong, Hong Kong, 2019.
181. Sills, G.J. and Rogawski, M.A., Mechanisms of action of currently used antiseizure drugs. *Neuropharmacology*, 168, 107966, 2020.
182. Löscher, W. and Klein, P., The pharmacology and clinical efficacy of antiseizure medications, from bromide salts to cenobamate and beyond. *CNS Drugs*, 35, 9, 935–63, 2021.
183. Kwan, P. and Brodie, M.J., Epilepsy after the first drug fails,substitution or add-on? *Seizure*, 9, 464–8, 2000.
184. Fattorusso, A., Matricardi, S., Mencaroni, E., Dell'Isola, G.B., Di Cara, G., Striano, P., Verrotti, A., The pharmacoresistant epilepsy: an overview on existant and new emerging therapies. *Front. Neurol.*, 12, 674483, 2021.
185. Brodie, M.J., Modern management of juvenile myoclonic epilepsy. *Expert Rev. Neurother.*, 16, 6, 681–8, 2016.
186. Cereghino, J.J., Brock, J.T., Van Meter, J.C., Penry, J.K., Smith, L.O., White, B.G., The efficacy of carbamazepine combinations in epilepsy. *Clin. Pharmacol. Ther.*, 18, 733–41, 1975.
187. Verrotti, A., Tambucci, R., Di Francesco, L., Pavone, P., Iapadre, G., Altobelli, E., Matricardi, S., Farello, G., Belcastro, V., The role of polytherapy in the management of epilepsy: suggestions for rational antiepileptic drug selection. *Expert Rev. Neurother.*, 20, 2, 167–73, 2020.
188. Stephen, L.J. and Brodie, M.J., Antiepileptic drug monotherapy versus polytherapy: pursuing seizure freedom and tolerability in adults. *Curr. Opin. Neurol.*, 25, 2, 164–72, 2012.
189. Stephen, L.J., Sills, G.J., Brodie, M.J., Lamotrigine and topiramate may be a useful combination. *Lancet*, 351, 958–9, 1998.

190. Brodie, M.J., Rosenfeld, W.E., Vasquez, B., Sachdeo, R., Perdomo, C., Mann, A. *et al.*, Rufinamide for the adjunctive treatment of partial seizures in adults and adolescents, a randomized placebo-controlled trial. *Epilepsia*, 50, 1899–909, 2009.
191. Sake´, J.K., Hebert, D., Isojarvi, J., Soty, P., De Bauker, M., Davies, K. *et al.*, A pooled analysis of lacosamide clinical trial data grouped by mechanisms of action of concomitant antiepileptic drugs. *CNS Drugs*, 24, 1055–68, 2010.
192. Barcs, G., Walker, E.B., Elger, C.E., Scaramelli, A., Stefan, H., Sturm, Y. *et al.*, Oxcarbazepine placebo-controlled, dose ranging trial in refractory partial epilepsy. *Epilepsia*, 41, 1597–607, 2000.
193. Wong, I.C.K., Mawer, G.E., Sander, J.W.A.S., Lhatoo, S.D., A pharmacoepidemiologic study of factors influencing the outcome of treatment with lamotrigine in chronic epilepsy. *Epilepsia*, 42, 1354–8, 2001.
194. Perucca, P. and Gilliam, F.G., Adverse effects of antiepileptic drugs. *Lancet Neurol.*, 11, 9, 792–802, 2012.
195. Poprac, P., Jomova, K., Simunkova, M., Kollar, V., Rhodes, C.J., Valko, M., Targeting free radicals in oxidative stress-related human diseases. *Trends Pharmacol. Sci.*, 38, 7, 592–607, 2017.
196. Patel, M., Targeting oxidative stress in central nervous system disorders. *Trends Pharmacol. Sci.*, 37, 9, 768–78, 2016.
197. Bhanu, P.S., Sagar, T.K., Zafer, R., Failure and successes of herbal medicines. *Indian Pharmacist*, 54, 2, 221–35, 2000.
198. Ritch, R., Potential role for Ginkgo biloba extract in the treatment of glaucoma. *Med. Hypotheses*, 54, 2, 221–35, 2000.
199. Altemimi, A., Watson, D.G., Kinsel, M., Lightfoot, D.A., Simultaneous extraction, optimization, and analysis of flavonoids and polyphenols from peach and pumpkin extracts using a TLC-densitometric method. *Chem. Cent. J.*, 9, 1, 1–5, 2015.
200. Mulinacci, N., Prucher, D., Peruzzi, M., Romani, A., Pinelli, P., Giaccherini, C., Vincieri, F.F., Commercial and laboratory extracts from artichoke leaves, estimation of caffeoyl esters and flavonoidic compounds content. *J. Pharm. Biomed. Anal.*, 34, 2, 349–57, 2004.
201. SaraJliJa, H., ČukelJ, N., Mršić, G.N., Brnčić, M., Ćurić, D., Preparation of flaxseed for lignan determination by gas chromatography-mass spectrometry method. *Czech J. Food Sci.*, 30, 1, 45–52, 2012.
202. Popova, I.E., Hall, C., Kubátová, A., Determination of lignans in flaxseed using liquid chromatography with time-of-flight mass spectrometry. *J. Chromatogr. A*, 1216, 2, 217–29, 2009.
203. Zhang, Z., Pang, X., Xuewu, D., Ji, Z., Jiang, Y., Role of peroxidase in anthocyanin degradation in litchi fruit pericarp. *Food Chem.*, 90, 1-2, 47–52, 2005.
204. Urbano, M., De Castro, M.D., Pérez, P.M., García-Olmo, J., Gomez-Nieto, M.A., Ultraviolet-visible spectroscopy and pattern recognition methods for differentiation and classification of wines. *Food Chem.*, 97, 1, 166–75, 2006.
205. Altemimi, A., Lakhssassi, N., Baharlouei, A., Watson, D.G., Lightfoot, D.A., Phytochemicals, Extraction, isolation, and identification of bioactive compounds from plant extracts. *Plants*, 6, 4, 42, 2017 Sep 22.
206. López-Lázaro, M., Distribution and biological activities of the flavonoid luteolin. *Mini-Rev. Med. Chem.*, 9, 1, 31–59, 2009 Jan 1.
207. Cherkaoui, A., Hibbs, J., Emonet, S., Tangomo, M., Girard, M., Francois, P., Schrenzel, J., Comparison of two matrix-assisted laser desorption ionization-time of flight mass spectrometry methods with conventional phenotypic identification for routine identification of bacteria to the species level. *J. Clin. Microbiol.*, 48, 4, 1169–75, 2010.
208. Beckey, H.D., 2–Field ionization sources, in: *International series of monographs on analytical chemistry, field ionization mass spectrometry*, vol. 42, pp. 11–7, 1971.

209. Beckey, H.D., 4-qualitative analyses with the fi mass spectrometer, in: *International series of monographs on analytical chemistry, field ionization mass spectrometry*, vol. 42, pp. 250–333, 1971.
210. Fisher, R.S., Acevedo, C., Arzimanoglou, A., Bogacz, A., Cross, J.H., Elger, C.E., Engel, J., Forsgren, L., French, J.A., Hesdorffer, D.C., Lee, B.I., How long for epilepsy remission in the ILAE definition. *Epilepsia*, 58, 8, 1486–7, 2017.
211. Duncan, J.S., Sander, J.W., Sisodiya, S.M., Walker, M.C., Adult epilepsy. *Lancet*, 367, 9516, 1087–100, 2006.
212. Perucca, E. and Tomson, T., The pharmacological treatment of epilepsy in adults. *Lancet Neurol.*, 10, 5, 446–56, 2011.
213. Iori, V., Aronica, E., Vezzani, A., Epigenetic control of epileptogenesis by miR-146a. *Oncotarget*, 8, 28, 45040, 2017.
214. Holmes, G.L., *Handbook of epilepsy*, Lippincott Williams & Wilkins, Baltimore, MD, 2008.
215. Striano, P. (Ed.), *Epilepsy towards the next decade, new trends and hopes in epileptology*, Springer, University of Genova, 2014.
216. Reid, C.A., Jackson, G.D., Berkovic, S.F., Petrou, S., New therapeutic opportunities in epilepsy, a genetic perspective. *Pharmacol. Ther.*, 128, 2, 274–80, 2010.
217. Møller, R.S., Dahl, H.A., Helbig, I., The contribution of next generation sequencing to epilepsy genetics. *Expert Rev. Mol. Diagn.*, 15, 12, 1531–8, 2015.
218. Covanis, A., Clinical management of epileptic encephalopathies of childhood and infancy. *Expert Rev. Neurother.*, 14, 6, 687–701, 2014.
219. Striano, P., Striano, S., Minetti, C., Zara, F., Refractory, life-threatening status epilepticus in a 3-year-old girl. *Lancet Neurol.*, 7, 3, 278–84, 2008.
220. Brunklaus, A. and Zuberi, S.M., Dravet syndrome—from epileptic encephalopathy to channelopathy. *Epilepsia.*, 55, 7, 979–84, 2014.
221. Lynch, J.M., Tate, S.K., Kinirons, P., Weale, M.E., Cavalleri, G.L., Depondt, C., Murphy, K., O'Rourke, D., Doherty, C.P., Shianna, K.V., Wood, N.W., No major role of common SV2A variation for predisposition or levetiracetam response in epilepsy. *Epilepsy Res.*, 83, 1, 44–51, 2009.
222. Piana, C., Antunes, N.D., Della Pasqua, O., Implications of pharmacogenetics for the therapeutic use of antiepileptic drugs. *Expert Opin. Drug Metab. Toxicol.* 10, 3, 341–58, 2014.
223. Balestrini, S. and Sisodiya, S.M., Pharmacogenomics in epilepsy. *Neurosci. Lett.*, 667, 27–39, 2018.

13

Antioxidants and Obesity

Parul Sood[1], Gagandeep Kaur[1], Komal Thapa[1], Kiran Sharma[2] and Rakesh K. Sindhu[3]*

[1]*School of Pharmacy, Chitkara University, Himachal Pradesh, Solan, India*
[2]*Amity Institute of Pharmacy, Amity University, Gurugram, Manesar, Panchgaon, Haryana, India*
[3]*School of Pharmacy, Sharda University, Greater Noida, Uttar Pradesh, India*

Abstract
Obesity, a global health epidemic, continues to impose significant burdens on healthcare systems and individual well-being. It is characterized by an excessive accumulation of adipose tissue resulting from an imbalance between energy intake and expenditure. The pathogenesis of obesity involves complex interplay between genetic, environmental, and lifestyle factors. As the prevalence of obesity rises, research into potential interventions and underlying mechanisms becomes increasingly crucial. Among the various factors implicated in obesity's development and progression, oxidative stress and its modulation by antioxidants have emerged as key players. Several examples of antioxidants belonging to chemical class including polyphenols, carotenoids, catechins, curcuminoids, isothiocyantes, etc. have shown promising effects in mitigating obesity. Resveratrol, a polyphenol found in grapes and red wine, has been shown to reduce body weight, improve insulin sensitivity, and regulate adipose tissue metabolism in animal and human studies. Green tea catechins, such as epigallocatechin gallate (EGCG), have also demonstrated anti-obesity effects by enhancing fat oxidation and reducing fat absorption. Moreover, vitamins C and E, as well as various phytochemicals found in fruits and vegetables, have exhibited antioxidant properties and potential benefits in obesity management. These antioxidants may exert their effects by modulating adipogenesis, inflammation, and metabolic pathways involved in energy homeostasis. In conclusion, this book chapter emphasizes the significant role of antioxidants in obesity management and highlights their therapeutic potential as adjunctive therapies for obesity prevention and management, thus promoting better health outcomes. Nonetheless, further research is warranted to elucidate the interaction between antioxidants and obesity-related pathways for devising targeted interventions to alleviate the burden of obesity-related comorbidities.

***Keywords*:** Antioxidant, obesity, oxidative stress, phenolic compounds, vitamins

13.1 Introduction

Obesity, a complex global health issue, is on the rise and results from a multifaceted interplay of several factors including metabolism, diet, physical activity, and genetics [1]. This condition is linked to severe complications like metabolic disorders such as, type 2 diabetes,

**Corresponding author*: drrakeshksindhu@gmail.com

Rakesh K. Sindhu, Inderbir Singh and M. Arockia Babu (eds.) Antioxidants: Nature's Defense Against Disease, (491–510)

cancer, stroke, hypertension, hyperlipidemia and cardiovascular dysfunction [2]. Reactive oxygen species (ROS), metabolic by-products, contribute to obesity's development and metabolic issues, influencing inflammation, adipocyte growth, and energy balance [3]. ROS serve as regulatory factors for mitochondrial activity. They alter inflammation-related molecule concentrations, associated with increased adipocyte count and size. Additionally, ROS promote adipogenesis, lipogenesis, and the differentiation of preadipocytes to mature ones. Notably, they also regulate energy balance in hypothalamic neurons controlling appetite [4, 5].

Recent nutrition research has ignited interest in leveraging natural substances to counteract obesity [2, 6]. With concerns over extended synthetic drug use and stringent approval criteria, attention has turned towards herbal remedies for clinical treatment reliability. The demand for these solutions has surged. Study of plant-derived phytochemicals, including alkaloids, terpenoids, glycosides, steroids, tannins, polyphenols, flavonoids, saponins, and proteins, has unveiled their pivotal role in addressing disorders [7]. These botanical derivatives contain dietary phytochemicals have proved its potential to promote health and prevent disease. Synergistic effects from combined phytochemical action offer advantages over single-agent therapies [8, 9]. Mechanisms behind anti-obesity effects encompass regulating pathways associated with lipid absorption, energy intake, expenditure, lipolysis, lipogenesis reduction, and preadipocyte modulation [10] as depicted in Figure 13.1. Various phytoconstituents like apigenin, caffeine, curcumin, ellagic acid, quercetin, resveratrol, genistein, guggulsterone, piperine, hydroxycitric acid, and have potential in combating obesity [11–13]. However, while some are used in anti-obesity drug formulations, comprehensive clinical investigation and validation are lacking. The potential of plant derivatives in treating obesity remains largely unexplored, promising secure and effective anti-obesity medications [14]. Phytochemicals as a source of new pharmaceuticals offer an expansive realm for inquiry, as only a fraction of known species have undergone evaluation [15]. Even medicinal plants lack scrutiny

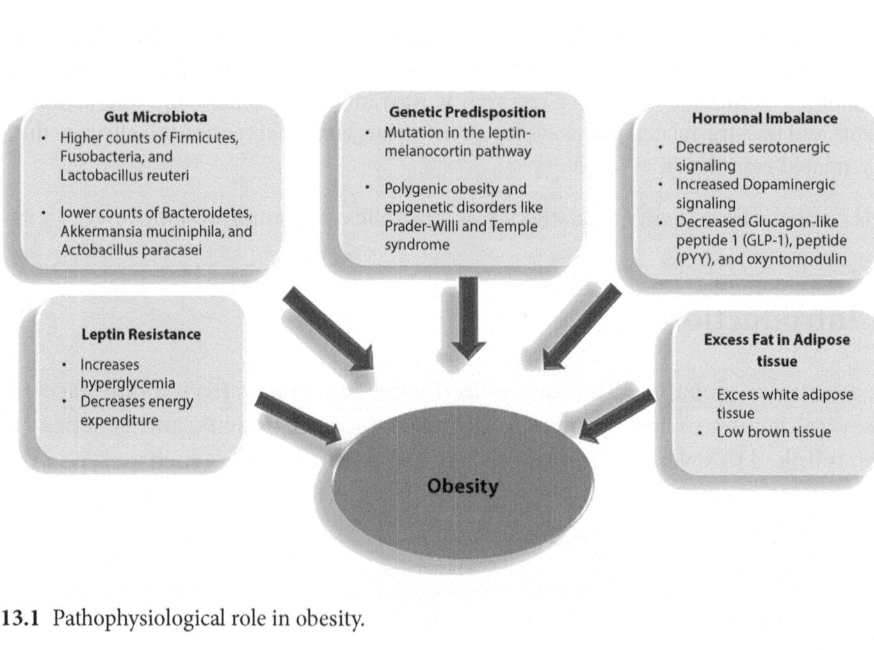

Figure 13.1 Pathophysiological role in obesity.

for effectiveness and safety, necessitating alternative therapies rooted in natural products with minimal side effects to replace existing approaches.

13.2 Pathological Role of Oxidative Stress in Obesity

Reactive oxygen species (ROS) and nitrogen species (RNS) such as superoxide (O2−), nitric oxide (NO), hydroxyl radical (OH), hydrogen peroxide (H2O2), hypochlorite (ClO−), and peroxynitrite (ONOO−) play significant roles in various biological processes [16]. The mitochondria serve as the main origin of ROS, resulting from electron leakage within the respiratory chain. However, other cellular components such as endoplasmic reticulum, peroxisomes, plasma membrane, lysosomes, and cytosolic enzymes also contribute in ROS generation. These molecules have dual effects, functioning both in assisting pathogenic defense and mediating intracellular signaling even present at low concentrations, but excessive levels can cause harm to DNA, proteins and lipids, leading to tissue damage and cell death [17]. The subsequent sections delve into ROS pathways and their implications for obesity.

13.2.1 ROS and Thermogenesis

Recent research has illuminated the multifaceted role of adipose tissue. Initially regarded as a passive calorie repository, it's now recognized as an active endocrine organ influencing energy equilibrium. Through the release of various molecules like cytokines, free fatty acids, and adipokines including adiponectin and leptin that play pivotal role in regulating food intake and insulin sensitivity, actively engaging in energy balance [18]. Adipose tissue classification includes two main types: white adipose tissue (WAT) which functions primarily for energy storage, and brown adipose tissue (BAT), responsible for thermogenic energy expenditure. In addition, intermediary beige adipocytes, emerging from WAT, has been identified [19] prompted to develop through external triggers like cold exposure, adrenergic stimulation, and long-term activation of peroxisome proliferator-activated receptor agonist [20]. It is noteworthy that, Both BAT and beige adipocytes feature by presence of substantial number of mitochondria and expression of uncoupling protein 1 (UCP-1). This dual characteristic not only regulates energy expenditure but also serves to mitigate adiposity and offer protection against diet-induced obesity [21]. Alongside UCP-1, brown adipocytes prompt other regulators like Cell death activator, type 2 iodothyronine deiodinase, Peroxisome proliferator-activated receptor gamma coactivator 1-α, and PR domain containing 16 (PRDM16), collectively influencing UCP1 transcription in BAT [22]. Recognizing BAT's potential for energy dissipation, current focus lies in pharmacological interventions promoting the browning of WAT to foster energy balance.

13.2.2 ROS and Mitochondrial Dysfunction

Aerobic organisms require oxygen for energy production and stability, yet this dependency also leads to generation of harmful free radicals, resulting in increased ROS levels [23]. These ROS emerge during cellular processes, primarily originating from the respiratory chain of mitochondria [24]. The foremost ROS is the superoxide anion (O^-), formed through oxygen molecule reduction due to the auto-oxidation of mitochondrial carriers [25]. The superoxide anion transforms into hydrogen peroxide (H_2O_2), catalyzed mainly by superoxide dismutase (SOD). Generation of O– radical is primarily localized in enzymes of complex I, II, and III [26]. The extent of mitochondrial ROS production varies with cellular conditions. Mitochondria employ an efficient antioxidant system to neutralize and transform synthesized ROS, minimizing their impact on the cytosol [27]. However, heightened ROS production, whether from mitochondria or other sources, not only disrupt the Krebs cycle and respiratory chain enzymes, but also subsequently affecting cellular activity and the entire organism [28]. This mitochondrial dysfunction is implicated in metabolic disorders leading to changes in ATP (Adenosine triphosphate) production, nutrient utilization, and apoptosis [29]. Elevated glucose levels further exacerbate ROS production, influencing mitochondrial enzymes, nutrient utilization, and contributing to the development of metabolic disorders [30]. In obesity, mitochondrial dysfunction affects adipocyte processes, energy generation, fatty acid oxidation, and apoptosis [31–33]. These mitochondrial anomalies result in altered lipid metabolism, increased lipid droplet formation, and a state of low-grade inflammation [34]. Damaged mitochondria undergo mitophagy, making way for replacement by new mitochondria, a process linked to ROS production [35].

13.2.3 ROS and Hormonal Imbalance

Neurobiological mechanisms underlie eating behavior irrespective of energy demand or hunger. These mechanisms involve brain regions such as the cortico-limbic system, hypothalamus, and hindbrain [36], which respond to various visual food cues tied to fasting, weight changes, overfeeding, exercise, hormones, and cognitive control [37]. Notably, key neurotransmitters like dopamine and serotonin play a significant role in regulating food intake. Serotonin elevation reduces consumption, while depletion leads to weight gain through overeating [38]. Similarly, decreased dopamine levels contribute to excessive eating [39]. Oxidative stress impacts crucial gut hormones involved in obesity regulation, where gastrointestinal cells release peptide (PYY), glucagon-like peptide 1 (GLP-1), and oxyntomodulin. These substances signal nutrient availability to the brain, effectively curbing appetite [40]. Leptin, sourced from adipose tissue, signals energy availability, thereby enhancing the feeling of satiety. Conversely, ghrelin, released during periods of food absence, heightens sensitivity to food intake [41].

13.2.3.1 ROS and Leptin Resistance

Leptin, an adipokine that play significant role in food intake, managing energy expenditure, and modulating the immune function reflects energy stores within adipose tissue [42]. It promotes energy expenditure, curbing food cravings when binding to brain receptors, while also countering insulin deficiency-induced hyperglycemia [43]. Although increasing circulating leptin levels was explored as an obesity treatment [44], obesity's progression disrupts leptin signaling causing resistance [45]. Despite high serum levels, leptin can't bind or mediate physiological activity, a common issue in obesity. This resistance renders leptin therapy ineffective for obese patients, highlighting the need for research into anti-obesity solutions targeting leptin resistance [41].

13.2.3.2 ROS and Decreased Level of Adiponectin

Adiponectin, composed of a globular domain resembling tumor necrosis factor-α (TNFα) and a collagen domain, demonstrates crucial functions. In the serum of obese individuals who also affecting with type-II diabetes, its concentration declines. This hormone enhances insulin sensitivity, promotes fatty acid oxidation, and reduces serum triglycerides. Its anti-inflammatory impact arises from decreased NF-κB levels, governing over 150 target genes, including pro-inflammatory elements [46]. Elevated adiponectin levels lead to reduced TNF-α, IL-6, and IL-8 production, as well as lowered expression of adhesion molecules. Moreover, interleukin 10 (IL-10) synthesis increases [47]. Interleukins IL-1, IL-6, and TNF-α, along with NF-κB, heighten lipolysis, fostering hepatic fatty acid production [48, 49]. Such associations relate to increased serum triglycerides and cholesterol. Adiponectin's benefits emerge when bolstered by fish oil, soy intake, or specific medications [50]. Elevated oxidative stress and decreased adiponectin contribute to conditions like obesity-related insulin resistance and elevate cardiovascular risk. Nuclear receptors like peroxisome proliferator-activated receptor gamma (PPARγ) govern adiponectin transcription, while mineralocorticoid receptor fosters oxidative stress due to aldosterone and glucocorticoids [51].

13.2.4 ROS and Altered Iron Metabolism in Obesity

Obesity leads to persistent mild inflammation, triggering the release of inflammatory cytokines that elevate OS and amplify hepcidin secretion in macrophages and hepatocytes [52]. Hepcidin, primarily synthesized mainly by liver play a crucial role in regulating iron metabolism [53]. This heightened OS contributes to obesity-related conditions like iron deficiency, where hepcidin plays a role [54]. Elevated serum hepcidin levels are observed in obese individuals [55]. Iron, an active metal, worsens OS, causing endoplasmic reticulum stress, inflammation, and adipose tissue dysfunction. Iron-related toxicity exacerbates obesity's progression by catalyzing the Fenton reaction and intensifying OS [56]. Ferritin is vital for liver fortification against oxidative damage. Dysregulated ferritin allows free iron to exert harmful effects. These events establish a feedback loop, leading to reactive oxygen species production and contributing to obesity's physiological changes. Hepcidin affects intestinal iron absorption and ferroportin function, crucial for iron homeostasis [57]. Excessive hepcidin may cause obesity-related hypoferremia [58], sequestering iron in cells and increasing local iron levels in adipose tissue, initiating adverse effects and iron overload.

13.2.5 ROS and Altered Gut Microbiota in Obese Patients

The gastrointestinal tract hosts a diverse array of microorganisms, encompassing bacteria, viruses, fungi, and protozoa, all of which play intricate roles in immune, metabolic, and neurological well-being [59]. These microorganisms engage in the fermentation of indigestible dietary fibers, producing short-chain fatty acids (SCFAs) such as butyrate, propionate, and acetate. These SCFAs have far-reaching effects on cholesterol metabolism, lipogenesis, and appetite regulation [60]. Disturbances in oxidative stress can upset the equilibrium and composition of the gut microbiota, leading to a condition known as dysbiosis. This is characterized by an imbalance between beneficial and harmful microorganisms and is evident in overweight individuals, where reduced microbial diversity is linked to impaired glucose control and mild inflammation [61]. The microbial makeup in obesity showcases elevated counts of *Fusobacteria, Firmicutes, Lactobacillus reuteri,* and, *Proteobacteria,* accompanied by decreased counts of *Akkermansia muciniphila, Bacteroidetes, Faecalibacter iumprausnitzii, Lactobacillus paracasei,* and *Lactobacillus plantarum* [62]. The balance between *Firmicutes* and *Bacteroidetes* is associated with obesity, displaying a contrast in weight loss-related changes [63]. Over the past two decades, research has emphasized the potential of altering gut microbiota composition through probiotic supplements to mitigate body weight and enhance glucose metabolism in rodents [64]. Dietary patterns also significantly shape the gut microbiome, favoring *Firmicutes* with a high-fat diet in animal models. Metabolites derived from the microbiome play significant role in modulating host metabolism, exemplified by conversion of bile acids promoting hormone secretion [65]. Additionally, choline metabolism influences microbiome composition, with the conversion of trimethylamine-N-oxide (TMAO) implicated in atherosclerosis and metabolic issues [66]. SCFAs influence insulin signaling, GLP-1 secretion, and gut immunity [67]. In a reciprocal manner, inflammation, and gut permeability, closely tied to adiposity, mutually amplify, leading to the leakage of metabolites into circulation. This fosters the mild inflammation characteristic of obesity and insulin resistance [68]. Additionally, it is important to note that proinflammatory cytokines also contribute significant role in disruption of the intestinal barrier [69].

13.3 Obesity Regulating Diverse Chemical Groups as Natural Antioxidants from Plant Sources

13.3.1 Alkaloids

Both sweet bell peppers and red-hot peppers contain alkaloids, specifically capsaicinoids and casinoids, which play important roles in physiological and pharmacologicalfunctions. These compounds have diverse effects, including anticancer, antioxidant, anti-obesity, and, anti-inflammatory properties [70]. The consumption of capsaicinoids has been linked to increased energy expenditure, lipid oxidation, suppressed appetite, and aiding in weight loss [71, 72]. Although the exact molecular mechanisms behind these effects are not completely understood, the activation of transient receptor potential vanilloid type-1 (TRPV1) appears to be a central factor in many of these benefits [73]. Additionally, capsaicin helps reduce obesity-induced inflammation by lowering levels of MCP-1, IL-6, IL-8, and TNF-α, while also increasing adiponectin levels, which are vital for insulin response [74].

13.3.2 Carotenoids

Carotenoids, vital antioxidant compounds, are present in plants, microorganisms, and in limited number of animals. Particularly, β-carotene, α-carotene, and β-cryptoxanthin exhibit pro-vitamin A activity. In terms of health, the connection between reduced inflammation, lower OS, improved endothelial function, decreased cardiovascular mortality, and decreased overall mortality among the elderly is associated with the intake of vegetables and fruits rich in carotenoids, such as apricots, broccoli, carrots, mandarins, pumpkin, spinach, and tomatoes [76]. This effect is also observed in the carotenoid Astaxanthin, sourced from microscopic algae, found in salmon, crabs, and lobster. It fortifies antioxidant defenses and curbs lipid peroxidation in overweight and obese adults [76]. Distinct carotenoids like β-cryptoxanthin have been shown to elevate adipokine levels in moderately obese post-menopausal women [77], while regular consumption of lycopene from tomatoes also mitigates OS in diabetic patients without reducing diabetes risk [78].

13.3.3 Catechins

Catechins, the primary flavonols in various sources like tea, cocoa, grapes, and red wine, play a vital role in reducing OS and inflammation. They achieve this by enhancing the expression of catalase and SOD, while also decreasing the activities of TNF-α, iNOS, Nox, and NF-κB [7]. One common catechin, epigallocatechin gallate, when supplemented, improves thermogenesis, enhances glucose tolerance, and increases PPARγ expression in rats fed a high-fat diet [79]. It also induces hypoglycemia and improves lipid profiles in streptozotocin-diabetic rats [80]. Epicatechin-enriched diets are linked to reduced IGF-1 levels and increased lifespan in diabetic mice and Drosophila melanogaster [81]. In clinical trial, catechins have shown the ability to lower obesity, blood pressure, LDL-cholesterol, and cardiovascular disease risk factors. Remarkably, beverages rich in catechins, such as green tea, are associated with decreased obesity and improved glucose levels among patients with type-II diabetes [82].

13.3.4 Flavonoids

Flavonoids, which are essential constituents of the human diet, are divided into distinct subgroups based on the oxidation levels of their defining rings. These subgroups encompass Flavonols (such as kaempferol and quercetin), flavanols (including epicatechin, EC, ECG, and EGCG), flavones (like apigenin), isoflavones (for instance, genistein), flavanones (such as hesperetin and naringenin), and anthocyanins (including cyanidin, delphidin, and malvidin). These compounds have shown potential therapeutic benefits against various conditions like atherosclerosis, cancer, myocardial infarction, congestive heart failure and, ischemic heart disease. They exert their effects through diverse biological mechanisms like scavenging reactive oxygen species (ROS), inducing apoptosis, and displaying antitumorigenic properties [83, 84]. This is attributed to their phenolic hydroxyl groups, which confer strong antioxidative properties capable of countering ROS-triggered lipid peroxidation initiation. Furthermore, some flavonoids act as soluble inhibitors, effectively neutralizing intermediate peroxyl and alkoxyl radicals that mediate peroxidation-induced molecular chain rupture. Noteworthy is the ability of specific flavonoids to hinder NF-κB–dependent gene expression [85].

13.3.5 Isoflavones

Isoflavones, such as daidzein, genistein, and glycitein, are primarily derived from soybeans and found in various grains, legumes, and vegetables. These compounds hold promise in combatting obesity and related conditions due to their anti-lipogenic and anti-adipogenic effects. Through their interactions with estrogen receptors and PPARγ, they influence adipose tissue growth, insulin sensitivity, and fatty acid metabolism while also mitigating oxidative stress and inflammation [86]. Genistein, for instance, showcases remarkable antioxidative abilities by countering ROS-induced damage, enhancing the GSH/GSSG ratio, and preserving mitochondrial membrane potential. Furthermore, it restores antioxidant enzymes, suppressing ROS, pro-inflammatory cytokines, iNOS, and eNOS, thus alleviating inflammation through actions like JNK activation and inhibition of NF-κB, TNF-α, and IL-6 [87]. Studies reveal that isoflavone supplementation contributes to reduced DNA oxidative damage and increased total plasma antioxidant capacity in healthy women [88]. In overweight diabetics, isoflavones not only boost antioxidant levels but also lower oxLDL and isoprostane 8-iso-PGF2α [89]. A significant finding is that supplementing with genistein in postmenopausal women leads to favorable reductions in body mass index, body fat mass, and waist size, coupled with elevated blood HDL levels [90].

13.3.6 Polyphenols

Polyphenols, abundant in foods like cranberries, red wine, and green teas, function as antioxidants contribute to addressing obesity-related comorbidities through various mechanisms, including influencing gut microbiota, promoting processes such as fatty acid oxidation, adipose tissue mobilization, and inducing apoptosis, satiety, thermogenesis and increased metabolism. These compounds also combat adipose differentiation and adipogenesis, mitigate the accumulation of triglycerides, and obesity-related inflammation [91]. Studies on cranberry extract's polyphenols in murine models reveal benefits such as reduced weight gain and visceral adipose tissue, lower hepatic triglyceride accumulation and plasma cholesterol levels, while enhancing insulin sensitivity [92]. Similarly, pomegranate extract polyphenols, combined with probiotics, exhibit promising outcomes in lowering triglycerides and intracellular lipids [92]. The red wine constituent resveratrol influences gut microbiota, improves endurance, and enhances insulin sensitivity in adipose tissue. It also boosts mitochondrial biogenesis, increases lipid utilization, and reduces glycolysis, ultimately elevating energy expenditure. Additionally, resveratrol's antioxidant function activates Nrf2, enhances SOD, and leads to reductions in BMI, blood pressure, and body weight [93].

Recent advancements highlight the beneficial role of curcumin, a polyphenol antioxidant, in alleviating obesity-related issues. Curcumin has ability to induce browning in white adipocytes by upregulating hormone-sensitive lipase and p-acyl-CoA carboxylase protein levels, thereby promoting lipolysis [94]. It is also associated with mitigating insulin resistance, hyperglycemia, hyperlipidemia, and obesity-linked comorbidities. Evidence suggests that curcumin can modulate several targets related to obesity and its related metabolic disorders, including the suppression of NF-κB and inflammatory cytokines' regulation [95]. Studies in murine models showcase dietary curcumin's effectiveness in lowering cholesterol, phospholipid and triglycerides, levels [96]. Additionally, bergamot's antioxidant properties contribute to improving obesity and related comorbidities. Rich in flavonoids and phenolic compounds, bergamot enhances dyslipidemia and reduces systemic inflammation in metabolic syndrome patients [97]. Bruteridin and melitidin isolated from Bergamot inhibit cholesterol synthesis by replacing HMG-CoA's substrate, leading to improved gut microbiota growth [98]. Preclinical evidence indicates that bergamot induction could enhance NASH features, reduce adipose tissue inflammation and cytokines like IL-6 and TNFα, alleviated yslipideia, hypoadiponectinemia and insulin resistance [99].

13.3.7 Vitamins

A variety of fruits and vegetables offer abundant vitamins, with vitamin C deficiency being particularly common among obese individuals. This deficiency shows a negative correlation with factors like waist-to-height ratio, body mass index and leptin concentrations. Vitamin C acts as both an antioxidant and electron donor, playing a vital role as a cofactor in fifteen enzymes. This function effectively combats free radicals, guarding tissues against oxidative stress and reducing inflammation. Its effects extend to inhibiting mature adipocyte formation, suppressing cell growth, blocking lipolysis, regulating glucocorticoid release, and modulating glucose metabolism and leptin secretion [100]. This comprehensive impact ultimately leads to reduced hyperglycemia and decreased glycosylation. Conversely, vitamin E's connection to obesity markers lacks clear confirmation. However, those with obesity-related metabolic syndrome require increased vitamin E intake due to elevated oxidative stress from weight-related concerns, even though these same concerns contribute to decreased vitamin E utilization [101]. Direct links exist between leptin levels and vitamin A, with higher vitamin A levels more common in non-obese women [102]. Additionally, an association between vitamin D deficiency and overweight has been observed [103]. Despite these findings, current data does not provide substantial support for using vitamin supplementation in obesity.

13.4 Nano Herbal Formulations with Anti-Obese Effects

Nanotherapy, a specialized field within therapeutics, harnesses nanotechnology's potential in drug delivery systems to achieve site-specific targeting and reduce dosage requirements. Recent years have witnessed a transformative impact of nanotechnology on drug delivery, fundamentally altering drug administration. This revolution is attributed to nanotechnology's unique characteristic of nano-sized dimensions (ranging from 1-100nm), enabling precise and targeted drug action. In the realm of herbal medicine, the fusion of nanotechnology and herbal knowledge, referred to as herbal nanotherapeutics, has significantly elevated treatment outcomes across diverse ailments like cancer, diabetes, and hypertension. This integration bolsters medical practices by enhancing herbal medicine stability, amplifying therapeutic effectiveness, and facilitating targeted delivery. Through the implementation of nanocarriers, the stability of active herbal components is augmented, leading to increased bioavailability, heightened pharmacological activity, reduced toxicity, and prevention of chemical degradation within the body. Moreover, these nanocarriers ensure sustained, controlled release, reducing the need for frequent administration and improving patient adherence. A multitude of nanocarrier types, including liposomes, nanoemulsions, polymeric nanoparticles, solid lipid nanoparticles, and phytosomes, are employed in crafting herbal nanotherapeutics.

13.4.1 Liposomes

Liposomes, consisting of closed spherical lipid-based bilayer systems, possess hydrophilic heads and hydrophobic tails, enabling them to effectively deliver both hydrophilic and lipophilic drugs. Implementing liposomes as nanocarriers for herbal medicine has demonstrated enhanced stability and tolerability of phytoconstituents, along with reduced toxicity. Research by Lacatusu *et al.* highlighted liposomes' utility in delivering capsaicin, an herbal anti-obesity component from red pepper extract [104]. This approach served as an alternative to traditional treatments, offering improved gastric tolerance and better targeting of adipose cells. Furthermore, treatment of Swiss Albino mice fed a high-fat diet with capsaicin-loaded liposomes exhibited a remarkable 15% reduction in body weight compared to the control [105]. Capsaicin at the dose of 10 mg/kg has proven its potential for the treatment of obesity in Sprague-Dawley rats. The proposed mechanism suggested that capsaicin downregulates Steap3 protein and heat shock protein 27 (Hsp27), while upregulating olfactory receptor (Olr1434) in obese white adipose tissue (WAT) [106].

In another study, resveratrol has been proved to be used as a potent anti-obesity agent, attributing its effects to the browning of white adipose tissue [107]. Although conventional resveratrol formulations showed promise, they encountered issues like low solubility, stability, and bioavailability [108]. Encapsulation of resveratrol in lipid nanocarriers found to be useful for boosting the pharmacological activity and maintaining chemical stability [109]. An assessment using a 3T3-L1 white adipocyte cell model further confirmed resveratrol's anti-obesity activity through WAT browning.

Silibinin, a significant constituent of silymarin extracted from milk thistle seeds, is utilized for treating obesity-associated chronic liver disease. However, its limited bioavailability and solubility hindered its use. Chen *et al.* addressed this issue by developing a lipid-based nanocarrier system for oral silibinin delivery, resulting in a 2.9-fold increase in bioavailability and improved therapeutic effects [110]. Preclinical testing on male Sprague-Dawley rats using silibinin-lipid nanocarriers has demonstrated its anti-obese effect at a dose of 300mg/kg evidenced by substantial reductions in collagen production and lipid droplet formation.

13.4.2 Solid Lipid Nanoparticles (SLNs)

Solid Lipid Nanoparticles (SLNs) are spherical shaped having diameter in the range of 50-1000 nm represent solid core lipid monolayer carrier system making them highly valuable in drug delivery due to their advantageous qualities including its non-toxic nature, stability, small size, large surface area, adaptability for large-scale production, high drug loading and its cost-effectiveness [111]. These SLNs play a crucial role in encapsulating hydrophobic herbal phytoconstituents, thereby boosting its bioavailability, increased stability, mitigating toxicity, and enabling precise control over the release of bioactive compounds [112]. For instance, in a study conducted by Ezhilarasi and its coworkers they developed SLNs containing hydroxycitric acid (HCA), a potent inhibitor sourced from *Garcinia cambogia* plants, known for its anti-obesity effects. This formulation demonstrated significantly improved bioavailability compared to non-encapsulated HCA, as evidenced by increased C_{max} in Wistar rats. Moreover, these SLNs not only heightened pharmacological efficacy but also facilitated targeted delivery to adipose tissues, effectively reducing HCA-related side effects like testicular atrophy and liver toxicity [113].

13.4.3 Polymeric and Magnetic Nanoparticles

Nanoparticles, having size in the range from 1-100nm, are extensively utilized in drug delivery systems due to their small size and precise action. Various types of nanoparticles, including polymeric, metal, and magnetic variants, contribute to effective drug delivery [114]. Integration of herbal medicine into nanoparticulate systems not only amplifies its therapeutic impact but also helps to reduce associated toxicities. Particularly, nanoparticles exhibit a unique capacity to transport anti-obesity bioactive constituents to targeted areas. Metal nanoparticles, such as silver, gold, and palladium, play a crucial role in targeted obesity treatment, accumulating around stored fat tissues [115]. This passive targeting mechanism, alongside an active approach involving specific ligand receptors, aids intracellular localization. Researchers are actively developing nanoparticulate herbal drug delivery systems for safer and alternative obesity treatments. An example is the formulation by Nallamuthu and its coworkers who formulated chitosan nanoparticles for targeted chlorogenic acid delivery, showcasing its anti-obesity effects. This approach led to a significant reduction in body weight, visceral fat mass, adipose tissue weight, plasma insulin and leptin level [116].

13.4.4 Nanosized Micelles

Nanosized micelles, which are clusters of surfactant molecules containing hydrophilic heads and hydrophobic tails, play a pivotal role in drug delivery systems, particularly significant for delivery the drugs, including bioactive compounds and phytoconstituents having poor solubility and low stability during delivery [117]. To overcome these limitations, researchers have harnessed the power of nanosized micelles to encapsulate these components. A compelling example is evidenced by study in which high fat diet-induced obese mice model shows that Micelles of capsaicin confers dramatic weight loss and adipose tissue browning, [118] demonstrating how this approach substantially enhances solubility and bioavailability. In addition, polymer PLA-co-PEG can be used as a carrier for orally administered fenofibrate drugs in the future for better water solubility and efficiency [119].

13.4.5 Gold Nanoparticles

In an investigation conducted by Ansari *et al.*, *in-vivo* anti-obesity potential of gold nanoparticles from *Smilax glabra* rhizome was examined in-vivo using male Wistar rats. The study revealed effective active uptake of these nanoparticles, showing their efficacy. Remarkably, treatment with these gold nanoparticles exhibited a positive impact on the restoration of internal cellular components such as membranes, nuclei, and cytoplasm [120]. Additionally, the administration of 50mg/kg of these nanoparticles led to a notable reduction in both body weight and body mass. Complementary research in the field includes, utilization of gold nanoparticles derived from *Salacia chinensis* [121], and from *Poria cocos*. Collectively, these studies underline the potential of gold nanoparticles as effective nanocarriers for herbal anti-obesity treatments [122].

13.4.6 Phytosomal Nanocarriers

Phytosomes represent a lipid-based nanocarrier system resulting from the intricate binding between bioactive phytoconstituents and a lipid layer. These chemical bonds significantly enhance the stability of the easily degradable phytoconstituents, ultimately improving bioavailability and therapeutic impact [123]. The integration of water-soluble herbal extracts with the lipophilic outer layer of phytosomes yields improved absorption, elevated bioavailability, and precise targeting when compared to conventional herbal extract forms. Menshawe *et al.* developed a novel nanosized thermogel infused with soybean phytosome exhibit properties to combat obesity. The anti-obesity effects of soybean, extracted from Glycine max of the Fabaceae family, are attributed to its active components such as proteins, saponins, and phosphatidylcholine. In preclinical trials, male Albino rats treated abdominally with a topical 2.5% phytosomal soybean formulation, weight gain increased by 24.570±3.35% in contrast to the control at 51.020±4.78%, together with reduced level of lipids. The nanosized phytosome particles demonstrated enhanced skin permeation, introducing an innovative topical avenue for obesity treatment [124].

13.5 Clinical Trials

Numerous clinical investigations have explored the potential of various herbal extracts and their active components in addressing obesity. For instance, the hydroxycitric acid in *Garcinia cambogia* inhibits lipogenesis and ATP citrate lyase, resulting in reduced serum triglycerides among obese women [125]. Additionally, *Curcuma longa*-derived curcumin [126], and *Salacia oblonga*-derived polyphenols [127] showcase lipid-lowering, glucose-reducing, and antioxidant properties in diabetic and obese patients. Moreover, *Salacia oblonga* tea extract demonstrates the ability to enhance lipid profiles and lower glucose levels [128]. Notably, *Portulaca oleracea* L. seed extract ameliorates lipid profiles among obese adolescents [129]. *Hibiscus sabdariffa* contributes to lowered body weight and fat [130], while *Phaseolus vulgaris* extracts inhibit α-amylase, resulting in reduced weight, fat mass, and waist/hip perimeters [131]. Furthermore, *C. annuum* stimulates catecholamine secretion, promoting fat oxidation and energy expenditure [132]. Green tea and coffee, containing catechins, are recognized for aiding in weight management. In the realm of effective obesity control, *Sorghum bicolor* L. stands out as a promising alternative [133]. The interest in harnessing herbal compounds for their safe and potent effects is driving research forward, highlighting the need for clinical studies that extend beyond *in vitro* and *in vivo* investigations.

13.6 Conclusion and Future Perspectives

To conclude, OS plays significant role in pathophysiology of obesity which alter mitochondrial activity regulators, inflammation-related molecules, and promotes lipogenesis. It also drives preadipocyte differentiation, adipogenesis, and regulates hypothalamic neurons controlling appetite and energy balance. As a result, natural antioxidants derived from plants play a vital role in obesity control. These antioxidants encompass diverse compounds like polyphenols (phenolic acids, flavonoids, stilbenes, etc.), carotenoids, capsaicinoids, isothiocyanates, and catechins. Curcumin, phloretin, quercetin, resveratrol, ferulic acid and other active phytoconstituents found in green tea such (-)-epicatechin, (-)-epigallocatechin, (-)-epicatechin-3-gallate, and (-)-epigallocatechin-3-gallate (EGCG) *etc.* are few examples of active phytoconstituents that hold potential against oxidative stress-linked diseases, including obesity. Their potency lies in ROS scavenging and NF-κB activation prevention, reducing target gene expression. Advancements and growing interest have expanded research in herbal medicine and natural products, driven by nanotechnology for innovative drug delivery. Nanocarriers like liposomes, polymeric nanoparticles, and gold nanoparticles offer promise, enhancing herbal bioactive delivery. However, challenges in scaling nanotechnology to therapy persist, given the complexities of nanomaterial interactions with biological systems. Rigorous clinical trials are crucial to understanding the impacts of herbal compounds and their nano-formulations in treating obesity effectively.

References

1. Furukawa, S., Fujita, T., Shimabukuro, M., Iwaki, M., Yamada, Y., Nakajima, Y., Nakayama, O., Makishima, M., Matsuda, M., Shimomura, I., Increased oxidative stress in obesity and its impact on metabolic syndrome. *J. Clin. Invest.*, 114, 1752, 2017.
2. Kumar, M., Kaushik, D., Kaur, J., Proestos, C., Oz, F., Oz, E., Gupta, P., Kundu, P., Kaur, A., Anisha, A., Ritika, R., A critical review on obesity: Herbal approach, bioactive compounds, and their mechanism. *Appl. Sci.*, 12, 8342, 2022.
3. Marrelli, M., Statti, G., Conforti, F., A review of biologically active natural products from Mediterranean wild edible plants: Benefits in the treatment of obesity and its related disorders. *Molecules*, 25, 649, 2020.
4. Savini, I., Catani, M.V., Evangelista, D., Gasperi, V., and Avigliano, L., Obesity-associated oxidative stress: strategies finalized to improve redox state. *Int. J. Mol. Sci.*, 14, 10497, 2013.
5. Boccellino, M. and D'Angelo, S., Anti-obesity effects of polyphenol intake: Current status and future possibilities. *Int. J. Mol. Sci.*, 21, 5642, 2020.
6. Singh, M., Thrimawithana, T., Shukla, R., Adhikari, B., Managing obesity through natural polyphenols: A review. *Future Foods*, 1, 100002, 2020.
7. Pérez-Torres, I., Castrejón-Téllez, V., Soto, M.E., Rubio-Ruiz, M.E., Manzano-Pech, L., Guarner-Lans, V., Oxidative stress, plant natural antioxidants, and obesity. *Int. J. Mol. Sci.*, 22, 1786, 2021.
8. Mohamed, G.A., Ibrahim, S.R., Elkhayat, E.S., El Dine, R.S., Natural anti-obesity agents. *Bull. Fac. Pharm. Cairo Univ.*, 52, 269, 2014.
9. Rodríguez-Pérez, C., Segura-Carretero, A., del Mar Contreras, M., Phenolic compounds as natural and multifunctional anti-obesity agents: A review. *Crit. Rev. Food Sci. Nutr.*, 59, 1212, 2019.
10. Gadde, K.M., Martin, C.K., Berthoud, H.R., Heymsfield, S.B., Obesity: pathophysiology and management. *J. Am. Coll. Cardiol.*, 71, 69, 2018.
11. Zhao, Y., Chen, B., Shen, J., Wan, L., Zhu, Y., Yi, T., Xiao, Z., The beneficial effects of quercetin, curcumin, and resveratrol in obesity. *Oxid. Med. Cell. Longev.*, 2017, 1–8, 2017.
12. Ghosh, S., Manchala, S., Raghunath, M., Sharma, G., Singh, A.K., Sinha, J.K., Role of phytomolecules in the treatment of obesity: targets, mechanisms and limitations. *Curr. Top. Med. Chem.*, 21, 863, 2021.
13. Saad, B. and Ghareeb, B., and Kmail, A., Metabolic and epigenetics action mechanisms of anti-obesity medicinal plants and phytochemicals. *Evid. Based Complementary Altern. Med*, 1, 2021, 2021.
14. Payab, M., Hasani-Ranjbar, S., Shahbal, N., Qorbani, M., Aletaha, A., Haghi-Aminjan, H., Soltani, A., Khatami, F., Nikfar, S., Hassani, S., Abdollahi, M., Effect of the herbal medicines in obesity and metabolic syndrome: a systematic review and meta-analysis of clinical trials. *Phytother. Res.*, 34, 526, 2020.
15. Mopuri, R. and Islam, M.S., Medicinal plants and phytochemicals with anti-obesogenic potentials: A review. *Biomed. Pharmacother.*, 89, 1442, 2017.
16. Nijhawan, P., Arora, S., Behl, T., Intricate role of oxidative stress in the progression of obesity. *Obes. Med.*, 15, 100125, 2019.

17. Čolak, E. and Pap, D., The role of oxidative stress in the development of obesity and obesity-related metabolic disorders. *J. Med. Biochem.*, 40, 1, 2021.
18. Rosen, E.D. and Spiegelman, B.M., Adipocytes as regulators of energy balance and glucose homeostasis. *Nature*, 444, 847, 2006.
19. Ikeda, K., Maretich, P., Kajimura, S., The common and distinct features of brown and beige adipocytes. *Trends Endocrinol. Metab.*, 29, 191, 2018.
20. Altshuler-Keylin, S., Shinoda, K., Hasegawa, Y., Ikeda, K., Hong, H., Kang, Q., Yang, Y., Perera, R.M., Debnath, J., Kajimura, S., Beige adipocyte maintenance is regulated by autophagy-induced mitochondrial clearance. *Cell Metab.*, 24, 402, 2016.
21. Cypess, A.M. and Kahn, C.R., Brown fat as a therapy for obesity and diabetes. *Curr. Opin. Endocrinol. Diabetes Obes.*, 17, 143, 2010.
22. Gesta, S. and Tseng, Y.H., and Kahn, C.R., Developmental origin of fat: tracking obesity to its source. *Cell*, 131, 242, 2007.
23. Lushchak, V.I., Free radicals, reactive oxygen species, oxidative stress and their classifications. *Ukr. Biochem. J.*, 87, 6, 11–8, 2015 Nov-Dec.
24. Hancock, J.T., Oxygen is instrumental for biological signaling: An overview. *Oxygen*, 1, 1, 3–15, 2021.
25. Lambert, A.J. and Brand, M.D., Reactive oxygen species production by mitochondria. *Mitochondrial DNA: Methods Protoc.*, 554, 165–181, 2009.
26. Tormos, K.V., Anso, E., Hamanaka, R.B., Eisenbart, J., Joseph, J., Kalyanaraman, B., Chandel, N.S., Mitochondrial complex III ROS regulate adipocyte differentiation. *Cell Metab.*, 14, 537, 2011.
27. Venditti, P., Di Stefano, L., Di Meo, S., Mitochondrial metabolism of reactive oxygen species. *Mitochondrion*, 13, 71, 2013.
28. Corkey, B.E., Reactive oxygen species: role in obesity and mitochondrial energy efficiency. *Philos. Trans. R. Soc. Lond., B, Biol. Sci.*, 378, 20220210, 2023.
29. Bhatti, J.S., Bhatti, G.K., Reddy, P.H., Mitochondrial dysfunction and oxidative stress in metabolic disorders—A step towards mitochondria based therapeutic strategies. *Biochim. Biophys. Acta (BBA)-Mol. Basis Dis.*, 1863, 1066, 2017.
30. Pintus, F., Floris, G., Rufini, A., Nutrient availability links mitochondria, apoptosis, and obesity. *Aging (Albany NY)*, 4, 734, 2012.
31. Cunarro, J., Casado, S., Lugilde, J., Tovar, S., Hypothalamic mitochondrial dysfunction as a target in obesity and metabolic disease. *Front. Endocrinol.*, 9, 283, 2018.
32. de Mello, A.H., Costa, A.B., Engel, J.D.G., Rezin, G.T., Mitochondrial dysfunction in obesity. *Life Sci.*, 192, 26, 2018.
33. Bournat, J.C. and Brown, C.W., Mitochondrial dysfunction in obesity. *Curr. Opin. Endocrinol. Diabetes Obes.*, 17, 446, 2010.
34. Serra, D., Mera, P., Malandrino, M.I., Mir, J.F., Herrero, L., Mitochondrial fatty acid oxidation in obesity. *Antioxid. Redox Signal.*, 19, 269, 2013.
35. Palikaras, K., Lionaki, E., Tavernarakis, N., Mechanisms of mitophagy in cellular homeostasis, physiology and pathology. *Nat. Cell Biol.*, 20, 1013, 2018.
36. Berthoud, H.R. and Morrison, C., The brain, appetite, and obesity. *Annu. Rev. Psychol.*, 59, 55, 2008.
37. Gibson, C.D., Carnell, S., Ochner, C.N., Geliebter, A., Neuroimaging, gut peptides and obesity: novel studies of the neurobiology of appetite. *J. Neuroendocrinol.*, 22, 833, 2010.
38. Lam, D.D., Garfield, A.S., Marston, O.J., Shaw, J., Heisler, L.K., Brain serotonin system in the coordination of food intake and body weight. *Pharmacol. Biochem. Behav.*, 97, 84, 2010.
39. Blum, K., Thanos, P.K., Gold, M.S., Dopamine and glucose, obesity, and reward deficiency syndrome. *Front. Psychol.*, 5, 919, 2014.

40. Di Domenico, M., Pinto, F., Quagliuolo, L., Contaldo, M., Settembre, G., Romano, A., Coppola, M., Ferati, K., Bexheti-Ferati, A., Sciarra, A., Nicoletti, G.F., The role of oxidative stress and hormones in controlling obesity. *Front. Endocrinol.*, 10, 540, 2019.
41. Obradovic, M., Sudar-Milovanovic, E., Soskic, S., Essack, M., Arya, S., Stewart, A.J., Gojobori, T., Isenovic, E.R., Leptin and obesity: role and clinical implication. *Front. Endocrinol.*, 12, 585887, 2021.
42. Dalamaga, M., Chou, S.H., Shields, K., Papageorgiou, P., Polyzos, S.A., Mantzoros, C.S., Leptin at the intersection of neuroendocrinology and metabolism: current evidence and therapeutic perspectives. *Cell Metab.*, 18, 29, 2013.
43. Kelesidis, T., Kelesidis, I., Chou, S., Mantzoros, C.S., Narrative review: the role of leptin in human physiology: emerging clinical applications. *Ann. Intern. Med.*, 152, 93, 2010.
44. Zhang, Y., Proenca, R., Maffei, M., Barone, M., Leopold, L., Friedman, J.M., Positional cloning of the mouse obese gene and its human homologue. *Nature*, 372, 425, 1994.
45. Myers, M.G., Heymsfield, S.B., Haft, C., Kahn, B.B., Laughlin, M., Leibel, R.L., Tschöp, M.H., Yanovski, J.A., Challenges and opportunities of defining clinical leptin resistance. *Cell Metab.*, 15, 150, 2012.
46. Nguyen, T.M.D., Adiponectin: role in physiology and pathophysiology. *Int. J. Prev. Med.*, 11, 1–9, 2020.
47. Liu, C., Feng, X., Li, Q., Wang, Y., Li, Q., Hua, M., Adiponectin, TNF-α and inflammatory cytokines and risk of type 2 diabetes: a systematic review and meta-analysis. *cytokine*, 86, 100, 2016.
48. Kern, P.A., Ranganathan, S., Li, C., Wood, L., Ranganathan, G., Adipose tissue tumor necrosis factor and interleukin-6 expression in human obesity and insulin resistance. *Am. J. Physiol. Endocrinol. Metab.*, 280, 5, E745–51, 2001.
49. Chen, Y.J., Zhang, L.Q., Wang, G.P., Zeng, H., Lü, B., Shen, X.L., Jiang, Z.P., Chen, F.P., Adiponectin inhibits tissue factor expression and enhances tissue factor pathway inhibitor expression in human endothelial cells. *Thromb. Haemost.*, 100, 291, 2008.
50. Kadowaki, T., Yamauchi, T., Kubota, N., Hara, K., Ueki, K., Tobe, K., Adiponectin and adiponectin receptors in insulin resistance, diabetes, and the metabolic syndrome. *J. Clin. Investig.*, 116, 1784, 2006.
51. Matsuda, M. and Shimomura, I., Roles of adiponectin and oxidative stress in obesity-associated metabolic and cardiovascular diseases. *Rev. Endocr. Metab. Disord.*, 15, 1, 2014.
52. Nicolas, G., Chauvet, C., Viatte, L., Danan, J.L., Bigard, X., Devaux, I., Beaumont, C., Kahn, A., Vaulont, S., The gene encoding the iron regulatory peptide hepcidin is regulated by anemia, hypoxia, and inflammation. *J. Clin. Investig.*, 110, 1037, 2002.
53. Nemeth, E. and Ganz, T., Regulation of iron metabolism by hepcidin. *Annu. Rev. Nutr.*, 26, 323, 2006.
54. Alshwaiyat, N.M., Ahmad, A., Wan Hassan, W.M.R., AlJamal, H.A.N., Association between obesity and iron deficiency. *Exp. Ther. Med.*, 22, 1, 2021.
55. Aeberli, I., Hurrell, R.F., Zimmermann, M.B., Overweight children have higher circulating hepcidin concentrations and lower iron status but have dietary iron intakes and bioavailability comparable with normal weight children. *Int. J. Obesity*, 33, 1111, 2009.
56. Kim, S.L., Shin, S., Yang, S.J., Iron Homeostasis and Energy Metabolism in Obesity. *Clin.Nutr. Res.*, 11, 316, 2022.
57. Menzie, C.M., Yanoff, L.B., Denkinger, B.I., McHugh, T., Sebring, N.G., Calis, K.A., Yanovski, J.A., Obesity-related hypoferremia is not explained by differences in reported intake of heme and nonheme iron or intake of dietary factors that can affect iron absorption. *J. Am. Diet. Assoc.*, 108, 145, 2008.
58. Nikonorov, A.A., Skalnaya, M.G., Tinkov, A.A., Skalny, A.V., Mutual interaction between iron homeostasis and obesity pathogenesis. *J. Trace Elem. Med. Bio.*, 30, 207, 2015.

59. Davis, C.D., The gut microbiome and its role in obesity. *Nutr. Today*, 51, 167, 2016.
60. Wong, J.M., De Souza, R., Kendall, C.W., Emam, A., Jenkins, D.J., Colonic health: fermentation and short chain fatty acids. *J. Clin. Gastroenterol.*, 40, 235, 2006.
61. Van Hul, M. and Cani, P.D., The gut microbiota in obesity and weight management: microbes as friends or foe? *Nat. Rev. Endocrinol.*, 1, 258–271, 2023.
62. Crovesy, L., Masterson, D., Rosado, E.L., Profile of the gut microbiota of adults with obesity: a systematic review. *Eur. J. Clin. Nutr.*, 74, 1251, 2020.
63. Ley, R.E., Bäckhed, F., Turnbaugh, P., Lozupone, C.A., Knight, R.D., Gordon, J.I., Obesity alters gut microbial ecology. *Proc. Natl. Acad. Sci. U.S.A.*, 102, 11070, 2005.
64. Cani, P.D., Dewever, C., Delzenne, N.M., Inulin-type fructans modulate gastrointestinal peptides involved in appetite regulation (glucagon-like peptide-1 and ghrelin) in rats. *Br. J. Nutr.*, 92, 521, 2004.
65. Wilson, A.S., Koller, K.R., Ramaboli, M.C., Nesengani, L.T., Ocvirk, S., Chen, C., Flanagan, C.A., Sapp, F.R., Merritt, Z.T., Bhatti, F., Thomas, T.K., Diet and the human gut microbiome: an international review. *Dig. Dis. Sci.*, 65, 723, 2020.
66. Wang, Z., Klipfell, E., Bennett, B.J., Koeth, R., Levison, B.S., DuGar, B., Feldstein, A.E., Britt, E.B., Fu, X., Chung, Y.M., Wu, Y., Gut flora metabolism of phosphatidylcholine promotes cardiovascular disease. *Nature*, 472, 57, 2011.
67. Tolhurst, G., Heffron, H., Lam, Y.S., Parker, H.E., Habib, A.M., Diakogiannaki, E., Cameron, J., Grosse, J., Reimann, F., Gribble, F.M., Short-chain fatty acids stimulate glucagon-like peptide-1 secretion via the G-protein–coupled receptor FFAR2. *Diabetes*, 61, 364, 2012.
68. Wondmkun, Y.T., Obesity, insulin resistance, and type 2 diabetes: associations and therapeutic implications. *Diabet., Metab. Synd. Ob.*, 13, 3611–3616, 2020.
69. Kurashima, Y., Yamamoto, D., Nelson, S., Uematsu, S., Ernst, P.B., Nakayama, T., Kiyono, H., Mucosal mesenchymal cells: secondary barrier and peripheral educator for the gut immune system. *Front. Immunol.*, 8, 1787, 2017.
70. Srinivasan, K., Biological activities of red pepper (*Capsicum annuum*) and its pungent principle capsaicin: a review. *Crit. Rev. Food Sci. Nutr.*, 56, 1488, 2016.
71. Shaik Mohamed Sayed, U.F., Moshawih, S., Goh, H.P., Kifli, N., Gupta, G., Singh, S.K., Chellappan, D.K., Dua, K., Hermansyah, A., Ser, H.L., Ming, L.C., Natural products as novel anti-obesity agents: insights into mechanisms of action and potential for therapeutic management. *Front. Pharmacol.*, 14, 1182937, 2023.
72. Ludy, M.J., Moore, G.E., Mattes, R.D., The effects of capsaicin and capsiate on energy balance: critical review and meta-analyses of studies in humans. *Chem. Senses.*, 37, 103, 2012.
73. Panchal, S.K., Bliss, E., Brown, L., Capsaicin in metabolic syndrome. *Nutrients*, 10, 630, 2018.
74. Kang, J.H., Kim, C.S., Han, I.S., Kawada, T., Yu, R., Capsaicin, a spicy component of hot peppers, modulates adipokine gene expression and protein release from obese-mouse adipose tissues and isolated adipocytes, and suppresses the inflammatory responses of adipose tissue macrophages. *FEBS Lett.*, 581, 4389, 2007.
75. Mounien, L., Tourniaire, F., Landrier, J.F., Anti-obesity effect of carotenoids: direct impact on adipose tissue and adipose tissue-driven indirect effects. *Nutrients*, 11, 1562, 2019.
76. Xia, W., Tang, N., Kord-Varkaneh, H., Low, T.Y., Tan, S.C., Wu, X., Zhu, Y., 161, The effects of astaxanthin supplementation on obesity, blood pressure, CRP, glycemic biomarkers, and lipid profile: A meta-analysis of randomized controlled trials. *Pharmacol. Res.*, 161, 105113, 2020.
77. Iwamoto, M., Imai, K., Ohta, H., Shirouchi, B., Sato, M., Supplementation of highly concentrated β-cryptoxanthin in a satsuma mandarin beverage improves adipocytokine profiles in obese Japanese women. *Lipids Health Dis.*, 11, 1, 2012.

78. Zhu, R., Chen, B., Bai, Y., Miao, T., Rui, L., Zhang, H., Xia, B., Li, Y., Gao, S., Wang, X.D., Zhang, D., Lycopene in protection against obesity and diabetes: A mechanistic review. *Pharmacol. Res.*, 159, 104966, 2020.
79. Roghani, M. and Baluchnejadmojarad, T., Hypoglycemic and hypolipidemic effect and antioxidant activity of chronic epigallocatechin-gallate in streptozotocin-diabetic rats. *Pathophysiol*, 17, 55–59, 2010.
80. Quine, S.D. and Raghu, P.S., Effects of (-)-epicatechin, a flavonoid on lipid peroxidation and antioxidants in streptozotocin-induced diabetic liver, kidney and heart. *Pharmacol. Rep.*, 57, 610, 2005.
81. Si, H., Fu, Z., Babu, P.V.A., Zhen, W., LeRoith, T., Meaney, M.P., Voelker, K.A., Jia, Z., Grange, R.W., Liu, D., Dietary epicatechin promotes survival of obese diabetic mice and Drosophila melanogaster. *J. Nutr.*, 141, 1095, 2011.
82. Hibi, M., Takase, H., Iwasaki, M., Osaki, N., Katsuragi, Y., Efficacy of tea catechin-rich beverages to reduce abdominal adiposity and metabolic syndrome risks in obese and overweight subjects: a pooled analysis of 6 human trials. *Nutr. Res.*, 55, 1, 2018.
83. Rufino, A.T., Costa, V.M., Carvalho, F., Fernandes, E., Flavonoids as antiobesity agents: A review. *Med. Res. Rev.*, 41, 556, 2021.
84. Nijhawan, P. and Behl, T., Nutraceuticals in the management of obesity. *Obesity Med.*, 17, 100168, 2020.
85. Park, Y.C., Rimbach, G., Saliou, C., Valacchi, G., Packer, L., Activity of monomeric, dimeric, and trimeric flavonoids on NO production, TNF-α secretion, and NF-κB-dependent gene expression in RAW 264.7 macrophages. *FEBS Lett.*, 465, 93, 2000.
86. Vitale, D.C., Piazza, C., Melilli, B., Drago, F., Salomone, S., Isoflavones: estrogenic activity, biological effect and bioavailability. *Eur. J. Drug Metab. Pharmacokinet.*, 38, 15, 2013.
87. Behloul, N. and Wu, G., Genistein: a promising therapeutic agent for obesity and diabetes treatment. *Eur. J. Pharmacol.*, 698, 31, 2013.
88. Erba, D., Casiraghi, M.C., Martinez-Conesa, C., Goi, G., Massaccesi, L., Isoflavone supplementation reduces DNA oxidative damage and increases O-β-N-acetyl-D-glucosaminidase activity in healthy women. *Nutr. Res.*, 32, 233, 2012.
89. Kuryłowicz, A., The role of isoflavones in type 2 diabetes prevention and treatment—A narrative review. *Int. J. Mol. Sci.*, 22, 218, 2020.
90. Goodman-Gruen, D. and Kritz-Silverstein, D., Usual dietary isoflavone intake is associated with cardiovascular disease risk factors in postmenopausal women. *J. Nutr.*, 131, 1202, 2001.
91. Aloo, S.O., Ofosu, F.K., Kim, N.H., Kilonzi, S.M., Oh, D.H., Insights on Dietary Polyphenols as Agents against Metabolic Disorders: Obesity as a Target Disease. *Antioxid.*, 12, 416, 2023.
92. Anhê, F.F., Roy, D., Pilon, G., Dudonné, S., Matamoros, S., Varin, T.V., Garofalo, C., Moine, Q., Desjardins, Y., Levy, E., Marette, A., A polyphenol-rich cranberry extract protects from diet-induced obesity, insulin resistance and intestinal inflammation in association with increased Akkermansia spp. population in the gut microbiota of mice. *Gut*, 64, 872, 2015.
93. Sorrenti, V., Randazzo, C.L., Caggia, C., Ballistreri, G., Romeo, F.V., Fabroni, S., Timpanaro, N., Raffaele, M., Vanella, L., Beneficial effects of pomegranate peel extract and probiotics on pre-adipocyte differentiation. *Front. Microbiol.*, 10, 660, 2019.
94. Mongioì, L.M., La Vignera, S., Cannarella, R., Cimino, L., Compagnone, M., Condorelli, R.A., Calogero, A.E., The role of resveratrol administration in human obesity. *Int. J. Mol. Sci.*, 22, 4362, 2021.
95. Lone, J., Choi, J.H., Kim, S.W., Yun, J.W., Curcumin induces brown fat-like phenotype in 3T3-L1 and primary white adipocytes. *J. Nutr. Biochem.*, 27, 193, 2016.
96. Bradford, P.G., Curcumin and obesity. *Biofactors*, 39, 78, 2013.

97. Musolino, V., Gliozzi, M., Nucera, S., Carresi, C., Maiuolo, J., Mollace, R., Paone, S., Bosco, F., Scarano, F., Scicchitano, M., Ruga, S., The effect of bergamot polyphenolic fraction on lipid transfer protein system and vascular oxidative stress in a rat model of hyperlipemia. *Lipids Health Dis.*, 18, 1, 115, 2019.
98. Leopoldini, M., Malaj, N., Toscano, M., Sindona, G., Russo, N., On the inhibitor effects of bergamot juice flavonoids binding to the 3-hydroxy-3-methylglutaryl-CoA reductase (HMGR) enzyme. *J. Agric. Food. Chem.*, 58, 10768, 2010.
99. Musolino, V., Gliozzi, M., Scarano, F., Bosco, F., Scicchitano, M., Nucera, S., Carresi, C., Ruga, S., Zito, M.C., Maiuolo, J., Macrì, R., Bergamot polyphenols improve dyslipidemia and pathophysiological features in a mouse model of non-alcoholic fatty liver disease. *Sci. Rep.*, 10, 2565, 2020.
100. Garcia-Diaz, D.F., Lopez-Legarrea, P., Quintero, P., Martinez, J.A., Vitamin C in the treatment and/or prevention of obesity. *J. Nutr. Sci. Vitaminol.*, 60, 367, 2014.
101. Niki, E., Evidence for beneficial effects of vitamin E. *Korean J. Intern. Med.*, 30, 571, 2015.
102. Mah, E., Sapper, T.N., Chitchumroonchokchai, C., Failla, M.L., Schill, K.E., Clinton, S.K., Bobe, G., Traber, M.G., Bruno, R.S., α-Tocopherol bioavailability is lower in adults with metabolic syndrome regardless of dairy fat co-ingestion: a randomized, double-blind, crossover trial. *Am. J. Clin. Nutr.*, 102, 1070, 2015.
103. Merino, O., Gregorio, B., Sampaio, F., Sanchez, R., Risopatrón, J., Role of Vitamin D in the Development of Obesity. *Int. J. Morphol.*, 35, 1568, 2017.
104. Lacatusu, I., Badea, N., Udeanu, D., Coc, L., Pop, A., Negut, C.C., Tanase, C., Stan, R., Meghea, A., Improved anti-obesity effect of herbal active and endogenous lipids co-loaded lipid nanocarriers: Preparation, *in vitro* and *in vivo* evaluation. *Mater. Sci. Eng., C*, 99, 12, 2019.
105. Joo, J.I., Kim, D.H., Choi, J.W., Yun, J.W., Proteomic analysis for antiobesity potential of capsaicin on white adipose tissue in rats fed with a high fat diet. *J. Proteome Res.*, 9, 2977, 2010.
106. Wang, S., Liang, X., Yang, Q., Fu, X., Rogers, C.J., Zhu, M., Rodgers, B.D., Jiang, Q., Dodson, M.V., Du, M., Resveratrol induces brown-like adipocyte formation in white fat through activation of AMP-activated protein kinase (AMPK) α1. *Int. J. Obes.*, 39, 967, 2015.
107. Walle, T., Hsieh, F., DeLegge, M.H., Oatis, J.E., Walle, U.K., High absorption but very low bioavailability of oral resveratrol in humans. *Drug Metab. Dispos.*, 32, 1377, 2004.
108. Zu, Y., Overby, H., Ren, G., Fan, Z., Zhao, L., Wang, S., Resveratrol liposomes and lipid nanocarriers: Comparison of characteristics and inducing browning of white adipocytes. *Colloids Surf. B. B.*, 164, 414, 2018.
109. Takke, A. and Shende, P., Nanotherapeutic silibinin: an insight of phytomedicine in healthcare reformation. *Nanomed. Nanotechnol. Biol. Med.*, 21, 102057, 2019.
110. Chen, C.H., Chen, C.J., Elzoghby, A.O., Yeh, T.S., Fang, J.Y., Self-assembly and directed assembly of lipid nanocarriers for prevention of liver fibrosis in obese rats: A comparison with the therapy of bariatric surgery. *Nanomed.*, 13, 1566, 2018.
111. Lingayat., V.J., Zarekar, N.S., Shendge, R.S., Solid lipid nanoparticles: a review. *Nanosci. Nanotechnol. Res.*, 4, 67, 2017.
112. Harish, V., Mohd, S., Tewari, D., Pandey, N.K., Vishwas, S., Babu, M.R., Salkini, M.A., ur Rehman, Z., Alotaibi, J.T., Alotaibi, R.F., Alrashed, F.A., Unravelling the role of solid lipid nanoparticles in drug delivery: Journey from laboratory to clinical trial. *J. Drug Deliv. Sci. Tech.*, 2023, 104616, 2023.
113. Ezhilarasi, P.N., Muthukumar, S.P., Anandharamakrishnan, C., Solid lipid nanoparticle enhances bioavailability of hydroxycitric acid compared to a microparticle delivery system. *RSC Adv.*, 6, 53784, 2016.

114. Zielińska, A., Carreiró, F., Oliveira, A.M., Neves, A., Pires, B., Venkatesh, D.N., Durazzo, A., Lucarini, M., Eder, P., Silva, A.M., Santini, A., Polymeric nanoparticles: production, characterization, toxicology and ecotoxicology. *Molecules*, 25, 3731, 2020.
115. Yaqoob, S.B., Adnan, R., Rameez Khan, R.M., Rashid, M., Gold, silver, and palladium nanoparticles: a chemical tool for biomedical applications. *Front. Chem.*, 8, 376, 2020.
116. Nallamuthu, I., Devi, A., Khanum, F., Chlorogenic acid loaded chitosan nanoparticles with sustained release property, retained antioxidant activity and enhanced bioavailability. *Asian J. Pharm. Sci.*, 10, 203, 2015.
117. Bose, A., Roy Burman, D., Sikdar, B., Patra, P., Nanomicelles: Types, properties and applications in drug delivery. *IET Nanobiotechnol.*, 15, 19–27, 2021.
118. Bao, C., Li, Z., Liang, S., Hu, Y., Wang, X., Fang, B., Wang, P., Chen, S., Li, Y., Microneedle Patch Delivery of Capsaicin-Containing α-Lactalbumin Nanomicelles to Adipocytes Achieves Potent Anti-Obesity Effects. *Adv. Funct. Mater.*, 31, 2011130, 2021.
119. Akib, A.A., Rumon, M.M.H., Moniruzzaman, M., Roy, C.K., Chowdury, A.N., 107, PLA-PEG Diblock Copolymer Micelle as Nanocarrier for Anti-Obesity Drug Delivery System. *ECS Trans.*, 1, 19031, 2022.
120. Ansari, S., Bari, A., Ullah, R., Mathanmohun, M., Veeraraghavan, V.P., Sun, Z., Gold nanoparticles synthesized with Smilax glabra rhizome modulates the anti-obesity parameters in high-fat diet and streptozotocin induced obese diabetes rat model. *J. Photochem. Photobiol. Biol.*, 201, 111643, 2019.
121. Gao, L., Hu, Y., Hu, D., Li, Y., Yang, S., Dong, X., Alharbi, S.A., Liu, H., Anti-obesity activity of gold nanoparticles synthesized from Salacia chinensis modulates the biochemical alterations in high-fat diet-induced obese rat model via AMPK signaling pathway. *Arab. J. Chem.*, 13, 6589, 2020.
122. Li, W., Wan, H., Yan, S., Yan, Z., Chen, Y., Guo, P., Ramesh, T., Cui, Y., Ning, L., Gold nanoparticles synthesized with Poria cocos modulates the anti-obesity parameters in high-fat diet and streptozotocin induced obese diabetes rat model. *Arab. J. Chem.*, 13, 5966, 2020.
123. Alharbi, W.S., Almughem, F.A., Almehmady, A.M., Jarallah, S.J., Alsharif, W.K., Alzahrani, N.M., Alshehri, A.A., Phytosomes as an emerging nanotechnology platform for the topical delivery of bioactive phytochemicals. *Pharmaceutics*, 13, 1475, 2021.
124. El-Menshawe, S.F., Ali, A.A., Rabeh, M.A., Khalil, N.M., Nanosized soy phytosome-based thermogel as topical anti-obesity formulation: an approach for acceptable level of evidence of an effective novel herbal weight loss product. *Int. J. Nanomed.*, 13, 307–318, 2018.
125. Heymsfield, S.B., Allison, D.B., Vasselli, J.R., Pietrobelli, A., Greenfield, D., Nunez, C., *Garcinia cambogia* (hydroxycitric acid) as a potential antiobesity agent: a randomized controlled trial. *Jama*, 280, 1596, 1998.
126. Akbari, M., Lankarani, K.B., Tabrizi, R., Ghayour-Mobarhan, M., Peymani, P., Ferns, G., Ghaderi, A., Asemi, Z., The effects of curcumin on weight loss among patients with metabolic syndrome and related disorders: a systematic review and meta-analysis of randomized controlled trials. *Front. Pharmacol.*, 10, 649, 2019.
127. Hao, L., Schlussel, Y., Fieselmann, K., Schneider, S.H., Shapses, S.A., Appetite and gut hormones response to a putative α-glucosidase inhibitor, *Salacia Chinensis*, in overweight/obese adults: A double blind randomized controlled trial. *Nutrients*, 9, 869, 2017.
128. Jeykodi, S., Deshpande, J., Juturu, V., *Salacia* extract improves postprandial glucose and insulin response: a randomized double-blind, placebo controlled, crossover study in healthy volunteers. *J. Diabetes Res.*, 2016, 1–9, 2016.
129. Ebrahimian, Z., Razavi, B.M., Mousavi Shaegh, S.A., Hosseinzadeh, H., Effects of *Portulaca oleracea* L. (purslane) on the metabolic syndrome: A review. *Iran J. Basic Med. Sci.*, 25, 1275–1285, 2022.

130. Chang, H.C., Peng, C.H., Yeh, D.M., Kao, E.S., Wang, C.J., *Hibiscus sabdariffa* extract inhibits obesity and fat accumulation, and improves liver steatosis in humans. *Food Funct.*, 5, 734, 2014.
131. Celleno, L., Tolaini, M.V., D'Amore, A., Perricone, N.V., Preuss, H.G., A dietary supplement containing standardized *Phaseolus vulgaris* extract influences body composition of overweight men and women. *Int. J. Med. Sci.*, 4, 45, 2007.
132. Jang, H.H., Lee, J., Lee, S.H., Lee, Y.M., Effects of *Capsicum annuum* supplementation on the components of metabolic syndrome: A systematic review and meta-analysis. *Sci. Rep.*, 10, 20912, 2020.
133. Anunciação, P.C., de Morais Cardoso, L., Alfenas, R.D.C.G., Queiroz, V.A.V., Carvalho, C.W.P., Martino, H.S.D., Pinheiro-Sant'Ana, H.M., xtruded sorghum consumption associated with a caloric restricted diet reduces body fat in overweight men: A randomized controlled trial. *Food Res. Int.*, 119, 693, 2019.

14

Antioxidants in Hypertension

Bharath Harohalli Byregowda[1], Farmiza Begum[2], Chaman Bala[3], Yogendra Nayak[1], Rakesh K. Sindhu[4] and Gautam Kumar[5]*

[1]*Department of Pharmacology, Manipal College of Pharmaceutical Sciences, Manipal, Karnataka, India*
[2]*Department of Pharmacology, Vaagdevi Pharmacy College, Bollikunta, Warangal, Telangana, India*
[3]*GNIT College of Pharmacy, Greater Noida, Knowledge Park II, Uttar Pradesh, India*
[4]*School of Pharmacy, Sharda University, Greater Noida, Uttar Pradesh, India*
[5]*Department of Pharmacy, Apeejay Stya University, Gurugram, Haryana, India*

Abstract

Millions of people worldwide suffer from hypertension, which is a significant risk factor for cardiovascular disease. Hypertension is responsible for various dysfunctions, including major organ tissue damage through oxidative mechanisms, lipid peroxidation, and myocardial dysfunction resulting from impairment in microvascular function. Oxidative stress, a powerful imbalance between the body's antioxidant defenses and the production of reactive oxygen species (ROS) and nitrogen species (RNS), is significant evidence of its complex pathogenesis. An abundance of ROS/RNS as destructive substances affects blood vessels through critical components such as endothelial cells and nitric oxide (NO) bioavailability, which ultimately leads to vascular dysfunction and hypertension. Apart from the endothelial cells and reduced NO, oxidative stress is also caused by redox signaling, insufficient antioxidants, and the neurohumoral system. Examining the complex interaction between antioxidants and hypertension, many studies explores how they may impact the onset of the disease and current and possible treatment approaches. On the other hand, endogenous antioxidants, which include substances like glutathione, catalase, and superoxide dismutase (SOD), function as cellular defenders and free radical scavengers, reducing the detrimental effects of ROS/RNS. Antioxidants play a significant role in the standardization process by lowering levels of superoxide, hydrogen peroxides, and organic hydroperoxides. In human and animal research, antioxidant treatment of hypertension has been found to lower blood pressure and enhance vascular function. Antioxidants may affect hypertension through various processes, inflammatory pathway modulations, neurohumoral regulation, endothelial repair, improved vascular tone, and rising blood pressure. Personalized antioxidant therapy is being made possible by recent research exploring the possibility of certain antioxidants focusing on certain ROS species, assembly of NADPH oxidase, signaling pathways (AMPK, TyrRS-PARP1, SIRT3, MAPK, TLR4, etc.), arterial stiffness, B and T cell proliferation, and activation of enzymes (nitric oxide synthase and Nrf2). More contemporary techniques, such as atrial natriuretic peptide (ANP) to block angiotensin II (Ang-II) and an antioxidant microinjection into the rostral ventrolateral medulla (RVLM) to directly shift ROS to NO, have been used to influence neurohumoral activity and manage hypertension. Many antioxidants under clinical trials have great

**Corresponding author*: drgautam9265@gmail.com; gautamkumar.phd.manipal@gmail.com

Rakesh K. Sindhu, Inderbir Singh and M. Arockia Babu (eds.) Antioxidants: Nature's Defense Against Disease, (511–536)

potential for managing hypertension, although there are still barriers to overcome. The review urges more investigation into the complexities of the antioxidant-hypertension axis to develop therapeutic strategies to fully utilize antioxidants in the fight against hypertension.

***Keywords*:** Hypertension, oxidative stress, antioxidants, clinical trials, new therapeutics

14.1 Introduction

As a primary cause of cardiovascular and cerebrovascular diseases with high morbidity and death rates worldwide, hypertension continues to be a prevalent risk factor and major health issue. Atrial fibrillation, coronary disease, and left ventricular hypertrophy are the major heart diseases associated with hypertension [1]. The two most common brain diseases linked to hypertension are dementia (due to reduced vasomotor activity, lowered cerebral blood flow, microbleeds, cerebral atrophy, and high cerebrovascular resistance) and stroke (due to direct transmission of high blood pressure from the large vessels to smaller resistant arteries) [2]. Blood pressure is defined as the lateral pressure of circulating blood exerted on the blood vessel walls [3]. Blood pressure of arteries is measured during the heartbeat (systolic), and the heart rests between the beats (diastolic); if there is a significant increase in either measurement, it refers to hypertension or high blood pressure. Systolic blood pressure greater than 140 mmHg and/or diastolic blood pressure of greater than 90 mmHg are considered hypertension. The changes were seen due to the failure of contraction and relaxation of vascular smooth muscle (VSMs) and poor vasodilation by vascular elasticity, reactivity, and caliber [4].

Modernization, a decrease in the amount of physical activity needed for daily life, consumption of alcohol and cigarettes, highly salted foods with higher sodium, saturated and trans fats, a poor intake of vegetables and fruits, and obesity have all contributed to a shift in lifestyle related to hypertension [5, 6]. There are many risk factors for hypertension, but lifestyle and population aging are the main ones. Compared to women, men have a slightly greater age-standardized (up to 50 years) prevalence of hypertension. Studies have shown that, in addition to sociocultural variables, the depletion of female sex hormones after menopause may cause a shift in the hypertriton rate in aged women (after 50 years) [7, 8]. Age-related arterial stiffness, which causes systolic and diastolic blood pressure spikes, aortal hyperplasia, and elastic lamellae fracture, all have a significant impact on the structural and functional alterations of the vascular system and significantly raise the risk of hypertension in elderly persons [9].

Molecular mechanisms in hypertension, particularly oxidative stress and inflammation, act as essential coordinators of various cellular processes and organ systems. Uncontrolled reactive oxygen species (ROS) production under pathological conditions causes oxidative stress, which is described as increased oxidant levels over the antioxidants produced within the system and causes an imbalance between them which results in molecular damage and redox signaling disturbance that exceed the resting-state level of normal cell homeostatic function [10]. Increased levels of oxidative stress during the etiology of hypertension play a crucial role. Pathogenic oxidative stress comes from the excessive generation of harmful reactive oxygen and nitrogen species (ROS and RNS) that exceed cellular antioxidant capacity. Studies on oxidative stress-induced hypertension have shown that it favors oxidation reactions by creating an imbalance of ROS/nitric oxide (NO) in the blood vessels, heart, brain, and kidney, where it regulates blood pressure [11].

Oxidative stress can be greatly elevated by interacting with several ROS sources and their effects on redox signaling. There should be a delicate balance between the generation of ROS and RNS and their elimination in normal physiological conditions. A deficiency in endogenous antioxidant levels will disrupt the balance. The antioxidants produced within the biological system have a scavenging effect on the reactive species that are released in normal metabolic pathways, including superoxide (O_2^-), peroxynitrite ($ONOO^-$), hydroxyl radical (·OH), hydrogen peroxide (H_2O_2), and hypochlorous acid (HOCl). During the pathological state, a supply of antioxidants (exogenous) maintains an oxido/redox balance to fight against oxidative stress and improve endogenous antioxidant levels [12]. On the other hand, antioxidants are essential for counteracting the pathological effects and neutralizing ROS or free radicals in the vascular system linked to inflammation and oxidative stress [13]. In addition to considering the part ROS plays in the pathophysiology of hypertension, a number of publications have proposed the possibility of antioxidants acting as emerging therapeutics to combat the rise in blood pressure of hypertensive patients [14]. Many studies covered antioxidants' role, possible antioxidative and antihypertensive properties, and molecular interactions. In both preclinical and clinical trials, the management of hypertension has demonstrated the effectiveness of several antioxidant therapies in lowering blood pressure, including superoxide dismutase (SOD) mimetics, vitamins, antioxidants, NADPH oxidase (NOX) inhibitors, and ROS scavengers. As new emerging treatments for hypertension, many new therapeutics (immunosuppressive, gp91ds peptide, atrial natriuretic peptide, Coenzyme Q10 (CoQ10), tempol, and vitamin D) have demonstrated lower blood pressure and oxidative stress due to their antioxidant properties.

14.2 Pathophysiology of Hypertension

The pathogenesis of hypertension is multifaceted and complex. The main biological mechanisms that cause vascular dysfunction and promote the emergence of hypertension are oxidative stress and changes in the signaling pathways of vascular endothelial cells and smooth muscle cells. Numerous factors, such as the production of ROS, redox signaling, endothelial dysfunction, and an antioxidant deficiency, contribute to oxidative stress.

14.2.1 Oxidative Stress-Induced Hypertension

Oxidative stress is an imbalanced redox state and ROS/NO production in cells, as represented in Figure 14.1, which influences mitochondrial dysfunction, endoplasmic reticulum (ER) stress, signal disruption, and protein oxidation, all of which lead to vascular dysfunction through a process of inflammation, proliferation, apoptosis, and migration [10]. Vascular dysfunction brings changes in endothelial function, inflammation, and remodeling of arteries. The excessive ROS production, if it is not offset by system antioxidants scavenging them, will cause ROS levels to increase above the threshold level due to the failure of antioxidant mechanism. ROS alters cell proteins through post-translational processes such as phosphorylation and oxidation, including glutathionylation, carbonylation, and

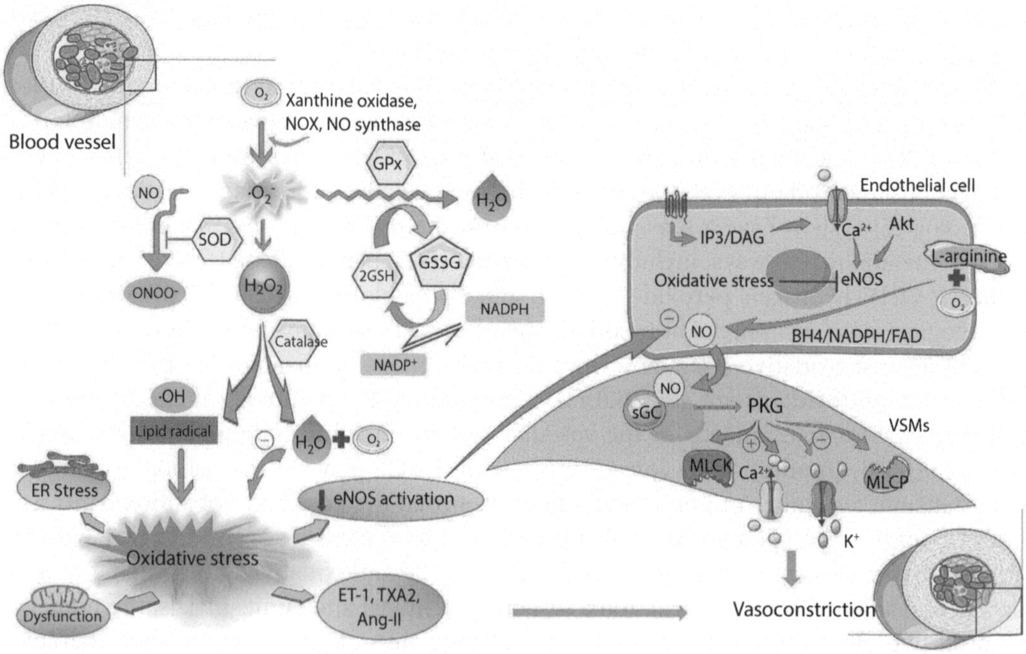

Figure 14.1 Antioxidant's role and pathophysiology of hypertension through oxidative stress, endothelial dysfunction, VSMs constriction, and activation of contraction factors. Increased xanthine oxidase, NOX, NO synthase activity, and lack of endogenous antioxidants enhance ROS generation. If ROS are not converted into water molecules by SOD, catalase, and glutathione, they intensify oxidative stress and reduce eNOS activation, which is required for NO production. Lowering NO levels and converting them to $ONOO^-$ species alters VSM function by increasing Ca^{2+} influx and MLCK activity, which is responsible for vasoconstriction. Oxidative stress also causes direct vasoconstriction by regulating the release of ET-1, TXA2, and angiotensin II (Ang-II). Antioxidant therapy that promotes endogenous antioxidant activity increases NO synthesis, which decreases VSM contraction factors and the release of vasoconstrictors in response to oxidative stress.

nitrosylation, to affect the cell function. Damage to the proteins, lipids, and DNA by ROS often leads to oxidative stress and induces pathology of vascular function [15].

Despite the lack of a ROS-generating mechanism from the mitochondria, producing ROS from isolated mitochondria is the subject of ongoing scientific discussion. Jensen claimed that because catalase was sensitive to the oxygen absorbed by submitochondrial particles when oxidizing NADH or succinate, only a small percentage of that oxygen was transformed to H_2O_2 [10, 16]. Unstable mitochondrial function regulates the production of ROS (byproducts of aerobic metabolism) by an assortment of mechanisms, including the ionic environment, the redox status of respiratory components, the redox buffer, and oxygen tension [17]. Enzymes are the sources of free-radical reactive species formation, and non-radical (H_2O_2) species are also responsible for ROS and RNS production. Superoxide free radicals of ROS are produced from enzyme sources like nicotinamide adenine dinucleotide phosphate (NADPH) oxidases, cyclooxygenases, xanthine oxidase, and mitochondrial respiration chain [18]. The catalytic center of NOX is responsible for the transfer of one electron from the oxygen molecule as it uses Fe^{2+}containing heme to form superoxide anion ($\cdot O_2^-$) radical [19]. ROS constitutes $\cdot O_2^-$, ·OH, and H_2O_2, and RNS comprises $ONOO^-$ and its reaction products, RNS are generated when superoxide and H_2O_2 react with the available NO in the system [20, 21]. Several NOX enzyme isoforms, which catalyze oxygen reduction by NADPH, an electron donor, to produce $\cdot O_2^-$, are involved in the control of ROS [22]. The signaling effects of ROS are influenced by the location of production inside the cell, local concentration, presence of antioxidants nearby, type of species produced, species half-life, and permeability of the cell membrane of ROS. When mitochondria are damaged, electrons leak from the electron transport chain at complexes I and III, which mostly results in post-translational modifications and the formation of ROS [23, 24]. Cyclophilin D, a protein found in mitochondria, is essential for opening the membrane transition pore, which involves depolarization and increases the generation of ROS [10].

Antioxidants are substances that can capture ROS; they may be able to lower blood pressure and oxidative damage. By eliminating free radical intermediates, antioxidants break the chain reactions of ROS and prevent further oxidation processes. In order to relax blood vessels from hypertensive-induced contraction, suppression of oxidative stress by antioxidants halts the activation of VSM contraction factors and mechanisms interfering with NO generation [25]. Antioxidants lower oxidative stress by inhibiting the xanthine oxidase enzyme, which is involved in purine metabolism and generates ROS [26]. The availability of two forms of xanthine oxidoreductase (xanthine dehydrogenase and xanthine oxidase) ratios plays a critical role in the generation of ROS. Preventing the conversion of xanthine dehydrogenase to xanthine oxidase should be controlled to keep oxidative stress under control [27]. From the same substrates, xanthine oxidase transfers electrons to oxygen to produce $\cdot O_2^-$ and H_2O_2. Supply of antioxidants which inhibits xanthine oxidase decrease the formation of superoxide radicals and maintain the balance with the xanthine dehydrogenase [28].

The formation of sulfenic form from the thiolate anion of cysteine residues during the redox signaling alters the protein function due to H_2O_2 mediated oxidation [29]. It can be reversible as it remains the first-degree oxidation, but further oxidation of thiolate with more peroxide forms irreversible sulfonic and sulfinic causing permanent damage to proteins (Oxidative stress) [30]. SOD I and II are present in the mitochondria and cytosol, which converts superoxides into H_2O_2 and prevents the accumulation (Figure 14.1). The presence of ferrous ions when the H_2O_2 is produced from the SOD undergoes a Fenton reaction to generate hydroxyl radicals, therefor maintenance of iron homeostasis of the cell plays a key role to suppress the hydroxyl radicals [17, 31]. Antioxidants that improves the SOD and catalase helps to inhibit the formation of sulfenic form by converting superoxide radicals to water molecule and eliminates H_2O_2 mediated oxidation.

14.2.2 Neurohumoral System-Induced Blood Pressure

Blood volume and cardiac output, or how much blood the heart pumps out per minute, are two cardiovascular system variables that affect blood pressure. The neurohumoral system is another factor that affects blood pressure by disturbing the balance of arterial tone with intravascular volume. Endothelium, immune system, natriuretic peptide, renin angiotensin aldosterone system (RAAS), and sympathetic nervous system are the various elements involved in the neurohumoral system [32].

Salt sensitivity is caused by endothelial dysfunction driven by chronic salt intake, environmental and genetic alterations. This endothelial dysfunction raises blood volume because of elevated serum Na^+ concentrations, which are responsible for fluid retention, and it also promotes oxidative stress by overproducing transforming growth factor beta (TGF-β). The regulation of blood pressure by RAAS depends on the renal perfusion pressure which regulates the reabsorption of Na^+ in the nephron proximal tubules, vascular injury, vascular constriction, salt sensitivity, and Na^+ retention [32, 33]. The amiloride-sensitive sodium channel is activated by aldosterone in order to promote sodium reabsorption in the cortical collecting duct, which results in vasoconstriction and hypertension [34]. The sympathetic nervous system controls the release of renin from the juxtaglomerular apparatus, which in turn causes peripheral vascular resistance by converting angiotensin I to Ang-II responsible for vasoconstriction and a rise in blood pressure [35]. Higher catecholamines were observed in the activated sympathetic nervous system associated with hypertension [36]. Natriuretic peptides responsible for vasodilation and Na^+ balance cause hypertension by vasoconstriction, when there is a deficiency of brain natriuretic peptide (BNP) and atrial natriuretic peptide (ANP) [37]. A spike in blood pressure is brought on by the immune system's increased sodium sensitivity as a result of macrophage infiltration [32].

14.2.3 Endothelial and Vascular Smooth Muscle in Hypertension

Endothelial cells consist of contraction and dilator factors of VSMs and regulate the vascular tone. When the balance of these factors is altered due to oxidative stress enhances hypertension, dilator factors like NO, prostacyclin (PGI2), and endothelium-derived hyperpolarizing factor (EDHF) cause a decrease in the intracellular Ca^{2+} levels responsible for the dilation of VSMs, whereas contraction factors like endothelin-1(ET-1), thromboxane (TXA2), Ang-II, and superoxide anion cause the influx of Ca^{2+}in the contraction process [38]. Decreased prostacyclin and elevated levels of thromboxane and ET-1 are involved in vasoconstriction and vascular resistance [39]. ET-1 is required to maintain the vascular tone, inflammation, immunity, blood flow, and neovascularization [15]. Antioxidants can be used to treat hypertension, which is caused by damage to the endothelium or VSMs that alters the normal physiology of vessel. Antioxidants may protect the endothelium from oxidative stress-induced damage [40]. Strong vasodilator NO levels are made more available by antioxidants, which lowers endothelial dysfunction. Through TXA2 suppression, which regulates ET-1-induced vascular contraction and ROS production, antioxidants can preserve VSM and endothelial function [41]. Antioxidants also control redox signaling to reduce ROS, preserve endothelial integrity, and support optimal vascular function [42].

14.2.4 Decreased Nitric Oxide Synthesis in Hypertension

NO is a signaling molecule for the relaxation of VSMs, synthesized from the oxidation of L-arginine catalyzed by endothelial nitric oxide synthase (eNOS) [32, 43]. When there is a dysfunction of endothelial cells due to oxidative stress as shown in Figure 14.1 alters the release of NO by decreasing the availability of soluble guanylyl cyclase (α & βsGC) required for the relaxation of VSMs, generation of ROS and RNS, and uncoupling of the nitric oxide synthase (NOS) [44]. When G-protein coupled receptors are activated, calmodulin is activated as a result of Ca^{2+} influx in the endothelial cell through the diacylglycerol (DAG) and inositol trisphosphate (IP3)signaling pathways and activates eNOS. Utilizing the L-arginine and oxygen molecules, the activated eNOS promotes oxidation in the presence of co-factors (NADPH/FAD/FMN/BH4) necessary for the production of NO [43].

Studies on the stimulation of eNOS in animals suggested that the kinase Protein kinase B(Akt1) is also involved [45]. Vascular endothelial cell dysfunction activates several ROS-generating enzymes, including NOX, xanthine oxidase, uncoupled eNOS, and mitochondrial respiratory chain enzymes [46]. This reduces bioactive NO levels by inhibiting NO synthesis and generating $ONOO^-$ in response to ROS. The release of $ONOO^-$ results in the oxidation of the cofactor BH4 to the BH3 radical, the dissociation of eNOS, and an increase in oxidative stress. As represented in the Figure 14.1, reduced NO levels enhance the VSMs ability to contract by reducing the amount of NO available to sGC, which is necessary for the stimulation of protein kinase G (PKG), which mediates the K+, Ca^{2+} channel, myosin light chain kinase (MLCK), and myosin light chain phosphatase (MLCP). Due to the negative regulation of PKG caused by a lack of NO, K^+ efflux and MLCP are inhibited. In the meantime, it promotes Ca^{2+} influx and activates MLCK to cause constriction of VSMs [43, 47, 48]. Antioxidants also control redox signaling to reduce ROS, preserve endothelial integrity, and support optimal vascular function. Through the redox control of transcription factors, which is important in defining the gene expression profile (Nuclear factor erythroid 2-related factor 2(Nrf2), Nuclear factor kappa B (NF-κB), cAMP-response element binding protein, and Activator protein 1) and cellular response to nitrosative stress, antioxidant enzymes play a crucial role in decreasing $ONOO^-$ levels. Transcriptional factor Nrf2, which controls the Keap1/Nrf2/ARE redox-sensitive signal and protects against nitrosative/oxidative stress, will be activated in order to initiate the activation of antioxidant enzymes [14, 49]. By inhibiting nitrosative stress ($ONOO^-$), antioxidant therapy that regulates NOS and Nrf2 reduces oxidation of BH4, a cofactor for NO synthase, and increases NO production (Figure 14.1) [50].

14.3 Contribution of Antioxidants in Managing Hypertension

14.3.1 Oxidative Stress Encounter

Numerous studies have revealed that antioxidants can prevent the propagation of free radicals or can prevent their generation, which can reduce oxidative stress and manage autoxidation. The oxidative stress generated in the body suppresses the efficiency of endogenous antioxidants (Table 14.1). Redox signaling is mediated by several enzymes and causes oxidation damage in hypertension [17]. SOD is a natural antioxidant defense mechanism produced in the body to fight against the oxidative stress caused by ROS formation. SOD converts the available superoxide anion ($\cdot O_2^-$) into the H_2O_2 and oxygen molecules to improve vasodilation and protective role in hypertension [51]. The correlation of increased levels of SOD with improved vascular structure and function in hypertensive patients showed the diminished effect of oxidative stress [52]. Catalase is another ubiquitous antioxidant enzyme that continues the process of SOD in the prevention of ROS generation (Figure 14.1). H_2O_2 produced by the SOD is converted by catalase into water and oxygen molecules [53].

Glutathione (GSH) is a major antioxidant that consists of glutathione reductase, glutathione peroxidase, and glutathione-S-transferase via enzymatic reactions to nullify the oxidant's effects. Whereas the thiol group of glutathione plays a role in ROS scavenging activity through non-enzymatic reaction [54]. Although SOD's dismutation of superoxide is the principal source of H_2O_2, mitochondria may additionally generate H_2O_2 through other processes.

Table 14.1 Important antioxidants involved in the scavenging of oxidative stress.

Sl. no.	Antioxidant enzymes	Non-enzymatic antioxidants
1	Superoxide dismutase	Vitamins
2	Catalase	Carotenoids
3	Glutathione peroxidase	Polyphenols
4	Glutathione reductase	Flavonoids
5	Thioredoxin	Minerals
6	Ceruloplasmin	Phenolic acid
7	Transferrin	Others

One of these processes involves the mitochondrial enzyme p66Shc's redox activity, which can produce H_2O_2 without the use of superoxide by oxidizing cytochrome C [10]. H_2O_2 functions as an oxidizing agent and can oxidize elements of the mitochondria, since H_2O_2 doesn't have any unpaired electrons, it should be highlighted that it is not regarded as a free radical. To help with H_2O_2 detoxification (Figure 14.1), mitochondria predominantly use the GSH redox system, which consists of glutathione peroxidases (Gpxs) and GSH reductases as well as peroxiredoxins (Prxs), which use the reducing equivalents of NADPH [55].

The series of glutathione enzymatic reactions during the process of ROS reduction in hypertension converts H_2O_2 into water. In the presence of glutathione peroxidase, glutathione is oxidized to the unstable oxidized glutathione (GSSG), and further glutathione reductase (NADPH-dependent) restores glutathione from GSSG. In order to produce GSSG, glutathione peroxidase needs glutathione as a co-substrate [56]. Additionally, glutathione has a propensity to interact with cysteine residues in proteins by producing mixed disulfides. However, glutathione S-transferase can readily return these unstable molecules to their normal state. The availability of NADPH determines the pace of H_2O_2 reduction by glutathione peroxidase and maintains the reduced state of mitochondria. Through glutathionylation of proteins, the uncontrolled production of GSSG is responsible for mitochondrial failure during oxidative stress and develops hypertension [57, 58]. By converting lipid hydroperoxides to protect the cell membrane and reducing $ONOO^-$ to nitrite, glutathione peroxidase prevents lipid peroxidation [19].

14.3.2 Antioxidative Strategies for Modulating Neurohumoral Activity in Hypertension

Antioxidants can shield the neurohumoral system from oxidative stress by scavenging ROS, preserving the system's normal function. The active sympathetic nervous system in the heart and kidney modulates pump function, renal tubular sodium reabsorption, glomerular filtration rate, and renin release to cause hypertensions [59]. Catecholamines on adrenergic receptors activate the sympathetic nervous system and increase blood pressure. Many hypertensive drugs with antioxidant activity have been proposed with improved endothelial activity by inhibiting sympathetic nervous system activation through beta-adrenergic receptors able to reduce superoxide anions and lipid peroxidation. The rostral ventrolateral medulla (RVLM) regulates the sympathetic nervous system responsible for the NO and ROS enzyme generation. The microinjection of antioxidants into RVLM decreases the sympathetic vasomotor tone by shifting the ROS to NO enzymes in RVLM and controls hypertension [60]. By lowering inflammation, antioxidants can indirectly affect the neurohumoral system and support normal blood pressure levels. Antioxidants with anti-inflammatory action and regulated neurotransmission may constitute a better therapeutic strategy for hypertension because they reduce salt sensitivity by lowering macrophage infiltrations.

14.3.3 Restoration of Endothelial Dysfunction

The activation of the mechanosensitive and mechanosensory systems corresponds to the release of vasodilator factors, such as NO, plasminogen activator, and prostacyclin, which regulate endothelial and vascular function [61]. One of the factors that contribute to endothelial dysfunction is oxidative stress. Several antioxidants enhance endothelial function by encouraging the generation of NO and reducing oxidative stress to preserve vascular tone [62]. SOD also plays an important role in reducing $ONOO^-$ formation by competing with the NO for the superoxide radical. With this SOD improves endothelial function and maintains VSMs contraction and relaxation by increasing the availability of NO [56]. The ability of natural antioxidants to speed up the reendothelialization process by increasing adhesion molecules and promoting accelerated endothelial cell development is now being investigated. Vitamins have recently been found to be an appealing new approach for directly converting stem or progenitor cells into endothelial cells [63].

14.3.4 Protection Against Vascular Damage

Regulation of vascular integrity and elasticity plays an important role in the protection of blood vessels and to reduce high blood pressure associated with cardiovascular disease risk and arterial stiffness. Vasculopathy, which affects arterial stiffness, contractility, and relaxation, is thought to have substantial side effects including vascular calcification and remodeling [64]. The inner layer of blood arteries, or endothelium, is crucial for preserving the functioning of the vascular system. Because oxidative stress impairs endothelial function, the amount of NO availability decreases and reduces vascular relaxation and dilation. To promote the elasticity and function of blood vessels, antioxidants are required to regulate oxidative stress which exerts on the vessels by increasing the synthesis of NO and preserving collagen and elastin structural activity. Collagen and elastin are two structural proteins needed to keep blood vessels flexible and strong, under oxidative stress proteins break down to cause vascular damage [65]. Antioxidant supplementation will aid in protecting collagen and elastin fibers, which can preserve blood vessel elasticity and resilience [66]. Expression of NOX in response to vascular resistance in hypertension and Ang-II-infused mice activates signalings like Rho kinase, MAPK (Mitogen-activated protein kinase), Src, and Ca^{2+} channels induce vascular remodeling, endothelial dysfunction, influx of inflammatory cells into the vascular walls, vascular hyperactivity, and increases vessel stiffness [67, 68]. Antioxidants were reported to interact with the pathways related to vascular damage in hypertension and decrease blood pressure by improving the vessel.

Vitamin C, an antioxidant, plays a significant role in collagen synthesis by assisting in the hydroxylation of proline and lysine in collagen [69, 70]. Additionally, it aids in preventing collagen from incurring oxidative damage, preserving the flexibility and integrity of the vascular system [71]. Vitamin E safeguards cell membranes by halting the lipid peroxidation that affects the collagen and elastin in blood vessels [72, 73]. It has been demonstrated that certain polyphenols, such as flavonoids and catechins, promote collagen synthesis and defend against collagen deterioration. Additionally, they can increase elastin synthesis and reduce the activity of enzymes that degrade elastin [74, 75].

14.3.5 Reducing Inflammation

Numerous signaling pathways involved in vascular remodeling, cell proliferation, and inflammatory reaction are stimulated by ROS. These pathways include the activation of STAT (signal transducers and activators of transcription), NF-κB, JAK-2 (Janus kinases), Pyk-2 (Proline-rich Tyrosine Kinase 2), MAPK, p21Ras, AKT, EGFR (Epidermal growth factor receptor), PDGFR (Platelet-derived growth factor receptor alpha), IGFR (Insulin-like growth factor 1), and activator Protein 1 (AP-1) and Hypoxia-inducible factor 1 (HIF-1), which regulate redox-sensitive transcriptor factors [76–79]. ROS activates a tumor necrosis factor (TNF-α) mediated pathway that decreases NO concentration by degrading the eNOS. Ang-II stimulates the aldosterone required to induce the inflammatory reaction by activating mineralocorticoid receptors. Anti-inflammatory enzymes like heme oxygenase-1 (HO-1) can interact with some antioxidants, including resveratrol, and increase their effectiveness. HO-1 maintains the vascular tone by reducing oxidative stress and inflammation [80]. Antioxidants like vitamins E and C reduce the chemokines and inflammatory cytokines with their immunomodulatory properties associated with vascular inflammation. Neutralizing high free radicals in the hypertensive case helps mitigate the inflammatory pathways during the initial stage of hypertension [81]. Interleukins produced from proinflammatory signaling activation by NLRP3 (nucleotide-binding and oligomerization domain-like receptor family pyrin domain-containing 3) inflammasome-induced cleavage of procaspase-1 and generate caspase-1 to induce pyroptosis [82]. It is becoming more widely understood that the inflammasome, which is significant in inflammatory, immune, VSMs, and endothelial cells, contributes to inflammation in blood vessels and enhances hypertension. Regulation of inflammasomes in hypertension could reduce the inflammatory reactions in blood vessels. Antioxidants interacting with the inflammatory pathways and NLRP3 inflammasomes associated with hypertension may contribute to regulating inflammatory markers and preventing blood pressure [83].

14.3.6 Antioxidants for Managing High Blood Pressure: Recent Research Findings

Much research is ongoing on antioxidants' role in treating oxidative stress-induced hypertension. The deficiency of CoQ10 endogenous antioxidants has been associated with a wide range of diseases, including hypertension, diabetes, congestive heart failure, Parkinson's disease, and coronary artery disease [84]. The potential role of CoQ10 as an antioxidant that is involved in energy synthesis can reduce oxidative stress-induced vascular hypertension by improving the levels of antioxidant enzymes in the body, such as SOD [85]. Supply of CoQ10 as microinjections has shown a greater neutralization of ROS [11]. Recent studies indicate that AMP-activated protein kinase (AMPK)signaling pathway dysfunction contributes to the rise of hypertension and cardiovascular diseases [86]. Many studies reported activating eNOS and Nrf2, AMPK controls the antioxidant system to prevent oxidative stress from damaging endothelial cells as a therapeutic approach. Quercetin treatment can activate the AMPK to increase the NO synthesis through eNOS positive regulation. Salvianolic acid B, a well-known naturally occurring antioxidant, lowers blood pressure by activating AMPK and Akt phosphorylation to increase NO synthesis [87]. A peptide gp91ds has been discovered that affects the assembly of NOX with antioxidant activity to reduce ROS and vascular resistance. The peptide treatment reduces the gradual and time-dependent rise in systolic blood pressure and markedly increases the bioavailability of NO [88]. As a ROS scavenger, lazaroid treatment lowers blood pressure and regulates eNOS and iNOS (inducible nitric oxide synthase) in blood vessels to increase NO availability [89]. Preliminary and observational studies of mycophenolate mofetil and immunosuppressive medications reduced arterial stiffness and B & T cell proliferation, which helped hypertensive patients' blood pressure control [90, 91]. Another recently discovered antioxidant, melatonin, inhibits ROS activity and lowers oxidative stress and vascular inflammation. Additionally, as a safer therapeutic approach for pregnant women, it improves the levels of Nrf-2 and HO-1 and has endothelium repair caused by hypertension. Melatonin prevents inflammatory activity by inhibiting IL-6, IL-8, and IL-1β inflammatory mediators involved in vascular inflammation and modulates the expression of HIF1α and eNOS genes [92, 93].

14.4 Antioxidant Therapy in Hypertension

Antioxidant defense mechanisms of cells and organisms depend on low molecular weight antioxidants. These antioxidants include carotenoids, polyphenols, minerals, cofactors, vitamins, and glutathione. Vitamin E and C are the two antioxidants that can be ingested externally. Some molecules will synthesize in the body and exert antioxidant effects (melatonin, lipoic acid, glutathione, melanins, uric acid, keto acids, taurine, and coenzyme Q). Consuming an antioxidant-rich diet has been linked to a reduction in hypertension by suppressing oxidative stress. Still, frequent chronic use of antioxidants might affect the response , although there isn't enough data to support this claim [94–96]. Antioxidant treatments explore many mechanistic approaches to suppress the oxidative stress that causes vasoconstriction, as shown in Figure 14.2 and Table 14.2.

14.4.1 Vitamins

The future outlook calls for the involvement of compounds with antioxidant activity that support healthy aging and combat oxidative stress [97]. In situations with a water-soluble environment, vitamin C functions directly to mitigate oxidative stress. Alpha-tocopherol, the active form of vitamin E, can be produced in lipids by reducing the tocopheroxyl radical created when exogenous oxidants interact with alpha-tocopherol. This reduces lipid peroxidation in the mitochondria, cell membrane, and ER to maintain cell integrity [98]. Vitamins E and C increase eNOS and downregulate NOX, a significant ROS generator in the vascular

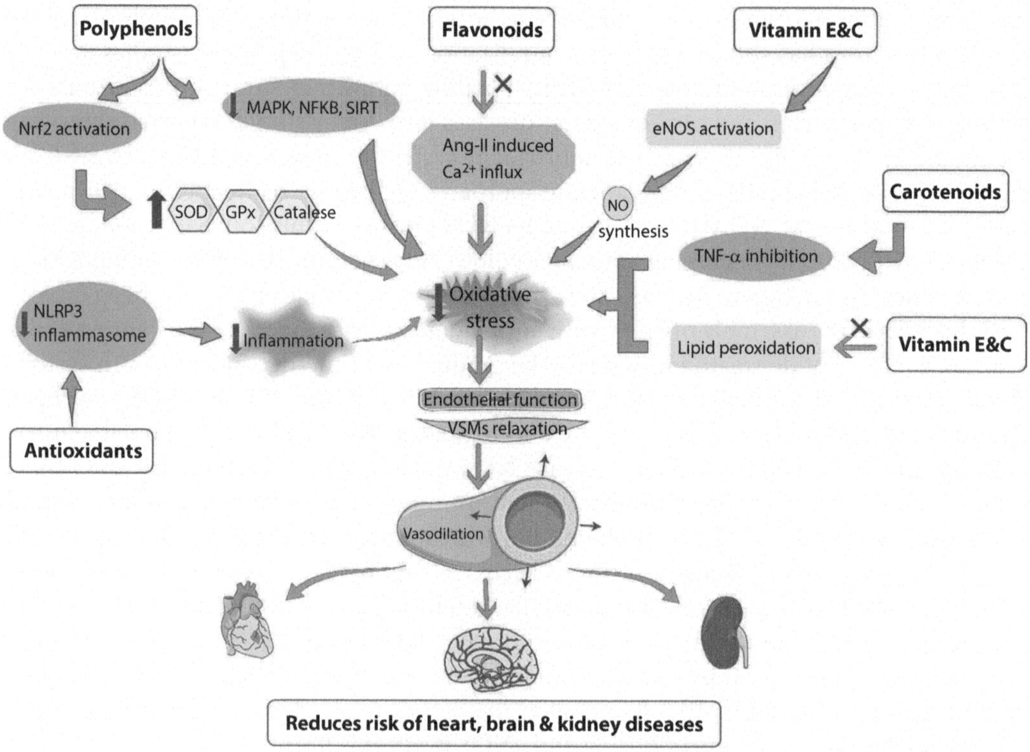

Figure 14.2 Antioxidant therapy in the prevention of hypertension. By activating eNOS and inhibiting lipid peroxidation, vitamins reduce oxidative stress in the body. Carotenoids significantly increase NO levels and decrease inflammatory responses in hypertension individuals, resulting in increased vasodilation. Polyphenols increase endothelial function by interfering with ROS generation via Nrf2 activation and signaling pathways, increasing the activity of the endogenous antioxidant. Flavonoids primarily inhibit ang-II-induced oxidative stress and inflammatory responses that result in elevated blood pressure. Various other antioxidants regulate many processes that lead to hypertension and lower the risk of cardiovascular, cerebrovascular, and renal dysfunction.

Table 14.2 Molecular mechanism of several antioxidant therapies to treat hypertension.

Sl. no.	Antioxidant	Mechanism
Vitamins		
1	Vitamin C	Increase eNOS and SOD and downregulate NOX, inhibit endothelin, restore GSH
2	Vitamin E	Tocopheroxyl radical, lipid peroxidation, maintains cell integrity, increases NOS and SOD, and downregulates NOX
Carotenoids		
3	β-carotene	Reduces TNF-mediated ROS, Improves SOD and catalase production to maintain NO levels

(*Continued*)

Table 14.2 Molecular mechanism of several antioxidant therapies to treat hypertension. (*Continued*)

Sl. no.	Antioxidant	Mechanism
4	Astaxanthin	Reduces malondialdehyde and improves SOD
5	Zeaxanthin	Improves NO bioavailability
6	Lycopene	Decreases TNF-mediated ROS production, endothelin, and inflammatory reaction
7	Lutein	Decreases release of endothelin and inflammatory reaction
Polyphenols		
8	Resveratrol	Enhances endothelial function via SIRT3 and TyrRS-PARP1 signaling
9	Red wine,	Inhibits VEGF and pro-angiogenic factors
10	Tert-butylhydroquinone	Activation of Nrf2 and increase in endogenous antioxidants
Flavonoids		
11	Quercetin	Inhibits Ang-II and HUVECs, Anti-inflammatory activity, Increases NO synthesis
12	Catechin	NO production by activating the eNOS
13	Flavonoids of genistein and legumes	Regulates TLR4 signaling
14	Luteolin	Improves ROS scavenger by inhibiting NF-κB mediated inflammation
Other antioxidants		
16	Melatonin	Improves the levels of Nrf-2 and HO-1, regulates endothelium function, inhibits inflammatory reaction, prevents lipid peroxidation
17	Allopurinol,	Xanthine oxidase inhibitor
18	N-acetylcysteine	Serves as a precursor for glutathione production
19	MCC950's	Inhibits NLRP3 inflammasome
20	Oridonin	Inhibits the NLRP3 and decreases inflammatory reaction
21	Columbianadin	Anti-inflammatory activity by inhibiting NF-κB p65 phosphorylation
22	Capsaicin	Decreases IL-6 and TNF- inflammatory markers release

(*Continued*)

Table 14.2 Molecular mechanism of several antioxidant therapies to treat hypertension. (*Continued*)

Sl. no.	Antioxidant	Mechanism
23	Okra seeds	Inhibits lipase, carbohydrate hydrolyzing enzymes, and ACE
24	1,8-Cineole	Activation of eNOS
25	Coenzyme Q10	Neutralization of ROS and improves endogenous antioxidants, lower oxidative stress by ROS to NO shift
26	gp91ds peptide	It affects the assembly of NOX, increases the bioavailability of NO
27	Lazaroid	Regulates eNOS and iNOS
28	Mycophenolate mofetil	Reduces arterial stiffness and B & T cell proliferation
29	Tempol	Lower oxidative stress by ROS to NO shift
30	Atrial natriuretic peptide	Maintains salt homeostasis and inhibits Ang-II
31	Carvedilol	Inhibits intracellular Ca^{2+} accumulation
32	Nebivilol	Reduce NOX expression, enhance eNOS and NOS activity
33	Celiprolol	Restores eNOS expression
34	Hydrochlorothiazide and exercise	Modulates autonomic mechanisms and inflammatory reactions, decreases MMP-2 activity and prevents vascular remodeling, inhibits NOX
35	Salvianolic acid B	Activates AMPK and Akt phosphorylation to increase NO synthesis

wall, reducing oxidative stress and lowering blood pressure [96]. By increasing eNOS and SOD activity and decreasing NOX in the aorta wall, vitamin C acts as an enzyme modulator in the vascular wall, protecting the endothelial cells. Vitamin C also regulates vasodilation by inhibiting the effects of ET-1 [99]. Natural antioxidants like -lipoic acid stop the Fenton-like radical process and restore GSH and vitamins C and E in their active state and prevent ROS generation [56]. During the vascular injury, antioxidants (vitamins E, C, and D) were reported to enhance vascular repair and reendothelialization [99, 100].

14.4.2 Carotenoids

Carotenoids are consumed in the form of green vegetables and fruits. It includes astaxanthin, β-carotene, zeaxanthin, lycopene, and lutein which protect the membrane from oxidative damage [101]. At the endothelium level, lycopene and beta-carotene are known to reduce TNF-mediated ROS production in blood vessels [102]. Carotenoids improve the production of SOD and catalase to increase the concentration of NO required to maintain vascular tone [103]. Astaxanthin decreases oxidative stress by reducing the malondialdehyde content and increasing the endogenous antioxidant activities. By inhibiting ET-1 and inflammatory response through the NF-κB signaling pathway, lutein and lycopene diminish oxidative stress [104].

14.4.3 Polyphenols

Phenolic components inhibit the nuclear transcription factor pathway (NF-κB) to activate vascular endothelial cells [105, 106]. Red wine polyphenols were reported to inhibit VEGF and pro-angiogenic factors due to the redox-sensitive activation of the MAPK pathway [107]. The polyphenol antioxidant resveratrol enhances endothelial function by raising the expression of Mfn1, Mfn2, and OPA1 through activation of the TyrRS-PARP1 signaling pathway. It also protects cell's mitochondrial quality control from oxidative damage. Resveratrol enhances endothelial function via silent mating type information regulation 2 homolog (SIRT3)signaling that increases Mn-SOD and glutathione peroxidase [108]. Nrf2 is indirectly involved in the expression of antioxidant genes: glutathione, glutathione-S-transferase, GPx, catalase, and SOD. Inhibition of Nrf2 decreases the binding of Nrf2 to antioxidant response elements (AREs) in the nucleus and reduces antioxidants by inhibiting the transcription of genes [109]. Tert-butylhydroquinone enhances antioxidants to prevent oxidative stress and preserves microvascular endothelial function in hypertension mice by activating the Nrf2 [110]. Polyphenols will improve the endothelial cell function of blood vessels in hypertension.

14.4.4 Flavonoids

Quercetin is an anti-inflammatory antioxidant that reduces blood pressure by inhibiting Ang-II and protects human umbilical vein endothelial cells (HUVECs) activities [108]. Baicalin is a flavonoid with antioxidant properties that have shown a vasorelaxant effect and reduced blood pressure by inhibiting the Ang-II-induced Ca^{2+} influx [111]. Catechin is another flavonoid that maintains endothelial cell NO production by activating the eNOS [112]. The vascular regeneration effect of flavonoids in genistein and legumes enhances antioxidant enzyme activities by regulating the TLR4 signaling pathway [113]. Luteolin is an antioxidant available in vegetables that s ROS and inflammatory reaction in hypertension through NF-κB-mediated inflammation cascade suppression [114].

14.4.5 Other Antioxidants

Melatonin prevents oxidative stress by protecting against the lipid peroxidation that causes membrane damage [115]. Allopurinol, purine-like a xanthine oxidase inhibitor, lowers the superoxide radical produced by the purine metabolism and suppresses diseases associated with the heart, including ischemic heart disease and heart failure [116]. As an antioxidant, N-acetylcysteine (NAC) reduces disulfide bonds, scavenges ROS, and serves as a precursor for the production of glutathione. In hypertension, NAC directly or indirectly enhances NO production, reduces oxidative stress, and regulate vasoactive molecules [96]. MCC950's inhibitory action on the NLRP3 inflammasome has been found to lower blood pressure in mice specifically produced with inflammatory hypertension by DOCA-salt [83]. Another antioxidant oridonin inhibits the NLRP3 to decrease inflammatory activity and might reduce hypertension [79, 117]. Columbianadin inhibits NF-κB p65 phosphorylation and decreased the expression of TNF-α and IL-1β [118]. By blocking SIRT1/MAPKs/NF-κB pathways, capsaicin can limit the release of IL-6 and TNF-inflammatory markers and lower blood pressure [119]. Okra seeds inhibited lipase, carbohydrate hydrolyzing enzymes, and ACE to suppress oxidative stress [120]. 1,8-Cineole improved the activation of eNOS by phosphoinositide 3-kinase (PI3K) agonistic effect reported in the cell line studies and reduced oxidative stress and malondialdehyde in rats [121].

14.5 Antioxidants to Treat Neurohumoral System-Induced Hypertension

Tempol, catalase, or CoQ10 antioxidant microinjections into the RVLM lower oxidative stress by ROS to NO shift [11]. Administration of ANP maintains salt homeostasis and inhibits Ang-II in hypertension to reduce the superoxide radicals. Tempol administrations also mimic the ANP effect in reducing oxidative stress [122, 123]. Long-acting ANP analogs (ANP has a high clearance rate) or inhibitors of ANP degradation will be a future perspective of hypertension treatment. Antihypertensive drugs with antioxidant activity include carvedilol, nebivilol, propanolol, and celiprolol, which inhibit catecholamines that cause hypertension-related oxidative stress [14]. Carvedilol lowers sympathetic activity in hypertension by inhibiting intracellular Ca^{2+} accumulation driven by 4-hydroxy-2-nonenal [124]. To reduce NOX expression and the production of ROS, nebivolol enhances NO, eNOS expression, and NOS activity [125]. Celiprolol restores eNOS expression by blocking beta receptors [126]. Blood pressure is regulated by sympathetic outflow, facilitated by antioxidants that preserve the balance of neurotransmitters involved in this process. The antioxidant effect of Hydrochlorothiazide (HCTZ) showed improved bioavailability of NO along with its inhibitory activity on NOX and ROS levels in blood vessels. In addition to its arterial dilatation impact, HCTZ lowers blood pressure in hypertensive individuals via decreased matrix metalloproteinases-2(MMP-2) activity and prevents vascular remodeling [127, 128]. In contrast, current therapy strategies for treating hypertension with HCTZ include regular exercise training, which may enhance the therapeutic efficacy by acting as a counter-regulatory mechanism for arterial pressure levels that HCTZ medication alone is unable to regulate. During HCTZ therapy, exercise modulates autonomic mechanisms and inflammatory reactions with oxidative stress to reduce blood pressure [129, 130].

14.6 Translating Antioxidant Research into Hypertension Management: Preclinical to Clinical

Clinical trials and preclinical research have demonstrated the significance of antioxidant therapy in treating hypertension. Numerous preclinical models were developed to use the antioxidant effect to treat hypertension; however, many of these models are not utilized in clinical studies. Many antioxidant therapies now undergoing clinical trials have been shown to have antihypertensive properties; some of these are included here. Ayurvedic medicine Tulsi combined with Wheatgrass as a formulation has been shown to alleviate free radical-induced oxidative stress as an antioxidant, anti-inflammatory, and immunomodulatory therapy in a clinical trial with the ID NCT05851040 [131]. A polyphenol-rich functional olive oil has undertaken a clinical trial (NCT05460221) to test its antihypertensive activity [132, 133]. A new treatment option for resistant hypertension in adult African Americans is allopurinol. This xanthine oxidase inhibitor lowers free radicals generated by purine metabolism and is currently undergoing clinical trials (NCT05888233) [134]. A small population with hypertension and cardiovascular disease is undergoing clinical studies using an antioxidant diet (NCT00756405), which is one of the supplements that comprises vitamin C, carotenoids, mixed tocopherol, and selenium [135]. Antioxidant-rich beetroot juice boosts endothelial cell activities to preserve normal blood vessel (relaxation) function in hypertensive pregnant women by enhancing the bioavailability of NO and is under clinical trials [136, 137]. A thorough and comprehensive study is necessary since a meta-analysis suggests that vitamin D has no impact on lowering blood pressure, despite a clinical study showing that vitamin D supplements significantly reduced both systolic and diastolic pressure to lower hypertension in vitamin D deficient patients [138, 139].

14.7 Future Scope and Conclusion

So far, the pathogenesis of hypertension is directly related to oxidative damage, contributing a higher percentage in increasing blood pressure by depleting the antioxidant enzymes, NO synthesis, and vascular dysfunction. Antioxidant therapy to treat diseases associated with the oxidative stress mechanism will be considered the most effective treatment in managing ROS formation. The functions of the heart and kidneys, which are susceptible to oxidative stress, may be improved by lowering hypertension with antioxidant therapy. Antioxidants, in addition to reducing ROS production, also increase NO bioavailability and minimize eNOS uncoupling, which helps to maintain vascular function and endothelial integrity. Although this viewpoint is supported by primary research and preclinical investigations, clinical studies continue to yield both positive and negative outcomes. Clinical experiments on antioxidant therapy have revealed findings that fall short of expectations in studies focused on the ROS in cell function of cardiovascular and neurodegenerative diseases administered with vitamin C or E [140, 141]. But clinical studies on a small number of subjects showed positive results, and in large trials, were shown unsatisfactory. Focusing on the other antioxidant therapies, except vitamin E or C, may show a good result. Clinical trials need additional research on specific treatments that reduce ROS/RNS formation and increase antioxidant enzymes. In contrast to studies on nonspecific oxidative stress markers, which are the focus of many clinical trials investigating antioxidants, studies on specific oxidative stress markers may result in an effective antioxidant therapy for the treatment of hypertension. Some studies have shown inhibiting specific enzymatic sources may be an effective antioxidant therapy to treat hypertension (NOX1 inhibition, xanthine oxidase, Nrf2 activation). Enhancing mitochondrial activity through the use of antioxidants that are mitochondria-targeted and activating sGC may be a key tactic to increase the efficacy of antioxidant therapy in clinical studies. Numerous documented preclinical researches have demonstrated the ability to lower blood pressure using novel approaches, and transferring these findings to clinical studies may serve as a trailblazer for the treatment of hypertension. Preclinical research has demonstrated the potential benefit of using antioxidants such as CoQ10, Salvianolic acid, quercetin, gp91ds peptide, and mycophenolate mofetil in the treatment of hypertension, which is a new therapeutic approach that has been developed recently. The other more recent treatment for hypertension involves microinjecting antioxidants into the RVLM. In order to regulate hypertension caused by the neurohumoral system, future research has to focus on long-acting ANP or inhibition of its degradation, which will help to reduce high blood pressure. Preclinical chronic administration of antioxidants in deoxycorticosterone acetate-salt hypertensive rats have shown blood pressure attenuation by decreasing NF-κB activation [120] and performing clinical studies will suggest good therapeutic options. More extensive clinical trials, in comparison to conventional hypertension treatment research, will be necessary to help advance these novel approaches as therapeutics. The most recent suggestions to manage blood pressure include regular exercise, maintaining a healthy weight, quitting smoking, reducing stress, and avoiding drinking too much alcohol. A thoughtful way to regulate hypertension is to consume less sodium and increase your intake of veggies, probiotics, fiber, meditation, and antioxidant therapy.

14.8 Acknowledgement

We want to acknowledge "smart servier medical art" as parts of the figure were drawn by using pictures from Servier Medical Art. Servier Medical Art by Servier is licensed under a Creative Commons Attribution 3.0 Unported License (https://creativecommons.org/licenses/by/3.0/).

References

1. Kjeldsen, S.E., Hypertension and cardiovascular risk: General aspects. *Pharmacol. Res.*, 129, 95–99, 2018.
2. Sidhu, N.S. and Kaur, S., Cerebrovascular Disease and Hypertension. *Cerebrovasc. Dis.*, 2021, https://doi.org/10.5772/intechopen.101180.
3. Ma, J., Li, Y., Yang, X. *et al.*, Signaling pathways in vascular function and hypertension: molecular mechanisms and therapeutic interventions. *Signal Transduct. Target Ther.*, 8, 168, 2023.
4. Augustin, H.G. and Koh, G.Y., Organotypic vasculature: From descriptive heterogeneity to functional pathophysiology. *Science*, 357, eaal2379, 2017, https://doi.org/10.1126/science.aal2379.
5. van Oort, S., Beulens, J.W.J., van Ballegooijen, A.J., Grobbee, D.E., Larsson, S.C., Association of Cardiovascular Risk Factors and Lifestyle Behaviors With Hypertension: A Mendelian Randomization Study. *Hypertens. (Dallas, Tex, 1979)*, 76, 1971–1979, 2020.
6. Shimbo, D., Dietary and lifestyle factors in hypertension. *J. Hum. Hypertens.*, 30, 571–572, 2016.
7. Mohanty, P., Patnaik, L., Nayak, G., Dutta, A., Gender difference in prevalence of hypertension among Indians across various age-groups: a report from multiple nationally representative samples. *BMC Public Health*, 22, 1524, 2022.
8. Wenger, N.K., Arnold, A., Bairey Merz, C.N. *et al.*, Hypertension Across a 'Woman's Life Cycle. *J. Am. Coll. Cardiol.*, 71, 1797–1813, 2018.

9. Oliveros, E., Patel, H., Kyung, S., Fugar, S., Goldberg, A., Madan, N., Williams, K.A., Hypertension in older adults: Assessment, management, and challenges. *Clin. Cardiol.*, 43, 99–107, 2020.
10. Zorov, D.B., Juhaszova, M., Sollott, S.J., Mitochondrial reactive oxygen species (ROS) and ROS-induced ROS release. *Physiol. Rev.*, 94, 909–950, 2014.
11. Tain, Y.-L. and Hsu, C.-N., Oxidative Stress-Induced Hypertension of Developmental Origins: Preventive Aspects of Antioxidant Therapy, *Antioxidants*, (Basel, Switzerland) 11, 3, 511, 2022, https://doi.org/10.3390/antiox11030511.
12. Lobo, V., Patil, A., Phatak, A., Chandra, N., Free radicals, antioxidants and functional foods: Impact on human health. *Pharmacogn. Rev.*, 4, 118–126, 2010.
13. Pizzino, G., Irrera, N., Cucinotta, M., Pallio, G., Mannino, F., Arcoraci, V., Squadrito, F., Altavilla, D., Bitto, A., Oxidative Stress: Harms and Benefits for Human Health. *Oxid. Med. Cell Longev.*, 2017, 8416763, 2017.
14. Sorriento, D., De Luca, N., Trimarco, B., Iaccarino, G., The Antioxidant Therapy: New Insights in the Treatment of Hypertension. *Front. Physiol.*, 9, 258, 2018.
15. Touyz, R.M., Rios, F.J., Alves-Lopes, R., Neves, K.B., Camargo, L.L., Montezano, A.C., Oxidative Stress: A Unifying Paradigm in Hypertension. *Can. J. Cardiol.*, 36, 659–670, 2020.
16. Boveris, A. and Chance, B., The mitochondrial generation of hydrogen peroxide. General properties and effect of hyperbaric oxygen. *Biochem. J.*, 134, 707–716, 1973.
17. Schieber, M. and Chandel, N.S., ROS function in redox signaling and oxidative stress. *Curr. Biol.*, 24, R453–62, 2014.
18. Brown, O.I., Bridge, K.I., Kearney, M.T., Nicotinamide Adenine Dinucleotide Phosphate Oxidases in Glucose Homeostasis and Diabetes-Related Endothelial Cell Dysfunction. *Cells*, 10, 9, 2315, 2021. https://doi.org/10.3390/cells10092315.
19. Loperena, R. and Harrison, D.G., Oxidative Stress and Hypertensive Diseases. *Med. Clin. North Am.*, 101, 169–193, 2017.
20. Martemucci, G., Costagliola, C., Mariano, M., D'andrea, L., Napolitano, P., D'Alessandro, A.G., Free Radical Properties, Source and Targets, Antioxidant Consumption and Health. *Oxygen*, 2, 48–78, 2022.
21. Alhasawi, A., Legendre, F., Jagadeesan, S., Appanna, V., Appanna, V., Chapter 10 - Biochemical Strategies to Counter Nitrosative Stress: Nanofactories for Value-Added Products, in: *Microb. Divers. Genomic Era*, S. Das and H.R. Das (eds), pp. 153–169, Academic Press, 2019.
22. Vermot, A., Petit-Härtlein, I., Smith, S.M.E., Fieschi, F., NADPH Oxidases (NOX): An Overview from Discovery, Molecular Mechanisms to Physiology and Pathology, *Antioxidants (Basel, Switzerland)*, 2021, https://doi.org/10.3390/antiox10060890.
23. Zhao, R.-Z., Jiang, S., Zhang, L., Yu, Z.-B., Mitochondrial electron transport chain, ROS generation and uncoupling (Review). *Int. J. Mol. Med.*, 44, 3–15, 2019.
24. Stram, A.R. and Payne, R.M., Post-translational modifications in mitochondria: protein signaling in the powerhouse. *Cell. Mol. Life Sci.*, 73, 4063–4073, 2016.
25. Cohen, R.A. and Tong, X., Vascular oxidative stress: the common link in hypertensive and diabetic vascular disease. *J. Cardiovasc. Pharmacol.*, 55, 308–316, 2010.
26. Liu, N., Xu, H., Sun, Q., Yu, X., Chen, W., Wei, H., Jiang, J., Xu, Y., Lu, W., The Role of Oxidative Stress in Hyperuricemia and Xanthine Oxidoreductase (XOR) Inhibitors. *Oxid. Med. Cell Longev.*, 2021, 1470380, 2021.
27. Chung, H.Y., Baek, B.S., Song, S.H., Kim, M.S., Huh, J.I., Shim, K.H., Kim, K.W., Lee, K.H., Xanthine dehydrogenase/xanthine oxidase and oxidative stress. *Age (Omaha)*, 20, 127–140, 1997.
28. Yoshida, S., Kurajoh, M., Fukumoto, S., Murase, T., Nakamura, T., Yoshida, H., Hirata, K., Inaba, M., Emoto, M., Association of plasma xanthine oxidoreductase activity with blood pressure

affected by oxidative stress level: MedCity21 health examination registry. *Sci. Rep.*, 10, 4437, 2020.
29. Winterbourn, C.C. and Hampton, M.B., Thiol chemistry and specificity in redox signaling. *Free Radic. Biol. Med.*, 45, 549–561, 2008.
30. Baba, S.P. and Bhatnagar, A., ROLE OF THIOLS IN OXIDATIVE STRESS. *Curr. Opin. Toxicol.*, 7, 133–139, 2018.
31. Lloyd, R.V., Hanna, P.M., Mason, R.P., The origin of the hydroxyl radical oxygen in the Fenton reaction. *Free Radic. Biol. Med.*, 22, 885–888, 1997.
32. Oparil, S., Acelajado, M.C., Bakris, G.L. *et al.*, Hypertension. *Nat. Rev. Dis. Prim.*, 4, 18014, 2018.
33. Ivy, J.R. and Bailey, M.A., Pressure natriuresis and the renal control of arterial blood pressure. *J. Physiol.*, 592, 3955–3967, 2014.
34. Tsilosani, A., Gao, C., Zhang, W., Aldosterone-Regulated Sodium Transport and Blood Pressure. *Front. Physiol.*, 13, 770375, 2022.
35. Delong C., Sharma S., *Physiology, Peripheral Vascular Resistance*, StatPearls [Internet], StatPearls Publishing, Treasure Island (FL), 2023 May 1, 2024 Jan–. PMID: 30855892.
36. Grisk, O. and Rettig, R., Interactions between the sympathetic nervous system and the kidneys in arterial hypertension. *Cardiovasc. Res.*, 61, 238–246, 2004.
37. Cannone, V., Burnett, J.C.J., Natriuretic Peptides and Blood Pressure Homeostasis a Novel Guanylyl Cyclase a Receptor Activator for Hypertension. *Front. Physiol.*, 12, 815796, 2021.
38. Sandoo, A., van Zanten, JJCSV, Metsios, G.S., Carroll, D., Kitas, G.D., The endothelium and its role in regulating vascular tone. *Open Cardiovasc. Med. J..*, 4, 302–312, 2010.
39. Cediel, E., Vázquez-Cruz, B., Navarro-Cid, J., de las Heras, N., Sanz-Rosa, D., Cachofeiro, V., Lahera, V., Role of endothelin-1 and thromboxane A2 in renal vasoconstriction induced by angiotensin II in diabetes and hypertension. *Kidney Int.*, 82, S2-7, 2002.
40. Varadharaj, S., Kelly, O.J., Khayat, R.N., Kumar, P.S., Ahmed, N., Zweier, J.L., Role of Dietary Antioxidants in the Preservation of Vascular Function and the Modulation of Health and Disease. *Front. Cardiovasc. Med.*, 4, 64, 2017.
41. Wilcox, C.S., Wang, C., Wang, D., Endothelin-1-Induced Microvascular ROS and Contractility in Angiotensin-II-Infused Mice Depend on COX and TP Receptors, *Antioxidants (Basel, Switzerland)*, 8, 6, 193, 2019, https://doi.org/10.3390/antiox8060193.
42. Alhayaza, R., Haque, E., Karbasiafshar, C., Sellke, F.W., Abid, M.R., The Relationship Between Reactive Oxygen Species and Endothelial Cell Metabolism. *Front. Chem.*, 8, 592688, 2020.
43. da Silva, G.M., da Silva, M.C., Nascimento, D.V.G., Lima Silva, E.M., Gouvêa, F.F.F., de França Lopes, L.G., Araújo, A.V., Ferraz Pereira, K.N., de Queiroz, T.M., Nitric Oxide as a Central Molecule in Hypertension: Focus on the Vasorelaxant Activity of New Nitric Oxide Donors, *Biology (Basel)*, 10, 10, 1041, 2021, https://doi.org/10.3390/biology10101041.
44. Schulz, E., Gori, T., Münzel, T., Oxidative stress and endothelial dysfunction in hypertension. *Hypertens. Res.*, 34, 665–673, 2011.
45. Schleicher, M., Yu, J., Murata, T. *et al.*, The Akt1-eNOS axis illustrates the specificity of kinase-substrate relationships *in vivo*. *Sci. Signal*, 2, ra41, 2009.
46. Förstermann, U., Xia, N., Li, H., Roles of Vascular Oxidative Stress and Nitric Oxide in the Pathogenesis of Atherosclerosis. *Circ. Res.*, 120, 713–735, 2017.
47. Förstermann, U. and Sessa, W.C., Nitric oxide synthases: regulation and function. *Eur. Heart J.*, 33, 829–37, 837a-837d, 2012.
48. Dominiczak, A.F. and Bohr, D.F., Nitric oxide and its putative role in hypertension. *Hypertens. (Dallas, Tex 1979)*, 25, 1202–1211, 1995.
49. Kurutas, E.B., The importance of antioxidants which play the role in cellular response against oxidative/nitrosative stress: current state. *Nutr. J.*, 15, 71, 2016.

50. Laursen, J.B., Somers, M., Kurz, S., McCann, L., Warnholtz, A., Freeman, B.A., Tarpey, M., Fukai, T., Harrison, D.G., Endothelial regulation of vasomotion in apoE-deficient mice: implications for interactions between peroxynitrite and tetrahydrobiopterin. *Circulation*, 103, 1282–1288, 2001.
51. Wang, Y., Branicky, R., Noë, A., Hekimi, S., Superoxide dismutases: Dual roles in controlling ROS damage and regulating ROS signaling. *J. Cell Biol.*, 217, 1915–1928, 2018.
52. Gómez-Marcos, M.A., Blázquez-Medela, A.M., Gamella-Pozuelo, L., Recio-Rodriguez, J.I., García-Ortiz, L., Martínez-Salgado, C., Serum Superoxide Dismutase Is Associated with Vascular Structure and Function in Hypertensive and Diabetic Patients. *Oxid. Med. Cell Longev.*, 2016, 9124676, 2016.
53. Alfonso-Prieto, M., Biarnés, X., Vidossich, P., Rovira, C., The molecular mechanism of the catalase reaction. *J. Am. Chem. Soc.*, 131, 11751–11761, 2009.
54. Kükürt, A., Gelen, V., Başer, ÖF, Deveci, H.A., Karapehlivan, M., Thiols: Role in Oxidative Stress-Related Disorders, in: *Accent Lipid Peroxidation*, 2021, https://doi.org/10.5772/intechopen.96682.
55. Ribas, V., García-Ruiz, C., Fernández-Checa, J.C., Glutathione and mitochondria. *Front. Pharmacol.*, 5, 151, 2014.
56. Sharifi-Rad, M., Anil Kumar, N.V., Zucca, P. *et al.*, Lifestyle, Oxidative Stress, and Antioxidants: Back and Forth in the Pathophysiology of Chronic Diseases. *Front. Physiol.*, 11, 694, 2020.
57. Tan, B.L., Norhaizan, M.E., Liew, W-P-P, Sulaiman Rahman, H., Antioxidant and Oxidative Stress: A Mutual Interplay in Age-Related Diseases. *Front. Pharmacol.*, 9, 1162, 2018.
58. Lubos, E., Loscalzo, J., Handy, D.E., Glutathione peroxidase-1 in health and disease: from molecular mechanisms to therapeutic opportunities. *Antioxid. Redox. Signal*, 15, 1957–1997, 2011.
59. Esler, M., The sympathetic system and hypertension. *Am. J. Hypertens.*, 13, 99S–105S, 2000.
60. Chan, S.H.H., Chan, J.Y.H., Brain stem NOS and ROS in neural mechanisms of hypertension. *Antioxid. Redox. Signal*, 20, 146–163, 2014.
61. Gallo, G., Volpe, M., Savoia, C., Endothelial Dysfunction in Hypertension: Current Concepts and Clinical Implications. *Front. Med.*, 8, 798958, 2021.
62. Lubos, E., Handy, D.E., Loscalzo, J., Role of oxidative stress and nitric oxide in atherothrombosis. *Front. Biosci.*, 13, 5323–5344, 2008.
63. Ulrich-Merzenich, G., Zeitler, H., Panek, D., Bokemeyer, D., Vetter, H., Vitamin C promotes human endothelial cell growth via the ERK-signaling pathway. *Eur. J. Nutr.*, 46, 87–94, 2007.
64. Touyz, R.M., Montezano, A.C., Rios, F., Widlansky, M.E., Liang, M., Redox Stress Defines the Small Artery Vasculopathy of Hypertension: How Do We Bridge the Bench-to-Bedside Gap? *Circ. Res.*, 120, 1721–1723, 2017.
65. Sena, C.M., Leandro, A., Azul, L., Seiça, R., Perry, G., Vascular Oxidative Stress: Impact and Therapeutic Approaches. *Front. Physiol.*, 9, 1668, 2018.
66. May, J.M. and Harrison, F.E., Role of vitamin C in the function of the vascular endothelium. *Antioxid. Redox. Signal*, 19, 2068–2083, 2013.
67. Touyz, R.M. and Schiffrin, E.L., Increased generation of superoxide by angiotensin II in smooth muscle cells from resistance arteries of hypertensive patients: role of phospholipase D-dependent NAD(P)H oxidase-sensitive pathways. *J. Hypertens.*, 19, 1245–1254, 2001.
68. Martínez-Revelles, S., García-Redondo, A.B., Avendaño, M.S. *et al.*, Lysyl Oxidase Induces Vascular Oxidative Stress and Contributes to Arterial Stiffness and Abnormal Elastin Structure in Hypertension: Role of p38MAPK. *Antioxid. Redox. Signal*, 27, 379–397, 2017.
69. Naidu, K.A., Vitamin C in human health and disease is still a mystery? An overview. *Nutr. J.*, 2, 7, 2003.

70. Pullar, J.M., Carr, A.C., Vissers, M.C.M., The Roles of Vitamin C in Skin Health. *Nutrients*, 9, 8, 866, 2017, https://doi.org/10.3390/nu9080866.
71. Rodrigo, R., Prat, H., Passalacqua, W., Araya, J., Bächler, J.P., Decrease in oxidative stress through supplementation of and E is associated with a reduction in blood pressure in patients with essential hypertension. *Clin. Sci. (Lond.)*, 114, 625–634, 2008.
72. Liao, S., Omage, S.O., Börmel, L., Kluge, S., Schubert, M., Wallert, M., Lorkowski, S., Vitamin E and Metabolic Health: Relevance of Interactions with Other Micronutrients, *Antioxidants (Basel, Switzerland)*, 11, 9, 1785, 2022, https://doi.org/10.3390/antiox11091785.
73. Traber, M.G., Stevens, J.F., , and E: beneficial effects from a mechanistic perspective. *Free Radic. Biol. Med.*, 51, 1000–1013, 2011.
74. De Bruyne, T., Steenput, B., Roth, L., De Meyer, G.R.Y., Santos Dos, C.N., Valentová, K., Dambrova, M., Hermans, N., Dietary Polyphenols Targeting Arterial Stiffness: Interplay of Contributing Mechanisms and Gut Microbiome-Related Metabolism. *Nutrients*, 11, 3, 578, 2019, https://doi.org/10.3390/nu11030578.
75. Ciumărnean, L., Milaciu, M.V., Runcan, O. *et al.*, The Effects of Flavonoids in Cardiovascular Diseases. *Molecules*, 25, 18, 4320, 2020, https://doi.org/10.3390/molecules25184320.
76. Dustin, C.M., Heppner, D.E., Lin, M.-C.J., van der Vliet, A., Redox regulation of tyrosine kinase signalling: more than meets the eye. *J. Biochem.*, 167, 151–163, 2020.
77. García-Redondo, A.B., Briones, A.M., Martínez-Revelles, S., Palao, T., Vila, L., Alonso, M.J., Salaices, M., c-Src, ERK1/2 and Rho kinase mediate hydrogen peroxide-induced vascular contraction in hypertension: role of TXA2, NAD(P)H oxidase and mitochondria. *J. Hypertens.*, 33, 77–87, 2015.
78. Lyle, A.N. and Griendling, K.K., Modulation of vascular smooth muscle signaling by reactive oxygen species. *Physiol. (Bethesda)*, 21, 269–280, 2006.
79. Griendling, K.K., Camargo, L.L., Rios, F.J., Alves-Lopes, R., Montezano, A.C., Touyz, R.M., Oxidative Stress and Hypertension. *Circ. Res.*, 128, 993–1020, 2021.
80. Son, Y., Lee, J.H., Chung, H.-T., Pae, H.-O., Therapeutic roles of heme oxygenase-1 in metabolic diseases: curcumin and resveratrol analogues as possible inducers of heme oxygenase-1. *Oxid. Med. Cell Longev.*, 2013, 639541, 2013.
81. Vasdev, S., Stuckless, J., Richardson, V., Role of the immune system in hypertension: modulation by dietary antioxidants. *Int. J. Angiol. Off. Publ. Int. Coll. Angiol. Inc.*, 20, 189–212, 2011.
82. Pasqua, T., Pagliaro, P., Rocca, C., Angelone, T., Penna, C., Role of NLRP-3 Inflammasome in Hypertension: A Potential Therapeutic Target. *Curr. Pharm. Biotechnol.*, 19, 708–714, 2018.
83. Krishnan, S.M., Ling, Y.H., Huuskes, B.M. *et al.*, Pharmacological inhibition of the NLRP3 inflammasome reduces blood pressure, renal damage, and dysfunction in salt-sensitive hypertension. *Cardiovasc. Res.*, 115, 776–787, 2019.
84. Niklowitz, P., Sonnenschein, A., Janetzky, B., Andler, W., Menke, T., Enrichment of coenzyme Q10 in plasma and blood cells: defense against oxidative damage. *Int. J. Biol. Sci.*, 3, 257–262, 2007.
85. Cordero, M.D., Cano-García, F.J., Alcocer-Gómez, E., De Miguel, M., Sánchez-Alcázar, J.A., Oxidative stress correlates with headache symptoms in fibromyalgia: coenzyme Q_{10} effect on clinical improvement. *PLoS One*, 7, e35677, 2012.
86. Tain, Y.-L. and Hsu, C.-N., AMP-Activated Protein Kinase as a Reprogramming Strategy for Hypertension and Kidney Disease of Developmental Origin. *Int. J. Mol. Sci.*, 19, 6, 1744, 2018, https://doi.org/10.3390/ijms19061744.
87. Heidary Moghaddam, R., Samimi, Z., Asgary, S., Mohammadi, P., Hozeifi, S., Hoseinzadeh-Chahkandak, F., Xu, S., Farzaei, M.H., Natural AMPK Activators in Cardiovascular Disease Prevention. *Front. Pharmacol.*, 12, 738420, 2021.

88. Greig, J.A., Shirley, R., Graham, D., Denby, L., Dominiczak, A.F., Work, L.M., Baker, A.H., Vascular-targeting antioxidant therapy in a model of hypertension and stroke. *J. Cardiovasc. Pharmacol.*, 56, 642–650, 2010.
89. Vaziri, N.D., Ni, Z., Oveisi, F., Trnavsky-Hobbs, D.L., Effect of antioxidant therapy on blood pressure and NO synthase expression in hypertensive rats. *Hypertens. (Dallas, Tex 1979)*, 36, 957–964, 2000.
90. Rodríguez-Iturbe, B., Pons, H., Quiroz, Y. *et al.*, Mycophenolate mofetil prevents salt-sensitive hypertension resulting from angiotensin II exposure. *Kidney Int.*, 59, 2222–2232, 2001.
91. Ferro, C.J., Edwards, N.C., Hutchison, C., Cockwell, P., Steeds, R.P., Savage, C.O., Townend, J.N., Harper, L., Does immunosuppressant medication lower blood pressure and arterial stiffness in patients with chronic kidney disease? An observational study. *Hypertens. Res.*, 34, 113–119, 2011.
92. Tobeiha, M., Jafari, A., Fadaei, S. *et al.*, Evidence for the Benefits of Melatonin in Cardiovascular Disease. *Front. Cardiovasc. Med.*, 9, 888319, 2022.
93. Valenzuela-Melgarejo, F.J., Lagunas, C., Carmona-Pastén, F., Jara-Medina, K., Delgado, G., Supraphysiological Role of Melatonin Over Vascular Dysfunction of Pregnancy, a New Therapeutic Agent? *Front. Physiol.*, 12, 767684, 2021.
94. Seyidoglu, N. and Aydin, C., Stress, Natural Antioxidants and Future Perspectives, in: *Heal Benefits Foods*, 2020, https://doi.org/10.5772/intechopen.91167.
95. Schiffrin, E.L., Antioxidants in hypertension and cardiovascular disease. *Mol. Interv.*, 10, 354–362, 2010.
96. Amponsah-Offeh, M., Diaba-Nuhoho, P., Speier, S., Morawietz, H., Oxidative Stress, Antioxidants and Hypertension, *Antioxidants (Basel, Switzerland)*, 12, 2, 281, 2023, https://doi.org/10.3390/antiox12020281.
97. Ulker, S., McKeown, P.P., Bayraktutan, U., Vitamins reverse endothelial dysfunction through regulation of eNOS and NAD(P)H oxidase activities. *Hypertens. (Dallas, Tex 1979)*, 41, 534–539, 2003.
98. Neuzil, J., Thomas, S.R., Stocker, R., Requirement for, promotion, or inhibition by alpha-tocopherol of radical-induced initiation of plasma lipoprotein lipid peroxidation. *Free Radic. Biol. Med.*, 22, 57–71, 1997.
99. Rodrigo, R., Guichard, C., Charles, R., Clinical pharmacology and therapeutic use of antioxidant vitamins. *Fundam. Clin. Pharmacol.*, 21, 111–127, 2007.
100. Wong, M.S.K., Leisegang, M.S., Kruse, C. *et al.*, Vitamin D promotes vascular regeneration. *Circulation*, 130, 976–986, 2014.
101. Milani, A., Basirnejad, M., Shahbazi, S., Bolhassani, A., Carotenoids: biochemistry, pharmacology and treatment. *Br. J. Pharmacol.*, 174, 1290–1324, 2017.
102. Di Tomo, P., Canali, R., Ciavardelli, D., Di Silvestre, S., De Marco, A., Giardinelli, A., Pipino, C., Di Pietro, N., Virgili, F., Pandolfi, A., β-Carotene and lycopene affect endothelial response to TNF-α reducing nitro-oxidative stress and interaction with monocytes. *Mol. Nutr. Food Res.*, 56, 217–227, 2012.
103. Rossoni-Júnior, J.V., Araújo, G.R., Pádua B da, C., Chaves, M.M., Pedrosa, M.L., Silva, M.E., Costa, D.C., Annato extract and β-carotene modulate the production of reactive oxygen species/nitric oxide in neutrophils from diabetic rats. *J. Clin. Biochem. Nutr.*, 50, 177–183, 2012.
104. Abbasian, F., Alavi, M.S., Roohbakhsh, A., Dietary carotenoids to improve hypertension. *Heliyon*, 9, e19399, 2023.
105. Linnewiel-Hermoni, K., Motro, Y., Miller, Y., Levy, J., Sharoni, Y., Carotenoid derivatives inhibit nuclear factor kappa B activity in bone and cancer cells by targeting key thiol groups. *Free Radic. Biol. Med.*, 75, 105–120, 2014.

106. Schubert, S.Y., Neeman, I., Resnick, N., A novel mechanism for the inhibition of NF-kappaB activation in vascular endothelial cells by natural antioxidants. *FASEB J. Off. Publ. Fed. Am. Soc. Exp. Biol.*, 16, 1931–1933, 2002.
107. Oak, M.-H., Chataigneau, M., Keravis, T., Chataigneau, T., Beretz, A., Andriantsitohaina, R., Stoclet, J.-C., Chang, S.-J., Schini-Kerth, V.B., Red wine polyphenolic compounds inhibit vascular endothelial growth factor expression in vascular smooth muscle cells by preventing the activation of the p38 mitogen-activated protein kinase pathway. *Arterioscler Thromb. Vasc. Biol.*, 23, 1001–1007, 2003.
108. Chang, X., Zhao, Z., Zhang, W., Liu, D., Ma, C., Zhang, T., Meng, Q., Yan, P., Zou, L., Zhang, M., Natural Antioxidants Improve the Vulnerability of Cardiomyocytes and Vascular Endothelial Cells under Stress Conditions: A Focus on Mitochondrial Quality Control. *Oxid. Med. Cell Longev.*, 2021, 6620677, 2021.
109. Tanase, D.M., Apostol, A.G., Costea, C.F., Tarniceriu, C.C., Tudorancea, I., Maranduca, M.A., Floria, M., Serban, I.L., Oxidative Stress in Arterial Hypertension (HTN): The Nuclear Factor Erythroid Factor 2-Related Factor 2 (Nrf2) Pathway, Implications and Future Perspectives. *Pharmaceutics*, 14, 3, 534, 2022, https://doi.org/10.3390/pharmaceutics14030534.
110. Wang, C., Luo, Z., Carter, G., Wellstein, A., Jose, P.A., Tomlinson, J., Leiper, J., Welch, W.J., Wilcox, C.S., Wang, D., NRF2 prevents hypertension, increased ADMA, microvascular oxidative stress, and dysfunction in mice with two weeks of ANG II infusion. *Am. J. Physiol. Regul. Integr. Comp. Physiol.*, 314, R399–R406, 2018.
111. Bajek-Bil, A., Chmiel, M., Włoch, A., Stompor-Gorący, M., Baicalin-Current Trends in Detection Methods and Health-Promoting Properties, *Pharmaceuticals (Basel)*, 16, 4, 570, 2023, https://doi.org/10.3390/ph16040570.
112. Basu, A. and Lucas, E.A., Mechanisms and effects of green tea on cardiovascular health. *Nutr. Rev.*, 65, 361–375, 2007.
113. Cong, L., Zhang, Y., Huang, H., Cao, J., Fu, X., DFMG reverses proliferation and migration of vascular smooth muscle cells induced by co-culture with injured vascular endothelial cells via suppression of the TLR4-mediated signaling pathway. *Mol. Med. Rep.*, 17, 5692–5699, 2018.
114. Gao, H.-L., Yu, X.-J., Feng, Y.-Q. *et al.*, Luteolin Attenuates Hypertension via Inhibiting NF-κB-Mediated Inflammation and PI3K/Akt Signaling Pathway in the Hypothalamic Paraventricular Nucleus. *Nutrients*, 15, 3, 502, 2023, https://doi.org/10.3390/nu15030502.
115. Morabito, R., Remigante, A., Marino, A., Melatonin Protects Band 3 Protein in Human Erythrocytes against H(2)O(2)-Induced Oxidative Stress. *Molecules*, 24, 15, 2741, 2019, https://doi.org/10.3390/molecules24152741.
116. Bredemeier, M., Lopes, L.M., Eisenreich, M.A., Hickmann, S., Bongiorno, G.K., Morsch, A.L.B., da Silva Stein, F., Campos, G.G.D., Xanthine oxidase inhibitors for prevention of cardiovascular events: a systematic review and meta-analysis of randomized controlled trials. *BMC Cardiovasc. Disord.*, 18, 24, 2018.
117. He, H., Jiang, H., Chen, Y. *et al.*, Oridonin is a covalent NLRP3 inhibitor with strong anti-inflammasome activity. *Nat. Commun.*, 9, 2550, 2018.
118. Jayakumar, T., Hou, S.-M., Chang, C.-C. *et al.*, Columbianadin Dampens *In Vitro* Inflammatory Actions and Inhibits Liver Injury via Inhibition of NF-κB/MAPKs: Impacts on (·)OH Radicals and HO-1 Expression, *Antioxidants (Basel, Switzerland)*, 2021, https://doi.org/10.3390/antiox10040553.
119. Jia, X.-Y., Jiang, D.-L., Jia, X.-T., Fu, L.-Y., Tian, H., Liu, K.-L., Qi, J., Kang, Y.-M., Yu, X.-J., Capsaicin improves hypertension and cardiac hypertrophy via SIRT1/NF-κB/MAPKs pathway in the hypothalamic paraventricular nucleus. *Phytomedicine*, 118, 154951, 2023.

120. Beswick, R.A., Zhang, H., Marable, D., Catravas, J.D., Hill, W.D., Webb, R.C., Long-term antioxidant administration attenuates mineralocorticoid hypertension and renal inflammatory response. *Hypertens. (Dallas, Tex 1979)*, 37, 781–786, 2001.
121. Chen, M., Hu, Q., Wang, S., Tao, L., Hu, X., Shen, X., 1,8-Cineole ameliorates endothelial injury and hypertension induced by L-NAME through regulation of autophagy via PI3K/mTOR signaling pathway. *Eur. J. Pharmacol.*, 954, 175863, 2023.
122. Laskowski, A., Woodman, O.L., Cao, A.H., Drummond, G.R., Marshall, T., Kaye, D.M., Ritchie, R.H., Antioxidant actions contribute to the antihypertrophic effects of atrial natriuretic peptide in neonatal rat cardiomyocytes. *Cardiovasc. Res.*, 72, 112–123, 2006.
123. Tan, A.C., Russel, F.G., Thien, T., Benraad, T.J., Atrial natriuretic peptide. An overview of clinical pharmacology and pharmacokinetics. *Clin. Pharmacokinet.*, 24, 28–45, 1993.
124. Nakamura, K., Murakami, M., Miura, D. *et al.*, Beta-Blockers and Oxidative Stress in Patients with Heart Failure. *Pharmaceuticals (Basel)*, 4, 1088–1100, 2011.
125. Wang, Y., Zhang, F., Liu, Y., Yin, S., Pang, X., Li, Z., Wei, Z., Nebivolol alleviates aortic remodeling through eNOS upregulation and inhibition of oxidative stress in l-NAME-induced hypertensive rats. *Clin. Exp. Hypertens.*, 39, 628–639, 2017.
126. Nishioka, S., Yoshioka, T., Nomura, A., Kato, R., Miyamura, M., Okada, Y., Ishizaka, N., Matsumura, Y., Hayashi, T., Celiprolol reduces oxidative stress and attenuates left ventricular remodeling induced by hypoxic stress in mice. *Hypertens. Res.*, 36, 934–939, 2013.
127. Griendling, K.K., Sorescu, D., Ushio-Fukai, M., NAD(P)H oxidase: role in cardiovascular biology and disease. *Circ. Res.*, 86, 494–501, 2000.
128. Ceron, C.S., Castro, M.M., Rizzi, E., Montenegro, M.F., Fontana, V., Salgado, M.C.O., Gerlach, R.F., Tanus-Santos, J.E., Spironolactone and hydrochlorothiazide exert antioxidant effects and reduce vascular matrix metalloproteinase-2 activity and expression in a model of renovascular hypertension. *Br. J. Pharmacol.*, 160, 77–87, 2010.
129. Lake, C.R., Ziegler, M.G., Coleman, M.D., Kopin, I.J., Hydrochlorothiazide-induced sympathetic hyperactivity in hypertensive patients. *Clin. Pharmacol. Ther.*, 26, 428–432, 1979.
130. Ferreira, M.J., Santos Ferreira Silva Dos, M.P., da Silva Dias, D., Bernardes, N., Irigoyen, M.C., De Angelis, K., Concurrent exercise training induces additional benefits to hydrochlorothiazide: Evidence for an improvement of autonomic control and oxidative stress in a model of hypertension and postmenopause. *PLoS One*, 18, e0289715, 2023.
131. Observe the Effect of Wheatgrass and Tulsi Formulation or Individuals Taken With Other Allopathic Drugs, ClinicalTrials.gov, (n.d.), https://clinicaltrials.gov/search?cond=Hypertension&term=Wheatgrass (accessed January 09, 2024).
132. Antihypertensive and Cardioprotective Evaluation of a Functional Olive Oil, ClinicalTrials.gov, (n.d.), https://clinicaltrials.gov/study/NCT05460221?cond=Hypertension&term=Olive%20Oil&rank=1 (accessed January 09, 2024).
133. Massaro, M., Scoditti, E., Carluccio, M.A., Calabriso, N., Santarpino, G., Verri, T., De Caterina, R., Effects of Olive Oil on Blood Pressure: Epidemiological, Clinical, and Mechanistic Evidence. *Nutrients*, 12, 6, 1548, 2020, https://doi.org/10.3390/nu12061548.
134. Allopurinol Improves Heart Function in African Americans With Resistant Hypertension (RESIST), ClinicalTrials.gov, (n.d.). https://clinicaltrials.gov/study/NCT05888233?cond=Hypertension&term=allopurinol&rank=2 (accessed January 09, 2024).
135. Effects of Dietary Antioxidants on Cardiovascular Risk Factors, ClinicalTrials.gov, (n.d.), https://clinicaltrials.gov/study/NCT00756405?cond=Hypertension&term=dietary%20antioxidants&rank=1 (accessed January 09, 2024).
136. Beetroot Juice Supplementation in Chronic Hypertensive Pregnancies (BEET_CH), ClinicalTrials.gov, (n.d.), https://clinicaltrials.gov/study/NCT06105775?cond=Hypertension&term=beetroot%20juice&rank=1 (accessed January 09, 2024).

137. Tropea, T., Renshall, L.J., Nihlen, C. *et al.*, Beetroot juice lowers blood pressure and improves endothelial function in pregnant eNOS(-/-) mice: importance of nitrate-independent effects. *J. Physiol.*, 598, 4079–4092, 2020.
138. Sheikh, V., Mozaianimonfared, A., Gharakhani, M., Poorolajal, J., Ph, D., Effect of vitamin D supplementation versus placebo on essential hypertension in patients with vitamin D deficiency: a double-blind randomized clinical trial. *J. Clin. Hypertens. (Greenwich)*, 22, 1867–1873, 2020.
139. Zhang, D., Cheng, C., Wang, Y., Sun, H., Yu, S., Xue, Y., Liu, Y., Li, W., Li, X., Effect of Vitamin D on Blood Pressure and Hypertension in the General Population: An Update Meta-Analysis of Cohort Studies and Randomized Controlled Trials. *Prev. Chronic. Dis.*, 17, E03, 2020.
140. Shekelle, P.G., Morton, S.C., Jungvig, L.K., Udani, J., Spar, M., Tu, W., J Suttorp, M., Coulter, I., Newberry, S.J., Hardy, M., Effect of supplemental vitamin E for the prevention and treatment of cardiovascular disease. *J. Gen. Intern. Med.*, 19, 380–389, 2004.
141. Juraschek, S.P., Guallar, E., Appel, L.J., Miller, E.R., Effects of vitamin C supplementation on blood pressure: a meta-analysis of randomized controlled trials. *Am. J. Clin. Nutr.*, 95, 1079–1088, 2012.

15

Antioxidants and Rheumatoid Arthritis

Chaman Bala[1], Bharath Harohalli Byregowda[2], Farmiza Begum[3], Yogendra Nayak[2], Rakesh K. Sindhu[4] and Gautam Kumar[5]*

[1]*GNIT College of Pharmacy, Greater Noida, Knowledge Park II, Uttar Pradesh, India*
[2]*Department of Pharmacology, Manipal College of Pharmaceutical Sciences, Manipal, Karnataka, India*
[3]*Department of Pharmacology, Vaagdevi Pharmacy College, Bollikunta, Warangal, Telangana, India*
[4]*School of Pharmacy, Sharda University, Greater Noida, Uttar Pradesh, India*
[5]*Department of Pharmacy, Apeejay Stya University, Gurugram, Haryana, India*

Abstract

An autoimmune condition known as rheumatoid arthritis (RA) affects a wide range of people. Reactive oxygen species (ROS), which are generated as a result of oxidative stress, are essential for the development of osteoarthritis and rheumatoid arthritis. In chondrocytes and synoviocytes, metabolic processes generate free radicals, ROS, and their derivatives. These harmful substances can build up in the synovial joint, where they can cause serious structural harm, inflammation, and cell death. Natural reducing agents known as antioxidants can scavenge free radicals, prevent the generation of ROS, and eliminate ROS derivatives. The immune system, cell differentiation, and the modulation of oxidative stress are all significantly influenced by antioxidant vitamins. The antioxidant properties of plant polyphenols, ascorbic acid, α-tocopherol, thiols (glutathione), and other nutrients can lessen oxidative stress in the body and neutralize ROS levels in joints. Understanding how antioxidants impact RA is essential because doing so may enable us to discover strategies for delaying the beginning and development of the ailment. With more emphasis on the significance of oxidative stress on rheumatoid arthritis, this chapter aims to investigate the biological roles of oxidants and naturally occurring antioxidants in the synovial joint at the cellular levels.

***Keywords*:** Oxidative stress, antioxidants, rheumatoid arthritis, vitamins, polyphenols

15.1 Introduction

Rheumatoid arthritis (RA) is an autoimmune and inflammatory disease of unknown cause that is characterized by immunological dysfunction and persistent inflammation which leads to deformation and destruction of synovial joints. Although the pathophysiology of RA is unclear, high levels of oxidants have been reported to have a link with RA [1, 2].

**Corresponding author*: drgautam9265@gmail.com; gautamkumar.phd.manipal@gmail.com

Rakesh K. Sindhu, Inderbir Singh and M. Arockia Babu (eds.) Antioxidants: Nature's Defense Against Disease, (537–550)

Common manifestations of RA include arthralgia, edema, redness, and restricted motion. Subcutaneous nodules, ulnar deviation, extra-articular symptoms, and swan neck deformity are additional signs of untreated and chronic RA [3]. Patients with RA had lower serum levels of several antioxidants, including zinc, selenium, β-carotene, vitamin C, and vitamin E, when compared to healthy individuals [1]. The main objective of RA treatment is to prevent joint damage and deformity, enhance joint function, and achieve sustained remission or low disease activity through the reduction of joint pain and inflammation [4].

One of the main barriers to the development of novel RA treatments is the inadequate knowledge of the disease's etiology. Conventional disease-modifying drugs for RA, nonsteroidal anti-inflammatory drugs, and corticosteroids are few of the drugs used to treat RA. However, many patients also experience side effects, including nephrotoxicity, hepatotoxicity, and cardiotoxicity [2, 4, 5]. According to data from a recent national survey conducted by the Arthritis Foundation (www.arthritis.org), 70% of RA patients receiving treatment still deal with pain, stiffness, and exhaustion on a daily basis and some studies also have revealed that these treatments have a poor prognosis and do not produce the expected results [6].

The role of nutrients in the context of RA has been studied for many years. It indicates the possibility of malnutrition in RA patients, due to food-drug interactions, that alters the absorption of nutrients leading to their poor bioavailability. In some patients' loss of appetite is also linked with RA. Similarly, studies on the utilization of natural remedies for the management of RA have been more focused in recent years. Due to their antioxidant and anti-inflammatory qualities, natural products have been investigated and documented for the potential role in the development of newer molecules for the management of disease [7, 8]. Natural remedies provide many anti-inflammatory, palliative, and antiarthritic qualities, making them an excellent option among complementary and alternative treatments for RA. It is now recognized that the pathophysiology of chronic inflammatory illnesses, such as RA, is influenced by oxidative stress and redox imbalance. Therefore, focusing on the underlying oxidative stress linked to RA may present a potential therapeutic strategy. We've outlined the functions of naturally occurring antioxidants for the management of RA in this chapter.

15.2 Oxidative Stress and Rheumatoid Arthritis

Upon activation, phagocytic cells—such as neutrophils and macrophages—produce copious amounts of reactive oxygen species (ROS) which destroy the foreign microorganisms. The NADPH oxidase system mediates this oxidative burst, which causes a significant rise in oxygen demand and the generation of superoxide (O_2-•) anions. NADPH is made up of several subunits that come together at the cytoplasmic membrane and join with the outer membrane or intracellular phagocytic vesicles. This enables the production and release of many subsequent oxidants. Superoxide dismutase (SOD), an enzyme that exists in two isoforms and is triggered by inflammatory cytokines like tumor necrosis factor-α (TNF-α), catalyzes the conversion of superoxide to hydrogen peroxide (H_2O_2), either spontaneously or quickly [9, 10]. Figure 15.1 demonstrates the normal vs. rheumatoid arthritis joint, where oxidative stress, hypoxia, cytokines, and immune cells damage the synovial membrane, and cartilage leads to inflammation, pain, and development of RA.

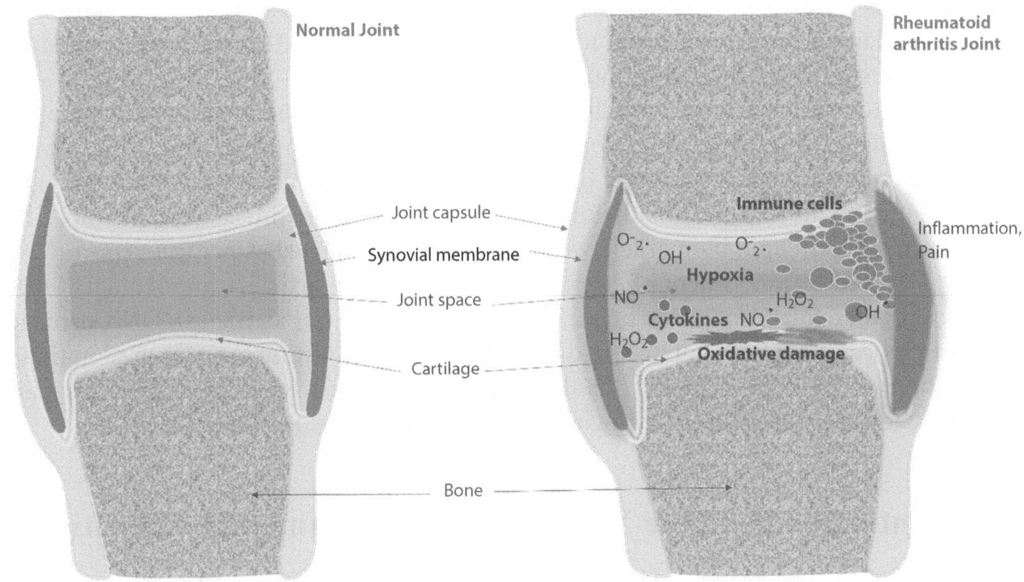

Figure 15.1 Normal knee Joint vs. rheumatoid arthritis knee joint.

H_2O_2 and O_2-• are transformed to highly reactive, water-soluble hydroxyl radicals (OH•) via the Fenton reaction, which is assumed to have a major role in the cellular toxicity associated with ROS, in the presence of transition metals including ferrous ions (Fe_2+). Myeloperoxidase, an enzyme found in neutrophils, is also capable of oxidizing halides like chloride (Cl^-) and transforming H_2O_2 into hypochlorous acid(HOCl), which can subsequently generate chloramines with the interaction of amino acids. Other halides like bromide (Br^-) & iodide (I^-) can cause similar reactions. Singlet oxygen is a kind of another harmful and highly reactive radical, produced when HOCl and H_2O_2 are further reacted. Aldehyde is produced when hypochlorous acid interacts with amino acids. Nitric oxide (NO), which is created when nitric oxide synthase (NOS) deaminates L-arginine, also, reacts with O_2-• to form the extremely reactive peroxy-nitrite radical ($ONOO^-$) [11].

Regulation of metabolic processes in cells, intracellular communication, and the functions of transcription factors like nuclear factor-κB (NF-κB), Hypoxia-Inducible Factor-1 (HIF-1), and Activator Protein 1 (AP-1) depends on the presence of ROS. They generate during the normal cellular metabolism in the presence of oxygen and play a significant role in sustaining redox status of the cell. Furthermore, ROS generated by phagocytes appears to play crucial role in stimulating the immune system. In murine pristane-induced arthritis, a functional mutation of Ncf1, a part of the NADPH oxidase complex, results in an increased sensitivity to and severity of arthritis as well as a decreased oxidative burst [12, 13]. In a report, when vitamin E was administered before arthritis induction, the NADPH complex was activated, which suggests that the Ncf1 polymorphism is implicated in the disease's immunological priming stage. According to previous reports, phagocytes' physiological generation of ROS in response to antigens influences interactions with T cell-antigen and may cause the death of arthritogenic T cells and suppress the autoimmune reactions. Ncf1 is redundant in humans, a kind of non-functional mutation is linked to a chronic granulomatous illness with a greater propensity for bacterial infections. The relationship between Ncf1 and other experimental autoimmune diseases imply that Ncf1 gene polymorphism may be significant for autoimmunity in general [14].

According to a systematic review published in 2016, 30 different markers of oxidants and/or antioxidants were investigated among the selected research investigations. They were divided into seven groups: (a) protein oxidation; (b) lipid peroxidation; (c) DNA damage; (d) enzymatic activity; (e) urate oxidation; (f) free radical/anions; and (g) antioxidants. The oxidation of different lipid biomarkers was extensively investigated as per this report (72% articles), followed by assessment of enzymatic reactions (68% articles), antioxidants (40% articles), protein oxidation (22% articles), free radical/anions (18% articles), DNA damage (4.5% articles), and oxidation of uric acid (4.5% articles). The testing samples used in these investigations were blood samples (whole blood, erythrocytes, lymphocytes, serum, and plasma), synovial tissue samples, synovial fluid samples, and urine samples [15].

15.3 Antioxidants in the Management of Rheumatoid Arthritis

To guard cellular structures from oxidative harm, several defensive mechanisms have developed. These include the sequestration of metal ion co-factors like iron and copper through protein interactions, thioredoxin reductase, catalase, glutathione peroxidase (GPX), and other peroxidases, as well as endogenous antioxidants. The less deadly H_2O_2 is produced when SOD speeds up the normally sluggish spontaneous breakdown of superoxide. This H_2O_2 can then interact with glutathione (GSH) to produce water and oxygen. There are two different types of SOD: constitutively produced SOD and an inducible form called MnSOD that resides in mitochondria. Via NF-κB, cytokines activate MnSOD. It may also require additional cofactors, such as the RNA-binding protein nucleolar phosphosmin [16]. The main mitochondrial defense against H_2O_2, GPX, is increased by p53 and hypoxia [17, 18]. Due to its absence from the mitochondria of the majority of cells, catalase breaks down hydrogen peroxide and is likely involved in cytosolic or extracellular defense against oxidants. Another crucial element of the physiological response to ROS is the thioredoxin reductase system, particularly in heart tissue [19]. Thioredoxin is induced by several stimuli, inclusive of oxidative stress and inflammatory cytokines. When active, thioredoxin controls the redox state of proteins and promotes interactions between proteins and DNA. Thioredoxin expression is increased in heart tissue when there is cyclic hypoxia and reperfusion. Additionally, it has been shown that RA synovial fluid and tissue express thioredoxin more strongly [20, 21].

Cellular systems are shielded against ROS and their harmful effects by endogenous antioxidants. Vitamin A (retinol and its metabolites), vitamin C (ascorbic acid), and vitamin E (α-tocopherol) are the three primary antioxidants. Provitamin A called carotene is a water-soluble antioxidant that inhibits the growth of reactive species and alters lipoxygenase activity. One of the first lines of defense against oxidative stress is vitamin C (ascorbic acid), which can stop peroxidation of lipids by catching peroxyl radicals ($ROO^{-\bullet}$) before they diffuse into lipid membranes. Vitamin C also responds to H_2O_2, $ROO^{-\bullet}$ and $O_2^{-\bullet}$ and plays a key role in recycling other antioxidants like vitamin E. Due to its lipid-soluble characteristics, vitamin E can function as a reagent to break the chain of lipid peroxidation [22]. Different types of antioxidants used for the management of RA are shown in Figure 15.2.

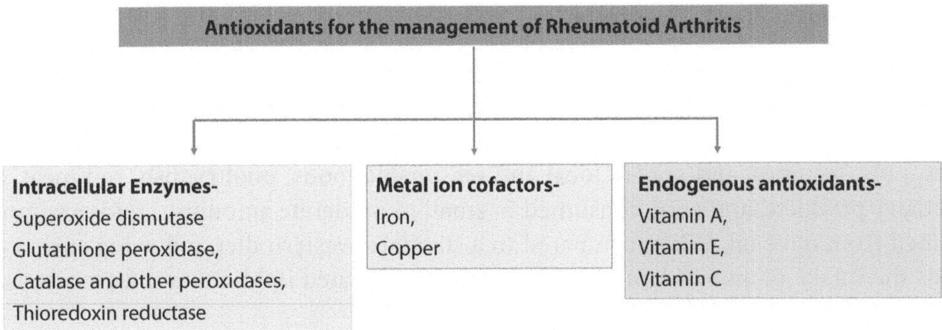

Figure 15.2 Types of antioxidants used for the management of RA.

15.3.1 Antioxidant Supplements

Despite relying on a relatively small number of publications, a recent comprehensive evaluation of clinical research found limited evidence for the effectiveness of antioxidant supplements in the treatment of RA. This is true, although evidence pointing to dietary measures, to treat antioxidant deficiency as a potential RA treatment option. Combining the data from five randomized clinical studies which found no conclusive evidence that vitamin E administration improved symptoms in RA patients. However, one of these trials did show that vitamin E was superior to a placebo in terms of reducing pain, but not inflammatory indicators [23, 24].

In line with this, five different clinical trials of selenium supplementation for the management of RA did not show any significant result in quantifiable indicators of disease activity as compared to placebo [24]. Few studies have shown that injecting an antioxidant like vitamin E truly increases the amount of vitamin E in the patient's tissues (e.g. in the knee joint) [23]. Importantly, only few research studies have looked at how antioxidant supplementation works on oxidative biomolecular damage using pharmacodynamic measures. As a result, it is unclear if therapeutic advancements, or even the lack thereof, are connected to any impact of supplementary antioxidants on oxidative damage.

As there are numerous antioxidant systems, uncertainty exists regarding the extent to which the administration of a particular antioxidant can revive the role of antioxidant systems that have been depleted. For instance, it has been hypothesized that ascorbic acid is essential for converting the α-tocopheryl radical (generated by the action of α-tocopherol) back to the reduced form. Without ascorbic acid to facilitate this recycling, α-tocopherol supplementation alone may not be adequate to restore normal antioxidant defenses. The fact that ROS influences physiological signaling processes and causes tissue damage at the same time adds another layer of complexity [25].

The possible negative effects of supplemental antioxidants, such as lung cancer in the smokers have become more common in patients who received supplementary β-carotene. It can be explained by the downregulation of these ROS-mediated cell-signaling pathways. So, it could be better to consume antioxidants through diet rather than consume high doses of commercially available pills [26].

15.3.2 Dietary Plant Materials

Numerous fruits and vegetables contain dietary antioxidants. The 'Mediterranean diet' is reflection of the eating customs of various nations in the Mediterranean region. The typical diet consists of an ampule quantity of plant foods (such as fruits, unrefined cereals, vegetables, nuts, legumes, and seeds), local and seasonable foods, poultry, fish, red meat, and with dairy products, and wine consumed in small to moderate amounts. Lipids are mostly obtained from olive oil. When compared to a standard western diet, following such a diet boosts the intake of antioxidant-rich foods and unsaturated lipid-containing foods, especially monounsaturated fatty acids [27, 28].

Clinical advantages for RA patients who follow a "Mediterranean type" diet have been documented. In clinical trials, RA patients who adhered to a modified conventional Cretan diet for three months revealed significant improvements in some quality of life markers as well as significant decrease in their Disease Activity Score using [28] joint counts and Health Assessment Questionnaire scores [29]. Further serological analysis, however, cast doubt on the cause of the clinical improvement since it showed that levels of antioxidants in plasma and malondialdehyde, a marker of oxidative stress, in urine were stable throughout the experiment. Women with RA who are from socially underprivileged locations in Glasgow participated in a 6-week instructional session with a focus on cooking and adhering to a "Mediterranean-type" diet as part of a study that employed a different technique. Only generic advice on a healthy diet was given to the controls. Females in the intervention group had noticed that they were eating better than the control group, and they showed substantial improvements in comparison to controls in global assessment at six months. Such research cannot pinpoint the precise mechanisms at play. It is plausible that consuming the omega-3 fatty acids contained in fish oil has an anti-inflammatory impact, however, the higher intake of dietary antioxidants and the probable rise in antioxidant activity remain more convincing explanation [30]. Several plant-based phytoconstituents are described here for the management of RA.

Resveratrol (Stilbenes), that is frequently found in natural products like grapes, berries, peanuts and is mostly known for its antioxidant capabilities, is one such polyphenol [31]. Previous studies have found strong support for this plant-derived compound's protective properties against autoimmune illnesses, and they have also noted a correlation between the disease physiology of RA and the level of oxidative stress at joints, mainly oxidative DNA damage. After two months of continuous use, it demonstrated a significant reduction in swelling of the knees as well as p65 expression mitigation and articular cartilage deterioration reduction [32].

Silibinin (Flavonolignans) is a naturally occurring polyphenolic flavonoid with well-established antioxidant, anticancer effects, and anti-inflammatory properties. According to studies, silibinin repressed the NF-κB pathway, lowered Sirtuin1, and blocked the generation of IL-6 and IL-1 that is caused by TNF-α. Furthermore, silibinin inhibited TNF-α mediated activation of NF-κB, p65 and IKKα/β in a concentration-dependent manner [33, 34].

Turmeric (*Curcuma longa*) contains the most active ingredient known as curcumin (Flavonoids), a polyphenol that has undergone extensive research. It has been found that proinflammatory cytokines and chemokines are expressed less frequently when the NF-κB signaling pathway is controlled [35].

Morin, a pentahydroxy flavone that is structurally related to 1-benzopyran phenylpropanoid, is frequently isolated from the leaves of *Psidium guajava* (guava), and *Maclurapomifera* (orange) and it has previously been linked to allergic reactions, despite being a member of the flavonoid family and having antioxidant, anti-inflammatory, and anti-microbial properties [36–38]. In addition to lowering inflammatory scores, due to decrease in cytokine levels and oxidative stress, a combination of indomethacin and morin in arthritic rats, also showed decreased paw edema, cartilage degradation, synovial hyperplasia, and bone collagen levels [39].

Epicatechin-3-O-d-Allopyranoside, extracted out from *Davallia formosana*, is a flavan-3-ol. According to reports, adding this flavonoid to the treatment group's rheumatoid arthritis mouse models was able to up the levels of IL-4 and IL-10 while decreasing the production of IL-17 and TNF-α [40].

Sinomenine or cocculine is a popular morphinan-derived alkaloid generally isolated from the root of *Sinomenium acutum*. In China, this plant has been used to treat RA for more than 2,000 years and is well-known in herbal therapy. Apart from improving arthritis scores in clinical settings, this alkaloid has been shown to decrease the production of IL-1a, IL-6, IL-1b, GMC-SF, Eotaxin-2, TNF-α. It has also been reported to decrease inflammatory cell infiltration and synovial hyperplasia, and in a dose-dependent manner, the scores for paw swelling, inflammation, and cartilage damage [41].

Taraxasterol, a Pentacyclic Triterpene obtained from *Taraxacum officinale* has been studied in the past for its protective effects, particularly those in anticipation of inflammation. Studies revealed that the level of IL-6, TNF-α, and IL-8 in the tissues of joints had significantly decreased [42–45].

Betulinic Acid is a naturally produced pentacyclic triterpenoid that is effective in treating arthritis and paw swelling in mice models, with results showing significantly lower scores from the end of the second week onward [46].

Madecassoside (Pentacyclic Triterpene), according to a recent study, is found in *Centella asiatica* plants, significantly reduced the histological lesions in rats with adjuvant-induced arthritis. Additionally, the administration of madecassoside decreased NF-κB translocation and phosphorylation, indicating that this medication had anti-RA effects by blocking the NF-κB/MMP-13 pathway [47].

Another significant bioactive derivative of Rhizome *Curcuma* is germacrone, a phenol lipid that belongs to the class of sesquiterpenoids. In the serum and synovial tissues of mice, the levels of the Th1 cytokines (TNF-α and IFN-γ) were partially reduced (almost decreased by 50%), and it has been demonstrated that the arthritis scores in the treated mice were almost cut in half compared to the untreated mice [48].

Triptolide (Diterpene Triepoxide), frequently found in *Tripterygium wilfordii Hook F*, a Chinese herb, successfully decreased the serum levels of IL-1 and IL-6 while reducing the severity of joint swelling [49].

Paeonol, which belongs to the category of methoxy benzene, is frequently present in *peonies* and has been freshly investigated in patients treated with IL-1β, which showed that it was able to effectively mitigate the generation of proinflammatory TNF-α, IL-6, IL-1β, and the levels of MMP-1/MMP-3 expression were also downregulated in a dose-dependent manner by the treatment and NF-κB, p65 activation as well as TLR4 expression was also inhibited [50].

Sulforaphane is a highly studied member of the isothiocyanate family of chemicals that are found in cruciferous vegetables, mainly broccoli and sprouts. These chemicals are the byproducts of the breakdown of glucosinolates [51]. A chemical with potential health-improvement qualities in defected tissues and low toxicity in healthy tissue is the subject of growing research. It works by boosting the GSH pathway in the cytoplasm of Nuclear factor (erythroid-derived 2)-like 2 (Nrf2), which enhances the generation of antioxidants in the brain [51, 52]. Histological findings like inflammation, degradation of cartilage and bone in the joints were all reported to have improved overall, along with the arthritis score. Apart from these, the anti-inflammatory potential of these treatments was linked to decreased expression of pro-inflammatory cytokines (IL-6, TNF-α, and IL-17) in the joints of RA mice in comparison to the diseased group [52, 53].

15.4 Mechanisms of Antioxidants in the Prevention of Rheumatoid Arthritis

Although the molecular processes by which food-based antioxidants may help prevent the beginning of RA are not well understood, they may include both hereditary and environmental variables. Somatic mutations may contribute to the development of RA, and ROS is considered as one of the important causes of DNA damage that may result in such mutations [54, 55].

The DNA damage caused by ROS produces the nucleoside adduct 8-hydroxydeoxyguanosine, which is more common in mononuclear cells of blood in RA patients. Antioxidants may also prevent some proteins from being altered through oxidative processes. This alteration may contribute to the development of neo-antigens, which results in autoimmunity [56]. Antioxidants including ascorbic acid, β-carotene, and α-tocopherol are known to inhibit NF-κB, a transcription factor associated with the overexpression of several 'proinflammatory' genes like tumor necrosis factor. The regulation of the inflammatory reactions connects several physiologically significant phosphorylation pathways, inclusive of those associated with the control of NF-κB activity. Additionally, since some protein tyrosine phosphatases are susceptible to oxidative inhibition, antioxidants can target these enzymes for beneficial effects in RA. The processes by which these complex cellular signaling pathways can be repaired may need a high dosage of antioxidant supplements, however, are difficult to identify because they are highly calibrated to maintain cellular function.

As a lipid-soluble antioxidant, vitamin E has anti-inflammatory properties. These properties may be associated with the inhibition of the metabolic pathways of arachidonic acid, which is involved in the generation of pro-inflammatory leukotrienes and prostaglandins [57]. It is also believed that α-tocopherol has bioactivities unrelated to its antioxidant action [58]. Pro-oxidants such as ascorbic acid and α-tocopherol also exist. Additionally, Hultqvist and associates hypothesized that RA might result from a diminished ability to create ROS. For instance, administering phytol to rats caused an increase in ROS production without causing arthritis to develop [59]. This perspective suggests that the protective effects seen in epidemiological research may be due to the chemicals' ability to increase ROS, such as ascorbic acid.

15.5 Conclusion and Future Perspectives

In the advanced phases of RA, new therapeutic modalities based on blocking of cytokine pathways are developing. Long-term use of these blocking medicines has revealed adverse effects despite their advantages. Exogenous (dietary) and endogenous (enzymatic and non-enzymatic) antioxidants work together to produce redox equilibrium in the body being a part of the antioxidant defense system. Downregulation of systemic oxidative stress is the main therapeutic advantage of antioxidant therapy. There is potential for using naturally occurring or plant-based products as many of their roots are used in the conventional treatment of RA. In many of the patients, low level of biomarkers like IL-6 and TNF-α indicates their anti-inflammatory activity, are effectively regulated regardless of the method used to prepare the plant extracts and because translating these effects to people are the ultimate goal, several factors have to be taken into consideration. For example, from the standpoint of nutrition and food, the efficacy evaluation of naturally occurring substances such as aqueous extracts of plant materials could offer insights for managing and preventing RA, despite its benefits, this cannot be said for the organic parts of the aqueous extract. In other words, the creation of new medications that will successfully identify and act on the disease-modifying agents depends on the evaluation of the different fractions of the extract as well as the discovery and isolation of pure substances. Given that research is intended to be carried out under extremely controlled conditions, it should come as no surprise that real-life variations relating to both human nature and the compound or product, such as cultivation conditions, genetics, methods of extraction, bioavailability, sex, age, comorbidities, and several other diversities, will necessitate more difficulty in problem-solving as we progress from the results of *in vitro* studies followed by animal models, and then to human trials. The proof is primarily based on outcomes in few patients or narrowly defined patient populations (based on sex, age, menopause status, comorbidities, etc.), which makes it difficult to make a convincing case for the value of naturally occurring substances as a complementary therapy. This is even though, in a few cases, as this chapter illustrates, the result of basic research has to be successfully transferred to the therapeutic environment. This shows a bottleneck in the research community's efforts to advance the investigation and use of environment friendly and sustainable pharmaceutical research. To develop an antioxidative therapeutic strategy for inflammatory-based disorders like RA, we propose that more clinical research with endogenous and exogenous antioxidants is required.

References

1. Bauerova, K., Kucharska, J., Ponist, S., Slovak, L., Svik, K., Jakus, V. *et al.*, Role of Endogenous Antioxidants in the Treatment of Experimental Arthritis, in: *Antioxidants*, 2019 Apr 6, Doi:10.5772/intechopen.85568.
2. Deligiannidou, G.E., Gougoula, V., Bezirtzoglou, E., Kontogiorgis, C., Constantinides, T.K., The Role of Natural Products in Rheumatoid Arthritis: Current Knowledge of Basic *In Vitro* and *In Vivo* Research. *Antioxidants*, 10, 4, 599, 2021 Apr 13. 2021.
3. Gulati, M., Farah, Z., Mouyis, M., Clinical features of rheumatoid arthritis. *Medicine*, 46, 4, 211–5, 2018 Apr 1.

4. Guo, Q., Wang, Y., Xu, D., Nossent, J., Pavlos, N.J., Xu, J., Rheumatoid arthritis: Pathological mechanisms and modern pharmacologic therapies, in: *Bone Research*, vol. 6, Sichuan University, China, 2018.
5. Burmester, G.R. and Pope, J.E., Novel treatment strategies in rheumatoid arthritis. *Lancet*, 389, 10086, 2338–48, 2017 Jun 10.
6. Mukhopadhyay, K., De, S., Kundu, S., Ghosh, P., Chatterjee, S., Chatterjee, M., Evaluation of levels of oxidative stress as a potential biomarker in patients with rheumatoid arthritis. *J. Fam. Med. Prim. Care*, 10, 5, 1981, 2021.
7. McCormack, B., A conceptual framework for person-centred practice with older people. *Int. J. Nurs. Pract.*, 9, 3, 202–9, 2003.
8. Hänninen, O., Kaartinen, K., Rauma, A.L., Nenonen, M., Törrönen, R., Häkkinen, S. *et al.*, Antioxidants in vegan diet and rheumatic disorders. *Toxicology*, 155, 1–3, 45–53, 2000 Nov 30.
9. Babior, B.M., Phagocytes and oxidative stress. *Am. J. Med.*, 109, 1, 33–44, 2000 Jul 28.
10. Weglicki, W.B., Mak, I.T., Dickens, B.F., Stafford, R.E., Komarov, A.M., Gibson, B. *et al.*, Neuropeptides, Free Radical Stress and Antioxidants in Models of Mg-Deficient Cardiomyopathy, in: *Magnesium: Curr. Status New Dev.*, pp. 169–78, 1997, https://doi.org/10.1007/978-94-009-0057-8_38.
11. Hitchon, C.A. and El-Gabalawy, H.S., Oxidation in rheumatoid arthritis. *Arthritis Res. Ther.*, 6, 6, 265–78, 2004 Dec 13.
12. Van de Loo, F.A.J., Bennink, M.B., Arntz, O.J., Smeets, R.L., Lubberts, E., Joosten, L.A.B. *et al.*, Deficiency of NADPH oxidase components p47phox and gp91phox caused granulomatous synovitis and increased connective tissue destruction in experimental arthritis models. *Am. J. Pathol.*, 163, 4, 1525–37, 2003 Oct 1.
13. Olofsson, P., Holmberg, J., Tordsson, J., Lu, S., Åkerström, B., Holmdahl, R., Positional identification of Ncf1 as a gene that regulates arthritis severity in rats. *Nat. Genet.*, 33, 1, 25–32, 2003 Jan 1.
14. van der Veen, R.C., Dietlin, T.A., Hofman, F.M., Pen, L., Segal, B.H., Holland, S.M., Superoxide prevents nitric oxide-mediated suppression of helper T lymphocytes: decreased autoimmune encephalomyelitis in nicotinamide adenine dinucleotide phosphate oxidase knockout mice. *J. Immunol.*, 164, 10, 5177–83, 2000 May 15.
15. Quinonez-Flores, C.M., Gonzalez-Chavez, S.A., Del Rio Najera, D., Pacheco-Tena, C., Oxidative Stress Relevance in the Pathogenesis of the Rheumatoid Arthritis: A Systematic Review. *Biomed. Res. Int.*, 2016, 6097417, 2016.
16. Dhar, S.K., Lynn, B.C., Daosukho, C., St. Clair, D.K., Identification of nucleophosmin as an NF-kappaB co-activator for the induction of the human SOD2 gene. *J. Biol. Chem.*, 279, 27, 28209–19, 2004 Jul 2.
17. Tan, M., Li, S., Swaroop, M., Guan, K., Oberley, L.W., Sun, Y., Transcriptional activation of the human glutathione peroxidase promoter by p53. *J. Biol. Chem.*, 274, 17, 12061–6, 1999 Apr 23.
18. Bierl, C., Voetsch, B., Jin, R.C., Handy, D.E., Loscalzo, J., Determinants of human plasma glutathione peroxidase (GPx-3) expression. *J. Biol. Chem.*, 279, 26, 26839–45, 2004 Jun 25.
19. Takimoto, E. and Kass, D.A., Role of Oxidative Stress in Cardiac Hypertrophy and Remodeling. *Hypertension*, 49, 2, 241–8, 2007 Feb 1.
20. Turoczi, T., Chang, V.W.H., Engelman, R.M., Maulik, N., Ho, Y.S., Das, D.K., Thioredoxin redox signaling in the ischemic heart: An insight with transgenic mice overexpressing Trx1. *J. Mol. Cell Cardiol.*, 35, 6, 695–704, 2003 Jun 1.
21. Das, D.K., Thioredoxin regulation of ischemic preconditioning. *Antioxid. Redox. Signal.*, 6, 2, 405–12, 2004.
22. Sowers, M.R. and Lachance, L.L., Vitamins and Arthritis: The Roles of Vitamins A, C, D and E. *Rheum. Dis. Clin. North Am.*, 25, 2, 315-332, 1999, doi:10.1016/s0889-857x(05)70070-3.

23. Edmonds, S.E., Winyard, P.G., Guo, R., Kidd, B., Merry, P., Langrish-Smith, A. *et al.*, Putative analgesic activity of repeated oral doses of vitamin E in the treatment of rheumatoid arthritis. Results of a prospective placebo controlled double blind trial. *Ann. Rheum. Dis.*, 56, 11, 649–55, 1997.
24. Canter, P.H., Wider, B., Ernst, E., The antioxidant vitamins A, C, E and selenium in the treatment of arthritis: a systematic review of randomized clinical trials. *Rheumatol. (Oxford)*, 46, 8, 1223–33, 2007 Aug.
25. Winrow, V.R., Winyard, P.G., Morris, C.J., Blake, D.R., Free radicals in inflammation: second messengers and mediators of tissue destruction. *Br. Med. Bull.*, 49, 3, 506–22, 1993.
26. Liu, Z., Ren, Z., Zhang, J., Chuang, C.C., Kandaswamy, E., Zhou, T. *et al.*, Role of ROS and Nutritional Antioxidants in Human Diseases. *Front. Physiol.*, 9, 477, 2018 May 17.
27. Dernini, S. and Berry, E.M., Mediterranean Diet: From a Healthy Diet to a Sustainable Dietary Pattern. *Front. Nutr.*, 2, 130610, 2015 May 7.
28. Shekhar, K.V., Pathak, M.M., Pisulkar, G., Diet and Lifestyle Impact on Rheumatoid Arthritis: A Comprehensive Review. *Cureus*, 15, 11, e48625, 2023 Nov 10, doi:10.7759/cureus.48625.
29. Sköldstam, L., Hagfors, L., Johansson, G., An experimental study of a Mediterranean diet intervention for patients with rheumatoid arthritis. *Ann. Rheum. Dis.*, 62, 3, 208, 2003 Mar 1.
30. McKellar, G., Morrison, E., McEntegart, A., Hampson, R., Tierney, A., Mackle, G. *et al.*, A pilot study of a Mediterranean-type diet intervention in female patients with rheumatoid arthritis living in areas of social deprivation in Glasgow. *Ann. Rheum. Dis.*, 66, 9, 1239–43, 2007 Sep.
31. Csaki, C., Keshishzadeh, N., Fischer, K., Shakibaei, M., Regulation of inflammation signalling by resveratrol in human chondrocytes *in vitro*. *Biochem. Pharmacol.*, 75, 3, 677–87, 2008 Feb 1.
32. Riveiro-Naveira, R.R., ValcáRcel-Ares, M.N., Almonte-Becerril, M., Vaamonde-García, C., Loureiro, J.S., Hermida-Carballo, L. *et al.*, Resveratrol lowers synovial hyperplasia, inflammatory markers and oxidative damage in an acute antigen-induced arthritis model. *Rheumatol. (Oxford)*, 55, 10, 1889–900, 2016 Oct.
33. Tong, W.W., Zhang, C., Hong, T., Liu, D.H., Wang, C., Li, J. *et al.*, Silibinin alleviates inflammation and induces apoptosis in human rheumatoid arthritis fibroblast-like synoviocytes and has a therapeutic effect on arthritis in rats. *Sci. Rep.*, 8, 1, 3241, 2018 Dec 1.
34. Anestopoulos, I., Sfakianos, A.P., Franco, R., Chlichlia, K., Panayiotidis, M.I., Kroll, D.J. *et al.*, A Novel Role of Silibinin as a Putative Epigenetic Modulator in Human Prostate Carcinoma. *Molecules*, 22, 1, 62, 2016 Jan 1.
35. Shoba, G., Joy, D., Joseph, T., Majeed, M., Rajendran, R., Srinivas, P.S.S.R., Influence of piperine on the pharmacokinetics of curcumin in animals and human volunteers. *Planta. Med.*, 64, 4, 353–6, 1998.
36. Green, A.E., Rowlands, R.S., Cooper, R.A., Maddocks, S.E., The effect of the flavonol morin on adhesion and aggregation of Streptococcus pyogenes. *FEMS Microbiol. Lett.*, 333, 1, 54–8, 2012 Aug 1.
37. Liu, S., Wu, N., Miao, J., Huang, Z., Li, X., Jia, P. *et al.*, Protective effect of morin on myocardial ischemia reperfusion injury in rats. *Int. J. Mol. Med.*, 42, 3, 1379–90, 2018 Sep 1.
38. Kandhare, A.D., Liu, Z., Mukherjee, A.A., Bodhankar, S.L., Therapeutic Potential of Morin in Ovalbumin-induced Allergic Asthma Via Modulation of SUMF2/IL-13 and BLT2/NF-kB Signaling Pathway. *Curr. Mol. Pharmacol.*, 12, 2, 122–38, 2019 Jan 3.
39. Sultana, F. and Rasool, M., A novel therapeutic approach targeting rheumatoid arthritis by combined administration of morin, a dietary flavanol and non-steroidal anti-inflammatory drug indomethacin with reference to pro-inflammatory cytokines, inflammatory enzymes, RANKL and transcription factors. *Chem. Biol. Interact.*, 230, 58–70, 2015 Mar 25.

40. Hsiao, H.B., Bin, W.J., Lin, W.C., Anti-arthritic and anti-inflammatory effects of (-)-Epicatechin-3-O-β-d-allopyranoside, a constituent of Davalliaformosana. *Phytomedicine*, 52, 12–22, 2019 Jan 1.
41. Liu, W., Zhang, Y., Zhu, W., Ma, C., Ruan, J., Long, H. *et al.*, Sinomenine Inhibits the Progression of Rheumatoid Arthritis by Regulating the Secretion of Inflammatory Cytokines and Monocyte/Macrophage Subsets. *Front. Immunol.*, 9, 2228, 2018 Sep 26.
42. Ma, C., Zhu, L., Wang, J., He, H., Chang, X., Gao, J. *et al.*, Anti-inflammatory effects of water extract of Taraxacummongolicum hand-Mazz on lipopolysaccharide-induced inflammation in acute lung injury by suppressing PI3K/Akt/mTOR signaling pathway. *J. Ethnopharmacol.*, 168, 349–55, 2015 Jul 20.
43. Jeon, D., Kim, S.J., Kim, H.S., Anti-inflammatory evaluation of the methanolic extract of Taraxacumofficinale in LPS-stimulated human umbilical vein endothelial cells. *BMC Complement. Altern. Med.*, 17, 1, 508, 2017 Nov 29.
44. Jeon, H.J., Kang, H.J., Jung, H.J., Kang, Y.S., Lim, C.J., Kim, Y.M. *et al.*, Anti-inflammatory activity of Taraxacumofficinale. *J. Ethnopharmacol.*, 115, 1, 82–8, 2008 Jan 4.
45. Chen, J., Wu, W., Zhang, M., Chen, C., Taraxasterol suppresses inflammation in IL-1β-induced rheumatoid arthritis fibroblast-like synoviocytes and rheumatoid arthritis progression in mice. *Int. Immunopharmacol.*, 70, 274–83, 2019 May 1.
46. Li, N., Gong, Z., Li, X., Ma, Q., Wu, M., Liu, D. *et al.*, Betulinic acid inhibits the migration and invasion of fibroblast-like synoviocytes from patients with rheumatoid arthritis. *Int. Immunopharmacol.*, 67, 186–93, 2019 Feb 1.
47. YU, W.G., SHEN, Y., WU, J.Z., GAO, Y.B., ZHANG, L.X., Madecassoside impedes invasion of rheumatoid fibroblast-like synoviocyte from adjuvant arthritis rats via inhibition of NF-κB-mediated matrix metalloproteinase-13 expression. *Chin. J. Nat. Med.*, 16, 5, 330–8, 2018.
48. Wang, Z., Zhuo, F., Chu, P., Yang, X., Zhao, G., Germacrone alleviates collagen-induced arthritis via regulating Th1/Th2 balance and NF-κB activation. *Biochem. Biophys. Res. Commun.*, 518, 3, 560–4, 2019 Oct 20.
49. Chen, G., Hao, B., Ju, D., Liu, M., Zhao, H., Du, Z. *et al.*, Pharmacokinetic and pharmacodynamic study of triptolide-loaded liposome hydrogel patch under microneedles on rats with collagen-induced arthritis. *Acta Pharm. Sin B*, 5, 6, 569–76, 2015 Nov 1.
50. Zhai, K., Duan, H., Luo, L., gen, C.W., kai, H.F., ling, S.L. *et al.*, Protective effects of paeonol on inflammatory response in IL-1β-induced human fibroblast-like synoviocytes and rheumatoid arthritis progression via modulating NF-κB pathway. *Inflammopharmacology*, 25, 5, 523–32, 2017 Oct 1.
51. Sita, G., Hrelia, P., Graziosi, A., Morroni, F., Sulforaphane from Cruciferous Vegetables: Recent Advances to Improve Glioblastoma Treatment. *Nutrients*, 10, 11, 1755, 2018 Nov 14.
52. Moon, S.J., Jhun, J., Ryu, J., Kwon, J.Y., Kim, S.Y., Jung, K.A. *et al.*, The anti-arthritis effect of sulforaphane, an activator of Nrf2, is associated with inhibition of both B cell differentiation and the production of inflammatory cytokines. *PLoS One*, 16, 8, ,e0256716 2021 Feb 1.
53. Healy, Z.R., Liu, H., Holtzclaw, W.D., Talalay, P., Inactivation of tautomerase activity of macrophage migration inhibitory factor by sulforaphane: a potential biomarker for anti-inflammatory intervention. *Cancer Epidemiol. Biomarkers Prev.*, 20, 7, 1516–23, 2011 Jul.
54. Souliotis, V.L., Vlachogiannis, N.I., Pappa, M., Argyriou, A., Ntouros, P.A., Sfikakis, P.P., DNA Damage Response and Oxidative Stress in Systemic Autoimmunity. *Int. J. Mol. Sci.*, 21, 1, 55, 2019 Dec 20.
55. Bashir, S., Harris, G., Denman, M.A., Blake, D.R., Winyard, P.G., Oxidative DNA damage and cellular sensitivity to oxidative stress in human autoimmune diseases. *Ann. Rheum. Dis.*, 52, 9, 659–66, 1993.

56. Eggleton, P., Haigh, R., Winyard, P.G., Consequence of neo-antigenicity of the "altered self". *Rheumatol. (Oxford)*, 47, 5, 567–71, 2008 May.
57. Reddanna, P., Krishna Rao, M., Channa Reddy, C., Inhibition of 5-lipoxygenase by vitamin E. *FEBS Lett.*, 193, 1, 39–43, 1985 Nov 25.
58. Rhee, S.J., Jeong, Y.C., Choi, J.H., Effects of Vitamin E on Phospholipase A2 Activity and Oxidative Damage to the Liver in Streptozotocin-Induced Diabetic Rats. *Ann. Nutr. Metab.*, 49, 6, 392–6, 2005 Dec 1.
59. Hultqvist, M., Olofssona, P., Gelderman, K.A., Holmbergb, J., Holmdahl, R., A new arthritis therapy with oxidative burst inducers. *PLoS Med.*, 3, 9, 1625–36, 2006.

16

Antioxidants in Skin Disorders

Satyender Kumar[1], Dil Prasad Subba[1], Seema[2], Sayed Mohammed Firdous[3], Oluwatoyin A. Odeku[4], Sandeep Kumar[5] and Rakesh K. Sindhu[1]*

[1]School of Pharmacy, Sharda University, Greater Noida, Uttar Pradesh, India
[2]HIMT College of Pharmacy, Greater Noida, Uttar Pradesh, India
[3]Department of Pharmacology, Calcutta Institute of Pharmaceutical Technology & AHS, Uluberia, Howrah, West Bengal, India
[4]Department of Pharmaceutics & Industrial Pharmacy, Faculty of Pharmacy, University of Ibadan, Ibadan, Nigeria
[5]Department of Regulatory Affairs, National Institute of Pharmaceutical Education and Research-Hyderabad, Balanagar, Hyderabad, Telangana, India

Abstract

Skin diseases provide serious risks to the health and wellbeing of the general population, ranging from dermatitis to illnesses linked to aging. Exploiting the potential of natural and plant-based antioxidants as therapeutic agents to treat a range of skin conditions has gained popularity in recent years. The wide range of naturally occurring and plant-derived antioxidants, their modes of action, and their effectiveness in reducing oxidative stress-related skin conditions are all examined in this comprehensive study. The findings gathered from numerous studies, including *in vitro*, *in vivo*, and clinical trials, provide a comprehensive grasp of the function of natural antioxidants in enhancing skin health. A detailed discussion has been held on the sources of natural antioxidants, with a focus on the abundance of phytochemicals present in fruits, vegetables, herbs, and other botanical extracts. The antioxidant potential of well-known compounds such as polyphenols, flavonoids, carotenoids, and vitamins is discussed, along with their mechanisms of action in combating oxidative stress in skin cells. A substantial portion of the chapter is dedicated to examining the impact of natural antioxidants on specific skin disorders. The role of these compounds in alleviating symptoms associated with dermatitis, psoriasis, acne, and photoaging is meticulously explored, drawing insights from both preclinical and clinical studies. Furthermore, the chapter addresses the challenges associated with the use of natural antioxidants, including issues of stability, bioavailability, and formulation. Strategies to enhance the efficacy of these antioxidants through innovative delivery systems and combination therapies are explored, highlighting recent advancements in the field. The chapter also discusses the importance of sustainable and ethically sourced natural ingredients in the cosmetic and skincare industry.

It sheds light on emerging trends in plant-based skincare formulations, emphasizing the shift towards eco-friendly and cruelty-free products. In conclusion, this chapter provides a valuable synthesis of the current knowledge on natural and plant-based antioxidants in the context of skin

**Corresponding author*: drrakeshksindhu@gmail.com

Rakesh K. Sindhu, Inderbir Singh and M. Arockia Babu (eds.) Antioxidants: Nature's Defense Against Disease, (551–572)

disorders. It not only underscores the therapeutic potential of these compounds but also identifies avenues for future research and development in the pursuit of safe and effective natural skincare solutions.

Keywords: Natural antioxidants, skin disorders, oxidative stress, polyphenols, flavonoids, carotenoids, vitamins, skincare formulations

16.1 Introduction

Oxidative stress is a condition where the body's tissues are unable to counteract reactive oxygen species (ROS), and it is strongly associated with aging and various health conditions. In the context of skin health, oxidative stress is impacted by environmental factors such as air pollution, water contamination, and exposure to light-emitting diodes and electromagnetic frequencies [1–3]. It has been found that oxidative stress plays a role in skin diseases like atopic dermatitis (AD) and chronic urticaria (CU), leading to protein oxidation and lipid peroxidation [1]. Additionally, oxidative stress exacerbates skin pigmentation and aging, resulting in uneven skin tone, pigmentary disorders, and wrinkles [4].

Oxidative stress in the scalp can have deleterious consequences for hair growth and retention [2]. Reactive oxygen species (ROS) generated in the skin contribute to ultra-weak photon emission and oxidation of biomolecules, leading to oxidative stress [1]. Melanocytes, which produce melanin, are particularly susceptible to oxidative stress and play a role in melanoma development. This stress triggers a cascade of reactions in the skin, leading to inflammation, collagen degradation, and impaired cellular function. Over time, chronic oxidative stress contributes to various skin disorders, including premature aging, acne, and conditions like eczema and psoriasis. Managing oxidative stress is crucial for maintaining skin integrity and preventing or mitigating these skin issues [5].

The body's organs and functions are negatively impacted by free radicals [Figure 16.1]. Hepatotoxicity, nephrotoxicity, and liver damage are among the problems they might induce

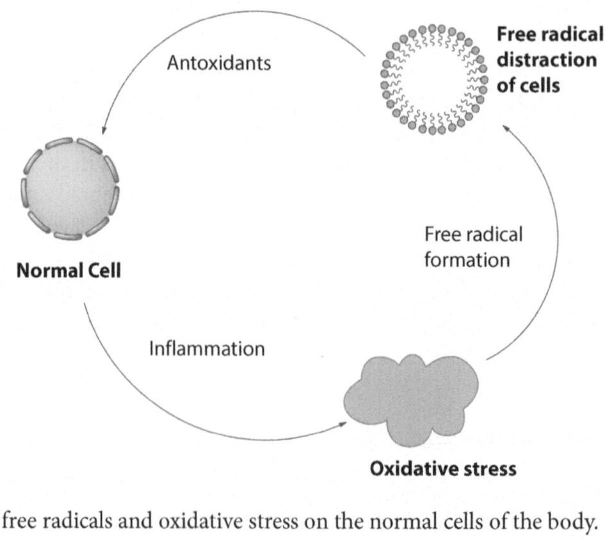

Figure 16.1 Effect of free radicals and oxidative stress on the normal cells of the body.

due to their extreme cellular damage and oxidative stress [6]. Furthermore, inflammatory illnesses, metabolic problems, and neurological deficits are pathophysiologically related to free radicals [7]. Sperm motility, DNA integrity, and protein synthesis may all be affected, which lowers fertility [8]. A number of chronic illnesses, including as diabetes, cancer, heart disease, and lung disorders, are linked to persistent inflammation, which is a condition that is also influenced by free radicals [9]. The impairment of pulmonary surfactant function resulting from oxidative damage causes elevated surface tension and respiratory distress syndrome [10]. In general, a broad variety of physiological processes are negatively impacted by free radicals, which also play a role in the onset and progression of several illnesses.

16.2 Importance of Antioxidants in Combating Free Radicals

Through a variety of processes, antioxidants demolish free radicals. Antioxidants have the ability to prevent the generation of free radicals by lowering their production [11]. Furthermore, as radical scavengers, antioxidants have the ability to absorb free radicals [12]. Thirdly, transition metal ions are complexable by antioxidants, and these ions could trigger prooxidative reactions [12]. Antioxidants also control enzymes linked to reactive oxygen species (ROS), which shields cells from harm caused by free radicals. Antioxidants may also seize products of free-radical reactions, which stops radicals from forming in live cells and preventing things like apoptosis from developing. Antioxidants work by preventing or limiting oxidative stress, which may result in a number of illnesses, including diabetes, immune system disorders, cancer, cardiovascular disease, and neurological diseases [13].

Plant polyphenols, such as curcumin and resveratrol, have therapeutic potential in combating oxidative stress in the skin [14, 15]. The use of antioxidants such as vitamin C, resveratrol, and ferulic acid has been explored as a strategy to mitigate oxidative stress-induced skin pigmentation. Overall, understanding oxidative stress and its impact on skin health is crucial for developing therapeutic methods and preventive strategies in dermatological diseases.

16.3 Types of Antioxidants

Antioxidants are categorized as follows:

1. Natural antioxidants,
2. Synthetic antioxidants.

16.3.1 Natural Antioxidants

such as vitamin E, flavonoids, and phenolic compounds, are obtained from plants and fruits [16, 17]. They work by preventing the formation of reactive oxygen species (ROS) and free radical oxidation processes in cells and tissues [18]. Natural antioxidants can scavenge harmful free radicals and reduce oxidative damage associated with various diseases, including cancer, cardiovascular diseases, and aging [19].

16.3.2 Synthetic Antioxidants

Antioxidants that are synthesized, such as butylated hydroxytoluene, stop oxygen radicals from being produced as a byproduct of electron transport inside the mitochondria. In order to prevent the production of reactive oxygen species (ROS), antioxidants work by blocking certain enzymes or chelating trace elements. Antioxidants, whether natural or synthetic, have been shown to have the ability to prevent illness and to assist prevent non-communicable diseases including aging, inflammatory processes, and neurological disorders.

Ascorbic acid, polyphenols, thiols, butylated hydroxytoluene, butylated hydroxyanisole, propyl gallate, and tert-butylhydroquinone are a few notable antioxidants [20, 22]. Resveratrol, lycopene, lutein, zeaxanthin, curcumin, and astaxanthin are other antioxidants that have been reported [21]. Important antioxidants include beta-carotene, lutein, vitamins A, C, E, and lycopene [22]. Fruits and vegetables include essential plant components called polyphenols, which have antioxidant properties [23]. Antioxidants of great importance include vitamins C and E. One major source of health-promoting antioxidants is plants, especially medicinal herbs.

16.4 Antioxidants Relevance to Skin Health

Antioxidants have various effects on skin care. They can improve skin health and youthfulness by combating oxidative stress and reducing the accumulation of free radicals [24, 25]. Antioxidants can also enhance the skin barrier function, increase hydration, and reduce inflammation and itching in skin conditions such as atopic dermatitis [26]. Additionally, antioxidants from marine sources have been found to be effective in treating skin disorders and improving skin health [27]. Incorporating antioxidants into skincare products can activate cells, scavenge free radicals, tighten skin, reduce wrinkles, promote growth of skin cells, and improve skin elasticity [28]. Overall, antioxidants play a crucial role in protecting the skin from damage, promoting its health and appearance, and preventing skin disorders.

Because antioxidants may neutralize highly reactive chemicals like free radicals, they are essential for skin protection against them. DNA, lipids, and proteins in cells can undergo damage due to free radicals such reactive oxygen species (ROS) and reactive nitrogen species (RNS).

Antioxidants are significant in this situation for the following reasons:

16.4.1 Neutralization of Free Radicals

Antioxidants stabilize unstable free radicals by giving them an electron, so halting more damage to DNA, proteins, and cells. Free radicals are neutralized by antioxidants by the donation of one electron, which stabilizes the free radical and lessens its destructive power [29]. Here, the antioxidant interrupts the chain reaction that the free radical started, a process known as a "chain-breaking mechanism" [12]. Reactive nitrogen species initiators and reactive oxygen species (ROS) may also be eliminated by antioxidants via the quenching of chain-initiating catalysts [14, 15]. Highly reactive molecules known as free radicals may develop in the body as a result of a number of different events, including chemical reactions, radiation exposure, and metabolic activities [30]. These free radicals have the ability to harm cells and play a role in the development of illnesses including diabetes, cancer, and cardiovascular disease. Antioxidants help shield cells from oxidative damage and preserve general health by scavenging free radicals. Eating a natural, well-balanced diet full of antioxidant-rich fruits and vegetables may help boost the body's defenses against disease.

Using skin care products that neutralize free radicals can provide several benefits for the skin. These products can help reduce inflammation and oxidative stress, which are known to contribute to skin diseases like atopic dermatitis (AD) [24, 31]. They can also enhance the skin barrier function, improving resilience against environmental factors and oxidative stress [32]. Additionally, these products can promote the growth of skin cells, accelerate cell metabolism, and enhance the activity of skin cells [33]. By scavenging free radicals and preventing their damage, these products can help improve skin elasticity, reduce wrinkles, and retard the aging processes of the skin [34]. It is important to note that the safety, associations, and mechanisms of action of these antioxidant skin care products should be considered for effective and safe intervention in oxidative processes. Some of the most effective ingredients for neutralizing free radicals in skin care products include natural compounds with antioxidant effects such as naringenin, nordihydroguaiaretic acid (NDGA), kaempferol, resveratrol, pomegranate, green tea, mango, amla, bearberry, ellagic acid, tetrahydrocurcuminoid, *Rhodiola rosea*, kakadu plum, apple peel extract, banana peel extract, orange peel extract, folium rosmarini officinalis extract, rose hip extract, and peony seed extract [27, 33, 35, 36]. These ingredients have been shown to have high antioxidant activities and can effectively scavenge free radicals, protecting the skin from oxidative stress and reducing the aging effects [37].

16.4.2 Cellular Protection

By neutralizing free radicals, antioxidants help protect skin cells from oxidative stress, preserving their structure and function. Antioxidants protect skin cells through various cellular mechanisms. Nuclear factor-erythroid 2-related factor-2 (NRF2), a master switch for antioxidant signaling, is activated in one manner [38]. Reactive oxygen and nitrogen species (ROS/RNS), which may harm cells and cell compartments, are neutralized by another method [39]. Additionally, antioxidants may mitigate UVB-induced damage by keeping human keratinocytes from going through apoptosis. These protective actions are observed with antioxidant compounds of natural origin, such as polyphenols found in virgin olive oil, and other food compounds like melatonin, creatine, and hydroxytyrosol [40]. Additionally, antioxidants can enhance the natural endogenous defenses of the skin, creating a reservoir for sustained protection against external environmental stress [41].

16.4.3 Reduction of Inflammation

Antioxidants in skin care products have various anti-inflammatory mechanisms. They can reduce the expression of inflammatory cytokines in keratinocytes and increase barrier-related molecules [24]. It has been shown that antioxidants, including xanthones, may alter inflammatory signaling pathways and pro- and anti-inflammatory cytokines [42]. Antioxidants can neutralize reactive oxygen and nitrogen species (ROS/RNS), which may prevent damage to cell compartments and tissues. ROS/RNS are important factors in aging and carcinogenesis [43]. In illustration, it has been shown that CoQ10 inhibits the activation of matrix-eroding enzymes and suppresses the inflammatory response in dermal fibroblasts [39]. It has been shown that the combination of carotenoids and CoQ10 improves the suppression of inflammation, offering improved defense against inflammation and sun-induced premature aging [44]. Human epidermal keratinocytes have been shown to produce less inflammatory cytokines and more molecules associated to barriers when exposed to antioxidants, such as those present in a natural extract made of curly kale, apple, and green tea [24]. Antioxidants like pyrrolidine dithiocarbamate (PDTC) and dimethylthiourea (DMTU) have been found to decrease the late rise in tissue vascular permeability and reduce neutrophil influx, possibly by altering neutrophil migration [45].

16.4.4 Collagen Preservation

Free radicals can break down collagen, a crucial protein for skin elasticity (Figure 16.2). Antioxidants help preserve collagen, maintaining skin firmness and reducing the signs of aging. Antioxidants in skin care help preserve collagen through various mechanisms. Hydrolyzed collagen (HC) is a popular ingredient with antioxidant properties due to the presence of hydrophobic amino acids [46]. Polyphenols, such as those found in collagen formulations, have been shown to have an amino oxidase-like activity that could contribute to collagen preservation [47]. Additionally, an antioxidant skin-care composition containing tuna collagen hydrolysate and grape seed extract has been developed to resist oxidation and supplement moisture loss [46].

In vitiligo lesions, collagen production is upregulated and collagen degeneration is attenuated, possibly due to the upregulation of the NRF2 signaling pathway, which is involved in defense against oxidative stress [48]. Antioxidant agents, particularly those of natural origin, have shown potential in maintaining the physiological balance of the skin and protecting against degradation caused by UV radiation and pollution [49]. The most effective antioxidants for collagen preservation in skin care include hydrolyzed collagen (HC) [46] [48]. Tuna collagen hydrolysate and grape seed extract [24]. Lysyl oxidases (LOX enzymes) and polyphenols [47]. and collagen peptides from Spanish mackerel skins [51]. HC, with its low molecular weight and hydrophobic amino acids, has been shown to improve skin properties and protect against UV melasma [51]. Tuna collagen hydrolysate and grape seed extract work together to resist oxidation, eliminate free radicals, and replenish moisture

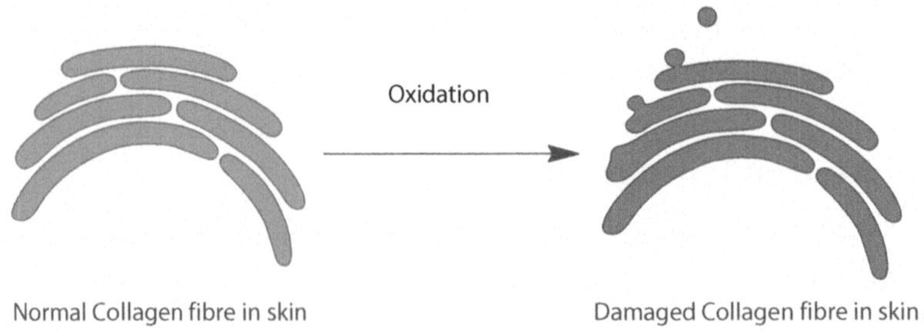

Figure 16.2 Effect of oxidation on collagen fibers in skin.

loss. LOX enzymes and polyphenols have an amino oxidase-like activity that promotes and maintains cross-linked collagen in the skin. Collagen peptides from Spanish mackerel skins have strong radical scavenging activities and can effectively reduce $Fe3+$ to $Fe2+$ and inhibit lipid peroxidation. These antioxidants offer potential for collagen preservation and skin care applications.

16.4.5 Defense Against Environmental Stressors

Exposure to UV radiation, pollution, and other environmental stressors increases free radical production. Vitamin C, resveratrol, and ferulic acid are well-known antioxidants that have been used in the treatment of hyperpigmentation, a common skin condition associated with oxidative stress. Furthermore, natural antioxidants have shown promise in preserving the skin's physiological equilibrium and guarding against pollution and UV ray damage. By acting as a barrier, antioxidants lessen the damaging effects of these outside variables on the skin. Antioxidants use a complex network of enzymatic and non-enzymatic systems to protect the body from environmental stresses. Reactive oxygen species (ROS) are reduced and scavenged by enzymatic antioxidants such as superoxide dismutase (SOD), catalase (CAT), peroxidases (POX), glutathione peroxidases (GPX), glutathione reductases (GR), monodehydroascorbate reductases (MDHAR), and dehydroascorbate reductases (DHAR) [52]. In addition to vitamins like ascorbate (AsA) and tocopherols, low molecular weight glutathione (GSH), polyphenols like flavonoids, and carotenoids, non-enzymatic antioxidants are also helpful in the quenching of dangerous ROS [53]. These antioxidants function alone or in combination to prevent the synthesis and anticipation of reactive oxygen species (ROS) and are distributed across different cellular compartments [54]. The transformation of ROS into less dangerous compounds is catalyzed by the enzymatic defense system, namely by glutathione reductase (GR), catalase (CAT), and superoxide dismutase (SOD). Overall, the antioxidant system in plants is crucial for maintaining cellular redox equilibrium and protecting against oxidative damage caused by environmental stressors.

16.4.6 Prevention of Premature Aging

By protecting against oxidative stress, antioxidants contribute to the prevention of premature aging, reducing the appearance of wrinkles and fine lines. Antioxidants have been shown to have beneficial effects on premature aging in skin care. They can protect the skin from damage caused by oxidation and prevent premature aging [55]. Antioxidants, such as polyphenols, have been found to have antioxidant, anticarcinogenic, and neuroprotective effects when taken into the body in appropriate amounts [39]. Reactive oxygen and nitrogen species (ROS/RNS), which contribute to aging processes, can be neutralized by antioxidants, thereby limiting damage to the skin [56]. Combining natural antioxidants in skin care products can have a synergistic effect, resulting in greater overall benefits for the skin [57]. In a clinical trial, the antioxidant SOD was found to significantly increase total antioxidant status (TAS), decrease transepidermal water loss (TEWL), and improve skin dryness in premature skin aging [58]. These findings suggest that antioxidants can play an important role in preventing and reducing premature aging in the skin.

16.4.7 Support for Skin Repair

Antioxidants aid in the repair of damaged skin cells, promoting the regeneration of healthy skin tissue. Antioxidants support skin repair through various mechanisms. Firstly, antioxidants help neutralize reactive oxygen species (ROS) and reactive molecule species (RMS) that can cause oxidative damage to the skin [38, 59].This helps in preventing skin aging, inflammation, and carcinogenesis [39]. Secondly, antioxidants can upregulate the expression of nuclear factor-erythroid 2-related factor-2 (NRF2), which is a master switch for antioxidant signaling, thereby enhancing the skin's ability to combat oxidative stress [60]. Thirdly, antioxidants can minimize the damage caused by external environmental irritants by restoring the oxidant and antioxidant balance in the skin [49]. Additionally, antioxidants can protect the skin's barrier function by preventing lipid peroxidation and maintaining the quality of the stratum corneum. Overall, antioxidants play a crucial role in supporting skin repair by reducing oxidative damage, enhancing antioxidant signaling, and maintaining the skin's physiological balance.

Certain antioxidants have been shown to be useful in promoting skin restoration. According to one research, antioxidants that target mitochondria reduced reactive oxygen species (ROS) and accelerated skin fibroblasts' ability to mend wounds [59]. With encouraging results from *in vitro*, *in vivo*, and clinical investigations, different research emphasized the function of naturally occurring antioxidant-incorporated biomaterials in fostering skin wound healing and tissue regeneration [61]. Furthermore, research revealed that the combination of partly lipophilic rutin and hydrophilic ascorbic acid provided a more robust defense against UV-induced alterations to membranes and the endocannabinoid system in keratinocytes and fibroblasts [62]. Furthermore, glutathione, a cellular antioxidant, was found to have a cytoprotective function in keratinocytes and played a role in maintaining skin integrity and efficient wound healing [63].

16.4.8 Overall Skin Health

Regular intake of antioxidants through diet and topical application in skincare routines contributes to maintaining overall skin health and resilience. Incorporating a variety of antioxidants, both through a balanced diet and skincare products, is essential for a comprehensive approach to combatting free radicals and promoting skin well-being are discussed following under:

Vitamin C (Ascorbic Acid): Vitamin C also acts as a redox buffer, reducing reactive oxygen species and neutralizing them. It is involved in various physiological processes, including cell division, growth, and immunological activation. Topical application of vitamin C can counteract oxidative stress induced by UVA, improving skin gloss, color, elasticity, and reducing wrinkle area. Different forms of vitamin C, such as ascorbic acid 2-glucoside (AA2G), have been developed to enhance stability and skin permeability. Vitamin C, a water-soluble antioxidant that promotes collagen formation and shields the skin from oxidative damage, is present in citrus fruits, berries, and vegetables.

Vitamin E (Tocopherols and Tocotrienols): Vitamin E is a group of eight different molecules, including tocopherols and tocotrienols, each with distinct biological activities and

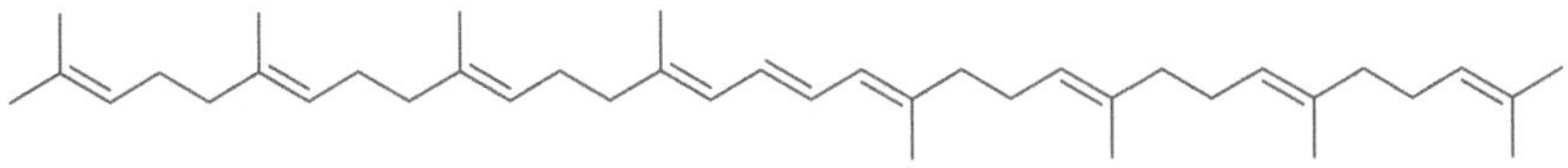

Phytoene (3 c.d.b.)

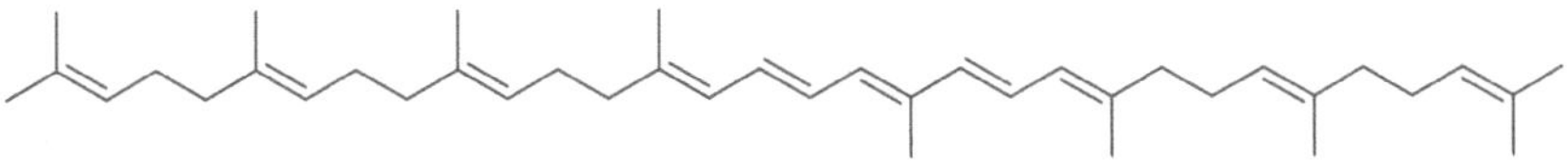

Phytofluene (5 c.d.b.)

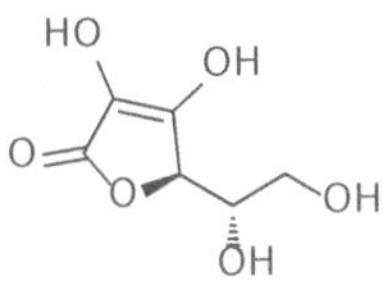

Ascorbic acid

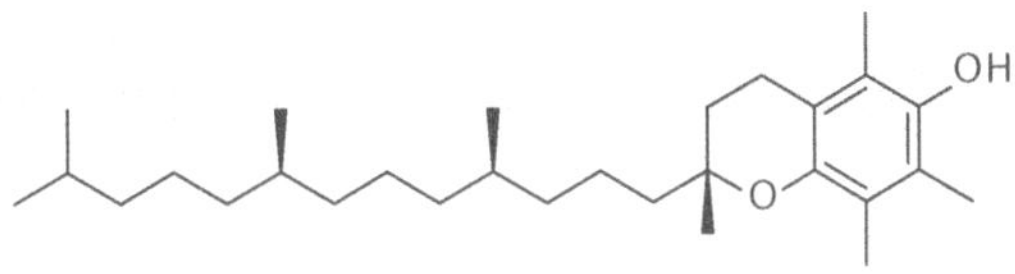

Vitamin E

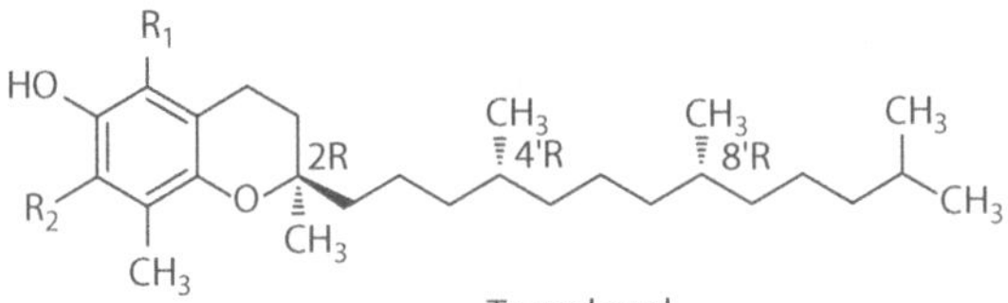

Tocopherol

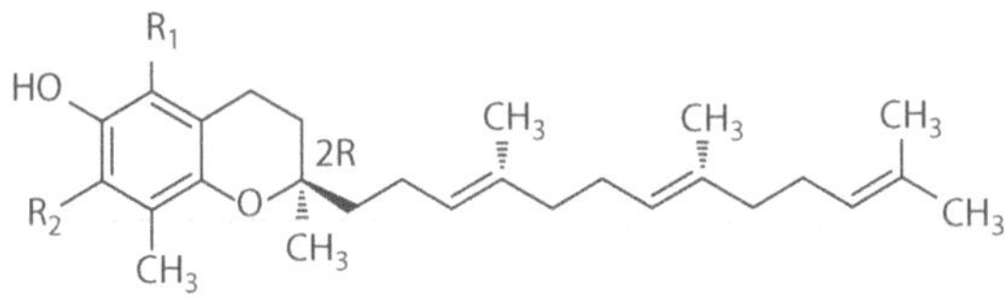

Tocotrienol

R1 = CH3 (alpha), CH3 (beta), H (gamma), H (delta)

R2 = CH3 (alpha), H (beta), CH3 (gamma), H (delta)

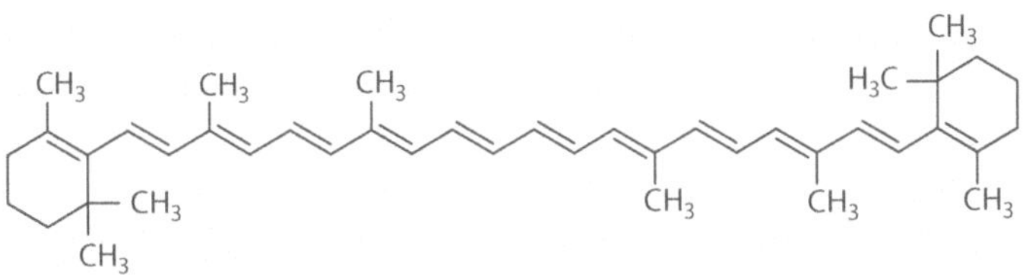

Beta-carotene

functions. While alpha-tocopherol has been extensively studied, other forms of vitamin E, such as tocotrienols, have been shown to possess unique properties, including neuroprotective, anticancer, and cholesterol-lowering effects. Vitamin E, a fat-soluble antioxidant that shields cell membranes from oxidative stress and supports skin health, is found in nuts, seeds, and oils.

Beta-Carotene: Found in colorful fruits and vegetables like carrots and sweet potatoes, beta-carotene is a precursor to vitamin A, playing a role in skin repair and maintaining a healthy complexion.

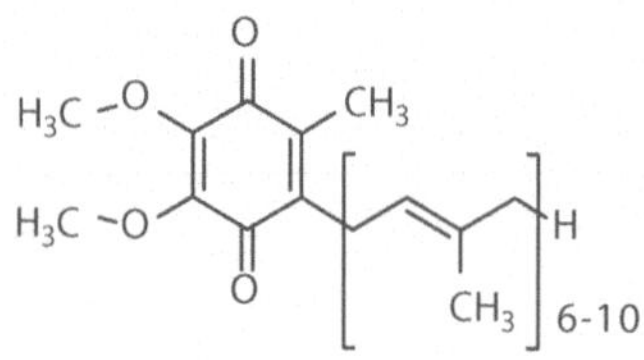

Coenzyme Q10

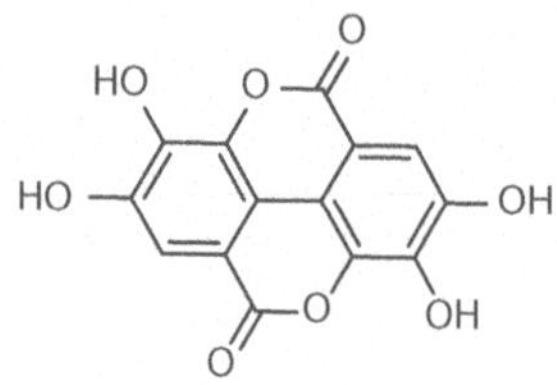

Polyphenols

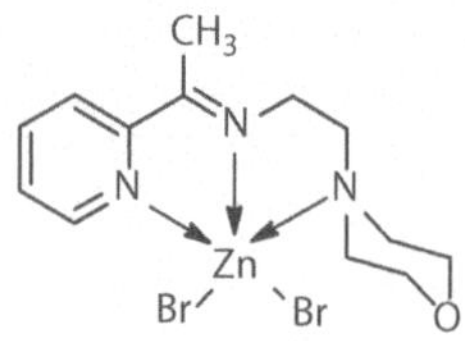

Zn Containing products

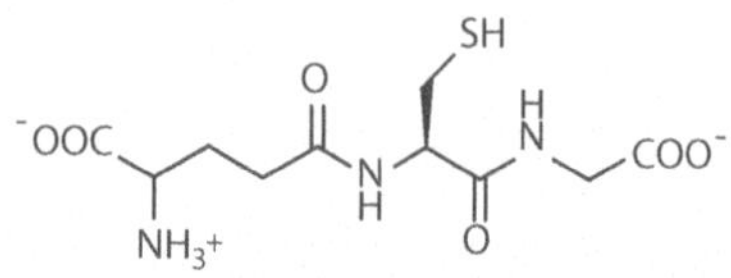

Glutathione

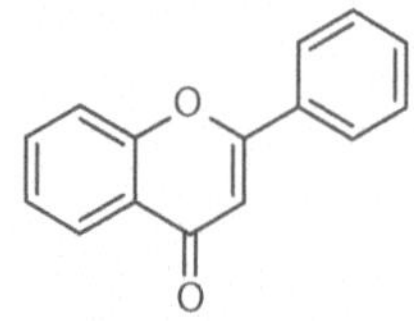

Flavonoids

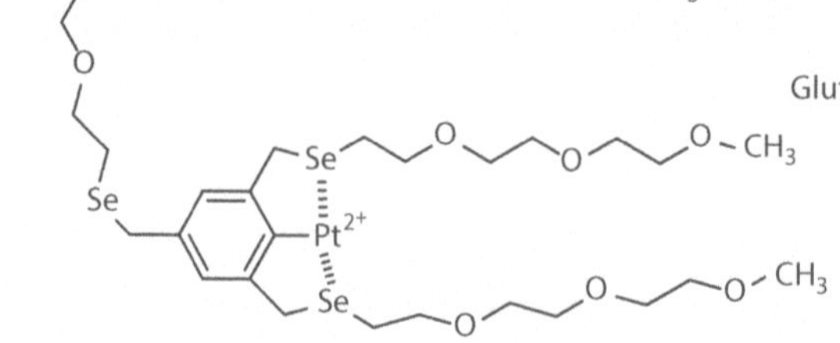

Selenium Products

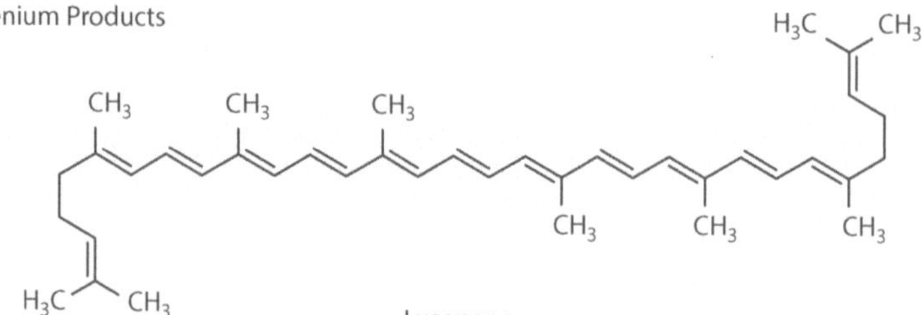

Lycopene

Coenzyme Q10 (Ubiquinone): Naturally occurring in the body, CoQ10 is an antioxidant involved in energy production. It helps protect the skin from oxidative damage and supports cellular function.

Polyphenols: Abundant in green tea, red wine, and dark chocolate, polyphenols have anti-inflammatory and antioxidant properties, contributing to skin health.

Selenium products: Found in nuts, seeds, and seafood, selenium is a trace element that supports the activity of antioxidant enzymes, helping protect the skin from oxidative stress.

Zinc: An essential mineral found in various foods, zinc supports the body's production of antioxidant enzymes, contributing to skin repair and maintenance.

Flavonoids: Present in fruits, vegetables, and teas, flavonoids have antioxidant and anti-inflammatory properties, aiding in skin protection and overall well-being.

Glutathione: A powerful antioxidant produced in the body, glutathione helps neutralize free radicals and supports the immune system.

Lycopene: Found in tomatoes, watermelon, and pink grapefruit, lycopene is a carotenoid with antioxidant properties that may help protect the skin from UV damage. Incorporating a variety of foods rich in these antioxidants into your diet and using skincare products containing them can contribute to a well-rounded approach to skin health.

16.5 Clinical Studies and Evidences

Numerous clinical studies provide evidence supporting the role of antioxidants in managing skin disorders and promoting skin health. Clinical studies and evidence suggest that antioxidants, such as vitamins C and E, have potential therapeutic benefits in various conditions. Vitamin C has been shown to have antioxidant effects at lower concentrations and has been effective in preventing respiratory diseases, improving immune response, and protecting the skin [64]. Vitamins C and E can reduce fasting blood glucose and HbA1c levels, although the clinical significance of these effects is limited [65]. Antioxidants, including carotenoids, anthocyanins, and vitamins, have shown promise in the treatment and prevention of ophthalmic diseases, such as age-related macular disease, glaucoma, and cataract [66]. In the context of male infertility, an antioxidant blend has been found to improve sperm count, semen parameters, and reduce DNA fragmentation [67]. In patients with autism spectrum disorder (ASD), antioxidant interventions, particularly N-acetylcysteine (NAC), have shown potential in improving symptoms such as irritability, communication disturbance, and hyperactivity [68]. However, more studies are needed to generate evidence-based guidelines and to further investigate the effectiveness of antioxidants in various conditions.

16.5.1 Green Tea Polyphenols in Acne

Research indicates that the polyphenols in green tea have anti-inflammatory and antioxidant properties, potentially benefiting individuals with acne by reducing inflammation and sebum production.

The polyphenols included in green tea, namely epigallocatechin-3-gallate (EGCG), have shown promise in the management of acne vulgaris. While ingesting green tea extract (GTE) orally has no benefits, topical application of GTE has been shown to significantly reduce the amount of inflammatory acne lesions [69]. Polyphenols, such EGCG, have anti-inflammatory and antibacterial qualities that make them useful in the treatment of inflammatory acne [70]. Nevertheless, there is little and inconsistent data supporting the effectiveness of green tea polyphenols in the treatment of acne [71]. For further evaluation of the efficacy of green tea polyphenols in the treatment of acne vulgaris, further excellent clinical studies are required [72].

16.5.2 Coenzyme Q10 in Wrinkle Reduction

Clinical trials suggest that CoQ10, when applied topically, may reduce the appearance of wrinkles and improve skin firmness, demonstrating its potential in addressing signs of aging. Coenzyme Q10 (CoQ10) has been studied for its effects on wrinkle reduction. CoQ10 supplementation has been shown in clinical studies to enhance mitochondrial activity and provide antioxidant protection to organs and tissues impacted by a range of pathophysiological diseases, including aging processes of the skin [73]. Research has also assessed the use of protransfersomal emulgel and nanostructured lipid carrier (NLC) formulations to improve CoQ10 stability and skin administration. These formulations have shown improved collagen density, fibroblast cell count, and anti-aging effectiveness in skin-aged mice caused by

UV radiation, suggesting that they may be able to reverse the aging process of the skin [74]. A comprehensive study also emphasized the use of CoQ10, plasma, and nutritional therapy in ocular aesthetic skin treatments like blepharoplasty. The analysis came to the conclusion that using CoQ10 in combination with plasma procedures and encouraging a healthy diet may improve skin look and slow down the aging process [75].

16.5.3 Selenium Products for Skin Care

Selenium has several skin care benefits. It plays a major role in antioxidant defense and helps protect against oxidative stress induced by environmental factors such as pollutants and UV radiation [76]. Supplementing with selenium has been shown to have a significant effect in preserving a decreased cellular environment, which prevents skin aging, and in the replicative life span of keratinocytes, the primary cells of the epidermis [77]. Because of their strong apoptotic properties, selenium nanoparticles have been discovered to be effective in treating psoriasis, a chronic inflammatory skin condition marked by hyperproliferation of keratinocytes [78]. In addition to its ability to affect immunological response, selenium also inhibits TNF-α levels in psoriasis patients [79]. Furthermore, dietary selenium enhances skin surface characteristics, guards against UV-induced damage and cancer, and is necessary for keratinocyte survival and proliferation [80]. Electrospun silk scaffolds may enhance their antibacterial activity while retaining the advantageous skin-healing qualities of silk by blending selenium nanoparticles into the fabric.

Products containing selenium have shown efficacy in treating a range of skin disorders. According to Bin Xie *et al.*, antiallergic selenium nanoparticles (LET-SeNPs) effectively suppressed mast cell activation and stimulated the Nrf2-Keap1 signaling pathway [81]. Selenium-rich yeast peptide fraction (SeP) was reported by Heng Guo *et al.* to have a possible therapeutic impact against psoriasis by suppressing the MAPK and NF-κB signaling pathways, which in turn reduced skin inflammation similar to psoriasis [82]. Selenium nanoparticles (SeNPs) were synthesized by Vinod Gangadevi *et al.* and shown strong apoptosis along with decreased inflammation and proliferation driven pathways, which made them a prime choice for psoriasis treatment [78]. An observational research, by Gabriela Turcu *et al.* found that a shampoo containing selenium disulfide (SeS2) was effective and well-tolerated in the treatment of dandruff and seborrheic dermatitis [83]. These trials demonstrate how well selenium-containing cosmetics work in treating psoriasis, seborrheic dermatitis, and allergic dermatitis.

16.5.4 Antioxidant Supplementation in Dermatitis

Some studies indicate that antioxidant supplementation, including vitamins C and E, may have a positive impact on inflammatory skin conditions like dermatitis, potentially reducing symptoms. Antioxidant supplementation has shown potential therapeutic value in the management of dermatitis. The benefits of natural antioxidants on the symptoms of dermatitis have been investigated. These include vitamin D, vitamin E, pyridoxine (vitamin B6), vitamin C, carotenoids, melatonin, iron, and zinc. Research on the relationship between vitamin D, E, C, and carotenoids levels and the severity of dermatitis has produced contradictory findings. Antioxidants, however, were linked to a substantial decrease in the severity of the condition in dermatitis patients, according to a systematic review and meta-analysis [84]. This was particularly true when oral vitamin D and topical vitamin B12 were added as supplements. Furthermore, an emollient medical device enhanced with antioxidant components decreased pruritus and enhanced skin condition in people with dermatitis, according to an open-label clinical investigation [84]. It has been shown that vitamin E and selenium antioxidant supplements hasten the clinical recovery of external parasitism in dogs, including dermatitis [85]. All things considered, antioxidant supplementation could be a secure and useful adjunctive therapy for dermatitis; nonetheless, further investigation is required to establish the ideal doses and course of care.

Studies in clinical settings have examined the effectiveness of antioxidant supplementation in treating dermatitis. Antioxidants were linked to statistically significant decreases in the severity of the illness in individuals with atopic dermatitis (AD), according to a comprehensive review and meta-analysis [84]. This was particularly true when oral vitamin D and topical vitamin B12 were added as supplements. In a different clinical study, people with mild-to-moderate dermatitis were assessed for the safety and therapeutic efficacy of an emollient medical device enhanced with antioxidant compounds. According to the research, most individuals' skin conditions were improved and pruritus was decreased by the medical device [33]. Nonetheless, there are few, inconsistent, and incompletely standardized trials on antioxidant treatments for dermatological conditions currently available. Antioxidants should be utilized in combination with conventional pharmacological therapies since they may not provide noticeable clinical effects on their own [86, 87]. Larger trials using standardized criteria are among the research required to have a better understanding of the effects of antioxidant supplements in dermatitis.

16.6 Incorporating Antioxidants in Skincare

There are many ways to include antioxidants into the routine of your skin care. Using substances that occur naturally with antioxidant properties, such as kaempferol, nordihydroguaiaretic acid (NDGA), and naringenin, is one method [27]. These substances may be incorporated into nanostructured lipid carriers (NLCs) for topical treatment. An additional choice is to use an antioxidant-rich natural extract, which has been shown to improve the skin barrier and lessen inflammation. Examples of such extracts include curly kale, apples, and green tea [24]. Furthermore, a composition of antioxidants including peony leaf extract, calendula stem cell extract, and rosemary stem extract has been created; this composition may be used in skincare products due to its potent antioxidant properties [25]. Additionally, it has been investigated if a low-molecular-weight gel prepared with Boc-L-DOPA(Bn)2-OH is an appropriate medium to host antioxidants and enhance their action [15, 67, 84]. The potential advantages of antioxidant protection for the skin may be realized by introducing antioxidants into your skincare regimen via these several techniques. Antioxidants have been found to be effective in skincare, with natural compounds showing promise in reducing oxidative stress and providing protection against free radicals [24]. Plant extracts rich in vitamins, flavonoids, and phenolic compounds have been shown to have antioxidant properties, although their efficacy *in vivo* is still not well-documented [30]. Lipid-based drug delivery systems, such as microemulsions and lyotropic liquid crystals, have been investigated for their potential in delivering antioxidants to the skin [27]. These systems have been found to enhance stability, antioxidant activity, and release of the incorporated antioxidants, making them a promising option for combined antioxidant therapy. Additionally, a study found that certain extracts, such as resveratrol, pomegranate, and green tea, exhibited high antioxidant activities in skincare formulations, although the efficacy of these extracts in the final product can be influenced by interactions with other compounds [87]. Overall, the latest research suggests that antioxidants have potential in skincare, but further studies are needed to fully understand their effectiveness and optimize their delivery.

16.7 Challenges and Future Directions

Antioxidants in skincare face several challenges and have future directions for improvement. One challenge is the need to identify specialized substances for treating specific skin conditions, such as aging, dehydration, and hyperpigmentation, to maximize effectiveness and minimize side effects [25]. The effectiveness of natural antioxidants *in vivo* presents another difficulty since, while having antioxidant advantages, their prooxidant qualities may cause oxidative damage [28]. Furthermore, topical antioxidants' stability and transdermal absorption should be taken into account for optimal formulation, administration, and efficacy [41]. Future research should look at using marine antioxidants from macroalgae, flora, and fauna in skincare products since they have shown encouraging benefits in enhancing skin health [88]. Additionally, the use of delivery methods including carbon nanotubes, lipidic systems, polymeric systems, and vesicular systems might boost the distribution of antioxidants to deeper layers of the skin and augment their anti-aging benefits [89].

16.7.1 Challenges

Certain antioxidants may be unstable, which might cause them to deteriorate over time. It might be difficult to guarantee product stability in skincare formulas. Many situations were studied in relation to antioxidant stability. Using sucrose as a reference, one research assessed the durability of freeze-dried prebiotic matrices in maintaining vitamin C. Another research that looked at the stability of vitamins A, E, and C in supplements after a year of storage discovered that the content of vitamin C had significantly decreased after a year. Tert-butylhydroxyquinone (TBHQ) shown the best efficiency in increasing thermal, storage, and oxidative stability in a distinct research that looked into the effects of synthetic antioxidants on the stability of biodiesel. Furthermore, it was discovered that an antioxidant composition comprising stabilized neohesperidin dihydrochalcone had unexpected stability and solubility, which made it useful for skin protection. Lastly, the stability of selenurea solutions was influenced by the concentration, acidity, and alkalinity of the medium, as well as the presence of various antioxidants.

16.7.1.1 Bioavailability

An effective absorption of antioxidants into the skin is a challenge, especially when considering the bioavailability of these compounds in different formulations. The effectiveness of antioxidants in skin care products may be impacted by their bioavailability. For a product to be successful, the active components must be able to maintain their antioxidant activity after being added to the formulation [31]. Certain antioxidants, such Trolox standard and amla extract, were shown to maintain their activities in the formulations, whilst the activities of other antioxidants, like green tea and pomegranate extracts, were found to decrease [24]. Furthermore, for antioxidants to have their protective benefits on skin, skin distribution is essential. Antioxidants administered topically, such I′-tocopheryl glucoside (DTG) and oxothiazolidine (OTZ), have been shown to provide both short-term and long-term protection against UV-induced oxidative stress [27]. The bioavailability of antioxidants, including their absorption and penetration into the skin, as well as their metabolism and conversion to active forms, can influence their ability to effectively scavenge reactive oxygen species and protect the skin [90].

16.7.1.2 Individual Variability

Responses to antioxidants vary among individuals, making it challenging to determine the optimal dosage or formulation for everyone. The individual variability of antioxidants in skin care products can affect their efficacy. Different antioxidants have different abilities to protect the skin from oxidative damage [24]. Some antioxidants, such as resveratrol, pomegranate, and green tea, exhibit high antioxidant activities [91]. However, the efficacy of antioxidants can be influenced by their interaction with other compounds in the product formulation [92]. For example, some antioxidants may retain their activities when mixed with the base cream, while others may experience a decrease in activities [93]. Additionally, the position and number of hydroxyl groups in phenol-based antioxidants can impact their capacity and reactivity [24]. Therefore, the choice of antioxidants and their formulation methods play a crucial role in enhancing their efficacy in skin care products.

16.7.1.3 Topical vs. Systemic Benefits

Topical administration of antioxidants in skin care products enhances their efficacy by creating a reservoir of high concentrations of antioxidants in the skin, which can provide sustained protection against external environmental stress [94]. This is because topical application allows for direct delivery of antioxidants to the skin, bypassing the digestive system and potential degradation that can occur with oral intake [92]. Additionally, the use of lipid-based carrier systems, such as nanostructured lipid carriers (NLCs) and lipid-based drug delivery systems, can further enhance the penetration of antioxidants into the skin and improve their stability [41, 95]. On the other hand, systemic administration of antioxidants through oral intake may not provide the same level of localized protection and may result in lower concentrations of antioxidants reaching the skin [92]. Therefore, topical administration of antioxidants in skin care products is more effective in improving skin health and combating oxidative stress.

16.7.2 Future Directions

16.7.2.1 Advanced Delivery Systems

Continued research into innovative delivery systems, such as nanotechnology or encapsulation, can enhance the penetration and stability of antioxidants in skincare products.

16.7.2.2 Personalized Skincare

Advancements in understanding genetic and environmental factors influencing skin health may lead to personalized skincare regimens tailored to individual needs.

16.7.2.3 Combination Therapies

Exploring synergies between antioxidants and other skincare ingredients, such as peptides or hyaluronic acid, for comprehensive skincare benefits.

Combination therapies that explore synergies between antioxidants and other skincare ingredients, such as peptides or hyaluronic acid, can provide comprehensive skincare benefits. Ascorbic acid (AA), an antioxidant, has been shown to promote collagen gene expression and maturation, which can help in reducing skin aging and photoaging.

16.8 Targeted Solutions for Specific Conditions

Developing antioxidant-rich formulations specifically designed for certain skin conditions, such as acne or rosacea, to address unique challenges associated with these disorders.

16.8.1 Robust Clinical Trials

Conducting well-designed and large-scale clinical trials to further substantiate the efficacy of antioxidants in skincare and their impact on various skin disorders.

16.8.2 Educational Initiatives

Increasing public awareness and education about the benefits and limitations of antioxidants in skincare to empower consumers in making informed choices.

16.8.3 Environmental Sustainability

Exploring sustainable sources for antioxidant-rich ingredients and adopting eco-friendly packaging practices in the skincare industry.

Navigating these challenges and embracing future directions in skincare research will contribute to refining the use of antioxidants and advancing their role in promoting skin health.

16.9 Conclusion

In conclusion, the incorporation of antioxidants into skincare represents a promising avenue for promoting skin health and addressing various skin disorders. Empirical research on antioxidants highlights their beneficial effects in reducing oxidative stress and its related side effects, including acne, inflammatory skin diseases, and early aging. Numerous advantages of incorporating antioxidants into skincare products have been shown. Antioxidants have the ability to reduce the build-up of free radicals and oxidative stress, two significant factors contributing to skin conditions including skin cancer. It has been discovered that antioxidants strengthen the skin's barrier function and decrease inflammation, which increases resistance to external stresses. They also slow down the aging process by neutralizing toxic chemicals produced during cellular senescence. In order to reduce skin photoaging and stop UV-induced oxidative damage, antioxidants are often included in sunscreen compositions. Antioxidants have also been shown to enhance the health and youth of the skin, and some compounds have been recommended for the treatment of certain skin disorders including hyperpigmentation, dehydration, and aging. All things considered, adding antioxidants to skincare products is a viable approach to safeguarding and enhancing the condition and look of the skin. Antioxidants, such as polyphenols, vitamin C, vitamin E, and Coenzyme Q10, are essential for preventing free radical damage, preserving cellular structures, and promoting healthy skin function. Antioxidants support a healthy complexion, preserve collagen, and lessen inflammation via their methods of action. Although there are obstacles to overcome, such as individual variability and stability, there are possibilities provided by continuing research and developments in skincare formulas. Personalized methods, cutting-edge delivery techniques, and a better understanding of the interactions between antioxidants and other skincare compounds are all necessary for the future of skincare. It is crucial to take into account elements like formulation compatibility, appropriate product layering, and maintaining a holistic approach to skincare as consumers adopt antioxidant-rich products. Campaigns for education and awareness enable people to make knowledgeable decisions, which promotes proactive and efficient skincare practices. Antioxidants are a crucial tactic for anybody looking to improve and preserve the health and vibrancy of their skin while navigating the ever-changing skincare environment.

References

1. Prasad, A., Duchová, H., Manoharan, R.R., Rathi, D., Pospíšil, P., Imaging and Characterization of Oxidative Protein Modifications in Skin. *Int. J. Mol. Sci.*, 24, 4, 3981, 2023.
2. Trüeb, R.M., Oxidative stress and its impact on skin, scalp and hair. *Int. J. Cosmet. Sci.*, 43 Suppl 1, S1, S9–13, 2021.
3. Galiniak, S., Mołoń, M., Biesiadecki, M., Bożek, A., Rachel, M., The Role of Oxidative Stress in Atopic Dermatitis and Chronic Urticaria. *Antioxidants*, 11, 8, 1590, 2022.
4. Nakai, K. and Tsuruta, D., What Are Reactive Oxygen Species, Free Radicals, and Oxidative Stress in Skin Diseases? *Int. J. Mol. Sci.*, 22, 19, 10799, 2021.
5. Emanuelli, M., Sartini, D., Molinelli, E., Campagna, R., Pozzi, V., Salvolini, E. *et al.*, The Double-Edged Sword of Oxidative Stress in Skin Damage and Melanoma: From Physiopathology to Therapeutical Approaches. *Antioxid. (Basel)*, 1, 4, 612, 2022.
6. Unsal, V., Cicek, M., Sabancilar, I., Toxicity of carbon tetrachloride, free radicals and role of antioxidants. *Rev. Environ. Health*, 36, 2, 279–95, 2020.
7. Tripathi, R., Gupta, R., Sahu, M., Srivastava, D., Das, A., Ambasta, R.K. *et al.*, Free radical biology in neurological manifestations: mechanisms to therapeutics interventions. *Environ. Sci. Pollut. Res. Int.*, 29, 41, 62160–62207, 2022.
8. Pourmasumi, S. and Sabeti, P., The effect of free radicals on sperm DNA and antioxidant protective role; an assessment and review. *Rev. Clin. Med.*, 7, 1, 37–42, 2020.

9. Biswas, S., Das, R., Banerjee, E.R., Role of free radicals in human inflammatory diseases. *AIMS Biophys.*, 4, 4, 596–614, 2017.
10. Al-Saiedy, M., Green, F., Amrein, M., The effects of free radicals on pulmonary surfactant lipids and proteins. *Oxid. Stress Lung Dis.*, 2, 3–24, 2019.
11. Miftode, A.M., Antioxidanţii: mecanisme deacţiuneşiaplicaţii înactulterapeutic. *Medicro.*, 306, 150, 8, 2022.
12. Sundaram Sanjay, S. and Shukla, A.K., Free Radicals Versus Antioxidants. *Potential Ther. App. Nano-Antioxidants*, 21, 26, 22, 1–17, 2021.
13. Izmailov, D.Y., Proskurnina, E.V., Shishkanov, S.A., Vladimirova, G.A., Vladimirov, Y.A., The effect of antioxidants on the formation of free radicals and primary products of the peroxidase reaction. *Biophys. (Russian Federation)*, 62, 4, 557–64, 2017.
14. Khaira, K., Menangkal Radikal BebasDengan Anti-Oksidan. *Sainstek: Jurnal Sains dan Teknologi.*, 2, 2, 183–7, 2016.
15. Liu, H.M., Cheng, M.Y., Xun, M.H., Zhao, Z.W., Zhang, Y., Tang, W. *et al.*, Possible Mechanisms of Oxidative Stress-Induced Skin Cellular Senescence, Inflammation, and Cancer and the Therapeutic Potential of Plant Polyphenols. *Int. J. Mol. Sci.*, 24, 4, 3755, 2023.
16. Shubham, D.L., Mahesh, S., Deepali, S., Vikas, Wamane, Vaibhav Jadhav. A Review on : Plant Based Antioxidant. *IJFMR-Int. J. Multi. Res.*, 4, 6, 1–12, 2022.
17. Sable, R., Sharma, S., Author, C., Antioxidants rich natural fruit and synthetic fruit products. *Pharma. Innovation J.*, 12, 5, 2026–35, 2026.
18. Nimbalkar, V.V., Jaggi, S.M., Shinde, G.P., Jogdand, S.S., Mhaismale, K.S., MS, R., Food Antioxidants. *Biochem. Biophys. Res. Commun.*, 2, 1, 386–93, 2022.
19. Swaroopa, G. and Srinath, D., Antioxidants and Their Roll In Human Health: A Review. *Int. J. Agri. Sci. Res.*, 7, 5, 9–18, 2017.
20. Islam, M. and Gracia, F., Antioxidants for Optimal Health. *Rev. Médico Científica.*, 26, 2, 3–9, 2013.
21. Sri, A. and Raj, P., An Overview of Antioxidants. *J. Food Sci. Res.*, 2, 2, 108, 2017.
22. Anwar, H., Hussain, G., Mustafa, I., Anwar, H., Hussain, G., Mustafa, I., Antioxidants from Natural Sources. *Antioxid. Foods App.*, 75961, 2018.
23. Soni, K.S. and Loonker, S., An Approach to Leading Antioxidant Activity of Different Plants and Food Material: A Review. *Orient. J. Chem.*, 38, 3, 663–70, 2022 Jun 30.
24. Zhang, Y., Heinemann, N., Rademacher, F., Darvin, M.E., Raab, C., Keck, C.M. *et al.*, Skin Care Product Rich in Antioxidants and Anti-Inflammatory Natural Compounds Reduces Itching and Inflammation in the Skin of Atopic Dermatitis Patients. *Antioxidants*, 11, 6, 1071, 2022.
25. Ngoc, L.T.N., Moon, J.Y., Lee, Y.C., Antioxidants for improved skin appearance: Intracellular mechanism, challenges and future strategies. *Int. J. Cosmet. Sci.*, 45, 3, 299–314, 2023.
26. Desai, P.S. and Kumar, M.S., Marine antioxidants and their role in improving skin health. *Mar. Antioxid.: Prep., Synth., App.*, 23, 2, 327–39, 2023 Jan 1.
27. Gonçalves, C., Ramalho, M.J., Silva, R., Silva, V., Marques-Oliveira, R., Silva, A.C. *et al.*, Lipid nanoparticles containing mixtures of antioxidants to improve skin care and cancer prevention. *Pharmaceutics*, 13, 12, 2042, 2021.
28. Hoang, H.T., Moon, J.Y., Lee, Y.C., Natural Antioxidants from Plant Extracts in Skincare Cosmetics: Recent Applications, Challenges and Perspectives. *Cosmetics*, 8, 106, 2021.
29. Imamović, M., Horozić, E., Salković, V., Pašalić, A., Hadžić, N., Metal complexes as potential agents for neutralizing free radicals: a review of *in vitro* studies. *Acta Med. Saliniana*, 52, 1–2, 18–23, 2023.
30. Suja, C., Shuhaib, B., Abdurahman, M., Khathoom, H., Simi, K., A Review on Dietary Antioxidants. *Res. J. Pharm. Technol.*, 9, 2, 196–202, 2016.

31. Ardizzone, A., Repici, A., Capra, A.P., De Gaetano, F., Bova, V., Casili, G. *et al.*, Efficacy of the Radical Scavenger, Tempol, to Reduce Inflammation and Oxidative Stress in a Murine Model of Atopic Dermatitis. *I*, 12, 6, 1278, 2023.
32. Martiniaková, S., Ácsová, A., Hojerová, J., Krepsová, Z., Kreps, F., Ceylon cinnamon and clove essential oils as promising free radical scavengers for skin care products. *Acta ChimicaSlovaca.*, 115, 1, 1–11, 2022.
33. Li Pomi, F., Papa, V., Borgia, F., Vaccaro, M., Allegra, A., Cicero, N. *et al.*, Rosmarinus officinalis and Skin: Antioxidant Activity and Possible Therapeutical Role in Cutaneous Diseases. *Antioxidants*, 12, 3, 680, 2023.
34. Addor, F.A.S., Antioxidants in dermatology. *An. Bras. Dermatol.*, 92, 3, 356, 2017.
35. Adeel, S., Habiba, M., Kiran, S., Iqbal, S., Abrar, S., Hassan, C.M., Utilization of Colored Extracts for the Formulation of Ecological Friendly Plant-Based Green Products. *Sustainability*, 14, 11758, 2022.
36. Nur, N., Zakaria, A., Okello, E.J., Idowu, O.C., Evaluation of the antioxidant efficacy of extracts/ingredients used in skin care products. *Int. J. Phytocosmetics Nat. Ingredients*, 5, 1, 7–7, 2018.
37. Estevez, A.Y., Pritchard, S., Harper, K., Aston, J.W., Lynch, A., Lucky, J.J. *et al.*, Nanoceria for the treatment of oxidative stress. *Free Radic. Biol. Med.*, 51, 6, 1155–63, 2013.
38. Furue, M., Uchi, H., Mitoma, C., Hashimoto-Hachiya, A., Chiba, T., Ito, T. *et al.*, Antioxidants for Healthy Skin: The Emerging Role of Aryl Hydrocarbon Receptors and Nuclear Factor-Erythroid 2-Related Factor-2. *Nutrients*, 9, 3, 223, 2017.
39. Wölfle, U., Seelinger, G., Bauer, G., Meinke, M.C., Lademann, J., Schempp, C.M., Reactive molecule species and antioxidative mechanisms in normal skin and skin aging. *Skin Pharmacol. Physiol.*, 27, 6, 316–32, 2014.
40. Salucci, S., Burattini, S., Curzi, D., Buontempo, F., Martelli, A.M., Zappia, G. *et al.*, Antioxidants in the prevention of UVB-induced keratynocyte apoptosis. *J. Photochem. Photobiol., B*, 141, 1–9, 2014.
41. Burke, K.E., Protection From Environmental Skin Damage With Topical Antioxidants. *Clin. Pharmacol. Ther.*, 105, 1, 36–8, 2019.
42. Fuller, B.B., Antioxidants and Anti-inflammatories. *Cosmet. Dermatol.: Prod. Proced.*, 3, 366–87, 2022.
43. Gunter, N.V., Teh, S.S., Lim, Y.M., Mah, S.H., Natural Xanthones and Skin Inflammatory Diseases: Multitargeting Mechanisms of Action and Potential Application. *Front. Pharmacol.*, 11, 594202, 2020.
44. Fuller, B., Smith, D., Howerton, A., Kern, D., Anti-inflammatory effects of CoQ10 and colorless carotenoids. *J. Cosmet. Dermatol.*, 5, 1, 30–8, 2006.
45. McGilvray, I.D. and Rotstein, O.D., Antioxidant modulation of skin inflammation: preventing inflammatory progression by inhibiting neutrophil influx. *Can. J. Surg.*, 42, 2, 109, 1999.
46. Aguirre-Cruz, G., León-López, A., Cruz-Gómez, V., Jiménez-Alvarado, R., Aguirre-Álvarez, G., Collagen Hydrolysates for Skin Protection: Oral Administration and Topical Formulation. *Antioxid. (Basel)*, 9, 2, 181, 2020.
47. Anazco, C., Rojas, A., Gonzalez, I., Castro, M.A., Robert, P., Oyarzun-Ampuero, F., Dermal Collagen Stabilization by Polyphenols and Spray Drying as an Encapsulation Strategy. *Curr. Top. Med. Chem.*, 18, 14, 1242–51, 2018.
48. Yokoi, K., Yasumizu, Y., Ohkura, N., Shinzawa, K., Okuzaki, D., Shimoda, N. *et al.*, Increased anti-oxidative action compensates for collagen tissue degeneration in vitiligo dermis. *Pigm. Cell Melanoma Res.*, 36, 5, 355–64, 2023.
49. Martins, T.E.A., Pinto CAS de, O., de Oliveira, A.C., Velasco, M.V.R., Guitiérrez, A.R.G., Rafael, M.F.C. *et al.*, Contribution of Topical Antioxidants to Maintain Healthy Skin—A Review. *Sci. Pharm.*, 88, 27, 2020.

50. Abdullah, A., Nurhayati, T., Nugroho, F.A., Antioxidant activity of trypsin and pepsin-hydrolyzed fish collagen. *IOP Conf. Ser. Earth Environ. Sci.*, 1033, 1, 012066, 2022.
51. Zhang, J.B., Zhao, Y.Q., Wang, Y.M., Chi, C.F., Wang, B., Eight Collagen Peptides from Hydrolysate Fraction of Spanish Mackerel Skins: Isolation, Identification, and *In Vitro* Antioxidant Activity Evaluation. *Mar. Drugs*, 17, 4, 2019.
52. Rajput, V.D., Harish, Singh, R.K., Verma, K.K., Sharma, L., Quiroz-Figueroa, F.R. *et al.*, Recent Developments in Enzymatic Antioxidant Defence Mechanism in Plants with Special Reference to Abiotic Stress. *Biol. (Basel)*, 10, 4, 267, 2021.
53. Rossnerova, A., Izzotti, A., Pulliero, A., Bast, A., Rattan, S.I.S., Rossner, P., The Molecular Mechanisms of Adaptive Response Related to Environmental Stress. *Int. J. Mol. Sci.*, 21, 19, 1–15, 2020.
54. Limón Pacheco, J.H., Carballo, M.A., Gonsebatt, M.E., Antioxidants Against Environmental Factor-Induced Oxidative Stress, in: *Nutritional Antioxidant Therapies: Treatments and Perspectives*, pp. 189–215, 2018.
55. MS, KARAGAC and CEYLAN, H., Effects of Natural Polyphenols Against Skin Aging. *Int. Conf. Recent Acad. Stud.*, 1, 1, 226–31, 2023.
56. Pouillot, A., Polla, L.L., Tacchini, P., Neequaye, A., Polla, A., Polla, B., Natural Antioxidants and their Effects on the Skin, in: *Formulating, Packaging, and Marketing of Natural Cosmetic Products*, pp. 239–57, 2011.
57. Svendsen, L., Rattan, S.I.S., Clark, B.F.C., Testing garlic for possible anti-ageing effects on long-term growth characteristics, morphology and macromolecular synthesis of human fibroblasts in culture. *J. Ethnopharmacol.*, 43, 2, 125–33, 1994.
58. Djawad, K. and Anggraini, D., The effectiveness of oral superoxide dismutase (sod) on total antioxidant status, transepidermal water loss (Tewl) and sebum concentration in photo aging skin. *Med. Sci. Ukraine (MSU)*, 17, 1, 47–56, 2021.
59. Zhu, X., Yuan, W., Li, Z., Lin, Y., Li, W., Ji, L., *et al.*, Progress of Research on Antioxidants and Carriers for Skin Wound Repair. *Processes*, 11, 7, 2069, 2023.
60. Xu, H., Zheng, Y.W., Liu, Q., Liu, L.P., Luo, F.L., Zhou, H.-C. *et al.*, Reactive Oxygen Species in Skin Repair, Regeneration, Aging, and Inflammation, in: *Reactive Oxygen Species (ROS) in Living Cells*, InTech, 23 May 2018. Crossref, doi:10.5772/intechopen.72747
61. Carney, J.M., Flateland, M.L., Hetland, B.L., Renquist, K.J., Owens, K.M., The Effect of Mitochondrially Targeted Antioxidants on Wound Healing of Human Dermal Fibroblasts. *FASEB J.*, 36, S1, R4276, 2022.
62. Fadilah, N.I.M., Phang, S.J., Kamaruzaman, N., Salleh, A., Zawani, M., Sanyal, A. *et al.*, Antioxidant Biomaterials in Cutaneous Wound Healing and Tissue Regeneration: A Critical Review. *Antioxid. (Basel)*, 12, 4, 787, 2023.
63. Gęgotek, A., Jastrząb, A., Dobrzyńska, M., Biernacki, M., Skrzydlewska, E., Exogenous Antioxidants Impact on UV-Induced Changes in Membrane Phospholipids and the Effectiveness of the Endocannabinoid System in Human Skin Cells. *Antioxid. (Basel)*, 10, 8, 1260, 2021.
64. Mullan, B.A., Young, I.S., Fee, H., McCance, D.R., Ascorbic Acid Reduces Blood Pressure and Arterial Stiffness in Type 2 Diabetes. *Hypertension*, 40, 6, 804–9, 2002.
65. Choo, P.P., Woi, P.J., Bastion, M.L.C., Omar, R., Mustapha, M., Md Din, N., Review of Evidence for the Usage of Antioxidants for Eye Aging. *Biomed Res. Int.*, 2022, 25810373, 2022.
66. Patki, A., Shelatkar, R., Singh, M., Agarwal, S., Umbardand, S., Reddy, A. *et al.*, Impact of antioxidants in improving semen parameters like count, motility and DNA fragmentation in sub-fertile males: a randomized, double-blind, placebo-controlled clinical trial. *Transl. Clin. Pharmacol.*, 31, 1, 28–39, 2023.

67. Liu, Y., Yang, Z., Du, Y., Shi, S., Cheng, Y., Antioxidant interventions in autism spectrum disorders: A meta-analysis. *Prog. Neuro-Psychopharmacol. Biol. Psychiatry*, 113, 110476, 2022.
68. Arhin, S.K., Ocansey, S., Barnes, P., Kwesi Botchey, C.P., Taylor-Adbulai, H.B., Efficacy of combined antioxidant therapy in male subfertility-A systematic review and meta-analysis. *Cell Mol. Biol. (Noisy-le-grand)*, 67, 4, 239–47, 2022.
69. Di Sotto, A., Gullì, M., Percaccio, E., Vitalone, A., Mazzanti, G., Di Giacomo, S., Efficacy and Safety of Oral Green Tea Preparations in Skin Ailments: A Systematic Review of Clinical Studies. *Nutrients*, 14, 15, 3149, 2022.
70. Kim, S., Park, T.H., Kim, W.I., Park, S., Kim, J.H., Cho, M.K., The effects of green tea on acne vulgaris: A systematic review and meta-analysis of randomized clinical trials. *Phytother. Res.*, 35, 1, 374–83, 2021.
71. Joshi, A., Elmasry, R., Shipp, Z., Is topical green tea extract an effective treatment of acne vulgaris? *Evidence-Based Pract.*, 26, 1, 9–10, 2023.
72. Hopkins, R., Natural Polyphenols for Acne Vulgaris: Clinical Evidence and Biological Mechanisms. *React. Oxygen Species*, 7, 21, 129–135, 2019.
73. Kostrzewska, P., Mandera, A., Nagi, K., Pawlikowska, A., Szuster, E., Coenzyme Q10 – a literature review on supplementation in chronic disorders and the anti-aging process. *Pediatriai. Medycyna. Rodzinna.*, 18, 2, 134–8, 2022.
74. Ayunin, Q., Miatmoko, A., Soeratri, W., Erawati, T., Susanto, J., Legowo, D., Improving the anti-ageing activity of coenzyme Q10 through protransfersome-loaded emulgel. *Sci. Rep.*, 12, 1, 1–13, 2022.
75. Shoviantari, F., Erawati, T., Soeratri, W., Coenzyme Q10 nanostructured lipid carriers as an inducer of the skin fibroblast cell and its irritability test in a mice model. *J. Basic. Clin. Physiol. Pharmacol.*, 30, 6, 31860468, 2019.
76. Jobeili, L., Rousselle, P., Béal, D., Blouin, E., Roussel, A.M., Damour, O. *et al.*, Selenium preserves keratinocyte stemness and delays senescence by maintaining epidermal adhesion. *Aging*, 9, 11, 2302–15, 2017.
77. Gangadevi, V., Thatikonda, S., Pooladanda, V., Devabattula, G., Godugu, C., Selenium nanoparticles produce a beneficial effect in psoriasis by reducing epidermal hyperproliferation and inflammation. *J. Nanobiotechnol.*, 19, 1, 101, 2021.
78. NazIroğlu, M., YIldIz, K., Tamtürk, B., Erturan, I., Flores-Arce, M., Selenium and psoriasis. *Biol. Trace Elem. Res.*, 150, 1–3, 3–9, 2012.
79. Sengupta, A., Lichti, U.F., Carlson, B.A., Ryscavage, A.O., Gladyshev, V.N., Yuspa, S.H. *et al.*, Selenoproteins are essential for proper keratinocyte function and skin development. *PLoS One*, 5, 8, e12249, 2010.
80. Chung, S., Ercan, B., Webster, T.J., Tran, P., Electrospun Silk Doped with Selenium for Antibacterial Skin Applications. *Proceedings of the IEEE Annual Northeast Bioengineering Conference, NEBEC*, 2014.
81. Xie, B., Zeng, D., Yang, M., Tang, Z., He, L., Chen, T., Translational Selenium Nanoparticles to Attenuate Allergic Dermatitis through Nrf2-Keap1-Driven Activation of Selenoproteins. *ACS Nano*, 17, 14, 14053–68, 2023.
82. Guo, H., Li, M., Liu, H., Selenium-Rich Yeast Peptide Fraction Ameliorates Imiquimod-Induced Psoriasis-like Dermatitis in Mice by Inhibiting Inflammation via MAPK and NF-κB Signaling Pathways. *Int. J. Mol. Sci.*, 23, 4, 2112, 2022.
83. Turcu, G., Artenie, C., Nowicka, D., Arenbergerová, M., Lazaridou, E., Khobzei, K. *et al.*, Selenium Disulfide-based shampoo applied for 4 weeks significantly improves dandruff and seborrheic dermatitis. *Eur. J. Dermatol.*, 33, S1, 19–23, 2023.

84. Yang, H., Chen, J., Luo, X., Wang, H., Efficacy and safety profile of antioxidants in the treatment of atopic dermatitis: A systematic review and meta-analysis of randomized controlled trials. *Dermatol. Ther.*, 35, 7, e15549, 2022.
85. Kubesy, A.A.M., Yehia, S.G., Salem, S.I., Rabah, M., Altered blood oxidative stress markers in association with antioxidant supplemented therapy for mange, tick, and flea allergic dermatitis of dogs. *Comp. Clin. Path.*, 29, 5, 937–43, 2020.
86. Vasques, L.I., Vendruscolo, C.W., Leonardi, G.R., Topical Application of Ascorbic Acid and its Derivatives: A Review Considering Clinical Trials. *Curr. Med. Chem.*, 30, 29, 3272–86, 2023.
87. Vitek, M., Gosenca Matjaž, M., Roškar, R., Gašperlin, M., Zvonar Pobirk, A., A comparative study of lipid-based drug delivery systems with different microstructure for combined dermal administration of antioxidant vitamins. *J. Dispers. Sci. Technol.*, 44, 9, 1711–24, 2023.
88. Hatem, S., Nasr, M., Elkheshen, S.A., Geneidi, A.S., Recent Advances in Antioxidant Cosmeceutical Topical Delivery. *Curr. Drug Deliv.*, 15, 7, 953–64, 2018.
89. Chen, L.L. and Wang, S.Q., From the bottle to the skin: challenges in evaluating antioxidants. *Photodermatol. Photoimmunol. Photomed.*, 28, 5, 228–34, 2012.
90. Jacques, C., Bacqueville, D., Jeanjean-Miquel, C., Génies, C., Noizet, M., Tourette, A. *et al.*, Sustained effect of two antioxidants (oxothiazolidine and δ-tocopheryl glucoside) for immediate and long-term sun protection in a sunscreen emulsion based on their different penetrating properties. *Int. J. Cosmet. Sci.*, 43, 4, 391–404, 2021.
91. Oresajo, C., Pillai, S., Manco, M., Yatskayer, M., McDaniel, D., Antioxidants and the skin: understanding formulation and efficacy. *Dermatol. Ther.*, 25, 3, 252–9, 2012.
92. D'Angelo Costa, G.M. and Maia Campos, P.M.B.G., Efficacy of topical antioxidants in the skin hyperpigmentation control: A clinical study by reflectance confocal microscopy. *J. Cosmet. Dermatol.*, 20, 2, 538–45, 2021.
93. Jung, K., Everson, R.J.C., Joshi, B., Bulsara, P.A., Upasani, R., Clarke, M.J., Structure-function relationship of phenolic antioxidants in topical skin health products. *Int. J. Cosmet. Sci.*, 39, 2, 217–23, 2017.
94. Samprasit, W., Vasarach, C., Opanasopit, P., Sriamornsak, P., Chamsai, B., Topical nanostructured lipid carriers of alpha-mangostin and resveratrol for synergistic antioxidant activity. *Pharm. Nanotechnol.*, 10, 3, 220–31, 2022.
95. Monteiro e Silva, S.A., de, S.JG., PS, M., Moreno, I.A.M., SM, A., Lopez, R.F.V. *et al.*, Target action of antioxidants using iontophoresis. *J. Cosmet. Dermatol.*, 20, 2, 664–76, 2021.

17

Antioxidants and Toxicity

Ritu Sanwal[1], Vichitra Kaushik[1]*, Brijendra Singh[2], Nittan Kumar[1] and Atul Kaushik[3]

[1]Shanti Niketan College of Pharmacy, Mandi, Himachal Pradesh, India
[2]Public Health Training Institute, SMS Hospital Campus, Jaipur, Rajasthan, India
[3]School of Applied Chemistry and Basic Sciences, SBS University, Balawala, Dehradun, Uttarakhand, India

Abstract

Antioxidants are widely used in a significant manner to attain and safeguard good health. Nutraceuticals and dietary supplements often include substantial quantities of antioxidants. Oxidative damage is a typical cellular process implicated in several illnesses and medication toxicities. Extensive study has been conducted to explore the potential of natural and newly created antioxidants in preventing or delaying oxidative damage. The initial enthusiasm about the possible health advantages of antioxidants has waned. Presently, there are assertions suggesting that antioxidants have the potential to elevate death rates. The oscillation of the antioxidant pendulum is seen to transition between beneficial and harmful effects, from being widely regarded as a cure-all to being considered an inconsequential component. Due to the divergent perspectives about antioxidants, the dietary guidance varies from advocating for the augmentation of plasma antioxidant levels to asserting the insignificance of such measurements. The aforementioned perspectives, which lack adequate scientific substantiation, give rise to misunderstandings that, from our perspective, restrict the logical utilization of dietary supplements and inhibit the formulation and advancement of novel antioxidant medications. Consequently, valuable possibilities may be inadvertently overlooked. The potential negative consequences of consuming large levels of antioxidants are not well acknowledged. Some researchers examine the toxicity of three antioxidants, namely vitamin E, β-carotene, and lipoic acid, in order to explore broad factors related to antioxidant toxicity. Based on the above instances, the following suggestions are proposed for the assessment of the harmful effects of antioxidants. (i) The use of conventional safety criteria is not recommended. There is a need to enhance our understanding of the mechanisms behind the effectiveness and potential adverse effects of antioxidants. (iii) The use of bio-kinetic/bio-efficacy modeling might potentially be advantageous in the optimization of dose. (iv) When antioxidant supplementation transitions into treatment, it becomes necessary to conduct a more precise analysis of the risks and benefits involved.

Keywords: Antioxidants, free-radicals, vitamin E, toxicity, safety, lipoic acid, β-carotene, glutathione

**Corresponding author*: vichitra.kaushik@gmail.com

Rakesh K. Sindhu, Inderbir Singh and M. Arockia Babu (eds.) Antioxidants: Nature's Defense Against Disease, (573–586)

17.1 Introduction

Antioxidants have emerged as a scientific topic that has elicited a multitude of contentious viewpoints. The subject matter is not just addressed within scholarly writing but also within popular media outlets. A search query on the Google search engine that combines the terms "antioxidant" and "health" yields a substantial number of results, exceeding 82 million, indicating the widespread interest in this particular topic. Within the ongoing dispute, it is evident that assertions pertaining to the efficacy of antioxidants may sometimes be excessively embellished and lacking in relevance. Furthermore, incorrect reasoning is occasionally used to support these assertions. An instance of a compound blend is commercially promoted as a 'life extension formula' due to the presence of antioxidants. It is said that the prolonged utilization of soap containing the antioxidant vitamin E is very beneficial for mitigating the appearance of stretch marks, blemishes, and wrinkles. The soap is furthermore appropriate for those with sensitive skin and offers advantages for those with psoriasis. It is not feasible to get a precise risk-benefit analysis of antioxidants by this approach.

An oxidant refers to a chemical substance that has the ability to accept electrons from another substance during a chemical reaction, resulting in the oxidation of that substance. From a chemical perspective, an oxidant is capable of accepting electrons. From a biological standpoint, it is observed that oxidants have the capability to cause damage to biomolecules [2]. Reactive species are derived from many biological processes. The oxidizing reactive species include the hydroxyl radical (OH), the superoxide anion radical (O2–), hydrogen peroxide (H2O2), and hypochlorous acid (HOCl). Reactive oxygen species (ROS) are produced as secondary products of mitochondrial respiration and are also synthesized by the activity of NADPH oxidase on phagocytes during inflammatory processes. Xanthine oxidase, nitric oxide synthase, the metabolic pathway of arachidonic acid, myeloperoxidase, and cytochrome P450 are further potential origins of reactive species. The process of drug biotransformation involves the generation of reactive species, as shown in the case of redox-cycling medications like doxorubicin and nitrofurantoin. An antioxidant refers to a substance that inhibits or counteracts the process of oxidation, which involves the transfer of electrons from one molecule to another. From a chemical perspective, an antioxidant refers to a substance that impedes or retards the process of oxidation in another material. Among the two chemicals, the molecule that undergoes oxidation serves as an antioxidant for the other compound. In this context, antioxidants do not possess exceptional characteristics. From a biological standpoint, it is essential for an antioxidant to exhibit activity in safeguarding physiological targets such as fatty acids, proteins, and DNA, even when present in relatively low quantities. Vitamins E and C are examples of dietary antioxidants. Substances such as the anesthetic propofol and the newly discovered antifibrotic drug pirfenidone have antioxidant properties. The compound N-acetylcysteine functions as a mucolytic agent by serving as a precursor for the naturally occurring antioxidant glutathione.

It is well acknowledged that antioxidants are an integral component of our daily dietary intake, found in many sources such as fruits, vegetables, drinks, spices, and herbs. Consequently, there has been a growing emphasis on the significance of antioxidant consumption. Recently, there has been a trend of fortifying designer meals with antioxidants. Supplementation with antioxidants is a prevalent practice among individuals, and the pharmaceutical industry offers medications that provide distinct antioxidant properties [1]. Over several decades, it has been observed that the perception of antioxidants has oscillated significantly, ranging from being seen as exclusively beneficial to being considered very detrimental and from favoring natural antioxidants as superior to questioning their efficacy altogether. The disputing factions seem to exhibit a lack of receptiveness towards other viewpoints. Inadequate assertions remain unaddressed, resulting in a fluctuation of the pendulum toward the opposing ends. This ultimately impedes research within the discipline and creates confusion among researchers and consumers. Consequently, there is a possibility that we may overlook potential instances where antioxidants might play a role in enhancing health. In this discourse, we want to address 10 stories and endeavor to rectify the prevailing imbalances.

1. Antioxidants are Effective Against any Illness

Antioxidants engage in a chemical reaction with the reactive oxygen species (ROS), therefore mitigating their chemical re-activity. It has been proposed that this process serves as a protective measure against the cellular damage caused by reactive oxygen species (ROS) [2]. As a result of these findings, it was postulated that disorders mediated by reactive oxygen species (ROS) may be amenable to therapeutic interventions [3]. Reactive oxygen species (ROS) have a significant role in the pathogenesis of several illnesses. Following this, there was a suggestion that antioxidants have the potential to mitigate and manage a wide range of illnesses, hence stimulating more investigation into the field of antioxidants. The importance of ROS in chronic obstructive pulmonary disease (COPD) is shown. Individuals diagnosed with chronic obstructive pulmonary disease (COPD) have been seen to expel the reactive oxygen species (ROS) known as hydrogen peroxide during exhalation. An additional illustration pertains to the proposed contributory function of reactive oxygen species (ROS) in the occurrence of ischemia-reperfusion injury in the brain, heart, and kidney. Atherosclerosis is a pathological condition characterized by the oxidation of low-density lipoprotein (LDL) and its subsequent uptake by macrophages, resulting in the production of foam cells and ultimately leading to cardiovascular problems. Substances such as nitrofurantoin and doxorubicin, which have the potential to generate oxygen radicals, may also elicit pulmonary and cardiac tissue injury, respectively [4, 5]. Numerous investigations have been carried out pertaining to the protective effects of antioxidants. The initial expectations placed upon antioxidants were too optimistic, leading to the realization that these substances had limitations in their ability to address all ailments. In addition, consumers and experts were frustrated by health promises that were deemed implausible. The initial excitement around antioxidants gradually transformed into skepticism, leading to doubts over the safety of some antioxidants, like vitamin C, which were even seen as potentially harmful [6]. The favorable perception of antioxidants' health benefits mostly stemmed from *in vitro* investigations. The *in vitro* demonstration of antioxidants' efficient chemical scavenging action has prompted the extrapolation of their possible beneficial effects *in vivo*. The assessment of antioxidant bioavailability has received less attention, despite the observed poor bioavailability of polyphenolic antioxidants [7]. Furthermore, it is important to take into account the concept of bio accessibility, which refers to the proportion of an antioxidant that becomes accessible for absorption. This entails the liberation of the antioxidant from the food matrix. Following the process of absorption, first-pass metabolism may exhibit a substantial degree of extent, prompting a recent focus on the biological significance of antioxidant metabolites. It has come to our understanding that the significant chemical reactivity shown by the parent molecule in laboratory settings does not provide definitive information about the substance's ability to effectively treat any diseases connected with reactive oxygen species (ROS). The current discourse around anti-oxidants has once again sparked significant interest due to the recent revelation of their impact on gene expression via pathways other than direct influence on the DNA sequence, namely through epigenetic processes [8]. The proposition of a novel comprehensive mechanism aimed at elucidating the many functions of antioxidants seems to follow a recurring pattern in scientific discourse.

2. Antioxidants Raise the Risk of Death

According to a recent meta-analysis conducted on a specific set of randomized clinical studies, it was shown that the addition of antioxidants resulted in an elevated rate of death across all causes [9]. Nevertheless, this assertion was contradicted subsequent to a thorough reassessment, which revealed that none of the aforementioned research had mortality as their primary focus [10]. Despite the presence of valid criticism, there persists a prevailing and unsupported belief that antioxidants may pose significant risks. Rather than adopting a polarized perspective that categorizes antioxidants as either beneficial or detrimental, it would be more suitable to use a high benefit-risk ratio when assessing the efficacy of antioxidants [11]. This ratio takes into consideration both the advantages and disadvantages, offering a more objective and equitable perspective. Previous research has shown that the use of vitamin E may have a beneficial impact on the prevention of coronary heart disease [12, 13]. Nevertheless, several extensive, randomized, placebo-controlled studies have shown unsatisfactory results regarding the impact of vitamin E on the likelihood of developing cardiovascular disease. A comprehensive assessment of these research indicated that a meticulous examination of the dietary patterns of the subjects might potentially provide an alternative inference. There was a suggestion that some individuals who seemed to be in good health had elevated levels of lipid peroxidation compared to others. people exhibiting greater levels of rancid compounds may potentially be at an increased risk of developing cardiovascular disease. Consequently, these people might potentially get extra health benefits from the use of antioxidants [14]. Likewise, the lack of significant impact of vitamin E on the overall incidence of strokes might perhaps be attributed to the inclusive categorization of strokes, which may not adequately account for the distinct pathophysiological mechanisms causing different types of ischemic and hemorrhagic episodes. The findings of a meta-analysis examining the impact of vitamin E on stroke indicate a 22% elevation in the risk of hemorrhagic stroke and a 10% reduction in the risk of ischemic stroke [15]. The authors expressed concern over the indiscriminate and widespread use of vitamin E due to the unfavorable risk-benefit ratio. This is attributed to the relatively modest risk decrease for the ischemic stroke and the typically more severe result associated with hemorrhagic stroke. The topic of antioxidants and mortality has garnered significant interest in academic literature. Recently, there has been a contention that the utilization of antioxidants is more prone to induce rather than mitigate the occurrence of cancer [16]. The possible risks associated with inhibiting the oxidant-induced apoptosis in cancer cells by some antioxidants were discussed. Given that antioxidants are not inherently required to have a positive impact, it becomes crucial to ascertain certain cohorts, such as those susceptible to oxidation, who may potentially get advantages from the utilization of antioxidants. Excessive reliance on imbalanced and sensationalized news should be avoided.

3. The Greater The Quantity, The More Advantageous It Is

Linus Pauling, widely recognized as the pioneer of orthomolecular medicine, pushed for the use of vitamin C as an antioxidant to enhance overall well-being [17]. His approach included administering dosages of 1000 mg, a quantity much beyond the recommended dietary limit of 75-90 mg per day [18]. However, there is a lack of convincing scientific evidence, supported by clinical research, for the presumed health advantage of mega dosages of vitamin C. Furthermore, it was observed by the Renaissance physician Paracelsus over five centuries ago that each substance has adverse consequences when administered in excessive quantities. This principle also applies to antioxidants. The potential correlation between the administration of large dosages of antioxidants and the observed instances of heightened toxicity may be elucidated. As an example, the administration of a 20 mg dosage of b-carotene to male individuals who smoke resulted in a rise of 18% in the occurrence of lung cancer [19]. It should be noted that the estimated average daily consumption of beta-carotene ranges from 2 to 7 mg [20, 21]. In retrospect, it is remarkable that throughout the period of conducting supplementation research, there was a notable absence of crucial knowledge on the biotransformation of beta-carotene. It is evident that the notion of "the more the better" is not applicable in this context. It is important to ascertain an ideal dosage that offers a favorable balance between benefits and risks, while also possessing a comprehensive understanding of the biotransformation processes associated with antioxidants [22].

4. At Elevated Concentrations, Antioxidants Have Pro-Oxidant Properties

Antioxidants possess the capacity to give electrons. The capacity to reduce is crucial for the neutralization of radicals and other highly reactive entities. The pro-oxidant effect may be induced by electron donation in the presence of transition metal ions. The impact of ascorbic acid on lipid peroxidation generated by iron in an *in vitro* setting serves as a very informative illustration of this phenomenon [23]. Iron has been shown to cause a moderate form of lipid peroxidation. Additionally, when combined with ascorbic acid, it further enhances the process of lipid peroxidation by lowering the transition metal ion. Nevertheless, when present at a sufficiently elevated quantity, ascorbic acid has the ability to impede the process of lipid peroxidation. Hence, it can be shown that ascorbic acid exhibits pro-oxidant properties at low concentrations, while transitioning to antioxidant behavior at higher concentrations [24]. This statement challenges the notion that a significant presence of an antioxidant consistently leads to a pro-oxidant outcome. The *in vivo* pro-oxidant action of vitamin C has been extensively examined in widely disseminated research investigations. The administration of a daily dose of 500 mg of vitamin C over a period of 6 weeks resulted in a significant increase of 60% in plasma vitamin C levels [25]. The authors proposed that vitamin C has pro-oxidant properties, as shown by their analysis of the levels of oxidized DNA bases measured in peripheral blood cells. One critique that was expressed was to the absence of evidence demonstrating *in vivo* DNA oxidation, since it was seen just during the process of DNA separation. Additionally, the verification of the projected increase in lymphocyte vitamin C concentrations was not conducted. Another study done revealed that vitamin C has the capacity to interact with lipid hydroperoxides, leading to the generation of agents that have the ability to induce DNA damage. The methodology utilized in the *in vitro* investigation did not take into account three pivotal factors: (i) the degradation of lipid hydroperoxides by glutathione peroxidases in living organisms, (ii) the excessively heightened levels of lipid peroxides employed in the experiment, and (iii) the observation that the production of lipid hydroperoxides in living organisms only transpires subsequent to the depletion of vitamin C. Carr *et al.* did a thorough evaluation that determined the existence of discrepancies and contradictions in the data on the correlation between DNA and vitamin C oxidation *in vivo*. However, it was observed that some discrepancies might be attributed to limitations in experimental design and technique. There seems to be a lack of clarity on the correlation between high dosage and toxicity as postulated by Paracelsus, as well as the concept of pro-oxidant behavior. However, this misperception continues to endure.

5. Any Antioxidant May Be Used

Antioxidants, in a general sense, operate by inhibiting or delaying the process of oxidation in diverse substances. Therefore, the term "antioxidant" is used as a comprehensive descriptive term owing to the presence of several chemical compounds that exhibit distinct methods of action, leading to diverse outcomes. Antioxidants include a diverse array of chemical substances, such as thiols, phenols, and amines, which may have either hydrophilic or lipophilic characteristics. The term "nutricophore" was coined as a counterpart to the term "pharmacophore" to describe the functional moiety inside a molecule and emphasize the diverse chemical properties of antioxidants [26, 27]. The presence of variety among antioxidants results in the development of distinct (bio-)chemical profiles, manifest in various sites of action or biological processes. Various antioxidants have distinct biological effects, necessitating the appropriate selection of an antioxidant for various pathological situations. The activity of the antioxidant hydroxytyrosol in protecting LDL from oxidation is attributed to its lipophilic and hydrophilic characteristics, which are both implicated in its function. This activity is seen at the interface between lipids and water.

6. In Theory, Antioxidants Are Incapable Of Exhibiting Their Characteristic Behavior.

This misperception arises due to the high reaction rate of radicals with endogenous biomolecules. This implies that in order to provide protection, antioxidants must exhibit enhanced reactivity towards radicals, a proposition that is contended to be unattainable. This statement accurately applies to highly reactive hydroxyl radicals. Nonetheless, it is worth noting that lipid peroxyl radicals present in membranes have a much longer half-life when compared to hydroxyl radicals. Nevertheless, it has been shown that the presence of vitamin E may effectively counteract the effects of these lipid peroxyl radicals [28]. Site-specific scavenging, such as the process of binding to iron and effectively neutralizing hydroxyl radicals at their site of generation, has the potential to provide protection. In essence, it is feasible to achieve primary protection, which entails the prevention of hydroxyl radical production. Antioxidants may also exert their effects indirectly. These compounds possess distinct properties, such as the ability to inhibit NADPH oxidase, regulate redox-sensitive pathways for signaling involving transcription variables, and inhibit poly (ADP-ribose) polymerase (PARP-1). In addition to direct scavenging, there may be an occurrence of indirect antioxidant action via activities such as anti-inflammatory effects or the development of protective factors against oxidative stress. Antioxidants are crucial for sustaining existence in an

Table 17.1 Biomarkers and their advantages and disadvantages.

Biomarker	Advantages	Disadvantages
Lipids [60–63]		
Lipids F2-isoprostanes-prostaglandin-like products of primarily arachidonic acid	Highly specific	Requires specialized analytical skills
Malondialdehyde (MDA)-reactive, breakdown product of PUFAs formed during lipid peroxidation	Frequently applied method	MDA is broken down by aldehyde dehydrogenases
Lipid dienes-product of PUFAs formed during lipid peroxidation	Simple	Low sensitivity and specificity
Ex vivo LDL oxidation-antioxidants prevent oxidation of LDL isolated after antioxidant supplementation	Semi-functional marker	Amphiphilic antioxidants are lost during LDL isolation
DNA [64, 65]		
8-Hydroxy-20-deoxyguanosine-damaged nucleotide formed by radical damage to DNA	Specific and sensitive	Prone to artifacts
Comet-DNA damage and strand breaks	Semi-functional marker	Well-standardized procedure required
Proteins [66–68]		
Protein carbonyls-protein oxidation products	Relatively stable	High background and poor selectivity
Nitrotyrosine-formed by the reaction of reactive nitrogen species with tyrosine	Specific	Nitration of tyrosine during sample preparation
Protein activity-the activity of many proteins is reduced after reaction with reactive species	Semi-functional marker	Low sensitivity

aerobic environment [29]. Antioxidants exhibit biological activity. The assessment of their impact on biomarkers indicative of oxidative damage has determined their course of action. The provided examples are shown in Table 17.1.

7. The Measurement Of Antioxidant Level Is Indicative Of An Individual's Health

Several methodologies, including the Trolox equivalent antioxidant capacity (TEAC), oxygen radical absorbance capacity (ORAC), and ferric reducing ability of plasma (FRAP), are used to evaluate the overall antioxidant status in biological and dietary specimens. The utilization of blood plasma's antioxidant status as a biomarker has been recommended owing to its capability to effectively reflect the comprehensive antioxidant capacity of a specific sample, encompassing all accessible antioxidants. The United States Department of Agriculture (USDA) has put up a suggested guideline for the daily recommended intake of ORAC units for dietary consumption. In 2010, the recommendation and the ORAC database for selected foods were revoked as a result of being deemed physiologically incorrect. This decision was made in light of the identification of the Oxygen Radical Absorbance Capacity (ORAC) of certain foods. It is imperative to recognize that the outcomes of varied tests measuring antioxidant capacity are subject to the effect of unique underlying mechanisms that use separate sources of radicals or oxidants, along with variations in experimental methodologies. The use of standardized methods is highly encouraged. The degree of precision in assays for measuring antioxidant levels is limited [30]. Nevertheless, this characteristic might potentially be advantageous since it allows for the assessment of the collective activity of many antioxidants simultaneously. The potential use of antioxidant status as a measure of disease severity is worth considering; nevertheless, it fails to consider the distinctive biochemical characteristics of individual antioxidants. This concept was further expounded upon in point 5.

8. After Being Used, Antioxidants Become Inert

It is reasonable to infer that the efficacy of an antioxidant diminishes with oxidation. However, this notion is not accurate. An illustrative instance within this framework is dihydrolipoic acid, which retains considerable efficacy as a scavenger of hypochlorous acid (HOCl) even after undergoing oxidation to lipoic acid [31]. Furthermore, it is important to note that antioxidants operate within a complex network, and the process of regenerating oxidized antioxidants is a common occurrence [32]. Furthermore, there is a growing body of research indicating that both glucuronidated and methylated intermediates of antioxidants possess biological activity. The enzyme b-Glucuronidase, found in macrophages or neutrophils, is capable of hydrolyzing glucuronidated flavonoids, resulting in the release of unconjugated phenolic compounds [33]. This phenomenon may manifest in a localized area of inflammation where these cells are found, and it seems to be a sophisticated approach for specifically targeting these chemicals [34].

9. Natural Antioxidants Have Exceptional Properties

Chemophobia is a prevalent phenomenon characterized by an aversion to chemicals, and there is a prevailing belief that naturally occurring antioxidants, sometimes labeled with the prefix "bio-", are considered to be more effective than their chemically manufactured counterparts [35]. For consumers, the term 'bio' connotes naturalness and is often seen as being connected with safety. The same ideological framework links the term "chemical" with potential hazards. The prevailing belief in this myth is extensive and serves as the foundation for organic farming, a practice that abstains from the use of pesticides, herbicides, and synthetic fertilizers. It is noteworthy that R, R, R-a-tocopherol, which represents the natural variant of a-tocopherol, has more efficacy in safeguarding against the peroxidation of membrane lipids compared to other enantiomers of a-tocopherol that are artificially synthesized. Nevertheless, the synthetic version of R, R, R-a-tocopherol is indistinguishable from its natural counterpart due to its similar molecular composition. The aforementioned statement holds true for bio-vitamin C, since it has bioequivalence to chemically manufacture L-ascorbic acid [36].

10. The Efficacy Of Antioxidant Medicines Is Limited

Numerous pharmaceutical substances possess antioxidant properties while lacking explicit design as antioxidants, hence potentially contributing to their therapeutic efficacy. The importance of this antioxidative effect is contingent upon both the specific molecule used and the particular disease under consideration. The current understanding of the possible antioxidant scavenging effect of the angiotensin-converting enzyme (ACE) inhibitor captopril and the dihydropyridine-type calcium antagonists is restricted owing to the insufficient administration of the required dosages to see this activity. The problem lies in establishing the direct and immediate effects of scavenging antioxidant molecules inside a living organism. The human physiological system has a very intricate network of antioxidants. The adequate assessment of the antioxidant effects of network-augmenting drugs, such as N-acetylcysteine (a precursor to glutathione) and propofol (an anesthetic with potential as a substitute for a-tocopherol), is currently lacking [37]. Despite the existing comprehensive biochemical comprehension of the antioxidant mechanism of action of N-acetylcysteine and its demonstrated effectiveness in decreasing exacerbation rates in patients diagnosed with chronic obstructive pulmonary disease (COPD), it is important to highlight that this therapeutic intervention is presently not incorporated in the treatment guidelines for this particular patient population. The observable effects of antioxidants are most pronounced in situations when the production of free radicals and resulting damage are also evident [38–40]. Research has shown that antioxidants have the ability to alleviate the toxicity caused by medications that produce radicals via redox cycling, such as the anti-tumor medication doxorubicin. The primary objective of this research was to undertake a thorough examination of the effects of 7-mono-hydroxyethylrutoside (monoHER), an iron chelator and antioxidant, on doxorubicin-induced cardiotoxicity. The effectiveness of monoHER in reducing cardiotoxicity is well shown in mouse models, however its effects on human individuals are still inconclusive. The observed differences between mice and humans in terms of monoHER bio-transformation may be attributable to a species-specific differential [41]. The findings of the research indicate that the inclusion of monoHER significantly improves the cytostatic efficacy of doxorubicin. Moreover, several techniques are used by alternative antioxidants to inhibit cellular proliferation. However, a considerable amount of scholarly evidence suggests that antioxidants can potentially impede the effectiveness of anticancer treatments [42–44].

It is noteworthy that while the physiological significance of the widely acknowledged antioxidant properties of certain compounds is often a subject of debate, there is a growing focus on the unsupported hypothesis that antioxidant dietary supplements may actually promote, rather than inhibit, cancer development by interfering with the process of oxidant-induced apoptosis in cancer cells [45, 46]. The paradox around the effectiveness of antioxidants arises from their inability to demonstrate antioxidant activity while simultaneously inducing toxicity and perhaps leading to early death via their antioxidant properties. This contradiction presents a challenge in understanding the true impact of antioxidants. A further mechanism through which antioxidants exert their effects is activating transcription factors, such as Nrf2. As a consequence, the activation of endogenous antioxidant enzymes takes place via the response of the antioxidant element, leading to the indirect impact of antioxidants [47].

In conclusion, it can be inferred that the aforementioned statements serve as a comprehensive summary of the main points discussed during this discourse- What strategies may be used to achieve an optimal balance of antioxidants? The toxicological risk associated with oxygen, similar to that of any other substance, is determined by multiplying the hazard with the exposure. Oxygen is widely acknowledged as being among the most harmful chemicals that have been identified. The inherent chemical reactivity of oxygen gives birth to its dangerous feature. Exposure is an enduring phenomenon that, if avoided, has inevitable repercussions for organisms reliant on aerobic respiration. In order to manage the presence of oxygen, the use of antioxidants is essential [48]. Hence, the endogenous defense mechanisms against oxidative stress exhibit a considerable level of efficacy, and several physiological processes use the reactivity of molecular oxygen. Antioxidants are not likely to elicit distinct pharmacological effects [49]. Adrien Albert has identified pharmacological reactions as a manifestation of selective toxicity. Pharmaceutical substances exert their effects by selectively interacting with certain biological entities, such as enzymes, receptors, or transporters. The optimal course of action for a medicine is characterized by specificity, meaning that it selectively targets a certain receptor or pathway and elicits a strong physiological response. Measuring the clinical effectiveness of a medicine is, therefore quite straightforward [50]. This stands in opposition to sustenance and sustenance-derived compounds such as dietary supplements. These compounds have a wide range of activities. The conduct in question lacks specificity, making it difficult to ascertain its precise impact on human health [51]. The positive impact of pharmaceuticals is associated with the conceptualization of health outlined in the original constitution of the World Health Organization (WHO), which defines health as "a state of complete physical, mental, and social well-being," rather than only the absence of sickness or disability [52]. The concept of health is open to revision since there is a growing recognition that health is a fluid condition that necessitates the capacity to adjust to changes instead of a state of flawless well-being. Antioxidants, minerals, and dietary supplements possess diverse mechanisms of action [53–55]. We agree with the proposal to replace the word "antioxidants" with "bioactives" to include its pleiotropic mode of action. The presence of bioactive substances may lead to nuanced adjustments in homeostasis, which in turn may enhance the capacity to adapt to various disturbances [56]. In order to get a deeper understanding of this phenomena, new linguistic concepts have emerged as modes of artistic representation. These include increasing metabolic flexibility, heterostasis, maximizing metabolic health via triage, fine-tuning metabolism, strengthening homeostasis, and augmenting elasticity [57]. The assessment of antioxidants' effectiveness is most effectively conducted by evaluating their capacity to preserve homeostasis. In this particular scenario, the identification of biomarkers pertaining to the impact of antioxidants may be achieved by the deliberate perturbation of homeostatic equilibrium. The identification of subtle illnesses may be achieved by the use of stress testing [58]. Similarly, there is a need for further exploration into novel methodologies for identifying the nuanced impacts of antioxidants. An illustrative instance of this phenomenon is the use of a cardiovascular health rating as a means to amalgamate various impacts of dietary supplement interventions on vascular health. The index was established by amalgamating the modifications observed in twenty-three discrete biomarkers. The observed substantial rise in the integrated biomarker indicates a positive impact of the dietary supplement on cardiovascular well-being. The process of oxidative stress resulting from the breakdown of aerobic life due to oxygen may contribute to the development of many diseases, which can potentially be mitigated by implementing suitable antioxidant interventions [59].

It would be unfortunate if the potential health benefits of antioxidant-rich meals or antioxidant nutritional supplements were overlooked. It would be regrettable if the advancement of novel antioxidant medications were to halt due to the prevalence of misunderstandings.

References

1. Bast, A., Antioxidant pharmacotherapy. *Drug News Perspect.*, 7, 465–472, 1994.
2. Halliwell, and Gutteridge, J.M.C., *Free Radicals in Biology and Medicine (4th edn)*, Oxford University Press, New York, 2007.
3. Willcox, J.K. *et al.*, Antioxidants and prevention of chronic disease. *Crit. Rev. Food Sci. Nutr.*, 44, 275–295, 2004.
4. Lee, S.H. *et al.*, Vitamin C-induced decomposition of lipid hydroperoxides to endogenous genotoxins. *Science*, 292, 2083–2086, 2001.
5. Podmore, I.D. *et al.*, Vitamin C exhibits pro-oxidant properties. *Nature*, 392, 559, 1998.
6. Anson, N.M. *et al.*, Bioprocessing of wheat bran improves *in vitro* bioaccessibility and colonic metabolism of phenolic compounds. *Food Chem.*, 57, 6148–6155, 2009.
7. Hollman, P.C. *et al.*, The biological relevance of direct antioxidant effects of polyphenols for cardiovascular health in humans is not established. *J. Nutr.*, 141, 989S–1009S, 2011.
8. Malireddy, S. *et al.*, Phytochemical antioxidants modulate mammalian cellular epigenome: implications in health and disease. *Antioxid. Redox Signal.*, 17, 327–339, 2012.
9. Bjelakovic, G. *et al.*, Mortality in randomized trials of antioxidant supplements for primary and secondary prevention: systematic review and meta-analysis. *J. Am. Med. Assoc.*, 297, 842–857, 2007.
10. Biesalski, H.K. *et al.*, Re-examination of a meta-analysis of the effect of antioxidant supplementation on mortality and health in randomized trials. *Nutrients*, 2, 929–949, 2010.
11. Stampfer, M.J. *et al.*, Vitamin-E consumption and the risk of coronary-disease in women. *N. Engl. J. Med.*, 328, 1444–1449, 1993.
12. Vivekananthan, D.P. *et al.*, Use of antioxidant vitamins for the prevention of cardiovascular disease: meta-analysis of randomised trials. *Lancet*, 361, 2017–2023, 2003.
13. Eidelman, R.S. *et al.*, Randomized trials of vitamin E in the treatment and prevention of cardiovascular disease. *Arch. Intern. Med.*, 164, 1552–1556, 2004.
14. Halliwell, B., The antioxidant paradox. *Lancet*, 355, 1179–1180, 2000.
15. Schurks, M. *et al.*, Effects of vitamin E on stroke subtypes: meta-analysis of randomised controlled trials. *Br. Med. J.*, 314, c5702, 2010.
16. Watson, J., Oxidants, antioxidants and the current incurability of metastatic cancers. *Open Biol.*, 3, 120144, 2013.
17. Pauling, L., *Vitamin C and the Common Cold*, W.H. Freeman, San Francisco, 1970.
18. *Institute of Medicine Dietary Reference Intakes for Vitamin C, Vitamin E, Selenium, and Carotenoids*, National Academy Press, Washington D.C., 2000.
19. The Alfa-Tocopherol, Beta Carotene Cancer Prevention Study Group, The effect of vitamin E and beta carotene on the incidence of lung cancer and other cancer in male smokers. *N. Engl. J. Med.*, 330, 1029–1035, 1994.
20. Kousik, A. *et al.*, Intake of the major carotenoids and the risk of epithelial ovarian cancer in a pooled analysis of 10 cohort studies. *Int. J. Cancer*, 119, 2148–2154, 2006.

21. EFSA Panel on Food Additives and Nutritional Sources Added to Food, Scientific opinion on the re-evaluation of mixed carotenes and beta-carotene as food additive. *EFSA J.*, 10, 2593, 2012.
22. Wang, X.D., Absorption and metabolism of beta-carotene. *J. Am. Coll. Nutr.*, 13, 314–325, 1994.
23. Bast, A. *et al.*, Oxidants and antioxidants: state of the art. *Am. J. Med.*, 91, 2S–13, 1991.
24. Halliwell, B., Vitamin C: antioxidant or pro-oxidant *in vivo*. *Free Radic. Res.*, 25, 439–454, 1996.
25. Carr, A. *et al.*, Does vitamin C act as a pro-oxidant under physiological conditions? *FASEB J.*, 13, 1007–1024, 1999.
26. Rezk, B.M. *et al.*, The antioxidant activity of ploretin: the disclosure of a new antioxidant pharmacophore in flavonoids. *Biochem. Biophys. Res. Commun.*, 295, 9–13, 2002.
27. Heijnen, C.G.M. *et al.*, Protection of flavonoids against lipid peroxidation: the structure activity relationship revisited. *Free Radic. Res.*, 36, 575–581, 2002.
28. Van Acker, S.A.B.E. *et al.*, Molecular pharmacology of vitamin E: structural aspects of antioxidant activity. *Free Radic. Biol. Med.*, 15, 311–328, 1993.
29. Weseler, A.R. *et al.*, Oxidative stress and vascular function: implications for pharmacologic treatment. *Curr. Hypertens. Rep.*, 12, 154–161, 2010.
30. Arts, M.J.T.J. *et al.*, A new approach to assess the total antioxidant capacity using the TEAC assay. *Food Chem.*, 88, 567–570, 2004.
31. Biewenga, G.P. *et al.*, The pharmacology of the antioxidant lipoic acid. *Gen. Pharmacol.*, 29, 315–33, 1997.
32. Jacobs, H. *et al.*, An essential difference between the flavonoids monoHER and quercetin in their interplay with the endogenous antioxidant network. *PLoS One*, 5, 11, e13880, 2010.
33. Chanet, A. *et al.*, Flavanone metabolites decrease monocyte adhesion to TNF-a-activated endothelial cells by modulating expression of atherosclerosis-related genes. *Br. J. Nutr.*, 110, 4, 587–98, 2013.
34. Bartholome, R. *et al.*, Deconjugation kinetics of glucuronidated phase II flavonoid metabolites by b-glucuronidase from neutrophils. *Drug Metab. Pharmacokinet.*, 25, 379–387, 2010.
35. Bast, A., The risk of eating: natural versus man-made toxins, in: *Chemistry of Crop Protection*, G. Voss, and G. Ramos (Eds.), pp. 63–68, Wiley-VCH, Germany, 2003.
36. Brigelius-Flohé, R. and Traber, M.G., Vitamin E: function and metabolism. *FASEB J.*, 13, 1145–1155, 1999.
37. Bast, A. *et al.*, A-Lipoic acid and N-acetylcysteine: new concepts for old drugs, in: *Biothiols in Health and Disease*, E. Cadenas, and L. Packer (Eds.), pp. 409–425, Marcel Dekker, New York/Basel/Hong Kong, 1995.
38. Stey, C. *et al.*, The effect of oral N-acetylcysteine in chronic bronchitis: a quantitative systematic review. *Eur. Respir. J.*, 16, 253–262, 2000.
39. Decramer, M. *et al.*, Effects of N-acetylcysteine on outcomes in chronic obstructive pulmonary disease (Bronchitis Randomized on NAC Cost-Utility Study, BRONCUS): a randomized placebo-controlled trial. *Lancet*, 365, 1552–1560, 2005.
40. Vestbo, J., Hurd, S.S., Agustí, A.G., Jones, P.W., Vogelmeier, C., Anzueto, A., Barnes, P.J., Fabbri, L.M., Martinez, F.J., Nishimura, M., Stockley, R.A., Sin, D.D., Rodriguez-Roisin, R., Global strategy for the diagnosis, management, and prevention of chronic obstructive pulmonary disease: GOLD executive summary. *Am. J. Respir. Crit. Care Med.*, 187, 4, 347–65, 2013.
41. Kappus, H. *et al.*, Toxic effects associated with oxygen metabolism: Redox cycling drugs. *Experientia*, 37, 1233–1241, 1981.
42. Bast, A. *et al.*, Protection by flavonoids against anthracyclinecardiotoxicity: from chemistry to clinical trials. *Cardiovasc. Toxicol.*, 7, 154–159, 2007.

43. Jacobs, H. *et al.*, Differences in pharmacological activities of the antioxidant flavonoid monoHER in humans and mice are caused by variations in its metabolic profile. *Clin. Pharmacol. Ther.*, 90, 852–859, 2011.
44. Jacobs, H. *et al.*, The semisynthetic flavonoid monoHER sensitizes human soft tissue sarcoma cells to doxorubicin-induced apoptosis via inhibition of nuclear factor-kB. *Br. J. Cancer*, 104, 437–440, 2011.
45. Bruynzeel, A.M. *et al.*, The effect of monohydroxyethylrutoside on doxorubicin-induced cardiotoxicity in patients treated for metastatic cancer in phase II study. *Br. J. Cancer*, 97, 1084–1089, 2007.
46. Sung, B. *et al.*, Fisetin, an inhibitor of cyclin-dependent kinase 6, down-regulates nuclear factor-kB-regulated cell proliferation, antiapoptotic and metastatic gene products through the suppression of TAK-1 and receptor-interacting protein-regulated IkBa kinase activation. *Mol. Pharmacol.*, 71, 1703–1714, 2007.
47. Reuland, D.J. *et al.*, The role of Nrf2 in the attenuation of cardiovascular disease. *Exerc. Sport Sci.*, 41, 3, 162–168, 2013.
48. Albert, A., Selective Toxicity, in: *The Physico-chemical Basis of Therapy*, 7th edn, Chapman & Hall, London, 1985.
49. Heaney, R.P., Nutrients, endpoints, and the problem of proof. *J. Nutr.*, 138, 1591–1595, 2008.
50. World Health Organisation, Preamble to the Constitution of the World Health Organisation. *International Health Conference*, WHO, 1946.
51. What is health? The ability to adapt. *Lancet*, 373, 781, 2009. Editorial.
52. Sies, H., Polyphenols and health: update and perspectives. *Arch. Biochem. Biophys.*, 501, 2–5, 2010.
53. Weseler, A., Pleiotropic-acting nutrients require integrative investigational approaches: the example of flavonoids. *J. Agric. Food. Chem.*, 60, 8941–8946, 2012.
54. Litvin, M. *et al.*, Recurrent hypoglycemia: boosting the brain's metabolic flexibility. *J. Clin. Invest.*, 123, 1922–1924, 2013.
55. Ames, B.N., Prevention of mutation, cancer, and other age-associated diseases by optimizing micronutrient intake. *J. Nucleic Acids*, 2010, 725071, 2010.
56. vanOmmen, B. *et al.*, Challenging homeostasis to define biomarkers for nutrition-related health. *Mol. Nutr. Food Res.*, 53, 795–804, 2009.
57. Biesalski, H.K. *et al.*, 26th Hohenheim Consensus Conference, September 11, 2010. Scientific substantiation of health claims: evidence-based nutrition. *Nutrition*, 27, S1–S20, 2011.
58. Weseler, A.R. *et al.*, Pleiotropic benefit of monomeric and oligomericflavanols on vascular health. A randomized controlled clinical pilot study. *PLoS One*, 6, e28460, 2011.
59. Halliwell, B., Free radicals and antioxidants: updating a personal view. *Nutr. Rev.*, 70, 257–265, 2012.
60. Morrow, J.D. *et al.*, Increase in circulating products of lipid peroxidation (F2-isoprostanes) in smokers – smoking as a cause of oxidative damage. *N. Engl. J. Med.*, 332, 1198–1203, 1995.
61. Nielsen, F. *et al.*, Plasma malondialdehyde as biomarker for oxidative stress: reference interval and effects of life-style factors. *Clin. Chem.*, 43, 1209–1214, 1997.
62. Guichardant, M. and Lagarde, M., Analysis of biomarkers from lipid peroxidation: a comparative study. *Eur. J. Lipid Sci. Technol.*, 111, 75–82, 2009.
63. Rietjens, S.J. *et al.*, New insights into controversies on the antioxidant potential of the olive oil antioxidant hydroxytyrosol. *J. Agric. Food. Chem.*, 55, 7609–7614, 2007.
64. Shigenaga, M.K. and Ames, B.N., Assays for 8-hydroxy-20- deoxyguanosine: a biomarker of *in vivo* oxidative DNA damage. *Free Radic. Biol. Med.*, 10, 211–216, 1991.
65. Collins, A.R., The comet assay for DNA damage and repair. *Mol. Biotechnol.*, 26, 249–261, 2004.

66. Dalle-Donne, I. *et al.*, Protein carbonyl groups as biomarkers of oxidative stress. *Clin. Chim. Acta*, 329, 23–38, 2003.
67. Van der Vliet, A. *et al.*, Nitrotyrosine as biomarker for reactive nitrogen species. *Methods Enzymol.*, 269, 175–184, 1996.
68. Aksenov, M. *et al.*, Oxidative modification of creatine kinase BB in Alzheimer's disease brain. *J. Neurochem.*, 74, 2520–2527, 2000.

18

Exploring the Therapeutic and Pharmaceutical Potential of Antioxidants

Sumanta Bhattacharya

Department of Textile Technology, Maulana Abul Kalam Azad University, West Bengal, India

Abstract

The health issues related to the oxidative stress is increasing rapidly due to the degradation of external environment. The pollution of environment and toxicity in food are some important reasons behind the increase of oxidative stress in human cells. The inherent enzyme based antioxidant mechanism of the body is not enough to maintain the redox balance of human body due to the increase in toxicity. The oxidative stress may lead to severe health issues like cancer and other genetic issues in human health. The use of antioxidant based pharmaceutical products will play significant role in dealing with this increasing rate of chronic health issues. Different types of antioxidants like nutrient antioxidants, enzyme antioxidants, metabolic oxidants etc. are used for mitigating the oxidative stress in the human cells and tissues. The efficiency of the antioxidant based drugs can be improved through different cutting edge technological modifications. The antioxidant chemicals play important role in the development of pharmaceutical industry and novel drug synthesis. The therapeutic potential of the antioxidants can be enhanced through the application of modern technologies like nanotechnology, biotechnology, genetic engineering etc. The farming of antioxidant based fruits and vegetables also needs to be prioritized for increasing the cultivation of antioxidant based herbal products which will expand the scope of Ayurveda in the treatment of oxidative stress related issues. The principals of green chemistry need to be adopted for the manufacturing of antioxidant based drugs for the achievement of sustainability.

***Keywords*:** Oxidative stress, antioxidant therapy, cancer treatment, pharmaceutical industry, drug synthesis

18.1 Introduction

When oxidative metabolism speeds up, it floods human cells with reactive oxygen species, damaging every part of the cell. Antioxidants are molecules that counteract the effects of oxidative stress and free radicals. Cancer, osteoporosis, diabetes, respiratory disorders, and other chronic conditions are all reduced in severity. Free radicals can cause chromosomal mutations, but antioxidants can stop that from happening by preventing DNA strand breaks. By suppressing cellular mutations, the chance of cancer is reduced.

Email: sumanta.21394@gmail.com; ORCID: 0000-0003-2563-2787

Rakesh K. Sindhu, Inderbir Singh and M. Arockia Babu (eds.) Antioxidants: Nature's Defense Against Disease, (587–612)

In addition to inhibiting pro-oncogenic signaling, antioxidants promote a wide variety of cell signaling pathways that limit cell growth and proliferation. Rather than directly targeting free radicals, certain enzymatic antioxidants, such as glucose-6-phosphate dehydrogenase, glutathione reductase, etc., operate as secondary antioxidants that react with cofactors such as glucose-6-phosphate, NADPH, GSH, etc. Pro-oxidant Reactive Oxygen Species are produced in large quantities by cancer cells. It promotes the growth and spread of cancer cells by activating other cells in the body to experience oxidative stress. Weak pro-oxidants, on the other hand, play a crucial role in antioxidant therapy by boosting the body's natural antioxidant capabilities. The synergistic effect of a diet high in antioxidants boosts the effectiveness of other nutritional components, increasing the overall anti-cancer effects of particular substances. In addition to lowering the risk of acute infections, it also lowers the effects of poisonous dietary additives. Chronic inflammation results from oxidative stress's disruption of cell signaling pathways, which triggers an inflammatory response in the absence of any external stimulus. Diseases like osteoporosis are directly attributable to it. It also accelerates cancer-causing cell division. Internal tissue damage caused by free radicals is another source of inflammation. Antioxidants stop free radicals in their tracks, decreasing your odds of developing osteoporosis and other chronic inflammatory disorders. In order to reduce the prevalence of chronic diseases that have become common in the modern era due to environmental deterioration, technological progress in the production of pharmaceutical goods rich in antioxidants is required. Better antioxidant medicines can be created with the help of in-depth pharmacokinetic and pharmacodynamic studies of the antioxidant components using cutting-edge biosensors. If antioxidants are to be used more effectively in therapy, synergistic effects with other medicinal chemicals need to be studied [1].

This requires technological advancements. Antioxidants may be found in the medicinal plants *Artemisiae annuae* Herba, *Cassia occidentalis, Cheilanthes albomarginata, Grateloupia lanceolata, Hibiscus sabdariffa, Houttuynia cordata, Inula japonica, Schisandra chinensis, Smilax glabra*, etc. Plant extracts like this can be used to make anti-oxidant drugs at a fraction of the manufacturing cost and with fewer side effects. Antioxidant medicine derived from plants will be produced in a way that adheres to green synthesis principles including waste minimization and the use of non-toxic chemicals. In addition to the industrial production of antioxidant medications, which would enhance the agricultural sector and the rural economy, the promotion of the antioxidant-rich diet is necessary [2].

18.2 Roles of Antioxidants in Medical Science

Reactive oxygen species generated in body through mitochondrial oxidative metabolism in response to the foreign invasion of xenobiotics, cytokines, and bacteria [1]. The reactive oxygen species are essential for normal physiological function like generation of ATP, various catabolic and anabolic processes, and the functioning of cell redox cycles as shown in Figure 18.1. However, the generation of excess reactive oxygen species (ROS) and reactive nitrogen species (RNS) cause chronic cellular damage, and affect homeostasis leads to oxidative stress as shown in Figure 18.2. The oxidative stress in the body also leads to the generation of cold stress, physical stress, chronic stress, nutritional stress, hypoxic stress etc.

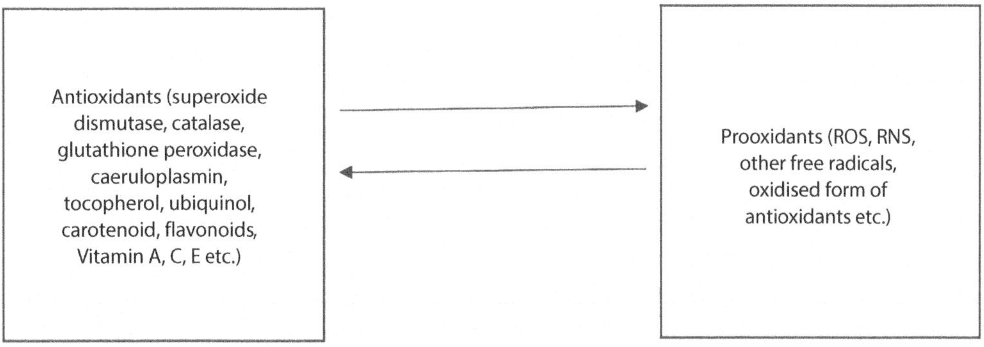

Figure 18.1 Redox equilibrium in cells.

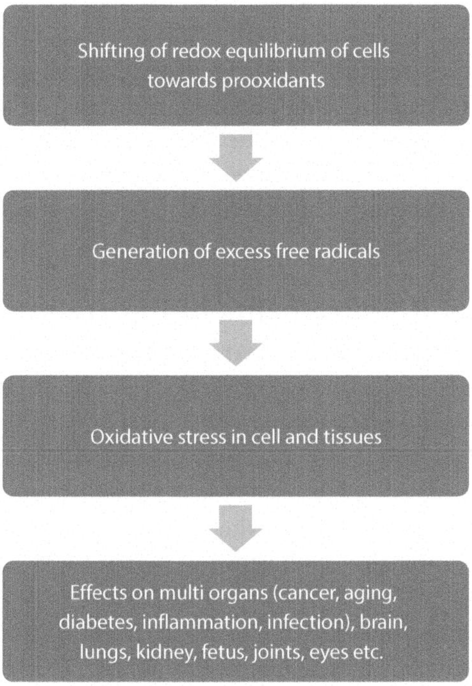

Figure 18.2 Causes of oxidative stress and its effects on health.

The accumulation of excessive ROS leads to the contractile dysfunction causing the weakness in muscle and fatigue. The oxidative stress causes the disruption of homeostasis mechanism and the affects the immunological response of the body. The uncontrolled rate of free radical generation leads to the injury of cells and damage in tissues. The ROS also inhibits the signaling of cytochrome c mediated apoptosis which leads to cancer. Several types of metabolic diseases like diabetes associate with the enhanced level of lipoperoxidation. The ROS produced in the tissues inflicts direct damage to the essential macromolecules like lipids, proteins, and nucleic acids [3]. The shifting of the balance between the antioxidants and prooxidants molecules in the human body towards the prooxidants leads to oxidative stress in human body. The antioxidants may produce in the body through internal mechanism due to the activities of the enzymes like superoxide dismutase, catalase, glutathione peroxidase, glutathione reductase etc. It may be introduced from outside through different food supplements containing minerals like manganese, copper, selenium etc. and vitamin A, C, and E. On the other hand, the prooxidant molecules produce in the body as byproducts of aerobic metabolism. The external factors like smoking, drinking, chewing tobacco, excessive sun exposure, prolonged exposure to metallic toxicity induce the production of antioxidants in the body [4].

The enzyme antioxidants like superoxide dismutase, catalase, glutathione peroxidase, caeruloplasmin etc. reduce the activities of the free radicals. The chain breaking antioxidants like tocopherol, ubiquinol, carotenoid, flavonoids etc. consume the free radicals during scavenging process to maintain the oxidative balance of the body. The metal binding protein based antioxidants like transferrin, ferritin, lactoferrin etc. bind with the metals like iron, copper etc. to prevent the generation of free radicals [5].

Antioxidants protect the human cells against the toxicity and prevent the disease. The external antioxidant supplements like carotene, ascorbic acid, tocopherol, sulphur amino acids, selenium, zinc, carotenoids, lutein, anthocyanins etc. enhances the inherent antioxidant mechanism of the human body. The effects of different antioxidants have been depicted in Table 18.1. Antioxidants also protect cells from premature and abnormal ageing and enhances the immune system of the body [6].

18.3 Technological Requirements in Improving the Therapeutic Potential of Antioxidants

Antioxidant is a sustainable alternative therapeutic mechanism for the treatment of chronic diseases emerges due to the loss of redox balance in the human cells. The application of modern science and technology will improve the therapeutic potential of the antioxidants. Moreover, the agricultural productivity of the antioxidant rich fruits and vegetables can be increased through the extensive application of agricultural science and technology as shown in Figure 18.3.

The application of nanotechnology will increase the bioavailability of the antioxidant based pharmaceutical products. The engineered nanomaterials like dendrimers, nanoemulsion, liposomes, graphene nanotubes, polymeric nanomaterials etc. will enable the successful targeted and non-targeted delivery of antioxidants to the desired destination. Moreover, the use of smart nanopills will also facilitate the pharmacokinetic study of the antioxidants that will enrich the research and development in the field [2].

The use of nanobiotechnology will enhance the agricultural outcome of the antioxidant rich crops. The technologies like nano priming, nano enabled organic fertilizers, nanophytoremediation, genetic engineering, bioinformatic study of the plant crops etc. will increase the sustainable productivity of the antioxidant based crops and herbs. The cultivation of antioxidant rich herbs will ensure the supply of high quality raw materials for the pharmaceutical industry [7].

Table 18.1 Effects of different antioxidants on human health.

Types of antioxidants	Name	Effects on health
Nutrient oxidants	Vitamin E	Protection against lipid peroxidation; Prevention against colon, prostate, and breast cancers; Treatment of cardiovascular diseases, ischemia, cataract, arthritis, and certain neurological disorders.
	Vitamin C	Reduction of stomach, lung, and colorectal cancer
	Beta carotene	Prevent the effects of prooxidants on eyes, increased risk of lung and prostate cancer
	Lycopene	Protect breast, prostate, and lung cancer
	Selenium	Anti-carcinogenic and immunomodulator
	Flavonoids	Prevent chronic and degenerative effects such as cancer, cardiovascular diseases, aging, cataract, memory loss, arthritis
	Omega-3 and omega-6 fatty acids	Reduce inflammation and prevent chronic diseases like cardiovascular diseases, memory loss, depression, cancer, cataract; improve diabetic neuropathy, osteoporosis, psoriasis, eczema.
Enzyme antioxidants	Superoxide dismutase, glutathione reductase,	Catalyse the conversion of superoxide radicals to hydrogen peroxide.
	Catalase and	Conversion of hydrogen peroxide to oxygen and water
	Glutathione peroxidase	Remove hydrogen peroxide by using it to oxidize reduced glutathione to oxidized glutathione, reduction of lipid or nonlipid hydroperoxides
Metabolic antioxidants	Lipoid acid, L-ariginine, glutathione, uric acid, transferrin, coenzyme Q10, melatonin,bilirubin, metal-chelating proteins	Suppress the formation of free radicals in various metabolic reactions at cellular level.

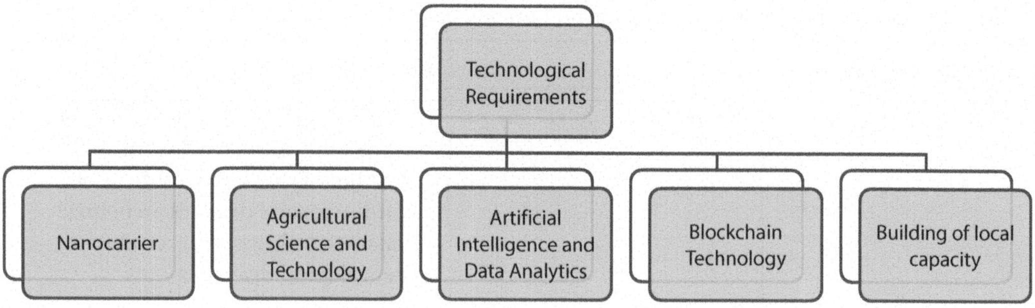

Figure 18.3 Technological requirements in improving the therapeutic potential of antioxidants.

Handling vast amounts of data generated by the pharmacokinetic monitoring systems antioxidant rich pharmaceutical products necessitates advanced analytics and artificial intelligence (AI). These technologies can identify patterns, predict potential issues, and optimize the performance of antioxidants in healthcare management. Understanding the dynamics of the pharmaceutical products in the body is crucial for the successful implementation of the antioxidants. Advanced modeling tools can simulate the understanding of the mechanism of medicines under different scenarios, aiding in the design and planning of pharmaceutical research [8].

Successful implementation of antioxidants relies on the involvement among different stakeholders like researchers, academicians, healthcare professionals, patients etc. Technology can facilitate this multistakeholder engagement through online platforms, mobile applications, and social media, fostering awareness, education, and collaboration in water management practices.

Blockchain technology can enhance transparency and accountability in healthcare governance and pharmaceutical research. Implementing blockchain in pharmaceutical research and technology can help the tracking of water usage, monitor the effectiveness of pharmaceutical projects, and ensure fair and equitable distribution of water resources [9].

Infrastructure supporting the healthcare professionals, researchers, and pharmaceutical industry must be designed with resilience in mind. This includes the use of climate resilient materials, adaptive design approaches, and the incorporation of smart technologies to ensure the longevity and effectiveness of antioxidant based pharmaceutical products.

The building of local capacity is crucial for the sustained success of the awareness against free radical toxicity. Online training platforms, virtual workshops, and knowledge-sharing systems can empower local communities, policymakers, and healthcare practitioners with the necessary skills and understanding of the impact of free radical toxicity and the necessity of antioxidant rich diet.

The achievement of a successful worldwide experience with antioxidant based treatment is the integration of advanced technologies across various domains. From remote sensing and AI to community engagement platforms and resilient infrastructure design, a holistic approach is necessary to harness the full potential of antioxidants for sustainable healthcare management on a global scale.

18.4 Economic Implication of the Therapeutic Potential of Antioxidants

The analysis of the therapeutic potential of antioxidants and their pharmaceutical applications has significant economic implications as shown in Figure 18.4. As research in this field progresses, it can lead to the development of novel drugs and treatments, impacting healthcare costs, pharmaceutical industries, and overall economic productivity.

The research and development of pharmaceuticals, including antioxidants, involve substantial investments. High-quality analysis can streamline the drug discovery process, potentially reducing costs associated with failed trials and accelerating successful outcomes. The application of modern technologies like nanotechnology improves the bioavailability of antioxidant drug molecules in the target site by avoiding the side effects. Moreover, the development of smart pills facilitates the pharmacokinetic study of the drug molecules that will lead to development of more effective drugs. Therefore, the application of the technology will reduce the necessity of trials due to the reduction of the possibilities of side effects. [10].

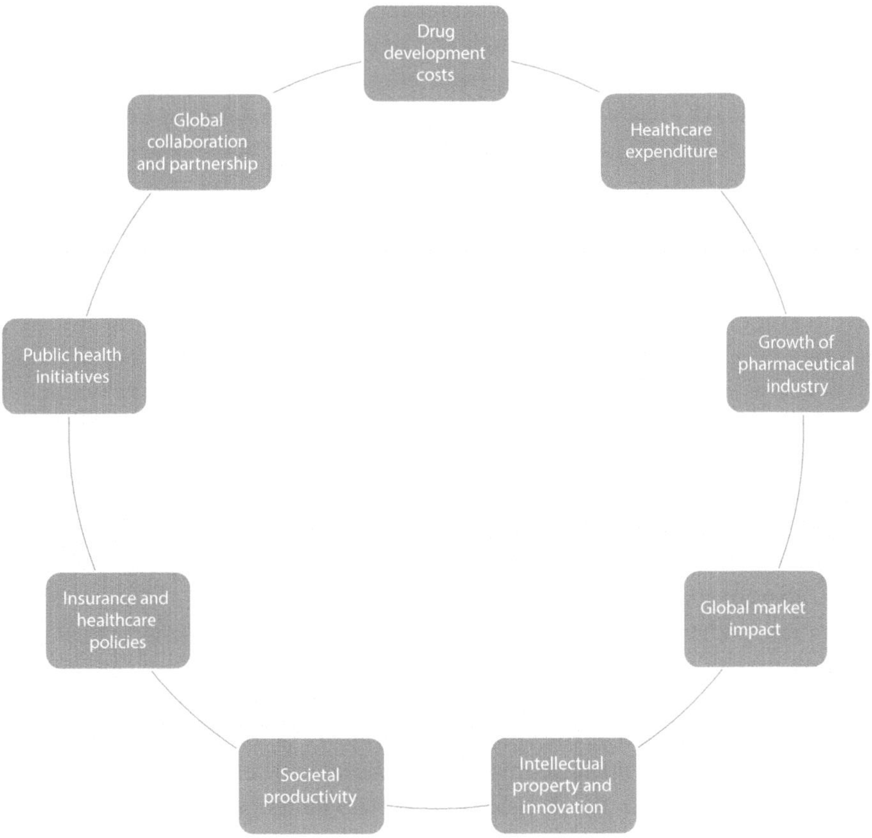

Figure 18.4 Economic implication of the therapeutic potential of antioxidants.

If antioxidants prove effective in preventing or treating diseases, there could be a subsequent reduction in healthcare expenditures. Diseases associated with oxidative stress, such as neurodegenerative disorders or cardiovascular diseases, may see a decline in prevalence, leading to lower healthcare costs. The consumption of antioxidant rich food items will reduce the expenditure in healthcare as it will mitigate the risk of attack by chronic diseases by maintaining the proper redox balance in the body cells effectively.

Positive findings in the therapeutic potential of antioxidants can stimulate growth in the pharmaceutical industry. Companies investing in antioxidant research and production may experience increased demand for their products, contributing to industry expansion and job creation. The application of modern science and technology in manufacturing the antioxidant rich pharmaceutical products will expand the market of the pharmaceutical industry.

As pharmaceutical companies develop and market antioxidant-based therapies, the global market for these products could witness substantial growth. This can have far-reaching economic effects, influencing trade balances and contributing to the GDP of countries involved in antioxidant production. Moreover, the market value of the antioxidant rich food items will also increase substantially due to the promotion of antioxidant rich diets which will facilitate the boost of agricultural sector. Successful analysis in this field may lead to the creation of valuable intellectual property, such as patents for new antioxidant compounds or therapeutic methods. This intellectual capital can drive innovation, attract investment, and foster a culture of research and development [11].

Improved health outcomes resulting from effective antioxidant-based treatments can have positive effects on societal productivity. A healthier population is generally more productive, leading to economic gains as people participate more actively in the workforce. The consumption of antioxidants will reduce the risk of the chronic diseases associated with the damage of cells and tissues and will enhance the productivity of the population. It will also prevent the aging and increases the efficiency of the human workforce which will lead to the economic growth of the society [12].

Positive therapeutic outcomes may influence insurance policies and healthcare regulations. Insurers may reevaluate coverage and premiums based on the potential reduction in disease risk and associated costs, impacting both individual and corporate budgets.

Governments might invest in public health initiatives promoting antioxidant-rich diets or supplements if the research supports their efficacy. This could influence agricultural practices, impacting the economy through changes in demand for certain crops or supplements. Government may promote the cultivation of antioxidant rich agricultural crops. The antioxidant rich diet may also be included in the public distribution schemes launched by central and different state governments.

Collaborations between countries, research institutions, and pharmaceutical companies can emerge as a result of shared interests in antioxidant research. These collaborations can foster knowledge exchange, technology transfer, and joint ventures, contributing to a more interconnected global economy. The cross border transfer of modern technology will boost the pharmaceutical industry and also improve the therapeutic efficiency of the antioxidants.

Economic impacts should also consider potential challenges, such as ethical concerns surrounding the marketing and accessibility of antioxidant-based therapies. Striking a balance between profitability and ethical practices is crucial for sustainable economic benefits. The economic impact of analyzing the therapeutic potential of antioxidants is multifaceted, ranging from direct effects on healthcare expenditures to broader influences on industries, innovation, and international collaborations. As research progresses, the full scope of these economic implications will become more apparent, shaping the landscape of healthcare and the pharmaceutical industry [4].

18.5 Environmental Implication of the Therapeutic Potential of Antioxidants

The analysis of the therapeutic potential of antioxidants and their pharmaceutical applications involves assessing both their benefits and environmental impacts as shown in Figure 18.5. Antioxidants, known for their ability to neutralize harmful free radicals, play a crucial role in various health treatments. However, the production and utilization of antioxidant-based pharmaceuticals can contribute to environmental concerns.

The manufacturing process of antioxidants often involves the use of chemicals and energy-intensive procedures, leading to carbon emissions and other pollutants. Additionally, the extraction of raw materials for antioxidant production may result in deforestation or habitat disruption. It's essential to consider these environmental costs when evaluating the overall sustainability of antioxidant-based therapies. Furthermore, the disposal of pharmaceutical products, including antioxidants, poses challenges. Improper disposal can lead to the release

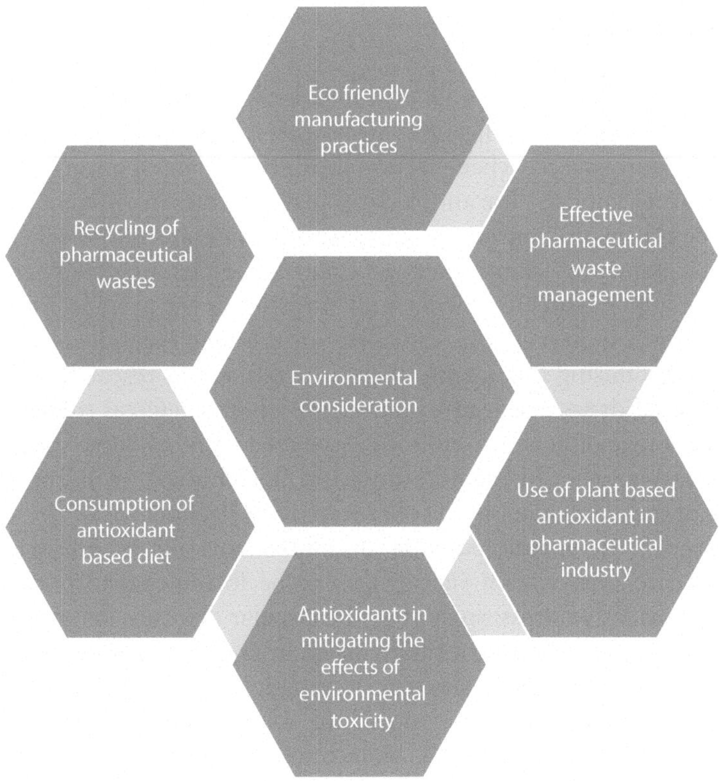

Figure 18.5 Environmental implication of the therapeutic potential of antioxidants.

of active compounds into the environment, affecting ecosystems and potentially harming aquatic life. Research into eco-friendly production methods and biodegradable formulations is crucial to mitigate these adverse effects. While antioxidants offer promising therapeutic potential, it's imperative to balance their medical benefits with the environmental consequences of their production and disposal. Sustainable practices and ongoing research are essential to minimize the ecological footprint associated with antioxidant-based pharmaceuticals [13].

The use of the plant based antioxidants in the manufacture of pharmaceutical products will improve the sustainability of the antioxidant based pharmaceutical industry. The effective management of the pharmaceutical wastes generated needs to be the priority of the industry. The pharmaceutical wastes are the potential sources of the products for other industries. Moreover, the use of antioxidant based diet will reduce the demand for the chemical based pharmaceutical products and mitigate the environmental implications.

The increase in the rate of consumption of fruits, vegetables, and certain dietary supplements can substantially enhance the protection against many common types of environmentally induced free radicals. The antioxidant based diets increase the oxidative stress caused by the exposure to environmental pollutants. The lung lining fluid antioxidants like ascorbate, glutathione, urate etc. provide protection against the particulate matter toxicity. The dietary uptake of tomato lycopene prevents nitrogen dioxide mediated damage [14].

18.6 Industrial Consideration of the Therapeutic Potential of Antioxidants

The analysis of the therapeutic potential of antioxidants and their pharmaceutical applications has significant industrial implications as shown in Figure 18.6. As research in this field advances, industries are poised to benefit from the development of novel antioxidant-based therapies and pharmaceuticals. The antioxidants play key roles in different industrial areas, viz. in drug development and innovation, increase of disease prevention efficiency, biotechnological analysis, regulatory compliance and standards, global market dynamics etc.

The analysis of antioxidants for therapeutic purposes opens avenues for drug development. Pharmaceutical industries can capitalize on the identification of specific antioxidants that exhibit promising therapeutic effects. Investment in research and development for antioxidant-based drugs can lead to innovative pharmaceutical solutions, potentially addressing a wide range of health conditions associated with oxidative stress. The application of modern technologies like nanotechnology, biotechnology will improve the therapeutic outcome of the antioxidant based pharmaceutical products [15].

Understanding the therapeutic potential of antioxidants allows industries to focus on preventive healthcare. Antioxidant-rich drugs can be formulated to prevent or manage diseases linked to oxidative stress, such as cardiovascular diseases, neurodegenerative disorders, and certain cancers. Industries may invest in creating formulations that enhance the body's natural antioxidant defenses, contributing to overall health and well-being. The investment in the commercial production of the antioxidant rich fruits and vegetables through the application microbiological, biotechnological, and genomic studies will also enhance the market potential of antioxidants.

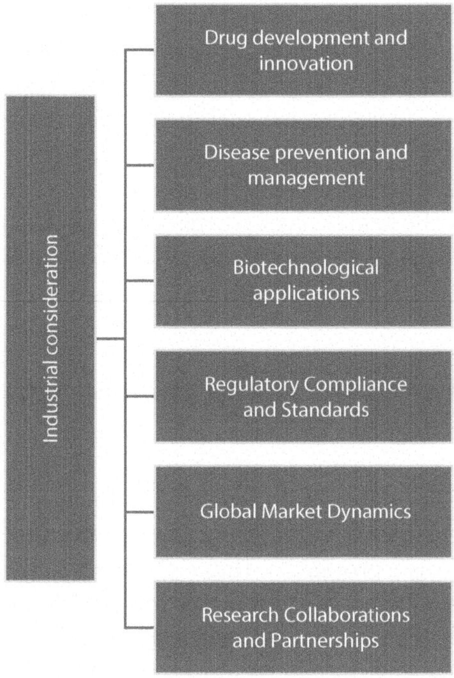

Figure 18.6 Industrial consideration of the therapeutic potential of antioxidants.

Antioxidants have applications beyond traditional pharmaceuticals. Biotechnological industries can explore the integration of antioxidants in various products, including skincare, dietary supplements, and functional foods. Incorporating antioxidants into biotechnological products can cater to consumer demands for health and wellness, leading to the development of a new range of marketable goods. Different plant tissue culture techniques like *in vitro* root cultures, micropropagation, callogenesis etc. facilitate the extraction of different antioxidant based bioactive components from the plant extracts which exhibit anti-ageing properties. It also enables the development of green label cosmeceutical products [16].

As antioxidant-based therapies emerge, regulatory bodies play a crucial role in setting standards for their production and use. Industries need to stay abreast of regulatory changes and ensure compliance with evolving guidelines. Meeting regulatory standards is essential for the successful development and commercialization of antioxidant-based pharmaceuticals, ensuring safety and efficacy in the market. The therapeutic potential of antioxidants can influence global market dynamics. Industries operating in the pharmaceutical and healthcare sectors may experience shifts in demand, with consumers seeking antioxidant-sbased treatments. Companies that strategically position themselves in this emerging market can gain a competitive advantage, driving economic growth within the pharmaceutical industry.

The analysis of antioxidant therapeutic potential fosters collaborations between pharmaceutical industries, research institutions, and academic entities. Such partnerships can accelerate the pace of discovery and development. Collaborative efforts can lead to shared resources, expertise, and a collective approach to addressing challenges associated with antioxidant-based drug development. The collaboration among different international bodies will enable the cross border exchange of technology and expertise which will benefit the developing economies technologically as well as financially [17].

The industrial impact of analyzing the therapeutic potential of antioxidants is multifaceted. From drug development and disease management to biotechnological applications and global market dynamics, industries have the opportunity to leverage this knowledge for innovation and growth. As the field evolves, a proactive approach to research, development, and collaboration will be essential for maximizing the industrial benefits of antioxidant-based therapies.

18.7 Quality Control in the Application of Antioxidants

The environmental impact of antioxidants-based drugs and their applications is an important consideration in the pharmaceutical industry as shown in Figure 18.7. The synthesis and manufacturing of antioxidants-based drugs can involve the use of various chemicals, solvents, and energy-intensive processes. These processes can generate waste materials, emissions, and consume resources. It is important for pharmaceutical companies to adopt sustainable manufacturing practices, such as green chemistry principles and waste management strategies, to minimize environmental impacts.

Proper disposal of pharmaceutical waste, including unused medications and manufacturing by-products, is crucial to prevent environmental contamination. Antioxidants-based drugs may have specific disposal requirements due to their chemical properties. It is important for pharmaceutical companies, healthcare facilities, and individuals to follow appropriate guidelines and regulations for Good Manufacturing Practices (GMP): Antioxidants-based drugs must be manufactured in accordance with GMP guidelines. GMP ensures that manufacturing processes are standardized, controlled, and consistent, minimizing the risk of contamination, errors, or variations that could impact product quality. GMP covers various aspects, including facility design, personnel training, equipment calibration and maintenance, documentation, and quality control testing [18].

Quality control starts with the selection and control of raw materials used in the production of antioxidants-based drugs. Pharmaceutical companies must establish stringent supplier qualification processes, conduct quality assessments of raw materials, and ensure their traceability. Raw material quality control includes testing for identity, purity, potency, and absence of contaminants. Robust analytical testing is performed at various stages of drug development and production to ensure the quality, identity, strength, and purity of antioxidants-based drugs. This includes testing of raw materials, in-process testing during

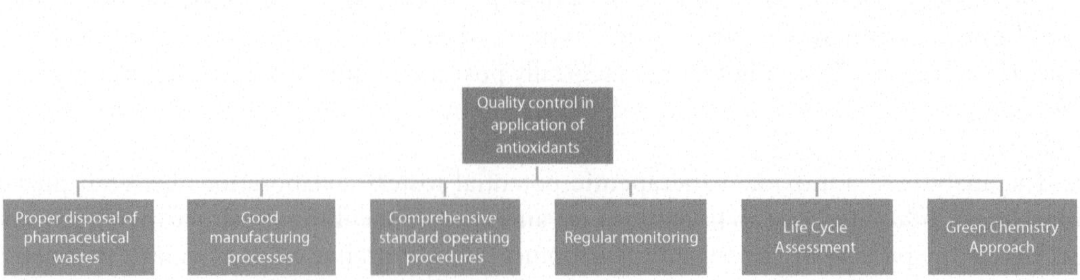

Figure 18.7 Quality control in the application of antioxidants.

manufacturing, and final product testing. Analytical techniques such as high-performance liquid chromatography (HPLC), mass spectrometry (MS), and spectroscopic methods are used to assess drug quality and detect any impurities or deviations from specifications [19].

Stability testing evaluates the shelf-life and storage conditions of antioxidants-based drugs. It involves subjecting the drugs to various environmental conditions (e.g., temperature, humidity) over a defined period to assess their stability, degradation, and potential changes in potency or quality. Stability testing helps determine appropriate storage and expiry date information for the drugs.

Comprehensive documentation is maintained to ensure traceability, accountability, and transparency in the quality control process. This includes standard operating procedures (SOPs), batch records, specifications, test methods, and quality control reports. Documentation provides a clear record of all activities and ensures compliance with regulatory requirements [20].

Regular internal and external quality audits and inspections are conducted to assess compliance with quality control standards, regulations, and guidelines. These audits may be performed by the company's quality control team, regulatory authorities, or independent third-party organizations. Audits help identify areas for improvement, ensure adherence to quality standards, and maintain regulatory compliance.

Quality control extends beyond the manufacturing process to include monitoring and reporting of adverse events associated with the use of antioxidants-based drugs. Robust systems are in place to collect, evaluate, and report any adverse reactions or safety concerns. Adverse event monitoring helps identify potential quality issues and enables timely risk management and corrective actions [21].

Quality control management in antioxidants-based drugs is guided by regulatory agencies and industry standards to ensure patient safety and product quality. Pharmaceutical companies dedicate resources to establishing quality control systems, conducting thorough testing, and implementing continuous improvement processes to meet or exceed regulatory requirements and deliver safe and effective drugs to patients [19].

Antioxidants-based drugs and their metabolites can potentially have ecotoxic effects if released into the environment. Depending on their properties, these compounds may pose risks to aquatic organisms, affecting aquatic ecosystems and biodiversity. It is essential to assess and mitigate these potential impacts through appropriate testing and environmental risk assessments during the drug development process [17].

Considering the environmental impact of antioxidants-based drugs involves conducting a life cycle assessment (LCA) that evaluates the environmental burdens associated with the entire life cycle of the drug, from raw material extraction to disposal. This assessment can help identify areas for improvement, such as reducing energy consumption, minimizing waste generation, and optimizing supply chain logistics. Antioxidants-based drugs often require raw materials derived from natural sources. Sustainable sourcing practices, such as ensuring responsible harvesting or cultivation of plant-based starting materials, can help minimize the impact on ecosystems and biodiversity [21].

Researchers and pharmaceutical companies are increasingly exploring green chemistry approaches to develop more sustainable and environmentally friendly synthesis routes for antioxidants-based drugs. These approaches focus on minimizing the use of hazardous chemicals, reducing waste, and maximizing resource efficiency. Regulatory bodies have implemented guidelines and regulations to address the environmental impact of pharmaceuticals. Compliance with these regulations, such as proper waste management and environmental risk assessment, is essential to ensure the safe and sustainable use of antioxidants-based drugs.

By considering the environmental impact throughout the life cycle of antioxidants-based drugs, pharmaceutical companies can work towards minimizing their ecological footprint and promoting sustainable practices in drug development and manufacturing.

18.8 Legal Considerations of the Application of Antioxidants

The analysis of the therapeutic potential of antioxidants and their pharmaceutical applications has significant implications for both law and policy as shown in Figure 18.8. This comprehensive examination encompasses various aspects, including regulatory frameworks, intellectual property rights, ethical considerations, and public health policies.

The introduction of antioxidant-based pharmaceuticals necessitates a thorough evaluation of existing regulatory frameworks. Health regulatory agencies play a pivotal role in ensuring the safety and efficacy of these therapeutic agents. Understanding the regulatory landscape is crucial for manufacturers, researchers, and policymakers to navigate the complex process of drug development, approval, and market entry [20].

The analysis of antioxidant pharmaceuticals also involves a consideration of intellectual property rights. Patents, trademarks, and other forms of intellectual property protection are essential for incentivizing innovation in the pharmaceutical industry. Exploring the balance

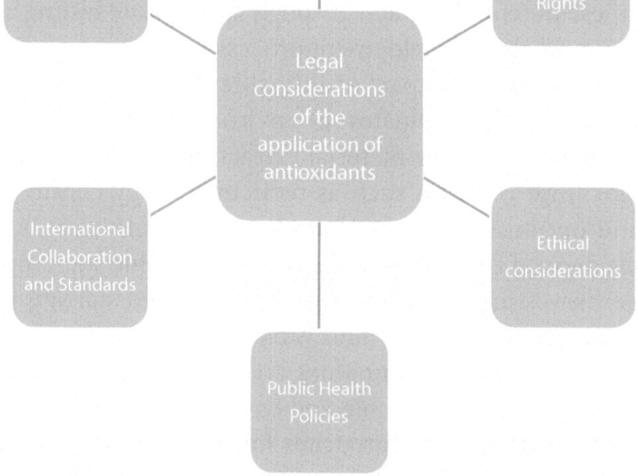

Figure 18.8 Legal considerations of the application of antioxidants.

between encouraging research and development through intellectual property rights and ensuring affordable access to these medicines is a critical aspect of this discussion. Different plants are the potential sources of antioxidants. Therefore, the farming, marketing, and effective utilization of the antioxidant rich herbs need to be ensured through the formulation of enforcement of proper legal framework [22].

The ethical dimensions of antioxidant-based therapies cannot be overlooked. Questions regarding equitable access to these treatments, informed consent, and potential conflicts of interest in research and development must be addressed. Policymakers must grapple with ethical dilemmas to create a framework that prioritizes patient welfare while promoting scientific progress. The marginal sections associated with the cultivation of antioxidant rich plant species needs to get benefitted effectively through the application of modern agricultural engineering and marketing of the products [23].

The broader impact of antioxidant pharmaceuticals on public health policies is another key area of exploration. Policymakers must assess the implications for healthcare systems, considering issues such as cost-effectiveness, reimbursement mechanisms, and the integration of these therapies into existing treatment protocols. Striking a balance between innovation and sustainability in healthcare delivery is paramount. The inclusion of the antioxidant rich diet in public distribution system needs to be considered for the improvement of public health [24].

The global nature of pharmaceutical research and development, international collaboration and the establishment of common standards are imperative. Harmonizing regulatory practices, sharing research findings, and fostering collaboration among countries can accelerate the integration of antioxidant therapies into global healthcare systems. The transfer of technology and expertise through multilateral collaboration among different nations will provide more effective outcome [17].

Anticipating future developments in antioxidant research and pharmaceutical applications is essential for effective policymaking. This includes monitoring emerging technologies, potential side effects, and adapting regulatory frameworks to keep pace with scientific advancements.

The analysis of the therapeutic potential of antioxidants and their pharmaceutical applications has far-reaching implications for law and policy. Addressing regulatory challenges, intellectual property considerations, ethical concerns, and the broader impact on public health policies requires a multidimensional and collaborative approach. Policymakers must navigate these complexities to foster innovation, ensure patient access, and promote the responsible integration of antioxidant-based therapies into healthcare systems worldwide.

18.9 Roles of Antioxidants in the Implication of Policies and Governance in Healthcare Sector

The analysis of the therapeutic potential of antioxidants and their pharmaceutical applications has significant implications for policies and governance in the healthcare sector as shown in Figure 18.9. The growing body of research in this field underscores the need for informed decision-making, regulatory frameworks, and strategic planning to harness the benefits of antioxidants effectively.

Figure 18.9 Roles of antioxidants in the implication of policies and governance in healthcare sector.

One key aspect of governance is the regulation of antioxidant-based pharmaceuticals. As the therapeutic potential of antioxidants is explored, policymakers must establish robust frameworks to ensure the safety, efficacy, and quality of these pharmaceutical products. This involves collaboration between regulatory bodies, research institutions, and pharmaceutical companies to establish and enforce standards that protect public health. Additionally, the economic impact of antioxidant-based therapies must be considered in healthcare policies. The introduction of new pharmaceutical applications may affect healthcare budgets, insurance policies, and accessibility to treatments. Policymakers need to strike a balance between promoting innovation and ensuring affordability and equitable access to antioxidant therapies for diverse populations [5].

Furthermore, the analysis of antioxidant potential raises ethical considerations. Governance frameworks should address issues such as the equitable distribution of benefits, transparency in research practices, and the inclusion of diverse populations in clinical trials. Ensuring that the benefits of antioxidant therapies reach all segments of society is a key aspect of ethical governance in healthcare [23].

The integration of antioxidant therapies into existing healthcare systems also requires policy adjustments. This includes updating medical guidelines, training healthcare professionals, and incorporating antioxidant treatments into standard protocols. Governance mechanisms should facilitate a smooth transition to the integration of antioxidants into mainstream healthcare, ensuring that patients receive the best available treatments [24].

Moreover, international collaboration and standardization are crucial aspects of governance in the context of antioxidant research. Policymakers need to engage in global discussions to harmonize regulations, share research findings, and promote a unified approach to the therapeutic use of antioxidants. This collaboration can lead to the development of international standards that enhance the quality and safety of antioxidant pharmaceuticals [22].

The analysis of the therapeutic potential of antioxidants and their pharmaceutical applications has far-reaching implications for policies and governance in healthcare. Policymakers must navigate regulatory, economic, ethical, and international considerations to ensure that the benefits of antioxidant therapies are maximized while minimizing potential risks. A well-informed and adaptive governance framework is essential to harness the full potential of antioxidants in improving health outcomes and advancing pharmaceutical innovation. This comprehensive analysis, spanning scientific, regulatory, economic, ethical, and global dimensions, will contribute valuable insights to inform healthcare policies and governance related to the therapeutic potential of antioxidants [22].

18.10 Roles of Stakeholder's Interest in Analyzing the Therapeutic Potential of Antioxidants

The analysis of the therapeutic potential of antioxidants and their pharmaceutical applications is a multifaceted task that involves understanding the impact on various stakeholders and investments as shown in Figure 18.10. The potential benefits of antioxidants in healthcare and medicine have garnered substantial interest from pharmaceutical companies, researchers, healthcare providers, and patients alike.

From a pharmaceutical perspective, investing in the research and development of antioxidant-based therapies requires a comprehensive analysis of the scientific landscape. Stakeholders in the pharmaceutical industry need to assess the current state of antioxidant research, identify potential drug candidates, and evaluate their efficacy and safety profiles. This analysis involves not only understanding the biochemical mechanisms of antioxidants but also navigating regulatory frameworks to ensure compliance with industry standards [25].

Healthcare providers are also key stakeholders in this analysis, as they play a pivotal role in the adoption and integration of antioxidant-based therapies into clinical practice. The impact on patient outcomes, treatment protocols, and overall healthcare costs must be considered. Assessing the feasibility and cost-effectiveness of incorporating antioxidant interventions into existing healthcare frameworks is crucial for stakeholders such as hospitals, clinics, and healthcare insurers [15].

Patients, as ultimate beneficiaries of therapeutic interventions, are deeply invested in the outcomes of antioxidant research. Understanding the potential benefits and risks of antioxidant-based treatments is essential for informed decision-making. Moreover, the accessibility and affordability of these therapies must be addressed to ensure equitable healthcare outcomes.

Researchers and scientists in academia, research institutions, and pharmaceutical companies are stakeholders with a keen interest in antioxidant-based drugs. They conduct

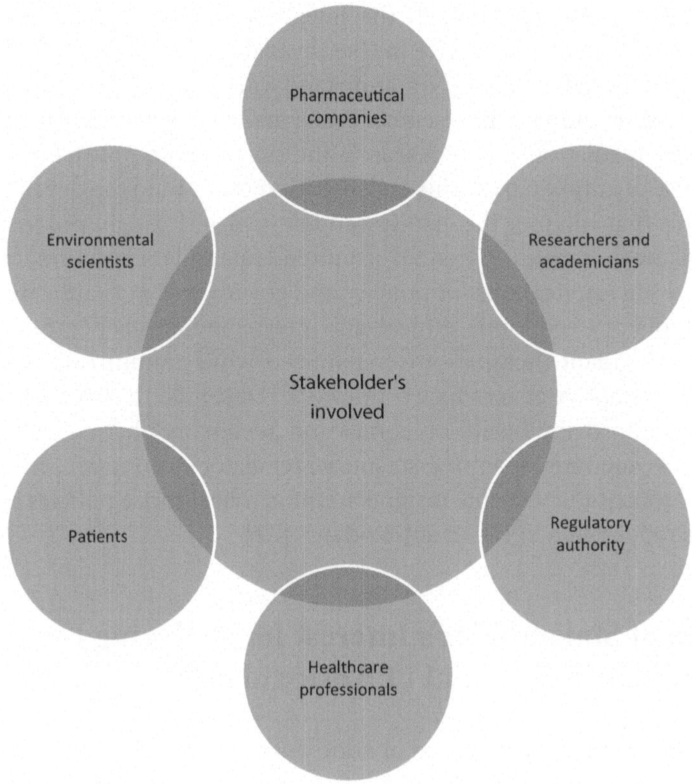

Figure 18.10 Stakeholder's involved in the analysis and use of antioxidants.

research to discover new compounds, study their mechanisms of action, assess their efficacy, and explore potential applications. Their work contributes to advancing the knowledge and understanding of antioxidant-based drugs.

Environmental and societal considerations are additional facets of the analysis, as the production and utilization of pharmaceuticals have broader implications. Stakeholders need to evaluate the environmental impact of manufacturing antioxidant-based drugs, considering factors such as raw material sourcing, waste disposal, and energy consumption. Additionally, societal aspects, including public awareness, education, and ethical considerations, play a role in shaping the overall landscape of antioxidant therapy [15].

Investors, including venture capitalists, private equity firms, and funding agencies, provide financial support for the research and development of antioxidant-based drugs. These stakeholders are interested in the commercial potential and return on investment in drug development projects. Their investments support the advancement of antioxidant-based drug research and enable their translation into viable products.

Society at large, as well as public health organizations, have an interest in the availability and accessibility of effective antioxidant-based drugs. These stakeholders focus on promoting public health, preventing diseases, and addressing healthcare inequalities. They advocate for policies that facilitate equitable access to these drugs and support initiatives that promote population health and well-being.

The analysis of the therapeutic potential of antioxidants and their pharmaceutical applications involves a comprehensive examination of scientific, economic, regulatory, and societal factors. Stakeholders, including pharmaceutical companies, healthcare providers, and patients, must collaborate to navigate the complexities of bringing antioxidant-based therapies to fruition. The investment in such research and development extends beyond financial considerations, encompassing the broader goal of advancing healthcare and improving patient outcomes [25].

18.11 Health and Safety Consideration of the Application of Antioxidants

Health and safety control is of utmost importance in the development, production, and use of antioxidant-based drugs. Before antioxidant-based drugs can progress to human trials, preclinical studies are conducted to assess their safety profile. These studies involve testing the drugs in laboratory settings, animal models, and *in vitro* systems to evaluate potential toxicities, adverse effects, and determine the appropriate dosage range as shown in Figure 18.11.

During clinical trials, the safety of antioxidant-based drugs is closely monitored. This involves rigorous evaluation of the drug's effects on human subjects, including the identification and reporting of any adverse events or side effects. Clinical trials are conducted in multiple phases to gradually assess safety and efficacy, with regular monitoring of

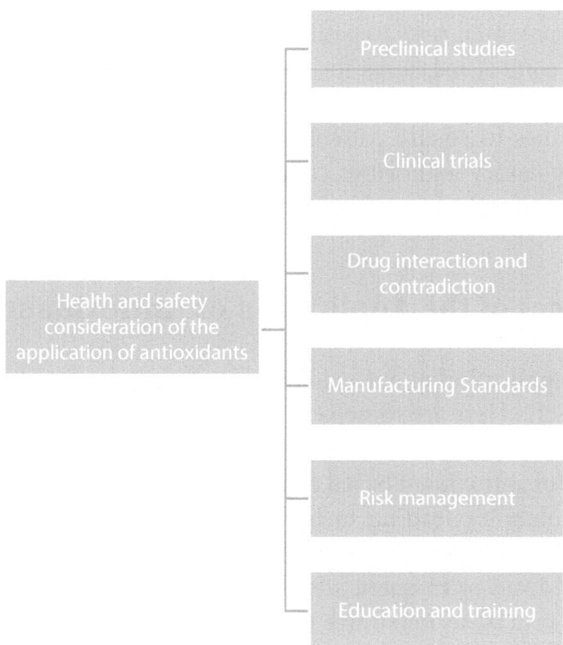

Figure 18.11 Health and safety consideration of the application of antioxidants.

participants. Once antioxidant-based drugs are approved and, on the market, pharmacovigilance systems are put in place to monitor and identify any unexpected adverse reactions or safety concerns. Healthcare professionals, patients, and other stakeholders are encouraged to report any observed or suspected side effects, which are then evaluated to ensure ongoing safety monitoring [26].

The manufacturing of antioxidant-based drugs must adhere to stringent quality control and Good Manufacturing Practices (GMP) guidelines. These guidelines ensure that the drugs are produced in a consistent, controlled, and safe manner, minimizing the risk of contamination or errors that could impact patient safety. The use of plant based materials for the manufacture of antioxidant based pharmaceutical products will increase the sustainability of the pharmaceutical products [23].

Accurate and comprehensive labeling and packaging of antioxidant-based drugs are crucial for ensuring their safe use. The labels provide information about dosage instructions, potential side effects, contraindications, warnings, and precautions to be taken. Clear instructions help healthcare professionals and patients use the drugs correctly and avoid potential risks. Antioxidant-based drugs may interact with other medications or have specific contraindications that need to be considered to ensure patient safety. Thorough evaluation of drug-drug interactions and identification of any contraindications help healthcare professionals make informed decisions when prescribing or administering these drugs [20].

Healthcare professionals involved in the prescribing, administration, and monitoring of antioxidant-based drugs require adequate education and training. This includes understanding the drug's mechanism of action, potential side effects, monitoring parameters, and appropriate patient counseling to ensure safe and effective use. Systems for reporting and managing adverse events associated with antioxidant-based drugs are in place to promptly address any safety concerns. These reports, along with risk management strategies, help identify and mitigate potential risks to patient health and safety [23].

Health and safety control measures for antioxidant-based drugs are guided by regulatory agencies, such as the FDA in the United States or the EMA in Europe, which establish standards and guidelines to ensure patient safety. Pharmaceutical companies, healthcare professionals, and regulatory authorities collaborate to maintain and improve health and safety practices throughout the lifecycle of these drugs [27].

18.12 Challenges of the Application of Antioxidants

The production and application of antioxidants-based drugs come with several challenges as shown in Figure 18.12. Here are some key challenges associated with antioxidants-based drugs.

The bioavailability of antioxidant based drugs is limited. Antioxidants-based drugs often exhibit poor water solubility, leading to limited bioavailability and reduced therapeutic effectiveness. Low bioavailability can result in suboptimal drug concentrations at the target site, requiring higher doses or frequent administration. Overcoming the challenge of limited bioavailability is crucial to maximize the therapeutic potential of antioxidants-based drugs. Moreover, antioxidants cannot survive in extreme thermal or oxidative stressed condition, osmotic shock and pH change [28].

Figure 18.12 Challenges of the application of antioxidants.

The stability of the antioxidant based drug molecules is another important challenge. Antioxidants-based drugs can be prone to instability and degradation, which can impact their shelf-life and efficacy. Factors such as light, temperature, pH, and oxidation can cause degradation and compromise drug stability [26]. Developing stable formulations and implementing appropriate storage conditions are essential to mitigate this challenge.

The effective delivery of drug molecules in the targeted needs to be ensured. Achieving targeted delivery of antioxidants-based drugs to specific cells, tissues, or organs remains a challenge. Effective targeting requires strategies such as functionalized nanoparticles, ligand-receptor interactions, or stimuli-responsive drug delivery systems. Designing and optimizing delivery systems that selectively deliver the drugs to the desired site while minimizing off-target effects are areas of ongoing research [29].

Drug resistance is a significant challenge in the treatment of various diseases, including cancer. Antioxidants-based drugs may encounter resistance mechanisms that reduce their efficacy over time. Understanding the mechanisms of resistance and developing strategies to overcome or prevent resistance are critical for improving the long-term effectiveness of antioxidants-based drugs [26].

The side effects of the application of drug molecules need to be mitigated. Antioxidants-based drugs can exhibit toxicities and side effects that limit their therapeutic use. Some antioxidants may have inherent toxicities, and their use must be carefully evaluated to balance the benefits and risks. Addressing the challenge of toxicity involves optimizing drug structures, dosage forms, and delivery systems to minimize adverse effects while maintaining therapeutic efficacy [29].

The protection of intellectual property rights is crucial for the commercialization of antioxidants-based drugs. However, challenges such as patent expiration, patent infringement, and the emergence of generic competitors can impact the profitability and exclusivity of these drugs. Pharmaceutical companies must navigate these challenges while balancing the need for innovation and affordable access to these drugs.

The regulatory approval process can be complex and time-consuming, presenting challenges in bringing antioxidants-based drugs to market. Complying with regulatory requirements, conducting extensive preclinical and clinical studies, and addressing safety and efficacy concerns are essential but demanding aspects of the drug development process [26].

The cost of developing, manufacturing, and accessing antioxidants-based drugs can be a barrier to their widespread use. Balancing the cost of drug development, manufacturing, and pricing with affordability and accessibility for patients and healthcare systems is a challenging task. Ensuring equitable access to these drugs while maintaining profitability and sustainability is an ongoing challenge [30].

Overcoming these challenges requires interdisciplinary collaboration, innovative research, advancements in drug delivery systems, regulatory support, and investment in infrastructure and technology. Addressing these challenges can enhance the development, production, and application of antioxidants-based drugs, ultimately improving patient outcomes and public health.

18.13 Future Perspective

The future research direction of antioxidant-based drugs production and applications is driven by the goal of enhancing therapeutic efficacy, improving drug delivery, enhancing the productivity of antioxidant rich crops, and addressing existing challenges as shown in Figure 18.13.

The drug delivery system needs to be improved in order to ensure the effective delivery of drug molecules. Researchers can focus on developing innovative drug delivery systems to improve the bioavailability and targeted delivery of antioxidant-based drugs. The use of antioxidant-based pharmaceutical products in combination therapies needs to be addressed in order to improve the therapeutic potential of drug molecules. Investigating the potential of antioxidant-based drugs in combination therapies is an important research direction. The research on the possibilities of the combination of smart pills with the antioxidant based drug molecules needs to be focused for understanding the mechanism of action and resistance of the antioxidant with the human cell and tissue. The exploration of the repurposing of existing antioxidant-based drugs for new indications can provide opportunities for expedited development and cost-effective solutions. Additionally, designing antioxidant-based drugs that can target multiple pathways or targets simultaneously can offer enhanced therapeutic benefits and address complex diseases. The improvement of the pharmacokinetics and pharmacodynamics study of the antioxidant based drugs is necessary in order to improve the therapeutic potential of antioxidant based drug products. Personalized medicine aims to tailor treatments based on individual patient characteristics. Research can focus on identifying biomarkers or genetic variations that influence the response to antioxidant-based drugs. Further research is needed to evaluate the safety and

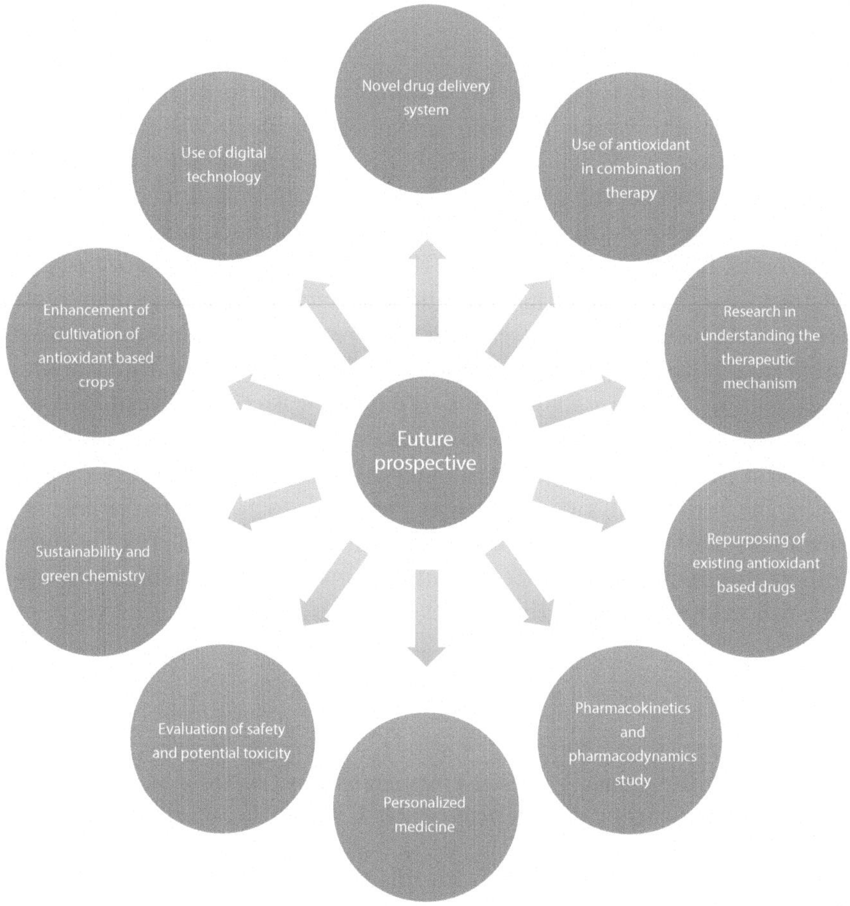

Figure 18.13 Future perspectives.

potential toxicity of antioxidant-based drugs. The adaptation of the principles of sustainable development and green chemistry is required for the development of antioxidant based medicinal products. The cultivation of the antioxidant rich medicinal herbs needs to be prioritized through the application of advanced agricultural science and technology. The research in the field of plant biotechnology needs to be prioritized in order to enhance the antioxidant contents in the crops and facilitates their use in pharmaceutical sector. Integration of digital technologies, such as artificial intelligence (AI), machine learning, and big data analytics, can accelerate drug discovery, optimize formulations, and improve treatment outcomes. Conducting patient-centered outcomes research is crucial to understand the real-world impact of antioxidant-based drugs on patients' lives. These research directions have the potential to advance the field of antioxidant-based drugs, leading to more effective therapies, improved patient outcomes, and enhanced healthcare delivery. Collaboration among researchers, pharmaceutical companies, healthcare providers, and regulatory agencies is essential to drive progress in these areas and translate research findings into clinical applications.

18.14 Conclusion

The demand of antioxidants is growing exponentially due to the increase in the rate of oxidative stress related health issues. The antioxidant rich products will enrich the healthcare and pharmaceutical sector in coming future. It also plays important role in mitigating the risk of cancer and other severe genetic issues by reducing the vulnerabilities of damage in the cells. It also improves the efficiency of the metabolic activity by reducing the risk of the loss of redox balance in human cells. The therapeutic efficiency of the antioxidants can also be increased through advanced technological modifications in drug synthesis and drug delivery mechanism. Further research is required in this field in order to expand the application of antioxidants in pharmaceutical and healthcare sectors. The awareness related to the increasing rate of vulnerabilities due to oxidative stress in human cells and tissues needs to be increased. The achievement of sustainable development goals and the reduction of environmental pollution is an important step of decreasing the risk of oxidative stress related issues. The antioxidant rich health diet also needs to be adopted largely by preventing the consumption of synthetically processed food items. The eco-friendly climate smart agricultural practices will also enhance the antioxidant related nutrients in the agricultural crops which will reduce the risk of these types of health issues.

References

1. Ray, P.D., Huang, B.W., Tsuji, Y., Reactive oxygen species (ROS) homeostasis and redox regulation in cellular signaling. *Cell. Signalling*, 24, 5, 980–990, 2012, https://doi.org/10.1016/j.cellsig.2012.01.008.
2. Cheng, H., Chen, L., McClements, D.J., Xu, H., Long, J., Zhao, J., Xu, Z., Meng, M., Jin, Z., Recent advances in the application of nanotechnology to create antioxidant active food packaging materials. *Taylor Francis, Crit. Rev. Food Sci. Nutr.*, 64, 1–16, 2022, https://doi.org/10.1080/10408398.2022.2128035.
3. Rahal, A., Kumar, A., Singh, V., Yadav, B., Tiwari, R., Chakraborty, S., Dhama, K., Oxidative Stress, Prooxidants, and Antioxidants: The Interplay, in: *Biomed Research International, Prooxidant Mechanisms in Toxicology*, Hindwai, Egypt, 2014, Article ID 761264, DOI: https://doi.org/10.1155/2014/761264.
4. Avcil, M., Role of Antioxidants and its Functions in Human Body. *Oxidants Antioxid. Med. Sci.*, 11, 7, 2022.
5. Singh, J., Khanna, S., Dwivedi, V., Pandey, S., Mishra, V.K., Role and Importance of Antioxidant in Medical Science- A Review. *Int. J. Pharma Bio. Sci.*, 11, 3, 129–134, 2020, DOI: http://dx.doi.org/10.22376/ijpbs.2020.11.2.b129-134.
6. Pham-Huy, L.A., He, H., Pham-Huy, C., Free Radicals, Antioxidants in Disease and Health. *Int. J. Biomed. Sci.*, 4, 2, 89–96, 2008.
7. Mishra, Y., Amin, H., II, Mishra, V., Vyas, M., Prabhakar, P.K., Gupta, M., Kanday, R., Sudhakar, K., Saini, S., Hromić-Jahjefendić, A., Alijabali, A.A.A., El-Tananani, M., Serrano-Aroca, A., Bakshi, H., Tambuwala, M.M., Application of nanotechnology to herbal antioxidants as improved phytomedicine: An expanding horizon. *Biomed. Pharmacother.*, 153, 113413, 2022, DOI: https://doi.org/10.1016/j.biopha.2022.113413.

8. Idowu, S.O. and Fatokun, A.A., Artificial Intelligence (AI) to the Rescue: Deploying Machine Learning to Bridge the Biorelevance Gap in Antioxidant Assays. *SLAS Technol.*, 26, 1, 16–25, 2021, DOI: https://doi.org/10.1177/2472630320962716.
9. Ghosh, P.K., Chakraborty, A., Hasan, M., Rashid, K., Siddique, A.H., Blockchain Application in Healthcare Systems: A Review. *MDPI, Sys.*, 11, 1, 2023, DOI: https://doi.org/10.3390/systems11010038.
10. Lauritano, C., Montuori, E., De Falco, G., Carrella, S., In Silico Methodologies to Improve Antioxidants' Characterization from Marine Organisms. *MDPI, Antioxid.*, 12, 3, 2023, DOI: https://doi.org/10.3390/antiox12030710.
11. Celestino, M.T., Magalhaes, U.O., Fraga, A.G.M., do Carmo, F.A., Lione, V., Castro, H.C., de Sousa, V.P., Rodrigues, C.R., Cabral, L.M., Rational use of antioxidants in solid oral pharmaceutical preparations. *Braz. J. Pharm. Sci.*, 48, 3, 2012, DOI: https://doi.org/10.1590/S1984-82502012000300007.
12. Wilson, D.W., Nash, P., Buttar, H.S., Griffiths, K., Singh, R., Meester, F.D., Horiuchi, R., Takahashi, T., The Role of Food Antioxidants, Benefits of Functional Foods, and Influence of Feeding Habits on the Health of the Older Person: An Overview. *MDPI, Antioxid.*, 6, 4, 81, 2017, DOI: https://doi.org/10.3390/antiox6040081.
13. Khounani, Z., Hosseinzadeh-Bandbafha, H., Nizami, A.S., Sulaiman, A., Goli, S.A.H., Tavassoli-Kafrani, E., Ghaffari, A., Rajaeifar, M.A., Kim, K.H., Talebi, A.F., Aghbashlo, M., Tabatabaei, M., Data on environmental analysis of natural antioxidant production from walnut husk by a solar photovoltaic-driven system as a replacement for potentially carcinogenic synthetic antioxidants. *Data Brief*, 28, 104933, 2020, DOI: https://doi.org/10.1016/j.dib.2019.104933.
14. Poljsak, B. and Fink, R., The Protective Role of Antioxidants in the Defence against ROS/RNS-Mediated Environmental Pollution, in: *Oxidative and Cellular Longevity*, Hindwai, Wisconsin, USA, 2014, Article ID 671539, DOI: https://doi.org/10.1155/2014/671539.
15. Egbujor, M.C., Egu, S.A., Okonkwo, V., II, Jacob, A.D., Egwautu, P., II, Amasiatu, I.S., Antioxidant Drug Design: Historical and Recent Developments. *J. Pharm. Res. Int.*, 32, 41, 36–56, 2021.
16. Bouzroud, S., Maaiden, E.E., Sobeh, M., Merghoub, N., Boukcim, H., Kouisni, L., Kharrassi, Y.E., Biotechnological Approaches to Producing Natural Antioxidants: Anti-Ageing and Skin Longevity Prospects. *MDPI, Int. J. Mol. Sci.*, 24, 2, 2023, DOI: https://doi.org/10.3390/ijms24021397.
17. Bresgen, N. and Siems, W., Peter Eckl: Research on the Pro-/Antioxidant Balance. *MDPI, Antioxid.*, 11, 6, 1079, 2022, DOI: https://doi.org/10.3390/antiox11061079.
18. Soliman, M.A., Pedersen, J.A., Park, H., Castaneda-Jimenez, A., Stenstrom, M.K., Suffet, I.H., Human Pharmaceuticals, Antioxidants, and Plasticizers in Wastewater Treatment Plant and Water Reclamation Plant Effluents. *Water Environ. Res.*, 79, 2, 156–167, 2007, DOI: https://doi.org/10.2175/106143006X111961.
19. Irina-Georgiana, M. and Apetrei, C., Analytical Methods Used in Determining Antioxidant Activity: A Review. *Int. J. Mol. Sci.*, 22, 7, 3380, 2021, DOI: https://doi.org/10.3390/ijms22073380.
20. Kebede, M. and Admassu, S., Application of Antioxidants in Food Processing Industry:Options to Improve the Extraction Yields and MarketValue of Natural Products. *Adv. Food Technol. Nutr. Sci.*, 5, 2, 38–49, 2019, DOI: 10.17140/AFTNSOJ-5-155.
21. Pérez-López, P., González-García, S., Jeffryes, C., Agathos, S.N., McHugh, E., Walsh, D., Murray, P., Moane, S., Feijoo, G., Moreira, M.T., Life cycle assessment of the production of the red anti-oxidantcarotenoid astaxanthin by microalgae: From lab to pilot scale. *HAL*, 64, 1, 322–344, 2016, https://hal.science/hal-01284775.
22. Zehiroglu, C. and Sarikaya, S.B.O., The importance of antioxidants and place in today's scientific and technological studies. *J. Food Sci. Technol.*, 56, 11, 4757–4774, 2019, DOI: 10.1007/s13197-019-03952-x.

23. Salehi, B., Martorell, M., Arbiser, J.L., Sureda, A., Martins, N., Maurya, P.K., Sharifi-Red, M., Kumar, P., Sharifi-Rad, J., Antioxidants: Positive or Negative Actors? *MDPI, Biomol.*, 8, 4, 124, 2018, DOI: https://doi.org/10.3390/biom8040124.
24. Lobo, V., Patil, A., Phatak, A., Chandra, N., Free radicals, antioxidants and functional foods: Impact on human health. *Pharmacogn. Rev.*, 4, 8, 118–126, 2010, DOI: 10.4103/0973-7847.70902.
25. Lourenço, S.C., Moldão-Martins, M., Alves, V., Antioxidants of Natural Plant Origins: From Sources to Food Industry Applications. *MDPI, Mol.*, 24, 22, 2019, DOI: https://doi.org/10.3390/molecules24224132.
26. Steinhubl, S.R., Why Have Antioxidants Failed in Clinical Trials? *Am. J. Cardiol.*, 101, 10Supplement, S14–S19, 2008, DOI: https://doi.org/10.1016/j.amjcard.2008.02.003.
27. Olsen, N., How can antioxidants benefit our health?, in: *Medical News Today*, 2023, https://www.medicalnewstoday.com/articles/301506.
28. Abourashed, E.A., Bioavailability of Plant-Derived Antioxidants. *MDPI, Antioxid.*, 2, 4, 2013, DOI: https://doi.org/10.3390/antiox2040309.
29. Xu, Y., Liu, H., Song, L., Novel drug delivery systems targeting oxidative stress in chronic obstructive pulmonary disease: a review. *J. Nanobiotechnol.*, 18, 2020, DOI: 10.3390/antiox2040309. Article Number 145.
30. Hermans, N., Cos, P., Maes, L., De Bruyne, T., Berghe, D.V., Vlietinck, A.J., Pieters, L., Challenges and pitfalls in antioxidant research. *Bentham Sci., Curr. Med. Chem.*, 14, 4, 2007, DOI: 10.2174/092986707779941005.

19

Regulatory Aspects of Antioxidants

Kiran Kumar Bellapu[1], Tejaswini Mergu[1], Nikhil Vinod Shirsath[1], Pooja Anil Shende[1], Pavankumar Yashvantrao Vasu[1], Deepak Jindal[2], Parul Grover[3], Saurabh Srivastava[4], Rakesh K. Sindhu[5] and Sandeep Kumar[1,6*]

[1]Department of Regulatory Affairs, National Institute of Pharmaceutical Education and Research- Hyderabad, Balanagar, Hyderabad, Telangana, India
[2]KIET School of Pharmacy, KIET Group of Institutions, Delhi-NCR, Ghaziabad, Uttar Pradesh, India
[3]Department of Pharmaceutical Sciences, Guru Jambheshwar University of Science & Technology, Hisar (Haryana), India
[4]Department of Pharmaceutics, National Institute of Pharmaceutical Education and Research- Hyderabad, Balanagar, Hyderabad, Telangana, India
[5]School of Pharmacy, Sharda University, Greater Noida, Uttar Pradesh, India
[6]Department of Pharmaceutics Nims Institute of Pharmacy Nims University Rajasthan, Jaipur, India

Abstract
To stop oxidative damage in the human body, antioxidants are frequently utilized in both food and pharmaceutical products. Oxidative damage is one of the root causes of many diseases, such as cardiovascular, Alzheimer's, cancer, etc. Antioxidants such as ascorbic acid, butylated hydroxytoluene, propyl gallate, butylated hydroxy anisole, and sodium metabisulphite are common excipients that are used in the finished product to reduce the oxidation of active substances. They could come from either natural or synthetic sources. Protecting food goods from oxidation-related deterioration, such as rancid fat, color changes, and nutritional loss, extends their shelf life. The approval and marketing of antioxidants based on the benefit-to-risk ratio are heavily influenced by regulatory bodies such as the European Economic Commission (EEC) in Europe, the Food and Drug Administration (FDA) in the United States, and the Food and Drug Regulations in Canada. The safety, quality, and efficacy information is essential to submit in a common technical document at the time of excipients approval. The amount of antioxidants in the food products is indicated by a nutrient content claims on the label. Moreover, inadequate doses and durations of therapy in clinical trials could explain the lack of positive outcomes.

***Keywords*:** Antioxidants, FDA, EMEA, oxidative stress, regulatory status

19.1 Introduction

Antioxidants are compounds that considerably reduce or completely prevent the oxidation of the substrate when given to meals at low concentrations relative to the amount of an

**Corresponding author*: Sandeep.kumar2@nimsuniversity.org; ssandeepkatarea@gmail.com; katariatniper@gmail.com

Rakesh K. Sindhu, Inderbir Singh and M. Arockia Babu (eds.) Antioxidants: Nature's Defense Against Disease, (613–628)

oxidizable substrate [1]. Antioxidants are compounds that considerably reduce or completely prevent the oxidation of the substrate when given to meals at low concentrations relative to the amount of an oxidizable substrate [2]. Electron-rich substances known as antioxidants have three rings, a double bond with a conjugation system, at least two different types of double bonds, and the capacity to stop free radicals from causing damage to proteins. Antioxidants, thus, prevent or slow down the process of oxidation of a therapeutic protein in aqueous formulations that are brought on by light exposure—such as UV, visible, and fluorescent light—as well as exposure to free radicals generated by other substances, such metal ions, peroxides, or superoxide anions. Antioxidants include both conjugated and aromatic molecules. Sunscreen and other ultraviolet (UV) protecting technologies, such as sulisobenzone, contain antioxidants. The Food and Drug Administration (FDA) and other national or regional regulatory bodies that have approved sunscreen active ingredients, such as dioxybenzone, homosalate, menthyl anthranilate, octocrylene, aminobenzoic acid, avobenzone, cinoxate, octyl methoxycinnamate, sulfonic acid, titanium dioxide, octyl salicylate, oxybenzone, padimate, phenylbenzimidazole trolamine salicylate, and zinc oxide, are among the other compounds considered in this class of antioxidants. Usually, an antioxidant that also serves as a preservative is used at a concentration lower than its Minimal Inhibitory Concentration (MIC), which can be as low as 5%, 10%, 15%, 20%, or 25% [3]. Antioxidants in this category include free radical scavengers, inactivators of peroxides, metal chelators, and other Reactive Oxygen Species (ROS), and quenchers of secondary lipid oxidation products that give off rotten smells [1].

19.2 Antioxidants Classification

Based on source:

1. Natural antioxidants: tocopherols, tocotrienols, ascorbic acid, carotenoids, etc.
2. Synthetic antioxidants: Ascorbyl palmitate, Butylated hydroxytoluene (BHT), Butylated hydroxyanisole (BHA), erythorbic acid, Tert-butylhydroquinone (TBHQ), etc.

Based on mechanism of action:

1. Primary antioxidants: Primary antioxidants are compounds that provide hydrogen to other molecules to create more stable radicals, thus preventing the oxidation chain reaction.
   Examples of important primary antioxidants include PG, BHA, BHT, and TBHQ.
2. Secondary antioxidants: These are the compounds that, among other things, chelate metals, regenerate primary antioxidants, break down hydroperoxides, and scavenge oxygen to limit the rate of oxidation.
   Examples include Citric acid, ascorbic acid, ascorbyl palmitate, lecithin, etc. [4].

Depending on activity:

1. Enzymatic antioxidants: Antioxidants that break down and eliminate free radicals are known as enzyme-based antioxidants. When cofactors like zinc, iron, copper, and manganese are available, the antioxidant enzymes convert hazardous oxidative products first to hydrogen peroxide (H2O2) and then to water in a multi-step process. These consist of glutathione systems, catalase, and superoxide dismutase, among others.
2. Non-enzymatic antioxidants: These compounds combat free radicals by halting their chain reactions. Various antioxidants that are not enzymatic consist of vitamin C, carotenoids vitamin E, plant polyphenols, and glutathione [3].

According to chemical class:

1. Antioxidants that break chains by reacting with peroxyl radicals, which are radicals that have weak O-H or N-H bonds include phenol, aminophenols, hydroquinone, aromatic amines, and naphthol.
2. Antioxidants such as quinones, iminoquinone, and nitrones react with alkyl radicals to break chains.
3. Hydroperoxide is broken down by antioxidants such as thiophosphate, phosphide, and sulfide.
4. Bifunctional compounds, hydroxyl acids, and diamines are examples of antioxidants that deactivate metals.
5. Compounds containing variable valence metals, nitroxyl radicals, and aromatic amines are examples of antioxidants that break cyclic chains.
6. Phenol sulfide is one of several antioxidants that have synergistic effects. The phenolic and sulfuric groups react with hydroperoxide and peroxyl radicals, respectively [3].

According to European Medicines Evaluation Agency:

1. True antioxidants: These are supposed to react with free radicals to prevent chain reactions. Ex: BHT.
2. Reducing agents: Compared to the medicine or excipient they are shielding, these have a lower redox potential. Ex: Ascorbic acid.
3. Antioxidant synergists: These strengthen the impact of antioxidants. Ex: Sodium edetate [5].

Mechanism of action of antioxidants:

1. Elimination of Oxygen.
2. Removing reactive nitrogen and oxygen ions.
3. Preventing ROS/RNS (Reactive nitrogen species) synthesis.
4. Binding the metal ions required for the ROS catalysis.
5. Increase in natural antioxidant defenses [6].

19.3 Importance of Regulations in the Approval of Antioxidants

The way that antioxidants are regulated varies from nation to nation. Regulatory agencies work to safeguard customers from any negative outcomes that may result from using antioxidants. Regulations are essential for the approval of antioxidants since they evaluate safety and efficacy, quality, and assist in the proper labeling of these chemicals. Regulations make it easier to conduct continuing monitoring in order to spot any unanticipated or unfavorable reactions that might occur during actual use. This enables regulatory bodies to take the proper action, such as product recalls or information updates, as needed. Government organizations in charge of policing pharmaceuticals, food, and other consumer goods frequently provide antioxidants regulatory certification in various nations. Examples of organizations that play a significant role in controlling the marketing of antioxidants and related products are the Food and Drug Administration (FDA) and the United States Department of Agriculture (USDA) in the United States; the Food Sanitation Law in Japan; the European Economic Community (EEC) in Europe; and the Food and Drug Regulations (National Health and Welfare) in Canada [1]. Consumers can have more faith in the safety and effectiveness of antioxidants by having a strong regulatory framework, which will promote overall health and well-being.

19.4 Regulatory Aspects of Antioxidants in Food

Foods are substances used as food or drink (or compounds used as food or drink components) that deliver taste, aroma, nutritional value, and technical influence by the Federal Food, Drug, and Cosmetic Act of 1938 [7].

Antioxidants are substances that extend the shelf life of food by protecting it from oxidation-related deterioration, which includes rancid fat, color changes, and nutritional loss. Because various types of antioxidants possess distinct chemical structures, their efficacy varies significantly when applied to different foods and in different processing and handling conditions. Selecting the optimal antioxidant or mix of antioxidants is made more difficult by the difficulty of forecasting how the additional antioxidant will function in the presence of pro-oxidants and antioxidants already present in the meal or formed during processing.

The use of antioxidants in food is controlled by national regulations or internal standards. While many substances—both natural and synthetic—have antioxidant qualities, only a small number of them have been approved by international organizations such as the Joint FAO/WHO Expert Committee on Food Additives (JECFA) and the European Community's Scientific Committee for Food (SCF) for inclusion in food products as "generally recognized as safe (GRAS)" substances. Anoxomer, stearyl and isopropyl esters, sodium and calcium salts, erythorbic acid, glycine, inox-100, thiodipropionic acid, dilauryl, trihydroxy butyrophenone, and octyl and dodecyl gallates are just a few of the antioxidants that are permitted in food in many countries Table 19.1 [8].

Antioxidants like, BHA, BHT, TBHQ and PG have been used in food products. The maximum level approved and regulatory status of antioxidants has been provided in Table 19.2.

Table 19.1 ADIs (Acceptable Daily Intake) of some antioxidants permitted in food by JECFA.

S. no.	Antioxidant's	ADI (mg/kg body weight)
1.	Propyl gallate	0 to 2.5
2.	BHT	0 to 0.125
3.	BHA	0 to 0.5
4.	Tert-butylhydroquinone	0 to 0.2
5.	Gum guaiac	0 to 2.5
6.	Tocopherols	0.15 to 2.0
7.	Phosphates	0 to 70.0
8.	Ethoxyquin	0 to 0.06
9.	Tartaric acid	0 to 30
10.	Lecithin	Not limited
11.	Ascorbic acid	Not limited
12.	Sulfites	0 to 0.7
13.	EDTA	2.5
14.	Citric acid	Not limited

BHA= Butylated hydroxyanisole, BHT= Butylated hydroxytoluene, EDTA= Ethylene Diamine Tetra Acetic acid, TBHQ= Tert-butylhydroquinone.

19.5 Toxicological Aspects

In addition to being thoroughly tested for their ability to cause a bitter taste in the food they are used in, antioxidants are also tested for their potential for cancer and other harmful effects in themselves, in their degraded forms, and their reaction products with food constituents. When determining the ADI limits and the safety of antioxidants, toxicological investigations are crucial. The ADI for widely used antioxidants, like gallates, BHA, and BHT, has changed throughout time as a result of their toxicological impacts on various species [9].

It is necessary to verify the antioxidant's safety. The Lethal Dose 50 (LD50) of an antioxidant must be more than 1000 mg/kg body weight, and in long-term studies, the antioxidant must not significantly affect the growth of the experimental animal at a level 100 times higher than that advised for human consumption to be deemed safe, according to Lehman *et al.* Before being permitted for use in food, an antioxidant's potential for mutational, teratogenic, and cancer-causing properties must also be thoroughly studied toxicologically.

Table 19.2 Food use, maximum usage level approved, regulatory approval status [1].

Name of the antioxidant	**Food use**	**Maximum level approved by regulatory agencies**	**Regulatory approval status**
BHA	Processed cheeses, Margarine, Dehydrated mashed potatoes, Spices, cereals, Essential oils, Beverage mixes, Dessert mixes, Nuts, Vitamins, Yeast, Vegetable oils, Animal fats, Bakery products, Meat products, and Chewing gum base.	For general usage, FDA 0.02% and USDA 0.01% of weight of fat.	UK and Europe: Approved • Australia/Pacific RIM and Japan: Approved • Canada: Approved
BHT	Potato chips, vegetable oils, butter, margarine, frozen seafood, chewing gum base, snack foods, Breakfast cereals, and baked goods.	FDA 0.02% and USDA 0.01% of weight of fat, respectively, for general use.	• UK and Europe: Approved • Australia/Pacific RIM and Japan: Approved • Canada: Approved
TBHQ	Potato chips, poultry, dried meats, sausages, beef patties, vegetable oils Dry cereals, edible fats, margarine, and pizza toppings.	For general usage, FDA – 0.02%, USDA – 0.01%, based on the lipid content of food	• Europe: allowed as a food additive • Canada: allowed for food use • Australia/Pacific RIM: Australia, New Zealand: allowed for food use • Japan: not allowed for food use
PG	Chewing gum base, non-alcoholic beverages, margarine, mixed nuts, fresh or dry sausages, pre-grilled beef patties, rendered animal fat, pizza toppings, and meatballs.	For general use, BHT or BHA by weight of fat proportion of food, FDA - 0.02% and USDA - 0.01% alone or in combination	• UK: Approved • Europe: listed • Canada: Approved • Australia/Pacific RIM and Japan: Approved

BHT= Butylated hydroxytoluene, TBHQ= Tert-butylhydroquinone, PG= Propyl gallate, BHA= Butylated hydroxyanisole.

Antioxidants are required to meet a variety of requirements before they can be used in food products. The antioxidant needs to be soluble in fats, not give the fat an odd color, smell, or flavor even after a long period of storage, effective for a minimum of a year at 25 to 30°C, resistant to thermal processing, and protecting the final product (carry-through effect), easy to incorporate, and effective at low concentrations [8]. The following substances are classified as antioxidants when they relocate from food-packaging material (limit of addition to food, 0.005%): BHA, BHT, Dilauryl thiodipropionate, Gum guaiac, Distearyl thiodipropionate, Propyl gallate, 2,4,5-Trihydroxy butyrophenone, Thiodipropionic acid, and gum guaiac [10].

19.6 Antioxidant Labeling

Claims for nutrient content that employ the word "antioxidant." A nutrient content claim can be made in the following situations, describing the amount of antioxidant nutrients listed on the food's label or in the labeling:

(1) There is a Recommended Daily Intake (RDI) for every nutrient.
(2) When there is strong scientific evidence that a nutrient is involved in physiological, biochemical, or cellular processes that deactivate free radicals or stop chemical reactions brought on by them, that nutrient is said to have accepted antioxidant activity.
(3) Each essential element of the claim has an adequate quantity to meet the requirements of 101.54(b), (c), or (e):

  (b) "High" assertions, The terms "high," "rich in," or "excellent source of" may be used on the label or in the labeling of foods, provided that the food consists of at least 20% of the RDI or the Daily Recommended Value (DRV) per reference the quantity typically consumed, with the exempt of food products defined in Section 101.13(l) and main a dish products as defined in Section 101.13(m).
  (c) "Good Source" implies something. Foods can employ the phrases "good source," "contains," or "provides" on the label and in labeling if the food contains 10 to 19% of the RDI or DRV per a reference the quantity typically being consumed, with the exception of meal products as specified in 101.13(l) and main dish products as stated in 101.13(m).
  (e) "More" claims: An equivalent claim utilizing the terms "more," "fortified," "enriched," "added," "extra," and "plus" may be used on a label or in labeling foods to indicate the level of protein, vitamins, minerals, dietary fiber, or potassium, in the exception of a meal product referred to in 101.13(l) and main dish products as defined in 101.13(m).

    (i) The item includes at least 10% more of the RDI for vitamins and minerals, or 10% more of the DRV (described as a percentage of the Daily Value) for potassium, dietary fiber, or protein, than a relevant reference food.

(ii) For claims based on nutrients added to the food, the fortification procedure conforms to section 104.20 of this chapter's policy on food fortification;
(iii) For pertinent claims, as mandated by 101.13(j) [2].

(4) The minerals that are the subject of the claim are named in the claim (e.g., "high in antioxidant vitamins C and E"). The term "antioxidant" or "antioxidants" (as in "high in antioxidants") can be linked to a nutrient content claim if it is used in conjunction with another symbol that appears on the same panel of a product label, and it should be followed by the name or names of the nutrients that have been found to have accepted antioxidant activity. The nutritional list needs to be displayed in uppercase letters with the type size of the most noticeable nutrient content claim being at least half the height of 1/16 inch [11].

19.7 Antioxidants Used as Excipients in Formulations

Excipients are utilized in almost all pharmaceutical preparations and are crucial to their effectiveness. Depending on how it is used in the formulation or manufacturing process, an excipient may serve a variety of functional objectives and possess a range of essential qualities. Antioxidants may be viewed as active components with complex molecules including specific metal ions when used for their biological action *in vivo*. Antioxidants prevent the complex oxidative reactions that could otherwise harm the drug material from starting and/or considerably slow them down. Antioxidants may also be used to safeguard flavors, essential oils, pegylated lipids, unsaturated oils, and other non-active ingredients. Antioxidants protect the dosage form's overall integrity from oxidative stress in this way. Antioxidants function best when they are incorporated in the formulation because they block free radicals and hydroperoxides from taking part in the cascading processes and prevent or delay the commencement of chain reactions. It is essential to comprehend oxidative processes and the kinds of by products they generate to apply antioxidants effectively and assess their efficacy.

Physical Properties: The antioxidant solubility must be at its peak at the formulation stage (oily, aqueous, or emulsion interface) because this is when the therapeutic ingredient is most soluble. The temperature at which an antioxidant breaks down is important for autoclaved preparations when antioxidant action may be lost. The antioxidant's stability must also be taken into account and may depend on the pH and the processing environment. PG and metal ions may combine to create colored complexes. Certain proteins and sodium salts may cause TBHQ to become discolored at alkaline pH levels.
Chemical characteristics: Activation energy, oxidation-reduction potential, stability at various formulations (such as pH), and processing conditions (such as heat) are crucial chemical attributes. If the projected shelf life of the dosage form depends on the antioxidant's ability to perform its job, the concentration must be taken into account and reevaluated on a regular basis to make sure that there is still enough antioxidant present without going overboard.

Table 19.3 List of USP general chapters used to assess the antioxidant properties of particular excipients.

Parameter	USP general chapter
Specific Surface Area	846
Crystallinity	695
Chromatography	621
Water Determination	921
Melting Range or Temperature	741
Iron	241

The USP general chapters which is used to evaluate specific antioxidant function has been provided in Table 19.3 [12].

19.8 Excipient Data Required to be Submitted in Regulatory Dossier

According to the European Medicine Agency (EMA), antioxidants are the excipients that are used in finished products to reduce the oxidation of active substances. Antioxidants need to be added in formulation because the presence of metallic and mineral contaminants, as well as exposure to light, increases oxidative deterioration. The use of antioxidants should be limited and justified. Regarding the request or modification of a relevant marketing authorization, EU Directive 2003/63/EC applies to all excipients in pharmaceutical goods intended for human consumption, including antioxidants and antimicrobial preservatives. Excipient data must be submitted in accordance with the Common Technical Document standard format.

If excipients are utilized to create sterile products, bioburden limitations must be defined. These limits should be made in accordance with the validation data that is currently available. Whether an excipient is included in the EU Pharmacopoeia or not, the EU directive stipulates the specifications for excipients used in formulation. While a third-country pharmacopoeia reference can be used if it is not in the EU pharmacopoeia. A suitable specification for the excipient must be provided if it is not listed in any pharmacopoeia by performing various tests:

- Identification Tests
- Purity tests, including limits for total or individual impurities.
- Other relevant tests including e.g. quantitative tests on parameters which have been determined to affect the performance of dosage form
- Physical characteristics
- Limit tests or assay

The specific application of the excipient is taken into consideration when justifying a specification. This will outline the characteristics that, in terms of the product's bioavailability, should be looked at during routine testing and that will be the subject of specific acceptance criteria. Excipients that are listed in European Pharmacopoeia as well as widely used excipients do not require a justification of specification. The final product release specifications should include limits and acceptance criteria in addition to identification and content determination tests. Batch data should be used to support release limitations when producing products containing antioxidants. In order to ensure that the product is preserved throughout the entirety of its shelf life and the recommended in-use period, controlled conditions and in-use stability testing should be used to justify the sufficiency of the established limitations.

The application for the drug product should adhere to the most recent CHMP/QWP stability recommendations and make sure that antioxidant limits are routinely measured all over the shelf-life.

The application for each antioxidant should include the following information:

1. The rationale for its inclusion.
2. Evidence of its safety and effectiveness.
3. The process of quality control in the final product.
4. Levels for the storage of broached and unbroached containers.
5. Information on the labeling of the finished product.

Bibliographic and/or experimental evidence should be used to support the antioxidant's safety.

For antioxidants, the applicant should demonstrate throughout the pharmaceutical development of the product:

A. The significance of adding an antioxidant to the final product at the chosen quantity.
B. The physical and chemical compatibility of the antioxidant with other components of the final product, the container, and closures.

The concentration that is being used must be explained in terms of effectiveness and safety to ensure that just the absolute minimal preservative quantity is being given to achieve the desired standard of efficacy. Even if the production process is optimized to reduce the risk for oxidation, antioxidants must still be used.

The composition, the point at which it is added to the finished product, the kind of container used, along with the formulation all affect how effective an antioxidant is. The effectiveness of antioxidants in the final product must be assessed under conditions that encourage real use, such as comparing the level of degradation in the product containing or without the antioxidant. Only if it is established that their usage cannot be avoided, antioxidants should be used in a formulation. This is relevant in situations where the production process is tailored to reduce the possibility of oxidation [13]. Table 19.4 describes the list of antioxidants used as excipients in the formulation [14].

Table 19.4 Antioxidants used as excipients in formulations.

Antioxidant	Route	Dosage form	Maximum daily exposure	CAS number
BHA	Intravenous	Injection	-	25013165
	Nasal	Solution		25013165
	Oral	Capsule	4mg	
	Oral	Liquid	8mg	25013165
	Oral	Syrup		25013165
	Oral	Tablet	3mg	25013165
	Oral	Suspension		25013165
BHT	Oral	Capsule	1mg	128370
Sodium metabisulphite	Intra-articular	Injection		7681574
	Intravenous	Injection		7681574
	Intracardiac	Injection		7681574
PG	Oral	Capsule		121799
	Oral	Tablet	7mg	121799
Cysteine	Intravenous	Solution		52904
Ascorbic acid	Intramuscular	Injection		50817
L-tocopherol	Intravenous	Injection	15 mg	

BHA= Butylated hydroxyanisole, BHT= Butylated hydroxytoluene, PG= Propyl gallate.

19.9 Antioxidants Used to Treat Diseases

Antioxidants have a variety of effects on different disease situations and aid in their prevention. An organism's natural antioxidants may be able to counteract the oxidative stress brought on by numerous physiological processes. These include antioxidant enzymes in the body, such as glutathione peroxidase, catalase, superoxide dismutase, and reduced glutathione. Furthermore, exogenous antioxidants from food—like vitamin C, vitamin A, and vitamin E—are ingested to protect against reactive oxygen species. One of the benefits of antioxidants is that, in addition to their capacity to scavenge free radicals, they also function in cell signaling. In endothelial cells, antioxidants like catechins have been identified to change the signaling that regulates transcription factors like AP-1 and NF-B [15]. Antioxidant combination therapy is one method for preserving the benefits of antioxidants and lowering the risk of disease aggravation. Additionally, targeting the antioxidants to

Table 19.5 List of antioxidants used to treat diseases.

Brand name	Name of the antioxidant	Mechanism of action/uses	Company name	Approved by
Evion 400mg Capsule	Vitamin E	As an antioxidant, vitamin E keeps free radicals from reacting with cell membranes. Although vitamin E has been demonstrated to have pro-oxidant effects in certain situations. Lipid peroxidation termination is one way by which vitamin E exerts its antioxidant effect.	Merck	US-FDA
Radicava	Edaravone	Authorized to treat Amyotrophic Lateral Sclerosis.	Mitsubishi Tanabe Pharma	JAPAN
ACETADOTE	Acetylcysteine	As well as to being an antioxidant in and of itself, it is deacetylated to cysteine, a process that helps in the production of glutathione, another antioxidant. Additionally, the antioxidant activity may change intracellular redox processes by reducing EGFR and MAPK phosphorylation, which in turn suppresses the transcription of the mucin-producing gene MUC5AC.	Cumberland	US-FDA
ASCOCID	Vitamin C	In the human body, ascorbic acid undergoes reversible oxidation to produce dehydroascorbic acid. Because it functions as a cofactor in the posttranslational formation of 4 hydroxyproline in -Xaa-Pro-Gly- sequences in collagens and other proteins, ascorbic acid from an external source is essential for the formation of collagen and tissue repair in humans.	Volchem	US-FDA
Dioptin	Alpha Lipoic Acid	Lipoic acid is a growth factor for specific microbes and is primarily involved in the oxidative decarboxylation of keto acids.	Novartis	EMA

(*Continued*)

Table 19.5 List of antioxidants used to treat diseases. (*Continued*)

Brand name	Name of the antioxidant	Mechanism of action/uses	Company name	Approved by
Genosyl	Nitric Oxide	Many body cells produce the chemical nitric oxide. By binding to the heme moiety of cytosolic guanylate cyclase, it activates the enzyme and raises intracellular levels of cyclic guanosine 3',5'-monophosphate, which causes vasodilation and relaxes vascular smooth muscle. Nitric oxide causes pulmonary vasodilation when inhaled.	VERO BIOTECH, INC.	US-FDA
Rimso- 50	-	Under some biological conditions, dimethyl sulfoxide has proven to have antioxidant properties. For instance, it is proposed that an antioxidant mechanism underlies the cardiovascular protective effect of dimethyl sulfoxide in rats deficient in copper. Additionally, antioxidant action is thought to be the cause of dimethyl sulfoxide's potential anti-inflammatory activity.	Mylan Institutional LLC	US-FDA
Allopurinol	-	The compound called xanthine oxidase, responsible for transforming hypoxanthine into xanthine and xanthine into uric acid, is blocked by allopurinol and its active metabolite. The actions of allopurinol are prompted by the inhibition of this enzyme	Vintage pharmaceuticals, LLC	US-FDA
Aquasol A	Vitamin A	In the absence of light, all-trans-retinol isomerizes to 11-cis-retinol, which is subsequently oxidized to 11-cis-retinal. Rhodopsin is created when 11-cis-retinal recombines with opsin.	Casper Pharma LLC	US-FDA
Carvedilol	-	Carvedilol's antioxidant effect stops low-density lipoprotein from oxidizing and from being absorbed into the coronary circulation.	Ranbaxy Inc.	US-FDA

(*Continued*)

Table 19.5 List of antioxidants used to treat diseases. (*Continued*)

Brand name	Name of the antioxidant	Mechanism of action/uses	Company name	Approved by
Pedmark Pedmarqsi Seacalphyx	Thiosulfuric acid	Cyanide exhibits a strong affinity for cytochrome a3, which is a constituent of the mitochondrial cytochrome c oxidase complex. Cyanide stops cells from using oxygen and forces anaerobic metabolism, which inhibits cellular respiration and causes lactic acidosis and cytotoxic hypoxia. Cyanide does this by binding to and inhibiting cytochrome a3.	Fennec Pharmaceuticals Inc. Fennec Pharmaceuticals (EU) Seaford Pharmaceuticals Inc.	US-FDA EMA CANADA
Catena Raxona	Idebenone	By means of its interaction with the electron transport chain, idebenone works to protect the lipid membrane and mitochondria from oxidative damage by decreasing free radicals, increasing ATP production that is necessary for mitochondrial function, and inhibiting lipid peroxidation.	Santhera Pharmaceuticals (Switzerland) Limited Santhera Pharmaceuticals (Switzerland) Limited	Canada EMA
Lorelco	Probucol	Probucol may prevent atherogenesis by preventing LDL cholesterol from oxidizing and depositing in tissues. It seems to prevent cellular lipid efflux that is mediated by ABCA1.	Sanofi Aventis	CANADA

particular areas linked to oxidative outbursts may be helpful to treat a variety of disorders linked to oxidative stress [16]. The list of antioxidants used to treat diseases has been presented in Table 19.5 [17].

19.10 Clinical Trial Status of Antioxidants

The use of antioxidant therapies, particularly vitamin E, to prevent the complications of atherosclerotic disease has been demonstrated repeatedly to be ineffective in hundreds of thousands of patients taking part in placebo-controlled trials. The observed results did not meet expectations in spite of strong supporting data, necessitating a reevaluation.

The studied antioxidant agents, such as carotene and vitamin E, may have untoward effects on lipids and HDL levels, possibly diminishing their benefits. Additionally, inadequate doses and durations of therapy in clinical trials could explain the lack of positive outcomes. Longer trials might be necessary to reveal any potential impact.

There is disagreement over the ideal antioxidant dosage. The optimal dosages are still unknown, although 800 units of vitamin E and 500 mg of vitamin C have been proposed as useful threshold doses. 400 IU/day doses have even been connected to higher death rates. The idea of individual variability in response to vitamin supplementation suggests that not everyone will experience the same antioxidant effect from vitamin E supplementation. Therefore, more investigation is required to establish the safest and most efficient antioxidant dosages for both therapeutic and preventive uses [18].

19.11 Antioxidant-Based Clinical Trials for Alzheimer's and Moderate Cognitive Impairment

The Prevention and Treatment of Alzheimer's Disease by Vitamin E and Selenium (PREADVISE) trial and the TEAM-AD trial, a randomized clinical study evaluating the combination of vitamin E and memantine in Alzheimer's disease, are the two clinical trials currently looking into the effects of vitamin E on the disease.

Thirty-six randomized clinical trials evaluating *Gingko biloba's* effects on Alzheimer and cognitive impairment were evaluated in a recent Cochrane review. The evaluation concentrated on nine six-month experiments (including patients from 2016), all of which used different doses of the standard *G. biloba* formulation, EGb 761. There were mixed outcomes on mood, depression, daily activities, cognition, and caregiver burden. In the end, the reviewers concluded that there is insufficient data to support *G. biloba's* consistently beneficial and clinically meaningful effects for patients suffering from dementia or cognitive impairment [19].

19.12 Conclusion

Antioxidants are widely used as a preservative in the food industry to store for longer time. In pharmaceutical industry it is used as excipient in formulation as well as to treat diseases. The significance of free radicals in diseases is becoming more understood, creating new opportunities for antioxidants potential in its prevention. They play a crucial role to treat various diseases by protecting them from the damaging effects of oxidative stress. The antioxidants like Vitamin E, Edaravone, Acetylcysteine, Vitamin C, Alpha Lipoic Acid, Nitric Oxide, Vitamin A, Thiosulfuric Acid, Idebenone, Probucol etc. have been approved by regulatory agencies. High quantities of the synthetic antioxidants might have a pro-oxidant impact. As a result, synthetic antioxidants in large concentrations have been associated with to safety issues. Food producers and pharmaceutical manufacturers must follow the acceptable daily intake in accordance with the relevant criteria controlled by the regulatory authorities' regulatory laws for approval and marketing. The researchers still focus on the pre-formulation, pharmacokinetic, and toxicological studies of antioxidant agents, which are promising health benefits in prevention.

References

1. Shahidi, F., Zhong, Y., Ambigaipalan, P., Antioxidants: regulatory status, in: *Bailey's industrial oil and fat products*, vol. 1, pp. 491–512, 2005.
2. Gulcin, I., Antioxidants and antioxidant methods: An updated overview. *Arch. Toxicol.*, 94, 3, 651–715, 2020.
3. Flora, S.J., Structural, chemical and biological aspects of antioxidants for strategies against metal and metalloid exposure. *Oxid. Med. Cell. Longev.*, 2, 191–206, 2009.
4. Elias, R.J. and Decker, E.A., Antioxidants and their mechanisms of action, in: *Food Lipids: Chemistry, Nutrition, and Biotechnology, Fourth Edition*, pp. 543–66, CRC Press, Taylor & Francis Group, Boca Raton, 2017.
5. EUROPEAN MEDICINES AGENCY, Excipients in the dossier for application for marketing authorisation of a medicinal product - Scientific guideline, 2007, [cited 2023 23 June], Available from: https://www.ema.europa.eu/en/excipients-dossier-application-marketing-authorisation-medicinal-product-scientific-guideline.
6. Halliwell, B., Antioxidants: the basics-what they are and how to evaluate them. *Adv. Pharmacol.*, 38, 3–20, 1996.
7. Force, A.T. (Ed.), *Intl Life Science Institute symposium*, International Life Sciences Institute Press, Bruxelles, Belgium, 1996.
8. Miková, K., *The regulation of antioxidants in food*, pp. 267–84, Woodhead Publishing Ltd, Cambridge, UK, 2001.
9. Madhavi, D.L., Deshpande, S., Salunkhe, D.K., *Food antioxidants: Technological: Toxicological and health perspectives*, CRC Press, Taylor & Francis Group, Boca Raton, 1995.
10. U.S. FOOD & DRUG ADMINISTRATION, Antioxidants, (Sec. 181.24) [cited 2023 19 August], Available from: https://www.accessdata.fda.gov/scripts/cdrh/cfdocs/cfcfr/CFRSearch.cfm?fr=181.24.
11. U.S. FOOD & DRUG ADMINISTRATION, Nutrient content claims for "good source," "high," "more," and "high potency, (Sec. 101.54) [cited 2023 15 August], Available from: https://www.accessdata.fda.gov/scripts/cdrh/cfdocs/cfcfr/cfrsearch.cfm?fr=101.54.
12. United State Pharmacopoiea- National Formulary, Excipient Performance <1059>, [cited 2023 15 May], Available from: https://www.uspnf.com/sites/default/files/usp_pdf/EN/USPNF/revisions/gc-1059-excipient-performance-pf-46-1.pdf.
13. EUROPEAN MEDICINES AGENCY, Excipients in the dossier for application for marketing authorisation of a medicinal product - Scientific guideline, [cited 2023 23 June], Available from: https://www.ema.europa.eu/en/excipients-dossier-application-marketing-authorisation-medicinal-product-scientific-guideline.
14. U.S. FOOD & DRUG ADMINISTRATION, Inactive Ingredient Search for Approved Drug Products, [cited 2023 07 July], Available from: https://www.accessdata.fda.gov/scripts/cder/iig/index.cfm?event=BasicSearch.page.
15. Jochmann, N., Baumann, G., Stangl, V., Green tea and cardiovascular disease: from molecular targets towards human health. *Curr. Op. Clin. Nut. Metab. Care*, 11, 6, 758–65, 2008.
16. Sarangarajan, R., Meera, S., Rukkumani, R., Sankar, P., Anuradha, G., Antioxidants: Friend or foe? *Asian Pac. J. Trop. Med. Hyg.*, 10, 12, 1111–6, 2017.
17. DRUG BANK Online, Antioxidants, [cited 2023 14 May], Available from: https://go.drugbank.com/categories/DBCAT000368.
18. Steinhubl, S.R., Why have antioxidants failed in clinical trials? *Am. J. Cardiol.*, 101, 10, S14–S9, 2008.
19. Mecocci, P. and Polidori, M.C., Antioxidant clinical trials in mild cognitive impairment and Alzheimer's disease. *Biochim. Biophys. Acta (BBA)-Mol. Basis Dis.*, 1822, 5, 631–8, 2012.

Index

Wiley The manufacturer's authorized representative according to the EU General Product Safety Regulation is Wiley-VCH GmbH, Boschstr. 12, 69469 Weinheim, Germany, e-mail: Product_Safety@wiley.com.

Printed and bound by CPI Group (UK) Ltd, Croydon, CR0 4YY
19/08/2026
02201086-0019